ADVANCES IN NUMERICAL SIMULATION OF NONLINEAR WATER WAVES

ADVANCES IN
COASTAL AND OCEAN ENGINEERING
VOLUME 11

ADVANCES IN NUMERICAL SIMULATION OF NONLINEAR WATER WAVES

Editor

Qingwei Ma

City University, London, UK

World Scientific

NEW JERSEY • LONDON • SINGAPORE • BEIJING • SHANGHAI • HONG KONG • TAIPEI • CHENNAI

Published by

World Scientific Publishing Co. Pte. Ltd.

5 Toh Tuck Link, Singapore 596224

USA office: 27 Warren Street, Suite 401-402, Hackensack, NJ 07601

UK office: 57 Shelton Street, Covent Garden, London WC2H 9HE

British Library Cataloguing-in-Publication Data
A catalogue record for this book is available from the British Library.

ADVANCES IN NUMERICAL SIMULATION OF NONLINEAR WATER WAVES

ISBN-13 978-981-283-649-6
ISBN-10 981-283-649-7

Printed in Singapore by B & Jo Enterprise Pte Ltd

PREFACE

Waves and wave–body interactions have been a research focus for many years and studied by a large number of researchers. Due to the compexity of problems of this kind, a lot of research work has been based on linear or other simplified theories such as second order perturbation methods, in which higher order terms may be ignored and boundary conditions satisfied at the initial position of the free and body surfaces. These analyses are valid only when the assumed conditions, such as small or moderate steepness of waves and/or small or moderate motion of bodies, are met. Beyond these conditions, fully nonlinear theory is necessary, in which all nonlinear terms are considered and boundary conditions are imposed on the instantaneous water and body surfaces. Two types of fully nonlinear models, i.e., the NS model (governed by the Navier–Stokes and the continuity equations) and the FNPT model (fully nonlinear potential theory model), may be employed. The latter is relatively simpler and needs relatively less computational resources than the former but cannot take into account viscosity. However, the FNPT model is normally considered to be sufficient if post-breaking of waves does not occur and wave forces rather than the viscous forces are of the main concern.

Unlike simplified theories, which are often amenable to analytical solution, approaches based on the fully nonlinear theories (NS model or FNPT model) rely heavily on numerical modeling. Many numerical methods have been developed over the past decades. These include the Finite Difference Method (FDM), Finite Volume Method (FVM), Finite Element Method (FEM), Boundary (Integral) Element Method (BEM) and Spectral Methods, all of which are mesh-based. More recently, meshless (or meshfree) methods have been proposed. These include the Smoothed Particle Hydrodynamics (SPH) method, Moving Particle Semi-implicit (MPS) method, Constrained Interpolation Profile (CIP) method, Method of Fundamental Solutions (MFS) and Meshless Local Petrov–Galerkin (MLPG) Method. Numerous solution techniques have been developed for the different mathematical models and formulations. Some of these have been devised for fully nonlinear potential theory, some are based on the higher order Boussinesq equations; some deal with the Navier–Stokes equations, some are suitable for single-fluid flow whilst others can cope with two or multiple fluids; some adopt a Lagrangian formulation; some are built on the familiar Eulerian formulation of mesh-based methods whilst others use a mixed Lagrangian–Eulerian formulation or Arbitrary Lagrangian–Eulerian (ALE) formulation.

There are many situations where a fully nonlinear model has to be applied. Typical examples are overturning waves, broken waves, waves generated by landslides, freak waves, solitary waves, tsunamis, violent sloshing waves, interaction of extreme waves with beaches, interaction of steep waves with fixed structures or with freely-responding floating structures.

This book comprises of 18 chapters, reviewing the pioneering work of the authors and their research teams over the past decades on modeling nonlinear waves. Altogether, they cover all numerical methods and all of the typical applications mentioned above. Chapters 1–6 present a review of research based on nonlinear potential theory implementing spectral methods, BEM and FEM, respectively with applications to nonbreaking freak waves, overturning waves, waves generated by landslides, interactions between steep waves and floating structures, and so on. Chapters 7 and 8 review the work on the higher order Boussinesq equations and simulation of various waves. Chapter 9 is also based on potential theory but adopts the meshless MFS method. Chapters 10 and 11 offer a review of numerical modeling based on the NS model using the FVM but adopting different techniques for meshing, for interface modeling and for velocity–pressure coupling. Chapters 12–15 present the development of various meshless methods (CIP, MPS, SPH, MLPG) based on the NS Model and their applications to steep or broken waves. When broken waves are involved, modeling turbulence becomes vital. Chapters 16 and 17 reflect research efforts on this aspect with the former chapter focusing on large eddy simulation and the latter discussing a range of turbulent models. Finally, Chapter 18 presents studies of freak (or rogue) waves and their interactions with structures.

Every chapter of this book has been reviewed by up to three anonymous referees. I would like to express my sincere thanks to those who have made contributions in the review process.

Qingwei Ma

CONTENTS

CHAPTER 1

MODEL FOR FULLY NONLINEAR OCEAN WAVE SIMULATIONS DERIVED USING FOURIER INVERSION OF INTEGRAL EQUATIONS IN 3D

John Grue* and Dorian Fructus

Mechanics Division, Department of Mathematics
University of Oslo, Norway
**johng@math.uio.no*

We describe the basics of the mathematical derivations and numerical implementation of a pseudo-spectral method for fully nonlinear-dispersive simulations of ocean surface waves in three dimensions, and illustrate the method through practical computations. Special features of the method include analytical inversion of the Laplace equation by application of Fourier transform, analytical time integration of the linear part of the prognostic equations using one long time-step, and time-integration of the genuinely nonlinear part of the prognostic equations using auto-adaptive time-step control. The dominant evaluation of the Laplace equation solver is obtained in a FFT fashion and is useful for highly accurate simulations of strongly nonlinear waves propagating over large domains, while any accuracy is obtained by including a remaining contribution which is highly nonlinear and highly local in physical space. The effect of a bottom topography that is variable in space and time is fully accounted for. Computations of three-dimensional wave patterns, extreme events in long-crested seas and short wave formation due to long tsunamis running into shallow water are performed.

1. Introduction

Improved methods for the computation of fully nonlinear ocean surface waves serve many purposes. Activities measuring ocean surface phenomenas use computer models as support for interpretation of the field data. Accurate phase resolved computations of nonlinear wave fields and wave induced orbital velocities are requested for subsequent statistical analysis and the modeling of long and short wave interaction. The wave fields in question typically take place in very large physical domains and over long time windows. Another need relates to the safety of maritime and offshore activity, and to development projects in the coastal area. Current research in the broad area of marine hydrodynamics represents a strong and highly active development on international scale and is documented by the publications that are available on e.g. www.iwwwfb.org. The research in marine hydrodynamics requires a continued improvement of models for the nonlinear wave motion that is input to advanced wave analysis of ships and offshore structures, and installations in the coastal area exposed to waves, e.g., clusters of fixed and floating wind turbines, active breakwaters for the protection of harbors, or other devices. Computation of very steep ocean waves represents a challenge in itself. The evolution of long wave fields and the effect of instability mechanisms have been focused in several studies following pioneering works by Benjamin and Feir[1] and Zakharov.[2] A review of the advances by the mid of 1980's can be found in Yuen and Lake.[3] Recent studies focus on instabilities, the formation of extreme wave events at sea – rogue waves, and how short-crested seas relax towards a steady state governed by the Tayfun distribution,[4] see e.g. Onorato et al.,[5] Dysthe et al.,[6] Socquet et al.[7]

The use of high-order spectral methods enables a relatively rapid modeling of the waves (West et al.,[8] Dommermuth and Yue,[9] Craig and Sulem[10]). Here, we describe a computational strategy in three dimensions that is based on extensive use of Fourier transform (Clamond and Grue,[11] Grue[12]) and with numerical implementation and convergence tested out (Fructus et al.,[13] Clamond et al.[14,15]). The method contains an auto-adaptive time integration procedure and fast converging iterative solution procedure of the Laplace equation. The dominant part of the solution is obtained by Fast Fourier Transform (FFT). The remaining highly nonlinear part have integrals with kernels that decay quickly in the space coordinate. In many applications involving the motion of moderately steep waves it suffices to evaluate the highly rapid FFT-part disregarding the slower part involving the integrals. The iterative method is so rapidly convergent that one iteration is sufficient for most practical applications. This means that the method is explicit. Any accuracy is obtained by a continued iteration, which has converged after three iterations.

The text is organized as follows, following the Introduction, section 2 provides the mathematical description of the method for wave motion in constant water depth and section 3 computations of three-dimensional patterns on the sea surface. Computations of extreme wave events with support from experimental results are compared to events in the field such as the Camille and Draupner waves in section 4. The theory is extended to include the effect of a sea bottom that varies in space and time in section 5, relevant for modeling tsunami waves, while section 6 describes model computations of how short waves and solitons may form in shallow seas, such as the Strait of Malacca, due to a disaster wave such as the Indian Ocean tsunami. Finally, concluding remarks are given in section 7.

2. Numerical Integration of the Wave Field

2.1. *Time integration procedure*

Derivations follow references[11–13,15] where we assume that the velocity field of the fluid is derived by the gradient of a velocity potential ϕ by $\mathbf{v} = \mathrm{grad}\,\phi$, where $\mathbf{v}$ denotes the fluid velocity. Cartesian coordinates are introduced by $(\mathbf{x}, y)$, where $\mathbf{x} = (x_1, x_2)$ denotes the horizontal coordinate vector and y the vertical. In what follows t denotes time. At the free surface we work with the variables

$$\eta(\mathbf{x}, t), \qquad \widetilde{\phi}(\mathbf{x}, t) = \phi(\mathbf{x}, y = \eta, t), \qquad V = \frac{\partial \phi}{\partial n}\sqrt{1 + |\nabla\eta|^2}, \tag{2.1}$$

where η denotes the elevation of the free surface, $\widetilde{\phi}$ the potential function at the free surface, V the normal velocity at the free surface multiplied by $\sqrt{1 + |\nabla\eta|^2}$, and $\nabla = \partial/\partial x_1, \partial/\partial x_2$ the horizontal gradient. The normal vector points out of the fluid. The functions η and $\widetilde{\phi}$ are integrated forward in time by the prognostic equations which result from the kinematic and dynamic boundary conditions, giving

$$\eta_t - V = 0, \quad \widetilde{\phi}_t + g\eta - \frac{|\nabla\widetilde{\phi}|^2 - V^2 - 2V\nabla\eta \cdot \nabla\widetilde{\phi} + |\nabla\eta \times \nabla\widetilde{\phi}|^2}{2 + 2\,\nabla\eta|^2} = 0, \tag{2.2}$$

where g denotes gravity. The time integration is performed in Fourier space, where the prognostic equations become

$$\mathcal{F}\{\eta\}_t - k\,\mathcal{F}\left\{\widetilde{\phi}\right\} = \mathcal{F}\{N_1\}, \quad \mathcal{F}\{\widetilde{\phi}\}_t + g\mathcal{F}\{\eta\} = \mathcal{F}\{N_2\}, \tag{2.3}$$

where $\mathcal{F}\{N_1\} = \mathcal{F}\{V\} - k\mathcal{F}\left\{\widetilde{\phi}\right\}$ and the expression for N_2 equals the fraction corresponding to the third term on the left of the second equation in (2.2). The variables $\mathcal{F}\{\eta\}$ and $\mathcal{F}\{\widetilde{\phi}\}$ are functions of time and wavenumber $\mathbf{k}$ in spectral space. We introduce the variables $\omega = \sqrt{gk\tanh kh}$, where h denotes the water depth and $k = |\mathbf{k}|$, and $\vec{Y} = [k\mathcal{F}\{\eta\}, k\omega\mathcal{F}\{\widetilde{\phi}\}/g]^T$, where

T denotes transpose, and a matrix A by

$$A = \begin{bmatrix} 0 & -\omega \\ \omega & 0 \end{bmatrix}. \tag{2.4}$$

The prognostic equations become

$$\vec{Y}_t + A\vec{Y} = \vec{N}, \qquad \vec{N} = [N_1, N_2]^T. \tag{2.5}$$

The linear part of the prognostic equations may be integrated analytically, which means that the linear part of the wave field is obtained at any time-instant t from the initial condition by using one (long) time-step. This is an advantage since the linear part of the equation contains the leading contribution to the motion while the nonlinear part is relatively smaller, and the applied time-step may be consequently be made larger. Another advantage relates to the stability of the integration. Thus we integrate instead

$$\vec{Y}(t) = \vec{Y}(t_0)\exp(-A(t-t_0)) + \int_{t_0}^{t} \exp A(s-t_0)\vec{N}(s)ds. \tag{2.6}$$

The second term of (2.6) is integrated using a Runge-Kutta (RK) scheme with a variable time-step where the latter is governed by a time-step control. This is done by estimating $\vec{Y}$ at the next time instant using both fifth and fourth order RK-schemes evaluating the difference between the two. The procedure continues until this difference is less than a small, prescribed value. An auto-adaptive method that identifies an optimal step-size speeds up the time integration procedure.[15] The time-step varies according to the strength of nonlinearity of the wave field. An example showing the effect of the time-step control is illustrated in figure 6. We have experienced that constant-step methods, such as symplectic integrators, are inefficient as compared to a well calibrated variable time-stepping method. All physical quantities are represented on a grid which is twice as fine as the desired one. Nonlinear terms up to cubic order are evaluated in physical space, and the product is transformed to spectral space where the highest half of the Fourier modes are padded to zero. This represents the anti-aliasing strategy of the nonlinear terms.

2.2. *Relation between η, $\widetilde{\phi}$ and $V = \phi_n\sqrt{1+|\nabla\eta|^2}$*

An additional relation between the variables η, $\widetilde{\phi}$ and V at the free surface is required in order to integrate (2.6) forward in time. This relation comes from the solution of the Laplace equation in the fluid domain, and is expressed here in terms of an integral equation involving the quantities at the free surface. In the first step we develop the integral equation for the simplest case, namely for water of infinite depth (the case of a finite, constant water depth is covered in the end of the paragraph, and the case of a sea bottom that may be variable both in space and time is covered in subsection 5), where we obtain

$$\int_S \frac{1}{r}\frac{\partial \phi'}{\partial n'}\mathrm{d}S' = 2\pi\widetilde{\phi} + \int_S \widetilde{\phi}'\,\frac{\partial}{\partial n'}\left(\frac{1}{r}\right)\mathrm{d}S'. \tag{2.7}$$

Here, S denotes the instantaneous free surface, n the normal at the free surface, pointing out of the fluid, and r the distance between the evaluation point, $(\mathbf{x}, y)$, and the integration point, $(\mathbf{x}', y')$, both at the free surface, i.e. $r = \sqrt{R^2 + (y - y')^2}$, where $R = |\mathbf{x}' - \mathbf{x}|$. For convenience we write $\eta = \eta(\mathbf{x}, t)$, $\eta' = \eta(\mathbf{x}', t)$, $\widetilde{\phi} = \widetilde{\phi}(\mathbf{x}, t)$, $\widetilde{\phi}' = \widetilde{\phi}(\mathbf{x}', t)$, and so on. Using that $\phi_n \mathrm{d}S = V\mathrm{d}\mathbf{x}$ the integral equation becomes

$$\int_S \frac{V'}{r}\mathrm{d}\mathbf{x}' = 2\pi\widetilde{\phi} + \int_S \widetilde{\phi}'\sqrt{1 + |\nabla\eta|^2}\,\frac{\partial}{\partial n'}\frac{1}{r}\mathrm{d}\mathbf{x}'. \tag{2.8}$$

We introduce the variable $D = (\eta' - \eta)/R$ corresponding to the difference in the wave elevation at two horizontal positions divided by the horizontal distance. In the case when $R \to 0$ the quantity D tends to $\partial D/\partial R$. Thus, the distance r is expressed in terms of R and D by $r = R[1 + D^2]^{1/2}$, meaning that $1/r = 1/R + 1/R[(1 + D^2)^{-\frac{1}{2}} - 1]$. A reorganization of the equation gives

$$\int \frac{V'}{R}\mathrm{d}\mathbf{x}' = 2\pi\widetilde{\phi} + ... - \int \frac{V'}{R}\left[(1 + D^2)^{-\frac{1}{2}} - 1\right]\mathrm{d}\mathbf{x}', \tag{2.9}$$

where intergration is performed over the horizontal plane. Now we apply Fourier transform to the equation, giving, for the l.h.s.

$$\mathcal{F}\left\{\int \frac{V'}{R}\,\mathrm{d}\mathbf{x}'\right\} = \frac{2\pi}{k}\int V'\,e^{-\mathrm{i}\mathbf{k}\cdot\mathbf{x}'}\,\mathrm{d}\mathbf{x}' = \frac{2\pi\mathcal{F}\{V\}}{k}, \tag{2.10}$$

where $\mathcal{F}$ denotes Fourier transform. This inverts the operator on the l.h.s. of the integral equation for V.[11,12] Introducing a decomposition of V by $V = V_1 + V_2 + V_3 + V_4$, we obtain the following expressions for the components:

$$\mathcal{F}\{V_1\} = k\mathcal{F}\left\{\widetilde{\phi}\right\}, \tag{2.11}$$

$$\mathcal{F}\{V_2\} = -k\mathcal{F}\{\eta V_1\} - \mathrm{i}\mathbf{k}\cdot\mathcal{F}\{\eta\nabla\widetilde{\phi}\}, \tag{2.12}$$

$$\mathcal{F}\{V_3\} = \frac{-k}{2\pi}\mathcal{F}\left\{\int \widetilde{\phi}'\left[(1+D^2)^{-\frac{3}{2}}-1\right]\nabla'\cdot\left[(\eta'-\eta)\nabla'\frac{1}{R}\right]\mathrm{d}\mathbf{x}'\right\}, \tag{2.13}$$

$$\mathcal{F}\{V_4\} = \frac{-k}{2\pi}\mathcal{F}\left\{\int \frac{V'}{R}\left[(1+D^2)^{-\frac{1}{2}}-1\right]\mathrm{d}\mathbf{x}'\right\}, \tag{2.14}$$

where we have used Gauss' theorem to arrive at the expression for V_2. While the contributions V_1, V_2, V_3 are explicit, the term V_4 is obtained in an implicit form. A good, leading approximation to the latter is obtained by introducing V_1+V_2 in the r.h.s. of the expression for V_4. This is the first step of an iterative procedure and is useful to obtain V_4. The procedure needs three iterations to be fully converged.[13]

It is of advantage to compute the leading contribution to V_4 in the form of a global evaluation using Fourier transform, i.e.

$$\begin{aligned}\mathcal{F}\{\hat{V}_4\} &= \frac{k}{2\pi}\mathcal{F}\left\{\int \frac{V'\frac{1}{2}D^2}{R}\mathrm{d}\mathbf{x}'\right\}\\ &= -\frac{1}{2}k\mathcal{F}\{\eta^2\mathcal{F}^{-1}[k\mathcal{F}(V)] - 2\eta\mathcal{F}^{-1}[k\mathcal{F}(\eta V)] + \mathcal{F}^{-1}[k\mathcal{F}(\eta^2 V)]\},\end{aligned} \tag{2.15}$$

where $\mathcal{F}^{-1}$ denotes inverse Fourier transform. In the computations the terms $V_1+V_2+\hat{V}_4$ are evaluated using FFT. This sum constitutes the fast FFT-part of the method. The remaining integrals determined by V_3 in eq. (2.13) above, and the difference,

$$\mathcal{F}\left\{V_4-\hat{V}_4\right\} = \frac{-k}{2\pi}\mathcal{F}\left\{\int\frac{V'}{R}\left[(1+D^2)^{-\frac{1}{2}}-1+\tfrac{1}{2}D^2\right]\mathrm{d}\mathbf{x}'\right\}, \tag{2.16}$$

are highly nonlinear and with kernels that decay rapidly with distance in the horizontal plane. These integrals have very local contributions and are evaluated by numerical integration over small domains in the horizontal plane, surrounding the field point $\mathbf{x}$.

In the case of a finite constant water depth, h, the Green function involved in the integral equation becomes $1/r+1/r_1$ where the latter is an image of the former with respect to the sea floor. The expression for $\mathcal{F}(V)$ becomes

$$\begin{aligned}\mathcal{F}(V) = {} & kE_1\mathcal{F}(\widetilde{\phi}) - kE_1\mathcal{F}(\eta V_1) - \mathrm{i}\mathbf{k}\cdot\mathcal{F}(\eta\nabla\widetilde{\phi})\\ & + kC_1[\mathcal{F}(T(\widetilde{\phi})+T_1(\widetilde{\phi})) + \mathcal{F}(N(V)+N_1(V))]\\ & + kC_1[e_1\mathcal{F}(\eta(V-V_1)) + \mathcal{F}(\eta\mathcal{F}^{-1}(e_1\mathcal{F}(V-V_1)))],\end{aligned} \tag{2.17}$$

where $\mathcal{F}(V_1) = E_1\mathcal{F}(\widetilde{\phi})$, $E_1 = \tanh kh$, $C_1 = 1/(1+e_1)$ and $e_1 = \exp(-2kh)$. The functions T, N, N_1 and T_1 are given in Grue,[12] eqs. (9), (10), (29), (30), respectively. This is the only modification to the time integration eq. (2.6) due to a finite constant water depth h.

2.3. *Note on global and local integration*

We note that, in cases when $D^2 < 1$, the term $-\left(1+D^2\right)^{-\frac{1}{2}}+1$ in (2.14) has a convergent series expansion of the form, $\frac{1}{2}D^2 - \frac{3}{8}D^4 + \frac{5}{16}D^6 + \cdots$, $D^2 < 1$, where the infinite expansion can formally be used to rewrite the integral V_4 obtaining a sum of convolutions. Such a procedure leads to the expansion suggested in the two-dimensional study by Craig and Sulem,[10] which, however, has essential drawbacks, including, among others: slow convergence; the convolution terms involving high-order derivatives lead to numerical instability; an inherent ill-conditioning due to cancellations. Our main conclusion is that the term is best divided into the sum $\frac{1}{2}D^2 - \left[\left(1+D^2\right)^{-\frac{1}{2}} - 1 + \frac{1}{2}D^2\right]$, where the first term $(\frac{1}{2}D^2)$ contributes to a global evaluation using FFT and the second to a local, truncated integration. This subdivision is found to be insensitive to the magnitude of D^2. We have tested out another subdivision of the form $\frac{1}{2}D^2 - \frac{3}{8}D^4 - \left[\left(1+D^2\right)^{-\frac{1}{2}} - 1 + \frac{1}{2}D^2 - \frac{3}{8}D^4\right]$, including also $-\frac{3}{8}D^4$ in the global evaluation, and the remaining part as an even stronger local, truncated integration, but this was found to lead to a numerically unstable procedure.[13]

2.4. *Comparison with other methods*

Numerical integration of the evolution of long wave fields over long time using the 2D counterpart of the present method[16] showed excellent comparison with the method by West et al.,[8] but that the formulation of Dommermuth and Yue[9] was inaccurate in the long term in the numerical simulation and is caused by a poor representation of the vertical free surface velocity in the model.

3. Computations of Three-Dimensional Wave Patterns[a]

Three-dimensional wave structures may occur at the sea surface in the form of crescent-shaped patterns. These play an important role in wave

[a]This section 3 is a reprint of section 4 in chapter 4 (pp. 191-202) of Grue and Trulsen,[17] with kind permission given by CISM, and is gratefully acknowledged.

breaking and cause transfer of momentum and energy between the ocean and atmosphere. The three-dimensional structures may result from spontaneous instabilities of a two-dimensional water wave field. Particularly those emerging naturally from the instability of Stokes waves are considered here. The stability of Stokes waves has been studied for many decades. Two-dimensional instability was studied by Lighthill[18] and Benjamin and Feir,[1] working with triads – or the side-band instability, while Zakharov[2] also included three-dimensional instabilities in his analysis, working with quartets in the wavenumber space. Later, McLean et al.[20] and McLean[21] discovered a new kind of three-dimensional instability which, prior to this, had been suggested in a two-dimensional study by Longuet-Higgins.[22] The new instability analysis involved the interaction between quintets and higher resonances in the wavenumber space. We study here 5-wave and higher interactions.

3.1. *The stability analysis by McLean et al.*[20]

This analysis was motivated by experimental measurements of three-dimensional wave patterns by Su, later documented in Su,[19] and by preliminary calculations of the patterns by Saffman and Yuen, see reference 12 in McLean et al.[20] The research group, including McLean, Ma, Martin, Saffman and Yuen at the Fluid Mechanics Department, TRW Defense and Space Systems Group in California, performed analysis and computations demonstrating that there were two distinct types of instabilities for gravity waves of finite amplitude on deep water. They solved the fully nonlinear equations for water waves, finding that one type is predominantly two-dimensional, relating to all the previously known results for special cases. The other predominantly three-dimensional instability becomes dominant when the wave steepness is large.

The set of equations that McLean et al. solved, working with the velocity potential, φ, was: $\nabla^2\varphi = 0$ for $-\infty < y < \eta(x_1, x_2, t)$; and $\varphi_t + \frac{1}{2}|\nabla\varphi|^2 + gy = 0$, $\eta_t + \nabla\varphi \cdot \eta - \varphi_y = 0$ on $y = \eta$. They calculated steady Stokes waves up to $2a/\lambda = 0.131$ (corresponding to $ak = 0.412$) on the form $\eta_s = \sum_0^\infty A_n \cos[2n\pi(x_1 - Ct)]$, where the Fourier coefficients A_n and the wave speed C were functions of $2a/\lambda$. In the computations λ was put to 2π (giving $k_0 = 1$) and $g = 1$. $2a$ denoted the wave height.

They perturbed the steady waves by infinitesimal perturbations η' of the form

$$\eta' = e^{\mathrm{i}[p(x_1 - Ct) + qx_2 - \sigma t]} \sum_{-\infty}^{\infty} a_n e^{\mathrm{i}n(x_1 - Ct)}, \tag{3.1}$$

where the perturbation wavenumbers p and q are real numbers. Eq. (3.1) is an eigenvector of the infinitesimal perturbations to the Stokes wave, with σ being the eigenvalue. Instability corresponds to $Im(\sigma) \neq 0$, since the perturbation also involves the complex conjugate. The task set by McLean et al. was to determine the eigenvalues, σ. This was done by inserting a truncated version of (3.1), and a corresponding perturbation of the velocity potential, φ', both including $2N+1$ modes, into the full equations. The kinematic and dynamic boundary conditions were satisfied at $2N+1$ nodes. The resulting homogeneous system of order $4N+2$ was solved as an eigenvalue problem for σ by means of standard methods. McLean et al. used $N = 20$ for $2a/\lambda < 0.1$ and $N = 50$ for the steepest case with $2a/\lambda = 0.131$.

In a frame of reference moving with the wave speed, C, the unperturbed wave, in the limit $2a/\lambda \to 0$, becomes:[21] $\eta_s = 0, \varphi_s = -x_1$, where C has been put to unity. In this case, the eigenfunctions and eigenvalues are

$$\eta' = e^{-\mathrm{i}\sigma_n t + \mathrm{i}[(p+n)x_1 + qx_2]}, \tag{3.2}$$

$$\sigma_n = -k_x \pm |\mathbf{k}|^{\frac{1}{2}}, \tag{3.3}$$

where the latter represents the linear dispersion relation on a unit current (of negative speed), and where $\mathbf{k} = (k_{x_1}, k_{x_2}) = (p+n, q)$. Eq. (3.3) gives

$$\sigma_n = -(p+n) \pm [(p+n)^2 + q^2]^{\frac{1}{4}}. \tag{3.4}$$

McLean et al. noted that the σ's were degenerate in the sense that $\sigma_n(p,q) = \sigma_{n+1}(p-1, q)$. In the case of finite amplitude Stokes waves, the eigenvalues may become complex, meaning instability, if

$$\sigma^{\pm}_{n_1}(p,q) = \sigma^{\pm}_{n_2}(p,q). \tag{3.5}$$

The corresponding eigenvectors have dominant wave vectors $\mathbf{k}_1 = (p+n_1, q)$ and $\mathbf{k}_2 = (p+n_2, q)$. For deep water waves, the solution to (3.5) was divided into two classes:

Class I:

$$\mathbf{k}_1 = (p+m, q),\ \mathbf{k}_2 = (p-m, q),\ \sigma^{+}_{m}(p,q) = \sigma^{-}_{-m}(p,q), \tag{3.6}$$

$$[(p+m)^2 + q^2]^{\frac{1}{4}} + [(p-m)^2 + q^2]^{\frac{1}{4}} = 2m,\ (m \geq 1). \tag{3.7}$$

Class II:

$$\mathbf{k}_1 = (p+m, q),\ \mathbf{k}_2 = (p-m-1, q),\ \sigma^{+}_{m}(p,q) = \sigma^{-}_{-m-1}(p,q), \tag{3.8}$$

$$[(p+m)^2 + q^2]^{\frac{1}{4}} + [(p-m-1)^2 + q^2]^{\frac{1}{4}} = 2m+1,\ (m \geq 1). \tag{3.9}$$

Class I curves are symmetric about $(p, q) = (0, 0)$. Class II curves are symmetric about $(p, q) = (\frac{1}{2}, 0)$.

The eigenvalues can alternatively be interpreted as resonance of two infinitesimal waves with the carrier wave, giving, in the fixed frame of reference:

$$\omega_1 = -\omega_2 + N\omega_0, \quad \mathbf{k}_1 = \mathbf{k}_2 + N\mathbf{k}_0. \tag{3.10}$$

Here, $\mathbf{k}_1 = (p' + N, q)$, $\mathbf{k}_2 = (p', q)$ (and $-\mathbf{k}_2 = (-p', -q)$), $\mathbf{k}_0 = (1, 0)$, and $\omega_i = |\mathbf{k}_i|^{\frac{1}{2}}$. Class I corresponds to N even, $m = \frac{1}{2}N$, $p' = p - m$. Class II corresponds to N odd, $m = \frac{1}{2}(N-1)$, $p' = p - m - 1$. $N = 2$ corresponds to 4-wave interaction, $N = 3$ to 5-wave interaction, $N = 4$ to 6-wave interaction, and so on. The resonance curves are visualized in figure 1.

For $p = \frac{1}{2}$ McLean et al. found that $Re(\sigma) = 0$, which means that the perturbation wave pattern co-propagates with the Stokes wave. They calculated the regions of nonlinear instability, and moreover, identified the growth rate of the most unstable perturbation (corresponding to maximum of $|Im(\sigma)|$). From McLean et al.,[20] figure 1, we obtain selected values of $(q, ak, |Im(\sigma)|_{\max})$ (and $p = \frac{1}{2}$):

Table 1. Growth rate of the most unstable perturbation of class II instability; 5-wave interaction. $p = \frac{1}{2}$. Extracted from McLean et al.,[20] figure 1.

q	ak	$\lvert Im(\sigma)\rvert_{\max}$
1.65	0.1	$6 \cdot 10^{-4}$
1.53	0.2	$5 \cdot 10^{-3}$
1.25	0.3	$2 \cdot 10^{-2}$
1.16	0.41	$1 \cdot 10^{-1}$

McLean[21] improved the accuracy of the stability calculations. In his table 2 and figure 3, more accurate values of the lateral wavenumber and growth rate were presented. Particularly for $ak = 0.3$, a new value of $q = 1.33$ was obtained, and is used here in the calculations that are discussed below (we use $q = \frac{4}{3}$). It is important to note that improved computations of the stability analysis were required, still. This was particularly true for waves with ak exceeding 0.412, for which no values were given by

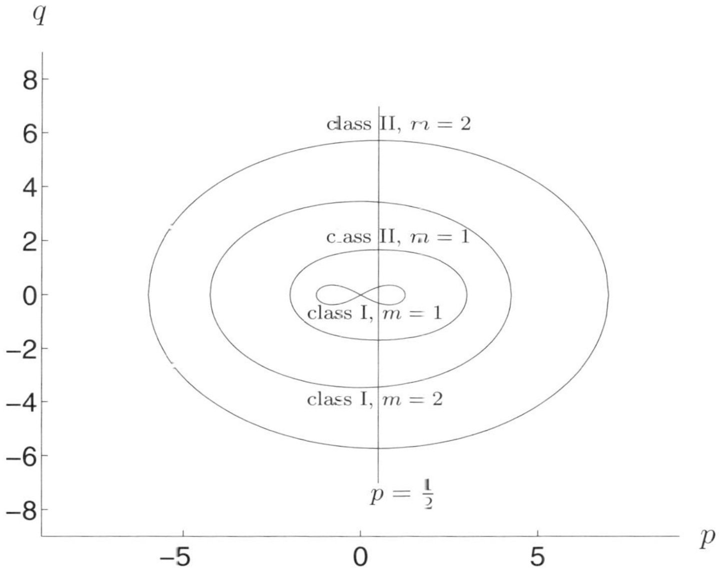

Fig. 1. Plot of resonance curves for $ak = 0$. Class I instabilities, eq. (3.7) with $m = 1$ (inner curve – including points $(\pm\frac{5}{4}, 0)$) and $m = 2$ (curve including points $(\pm\frac{17}{4}, 0)$). Class II instabilities, eq. (3.9) with $m = 2$ (outer curve – including points $(-6, 0)$ and $(7, 0)$) and $m = 1$ (including points $(-2, 0)$ and $(3, 0)$). Class II instability curves are symmetric with respect to $p = \frac{1}{2}$.

McLean.[21] This was the motivation for the studies by Kharif and Ramamonjiarisoa.[23,24]

3.2. *Computations of the classical horseshoe pattern*

The stability analysis of McLean et al.[20] was motivated by the three-dimensional patterns observed experimentally by Su, published in 1982, as mentioned. The waves were also investigated experimentally by Melville,[25] Kusuba and Mitsuyasu[26] and Collard and Caulliez.[27] The experimental observations of the waves are here numerically reproduced in the sense that we compute the growing wave instabilities up to the point of breaking. In the original experiments the Stokes waves had an amplitude corresponding to $(ak)_0 = 0.33$. For such steep waves the class II instability is stronger than the class I instability. We here illustrate numerically the growth of the class II instability, where in the computations the class I instability

Table 2. Characteristics of three-dimensional wave patterns. Numerical computations prior to breaking for $(ak)_0 = 0.2985$ and 0.33 compared to experiments of steady pattern by Su, $(ak)_0 = 0.33$. $S_x = \partial\eta/\partial x$. The numerical waves grow up to breaking during 18 wave periods ($(ak)_0 = 0.2985$) and 11 wave periods ($(ak)_0 = 0.33$).

$(ak)_0$	$\frac{\lambda_2}{\lambda_1}$	$\frac{h_{11}}{h_{12}}$	$\frac{h_{21}}{h_{22}}$	$\frac{h_{11}}{h_{21}}$	$\frac{h_{22}}{h_{11}}$	$\frac{h_{11}+h_{12}}{h_{22}+h_{21}}$	S_x,max
Su; 0.33	1.28	1.10	0.88	1.66	0.68	1.49	0.65
Sim, 0.2985	1.28	1.11	0.88	1.56	0.73	1.38	0.66
Sim, 0.33	1.28	1.12	0.85	1.53	0.76	1.33	0.67

is suppressed. Description follows Fructus et al.[28] The waves were also studied numerically by Xue et al.[29] and Fuhrman et al.[30]

Stokes waves with frequency $\omega_0 = \omega(k_0)$ and wave slope $(ak)_0$ were computed using the procedure of Fenton.[31] To this wave train, a small perturbation, taking the form

$$\hat{\eta} = \epsilon a_0 \sin((1+p)k_0 x_1)\cos(qk_0 x_2) \tag{3.11}$$

was superposed. Here, ϵ is a small number, making the amplitude of the initial perturbation field a fraction of the Stokes wave field, and $(1+p,q)k_0$ denotes the directional wavenumber. Computations performed with $(ak)_0 = 0.2985$ and 0.33, have perturbation at $p = \frac{1}{2}$, $q = \frac{4}{3}$ and value of ϵ of 0.05.

Four periods of Stokes waves in the longitudinal direction and three periods of perturbation in the lateral direction were resolved by 128×64 nodes. The magnitude of the Fourier transform of the perturbation field, $|\mathcal{F}(\hat{\eta})|$, visualizes the growth of the modes with wavenumbers $(\frac{3}{2}, \pm\frac{4}{3})k_0$ up to the point of breaking, see figure 2a. The sum of the wave vectors of the satellites is $(3,0)k_0$. The initial growth rate in the nonlinear calculations is 0.0194, slightly smaller than the analytical value of 0.021 computed by perturbation theory, when $(ak)_0 = 0.2985$. The frequency spectrum, $\omega(k_{x_1}, k_{x_2}, t)$, obtained from the Fourier transformed perturbation field by $\mathcal{F}(\hat{\eta}) = |\mathcal{F}(\hat{\eta})|\exp(\mathrm{i}\chi)$ and $\omega(k_{x_1}, k_{x_2}, t) = \partial\chi/\partial t$, shows that $\omega(\frac{3}{2}k_0, \pm\frac{4}{3}k_0) = \frac{3}{2}\omega_0(1+\hat{\epsilon})$, where $\hat{\epsilon}$ represents a very small variation (caused by interactions with modes of higher wavenumber) around zero and is visualized in figure 2b. Indeed, the figure illustrates that a quintet interaction is evident. The resulting wave field prior to breaking is visualized in figure 4a, and the wave frequency spectrum in figure 4c. The latter exhibits peaks at ω_0 – the fundamental frequency, and $n\omega_0$, $(n > 1)$ – the locked components of the Stokes wave. The dominant peaks corresponding

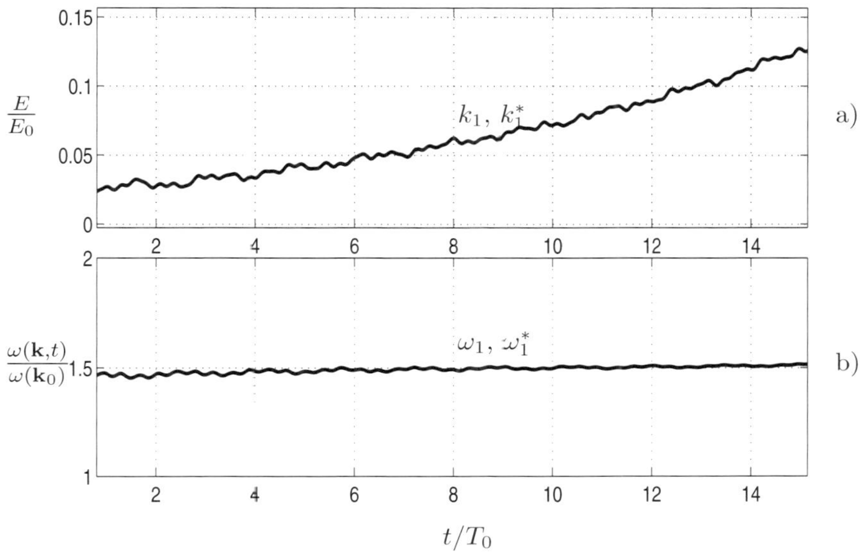

Fig. 2. a) Energy evolution, E, relative to initial energy, $E_0 = \frac{1}{2}\rho g a_0^2$, of the dominant modes: $\mathbf{k}_1 = (\frac{3}{2}, \frac{4}{3})k_0$ and $\mathbf{k}_1^* = (\frac{3}{2}, -\frac{4}{3})k_0$. b) Evolution of frequency. Initial perturbation with $(p, q) = (\frac{1}{2}, \frac{4}{3})$. $(ak)_0 = 0.2985$, $\epsilon = 0.05$. Adapted from Fructus et al.[28] Reproduced by permission by J. Fluid Mech.

to the satellites is observed at $\frac{3}{2}\omega_0$. Higher order motions of the satellites have small peaks at $\frac{1}{2}\omega_0$, $\frac{5}{2}\omega_0$, $\frac{7}{2}\omega_0$, ... The number of time steps in the computation of the waves is indicated in figure 2, 4a,c as visualized in figure 5.

The wave breaking appears in the computations by a rapid growth of the high wavenumbers of the spectrum. The computational wave breaking – when it is independent of the resolution – corresponds to breaking of the physical waves. The computed waves right before breaking resemble the steady state of the three-dimensional wave field in the experiments. Wave characteristics are defined in figure 3. The numerical waves compare very well to the waves in steady state, observed by Su,[19] see table 5. Su termed the pattern by L_2 pattern, since it repeated itself once per two wavelengths of the fundamental train. More generally, L_n patterns correspond to interactions between satellites $\mathbf{k}_1 = (1 + 1/n, q)k_0$ and $\mathbf{k}_2 = (2 - 1/n, -q)k_0$, meaning that $p = 1/n$, where the integer $n \geq 2$. The definition can also be used for $n = 1$ (L_1 pattern) corresponding to $p = 0$.

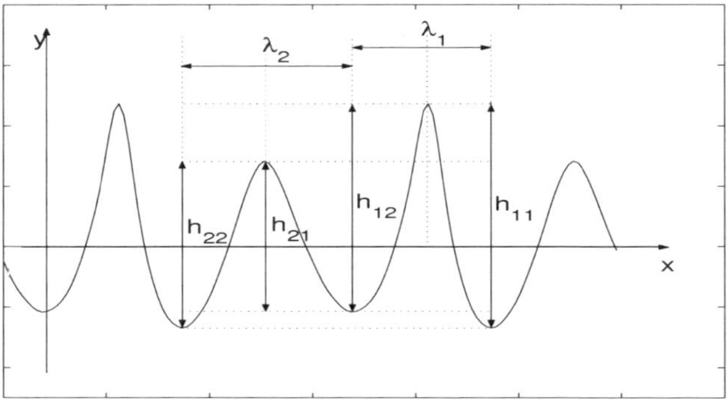

Fig. 3. Definition of Su's geometric parameters.

3.3. *Oscillating horseshoe pattern. Computations of the experiments by Collard and Caulliez*

A new surface wave pattern was experimentally observed by Collard and Caulliez.[27] They performed water wave experiments in a long, relatively wide indoor tank, finding an oscillatory, crescent formed, horse-shoe like pattern, riding on top of long-crested steep Stokes waves with wavenumber k_0. The crescents of the sideways oscillating pattern were always aligned, both in the longitudinal and transversal directions, see figure 4b for illustration. The wave patterns appeared due to an instability in the form of a quintet resonant interaction: wave satellites had longitudinal wavenumber k_0 and transversal wavenumber qk_0, and the other wavenumber $2k_0$ (longitudinal) and $-qk_0$ (transversal). These summed up to three times the wavenumber of the fundamental Stokes wave. The observed frequencies (in the experiments) were $1.36\omega_0$, $1.64\omega_0$, adding up to three times the fundamental frequency, ω_0 of the Stokes wave, satisfying also a quintet in the frequencies. Along the propagation direction, the pattern repeated itself once per fundamental wave length, and is so an L_1 pattern. The stability analysis above, with $N = 3$ and $p = 0$, supports the observations of the three-dimensional pattern.

This aligned L_1 pattern was numerically evidenced by Skandrani[32] and was also computed by Fuhrman et al.[30] Here we follow Fructus et al.[28] who performed simulations with an initial perturbation corresponding to $(p, q) = (0, \frac{4}{3})$ and $\epsilon = 0.05$, for Stokes waves with $(ak)_0 = 0.2985$. The wave field was initially perturbed by $\mathbf{k}_1 = (1, \frac{4}{3})k_0$. This perturbation generated a

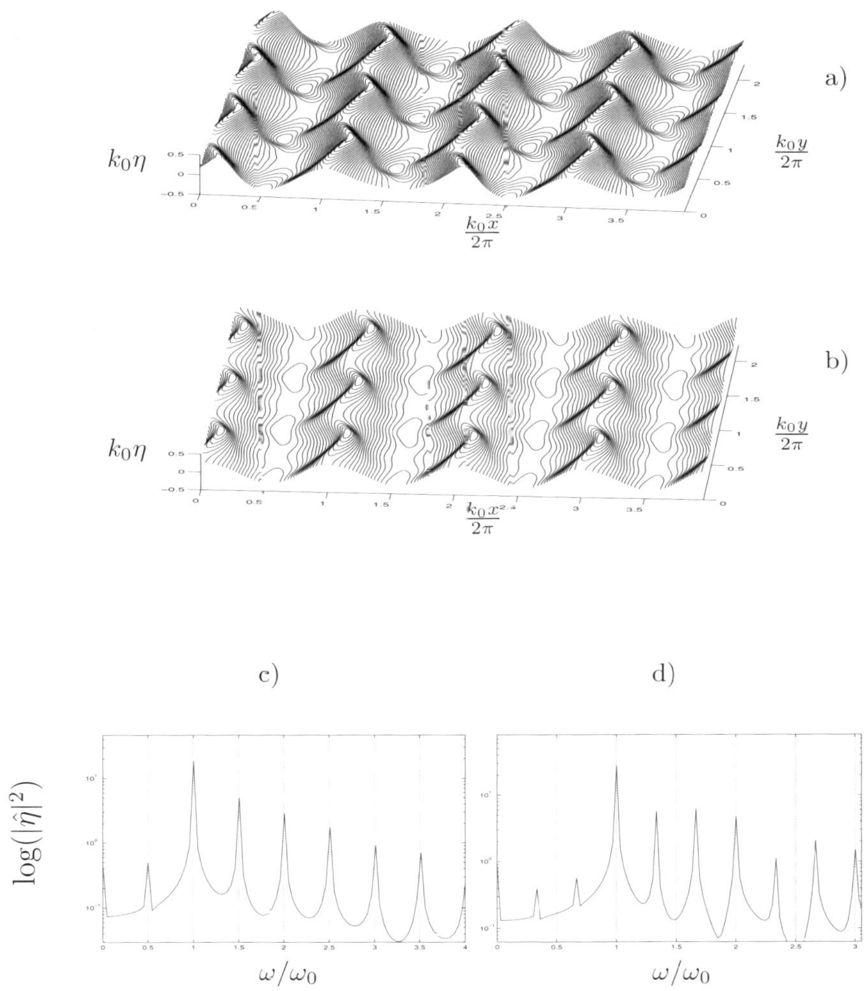

Fig. 4. Three-dimensional wave patterns on Stokes waves with wave slope 0.30 of the fundamental mode. a) Steady (most unstable) horse shoe pattern, observed experimentally by Su in 1982. Snapshot at $t/T_0 = 16$ (T_0–wave period). b) Unsteady horse shoe pattern, observed experimentally by Collard and Caulliez in 1999 ($t/T_0 = 23$). c) Wave frequency spectrum corresponding to a). d) Wave frequency spectrum corresponding to b). From Fructus et al.[13]

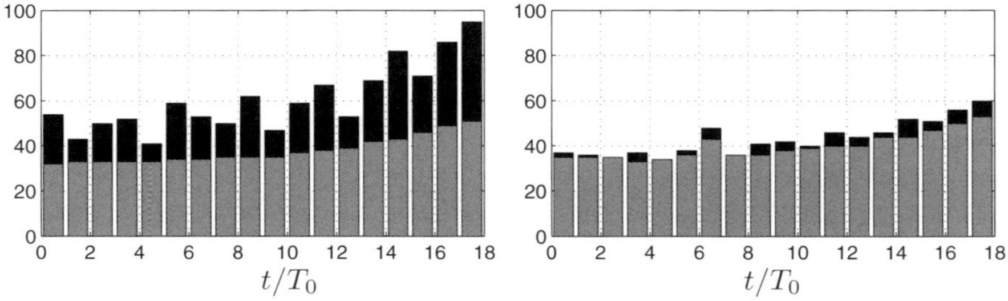

Fig. 5. Number of time steps per wave period for the computation in figure 2. Time integration without (right) and with (left) stabilizing step size control. Number of accepted (grey) and rejected (black) steps per wave period. From Clamond et al.[15]

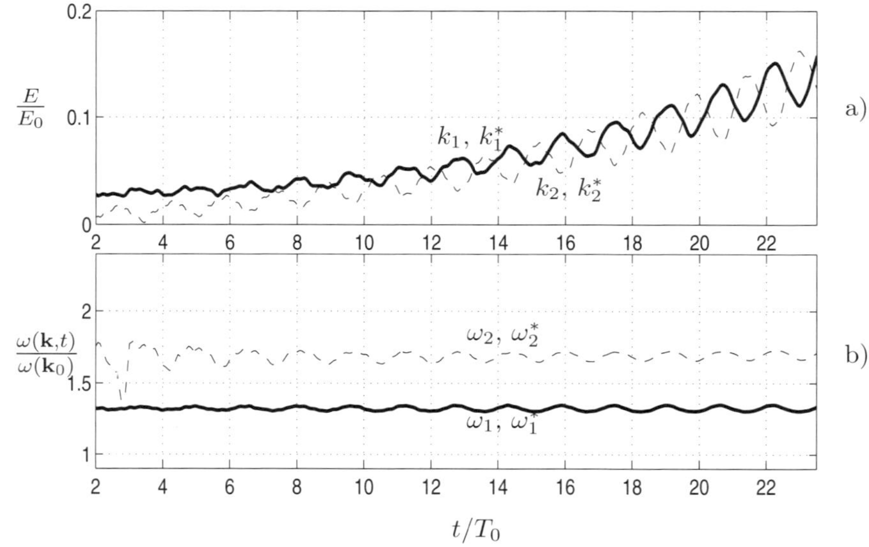

Fig. 6. a) Energy evolution, E, relative to initial energy, $E_0 = \frac{1}{2}\rho g a_0^2$, of the dominant modes: $\mathbf{k}_1 = (1, \frac{4}{3})k_0$, $\mathbf{k}_1^* = (1, -\frac{4}{3})k_0$, and $\mathbf{k}_2 = (2, \frac{4}{3})k_0$, $\mathbf{k}_2^* = (2, -\frac{4}{3})k_0$. b) Evolution of frequency. Initial perturbation with $(p, q) = (0, \frac{4}{3})$. $(ak)_0 = 0.2985$, $\epsilon = 0.05$. Adapted from Fructus et al.[28] Reproduced by permission by J. Fluid Mech.

motion with energy at wavenumber $\mathbf{k}_2^* = (2, -\frac{4}{3})k_0$. An interaction between the two modes then became evident. The pair of modes with wave vectors $\mathbf{k}_1 = (1, \frac{4}{3})k_0$, $\mathbf{k}_2^* = (2, -\frac{4}{3})k_0$ satisfied $\mathbf{k}_1 + \mathbf{k}_2^* = 3\mathbf{k}_0$. This growth is visualized in figure 6a. The growth is somewhat weaker than for the classical horseshoe pattern, see figure 2a. The resulting wave field prior to breaking is illustrated in figure 4.

The dominant frequencies of the satellites are visualized in figure 6b. The figure shows that $\omega(\mathbf{k}_1)+\omega(\mathbf{k}_2^*) = 3\omega_0(1+\hat{\epsilon})$ where $\hat{\epsilon}$ represents a small oscillation around zero. The oscillation has period $\frac{11}{7}$, approximately, of the fundamental period, T_0. The dominant frequencies are further illustrated in figure 4d, where peaks in the wave frequency spectrum is observed pair-wise, at $(n+a)\omega_0$, $(n+1-a)\omega_0$, for $n = 0, 1, 2, \ldots$, where $a = 0.33 \pm 0.02$. (Note that the value of $(ak)_0$ is different in the computation and the experiment by Collard and Caulliez, and thus makes a small difference in the value of the peak frequencies of the satellites.) The wave frequencies of the Stokes waves at $n\omega_0$, $n \geq 1$, are also present.

The computations by Fructus et al.[28] further showed that the modes ($\mathbf{k}_1 = (1, \frac{4}{3})k_0$, $\mathbf{k}_2^* = (2, -\frac{4}{3})k_0$) and ($\mathbf{k}_1^* = (1, -\frac{4}{3})k_0$, $\mathbf{k}_2 = (2, \frac{4}{3})k_0$) had the same development. A resonant interaction between the two pairs of modes was observed. The triggering mechanism of the waves was investigated, finding that the waves could be generated by parametric resonance due to the wave-maker. This L_1 pattern should be observed in coastal waters where the modulational instability becomes weaker.

3.4. *Other features of class II instability*

3.4.1. *Class I instability may restabilize class II instability;* $(ak)_0 = 0.10$.

For moderate $(ak)_0$ the class II instability may be restabilized by class I instability. In an example, shown in figure 7a, the Stokes wave with $(ak)_0 = 0.10$ was perturbed by the most unstable modes of the class I and class II instabilities, corresponding to $p = \frac{1}{6}$ and $p = \frac{1}{2}$, with wave vectors $\mathbf{k}_1 = (\frac{5}{6}, 0)k_0$, $\mathbf{k}_2 = (\frac{7}{6}, 0)k_0$, $\mathbf{k}_3 = (\frac{3}{2}, 1.645)k_0$, $\mathbf{k}_3^* = (\frac{3}{2}, -1.645)k_0$. The explanation for the restabilization of the class II instability is that the modulational instability moves the very narrow class II instability region slightly (when $(ak)_0 = 0.1$), so that the perturbations at the wave vectors $\mathbf{k}_3 = (\frac{3}{2}, 1.645)k_0$, $\mathbf{k}_3^* = (\frac{3}{2}, -1.645)k_0$ fall outside the instability region. No breaking was observed in the simulation.

3.4.2. *Class II instability may trigger class I instability, leading to breaking;* $(ak)_0 \geq 0.10$.

If the wave field is perturbed by class II instability only, the side-band instability becomes triggered. If $(ak)_0 \geq 0.10$, the interaction between

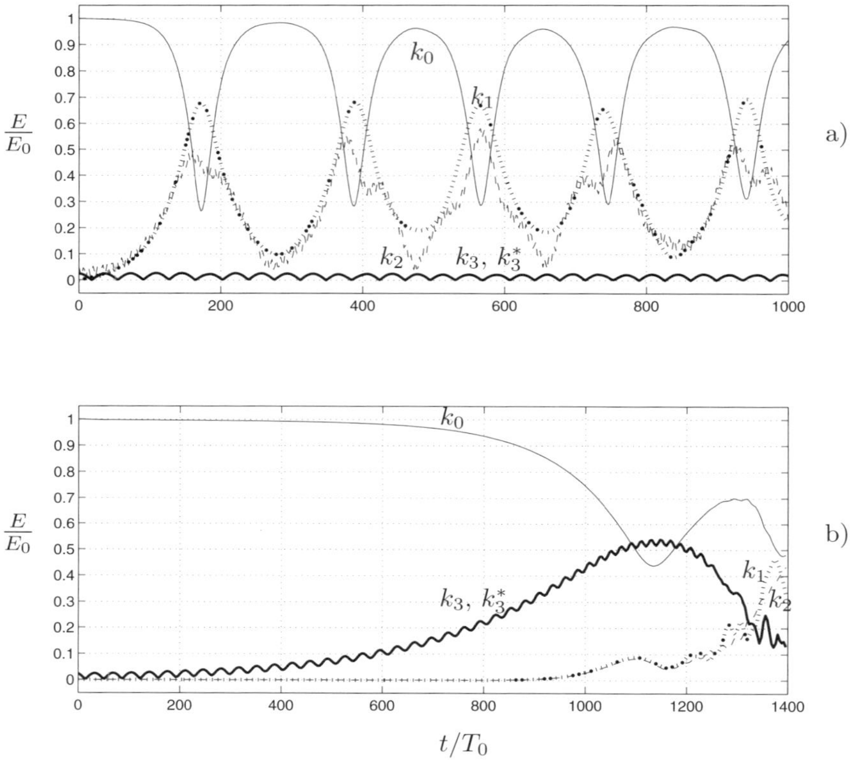

Fig. 7. Energy evolution, E, relative to initial energy, $E_0 = \frac{1}{2}\rho g a_0^2$, of the main modes: $\mathbf{k}_0 = (1, 0)k_0$, $\mathbf{k}_1 = (\frac{5}{6}, 0)k_0$, $\mathbf{k}_2 = (\frac{7}{6}, 0)k_0$, $\mathbf{k}_3 = (\frac{3}{2}, 1.645)k_0$, $\mathbf{k}_3^* = (\frac{3}{2}, -1.645)k_0$. Stokes wave with $(ak)_0 = 0.1$ perturbed initually by a) $\mathbf{k}_1$, $\mathbf{k}_2$, $\mathbf{k}_3$, $\mathbf{k}_3^*$, and b) $\mathbf{k}_3$, $\mathbf{k}_3^*$ only. Adapted from Fructus et al.[28] Reproduced by permission by J. Fluid Mech.

the class II and class I instabilities and the fundamental wave train leads to breaking. An example with the initial unstable perturbation corresponding to the phase-locked crescent-shaped pattern, with wave satellites $\mathbf{k}_3 = (\frac{3}{2}, 1.645)k_0$, $\mathbf{k}_3^* = (\frac{3}{2}, -1.645)k_0$, is visualized in figure 7b.

3.4.3. *Class I instability may trigger class II instability, leading to breaking;* $(ak)_0 > 0.12$.

Numerical simulations taking into account both class I and class II instabilities show that for moderately steep waves, namely $(ak)_0 > 0.12$, their nonlinear coupling (involving the fundamental of the Stokes wave) results in breaking of the wave when in the initial condition only the modulational instability was considered. This result is in agreement with the experiments conducted by Su and Green.[33]

3.4.4. *Class II leading to breaking; $(ak)_0 > 0.17$.*

For steeper waves (in deep water), the strength of class II instability alone is found sufficient to trigger breaking of the wave. It is shown that the nonlinear dynamics of the most unstable class II perturbation lead to breaking when $(ak)_0 > 0.17$. For very steep waves (in deep water), with $(ak)_0 > 0.31$, class II instability dominates over class I instability, being the primary source of breaking of the waves.

3.4.5. *Predominance of class I and class II instabilities. Recurrence versus breaking. Wave slope thresholds*

Computations show that class I and class II instabilities are equally strong for $(ak)_0 = 0.314$, when the water depth is infinite. Below this wave slope the modulational instability is the strongest, while above class II is the strongest. The threshold wave slope becomes reduced when the water depth is finite. For $kh = 1$, the value is $(ak)_0 = 0.1$, for example. The class II instability exhibits recurrence for $(ak)_0 < 0.17$ and breaking above this value (when $h = \infty$), while in finite water depth, for $kh = 1$, recurrence occurs when $(ak)_0$ is less than 0.13, approximately. The thresholds are summarized in figure 8. In general, higher order instabilities in the form of 5-, 6- and higher order wave interactions become more important for

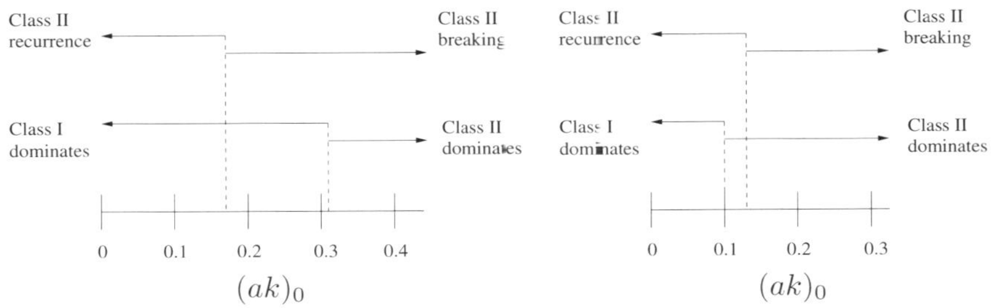

Fig. 8. Predominance of class I and class II instabilities. Recurrence vs. breaking. Infinite depth and $kh = 1$. Adapted from Fructus et al.[28] Reproduced by permission by J. Fluid Mech.

shallower water than in deep water, see also Francius and Kharif,[34] Kristiansen et al.[35]

4. Computations of Extreme Wave Events

4.1. *Effect of sideband instability*

Computations of side-band instability are preformed in the following way: A Stokes wave train of initial amplitude a_0 and wavenumber k_0 propagating along the x_1-direction is multiplied by an amplitude function given by $\mathrm{sech}[\epsilon_1\sqrt{2}a_0k_0^2x_1]$ where in the present computations $(a_0k_0, \epsilon_1) = (0.13, 0.263)$ in figure 9 and (0.1,0.25) in figure 10. The values of a_0k_0 mean that the initial waveslope is relatively moderate, while the values of ϵ_1 mean that the wavegroup will split into several envelope solitons and form local extreme events during the interaction between the group solitons and the inherent effect of the sideband instability.[36] In the case when $\epsilon_1 = 1$ the initial condition represents that of an envelope soliton of permanent shape and no large wave events will occur. Figure 9 shows the three-dimensional calculation of the group on a grid that is 32 wavelengths long and four wavelengths wide, where the wavelength is $\lambda_0 = 2\pi/k_0$. A resolution of 32 nodes per wavelength in each horizontal direction gives totally $1.3 \cdot 10^5$ nodes in the calculation where all nonlinear terms are included. The effect of sideband instability is triggered because of the imperfect shape of the amplitude function. The sideband instability moves energy from the fundamental wavenumber to both longer and shorter waves and is visualized in figure 10b. The sideband instability causes a local extreme wave event in the middle of the wave group after a time corresponding to 70 wave periods in figure 9b and 120 wave periods in figure 10a. The largest events are non-breaking. The wave in figure 10 reaches a maximum in $k\eta_m$ of 0.325 where η_m denotes the maximum elevation above $y = 0$ and $k = 2\pi/\lambda$, where λ is the local trough-to-trough wavelength.

4.2. *Events similar to the Camille and Draupner waves*

Very large waves on the ocean have several causes. Here we discuss large waves resulting from nonlinear focusing and linear superposition. Other causes include waves interacting with currents such as the Aghulas current, and wind. Recent reviews of the phenomenon of rogue waves is provided by Kharif and Pelinovsky,[39] see also Grue and Trulsen.[17]

a)

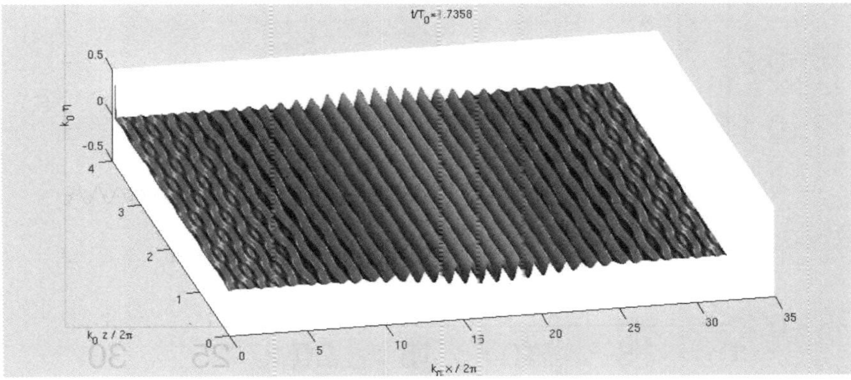

b)

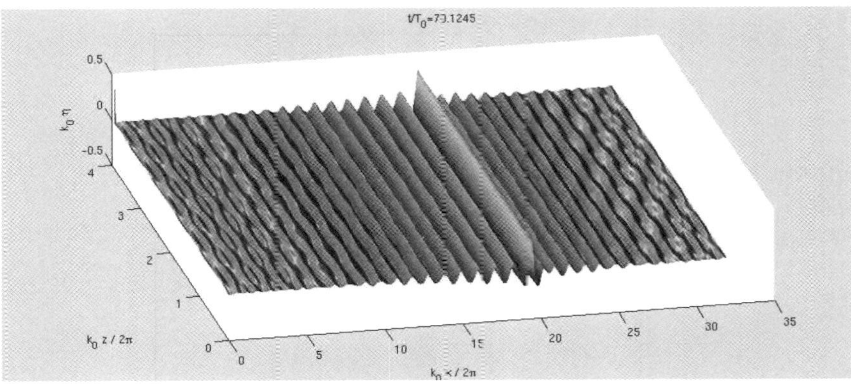

Fig. 9. Formation of an extreme wave event in a modulated long wave group. Initial condition: Stokes wave of amplitude a_0 and wavenumber k_0 multiplied by amplitude function $\mathrm{sech}[\epsilon_1 \sqrt{2} a_0 k_0^2 x_1]$, with $a_0 k_0 = 0.13$, $\epsilon_1 = 0.263$. a) elevation after 1.73 wave periods, and b) after 70 wave periods.

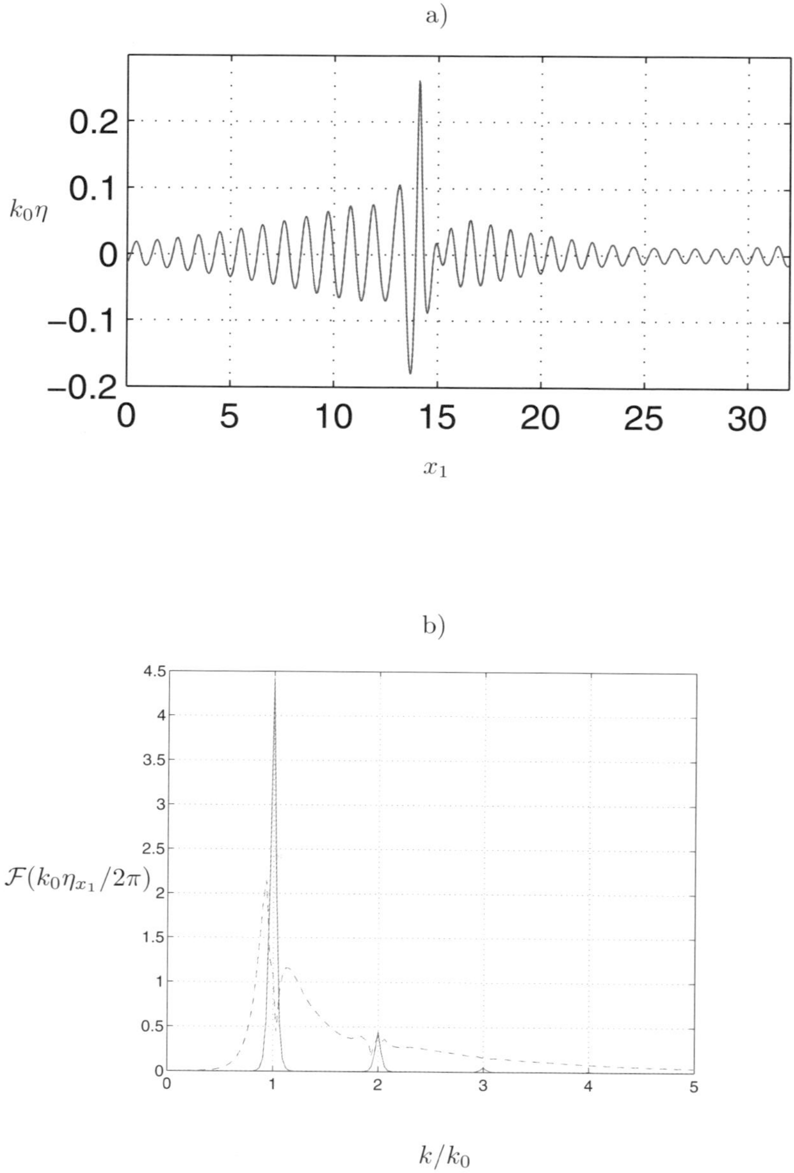

Fig. 10. Same as figure 10, but $a_0k_0 = 0.1$ and $\epsilon_1 = 0.25$. a) elevation after 120 wave periods, and b) wave spectrum, $\mathcal{F}(\eta_{x_1})$, at t=0 (solid line) and after 120 wave periods (dashed line).

In order to bridge numerical modeling and physical recordings we use a scaling of the waves, following the documentation provided by laboratory measurements of several large wave events.[40,41] From experimental wave record at a fixed point, the local trough-to-trough period, T_{TT}, and the maximal elevation above mean water level, η_m, of an individual steep wave event were obtained. The local wavenumber, k, an estimate of the wave slope, ϵ, and a reference velocity were defined by, respectively,

$$\omega^2/gk = 1 + \epsilon^2, \qquad k\eta_m = \epsilon + \frac{1}{2}\epsilon^2 + \frac{1}{2}\epsilon^3, \qquad \epsilon\sqrt{g/k}, \tag{4.1}$$

where $\omega = 2\pi/T_{TT}$.

We compute wave events that match the very large Camille and Draupner waves measured in the field using the scaling (4.1). These very large events at sea are much referred to because of their unique documentation. The time records of the waves provide unique references for how very large ocean waves look like, and for how large and steep they may become. The time series of the Camille and Draupner waves may be found in Ochi[37] figure 8, p. 218, and Trulsen,[38] respectively. Estimates using the scaling (4.1) gives $\epsilon \simeq 0.38$, $k\eta_m \simeq 0.49$ for the Camille wave and $\epsilon \simeq 0.39$, $k\eta_m \simeq 0.49$ for the Draupner wave, and is about 50 per cent above the computations shown in figures 9–10. The non-dimensional wave phase velocity estimated from (4.1) becomes $[\omega/k]/\epsilon\sqrt{g/k}] = \sqrt{1+\epsilon^2}/\epsilon \simeq 2.75$, and $u/[\omega/k] \simeq 0.67$, for the Draupner wave.

4.2.1. *Non-breaking wave events*

Large waves developing from a perturbed Stokes wave train moving in a periodic five wavelength's long numerical tank are computed. While these simulations are classical and follow earlier studies[42,43] less emphasis has been given to the wave kinematics, however. The Stokes waves with initial wave slope a_0k_0 are perturbed by

$$\epsilon_0 a_0 \cos\left(\frac{n+m}{n}k_0x_1\right) + \epsilon_0 a_0 \cos\left(\frac{n-m}{n}k_0x_1\right) \tag{4.2}$$

where the value of ϵ_0 is put to 0.1, and $n = 5$, $m = 1$. The latter means that pertubations are initiated at wavenumbers 0.8 and 1.2 times k_0. The first simulation with $a_0k_0 = 0.11$ develops recurrence and does not break.[43] Large events develop, as illustrated in figure 11a, with trough-to-trough wavenumber, maximal elevation and value of ϵ extracted using (4.1), giving that $\epsilon = 0.32$, $k\eta_m = 0.39$, u/c up to 0.45 after 83 wave periods (the largest event), while $\epsilon = 0.29$, $k\eta_m = 0.34$, u/c up to 0.39 after 83.5 wave periods, and $\epsilon = 0.23$ after 84.5 wave periods.

Baldock et al.[44] developed non-breaking events in physical wave tank using linear focusing technique (superposition), measuring both surface elevation and wave induced velocities using Laser Doppler Anemometry. A recalculation using (4.1) gives $\epsilon = 0.29$ and $k\eta_m = 0.34$ for their largest wave event. We have evaluated the wave induced velocity profile below crest of the large events shown in figures 11a and c. The velocity profiles for $\epsilon = 0.29$, divided by the reference velocity $\epsilon\sqrt{g/k}$, are plotted in figure 12a together with experimental data, also with $\epsilon = 0.29$. The figure illustrates that the non-dimensional velocity profile below crest is almost the same for the three large wave events, even though they have rather different origin, being, in the theoretical computations, the result of nonlinear focusing in a short wave group (figure 11a) and long wave group (figure 11c), and linear superposition in experiment.[44] We may conclude that gravity waves with $\epsilon < 0.32$, $k\eta_m < 0.39$ and $u/c < 0.45$ are non-breaking, while waves with $\epsilon > 0.32$, $k\eta_m > 0.39$ and $u/c > 0.45$ develop breaking.

4.2.2. *Very large wave events; waves developing breaking*

In the next set of simulations we put $a_0k_0 = 0.1125$ in the initial perturbation (4.2) leading to breaking.[43] Resulting waves exhibit $\epsilon = 0.32$, $k\eta_m = 0.39$ after 82.9 wave periods; $\epsilon = 0.36$, $k\eta_m = 0.45$ after 85.2 periods; and $\epsilon = 0.38$, $k\eta_m = 0.49$ after 85.4 periods (figure 11b). The numerical waves develop steepening towards breaking for subsequent time and the computations are stopped. Computational velocity profile below crest for the largest wave with $\epsilon = 0.38$ compares well with measured velocity profiles of the six largest among 122 wave events[40,41] with value of ϵ in the range $0.40 - 0.46$, obtained using Particle Image Velocimetry (PIV), see figure 12b. There is a strong match between computation and experiment and is true even though the large computational wave was produced by nonlinear focusing resulting from side band instability, while the experimental waves were produced by linear superposition.

Computational velocity vectors along the free surface obtained for the two largest waves with $\epsilon = 0.36, 0.38$ (figure 13a) are compared to experimental velocity vectors in a non-overturning event with $\epsilon = 0.40$ (figure 13b) and a moderately overturning event with $\epsilon = 0.41$ (figure 13c). The computational and experimental velocity vectors have the following main features: the maximal horizontal velocities have non-dimensional values up

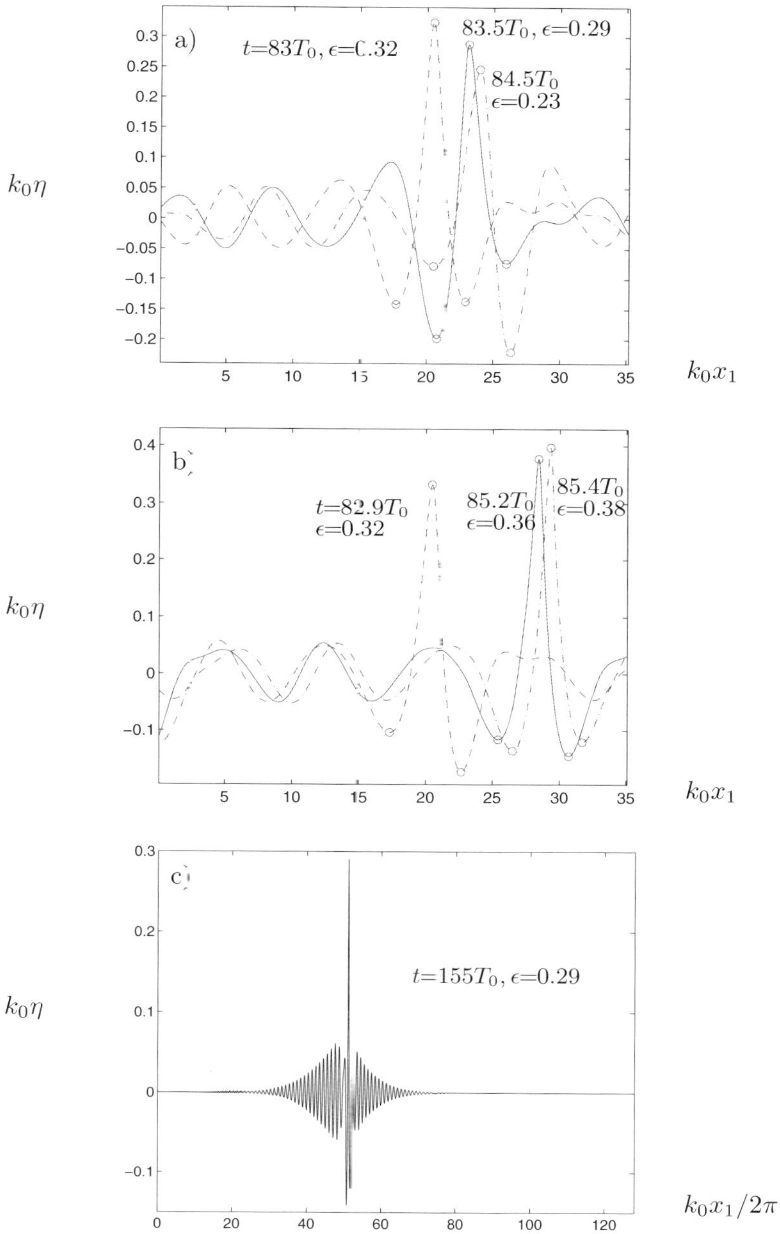

Fig. 11. Large wave events in periodic wave tank. a) Non-Breaking case. b) Breaking case. c) Non-Breaking large wave event during the split-up of a long wave group.[36]

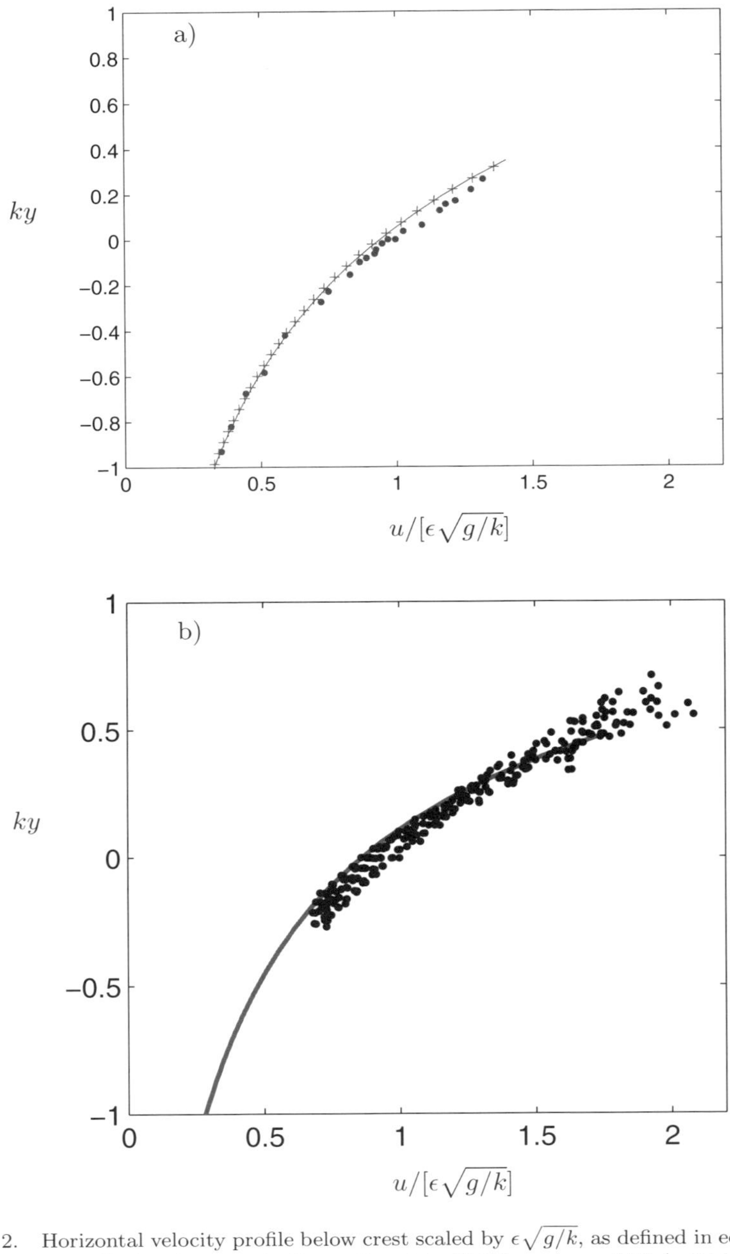

Fig. 12. Horizontal velocity profile below crest scaled by $\epsilon\sqrt{g/k}$, as defined in eq. (4.1). a) Velocity profile for the wave in figure 11a at t=83.5T_0 and $\epsilon = 0.29$ (solid line), for the wave in figure 11c and $\epsilon = 0.29$ (+ + +), experiments by Baldock et al.[44] (dots). b) Velocity profile for the wave in figure 11b at t=85.4T_0 and $\epsilon = 0.38$ (solid line) and data from experiments[40,41] and $\epsilon = 0.40 - 46$ (dots).

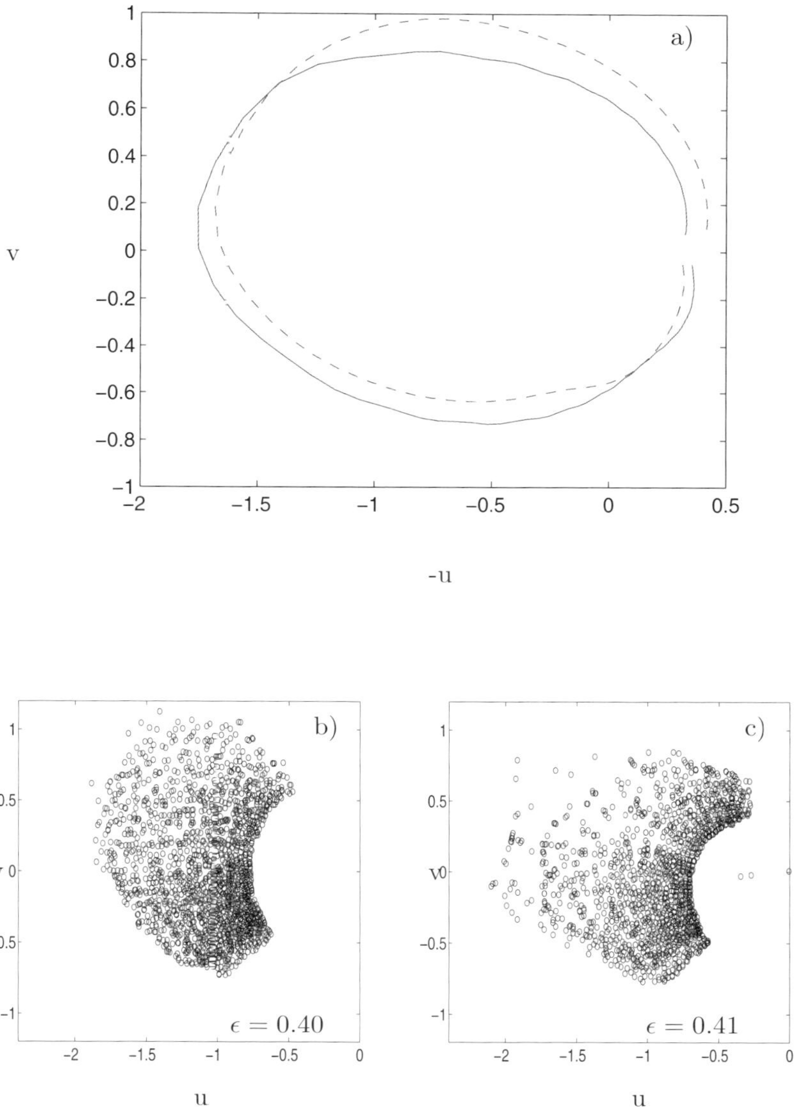

Fig. 13. Non-dimensional velocity plane vectors $(\mathrm{u}, \mathrm{v}) = (u, v)/\epsilon\sqrt{g/k}$. a) Computed velocity vectors along the surface of the wave, with $\epsilon = 0.36$ (dashed line) and $\epsilon = 0.38$ (solid line). b)–c): Experimental velocity plane vectors with $\epsilon = 0.40$ (non-breaking) and $\epsilon = 0.41$ (breaking).[41]

to, about, 1.75 for $\epsilon = 0.38$; 1.85 for $\epsilon = 0.40$; and 2 for $\epsilon = 0.41$ (overturning). (The real velocity is obtained multiplying by $\epsilon\sqrt{g/k}$.) The velocity plot is asymmetric with larger velocities in the front part of the wave than in the rear. When the wave is about to break, the horizontal velocities exhibit increase, while the vertical velocities become reduced. The results shown here, with ϵ in the range about 0.38–0.4, and $k\eta_m$ up to about 0.49, document that the Camille and Draupner waves eventually developed into breaking.

The non-dimensional wave phase velocity is estimated using (4.1) giving $[\omega/k]/[\epsilon\sqrt{g/k}] = \sqrt{1+\epsilon^2}/\epsilon$. The numerical values of this quantity are 2.8 for $\epsilon = 0.38$ (figure 13a); 2.6 for $\epsilon = 0.40$ (figure 13b); and 2.5 for $\epsilon = 0.41$ (figure 13c). It can be concluded that the fluid velocity is significantly smaller than the wave phase velocity in the examples discussed.

5. Generalization to Variable Bottom Topography

We now extend the fully nonlinear and fully dispersive method accounting for a sea floor that is varying in both space and time, following derivations by Fructus and Grue.[45] The sea floor is represented by the function $y = -h + \delta(\mathbf{x}, t)$ where h is constant. A scaled normal velocity of the moving sea bottom is introduced by $V_b = \partial\delta/\partial t = \partial\phi/\partial n\sqrt{1 + |\nabla\delta|^2}$ where the normal vector at the sea floor points into the fluid.

The prognostic equation (2.6) and the time integration procedures discussed below (2.6) are used to integrate η and $\widetilde{\phi}$ forward in time once the (scaled) normal velocity at the free surface, V, and the velocity potential at the bottom topography, ϕ_b are found, see below. At each time step V has to be evaluated, given η, $\widetilde{\phi}$ and the shape and motion of the sea bottom. This implies the solution of a coupled set of integral equations for V and ϕ_b.

5.1. *Laplace equation solver. Field point on free surface*

Solution of the Laplace equation is obtained by applying Green's theorem to the velocity potential ϕ and a suitable source Green function. For an evaluation point that is on the free surface we obtain

$$\int_{S+B} \left(\frac{1}{r} + \frac{1}{r_{\mathrm{B}}}\right) \frac{\partial\phi'}{\partial n_1'}\, \mathrm{d}S' = 2\pi\phi + \int_{S+B} \phi' \frac{\partial}{\partial n_1'} \left(\frac{1}{r} + \frac{1}{r_{\mathrm{B}}}\right) \mathrm{d}S', \quad (5.1)$$

where $r^2 = R^2 + (y' - y)^2$, $r_{\mathrm{B}}^2 = R^2 + (y' + y + 2h)^2$, $\mathbf{R} = \mathbf{x}' - \mathbf{x}$, and $R = |\mathbf{x}' - \mathbf{x}|$. Further, S denotes the instantaneous free surface and B

the instantaneous sea bottom. The normal n_1 is pointing out of the fluid ($n = n_1$ at the free surface, and $n = -n_1$ at the sea-floor). We introduce $\mathrm{d}S = \sqrt{1+|\nabla\eta|^2}\mathrm{d}\mathbf{x}$ at the free surface S and $\mathrm{d}S = \sqrt{1+|\nabla\delta|^2}\mathrm{d}\mathbf{x}$ at the sea bottom B. For notation we use (as before) $\eta' = \eta(\mathbf{x}', t)$, $\eta = \eta(\mathbf{x}, t)$, $\delta' = \delta(\mathbf{x}', t)$, $\delta = \delta(\mathbf{x}, t)$ and so on. The integral equation may be expressed on the form

$$\int_S \left(\frac{1}{r} + \frac{1}{r_{\mathrm{B}}}\right) V' \, \mathrm{d}\mathbf{x}' = 2\pi\widetilde{\phi} + \int_S \widetilde{\phi}' \frac{\partial}{\partial n'} \left(\frac{1}{r} + \frac{1}{r_{\mathrm{B}}}\right) \sqrt{1+|\nabla'\eta'|^2} \, \mathrm{d}\mathbf{x}'$$
$$+ \int_B \left(\frac{1}{r} + \frac{1}{r_{\mathrm{B}}}\right) V_b' \, \mathrm{d}\mathbf{x}' - \int_B \phi_b' \frac{\partial}{\partial n'} \left(\frac{1}{r} + \frac{1}{r_{\mathrm{B}}}\right) \sqrt{1+|\nabla'\delta'|^2} \, \mathrm{d}\mathbf{x}', \tag{5.2}$$

which is valid on the free surface at $y = \eta(\mathbf{x}, t)$. We evaluate the terms

$$\int_B \left(\frac{1}{r} + \frac{1}{r_{\mathrm{B}}}\right) V_b' \, \mathrm{d}\mathbf{x}', \quad - \int_B \phi_b' \frac{\partial}{\partial n'} \left(\frac{1}{r} + \frac{1}{r_{\mathrm{B}}}\right) \sqrt{1+|\nabla'\delta'|^2} \, \mathrm{d}\mathbf{x}'. \tag{5.3}$$

We note that

$$\frac{1}{r} + \frac{1}{r_{\mathrm{B}}} = \frac{2}{R_2} + 2\eta \frac{\partial}{\partial h} \frac{1}{R_2} + (\delta'^2 + \eta^2) \frac{\partial^2}{\partial h^2} \frac{1}{R_2} + \frac{1}{\mathcal{R}_1}, \tag{5.4}$$

where $R_2^2 = R^2 + h^2$ (and $R = |\mathbf{x}' - \mathbf{x}|$). The first three terms on the right of (5.4) are the leading terms of the sources on the left of the equation. The remaining part is defined by the difference, i.e.

$$\frac{1}{\mathcal{R}_1} = \frac{1}{r} + \frac{1}{r_{\mathrm{B}}} - \frac{2}{R_2} - 2\eta \frac{\partial}{\partial h} \frac{1}{R_2} - (\delta'^2 + \eta^2) \frac{\partial^2}{\partial h^2} \frac{1}{R_2},$$

and is a small quantity, decaying rapidly with R. Now, we use that

$$\frac{\partial^n}{\partial h^n} \frac{1}{R_2} = \mathcal{F}^{-1} \left\{ \frac{2\pi(-k)^n}{k} e^{-\mathrm{i}\mathbf{k}\cdot\mathbf{x}' - kh} \right\}. \tag{5.5}$$

($\mathcal{F}^{-1}$ denotes inverse Fourier transform.) We evaluate the integral, giving

$$\int_B \left(\frac{1}{r} + \frac{1}{r_{\mathrm{B}}}\right) V_b' \mathrm{d}\mathbf{x}' = \mathcal{F}^{-1}\{4\pi e^{-kh} \mathcal{F}(V_b)/k\} - \eta \mathcal{F}^{-1}\{4\pi e^{-kh} \mathcal{F}(V_b)\}$$
$$+ \mathcal{F}^{-1}\{2\pi k e^{-kh} \mathcal{F}(\delta^2 V_b)\} + \eta^2 \mathcal{F}^{-1}\{2\pi k e^{-kh} \mathcal{F}(V_b)\} + \int_B \left(\frac{1}{\mathcal{R}_1}\right) V_b' \, \mathrm{d}\mathbf{x}'. \tag{5.6}$$

To evaluate the second integral in (5.3) we first note that

$$\sqrt{1+|\nabla'\delta'|^2} \frac{\partial}{\partial n'} \left(\frac{1}{r} + \frac{1}{r_{\mathrm{B}}}\right) = \frac{\mathbf{R} \cdot \nabla'\delta' - y' + y}{r^3} + \frac{\mathbf{R} \cdot \nabla'\delta' - y' - y - 2h}{r_{\mathrm{B}}^3},$$

where $y' = -h + \delta'$ on the sea floor and $y = \eta$ on the sea surface. The r.h.s. of the equation may be developed by

$$\sqrt{1+|\nabla'\delta'|^2}\frac{\partial}{\partial n'}\left(\frac{1}{r}+\frac{1}{r_{\rm B}}\right) = -2\nabla'\cdot\left(\delta'\nabla'\frac{1}{R_2}\right) - 2\eta\nabla'\cdot\left(\delta'\nabla'\frac{\partial}{\partial h}\frac{1}{R_2}\right) + \frac{1}{\mathcal{R}_2},$$

where the two first terms on the right give the leading contributions to the left hand side and $1/\mathcal{R}_2$ is a small remainder. We obtain for the integral

$$-\int_B \phi_b{}'\frac{\partial}{\partial n}\left(\frac{1}{r}+\frac{1}{r_{\rm B}}\right)\sqrt{1+|\nabla'\delta'|^2}\,{\rm d}\mathbf{x}' = \mathcal{F}^{-1}\{4\pi{\rm i}e^{-kh}(\mathbf{k}/k)\cdot\mathcal{F}(\delta\nabla\phi_b)\}$$
$$-\eta\mathcal{F}^{-1}\{4\pi{\rm i}e^{-kh}\mathbf{k}\cdot\mathcal{F}(\delta\nabla\phi_b)\} - \int_B\left(\frac{1}{\mathcal{R}_2}\right)\phi_b{}'{\rm d}\mathbf{x}'. \quad (5.7)$$

Adding contributions (5.6) and (5.7) we obtain

$$\mathcal{F}^{-1}\left\{\frac{1+e_h}{k}[\mathcal{F}(V)+{\rm i}\mathbf{k}\cdot\mathcal{F}(\eta\nabla\widetilde{\phi})]\right\} = \mathcal{F}^{-1}\left\{(1-e_h)\mathcal{F}(\widetilde{\phi}-\eta V^{(1)})\right\}$$
$$+\mathcal{F}^{-1}\{e_h\mathcal{F}(\eta(V-V^{(1)}))\} + T(\widetilde{\phi}) + T_1(\widetilde{\phi}) + N(V) + N_1(V)$$
$$+\eta\mathcal{F}^{-1}(e_h\mathcal{F}(V-V^{(1)})) + \mathcal{F}^{-1}\left\{\frac{2\sqrt{e_h}\mathcal{F}(V_b)}{k}\right\}$$
$$-\eta\mathcal{F}^{-1}\{2\sqrt{e_h}\mathcal{F}(V_b)\} + \mathcal{F}^{-1}\{k\sqrt{e_h}\mathcal{F}(\delta^2 V_b)\} + \eta^2\mathcal{F}^{-1}\{k\sqrt{e_h}\mathcal{F}(V_b)\}$$
$$+\mathcal{F}^{-1}\left\{\frac{{\rm i}\mathbf{k}}{k}\cdot 2\sqrt{e_h}\mathcal{F}(\delta\nabla\phi_b)\right\} - \eta\mathcal{F}^{-1}\{2{\rm i}\sqrt{e_h}\mathbf{k}\cdot\mathcal{F}(\delta\nabla\phi_b)\}$$
$$+\int_B\left(\frac{1}{\mathcal{R}_1}\right)V_b{}'\,{\rm d}\mathbf{x}' - \int_B\left(\frac{1}{\mathcal{R}_2}\right)\phi_b{}'{\rm d}\mathbf{x}', \quad (5.8)$$

where $e_h = e^{-2kh}$ and $V^{(1)}$ is given in eq. (5.19) below. The functions $T(\widetilde{\phi})$, $N(V)$, $N_1(V)$ and $T_1(\widetilde{\phi})$ are the same as in eq. (2.17). In (5.8) we have divided by a factor of 2π.

5.2. *Field point on bottom surface*

An additional equation for the velocity potential ϕ_b at the uneven sea bottom is required in order to solve (5.2), and is obtained by the application of Green's theorem as well. In the case when the field point is on the sea bottom, we employ $1/r + 1/r_{\rm C}$ where $r^2 = R^2 + (y'-y)^2$, and $r_{\rm C}^2 = R^2 + (y'+y)^2$ is the image with respect with $y = 0$. The resulting equation becomes

$$2\pi\phi_b = \int_S\left(\frac{1}{r}+\frac{1}{r_{\rm C}}\right)V'\,{\rm d}\mathbf{x}' - \int_S\widetilde{\phi}'\frac{\partial}{\partial n}\left(\frac{1}{r}+\frac{1}{r_{\rm C}}\right)\sqrt{1+|\nabla'\eta'|^2}\,{\rm d}\mathbf{x}'$$
$$-\int_B\left(\frac{1}{r}+\frac{1}{r_{\rm C}}\right)V_b{}'\,{\rm d}\mathbf{x}' + \int_B\phi_b{}'\frac{\partial}{\partial n}\left(\frac{1}{r}+\frac{1}{r_{\rm C}}\right)\sqrt{1+|\nabla'\delta'|^2}\,{\rm d}\mathbf{x}', \quad (5.9)$$

and is valid for $y = -h + \delta(\mathbf{x}, t)$. For the integration over B we note that

$$\frac{1}{r} + \frac{1}{r_C} = \frac{1}{R} + \frac{1}{R_1} - (\delta' + \delta)\frac{\partial}{\partial(2h)}\frac{1}{R_1} + \frac{1}{\mathcal{R}_3}, \tag{5.10}$$

where the first three terms on the right capture the leading behaviour of the function on the left, $R_1^2 = R^2 + (2h)^2$, and $1/\mathcal{R}_3$ is a small remainder. Further,

$$\sqrt{1+|\nabla'\delta'|^2}\frac{\partial}{\partial n'}\left(\frac{1}{r} - \frac{1}{r_C}\right)$$
$$= -\nabla' \cdot \left((\delta' - \delta)\nabla'\frac{1}{R}\right) - \nabla' \cdot \left((\delta' + \delta)\nabla'\frac{1}{R_1}\right) - \frac{\partial}{\partial(2h)}\frac{1}{R_1} + \frac{1}{\mathcal{R}_4}. \tag{5.11}$$

For the integration over S we obtain

$$\frac{1}{r} + \frac{1}{r_C} = \frac{2}{R_2} - 2\delta\frac{\partial}{\partial h}\frac{1}{R_2} + \frac{1}{\mathcal{R}_5}, \tag{5.12}$$

$$\sqrt{1+|\nabla'\eta'|^2}\frac{\partial}{\partial n'}\left(\frac{1}{r} + \frac{1}{r_C}\right) = -\nabla' \cdot \left(2\eta'\nabla'\frac{1}{R_2}\right) + \frac{1}{\mathcal{R}_6}. \tag{5.13}$$

Now, using (5.5) and

$$\frac{1}{R} = \mathcal{F}^{-1}\left\{\frac{2\pi}{k}e^{-i\mathbf{k}\cdot\mathbf{x}'}\right\}, \qquad \frac{1}{R_1} = \mathcal{F}^{-1}\left\{\frac{2\pi}{k}e^{-i\mathbf{k}\cdot\mathbf{x}'-2kh}\right\}, \tag{5.14}$$

we may evaluate the integrals in (5.9). The equation becomes:

$$2\pi\phi_b = 2\pi\mathcal{F}^{-1}(\mathcal{A})$$
$$+\int_S \left(\frac{1}{\mathcal{R}_5}\right) V\mathrm{d}\mathbf{x}' - \int_B \left(\frac{1}{\mathcal{R}_3}\right) V_b\mathrm{d}\mathbf{x}' - \int_S \left(\frac{1}{\mathcal{R}_6}\right) \widetilde{\phi}\mathrm{d}\mathbf{x}' + \int_B \left(\frac{1}{\mathcal{R}_4}\right) V\mathrm{d}\mathbf{x}' \tag{5.15}$$

where

$$\mathcal{A} = \frac{2\sqrt{e_h}\mathcal{F}(V)}{k} + 2\delta\sqrt{e_h}\mathcal{F}(V) - \frac{(1+e_h)\mathcal{F}(V_b)}{k} - \delta e_h\mathcal{F}(V_b) - e_h\mathcal{F}(\delta V_b)$$
$$+\frac{2i\mathbf{k}}{k}\cdot\sqrt{e_h}\mathcal{F}(\eta\nabla\widetilde{\phi}) + e_h\mathcal{F}(\phi_b) - \delta(1-e_h)k\mathcal{F}(\phi_b) - \frac{i\mathbf{k}}{k}\cdot(1+e_h)\mathcal{F}(\delta\nabla\phi_b). \tag{5.16}$$

By taking the Fourier transform on both sides of (5.15) we obtain

$$\mathcal{F}(\phi_b) = \mathcal{F}({\phi_b}^{(1)}) + \mathcal{F}(\delta V_b) + \frac{\mathbf{i}\mathbf{k}\cdot\mathcal{F}(\eta\nabla\widetilde{\phi})}{k\sinh kh} - \frac{\mathbf{i}\mathbf{k}\cdot\mathcal{F}(\delta\nabla\phi_b)}{k\tanh kh}$$
$$-\frac{1}{1-e_h}\mathcal{F}(\delta\mathcal{F}^{-1}\{k(1-e_h)[\mathcal{F}(\phi_b) - \mathcal{F}({\phi_b}^{(1)})]\})$$
$$+\int_S\left(\frac{1}{\mathcal{R}_5}\right)V\mathrm{d}\mathbf{x}' - \int_B\left(\frac{1}{\mathcal{R}_3}\right)V_b\mathrm{d}\mathbf{x}' - \int_S\left(\frac{1}{\mathcal{R}_6}\right)\widetilde{\phi}\mathrm{d}\mathbf{x}' + \int_B\left(\frac{1}{\mathcal{R}_4}\right)V\mathrm{d}\mathbf{x}', \tag{5.17}$$

where

$$\mathcal{F}({\phi_b}^{(1)}) = \frac{\mathcal{F}(V)}{k\sinh kh} - \frac{\mathcal{F}(V_b)}{k\tanh kh} = \frac{\mathcal{F}(\widetilde{\phi})}{\cosh kh} - \frac{\tanh kh\mathcal{F}(V_b)}{k}. \tag{5.18}$$

Similarly,

$$\mathcal{F}(V^{(1)}) = k\tanh hk\mathcal{F}(\widetilde{\phi}) + \frac{\mathcal{F}(V_b)}{\cosh hk}. \tag{5.19}$$

The linear approximation (5.18)–(5.19) is a variant of previous solutions.[46]

5.3. *Successive approximations*

Equations (5.18)–(5.19) represent a first step in a series of successive approximations to the full solution. An improved approximation to the nonlinear wave field is obtained by keeping terms that are linear in η, η', δ and δ', giving

$$\mathcal{F}(V^{(2)}) = \mathcal{F}(V^{(1)}) - k\tanh hk\mathcal{F}(\eta V^{(1)}) + \mathrm{i}\mathbf{k}\cdot\mathcal{F}\{\eta\nabla\widetilde{\phi}\} + \frac{\mathrm{i}\mathbf{k}\cdot\mathcal{F}\{\delta\nabla{\phi_b}^{(1)}\}}{\cosh hk}, \tag{5.20}$$

$$\mathcal{F}({\phi_b}^{(2)}) = \mathcal{F}({\phi_b}^{(1)}) - \frac{\mathcal{F}(\eta V^{(1)})}{\cosh hk} - \frac{\mathrm{i}\tanh hk}{h}\mathbf{k}\cdot\mathcal{F}\{\delta\nabla{\phi_b}^{(1)}\} + \mathcal{F}(\delta V_b), \tag{5.21}$$

where ${\phi_b}^{(1)}$ and $V^{(1)}$ are obtained from (5.18)–(5.19). By performing another analytical iteration, keeping terms in the kernels of the integral equations that are quadratic in η, η', δ and δ', we obtain

$$\mathcal{F}(V^{(3)}) = \mathcal{F}(V^{(2)}) - [k \tanh hk\mathcal{F}(\eta(V^{(2)} - V^{(1)})) - \frac{k^2}{2}\mathcal{F}(\eta^2 V^{(1)})]$$
$$+\frac{k}{1+e_h}\left\{\mathcal{F}(\eta\mathcal{F}^{-1}[-\mathrm{i}\mathbf{k}\cdot\mathcal{F}\{\eta\nabla\widetilde{\phi}\}]) + \mathrm{i}\frac{e_h}{2}\mathbf{k}\cdot\mathcal{F}\{\eta^2\nabla\widetilde{\phi}\} - \mathcal{F}\{\eta^2\mathcal{F}^{-1}[\frac{k^2}{2}\widetilde{\phi}]\}\right\}$$
$$+\frac{k}{1+e_h}\left\{k\sqrt{e_h}\mathcal{F}\{\delta^2 V_b\} + 2\mathrm{i}\frac{\sqrt{e_h}}{k}\mathbf{k}\cdot\mathcal{F}\{\delta\nabla({\phi_b}^{(2)} - {\phi_b}^{(1)})\}\right\}, \qquad (5.22)$$

where ${\phi_b}^{(1)}$ and $V^{(1)}$ are obtained from (5.18)–(5.19), $V^{(2)}$ and ${\phi_b}^{(2)}$ from (5.20)–(5.21) and $e_h = \exp(-2kh)$. The equations for $V^{(1)}$, $V^{(2)}$, $V^{(3)}$ are valid on the exact position of the free surface. Similary, the potential and normal velocity at the sea floor are evaluated on its exact position.

In Fructus and Grue[45] they studied the full expression for V and computed the difference in the resulting wave field using the full V and the approximation provided by $V^{(3)}$. For very strong tsunami waves propagating over long distance, they found (in that investigation) that the relative error in amplitude and phase at maximum was 0.0014 and 0.07×10^{-4}, respectively, and thus that $V^{(3)}$ represents a most useful approximation for the practical evaluation of highly nonlinear and highly dispersive phenomenons relating to tsunami waves. The formulation is valid up to the physical beach where numerical boundary conditions are applied.[14]

6. Short Wave Formation by Tsunami in Shallow Water

The natural disaster caused by the Indian Ocean tsunami on December 26, 2004 led to large number of recent publications on tectonics, earthquakes, tectonically generated water waves and wave run-up along the coast line.[47–50] We shall here not model the tsunami itself but rather simulate the motion that was observed when the long tsunami ran into the Strait of Malacca, generating undular bores in the shallow sea. We use equations (2.6) and (5.22) to model the nonlinear-dispersive process taking place during the formation of the short waves. The input wave to the simulation starting at the entrance of the Strait corresponds to a model simulation of the Indian Ocean tsunami.

The computations presented here represent a short summary of a fuller investigation where fully nonlinear-dispersive computations and a variant of the KdV equation were compared.[51] We perform a modeling along a 2D section that is midways between the Sumatra Island and Malaysian Peninsula. The depth profile of the sea ranges from 160 m at the starting

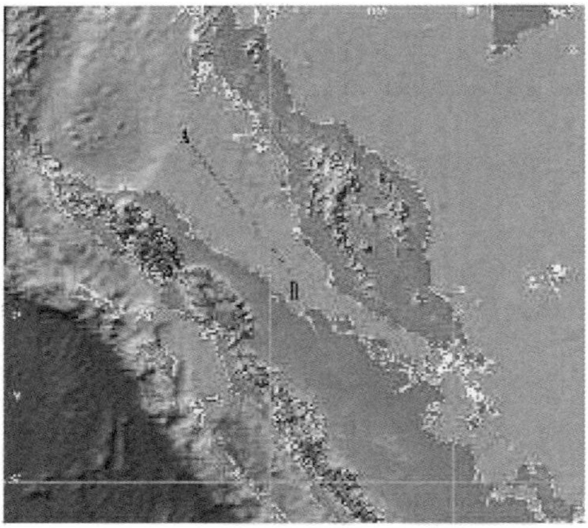

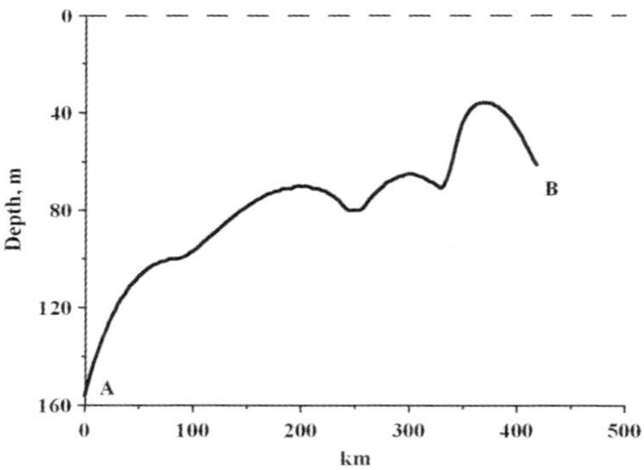

Fig. 14. Upper: computational section A–B in the Strait of Malacca. Lower: depth profile along section.

position and reduces to 80 m after 200 km, and to 37 m after 375 km (figure 14). The input wave has an initial depression period of about 1500 s, a maximal depression of -2.4 m and a subsequent elevation of 2.8 m, see figure 16a. The numerical tank is 1430 km long. Two resolutions of 20 m and 40 m were used, meaning up to $1.4 \cdot 10^5$ nodes, since there are always a minimum of two nodes in the lateral direction. An estimated wavelength of the train of short waves is about 400 m (see below) giving (about) 20 nodes per wavelength with the fine resolution and 10 with coarse, both producing same result.

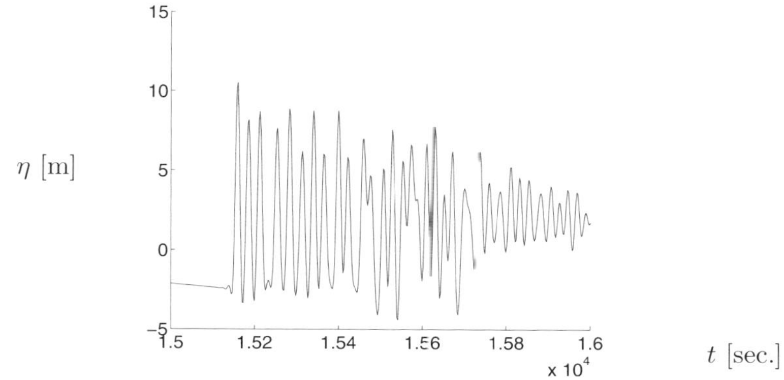

Fig. 15. Elevation vs. time at position 375 km from entrance of the Strait. Local water depth 37 m.

The back face of the leading depression wave undergoes a significant steepening during the propagation in the shallow strait, where the value of the vertical velocity has a local maximum of 0.008 m/s at the entrance, and reaches a local maximum of 0.4 m/s at the position at 250 km. A train of short waves then develops in the subsequent motion (figure 16c). The crest to crest period is in the beginning about half a minute, corresponding to peak frequency of $\omega_p = 10^{-0.631}$ s^{-1} in figure 16b, at 250 km.

The energy supply to the short waves comes from the steep back of the long, leading depression wave. In the beginning, the group of short waves has very small amplitude and behaves like a linear dispersive wave train. This fact is illustrated by plotting $\partial\eta/\partial t$ as function of time (figure 16d). A close examination of the train of short waves supports that the local wave length during the generation phase is governed by the linear dispersion relation $\omega^2 = gk \tanh kh$, h local depth. While the front of the train of short waves moves with the shallow water speed, $\sqrt{gh}$, the tail moves with the slower group velocity $\partial\omega/\partial k$. The train extends behind the steep front since $\sqrt{gh} > \partial\omega/\partial k$.

At the third bottom peak at 375 km and depth of 37 m, nondimensional wavenumber increases to $kh \simeq 0.6$ ($2\pi/k \simeq 400$ m) and wave height to depth ratio to 0.41, and the motion falls outside the range of validity of weakly nonlinear-dispersive theory. The train of short waves continues to grow in length and strength until the local wave height is about twice that of the incoming wave. The wave train develops into a sequence of solitary waves that propagate along the leading depression wave (figure 15). The

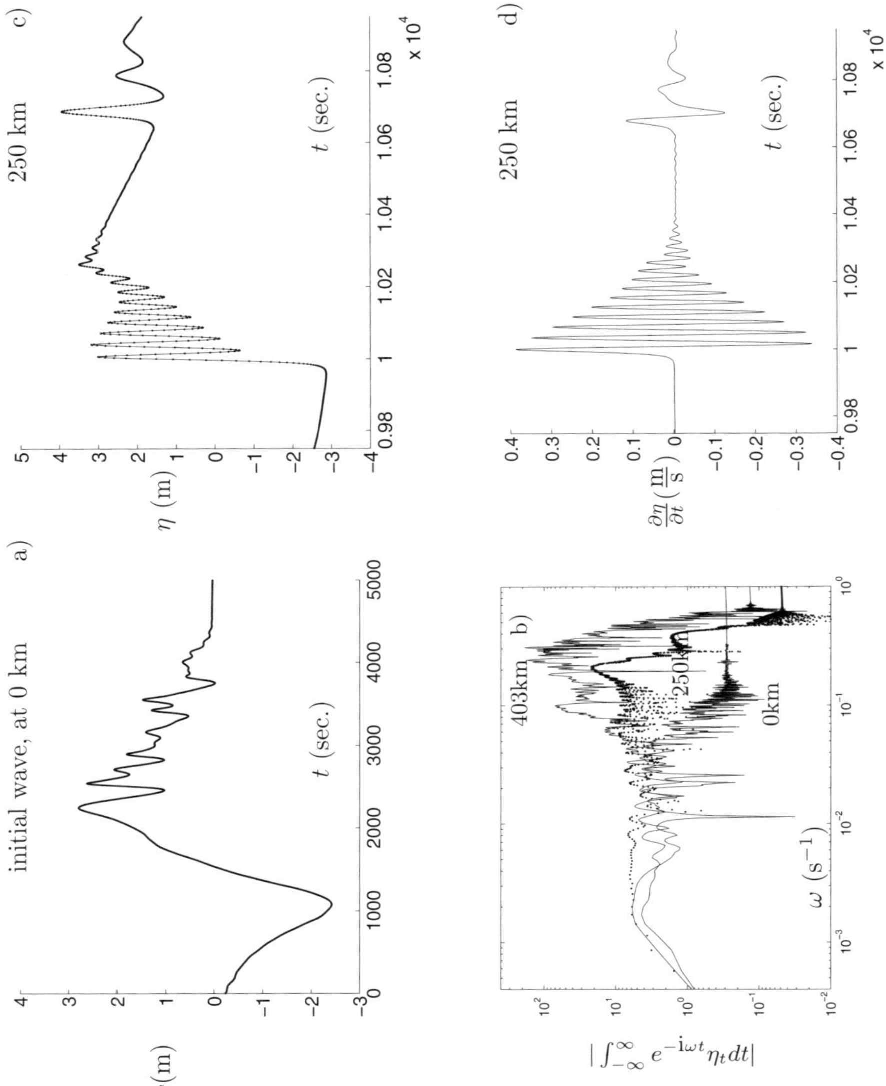

Fig. 16. a) Time history of surface elevation at the entrance of the Strait. b) Up-shift of energy by fully nonlinear-dispersive computations: $|\int_{-\infty}^{\infty} e^{-i\omega t}\eta_t dt|$ vs. ω. for 0 km (initial), 250 km, 403 km. c) Time history of η at 250 km and d) of $\partial\eta/\partial t$ at 250 km.

transfer of energy from long to short modes shows a peak frequency at the position at 403 km of $\omega_p = 10^{-0.514}$ s^{-1}, corresponding to a wave period of 20.5 s (figure 16b). The motion due to an input wave of half amplitude exhibits a similar energy transfer taking place at the third bottom ridge. Finally, simulations using the fully nonlinear-dispersive formulation and a KdV model both support the formation of solitons in the shallow Strait of Malacca due to a disaster wave such as the Indian Ocean tsunami, but show that KdV is inferior in representing the short waves.[51]

7. Concluding Remarks

We have presented the mathematical description and numerical implementation of a fully nonlinear and dispersive model in three-dimensions. The essential parts of the method include a trimmed time-integration procedure of the free surface variables, the elevation, η, and wave potential at the free surface, $\widetilde{\phi}$, where the quantities are evaluated in spectral space. In this integration, the linear contribution is obtained analytically, corresponding to one long time-step using the initial condition as input. The genuinely nonlinear motion is stepped forward in time using a RK-54 method with auto-adaptive time-step selection. Moreover, an anti-alising strategy has been implemented for multiplications up to cubic nonlinearities, where the spectra of the variables are doubled and energy at the highest half of the wavenumber nodes padded to zero.

The other important part is the analytical inversion of the Laplace equation expressing the normal velocity at the free surface in terms of η and $\widetilde{\phi}$. This defines an analytical iteration method to obtain the fully nonlinear solution, with rapid convergence. The numerical counterpart of the method provides a highly efficient computational strategy where the dominant part of the evaluations are performed using FFT. A remaining part of the Laplace equation solver is obtained by highly nonlinear and local integrals where the integrands decay rapidly and are evaluated by ordinary integration over squares in the horizonal plane, with sides one characteristic wavelength. The method has been fully numerically implemented, convergence properties documented,[13] with wave generation and absorption procedures embedded.[14] Numerical integration of the evolution of long wave fields over long time using the two-dimensional counterpart of the method[16] showed excellent comparison with the high-order spectral method by West et al.,[8] but that the formulation by Dommermuth and Yue[9] was inaccurate at large times of the simulation, due to poor representation of the vertical velocity at the free surface. The present method, particularly the pure FFT-part, is faster than other related methods.

Various wave phenomenas are computed, like formation of three-dimensional surface wave patterns, very large wave events caused by linear and nonlinear focusing, and short wave formation due to a long tsunami propagating into shallow water. The computations are illustrated in this chapter. We point to particularly three highlights. Firstly, a main objective with the present computations of three-dimensional patterns was to model the relatively obscure L_1-pattern – the oscillating horse shoes – observed by Collard and Caulliez[27] in a wave tank, which, until recently, was not given a theoretical explanation.[28,30] A reason why they observed the pattern was, first, that they suppressed the class I instability – the sideband instability – by putting in the wave tank a thin sheet of plastic-film on the water surface, and, secondly, the L_1-pattern was triggered because of parametric resonance taking place in the lateral motion in the wave tank, determining the value of the lateral wavenumber, qk_0. This L_1-pattern should be observed in coastal waters where the modulational instability becomes relatively weaker. For steep waves in deep water the strength of class II instability alone is sufficient to trigger breaking of the wave. In water of moderate depth, with $kh = 1$, class II instability dominates class I instability when $(ak)_0 > 0.1$, exhibits recurrence for $0.1 < (ak)_0 < 0.13$, and class II breaking for $(ak)_0 > 0.13$.

A second main result is the recomputation and experimental investigation of extremely large wave events corresponding to the largest waves that are documented at sea, namely the Camille and Draupner waves. Both are characterized by a maximal elevation above mean sea level, η_m, times an estimated local wavenumber of the event, k, with product $k\eta_m = 0.49$, for the waves in the field, where the procedure of how to obtain the local wavenumber is detailed in the text prior to equation (4.1). Recomputations of the high waves compare favorably with available sets of laboratory measurements using Laser Doppler Anemometry and Particle Image Velocimetry. Based on the simulations presented here one may conclude that the local properties of waves such as the large Camille and Draupner waves share a common kinematics and is obtained by simulation and experiment. Moreover, the results indicate that both the Camille and Draupner waves eventually developed into breaking at a later stage loosing a substantial fraction of momentum and energy. (Waves with $k\eta_m < 0.39$ and $u/c < 0.45$ are recurrent and do not break.)

The third important point relates to the extension of the mathematical model and numerical implementation taking into account the effect of a general variable bottom in space and time. This expresses the free surface normal velocity and bottom potential in an FFT fashion of the involved variables. The formulation has an unexplored potential for further use in computation, application and interpretation.

The method requires $\mathcal{O}(N \log N)$ operations per time-step, N the number of computational nodes. Computations so far are carried out on lap-tops using a typical resolution of 32 nodes per characteristic wavelength in both directions, which means 10^3 nodes per wavelength squared. In the largest computation of formation of extreme events, using the full method, a wave tank of 32 by 4 wavelengths was used, corresponding to a total number of nodes of $N = 1.3 \cdot 10^5$. In the computations of the short wave formation due to the tsunamis, the number of nodes was $N = 1.4 \cdot 10^5$ in the finest computation, which involved only the rapid FFT-part of the formulation, since in this application the remaing integrals are vanishingly small. In the tsunami-wave application the number of nodes per wavelength was (about) 20 with the fine resolution and 10 with coarse. The wave propagation period was $2 \cdot 10^4$ seconds corresponding to slightly less than 6 hours.

The code is implemented with version FFTW2.5 of the Fourier transform and with parallel architecture that awaits to be tested. An upgrade of the program using FFTW3 will divide the applied number of nodes by a factor of two. We believe that an upgrade of the code for runs on clusters represents a next step. From the application side the code should be used as support for interpretation of field data computing wave fields over areas, at least, 100 by 100 typical wavelengths, or more desireably, 1000 by 1000 wavelengths. The former means a number of $N = 10^7$ nodes while the latter means 10^9 nodes, and is a factor $10^2 - 10^4$ higher than in the computations pursued so far. This should be feasible, given available computer resources.

Acknowledgements

We acknowledge the collaboration with Drs. Didier Clamond and Atle Jensen. The text was written when J.G. was on sabbatical leave at Scripps Institution of Oceanography in La Jolla, San Diego, USA. He expresses his gratitude to Professor W. Kendall Melville and his group for making the visit possible and most beneficial. Development of the method was funded by the Research Council of Norway through: NFR 146526/420 *General analysis of Realistic Ocean Waves (GROW)* 1998-2002,

NFR 121078/420 *Modelling of currents and waves for sea structures* 2002-6, and BeMatA-program *Computational methods for stratified flows involving internal waves.*

References

1. Benjamin, T. B. and Feir, J. E. (1967). The disintegration of wave trains on deep water. *J. Fluid Mech.*, **27**, pp. 417–430.
2. Zakharov, V. E. (1968). Stability of periodic wave of finite amplitude on the surface of a deep fluid. *J. Appl. Mech. Phys., Engl. Transl*, **2**, p. 190.
3. Yuen, H. C. and Lake, B. M. (1982). Nonlinear dynamics of deep-water gravity waves. *Adv. Appl. Mech.*, **22**, pp. 67–229.
4. Tayfun, M. (1980). Narrow-band nonlinear sea waves. *J. Geophys. Res.*, **85**, C3, pp. 1548-1552.
5. Onorato, M., Osborne, A. R. and Serio, M. (2002). Extreme wave events in directional, random oceanic sea states. *Phys. Fluids*, **14**, pp. L25–L28.
6. Dysthe, K. B., Trulsen, K., Krogstad, H. E. and Socquet-Juglard, H. (2003). Evolution of a narrow-band spectrum of random surface gravity waves. *J. Fluid Mech.*, **478**, pp. 1–10.
7. Socquet-Juglard, H., Dysthe, K. B., Trulsen, K., Krogstad, H. E. and Liu, J. (2005). Distribution of surface gravity waves during spectral changes. *J. Fluid Mech.*, **542**, pp. 195–216.
8. West, B. J., Brueckner, K. A., Janda, R. S., Milder, D. M. and Milton, R. L. (1987). A new numerical method for surface hydrodynamics. *J. Geophys. Res.* **92**, 11, pp. 11803–11824.
9. Dommermuth, D. G. and Yue, D. K. P. (1987). A high-order spectral method for the study of nonlinear gravity waves. *J. Fluid Mech.*, **184**, pp. 267–288.
10. Craig, W. and Sulem, C. (1993). Numerical simulation of gravity waves. *J. Comp. Phys.* **108**, pp. 73–83.
11. Clamond, D. and Grue, J. (2001). A fast method for fully nonlinear water-wave computations. *J. Fluid Mech.*, **447**, pp. 337–355.
12. Grue, J. (2002). On four highly nonlinear phenomena in wave theory and marine hydrodynamics. *Appl. Ocean Res.*, **24**, pp. 261–274.
13. Fructus, D., Clamond, D., Grue, J. and Kristiansen, Ø. (2005). An efficient model for three-dimensional surface wave simulations. Part I. Free space problems. *J. Comp. Phys.* **205**, pp. 665–685.
14. Clamond, D., Fructus, D., Grue, J. and Kristiansen, Ø. (2005). An efficient model for three-dimensional surface wave simulations. Part II: Generation and absorption. *J. Comp. Phys.*, **205**, pp. 686–705.
15. Clamond, D., Fructus, D. and Grue, J. (2007). A note on time integrators in water-wave simulations. *J. Eng. Math.*, **58**, pp. 149–156. Special issue in honor of Prof. J. N. Newman.
16. Clamond, D., Francius, M., Grue, J. and Kharif, Ch. (2006). Long time interaction of envelope solitons and freak wave formations. *Eur. J. Mech. B/Fluids*, **25**, pp. 536-553.

17. Grue, J. and Trulsen, K. (eds.) (2006). Waves in Geophysical Fluids; Tsunamis, Rogue Waves, Internal Waves and Internal Tides. CISM lectures, Vol. 489, Springer 332 pp.
18. Lighthill, M. J. (1965). Contributions to the theory of of waves in nonlinear dispersive systems. *J. Fluid Mech.*, **183**, pp. 451–465.
19. Su, M. Y. (1982). Three-dimensional deep water waves. Part 1. Experimental measurements of of skew and symmetric wave patterns. *J. Fluid Mech.*, **124**, pp. 73-108.
20. McLean, J. W., Ma, Y. C., Martin, D. U., Saffman, P. G. and Yuen, H. C. (1981). Three dimensional instability of finite amplitude water waves. *Phys. Rev. Letters*, **46**, pp. 817-820.
21. McLean, J. W. (1982). Instabilities of finite-amplitude water waves. *J. Fluid Mech.*, **114**, pp. 315-330.
22. Longuet-Higgins, M. S. (1978). The instabilities of gravity waves of finite amplitude in deep water. II. Subharmonics. *Proc. R. Soc. Lond. A*, **360**, pp. 489-505.
23. Kharif, C. and Ramamonjiarisoa, A. (1988). Deep-water gravity waves instabilities at large steepness. *Phys. Fluids*, **31**, pp. 1286-1288.
24. Kharif, C. and Ramamonjiarisoa, A. (1990). On the stability of gravity waves on deep water. *J. Fluid Mech.*, **218**, pp. 163-170.
25. Melville, W. K. (1982). The instability and breaking of deep-water waves. *J. Fluid Mech.*, **115**, pp. 165-185.
26. Kusuba, T. and Mitsuyasu, M. (1986). Nonlinear instability and evolution of steep water waves under wind action. *Rep. Res. Inst. Appl. Mech. Kyushu Univ.* **33**, (101), pp. 33-64.
27. Collard, F. and Caulliez, G., 1999. Oscillating crescent-shaped water wave patterns. *Phys. Fluids, Letters*, **11**, pp. 3195–3197.
28. Fructus, D., Kharif, Ch., Francius, M., Kristiansen, Ø, Clamond, D. and Grue, J. (2005). Dynamics of crescent water wave patterns. *J. Fluid Mech.*, **537**, pp. 155-186.
29. Xue, M., Xü, H., Liu Y. M. and Yue, D. K. P. (2001). Computations of fully nonlinear three-dimensional wave-wave and wave-body interactions. Part 1. Dynamics of steep three-dimensional waves. *J. Fluid Mech.*, **438**, pp. 11-39.
30. Fuhrman, D. R., Madsen P. A. and Bingham H. B. (2004). A numerical study of crescent waves. *J. Fluid Mech.*, **513**, pp. 309-342.
31. Fenton, J. D. (1988). The numerical solution of steady water wave problems. *Computers Geosci.* **14**, pp. 357-368.
32. Skandrani, C. (1996). Contribution á l'étude de la dynamique non linéaire de champs de vagues tridimensionnels en profondeur infinie. PhD thesis, Univ. de la Méditerranée.
33. Su, M. Y. and Green, A. W. (1984). Coupled two- and three-dimensional instabilities of surface gravity waves. *Phys. Fluids*, **27**, pp. 2595-2597.
34. M. Francius and C. Kharif (2006). Three-dimensional instabilities of periodic gravity waves in shallow water. *J. Fluid Mech.*, **561**, pp. 417-437.
35. Kristiansen, Ø., Fructus, D., Clamond, D. and Grue, J. (2005). Simulations of of crescent water wave patterns on finite depth. *Phys. Fluids.*, **17**, 064101-1-15.

36. Clamond, D. and Grue, J. (2002). Interaction between envelope solitons as a model for freak wave formations. Part I: Long time interaction. *C. R. Mecanique*, **330**, pp. 575–580.
37. Ochi, M. K. (1998). Ocean waves - the stochasitc approach. Camb. Univ. Press. 320 pp.
38. Trulsen, K. (1999). *Trans. ASME*, **121**, pp. 126-130.
39. Kharif, C. and Pelinovsky, E. (2003). Physical mechanisms of the rogue wave phenomenon. *Eur. J. Mech. B/Fluids*, **22**, pp. 603-634.
40. Grue, J., Clamond, D., Huseby, M. and Jensen, A. (2003). Kinematics of extreme waves in deep water. *Appl. Ocean Res.*, **25**, pp. 355-366.
41. Grue, J. and Jensen, A. (2006). Experimental velocities and accelerations in very steep wave events in deep water. *Eur. J. Mech. B/Fluids*, **25**, pp. 554-564.
42. Dold, J. W. & Peregrine, D. H., 1986. Water-wave modulation. *Proc. 20th Int. Conf. on Coastal Engineering*, American Society of Civil Engineers, Taipei, 10-14 Nov. 1986, 163–175.
43. Banner, M. L. and Tian, X. (1998). On the determination of the onset of breaking for modulating surface water waves. *J. Fluid Mech.*, **367**, pp. 107-137.
44. Baldock, T. E., Swan, C. and Taylor, P. H. (1996). A laboratory study of nonlinear surface waves on water. Phil. Trans. R. Soc. A, Vol. 354, *Phil. Trans. R. Soc. A*, **354**, pp. 649-676.
45. Fructus, D. and Grue, J. (2007). An explicit method for the nonlinear interaction between water waves and variable and moving bottom topography. *J. Comp. Phys.*, **222**, pp. 720-739.
46. Hammack, J. (1973). A note on tsunamis: their generation and propagation in an ocean of uniform depth. *J. Fluid Mech.*, **60**, pp. 769–700.
47. Liu, P. L.-F. (2005). Tsunami simulations and numerical models. *The Bridge (National Academy of Sciences, USA)*, **35**, (2), pp. 14-20.
48. Dalrymple, R. A., Grilli, S. T. and Kirby, J. T. (2006). Tsunamis and challenges for accurate modeling. *Oceanography I*, **19**, (1), pp. 142-151.
49. Glimsdal, S., Pedersen, G. K., Atakan, K., Harbitz, C. B., Langtangen, H. P. and Løvholt, F. (2006). Propagation of the Dec. 26, 2004, Indian Ocean Tsunami: Effects of Dispersion and Source Characteristics. *Int. J. Fluid Mech. Res.* **33**, (1), pp. 15-43.
50. Pelinovsky, E. N. (2006). Hydrodynamics of tsunami waves. In: Waves in geophysical fluids. Tsunamis, rogue waves, internal waves and internal tides, J. Grue and K. Trulsen (Eds.), CISM courses and lecture series No. 489, Springer, 2006, pp. 1-48.
51. Grue, J., Pelinovsky, E. N., Fructus, D., Talipova, T. and Kharif, Ch. (2008). Formation of undular bores and solitary waves in the Strait of Malacca caused by the Dec. 26, 2004, Indian Ocean tsunami. *J. Geophys. Res.* (in press).

CHAPTER 2

TWO-DIMENSIONAL DIRECT NUMERICAL SIMULATIONS OF THE DYNAMICS OF ROGUE WAVES UNDER WIND ACTION

Julien Touboul

Laboratoire de Sondages Electromagnétiques
de l'Environnement Terrestre
Université de Toulon et du Var, BP 20132
83957 La Garde Cedex, France
julien.touboul@univ-tln.fr

Christian Kharif

Institut de Recherche sur les Phénomènes Hors Equilibre
49 rue F. Joliot Curie, Technopôle de Château Gombert
13384 Marseille Cedex 13, France
kharif@irphe.univ-mrs.fr

The understanding of the role of water waves in air–sea interaction is of prime importance to improve approximate wave models predicting the sea state, namely the action balance equation which is forced by wind input, wave–wave interactions and dissipation. In this equation the physics is embodied in a set of source functions whose modelling can be improved through a more detailed analysis based on the numerical simulation of the fully nonlinear equations of water waves. The knowledge of velocity, acceleration and pressure fields of extreme wave events such as rogue waves is crucial for computing loads on structures, ship routing and human safety. The dynamics of these huge waves which are strongly nonlinear phenomena can be properly determined only by integrating numerically the fully nonlinear governing equations.

Two numerical methods are presented and used to study the dynamics of rogue waves in the presence of wind: A High-Order Spectral Method (HOSM) and a Boundary Integral Equation Method (BIEM). For both numerical methods the convergence and accuracy of the models are tested using exact solutions or experimental data.

1. Introduction

Among dangerous phenomena for human activity at sea, freak, or rogue waves are of special interest. They have been part of the marine folklore for centuries (Lawton[1]), while oceanographers did not believe in their existence. They correspond to giant waves, abnormally high compared to the sea state. Mallory[2] was the first to provide a discussion about giant waves observed in the Agulhas current, between 1952 and 1973. Nowadays, some of these waves have been observed, photographed and measured, the most famous of them being the new year wave, measured in 1995 in the North Sea (Haver[3]).

For the last twenty years, rogue waves have been widely studied. A first approach is to understand the physical processes generating these huge waves. Among these various mechanisms, one may cite the spatial focusing, that can be achieved by refraction of waves propagating in inhomogeneous media (e. g. variable bottom topography, or variable current), and leading to the formation of caustics. This mechanism provides an explanation to the formation of extreme waves in the Agulhas current (Peregrine,[4] Lavrenov[5]). Another mechanism is the dispersive focusing, corresponding to the focusing of energy in time and space due to the dispersive behavior of water waves. Natural existence of chirped wave packet amenable to produce giant waves is still subject to controversy, but this mechanism remains extremely popular for experimental approaches (Baldock *et al.*,[6] Kharif *et al.*[7]). The mechanism the most amenable to produce freak waves in field is the nonlinear focusing. This process corresponds to the focusing of energy in time and space due to the unstable nature of periodic wave fields on sea surface, as discovered by Benjamin & Feir.[8] Clamond & Grue[9] and Clamond *et al.*[10] showed how interacting solitary wave envelopes that emerge from long wave packet can produce extreme wave events. Later, Dyachenko & Zakharov[11] and Zakharov *et al.*[12] claimed that rogue waves are due to solitonic turbulence generated from modulational instability of Stokes wave trains. A detailed review of these mechanisms can be found in Kharif & Pelinovsky.[13]

From a practical point of view, it is more useful to investigate the statistical behaviour of rogue waves. Classical linear wave model used in ocean engineering and naval architecture consider wave fields as a linear superposition of sine waves. Elevations observed in such a wave field follow a Gaussian distribution (Dean & Dalrymple[14]). Within the framework of a free surface description up to second order in wave steepness, elevations in narrow banded wave fields are shown to follow a Tayfun distribution

(Tayfun[15]). None of these distributions take into account the nonlinear focusing, leading to an increase of freak waves occurrence in free wave fields. Onorato *et al.*[16] and Janssen[17] introduced a dimensionless number, known as the Benjamin Feir Index (BFI), and studied the probability of occurrence of rogue waves in random sea states. They observed a correlation between the BFI and freak waves occurrence. More recently, Gramstad & Trulsen[18] observed that this dependence was strongly depending on the crest length of the wave field. While they confirmed previous results for long crested wave fields, they observed that elevations in short crested wave fields were behaving as a Gaussian process, whatever was the BFI. A careful review of the state of the art in rogue waves research was completed recently by Dysthe *et al.*[19]

Freak waves observation being extremely rare, the numerical tool is essential for improving the practical and theoretical knowledge in the field. Given the variety of research approaches, one can understand the need of a large number of numerical simulations in the domain. The study of processes requires the use of highly accurate numerical methods, while the stochastic approach implies to achieve a large number of simulations as fast as possible. For the latter argument, model equations such as the Nonlinear Schrödinger equation, Dysthe equation, or Zakharov equation were often used. However, the order of nonlinearity being intrinsically limited in such approximate equations, it became a key point to improve the numerical treatment of the exact equations.

To achieve this purpose, exact methods based on potential flow theory were developed. Longuet-Higgins & Cokelet[20] first introduced a numerical technique providing a solution for the fully nonlinear potential equations. This technique was widely used, and improved, in the framework of nonlinear water waves. One should cite Vinje & Brevig[21] who reformulated this approach into Cauchy's integral formulation, Grilli *et al.*[22] or Cointe[23] who introduced significant improvements in boundary conditions formulation. A comprehensive review of those methods was performed by Tsai & Yue.[24] Nowadays, this technique is known as Boundary Integral Elements Methods (BIEM). Dommermuth *et al.*[25] used this technique to propagate chirped wave packets, producing two-dimensional deep water plunging breakers, and achieving the simulation up to overturning of the crest. More recently, Fochesato *et al.*[26] studied the three dimensional shape and kinematics of extreme waves produced by directional focusing.

Spectral methods are highly successful in nonlinear waves. A major advances was the pseudo-spectral approach of Kriess & Oliger,[27] which like most other spectral methods benefited greatly from the fast Fourier transform (FFT) algorithm of Cooley & Turkey.[28] Later, Fornberg & Whitham[29] introduced the pseudo-spectral approximation to study nonlinear wave dynamics. The simulation of long time evolution of wave groups for studying modulational instability needs to use numerical methods presenting better computational ability such as the spectral methods (errors typically decays at exponential rates). Dommermuth & Yue[30] and West *et al.* used a highly accurate technique to solve the potential water wave problem, based on a spectral treatment of the equations. This method is known as High-Order Spectral Method (HOSM). However, the representation and the treatment of the vertical velocity in the HOSM formulations derived by these authors are different. The feature was emphasized by Clamond *et al.*[10] Brandini,[31] and Brandini & Grilli[32] investigated numerically by using this technique the formation of rogue waves due to modulational instability of classes I and II. The latter is dominated by three-dimensional instabilities. Several authors, including Tanaka[33] and Ducrozet *et al.*,[34] also used this approach to compute statistics of rogue waves occurrence. New approaches were recently developed, taking advantages of the two methods. Among them, one may cite the efficient methods derived by Clamond & Grue[9] and Dyachenko & Zakharov,[11] respectively.

To study the wind effect on the extreme water wave events, an additional term is introduced in the dynamic boundary condition. The general equations are presented in section 2 whereas sections 3 and 4 are devoted to the numerical description of both methods. Finally, a wind model is introduced, and the dynamics of rogue waves due to both dispersive and nonlinear focusing in the presence of wind is numerically simulated.

2. General Equations of Potential Flow Theory

2.1. *Governing equations*

Water wave equations are based on conservation laws. Among them, the continuity equation describes the mass conservation of the fluid.

$$\frac{d\rho}{dt} + \rho\vec{\nabla}.\vec{U} = 0, \tag{2.1}$$

where t is the time, ρ is the water density, $\vec{U}$ is the fluid particle velocity and $d()/dt = \partial()/\partial t + \vec{u}.\vec{\nabla}()$ is the material derivative. The operator $\vec{\nabla}\cdot$ is the divergence.

By assuming negligible the variations of temperature in the fluid, the Navier-Stokes equations expressing the fundamental principle of the dynamics applied to the fluid particles writes

$$\rho(\frac{\partial \vec{U}}{\partial t} + \vec{U}.\vec{\nabla}\vec{U}) = \vec{\nabla}.(2\mu \underline{\underline{D}}) - \rho g \vec{e_z} - \vec{\nabla} p, \tag{2.2}$$

where g is the acceleration due to gravity, μ is dynamic viscosity and where $\underline{\underline{D}}$ is the rate of strain tensor, defined as

$$D_{ij} = \frac{1}{2}(\partial_i u_j + \partial_j u_i). \tag{2.3}$$

The operator ∂_i means spatial partial derivative and u_j is the velocity component. Here, the operator $\vec{\nabla}$ is the gradient.

These equations have to be satisfied at any point of the fluid domain. Upon several assumptions suitable for the problem under consideration, equations (2.1) and (2.2) can be simplified.

By assuming the water to be inviscid and incompressible, and the motion irrotational, several simplifications of the above equations can be made. The water being incompressible, we have $d\rho/dt = 0$ leading ρ to be constant in the water, hence equation (2.1) can be written as

$$\vec{\nabla}.\vec{U} = 0. \tag{2.4}$$

Assuming that water is a weakly viscous fluid and waves are born from the state at rest, one can consider the motion globally irrotational which means that we can write $\vec{U} = \vec{\nabla}\Phi$, where Φ is the velocity potential. Introducing this expression in equation (2.4) yields to the Laplace's equation

$$\Delta\Phi = 0. \tag{2.5}$$

The Navier-Stokes equation reduces to

$$\frac{\partial \vec{\nabla}\Phi}{\partial t} + \vec{\nabla}\Phi.\vec{\nabla}\vec{\nabla}\Phi = -g\vec{e_z} - \frac{1}{\rho}\vec{\nabla}p \tag{2.6}$$

that can be integrated in space to give

$$\frac{\partial \Phi}{\partial t} + \frac{\vec{\nabla}\Phi^2}{2} = -gz - \frac{p}{\rho}, \tag{2.7}$$

which is the Bernoulli's equation. Note that the arbitrary time dependent function due to space integration has been included in the velocity potential. The Laplace equation (2.5) has to be solved at any time in the fluid domain.

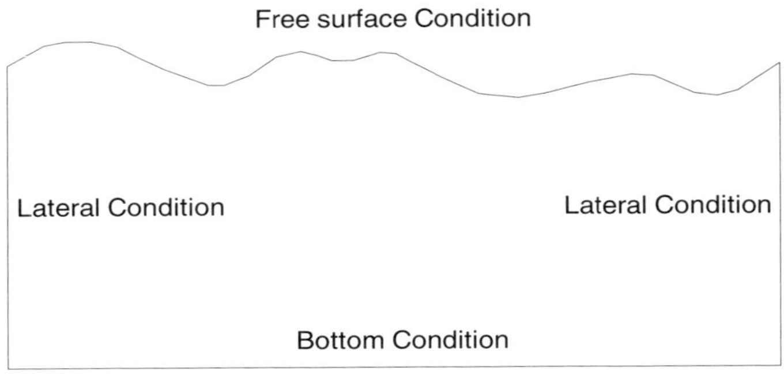

Fig. 1. Geometry of the problem.

The time dependency of the problem is introduced through the boundary conditions.

2.2. *Boundary conditions*

To solve this partial differential equation, the boundary conditions have to be fixed on the domain boundaries. The domain considered is shown in Figure 1, enclosed by the free surface, two lateral conditions, and a bottom condition. Boundary conditions have to be set on each of those surfaces. Several conditions can be used, depending on the physical problem that has to be solved.

2.2.1. *Free surface condition*

Among these boundaries, the free surface is of major importance for water waves problems. From the Bernoulli equation (2.7) and pressure continuity at the interface the following condition can be obtained

$$\frac{\partial \Phi}{\partial t} + \frac{1}{2}\vec{\nabla}\Phi^2 + g\eta = -\frac{p_a}{\rho} \quad \text{on} \quad z = \eta(x, y, t), \tag{2.8}$$

where p_a is the atmospheric pressure, η the free surface elevation, and where x, y and z are the spatial coordinates. This condition is known as the dynamic condition.

The free surface location is not known *a priori*, hence a second boundary condition is needed. The kinematic boundary condition states that any particle on the free surface will remain on it. This condition corresponds to a non-penetration condition. The kinematic problem can be expressed from both Eulerian or Lagrangian point of view. In an Eulerian description the free surface evolution is given by the following equation

$$\frac{\partial \eta}{\partial t} = -\frac{\partial \Phi}{\partial x}\frac{\partial \eta}{\partial x} - \frac{\partial \Phi}{\partial y}\frac{\partial \eta}{\partial y} + \frac{\partial \Phi}{\partial z} \quad \text{on} \quad z = \eta(x, y, t) \tag{2.9}$$

while in a Lagrangian description the kinematic condition writes

$$\begin{aligned} \frac{dx}{dt} &= \frac{\partial \Phi}{\partial x} \\ \frac{dy}{dt} &= \frac{\partial \Phi}{\partial y} \\ \frac{dz}{dt} &= \frac{\partial \Phi}{\partial z} \end{aligned} \tag{2.10}$$

where $d/dt = \partial/\partial t + \vec{\nabla}\Phi.\vec{\nabla}$ corresponds to the material derivative. These formulations are equivalent, even if they will lead to different properties of the method, as discussed in the following sections.

2.2.2. *Bottom condition*

Depending on the problem to solve, two conditions can be considered for the bottom boundary condition. In infinite depth the bottom condition is

$$\nabla \Phi \to 0 \quad \text{when} \quad z \to -\infty, \tag{2.11}$$

while in finite depth this condition writes

$$\vec{\nabla}\Phi.\vec{n} = \frac{\partial \Phi}{\partial n} = 0 \quad \text{on} \quad z = -h(x, y) \tag{2.12}$$

where $z = -h(x, y)$ is the bottom equation.

2.2.3. *Lateral condition*

The choice of the lateral boundary condition also depends on the physical problem considered. In a first case one can consider a closed geometry. The lateral surfaces are then understood as walls, and a numerical wave tank can be modeled. To do so, an impermeability condition needs to be set,

$$\vec{\nabla}\Phi.\vec{n} = \frac{\partial \Phi}{\partial n} = \vec{V}.\vec{n}, \tag{2.13}$$

where $\vec{V}$ refers to the wall velocity. This velocity can be taken equal to zero for a fixed wall, or can obey some time dependent law to simulate a wavemaker.

Another case concerns periodic flows which need the use of periodic boundary conditions. The potential and its normal derivative on the left and right lateral surfaces are equal.

$$\Phi_{Left} = \Phi_{Right}$$
$$\left(\frac{\partial \Phi}{\partial n}\right)_{Left} = \left(\frac{\partial \Phi}{\partial n}\right)_{Right}. \tag{2.14}$$

Concerning some local problems, another condition which will not be used in the following can be enforced. A radiation condition can be applied on the vertical control surfaces. This condition states that the flow disturbances should vanish far away from it, namely $\vec{\nabla}\Phi \to 0$ when $\vec{X} \to \infty$.

3. Boundary Integral Element Method (BIEM)

This section describes, within the framework of two-dimensional flows, a numerical method initially introduced by Longuet-Higgins & Cokelet[20] to solve the partial differential equations stated in the previous section. The method is a Mixed Eulerian-Lagrangian method, meaning that the description of the free surface is Lagrangian. The free surface boundary, which corresponds to a Dirichlet boundary condition, is denoted $\partial\Omega_D$. The remaining boundaries fulfil the impermeability condition described in the previous section. They correspond to Neumann conditions and are denoted $\partial\Omega_N$. This method is used to study rogue waves generated by dispersive focusing of wave energy.

3.1. *Boundary integral formulation*

The problem described in the previous section can be rewritten in the following way

$$\begin{cases} \Delta\Phi = 0 \text{ in the fluid domain } \Omega \\ \Phi \qquad \text{known on } \partial\Omega_D \\ \dfrac{\partial \Phi}{\partial n} \qquad \text{known on } \partial\Omega_N, \end{cases}$$

at each time step. In terms of Green's second identity, it can be rewritten

$$\int_{\partial\Omega} \Phi(P)\frac{\partial G}{\partial n}(P,Q)dl - \int_{\partial\Omega} \frac{\partial \Phi}{\partial n}(P)G(P,Q)dl = c(Q)\Phi(Q) \tag{3.1}$$

for points P and Q belonging to the domain Ω. Here G is the free space Green's function defined by

$$G(\vec{P},\vec{Q}) = ln(|\vec{Q}-\vec{P}|) \tag{3.2}$$

and the normal vector $\vec{n}$ points outside the fluid domain. The angle c(Q) is defined as

$$c(Q) = \begin{cases} 0 & \text{if Q is outside the fluid domain} \\ \alpha & \text{if Q is on the boundary} \\ -2\pi & \text{if Q is inside the fluid domain} \end{cases}$$

where α is the inner angle relative to the fluid domain at point Q along the boundary. When we apply this representation at points Q along the boundary $\partial\Omega$, the equations necessary to solve the problem are obtained. In more detail, we get the following system of equations

$$\begin{aligned} &\alpha\Phi(Q) - \int_{\partial\Omega_D} \Phi(P)\frac{\partial G}{\partial n}(P,Q)dl + \int_{\partial\Omega_N} \frac{\partial \Phi}{\partial n}(P)G(P,Q)dl \\ &= \int_{\partial\Omega_N} \Phi(P)\frac{\partial G}{\partial n}(P,Q)dl - \int_{\partial\Omega_D} \frac{\partial \Phi}{\partial n}(P)G(P,Q)dl \end{aligned} \tag{3.3}$$

when Q belongs to the Dirichlet boundary $\partial\Omega_D$, and

$$\begin{aligned} &\int_{\partial\Omega_D} \Phi(P)\frac{\partial G}{\partial n}(P,Q)dl - \int_{\partial\Omega_N} \frac{\partial \Phi}{\partial n}(P)G(P,Q)dl \\ &= \alpha\Phi(Q) - \int_{\partial\Omega_N} \Phi(P)\frac{\partial G}{\partial n}(P,Q)dl + \int_{\partial\Omega_D} \frac{\partial \Phi}{\partial n}(P)G(P,Q)dl \end{aligned} \tag{3.4}$$

when Q belongs to the Neumann boundary $\partial\Omega_N$. The unknowns are $\partial\Phi/\partial n$ on $\partial\Omega_D$ and Φ on $\partial\Omega_N$. To solve numerically this system, a linear system of algebraic equations for a finite number of unknowns is derived by discretizing the boundaries of the domain.

3.2. *Numerical modelling*

The method used here is similar to that developed by Greco[35] and Faltinsen *et al.*[36] Equations (3.3) and (3.4) are solved numerically. The boundary of domain is discretized by introducing N_N elements on the Neumann boundary, and N_D elements on the Dirichlet boundary. Then, the boundary integral equations (3.3) and (3.4) are discretized at the ends of these elements.

Even if higher-order methods were suggested (see Sen[37]), the integrals are approximated as integrals along the elements where a linear variation of Φ and $\partial\Phi/\partial n$ is assumed. The following equations

$$\alpha\Phi_i - \sum_{j=1}^{N_D}\int_j \Phi\frac{\partial G}{\partial n}(i,j)ds + \sum_{j=1}^{N_N}\int_j \frac{\partial \Phi}{\partial n}G(i,j)ds$$
$$= \sum_{j=1}^{N_N}\int_j \Phi\frac{\partial G}{\partial n}(i,j)ds - \sum_{j=1}^{N_D}\int_j \frac{\partial \Phi}{\partial n}G(i,j)ds \qquad (3.5)$$

are obtained for any point i of the Dirichlet boundary, and

$$\sum_{j=1}^{N_D}\int_j \Phi\frac{\partial G}{\partial n}(i,j)ds - \sum_{j=1}^{N_N}\int_j \frac{\partial \Phi}{\partial n}G(i,j)ds$$
$$= \alpha\Phi_i - \sum_{j=1}^{N_N}\int_j \Phi\frac{\partial G}{\partial n}(i,j)ds + \sum_{j=1}^{N_D}\int_j \frac{\partial \Phi}{\partial n}G(i,j)ds \qquad (3.6)$$

for any point i of the Neumann boundary. With Green's function defined by equation (3.2), it comes, in the pannel's coordinates (ξ, ϑ)

$$G(\vec{P},\vec{Q}) = \ln(\sqrt{\xi^2+\vartheta^2}) \quad \text{and} \quad \frac{\partial G}{\partial n}(\vec{P},\vec{Q}) = \frac{\vartheta}{\xi^2+\vartheta^2}. \qquad (3.7)$$

Thus, equations (3.5) and (3.6) can be written respectively

$$\alpha\Phi_i - \sum_{j=1}^{N_D}\left\{\Phi_{j+1}\frac{I_4-\xi_j I_2}{\xi_{j+1}-\xi_j} + \Phi_j\frac{I_2\xi_{j+1}-I_4}{\xi_{j+1}-\xi_j}\right\}$$
$$+ \sum_{j=1}^{N_N}\left\{\Psi_{j+1}\frac{I_3-\xi_j I_1}{\xi_{j+1}-\xi_j} + \Psi_j\frac{I_1\xi_{j+1}-I_3}{\xi_{j+1}-\xi_j}\right\}$$
$$= \sum_{j=1}^{N_N}\left\{\Phi_{j+1}\frac{I_4-\xi_j I_2}{\xi_{j+1}-\xi_j} + \Phi_j\frac{I_2\xi_{j+1}-I_4}{\xi_{j+1}-\xi_j}\right\}$$
$$- \sum_{j=1}^{N_D}\left\{\Psi_{j+1}\frac{I_3-\xi_j I_1}{\xi_{j+1}-\xi_j} + \Psi_j\frac{I_1\xi_{j+1}-I_3}{\xi_{j+1}-\xi_j}\right\}, \qquad (3.8)$$

and

$$\sum_{j=1}^{N_D}\left\{\Phi_{j+1}\frac{I_4-\xi_j I_2}{\xi_{j+1}-\xi_j}+\Phi_j\frac{I_2\xi_{j+1}-I_4}{\xi_{j+1}-\xi_j}\right\}$$
$$-\sum_{j=1}^{N_N}\left\{\Psi_{j+1}\frac{I_3-\xi_j I_1}{\xi_{j+1}-\xi_j}+\Psi_j\frac{I_1\xi_{j+1}-I_3}{\xi_{j+1}-\xi_j}\right\}$$
$$=\alpha\Phi_i-\sum_{j=1}^{N_N}\left\{\Phi_{j+1}\frac{I_4-\xi_j I_2}{\xi_{j+1}-\xi_j}+\Phi_j\frac{I_2\xi_{j+1}-I_4}{\xi_{j+1}-\xi_j}\right\}$$
$$+\sum_{j=1}^{N_D}\left\{\Psi_{j+1}\frac{I_3-\xi_j I_1}{\xi_{j+1}-\xi_j}+\Psi_j\frac{I_1\xi_{j+1}-I_3}{\xi_{j+1}-\xi_j}\right\}, \tag{3.9}$$

where $\Psi = \partial\Phi/\partial n$, and I_1, I_2, I_3 and I_4 are

$$\begin{aligned}
I_1 &= \int_{\xi_1}^{\xi_2} \ln(\sqrt{x^2+\vartheta^2})dx \\
&= \frac{1}{2}\xi_2\ln(\xi_2^2+\vartheta^2)-\xi_2-\vartheta\arctan(\frac{\xi_2}{\vartheta})- \\
&\quad \frac{1}{2}\xi_1\ln(\xi_1^2+\vartheta^2)+\xi_1-\vartheta\arctan(\frac{\xi_1}{\vartheta}) \\
I_2 &= \int_{\xi_1}^{\xi_2} \frac{\vartheta}{x^2+\vartheta^2}dx \\
&= \arctan(\frac{\xi_2}{\vartheta})-\arctan(\frac{\xi_1}{\vartheta}) \\
I_3 &= \int_{\xi_1}^{\xi_2} x.\ln(\sqrt{x^2+\vartheta^2})dx \\
&= \frac{1}{4}\xi_2^2\ln(\xi_2^2+\vartheta^2)+\frac{1}{4}\vartheta^2\ln(\xi_2^2+\vartheta^2)-\frac{1}{4}\xi_2^2- \\
&\quad \frac{1}{4}\xi_1^2\ln(\xi_1^2+\vartheta^2)-\frac{1}{4}\vartheta^2\ln(\xi_1^2+\vartheta^2)+\frac{1}{4}\xi_1^2 \\
I_4 &= \int_{\xi_1}^{\xi_2} \frac{x.\vartheta}{x^2+\vartheta^2}dx \\
&= \frac{1}{2}\vartheta\ln(\xi_2^2+\vartheta^2)-\frac{1}{2}\vartheta\ln(\xi_1^2+\vartheta^2).
\end{aligned} \tag{3.10}$$

In this way, we have $N_N + N_D + 2$ equations for the $N_N + N_D + 2$ unknowns of the problem. In equations (3.8) and (3.9), all the unknowns are on the right hand side, while the known terms are on the left hand side. When we rewrite the known terms of equation (3.8) with f_i, and g_i known terms of equation (3.9), then, the system can formally be written as

$$\begin{bmatrix} A_{ij} & B_{ij} \\ C_{ij} & D_{ij} \end{bmatrix} \begin{pmatrix} \partial\Phi/\partial n_i \\ \Phi_i \end{pmatrix} = \begin{pmatrix} f_i \\ g_i \end{pmatrix}.$$

Hence, Φ and its normal derivative can be computed on all the boundaries.

4. High Order Spectral Method (HOSM)

Within the framework of two-dimensional flows, a High-Order Spectral Method is used to solve numerically the basic partial differential equations. The lateral conditions in Figure 1 correspond here to space-periodic conditions. The horizontal bottom condition can correspond either to infinite or finite depth. The velocity potential is expanded in a series of eigenfunctions fulfilling periodic conditions. A spectral treatment is well adapted to investigate numerically rogue waves occurrence due to periodic modulational instability.

4.1. *Mathematical formulation*

We first consider the case of infinite depth and introduce the following dimensionless variables into equations (2.5), (2.8), (2.9) and (2.11): $x = k_0 X$, $z = k_0 Z$, $t = \sqrt{gk_0}T$, $\zeta = k_0\eta$, $\phi = \Phi/\sqrt{g/k_0^3}$ and $p = P/(\rho g/k_0)$, where X, Z, T, η, Φ and P are dimensional variables, and where k_0 is a reference wavenumber. Hence, the kinematic and dynamic boundary conditions become

$$\frac{\partial\zeta}{\partial t} + \frac{\partial\phi}{\partial x}\frac{\partial\zeta}{\partial x} - \frac{\partial\phi}{\partial z} = 0 \quad \text{on} \quad z = \zeta \tag{4.1}$$

$$\frac{\partial\phi}{\partial t} + \frac{1}{2}\nabla\phi\cdot\nabla\phi + p_a + z = 0 \quad \text{on} \quad z = \zeta. \tag{4.2}$$

Following Zakharov,[38] we introduce the velocity potential at the free surface $\phi^s(x,t) = \phi(x, z = \zeta(x,t), t)$ into equations (4.1) and (4.2)

$$\frac{\partial\zeta}{\partial t} = -\nabla\phi^s\cdot\nabla\zeta + w[1 + (\nabla\zeta)^2] \tag{4.3}$$

$$\frac{\partial \phi^s}{\partial t} = -\zeta - \frac{1}{2}\nabla\phi^s \cdot \nabla\phi^s + \frac{1}{2}w^2[1 + (\nabla\zeta)^2] - p_a \tag{4.4}$$

with

$$w = \frac{\partial \phi}{\partial z}(x, z = \zeta(x,t), t). \tag{4.5}$$

The main difficulty is the computation of the vertical velocity at the free surface, w. Following Dommermuth & Yue,[30] the potential $\phi(x, z, t)$ is written in a finite perturbation series up to a given order M

$$\phi(x, z, t) = \sum_{m=1}^{M} \phi^{(m)}(x, z, t).$$

The term $\phi^{(m)}$ is of $\mathcal{O}(\varepsilon^m)$ where ε, a small parameter, is a measure of the wave steepness. Then expanding each $\phi^{(m)}$ evaluated on $z = \zeta$ in a Taylor series about $z = 0$, we obtain

$$\phi^s(x, t) = \sum_{m=1}^{M} \sum_{l=0}^{M-m} \frac{\zeta^l}{l!} \frac{\partial^l}{\partial z^l} \phi^{(m)}(x, z = 0, t). \tag{4.6}$$

At a given instant of time, ϕ^s and ζ are known so that from (4.6), at each order we can calculate $\phi^{(m)}$:

$\mathcal{O}(1)$

$$\phi^{(1)}(x, z = 0, t) = \phi^s(x, t) \tag{4.7}$$

$\mathcal{O}(m)$

$$\phi^{(m)}(x, z = 0, t) = -\sum_{l=1}^{m-1} \frac{\zeta^l}{l!} \frac{\partial^l}{\partial z^l} \phi^{(m-l)}(x, z = 0, t), \quad m \geq 2. \tag{4.8}$$

These boundary conditions, with Laplace equations $\nabla\phi^{(m)}(x, z, t) = 0$, define a series of Dirichlet problems for $\phi^{(m)}$.

For $2\pi-$periodic conditions in x, say, $\phi^{(m)}$ can be written as follows in deep water

$$\phi^{(m)}(x, z, t) = \sum_{j=0}^{\infty} \phi_j^{(m)}(t) \exp(\kappa_j z) \exp(ijx) \tag{4.9}$$

where $\kappa_j = j$. Note that $\phi^{(m)}(x, z, t)$ satisfies automatically the Laplace equation and the condition $\nabla\phi^{(m)}(x, z, t) \to 0$ as $z \to -\infty$ (2.11). An alternative decomposition can be used in the case of finite depth h. By replacing projection (4.9) with the following expression,

$$\phi^{(m)}(x,z,t)=\sum_{j=0}^{\infty}\phi_j^{(m)}\frac{\cosh(\kappa_j(z+h))}{\cosh(\kappa_j h)}\exp(ijx) \tag{4.10}$$

$\phi^{(m)}(x,z,t)$ satisfies automatically the Laplace equation and the bottom condition (2.12).

4.2. *Computation of the vertical velocity*

Substitution of (4.9) into the set of equations (4.7) and (4.8) gives the modes $\phi_j^{(m)}(t)$. The vertical velocity at the free surface is then

$$w=\sum_{m=1}^{M}\sum_{l=0}^{M-m}\frac{\zeta^l}{l!}\frac{\partial^{l+1}}{\partial z^{l+1}}\phi^{(m)}(x,z=0,t). \tag{4.11}$$

Substitution of (4.11) into the boundary conditions (4.3) and (4.4) yields the evolution equations for ϕ^s and ζ. Another version of HOSM developed by West *et al.*[39] can be used. The difference between both methods lies in the way of computing w from $\phi^{(m)}$. West *et al.*[39] assume a power series for w as

$$w(x,t)=\sum_{m=1}^{M}w^{(m)} \tag{4.12}$$

where

$$w^{(m)}=\sum_{l=0}^{m-1}\frac{\zeta^l}{l!}\frac{\partial^{l+1}}{\partial z^{l+1}}\phi^{(m-l)}(x,z=0,t). \tag{4.13}$$

In fact, the version of West *et al.*[39] differs from the version of Dommermuth & Yue[30] not only in the expression of the approximated vertical velocity at the surface, but also in its subsequent treatment in the free surface equations. According to West *et al.*, the surface equations must be truncated at consistent nonlinear order if they are to simulate a conservative Hamiltonian system. This requires to treat carefully all nonlinear terms containing w in the prognostic equations. In contrast to the series used by Dommermuth & Yue, those used by West *et al.*[39] are naturally ordered with respect to the nonlinear parameter ε. The Dommermuth & Yue formulation is not consistent, after truncation, with the underlying Hamiltonian structure of the canonical pair of free-surface equations. Thus, the formulation of West *et al.*[39] preserves the Hamiltonian structure of the prognostic equations.

The treatment of nonlinear terms in the HOSM method is useful for comparisons between truncated fully nonlinear equation and approximate models such as Zakharov equation. However, Clamond *et al.*[10] noticed differences in the inter-comparison of both HOSM formulations. For the specific case $M = 3$, the HOSM is supposed to be an extension of the third order Zakharov equation. However, Clamond *et al.*[10] noticed that the HOSM formulation by Dommermuth & Yue[30] was not able to predict long-time evolution of nonlinear wave packets while the formulation by West *et al.*[39] was. The observed behavior of the two approaches was only found to be similar for $M \geq 5$.

4.3. *Numerical treatment*

The numerical method used to solve the evolution equations (4.3) and (4.4) is the same to that developed by Dommermuth & Yue,[30] but using West *et al.*[39] approach to compute vertical velocity. Equations (4.3) and (4.4) are integrated using a pseudo-spectral treatment with N wave modes and retaining nonlinear terms up to order M. Once known the surface elevation $\zeta(x,t)$ and the potential at the free surface $\phi^s(x,z,t)$ at time t, the modal amplitudes may be computed. The spatial derivatives of $\phi^{(m)}$, ϕ^s, ζ and w are calculated in the spectral space, while nonlinear terms are evaluated in the physical space at a discrete set of collocation points x_j. Fast Fourier Transforms are used to link the spectral and physical spaces. Equations (4.7) and (4.8) are solved in the spectral space. Evolution equations for ϕ^s and ζ are integrated in time using a fourth-order Runge-Kutta integrator with constant time step. The calculation accuracy depends on several sources of errors due to truncation in the number of modes N, and order M, amplification of round-off error, aliasing phenomenon, or numerical time integration. Errors due to truncation correspond to the value of the residuals after truncation to an order M. For sufficiently smooth functions, error after truncation of order M are $\mathcal{O}(\varepsilon^{M+1})$, and converges exponentially as M increases. Amplification of round-off error is due to the summation of numerical errors made on each Fourier component before summation. One should note that such error does not appear in BIEM methods. To avoid it, a low pass filter needs to be used. In a pseudo-spectral method the nonlinear terms are computed in the physical space instead of spectral space. To avoid aliasing errors inherent in this numerical treatment the number of collocation points in x-direction has to be greater than $(M+1)N$. Numerical convergence tests can be found in the papers by Dommermuth & Yue[30] and Skandrani *et al.*[40]

5. Application to the Interaction of Wind over Rogue Waves

This section is devoted to numerical simulations of frequency modulated wave trains with and without wind forcing. Two kinds of modulations are used to produce focus of wave energy corresponding to rogue wave occurrence. In a first series of numerical experiments the extreme wave event is due to the dispersive nature of the water waves while in a second series it is due to the modulational instability. The dispersive focusing occurs when the leading waves of a group have a higher frequency than the trailing waves while the modulational instability of uniform wave trains may evolve to extreme wave events. The first and second mechanisms are sometimes called linear focusing and nonlinear focusing respectively. These mechanisms are investigated numerically by using the BIEM and HOSM approaches with and without wind forcing.

The question is to know how extreme wave events due to dispersive focusing or modulational instability evolve under wind action. How are modified the amplification and time duration of these waves under wind effect? In the following subsection is presented the wind modelling.

5.1. *Wind modelling: The modified Jeffreys' sheltering theory*

In previous works, Touboul *et al.*,[41] and Kharif *et al.*,[7] investigated experimentally, and numerically the structure of air flow above extremely steep waves. From the experimental point of view, chirped wave packets were propagated in the large air-sea interaction facility of IRPHE, and their dynamics with and without wind were compared. A significant persistence of the steepest stages of the wave packet evolution was found. These experiments are the one that will be referred to later. A further investigation of the structure of the air flow demonstrated experimentally for a wind velocity $U = 4m/s$ that steep wave events occurring in water wave groups are accompanied by air flow separation. This assumption was confirmed numerically through the approach presented here. Numerically, the Jeffreys' sheltering mechanism which was introduced by Jeffreys[42] could be used as wind model. Since air flow separation only occurs over steep waves, the

Jeffreys' sheltering mechanism has to be applied locally in time and space and not permanently over the whole wave field. It is well known that this mechanism cannot be applied continuously over water waves. This mechanism is working only when air flow separation occurs over steep waves (Banner & Melville,[43] Kawai[44]).

Previous works on rogue waves have not considered the direct effect of wind on their dynamics. It was assumed that they occur independently of wind action, that is, far away from storm areas where wind wave fields are formed. Herein the Jeffreys' theory is invoked for the modelling of the pressure, p_a. Jeffreys proposed a plausible mechanism to explain the phase shift of the atmospheric pressure, p_a, needed for an energy transfer from wind to the water waves. He suggested that the energy transfer was due to the form drag associated with the flow separation occurring on the leeward side of the crests. The air flow separation would cause a pressure asymmetry with respect to the wave crest resulting in a wave growth. This mechanism can be invoked only if the waves are sufficiently steep to produce air flow separation. Banner & Melville[43] have shown that separation occurs over breaking waves. For weak or moderate steepness of the waves this phenomenon cannot apply and the Jeffreys' sheltering mechanism becomes irrelevant.

Following Jeffreys,[42] the pressure at the interface $z = \eta(x,t)$ is related to the local wave slope according to the following expression

$$p_a = \rho_a s(U - c)^2 \frac{\partial \eta}{\partial x} \tag{5.1}$$

where the constant s is termed the sheltering coefficient, U is the wind speed, c is the wave phase velocity and ρ_a is atmospheric density. The value of the sheltering coefficient s was initially suggested by Jeffreys, and taken such as $s = 0.5$. We checked this value experimentally, using air flow measurements presented in Kharif *et al.*[7] However, the accuracy of this value remains perfectible, since the air flow measurement becomes a tricky task in the vicinity of the air-sea interface.

In order to apply the relation (5.1) for only steep waves we introduce a threshold value for the slope $(\partial\eta/\partial x)_c$. When the local slope of the waves becomes larger than this critical value, the pressure is given by equation (5.1) otherwise the pressure at the interface is taken equal to a constant which is chosen equal to zero without loss of generality. This means that wind forcing is applied locally in time and space. Figure 2 shows the pressure distribution at the interface in the vicinity of the crest, given by equation (5.1), for a threshold value close to the slope corresponding to the

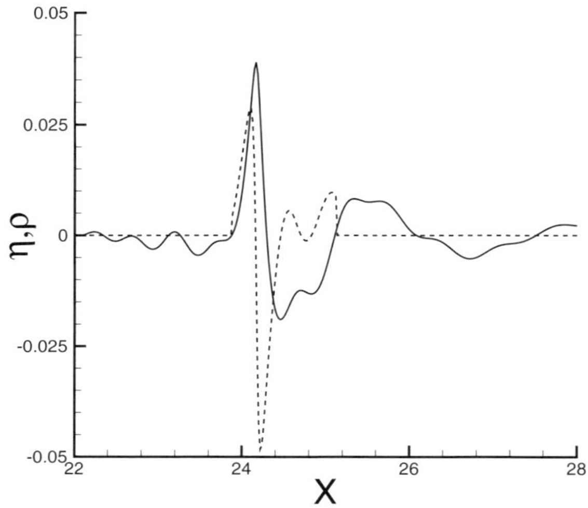

Fig. 2. Pressure at the interface given in $10^{-1}HPa$ (dashed line) and surface elevation given in m (solid line) as a function of x.

Stokes' corner, which is the limiting Stokes' wave presenting an angle of 120° at the crest. According to the experiments, the critical value of the slope, $(\partial\eta/\partial x)_c$, is chosen close to 0.35, in the range (0.30-0.40) for the spatio-temporal focusing. For the nonlinear focusing due to modulational instability we used higher values to avoid a rapid evolution towards breaking. When the critical value is low, the transfer of energy from the wind to the waves leads the wave group to breaking and when it is too high this transfer becomes negligible to influence the wave dynamics. The choice of the value of the sheltering coefficient is also of importance. This coefficient has been computed experimentally. We have not performed a systematic study on the influence of $(\partial\eta/\partial x)_c$ and s on the wind-wave coupling. Our main purpose is to show that the application of the modified Jeffreys mechanism could explain simply some features of the interaction between wind and strongly modulated water wave groups.

5.2. *The spatio-temporal focusing*

A focusing wave train is generated by the piston wave maker, leading during the focusing stage to the generation of a extreme wave followed by a

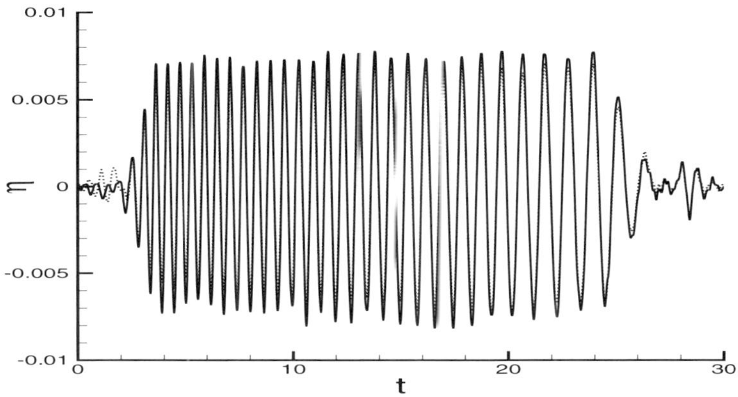

Fig. 3. Surface elevation (m) as a function of time (s) at fetch $x = 1m$: Experiments (solid line) and numerical simulation (dotted line) within the framework of the spatio-temporal focusing.

defocusing stage. The water surface and the solid boundaries (downstream wall, bottom and wave maker) are discretised by 2000 and 1000 meshes respectively, uniformly distributed. The time integration is performed using a fourth-order Runge & Kutta scheme with a constant time step of 0.01 s. Figure 3 displays the experimental and computed surface elevation $\eta(t)$ at fetch $x = 1m$ while Figure 4 shows the surface elevation at several fetches, measured experimentally and computed numerically. The origin of the surface elevation corresponding to fetches $x = 18m$ and $x = 21m$ are located at 0.05 and 0.1 respectively, for sake of clarity. This set of figures illustrate perfectly the dispersive focusing principle. One will notice from figure 3 that the frequency of waves present in the train is modulated, short waves being emitted before long waves.

Figure 4 shows the evolution of this frequency modulated wave packet while propagating downstream. One will notice that the packet focuses at fetch $x = 18m$. Beyond the focus point a defocusing is observed corresponding to long waves ahead of short waves.

Furthermore, numerical and experimental data at fetch $x = 1m$ are in excellent agreement while discrepancies observed for steep waves at fetches $x = 11m$, $18m$ and $21m$ are possibly due to local breaking. Nevertheless the phases of the numerical and experimental wave trains are the same demonstrating the efficiency of the numerical code to reproduce correctly

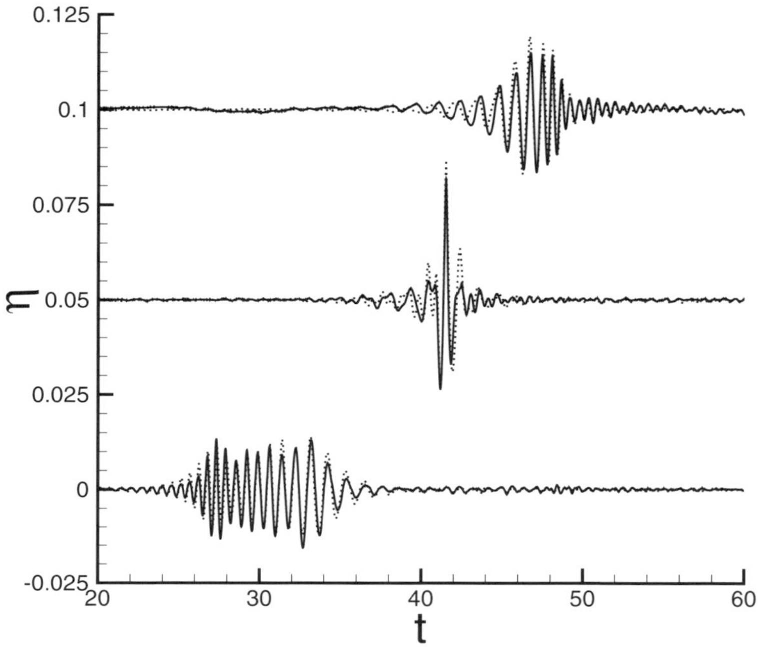

Fig. 4. Surface elevation (m) as a function of time (s) at fetches $x = 21m$ (top), $x = 18m$ (middle) and $x = 11m$ (bottom): Experiments (solid line) and numerical simulation (dotted line) within the framework of the spatio-temporal focusing.

the nonlinear evolution of water wave groups during the focusing-defocusing cycle.

The focusing mechanism is investigated with and without wind as well. A series of numerical simulations has been run for two values of the wind velocity: $U = 0m/s$ and 6 m/s. Let us introduce the amplification factor

$$A(x) = \frac{\max|\eta(x,t)|}{\max|\eta(1,t)|}, \tag{5.2}$$

corresponding to the maximum elevation reached at a given fetch, normalized by its initial value. Figure 5 shows the experimental and numerical amplification factors as a function of the normalized fetch x/x_{f} where x_{f} is the abscissa of the focusing point without wind. We can observe an excellent agreement between the experimental and numerical results. The

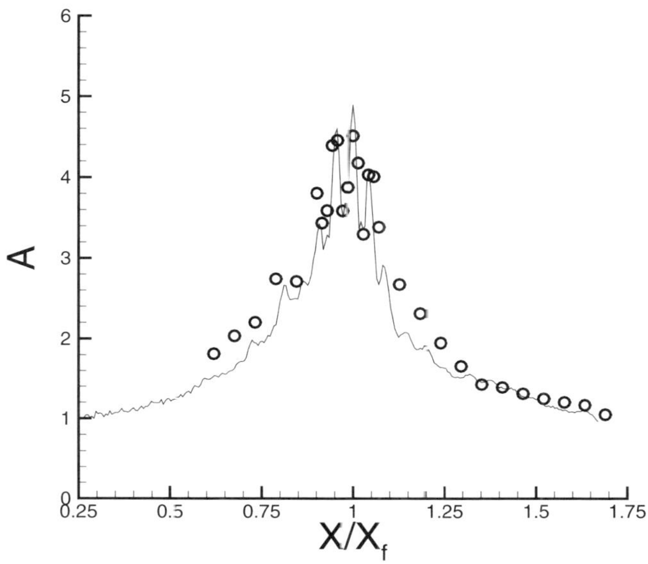

Fig. 5. Numerical (solid line) and experimental (circle) amplification factor $A(X/X_{\rm f}, U)$ as a function of the normalized distance without wind within the framework of the spatio-temporal focusing.

experimental and numerical values of the abscissa of the focus point, $x_{\rm f}$, and amplification factor, A, well correspond.

Figure 6(a) describes the spatial evolution of the amplification factor computed numerically, in the presence of wind of velocity $U = 6m/s$, plotted with experimental results. It demonstrates that the numerical and experimental amplification factors disagree beyond the focus point. For $(\partial\eta/\partial x)_c = 0.3$ a blow-up of the numerical simulation occurs due to the onset of breaking. This threshold value is low, and overcame for a long time. As a result, the transfer of energy from the wind to the steep waves lasts for a long time. The total amount of energy transferred from wind to waves leads to develop wave breaking. The threshold value of the slope beyond which the wind forcing is applied has been increased and is $(\partial\eta/\partial x)_c = 0.4$. This value corresponds to a wave close to the limiting form for which the modified Jeffreys' theory applies. The Jeffreys' sheltering mechanism is not effective enough in the present case.

Wind waves are generally propagating in the presence of current, which is due to the shear of wind at the surface. Figure 6(b) corresponds to

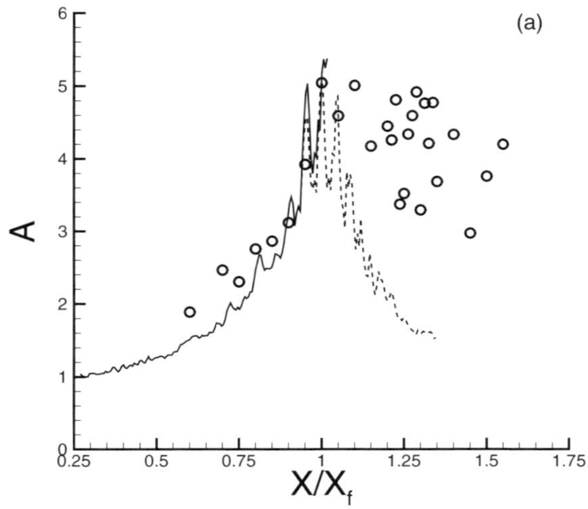

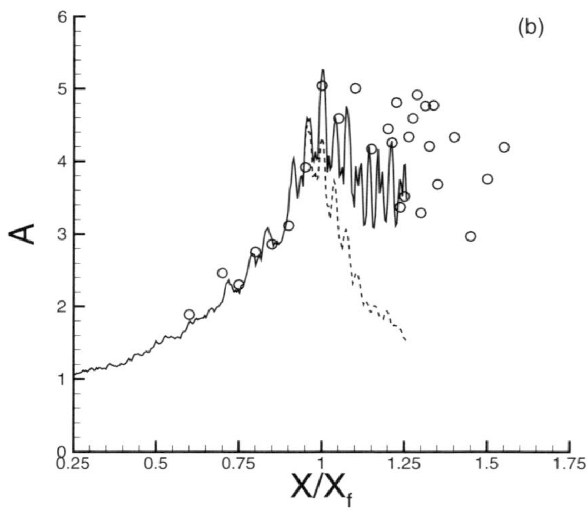

Fig. 6. (a) Numerical (solid and dashed lines) and experimental (circle) amplification factor $A(X/X_{\rm f}, U)$ as a function of the normalized distance with wind ($U = 6m/s$) for $(\partial\eta/\partial x)_c = 0.3$ (solid line) and $(\partial\eta/\partial x)_c = 0.4$ (dashed line) within the framework of the spatio-temporal focusing. (b) Amplification factor obtained in the presence of a following current.

the spatio-temporal focusing in the presence of wind and current with $(\partial\eta/\partial x)_c = 0.3$. The wind velocity is $U = 6m/s$ and a uniform following current corresponding to 2% of U has been introduced to have the numerical value of the focus point equal to the experimental value. Generally, the current induced by wind is taken equal to 3% of the wind velocity (Large & Pond[45]). More information about the introduction of a current in the model can be found in the paper by Touboul *et al.*[46] In this work, chirped wave packet were propagated in the presence of current, and compared to the linear theory. The role of nonlinearity on the kinematics and the dynamics of the process was emphasized. The current was introduced in the BIEM model by considering the velocity potential $\Phi = U.x + \varphi$, and by solving the equations obtained for φ

Here, the introduction of the following current prevents the onset of breaking. During extreme wave events the wind-driven current may play a significant role in the wind-wave interaction. The combined action of the Jeffreys' sheltering mechanism and wind-driven current may sustain longer extreme wave events. We can observe a better agreement between the numerical simulation and experiment. The steep wave event is propagating over a longer distance (or period of time) in the numerical simulations and experiments as well. The observed asymmetry between the focusing and defocusing regimes can be explained as follows. Without wind the amplitude of the extreme wave is decreasing during defocusing. In the presence of wind, the modified Jeffreys' mechanism, which is acting locally in time and space, amplifies only the highest waves and hence delays their amplitude decrease during the very beginning of the defocusing stage. The competition between the dispersive nature of the water waves and the local transfer of energy from the wind to the extreme wave event leads to a balance of these effects at the maximum of modulation. This asymmetry results in an increase of the life time of the steep wave event, which increases with the wind velocity. The main effect of Jeffreys' sheltering mechanism is to sustain the coherence of the short group involving the steep wave event.

5.3. *Focusing due to modulational instability*

Besides the focusing due to dispersion of a chirped wave group, another mechanism, the modulational instability or Benjamin-Feir instability (see the paper by Benjamin & Feir[8]) of uniform wave trains, can generate extreme wave events. Indeed this instability was discovered at the same time by different researchers. Lightill[47] provided a geometric condition for wave instability in deep water. Later, Benjamin & Feir[8] demonstrated the result analytically. Using a Hamiltonian approach, Zakharov[38] derived the same instability result. It would have been more appropriate to call this instability the BFLZ instability instead of BF instability. This periodic phenomenon is investigated numerically using a high-order spectral method (HOSM) without experimental counterpart. The question is to know how evolve extreme wave events due to modulational instability under wind action. How are modified their amplification and time duration under wind effect? Are these effects similar or different from those observed in the case of extreme wave due to dispersive focusing? A presentation of the different classes of instability of Stokes waves is given in the review paper by Dias & Kharif.[48]

The procedure used to calculate the linear stability of Stokes waves is similar to the method described by Kharif & Ramamonjiarisoa.[49] McLean *et al.*[50] and McLean[51] showed that the dominant instability of a uniformly-traveling train of Stokes' waves in deep water is the two-dimensional modulational instability (class I) provided its steepness is less than $\varepsilon = 0.30$. For higher values of the wave steepness three-dimensional instabilities (class II) become dominant, phase locked to the unperturbed wave. Herein we shall focus on the two-dimensional nonlinear evolution of a Stokes' wave train suffering modulational instability with and without wind action.

We consider the case of wave trains of five waves. The initial condition is a Stokes wave of steepness $\varepsilon = 0.11$, disturbed by its most linearly unstable perturbation which corresponds to $p \simeq 0.20 = 1/5$. The fundamental wavenumber of the Stokes wave is chosen so that integral numbers of the sidebands perturbation (satellites) can be fitted into the computational domain. For $p = 1/5$ the fundamental wave harmonic of the Stokes wave is $k_0 = 5$ and the dominant sidebands are $k_1 = 4$ and $k_2 = 6$ for subharmonic and superharmonic part of the perturbation respectively. The wave parameters have been re-scaled to have the wavelength of the perturbation equal to 2π. There exists higher harmonics present in the interactions which are not presented here. The normalized amplitude of the perturbation relative to Stokes wave amplitude is initially taken equal to 10^{-3}. The order of nonlinearity is $M = 6$, the number of mesh points is greater than $(M+1)N$ where N is the highest harmonic taken into account in the simulation. The latter criterion concerning N is introduced to avoid aliasing errors. To compute the long time evolution of the wave train the time step Δt is chosen equal to $T/100$ where T is the fundamental period of the basic wave. This temporal discretisation satisfies the CFL condition.

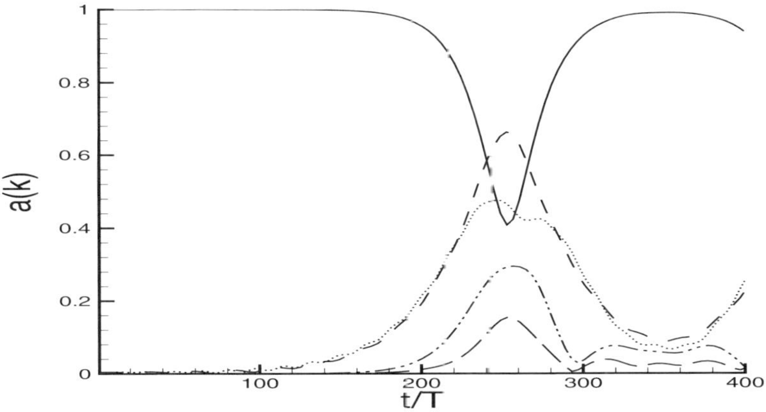

Fig. 7. Time histories of the amplitude of the fundamental mode, $k_0 = 5$ (solid line), subharmonic mode, $k_1 = 4$ (dashed line), and superharmonic mode, $k_2 = 6$ (dotted line), for an evolving perturbed Stokes wave of initial wave steepness $\epsilon = 0.11$ and fundamental wave period T, without wind action. The two lowest curves (dot-dot-dashed and dot-dashed lines) correspond to the modes $k_3 = 3$ and $k_4 = 7$.

For the case without wind, the time histories of the normalized amplitude of the carrier, lower sideband and upper sideband of the most unstable perturbation are plotted in Figure 7. Another perturbation which was initially linearly stable becomes unstable in the vicinity of maximum of modulation resulting in the growth of the sidebands $k_3 = 3$ and $k_4 = 7$. The nonlinear evolution of the two-dimensional wave train exhibits the Fermi-Pasta-Ulam recurrence phenomenon. This phenomenon is characterized by a series of modulation-demodulation cycles in which initially uniform wave trains become modulated and then demodulated until they are again uniform. Herein one cycle is reported. At $t \approx 360T$ the initial condition is more or less recovered. At the maximum of modulation $t = 260T$, one can observe a temporary frequency (and wavenumber) downshifting since the subharmonic mode $k_1 = 4$ is dominant. At this stage a very steep wave occurs in the group as it can be seen in Figure 8(a). Note that the solid line represents the free surface without wind effect while the dotted line corresponds to the case with wind effect that is discussed below. Figures 8(b), 8(c) and 8(d) show the free surface profiles at several instants of time. The solid lines correspond to the case without wind action. We can emphasize that no breaking occurs during the numerical simulation.

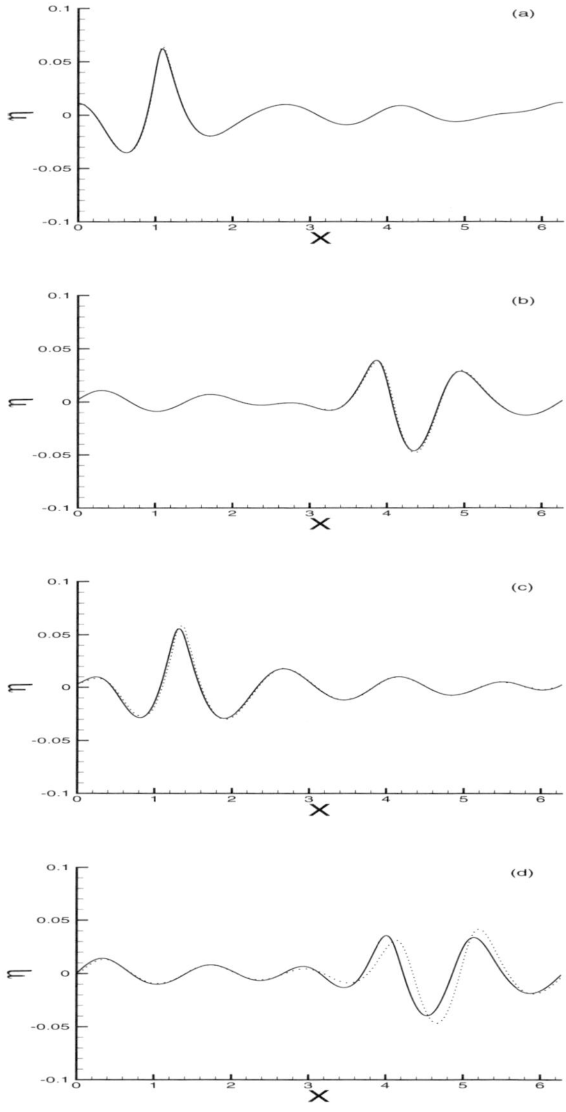

Fig. 8. Surface wave profile at (a) $t = 260T$, (b) $t = 265T$, (c) $t = 270T$, (d) $t = 275T$: without wind (solid line) and with wind (dotted line).

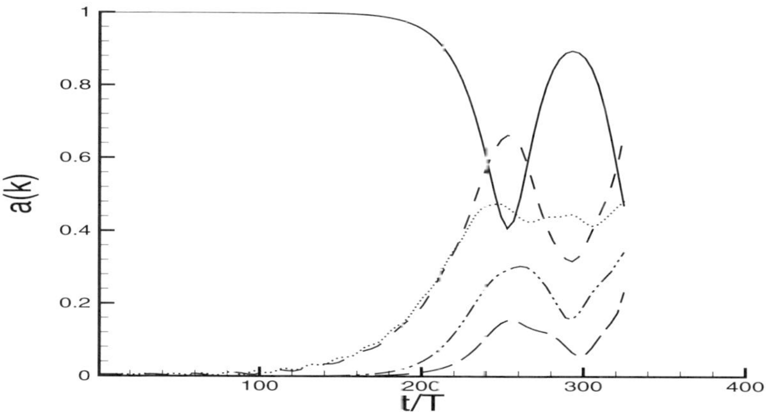

Fig. 9. Time histories of the amplitude of the fundamental mode, $k_0 = 5$ (solid line), subharmonic mode, $k_1 = 4$ (dashed line), and superharmonic mode, $k_2 = 6$ (dotted line), for an evolving perturbed Stokes wave of initial wave steepness $\epsilon = 0.11$ and fundamental wave period T, with wind action. The two lowest curves (dot-dot-dashed and dot-dashed lines) correspond to the modes $k_3 = 3$ and $k_4 = 7$.

Dold & Peregrine[52] have studied numerically the nonlinear evolution of different modulating wave trains towards breaking or recurrence. For a given number of waves in the wave trains breaking always occurs above a critical initial steepness, and below a recurrence towards the initial wave group is observed. This problem was revisited by Banner & Tian[53] who however did not consider the excitation at the maximum of modulation of the perturbation corresponding to $p = 2/5$.

Figure 9 is similar to Figure 7, except that now water waves evolve under wind action. Wind forcing is applied over crests of the group of slopes larger than $(\partial\eta/\partial x)_c = 0.405$. This condition is satisfied for $256T < t < 270T$, that is during the maximum of modulation which corresponds to the formation of the extreme wave event. When the values of the wind velocity are too high numerical simulations fail during the formation of the extreme wave event, due to breaking. During breaking wave process the slope of the surface becomes infinite, leading numerically to a spread of energy into high wavenumbers. This local steepening is characterized by a numerical blow-up (for methods dealing with an Eulerian description of the flow). To avoid a too early breaking wave, the wind velocity U is fixed close to $1.75c$. Owing to the weak effect of the wind on the kinematics

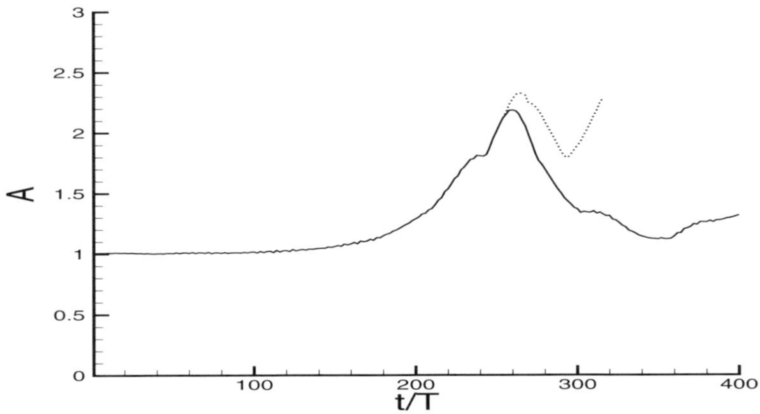

Fig. 10. Numerical amplification factor as a function of time for a wave train of five waves without wind (solid line) and with wind (dotted line) for $U = 1.75c$.

of the crests on which it acts, the phase velocity, c, is computed without wind. The effect of the wind reduces significantly the demodulation cycle and thus sustains the extreme wave event. This feature is clearly shown in Figure 10. The amplification factor is stronger in the presence of wind and the rogue wave criterion given by $A > 2.2$, is satisfied during a longer period of time. In the presence of wind forcing extreme waves evolve into breaking waves at $t \approx 330T$. Figures 8(a), 8(b), 8(c) and 8(d) display water wave profiles at different instants of time in the vicinity of the maximum of modulation with and without wind. The solid lines corresponds to waves propagating without wind while the dotted lines represent the wave profiles under wind action. These Figures show that the wind does not modify the phase velocity of the very steep waves while it increases their height and their duration.

6. Conclusions

This paper reports on application of both Boundary Integral Element Method (BIEM) and High-Order Spectral Method (HOSM) to investigate wind effect on rogue waves. Both methods are used to propagate nonlinear wave groups producing extreme wave events. Following Jeffreys, the influence of wind is introduced through a pressure model describing the air flow separation over the steepest crests.

The numerical methods are found to be efficient and well adapted to the physical purposes considered. Within the framework of BIEM, a boundary value problem is derived through the modelling of a wavemaker's motion producing aperiodic wave groups. This method is suitable to investigate the linear focusing of a chirped wave packet and to provide comparison with experiments in wave tank. The second approach (HOSM) is more accurate to propagate periodic nonlinear wave groups on longer distances. The characteristic time scale of the modulational instability being quite important, it is desirable to use fast numerical methods such as HOSM.

Furthermore, extreme wave events generated by two different mechanisms are found to exhibit the same behavior in the presence of wind. The good agreement obtained between numerical and experimental results justifies the use of the modified Jeffreys'mechanism as a simple approximate model of wind effect on steep wave groups. Generally speaking, this emphasizes the ability of potential theory to simulate wind wave interaction, as soon as a wind model is known.

References

1. G. Lawton, Monsters of the deep (the perfect wave), *New Scientist.* **170 (2297)**, 28–32, (2001).
2. J. K. Mallory, Abnormal waves on the south-east africa, *Int. Hydrog. Rev.* **51**, 89–129, (1974).
3. S. Haver, A possible freak wave event measured at the Draupner Jacket January 1 1995. In *Proc. Rogue Waves.* IFREMER, Brest, (2004).
4. D. H. Peregrine, Interaction of water waves and current, *Adv. Appl. Mech.* **16**, 9–117, (1976).
5. I. V. Lavrenov, The wave energy concentration at the aguhlas current of south africa, *Natural hazards.* **17**, 117–127, (1998).
6. T. Baldock, C. Swan, and P. Taylor, A laboratory study of surface waves on water, *Phil. Trans. R. Soc. Lond. A.* **354**, 649–676, (1996).
7. C. Kharif, J.-P. Giovanangeli, J. Touboul, L. Grare, and E. Pelinovsky, Influence of wind on extreme wave events: Experimental and numerical approaches, *J. Fluid Mech.* **594**, 209–247, (2008).
8. T. B. Benjamin and J. E. Feir, The desintegration of wave trains on deep water. part 1. theory., *J. Fluid Mech.* **27**, 417–430, (1967).
9. D. Clamond and J. Grue, A fast method for fully nonlinear water wave computations, *J. Fluid Mech.* **447**, 337–355, (2001).
10. D. Clamond, M. Francius, J. Grue, and C. Kharif, Strong interaction between envelope solitary surface gravity waves, *Eur. J. Mech. B/Fluids.* **25 (5)**, 536–553, (2006).

11. A. Dyachenko and V. Zakharov, Modulational instability of stokes wave $\rightarrow$ freak wave, *J. Exp. Theor. Phys.* **81 (6)**, 318–322, (2005).
12. V. Zakharov, A. Dyachenko, and A. Prokofiev, Freak waves as nonlinear stage of stokes wave modulation instability, *Eur. J. Mech. B/Fluids.* **25 (5)**, 667–692, (2006).
13. C. Kharif and E. Pelinovsky, Physical mechanisms of the rogue wave phenomenon, *Eur. J. Mech. B/Fluids.* **22**, 603–634, (2003).
14. R. G. Dean and R. A. Darlymple, Water waves mechanics for engineers and scientists. In *Proc. 20th Int. Conf. Coastal Engng.*, p. 368. Word Sci., Singapore, (1991).
15. M. A. Tayfun, Narrow band nonlinear sea waves, *J. Geophys. Res.* **85**, 1548–1552, (1980).
16. M. Onorato, A. R. Osborne, M. Serio, and S. Bertone, Freak waves in random oceanic sea states, *Phys. Rev. Lett.* **86**, 1–4, (2001).
17. P. Janssen, Nonlinear four-wave interactions and freak waves, *J. Phys. Oceanogr.* **33**, 863–884, (2003).
18. O. Gramstad and K. Trulsen, Influence of crest and group length on the occurrence of freak waves, *Journ. Fluid Mech.* **582**, 463–472, (2007).
19. D. K. Dysthe, H. E. Krogstad, and P. Müller, Oceanic rogue waves, *Ann. Rev. Fluid Mech.* **40**, 287–310, (2008).
20. M. S. Longuet-Higgins and E. Cokelet, The deformation of steep surface waves on water, *Proc. Roy. Soc. Ser. A.* **350**, 1–26, (1976).
21. T. Vinje and P. Brevig, Breaking waves on finite depth: a numerical study. Technical Report R-118-81, Ship Res. Inst. Norway, (1981).
22. S. T. Grilli, J. Skourup, and I. A. Svendsen, An efficient boundary element method for nonlinear waves, *Eng. Anal. Bound. Elem.* **6**, 97–107, (1989).
23. R. Cointe, Numerical simulation of a wave channel, *Eng. Anal. Bound. Elem.* **7**, 167–177, (1990).
24. W. Tsai and D. K. P. Yue, Computation of nonlinear free surface flows, *Annu. Rev. Fluid Mech.* **28**, 249–278, (1996).
25. D. G. Dommermuth, D. K. P. Yue, W. M. Lin, R. J. Rapp, E. S. Chan, and W. K. Melville, Deep-water plunging breakers: a comparison between potential theory and experiments, *J. Fluid Mech.* **189**, 423–442, (1988).
26. C. Fochesato, S. T. Grilli, and F. Dias, Numerical modeling of extreme rogue waves generated by directional energy focusing, *Wave Mot.* **44**, 395–416, (2007).
27. H.-O. Kreiss and J. Oliger, Comparison of accurate methods for the integration of hyperbolic equations, *Tellus.* **24**, 703–714, (1972).
28. J. W. Cooley and J. W. Turkey, An algorithm for the machine calculation of the fourier series, *Math. Comput.* **19**, 297–301, (1965).
29. B. Fornberg and G. B. Whitham, A numerical and theoretical study of certain nonlinear wave phenomena, *Phil. Trans. Roy Soc. Lond. A.* **289**, 373–404, (1978).
30. D. G. Dommermuth and D. K. P. Yue, A high-order spectral method for the study of nonlinear gravity waves, *J. Fluid Mech.* **184**, 267–288, (1987).

31. C. Brandini, *Nonlinear interaction processes in extreme wave dynamics*. PhD thesis, University of Firenze, (2001).
32. C. Brandini and S. Grilli, Evolution of three-dimensional unsteady wave modulations.
33. M. Tanaka, A method of studying nonlinear random field of surface gravity waves by direct numerical simulation, *Fluid Dyn. Res.* **28(1)**, 41–60, (2001).
34. G. Ducrozet, F. Bonnefoy, D. L. Touzé, and P. Ferrant, 3-d hos simulations of extreme waves in open seas, *Nat. Hazards Earth Syst. Sci.* **7 (1)**, 109–122, (2007).
35. M. Greco, *A two-dimensional study of green water loading*. PhD thesis, Dept. Marine Hydrodynamics, NTNU, Trondheim, (2001).
36. O. M. Faltinsen, M. Greco, and M. Landrini, Green water loading on a fpso, *J. Offshore Mech. Art. Eng.* **124**, 97–103, (2002).
37. D. Sen, A cubic-spline boundary integral method for two-dimensional free-surface flow problems, *Int. Journ. Nume. Meth. Eng.* **38(11)**, 1809–1830, (1995).
38. V. E. Zakharov, Stability of periodic waves of finite amplitude on the surface of a deep fluid, *J. Appl. Tech. Phys.* **9**, 190–194, (1968).
39. B. J. West, K. A. Brueckner, R. S. Janda, M. Milder, and R. L. Milton, A new numerical method for surface hydrodynamics, *J. Geophys. Res.* **92**, 11803–11824, (1987).
40. C. Skandrani, C. Kharif, and J. Poitevin, Nonlinear evolution of water surface waves: The frequency downshifting phenomenon, *Contemp. Math.* **200**, 157–171, (1996).
41. J. Touboul, J.-P. Giovanangeli, C. Kharif, and E. Pelinovsky, Freak waves under the action of wind: Experiments and simulations, *Eur. J. Mech. B/Fluids.* **25 (5)**, 662–676, (2006).
42. H. Jeffreys, On the formation of waves by wind, *Proc. Roy. Soc. A.* **107**, 189–206, (1925).
43. M. I. Banner and W. K. Melville, On the separation of air flow over water waves, *J. Fluid Mech.* **77**, 825–842, (1976).
44. S. Kawai, Structure of air flow separation over wind wave crests, *Boundary-Layer Mteoro.* **23 (4)**, 503–521, (1982).
45. W. P. Large and S. Pond, Open ocean momentum flux measurements in moderate to strong winds, *J. Phys. Oce.* **11**, 324–336, (1981).
46. J. Touboul, E. Pelinovsly, and C. Kharif, Nonlinear focusing wave groups on current, *J. Kor. Soc. Cost. Oc. Eng.* **3**, 222–227, (2007).
47. M. Lightill, Contribution to the theory of waves in nonlinear dispersive systems, *J. Inst. Math. Appl.* **1**, 269–306, (1965).
48. F. Dias and C. Kharif, Nonlinear gravity and capillary-gravity waves, *Annu. Rev. Fluid Mech.* **31**, 301–346, (1999).
49. C. Kharif and A. Ramamonjiarisoa, Deep water gravity wave instabilities at large steepness, *Phys. Fluids.* **31**, 1286–1288, (1988).
50. J. W. McLean, Y. C. Ma, D. U. Martin, P. G. Saffman, and H. C. Yuen, Three-dimensional instability of finite-amplitude water waves, *Phys. Rev. Lett.* **46**, 817–820, (1981).

51. J. W. McLean, Instabilities of finite-amplitude water waves, *J. Fluid Mech.* **114**, 315–330, (1982).
52. J. W. Dold and D. H. Peregrine, Water wave modulation. In *Proc. 20th Int. Conf. Coastal Engng.*, vol. 1, pp. 163–175. ASCE, Taipei, (1986).
53. M. I. Banner and X. Tian, On the determination of the onset of breaking for modulating surface gravity water waves, *J. Fluid Mech.* **367**, 107–137, (1998).

CHAPTER 3

PROGRESS IN FULLY NONLINEAR POTENTIAL FLOW MODELING OF 3D EXTREME OCEAN WAVES

Stephan T. Grilli*

Department of Ocean Engineering, University of Rhode Island Narragansett, RI 02882, USA

grilli@oce.uri.edu

Frédéric Dias†

CMLA, ENS Cachan, CNRS, PRES UniverSud, 61 Av. President Wilson, F-94230 Cachan, France

Philippe Guyenne‡

Department of Mathematical Sciences, University of Delaware, Newark DE 19716, USA

Christophe Fochesato§

CMLA, ENS Cachan, CNRS, PRES UniverSud, 61 Av. President Wilson, F-94230 Cachan, France

François Enet¶

Alkyon Hydraulic Consultancy and research, 8316PT Marknesse The Netherlands

---

*SG and FE acknowledge the US National Science Foundation (NSF), under grant CMS-0100223 of "the Engineering/Earthquake, Hazards and Mitigation Program", and SG also acknowledges the US Office of Naval Research, under grant N000140510068, for supporting part of this work. Jeff Harris' help is gratefully acknowledged for providing original data for the last two figures.

†frederic.dias@cmla.ens-cachan.fr.

‡guyenne@math.udel.edu. PG acknowledges support from the University of Delaware Research Foundation and the US NSF under grant DMS-0625931.

§fochesat@cmla.ens-cachan.fr.

¶enet@alkyon.nl.

This article reviews recent research progress by the authors and co-workers, in the application of three-dimensional (3D) Numerical Wave Tanks (NWT), based on Fully Nonlinear Potential Flow theory (FNPF), to the modeling of extreme, overturning, ocean waves and of their properties, in both deep and shallow water. Details of the model equations and numerical methods are presented. Applications are then presented for the shoaling and 3D overturning in shallow water of solitary waves over a sloping ridge, for the generation of extreme deep and intermediate water waves, often referred to as "rogue" waves, by directional energy focusing, for the generation of tsunamis by solid underwater landslides, and for the generation of surface waves by a moving pressure disturbance. In all cases, physical and numerical aspects are presented and properties of generated waves are discussed at the breaking point. Aspects of numerical methods influencing the accuracy and the efficiency of the NWT solution are detailed in the article. Specifically, the 3D-NWT equations are expressed in a mixed Eulerian-Lagrangian formulation (or pseudo-Lagrangian in one case) and solved based on a higher-order Boundary Element Method (BEM), for the spatial solution, and using explicit higher-order Taylor series expansions for the time integration. Direct and iterative solutions of the governing equations are discussed, as well as results of a recent application of the Fast Multipole Algorithm. Detailed aspects of the model such as the treatment of surface piercing solid boundaries are discussed as well.

1. Introduction

Over the last three decades, many studies have been carried out to achieve a better modeling and understanding of the generation of extreme waves in the deep ocean, as well as the propagation, shoaling, and breaking of ocean waves over a sloping nearshore topography. Such work was motivated by a variety of fundamental and practical ocean engineering problems. For instance, a description of the dynamics of breaking waves is necessary to explain the mechanisms of air–sea interactions, such as energy and momentum transfer from wind to water and from waves to currents, and the generation of turbulence in the upper ocean. The interaction and impact of extreme ocean waves (often referred to as rogue waves) with fixed or floating structures in deep water is thought to represent one of the highest hazard for the design of such structures. In nearshore areas, breaking wave induced currents are the driving mechanism for sediment transport, which leads to beach erosion and accretion, and also represent the design load for coastal structures used for shoreline and harbor protection. Despite significant progress, due to its complexity, the process of wave breaking has not yet been fully explained. Reviews of wave breaking phenomena in shallow and deep water can be found in Refs. 1 and 2, respectively.

In this chapter, we report on our experience with three-dimensional (3D) numerical simulations aimed at very accurately describing the early stages of wave breaking induced by directional energy focusing in deep water and changes in topography in shallow water, namely the phenomenon of wave overturning. Additionally, we report on additional applications of the same modeling approach, to the generation of tsunamis by underwater landslides and the generation of waves by a moving free surface pressure disturbance representing the air cushion of a fast surface effect ship.

We concentrate on cases in which 3D effects are induced in the wave flow, and, at the final stages of wave overturning, we pay particular attention to large size plunging breakers which are characterized by the formation of a more prominent jet (rather than smaller size spilling breakers). Such large jets would create the strongest wave impact on ocean structures, as well as cause the larger disturbances in beach sediment. A high-order 3D numerical model, solving Fully Nonlinear Potential Flow (FNPF) equations is developed and used in this work. The potential flow approximation is justified for initially irrotational waves or flows starting from rest, considering the slow diffusion of vorticity from boundaries until the breaker jet touches down. In fact, comparisons of two-dimensional (2D) numerical results with laboratory experiments have consistently shown that FNPF theory accurately predicts the characteristics of wave overturning, in deep water (e.g., Refs. 3 and 4), as well as wave shoaling and overturning over slopes (e.g., Refs. 5 and 6). In the latter work, the model predicts the shape and kinematics of shoaling solitary waves over mild slopes, within 2% of experimental measurements in a precision wavetank, up to the breaking point. Other experiments showed that the shape of such overturning waves is then accurately modeled up to touch down.[7,8] Beyond touch down of the breaker jet, strong vorticity and energy dissipation occur and a full Navier-Stokes (NS) model must be used. Lin and Liu,[9] Chen et al.[10] and Christensen and Deigaard,[11] for instance, studied the breaking and post-breaking of solitary waves using a 2D-NS model. The latter authors also calculated 3D turbulent fields using a LES model and Lubin[12] similarly performed 3D LES simulations of plunging breaking waves. Guignard et al.[13] and Lachaume et al.[14] proposed a coupled model that combines both the accuracy and efficiency of a 2D-FNPF model, used during the shoaling and overturning phases of wave propagation, and a 2D-NS modeling of the surfzone.

Due to computer limitations, detailed numerical studies of wave breaking with FNPF models have initially focused on 2D problems. In this respect, significant contributions in the numerical simulation of steep fully nonlinear waves were made by Longuet-Higgins and Cokelet,[15] who first proposed a mixed Eulerian–Lagrangian (MEL) approach for the time updating, combined with a Boundary Integral Equation (BIE) formulation, in a deep water space-periodic domain. Their computations were able to reproduce overturning waves by specifying a localized surface pressure. Similar methods were adopted in subsequent works, notably, by Vinje and Brevig[16] and Baker et al.,[17] who considered the case of finite depth. Results obtained by New et al.[18,19] for plunging waves over constant depth, greatly contributed to our understanding of breaking wave kinematics. These authors carried out high-resolution computations for various types of breakers and analyzed in detail the overturning motions, by following fluid particle trajectories in the space, velocity, and acceleration planes. More recent 2D-FNPF models can accommodate both arbitrary waves and complex bottom topography, as well as surface-piercing moving boundaries such as wavemakers. These models are directly implemented in a physical space region, where incident waves can be generated at one extremity and reflected, absorbed or radiated at the other extremity (e.g., Refs. 20–23). For these reasons, they are often referred to as Numerical Wave Tanks (NWT).

A comparatively smaller number of works have addressed, essentially non-breaking, 3D-FNPF wave simulations, due to the more difficult geometric representation as well as the more demanding computational problems (e.g., Refs. 24, 25, 28–33). In particular, the problem of strongly nonlinear waves requires very accurate and stable numerical methods, and this consequently leads to an increase of the computational cost. Xü and Yue[34] and Xue et al.[35] calculated 3D overturning waves in a doubly periodic domain with infinite depth (i.e. only the free surface is discretized). They used a quadratic Boundary Element Method (BEM) to solve the equations in a MEL formulation. As in Ref. 15, the initial conditions were progressive Stokes waves and a localized surface pressure was applied to make waves break. These authors performed a detailed analysis of the kinematics of plunging waves and quantified the three-dimensional effects on the flow,[26,27] developed a similar method, for a non-periodic domain with finite depth. They were also able to produce the initial stages of wave overturning over a bottom obstacle. More recently, Grilli et al.[36] proposed an accurate 3D NWT model, for the description of strongly nonlinear wave generation and propagation over complex bottom topography. This NWT is based on a MEL explicit time stepping and a high-order BEM with third-order spatial discretization, ensuring local continuity of the inter-element slopes. Arbitrary waves can be generated in this NWT and, if needed, absorbing conditions can be specified on lateral boundaries. Although an application to the shoaling of a solitary wave up to overturning was shown, this initial paper focused more on the derivation and validation of the numerical model and methods rather than on the physical implications of results. Other applications and extensions of this NWT to other nonlinear wave processes were done for the modelling of: (i) wave impact on a vertical wall;[37] (ii) freak wave generation due to directional wave focusing;[38–40] (iii) tsunami generation by submarine mass failure;[41–43] (iv) the generation of waves by a moving surface pressure disturbance;[44–46] and (v) the simulation of co-seismic tsunami generation by a specified bottom motion.[47] The reader interested in boundary integral methods, especially in applications to 3D free surface flows, is also referred to Ref. 48.

In this chapter, we first detail equations and numerical methods for the 3D-FNPF NWT model initially developed by Ref. 36, and subsequently extended by Refs. 38, 41, 45, 50 and 49, concentrating on the different types of boundary conditions, free surface updating, boundary representation, and fast equation solver. We then present a number of typical applications of the model to the breaking of solitary waves over a sloping ridge, extreme overturning waves created by directional energy focusing using a wavemaker, landslide tsunami generation, and surface wave generation by a moving disturbance. Due to the complexity of the problem and the high computational cost, our studies of 3D breakers are typically restricted to fairly small spatial domains. Beside wave shape, various results are presented for the velocity and acceleration fields before and during wave overturning, both on the free surface and within the flow. We stress that no smoothing of the solution is required, e.g., to suppress spurious waves at any time in the computations, as experienced in most of the other proposed models.

2. Mathematical Formulation

2.1. *Governing equations and boundary conditions*

We assume an incompressible inviscid fluid, with irrotational motion described by the velocity potential $\phi(\boldsymbol{x}, t)$, in a Cartesian coordinate system $\boldsymbol{x} = (x, y, z)$ (with z the vertical upward direction and $z = 0$ at the undisturbed free surface).

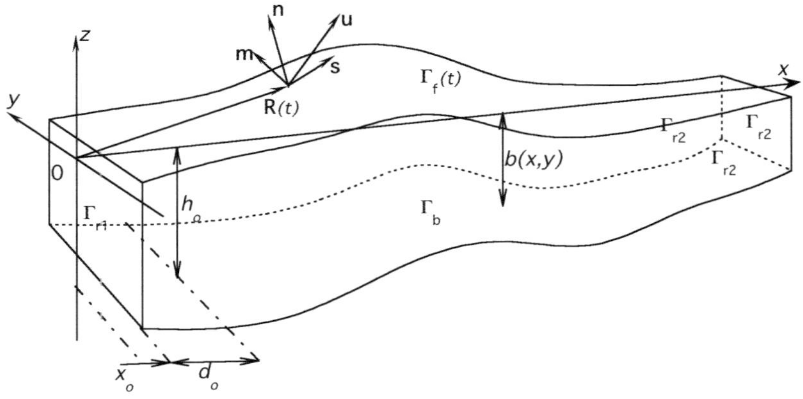

Fig. 1. Sketch of typical computational domain for 3D-BEM solution of FNPF equations. Tangential vectors at point $\boldsymbol{R}(t)$ on the free surface $\Gamma_f(t)$ are defined as $(\boldsymbol{s}, \boldsymbol{m})$ and outward normal vector as $\boldsymbol{n}$.

The fluid velocity is thus defined as $\boldsymbol{u} = \boldsymbol{\nabla}\phi = (u, v, w)$ and mass conservation is Laplace's equation for the potential, which in the fluid domain $\Omega(t)$ with boundary $\Gamma(t)$ reads (Fig. 1),

$$\nabla^2 \phi = 0 \,. \tag{2.1}$$

Applying Green's second identity, Eq. (2.1) transforms into the BIE,

$$\alpha(\boldsymbol{x}_l)\phi(\boldsymbol{x}_l) = \int_\Gamma \left\{ \frac{\partial \phi}{\partial n}(\boldsymbol{x}) G(\boldsymbol{x}, \boldsymbol{x}_l) - \phi(\boldsymbol{x}) \frac{\partial G}{\partial n}(\boldsymbol{x}, \boldsymbol{x}_l) \right\} \mathrm{d}\Gamma \tag{2.2}$$

where the field point $\boldsymbol{x}_l$ and the source point $\boldsymbol{x}$ are both on the boundary, $\alpha(\boldsymbol{x}_l) = \frac{1}{4\pi}\theta_l$, with θ_l the exterior solid angle at point $\boldsymbol{x}_l$, and the three-dimensional (3D) free-space Green's function is defined as,

$$G(\boldsymbol{x}, \boldsymbol{x}_l) = \frac{1}{4\pi r} \quad \text{with} \quad \frac{\partial G}{\partial n}(\boldsymbol{x}, \boldsymbol{x}_l) = -\frac{1}{4\pi} \frac{\boldsymbol{r} \cdot \boldsymbol{n}}{r^3} \tag{2.3}$$

where $r = |\boldsymbol{r}| = |\boldsymbol{x} - \boldsymbol{x}_l|$ and $\boldsymbol{n}$ is the outward unit vector normal to the boundary at point $\boldsymbol{x}$.

The boundary is divided into various parts with different boundary conditions (Fig. 1). On the free surface $\Gamma_f(t)$, ϕ satisfies the nonlinear kinematic and dynamic boundary conditions, which in the MEL formulation read,

$$\frac{\mathrm{D}\boldsymbol{R}}{\mathrm{D}t} = \boldsymbol{u} \tag{2.4}$$

$$\frac{\mathrm{D}\phi}{\mathrm{D}t} = -gz + \tfrac{1}{2}\boldsymbol{\nabla}\phi\cdot\boldsymbol{\nabla}\phi - \frac{p_a}{\rho} \tag{2.5}$$

respectively, with $\boldsymbol{R}$ the position vector of a fluid particle on the free surface, g the gravitational acceleration, p_a the atmospheric pressure, ρ the fluid density and $\mathrm{D}/\mathrm{D}t = \partial/\partial t + \boldsymbol{\nabla}\phi\cdot\boldsymbol{\nabla}$ the Lagrangian (or material) time derivative. Effects of surface tension are neglected considering the large scale waves modeled here (however, this could be easily added to Eq. (2.5)).

In earlier applications of the model, waves have been generated in various ways: (i) by directly specifying wave elevation and potential (such as a solitary wave; e.g., Ref. 51) on the free surface at $t = 0$;[36,37,52,53] (ii) by simulating a (solid) wavemaker motion at the 'open sea' boundary side of the model Γ_{r1};[38–40] (iii) by specifying the motion of a solid underwater landslide on the bottom boundary Γ_b;[41–43] or (iv) by a moving pressure disturbance on the free surface.[44–46,54]

Over moving boundaries such as wavemakers (or landslides), both the boundary geometry $\boldsymbol{x}_p(t)$ and normal fluid velocity are specified as,

$$\overline{\boldsymbol{x}} = \boldsymbol{x}_p \quad \text{and} \quad \overline{\frac{\partial\phi}{\partial n}} = \boldsymbol{u}_p\cdot\boldsymbol{n} \tag{2.6}$$

where overbars denote specified values and $\boldsymbol{u}_p(\boldsymbol{x}_p, t)$ is the boundary velocity.

In all cases, one (or two in case (iv)) open boundary conditions can be specified on some vertical sections $\Gamma_{r2}(t)$, of the model boundary. Following,[38,41] the open boundary is modeled as a pressure sensitive 'snake' flap- or piston-type absorbing wavemaker. For a piston-type boundary in depth h_o, for instance, the piston normal velocity is specified as,

$$\overline{\frac{\partial\phi}{\partial n}} = u_{ap}(\sigma, t) \qquad \text{on } \Gamma_{r2}(t), \text{ with,} \tag{2.7}$$

$$u_{ap}(\sigma, t) = \frac{1}{\rho h_o\sqrt{gh_o}}\int_{-h_o}^{\eta_{ap}(\sigma,t)} p_D(\sigma, z, t)\,\mathrm{d}z \tag{2.8}$$

calculated at the curvilinear abscissa σ, horizontally measured along the piston boundary, where η_{ap} is the surface elevation at the piston location and $p_D = -\rho_w\{\frac{\partial\phi}{\partial t} + \frac{1}{2}\nabla\phi\cdot\nabla\phi\}$ denotes the dynamic pressure. The integral in Eq. (2.8) represents the horizontal hydrodynamic force $F_D(\sigma, t)$ acting on the piston at time t, as a function of σ. For 2D problems, Grilli and Horrillo[23] showed that this type of boundary condition absorbs well energy from long incident waves. Shorter waves, however, will be reflected to a greater extent by the boundary, but can directly be damped on the free surface using an 'Absorbing Beach' (AB). Such a pressure always works against the waves and hence is equivalent to an energy dissipation over time. For 2D problems, Grilli and Horrillo[23] implemented the AB over a section of the free surface by specifying the pressure in the dynamic boundary condition Eq. (2.5) such as to be opposite and proportional to the normal fluid velocity,

$$p_a = \nu(\boldsymbol{x}, t) \frac{\partial \phi}{\partial n}(\eta(\boldsymbol{x}, t)) \tag{2.9}$$

in which ν, the AB absorption function, is smoothly varied along the AB and η refers to nodes on the free surface of elevation η. This approach was extended by Grilli et al.[55] to model the effect of energy dissipation due to bottom friction on the shoaling of 2D periodic waves. In this case, the coefficient $\nu(\boldsymbol{x}, t)$ was found at each time step by expressing a balance between the time averaged energy dissipation on the bottom and the free surface pressure. In some earlier applications of the 3D model,[44–46] a similar AB was specified over an area of the free surface to damp shorter waves, in combination with the absorbing wavemaker boundary. Note, recently, Dias et al.[56] derived new free surface boundary conditions to account for energy dissipation due to viscosity in potential flow equations. In these equations, a correction term is added to both the kinematic and dynamic boundary conditions, Eqs. (2.4) and (2.5), respectively. It should be of interest to implement this new set of equations in the NWT.

Except in case (iii), a no-flow condition is specified on the bottom Γ_b and other fixed parts of the boundary referred to as Γ_{r2} as,

$$\overline{\frac{\partial \phi}{\partial n}} = 0\,. \tag{2.10}$$

2.2. *Internal velocity and acceleration*

Once the BIE Eq. (2.2) is solved, the solution within the domain can be evaluated from the boundary values. Using Eq. (2.2), the internal velocity is given by,

$$\boldsymbol{u}(\boldsymbol{x}_i) = \boldsymbol{\nabla}\phi(\boldsymbol{x}_i) = \int_\Gamma \left[\frac{\partial \phi}{\partial n}(\boldsymbol{x})\boldsymbol{Q}(\boldsymbol{x}, \boldsymbol{x}_i) - \phi(\boldsymbol{x})\frac{\partial \boldsymbol{Q}}{\partial n}(\boldsymbol{x}, \boldsymbol{x}_i)\right] \mathrm{d}\Gamma \tag{2.11}$$

with,

$$Q(x, x_i) = \frac{1}{4\pi r^3} r\,, \quad \frac{\partial Q}{\partial n}(x, x_i) = \frac{1}{4\pi r^3}\left[n - 3(r \cdot n)\frac{r}{r^2} \right] \tag{2.12}$$

and r denoting the distance from the boundary point x to the interior point x_i. Note that the coefficient $\alpha(x_i)$ is unity for interior points.

Similarly, we express the internal Lagrangian acceleration as,

$$\frac{\mathrm{D}u}{\mathrm{D}t} = \frac{\mathrm{D}}{\mathrm{D}t}\nabla\phi = \frac{\partial}{\partial t}\nabla\phi + (\nabla\phi \cdot \nabla)\nabla\phi \tag{2.13}$$

where the first term on the right-hand side, corresponding to the local acceleration, is given by,

$$\nabla\frac{\partial\phi}{\partial t}(x_i) = \int_\Gamma \left[\frac{\partial^2\phi}{\partial t\partial n}(x) Q(x, x_i) - \frac{\partial\phi}{\partial t}(x)\frac{\partial Q}{\partial n}(x, x_i) \right] \mathrm{d}\Gamma \tag{2.14}$$

and the second term is computed using Eq. (2.11) and differentiating $\nabla\phi$. This requires calculating the spatial derivatives of all components of Q and $\partial Q/\partial n$ as, (note, the index summation convention does not apply for the two following equations)

$$\frac{\partial Q_k}{\partial x_j} = \begin{cases} \frac{3}{4\pi r^5} r_k r_j\,, & k \neq j \\ \frac{1}{4\pi r^3}\left(\frac{3}{r^2} r_k^2 - 1\right)\,, & k = j \end{cases} \tag{2.15}$$

$$\frac{\partial}{\partial x_j}\left(\frac{\partial Q_k}{\partial n}\right) = \begin{cases} \frac{3}{4\pi r^5}\left[r_j n_k + r_k n_j - \frac{5}{r^2}(r \cdot n) r_k r_j \right]\,, & k \neq j \\ \frac{3}{4\pi r^5}\left[r \cdot n + 2 r_k n_k - \frac{5}{r^2}(r \cdot n) r_k^2 \right]\,, & k = j\,, \end{cases} \tag{2.16}$$

where k, j refer to the spatial dimensions and r_k stands for the k-th component of r.

The boundary quantities $\partial\phi/\partial t$ and $\partial^2\phi/\partial t\partial n$ in Eq.(2.14) also satisfy a BIE similar to Eq. (2.2), for ϕ and $\partial\phi/\partial n$. In fact, this second BIE is solved at each time step to compute these fields, which are necessary to perform the second-order time updating of free surface nodes; this is detailed in Section 3.2.

2.3. *Boundary velocity and acceleration*

As will be detailed in Section 3.2, particle velocity u and acceleration $\mathrm{D}u/\mathrm{D}t$ are used to perform the second-order Lagrangian time updating of collocation nodes on the free surface. Similar terms are also needed for expressing boundary conditions on other moving or deforming parts of the boundary, such as wavemakers, open boundaries, or underwater landslides.

At boundary nodes, geometry, field variables and their derivatives are expressed in a local curvilinear coordinate system $(\boldsymbol{s}, \boldsymbol{m}, \boldsymbol{n})$ (Fig. 2) defined within a single reference element, which for practical reasons is different from that of the BEM discretization that will be detailed in Section 3.1. Thus, at each collocation point $\boldsymbol{x}_l$ on the boundary, the unit tangential vectors are defined as

$$\boldsymbol{s} = \frac{1}{h_1}\frac{\partial \boldsymbol{x}_l}{\partial \xi} \quad \text{and} \quad \boldsymbol{m} = \frac{1}{h_2}\frac{\partial \boldsymbol{x}_l}{\partial \eta} \tag{2.17}$$

where

$$h_1 = \left|\frac{\partial \boldsymbol{x}_l}{\partial \xi}\right|, \quad h_2 = \left|\frac{\partial \boldsymbol{x}_l}{\partial \eta}\right| \tag{2.18}$$

and $-1 \leq (\xi, \eta) \leq 1$ denote the intrinsic coordinates of the reference element (this aspect will also be detailed later). A third unit vector in the normal direction is then defined as $\boldsymbol{n} = (\boldsymbol{s} \times \boldsymbol{m})/\sin\widehat{(\boldsymbol{s}, \boldsymbol{m})}$.

For the initial typically orthogonal discretization in the model, the local coordinate system $(\boldsymbol{s}, \boldsymbol{m}, \boldsymbol{n})$ is also orthogonal and we simply have $\boldsymbol{n} = \boldsymbol{s} \times \boldsymbol{m}$. Unlike initially assumed,[36] however, the orthogonality of $\boldsymbol{s}$ and $\boldsymbol{m}$ does not strictly hold when the boundary elements are distorted, such as in regions of large surface deformations. Fochesato et al.[50] extended the expressions of tangential derivatives to the general case when $\boldsymbol{s}$ and $\boldsymbol{m}$ are not orthogonal; this is summarized below.

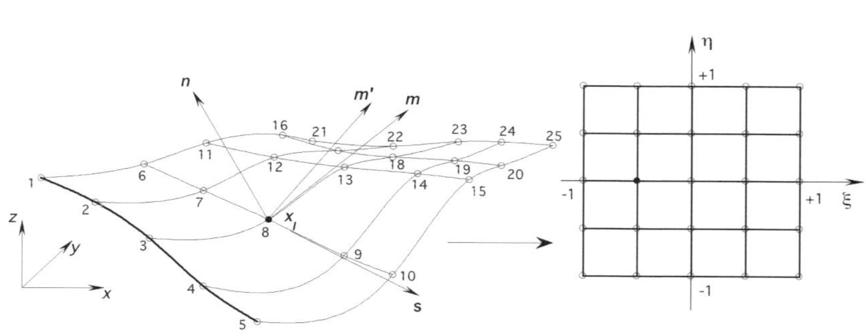

Fig. 2. Sketch of local interpolation by fourth-order two-dimensional sliding polynomial element of (ξ, η), for calculating tangential derivatives in orthogonal axes $(\boldsymbol{s}, \boldsymbol{m}', \boldsymbol{n})$ at collocation point $\boldsymbol{x}_l$ on the boundary.

Using $\boldsymbol{s}$ and $\boldsymbol{m}$ in Eq. (2.17), one can define a new unit tangential vector as (Fig. 2)

$$\boldsymbol{m}' = \frac{1}{\sqrt{1-\kappa^2}}\boldsymbol{m} - \frac{\kappa}{\sqrt{1-\kappa^2}}\boldsymbol{s} \tag{2.19}$$

so that $\boldsymbol{s}$ and $\boldsymbol{m}'$ are orthogonal (i.e., $\boldsymbol{s}\cdot\boldsymbol{m}' = 0$), with $\kappa = \boldsymbol{s}\cdot\boldsymbol{m} = \cos\widehat{(\boldsymbol{s},\boldsymbol{m})}$ (Fig. 2). This implies that $-1 < \kappa < 1$. The unit normal vector now takes the form

$$\boldsymbol{n} = \boldsymbol{s}\times\boldsymbol{m}' = \frac{1}{\sqrt{1-\kappa^2}}\boldsymbol{s}\times\boldsymbol{m} \tag{2.20}$$

(which does yield $\boldsymbol{m}' = \boldsymbol{m}$ and $\boldsymbol{s}\cdot\boldsymbol{m} = 0$ only when $\kappa = 0$). For clarity, let us introduce the following notations

$$(\,)_s \equiv \frac{\partial}{\partial s} = \frac{1}{h_1}\frac{\partial}{\partial \xi}\,, \quad (\,)_m \equiv \frac{\partial}{\partial m} = \frac{1}{h_2}\frac{\partial}{\partial \eta}\,, \quad (\,)_n \equiv \frac{\partial}{\partial n} \tag{2.21}$$

and

$$(\,)_{ss} \equiv \frac{1}{h_1^2}\frac{\partial^2}{\partial \xi^2}\,, \quad (\,)_{sm} \equiv \frac{1}{h_1 h_2}\frac{\partial^2}{\partial \xi \partial \eta}\,, \quad (\,)_{mm} \equiv \frac{1}{h_2^2}\frac{\partial^2}{\partial \eta^2}\,. \tag{2.22}$$

In the orthonormal coordinate system $(\boldsymbol{s}, \boldsymbol{m}', \boldsymbol{n})$, the particle velocity on the boundary is expressed as

$$\boldsymbol{u} = \boldsymbol{\nabla}\phi = \phi_s\boldsymbol{s} + \phi_{m'}\boldsymbol{m}' + \phi_n\boldsymbol{n} \tag{2.23}$$

where ϕ denotes the velocity potential. In the general coordinate system $(\boldsymbol{s}, \boldsymbol{m}, \boldsymbol{n})$, this equation thus becomes

$$\boldsymbol{u} = \frac{1}{1-\kappa^2}(\phi_s - \kappa\phi_m)\boldsymbol{s} + \frac{1}{1-\kappa^2}(\phi_m - \kappa\phi_s)\boldsymbol{m} + \phi_n\boldsymbol{n} \tag{2.24}$$

after using Eq. (2.19) and the fact that

$$\phi_{m'} = \frac{1}{\sqrt{1-\kappa^2}}\phi_m - \frac{\kappa}{\sqrt{1-\kappa^2}}\phi_s\,. \tag{2.25}$$

Laplace's equation $\boldsymbol{\nabla}^2\phi = 0$ can be similarly expressed on the boundary, as well as particle accelerations (by applying the material derivative to Eq. (2.24)), which are both required to calculate second-order terms in the time updating method (see Section 3.2). These expressions are given in Appendix A.1.

3. Numerical Methods

Many different numerical methods have been proposed for solving FNPF equations for water waves in 2D or 3D (see, e.g., Ref. 57 for a recent review). Here we solve FNPF equations using the BIE formulation outlined above, which was initially implemented[36] in this 3D model, based on initial 3D-BEM modeling of linear waves[58] and as an extension of earlier 2D-FNPF models.[20,22,23,59,60] The model benefits from recent improvements in the numerical formulations and solution.[49,50] Numerical methods are briefly summarized below and detailed in the following subsections.

The model consists in a time stepping algorithm in which the position vector and velocity potential on the free surface are updated, based on second-order Taylor series expansions. At each time step, the BIE Eq. (2.2) is expressed for N_Γ collocation nodes, defining the domain boundary Γ, and solved with a BEM, in which boundary elements are specified in between nodes, to locally interpolate both the boundary geometry and field variables, using bi-cubic polynomial shape functions. A local change of variables is defined to express the BIE integrals on a single curvilinear reference element, and compute these using a Gauss-Legendre quadrature and other appropriate techniques removing the weak singularities of the Green's function (based on polar coordinate transformations). The number of discretization nodes yields the assembling phase of the system matrix, resulting in an algebraic system of equations. The rigid mode technique is applied to directly compute external angles α_l and diagonal terms in the algebraic system matrix, which would normally require evaluating strongly singular integrals involving the normal derivative of the Green's function. Multiple nodes are specified on domain edges and corners, in order to easily express different normal directions on different sides of the boundary. Additional equations derived for enforcing continuity of the potential at these nodes also lead to modifications of the algebraic system matrix. The velocity potential (or its normal derivative depending on the boundary condition) is obtained as a solution of the linear system of equations. Since the system matrix is typically fully populated and non-symmetric, the method has, at best, a computational complexity of $O(N_\Gamma^2)$, when using an iterative, optimized conjugate gradient method such as GMRES (see, e.g., Refs. 34 and 35). Thus the spatial solution at each time step is of the same complexity as the numerical cost of assembling of the system matrix. The Fast Multipole Algorithm (FMA) is implemented in the model to reduce this complexity to $O(N_\Gamma \log N_\Gamma)$.

3.1. *Boundary discretization*

A BEM is used to solve the BIE (2.2) for ϕ, and due to the second-order time stepping, a similar equation for $\frac{\partial \phi}{\partial t}$. Thus, the boundary is discretized into N_Γ collocation nodes and M_Γ high-order boundary elements are defined for the local interpolation of both the geometry and field variables in between these nodes. While standard isoparametric elements based on polynomial shape functions can provide a high-order approximation within their area of definition, they only offer $\mathcal{C}^0$ continuity at nodes located in between elements (e.g., Ref. 61). In their 2D work, Grilli et al.[20,22] showed that such discontinuities in slope and curvature can be a source of inaccuracy that, through time updating, may trigger sawtooth instabilities in the model near the crest of steep waves or within overturning jets. These authors showed that a robust treatment requires defining elements, which are both of high-order within their area of definition and at least locally $\mathcal{C}^2$ continuous at their edges. Among various types of approximations, they proposed using so-called middle-interval-interpolation (MII) elements, which in 2D are 4-nodes cubic isoparametric elements, in which only the interval between the middle two nodes is used for the interpolation. In the initially developed 3D model,[36] we used an extension of the MII by defining boundary elements that are 4×4-node quadrilaterals associated with bi-cubic shape functions. Only one out of the nine sub-quadrilaterals so defined is used for the interpolation, typically the central one, but any other is used for elements located at the intersections between different boundary sections, depending on the location (Fig. 3).

Specifically, the boundary geometry and field variables (denoted hereafter by $u \equiv \phi$ or $\frac{\partial \phi}{\partial t}$ and $q \equiv \frac{\partial \phi}{\partial n}$ or $\frac{\partial^2 \phi}{\partial t \partial n}$ for simplicity) are represented within each MII element using shape functions N_j as,

$$\boldsymbol{x}(\xi, \eta) = N_j(\xi, \eta)\, \boldsymbol{x}_j^k \tag{3.1}$$

$$u(\xi, \eta) = N_j(\xi, \eta)\, u_j^k \quad \text{and} \quad q(\xi, \eta) = N_j(\xi, \eta)\, q_j^k \tag{3.2}$$

where $\boldsymbol{x}_j^k$, u_j^k and q_j^k, are nodal values of geometry and field variables, respectively, for $j = 1, \ldots, m$, locally numbered nodes within each element Γ_e^k, $k = 1, \ldots, M_\Gamma$, and the summation convention is applied to repeated subscripts. Here, shape functions are analytically defined as bi-cubic polynomials over a single reference element $\Gamma_{\xi,\eta}$, to which the M_Γ "Cartesian" elements of arbitrary shape are transformed by a curvilinear change of variables, defined by the Jacobian matrix $\mathbf{J}^k$. The intrinsic coordinates on the reference element are denoted by $(\xi, \eta) \in [-1, 1]$. The shape function polynomial coefficients are simply found analytically by requiring that $u(\xi, \eta)$ take the value u_i^k at node $\boldsymbol{x}_i^k$, that is in Eq. (3.2),

$$u(\xi(\boldsymbol{x}_i^k), \eta(\boldsymbol{x}_i^k)) = N_j(\xi_i, \eta_i)\, u_j^k = u_i^k \,.$$

Hence, for the i-th node of an m-node reference element, shape functions must satisfy,

$$N_j(\xi_i, \eta_i) = \delta_{ij} \quad \text{with} \quad i, j = 1, \ldots, m \ \text{ on } \Gamma_{\xi,\eta} \tag{3.3}$$

and δ_{ij} the Kronecker symbol.

We further define the shape functions as the product of two one-dimensional cubic shape functions $N_c'(\mu)$, with $c = 1, \ldots, 4$ and $\mu \in [-1, 1]$, i.e.

$$N_j(\xi, \eta) = N'_{b(j)}(\mu(\xi, \xi_o))\, N'_{d(j)}(\mu(\eta, \eta_o)) \tag{3.4}$$

with b and $d = 1, \ldots, 4$; $j = 4\,(d-1) + b$, for which the property (3.3) implies,

$$N_c'(\mu_i) = \delta_{ic} \quad \text{with} \quad \mu_i = (2i - 5)/3 \tag{3.5}$$

for $i = 1, \ldots, 4$. Hence, solving Eq. (3.5) we find

$$\begin{aligned} N_1'(\mu) &= \frac{1}{16}\,(1-\mu)\,(9\mu^2 - 1) \ ; \ N_2'(\mu) = \frac{9}{16}\,(1-\mu^2)\,(1-3\mu) \\ N_3'(\mu) &= \frac{9}{16}\,(1-\mu^2)\,(1+3\mu) \ ; \ N_4'(\mu) = \frac{1}{16}\,(1+\mu)\,(9\mu^2 - 1)\,. \end{aligned} \tag{3.6}$$

For the MII method, the additional transformation from μ to the intrinsic coordinates (ξ, η) on the reference element is formally expressed as,

$$\mu(\chi, \chi_o) = \chi_o + \frac{1}{3}\,(1 + \chi) \tag{3.7}$$

with $\chi = \xi$ or η, and $\chi_o = \xi_o$ or $\eta_o = -1$, $-1/3$ or $1/3$, depending on which of the 9 quadrilaterals defined by the $m = 16$ nodes is selected as a function of the location of the Cartesian element with respect to intersections between various parts of the boundary (Fig. 3).

Discretized BIEs

Integrals in Eq. (2.2) are transformed into a sum of integrals over the boundary elements Γ_e^k, $k = 1, \ldots, M_\Gamma$. Each of these integrals is calculated within the reference element $\Gamma_{\xi,\eta}$ by applying the curvilinear change of variables discussed above: $[\boldsymbol{x} \to (\xi, \eta)]$, for which the Jacobian matrix is

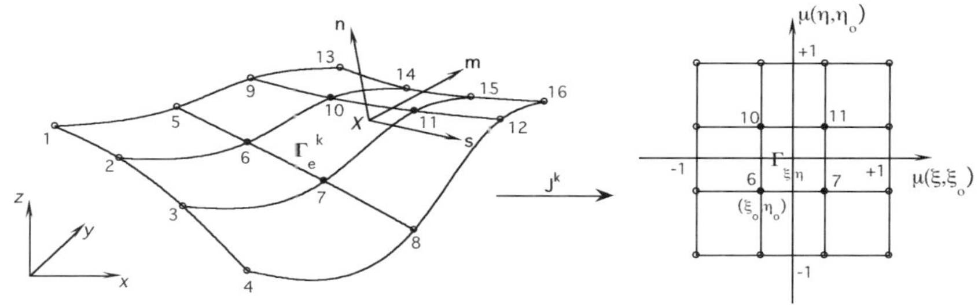

Fig. 3. Sketch of 16-node cubic 3D-MII Cartesian element Γ_e^k and corresponding reference element $\Gamma_{\xi,\eta}$. Quadrilateral element nodes are indicated by symbols (•), and additional nodes by symbols (∘). The curvilinear coordinate system $(\boldsymbol{s}, \boldsymbol{m}, \boldsymbol{n})$ has been marked at point $\boldsymbol{r}$ of the element. (ξ_o, η_o) marks the bottom left node of the quadrilateral, transformed as part of the reference element by Jacobian $\mathbf{J}^k$.

obtained as follows. Using Eqs. (3.2) and (3.3), two tangential vectors are defined at arbitrary boundary point $\boldsymbol{x}(\xi, \eta)$ as,

$$\begin{aligned}
\frac{\partial \boldsymbol{x}}{\partial \xi} &= \frac{\partial N_j(\xi,\eta)}{\partial \xi}\, \boldsymbol{x}_j^k = \left\{ \frac{\partial N'_{b(j)}(\mu(\xi,\xi_o))}{\partial \mu}\, N'_{d(j)}(\mu(\eta,\eta_o))\, \frac{\partial \mu}{\partial \xi} \right\} \boldsymbol{x}_j^k \\
\frac{\partial \boldsymbol{x}}{\partial \eta} &= \frac{\partial N_j(\xi,\eta)}{\partial \eta}\, \boldsymbol{x}_j^k = \left\{ N'_{b(j)}(\mu(\xi,\xi_o))\, \frac{\partial N'_{d(j)}(\mu(\eta,\eta_o))}{\partial \mu}\, \frac{\partial \mu}{\partial \eta} \right\} \boldsymbol{x}_j^k
\end{aligned} \tag{3.8}$$

with $j = 1, \ldots, m$ on Γ_e^k $(k = 1, \ldots, M_\Gamma)$ and, by applying Eq. (3.7), $\frac{\partial \mu}{\partial \xi} = \frac{\partial \mu}{\partial \eta} = 1/3$. The corresponding tangential unit vectors $\boldsymbol{s}$ and $\boldsymbol{m}$ are further defined similar to Eqs. (2.17) and (2.18) for nodal points. As discussed before, these vectors are in general non-orthogonal. A local normal vector is defined based on these as,

$$\frac{\partial \boldsymbol{x}}{\partial \zeta} = \frac{\partial \boldsymbol{x}}{\partial \xi} \times \frac{\partial \boldsymbol{x}}{\partial \eta} \tag{3.9}$$

where the corresponding unit normal vector is thus defined as,

$$\boldsymbol{n}(\xi,\eta) = \frac{1}{h_1\, h_2} \frac{\partial \boldsymbol{x}}{\partial \zeta} \qquad \text{with} \qquad \left| \frac{\partial \boldsymbol{x}}{\partial \zeta} \right| = h_1\, h_2\,. \tag{3.10}$$

Vector $\boldsymbol{n}$ will be pointing in the outward direction with respect to the domain if vectors $(\boldsymbol{s}, \boldsymbol{m})$ are such that their cross product is outward oriented (this is only a matter of definition of the considered element nodes numbering).

The Jacobian matrix is then defined as,

$$\mathbf{J}^k = \left\{ \frac{\partial \boldsymbol{x}}{\partial \xi}, \frac{\partial \boldsymbol{x}}{\partial \eta}, \boldsymbol{n} \right\}^T$$

and the determinant of the Jacobian matrix, to be used in boundary integrals for the k-th element is thus given by

$$\mid \mathbf{J}^k(\xi, \eta) \mid = h_1 \, h_2 \qquad \text{for} \;\; k = 1, \ldots, M_\Gamma \text{ on } \Gamma \tag{3.11}$$

which can be analytically calculated at any point of a given element by using Eqs. (3.8) with the definitions Eqs. (3.6) of shape functions.

After elementary transformation, the discretized expressions of the integrals in Eq. (2.2) are obtained as,

$$\int_{\Gamma(\boldsymbol{x})} \frac{\partial \phi}{\partial n} G_l \, \mathrm{d}\Gamma = \sum_{j=1}^{N_\Gamma} \left\{ \sum_{k=1}^{M_\Gamma} \int_{\Gamma_{\xi,\eta}} N_j(\xi,\eta) G(\boldsymbol{x}(\xi,\eta), \boldsymbol{x}_l) \mid \mathbf{J}^k(\xi,\eta) \mid \mathrm{d}\xi \mathrm{d}\eta \right\}$$
$$\frac{\partial \phi}{\partial n}(\boldsymbol{x}_j) = \sum_{j=1}^{N_\Gamma} \left\{ \sum_{k=1}^{M_\Gamma} D_{lj}^k \right\} \frac{\partial \phi_j}{\partial n} = \sum_{j=1}^{N_\Gamma} K_{lj}^d \, \frac{\partial \phi_j}{\partial n} \tag{3.12}$$

$$\int_{\Gamma(\boldsymbol{x})} \phi \, \frac{\partial G_l}{\partial n} \, d\Gamma = \sum_{j=1}^{N_\Gamma} \left\{ \sum_{k=1}^{M_\Gamma} \int_{\Gamma_{\xi,\eta}} N_j(\xi,\eta) \frac{\partial G(\boldsymbol{x}(\xi,\eta), \boldsymbol{x}_l)}{\partial n} \mid \mathbf{J}^k(\xi,\eta) \mid \mathrm{d}\xi \mathrm{d}\eta \right\}$$
$$\phi(\boldsymbol{x}_j) = \sum_{j=1}^{N_\Gamma} \left\{ \sum_{k=1}^{M_\Gamma} E_{lj}^k \right\} \phi_j = \sum_{j=1}^{N_\Gamma} K_{lj}^n \, \phi_j \tag{3.13}$$

in which, $l = 1, \ldots, N_\Gamma$, for nodes on the boundary, and $\mathbf{D}^k$ and $\mathbf{K}^d$ denote so-called local (i.e., for element k) and global (i.e., assembled for the entire discretization) Dirichlet matrices, and $\mathbf{E}^k$ and $\mathbf{K}^n$, Neumann matrices, respectively. Note, here j follows the global node numbering convention on the boundary and refers to nodal values of element k. Expressions for the Green's and shape functions, to be used in these equations are given by Eqs. (2.3) and Eqs. (3.6), respectively.

With Eqs. (3.12) and (3.13), the discretized form of the BIE (2.2) (and the equivalent BIE for $(\frac{\partial \phi}{\partial t}, \frac{\partial^2 \phi}{\partial t \partial n})$) finally reads,

$$\alpha_l \, u_l = \sum_{j=1}^{N_\Gamma} \{ \, K_{lj}^d \, q_j - K_{lj}^n \, u_j \, \} \tag{3.14}$$

in which, $l = 1, \ldots, N_\Gamma$.

Boundary conditions are directly specified in Eq. (3.14); these are: (i) Dirichlet conditions for $u = \phi$ or $\frac{\partial \phi}{\partial t}$; and (ii) Neumann conditions for

$q = \frac{\partial \phi}{\partial n}$ or $\frac{\partial^2 \phi}{\partial t \partial n}$. The final algebraic system is assembled by moving nodal unknowns to the left hand side and keeping specified terms in the right hand side,

$$\{ C_{pl} + K^n_{pl} \} u_p - K^d_{gl} q_g = K^d_{pl} \overline{q_p} - \{ C_{gl} + K^n_{gl} \} \overline{u_g} \qquad (3.15)$$

where $l = 1, \ldots, N_\Gamma$; $g = 1, \ldots, N_g$, refers to nodes with a Dirichlet condition on boundary Γ_f and $p = 1, \ldots, N_p$, refers to nodes with a Neumann condition on other parts of the boundary; **C** is a diagonal matrix made of coefficients α_l.

3.1.1. *Rigid mode method*

Diagonal terms of the system matrix in Eq. (3.15) $\{C_{ll} + K^n_{ll}\}$ include both C_{ll} coefficients that can be obtained through a direct, purely geometric, calculation of solid angles θ_l at nodes of the discretized boundary, and Neumann matrix diagonal terms K^n_{ll}. As can be seen in Eq. (3.13), the latter terms contain integrals, which although regularized by the separate evaluation of their strongly singular part[61] in the form of coefficients C_{ll}, still contain a highly varying kernel that should be integrated with great care and accuracy.

Rather than directly computing these coefficients one can derive them by applying a straightforward and overall more accurate method referred to as "rigid mode"[61] by analogy with structural analysis problems. By considering a homogeneous Dirichlet problem where a uniform field, $\overline{u} = cst \neq 0$, is specified over the entire boundary Γ (hence, $N_\Gamma = N_g$), potential flow theory implies that normal gradients q must vanish at each node. Thus, for these particular boundary conditions, Eq. (3.15) simplifies to,

$$\{ C_{jl} + K^n_{jl} \} \overline{u_j} = 0 \qquad (3.16)$$

which requires that the summation in curly brackets vanish for all l. Thus, by isolating the diagonal terms in the left-hand-side, we get,

$$\{ C_{ll} + K^n_{ll} \} = - \sum_{j(\neq l)=1}^{N_\Gamma} K^n_{jl} \qquad l = 1, \ldots, N_\Gamma \qquad (3.17)$$

which yields the value of the diagonal term of each row of Eq. (3.16) as minus the sum of its off-diagonal coefficients. These diagonal terms are directly substituted in the discretized system Eq. (3.15).

This method was earlier shown[20] to significantly improve the conditioning of algebraic systems such as Eq. (3.15), and thus the accuracy of their numerical solution (particularly for iterative methods such as GMRES; see below).

3.1.2. *Multiple nodes*

Boundary conditions and normal directions are in general different on intersecting parts of the boundary, such as between the free surface or the bottom, and the lateral boundary of the computational domain (Fig. 1). Such intersections are referred to as 'edges', and corresponding discretization nodes as 'corners'. To be able to specify such differences, corners are represented by double-nodes, for which coordinates are identical but normal vectors are different. This double or multiple node method is essentially an extension of earlier 2D treatments.[20,59] Thus, different discretized BIE's (Eq. (3.15)) are expressed for each node of a corner multiple-node.

For Dirichlet-Neumann boundary conditions we have, for instance, equations: (i) for $l = p$, on a wavemaker boundary Γ_{r1}; and (ii) for $l = f$, on the free surface Γ_f. Since the potential must be unique at a given location, one of these two BIE's is modified in the final discretized system, to explicitly satisfy, $\phi_p = \overline{\phi_f}$ (i.e., "continuity of the potential"), where the overline indicates that the potential is specified on the free surface. For Neumann-Neumann boundary conditions at corners we have, for instance, equations: (i) for $l = p$, on a wavemaker boundary Γ_{r1}; and (ii) for $l = b$, on the bottom Γ_b. The potential continuity equation for this case reads, $\phi_p - \phi_b = 0$, both of these being unknown. Similar continuity relationships are expressed for $\frac{\partial \phi}{\partial t}$ at corners, in the corresponding BIE.

Note, at the intersection between three boundaries, triple-nodes are specified for which three BIE equations are expressed, two of which are replaced in the final algebraic system by equations specifying continuity of the potential (and of $\frac{\partial \phi}{\partial t}$).

The use of this multiple node technique makes this model especially suitable for problems involving surface piercing bodies such as wavemakers, slopes, or ships.

3.1.3. *Numerical integrations*

The discretized boundary integrals D^k_{lj} and E^k_{lj} in Eqs. (3.12) and (3.13) are evaluated for each collocation node $\boldsymbol{x}_l$ by numerical integration. When the collocation node l does not belong to the integrated element k, a standard Gauss–Legendre quadrature is applied. When l belongs to element k, the distance r in the Green's function G and in its normal gradient vanishes at one of the nodes of the element. For such singular situations, a method of singularity extraction is used based on polar coordinate transformations. This method was first developed for a 3D-BEM using higher-order elements[58] and later applied and extended to higher-order Green's function.[62] Grilli et al.[36] optimized this method for the bi-cubic MII elements used in this model; the reader is referred to the latter reference for details.

Due to the form of Green's function Eq. (2.3), non-singular integrals may still have a highly varying kernel when the distance r becomes small, albeit non-zero, in the neighborhood of a collocation point $\boldsymbol{x}_l$. Such situations may occur near intersections of boundary parts (e.g., such as between the free surface and lateral boundaries) or in other regions of the free surface, such as overturning breaker jets, where nodes are close to elements on different parts of the boundary. In such cases, a standard Gauss quadrature, with a fixed number of integration points, may fail to accurately calculate such integrals. One can refer to such cases as "almost" or "quasi-singular" integrals. Grilli et al.,[60] for instance, showed for 2D problems, that the loss of accuracy of Gauss integrations (with ten integration points) for such quasi-singular integrals may be several orders of magnitudes, when the distance to the collocation node becomes very small. For such 2D cases, Grilli and Svendsen[59] developed an adaptive integration scheme based on a binary subdivision of the reference element and obtained almost arbitrary accuracy for the quasi-singular integrals, when increasing the number of subdivisions. This method, however, can be computationally expensive and Grilli and Subramanya[22] developed a more efficient method that essentially redistributes integration points around the location of the quasi-singularity (point of minimum distance from an element k to the nearest collocation node, $\boldsymbol{x}_l$). A method similar to Grilli and Svendsen's, but applicable to 3D problems, was implemented in the model[36] and showed to be both accurate and efficient in applications; the reader is referred to the latter reference for details.

3.1.4. *Solution of the algebraic system of equations*

The linear algebraic system Eq. (3.15) is in general dense and non-symmetric. Since the total number of nodes N_Γ can be very large in 3D applications, the solution by a direct method, of order N_Γ^3, as was done in 2D applications,[20,22] and in initial 3D applications of the model,[36] becomes prohibitive. Hence, for large 3D applications, an iterative solver[35] was implemented[36,53] to solve the linear system in the model, based on a generalized minimal residual (GMRES) algorithm with preconditioning (typically the "Symmetric Successive Overrelaxation" (SSOR) method with relaxation parameter equal to 0.6), and initial solution equal to that of the earlier time step. GMRES yields a more favorable N_Γ^2 numerical complexity, similar to that of the assembling of the system matrix and is amenable to vectorization on a supercomputer. The downside with an iterative method, however, is that for the second-order time stepping scheme used here, two full systems of equations must be solved at each time step (one for ϕ and one for $\frac{\partial\varphi}{\partial t}$) whereas with a direct method, the solution of the second system takes only a few percent of the time needed to solve the first system. Nevertheless, results on a CRAY-C90 showed that, for more than 2,000 nodes and a similar accuracy, the GMRES-SSOR method became faster when used in the model than the direct solution.

To tackle very large problems with a reasonable computational time, we more recently implemented the Fast Multipole Algorithm (FMA) in the model, which reduced this complexity to $O(N_\Gamma \log N_\Gamma)$. First developed by Greengard and Rokhlin[63] for the N-body problem, the FMA allows for a faster computation of all pairwise interactions in a system of N particles, and in particular, interactions governed by Laplace's equation. Hence, it is well suited to our problem. The basis of the algorithm is that due to the 3D Green's function, the interaction strength decreases with distance, so that points that are far away on the boundary can be grouped together to contribute to one distant collocation point. A hierarchical subdivision of space automatically verifies distance criteria and distinguishes near interactions from far ones. The FMA can be directly used to solve Laplace's equation, but it can also be combined with a BIE representation of this equation, whose discretization then leads to a linear system, as detailed above. Matrix-vector products in the system can be evaluated as part of an iterative solver (such as GMRES), that can be accelerated using the FMA. Rokhlin[64] applied this idea to the equations of potential theory. A review of the application of this algorithm to BIE methods can be found in Nishimura.[65] Korsmeyer et al.[66] combined the FMA with a BEM, through a Krylov-subspace iterative algorithm, for water wave computations. Following Rokhlin's ideas, they designed a modified multipole algorithm for the equations of potential theory. First developed for electrostatic analysis, their code was generalized to become a fast Laplace solver, which subsequently has been used for potential fluid flows. Their model was efficient but its global accuracy was limited by the use of low order boundary elements. Scorpio and Beck[67] studied wave forces on bodies with a multipole-accelerated desingularized method, and thus did not use boundary elements to discretize the problem. Neither did Graziani and Landrini,[68] who used the Euler-McLaurin quadrature formula in their 2D model. Srisupattarawanit et al.[69] also used a fast multipole solver to study waves coupled with elastic structures. We show briefly below how the FMA can be combined with our model to yield a more efficient numerical tool. Details and validation of this implementation of the FMA can be found in Refs. 70 and 49. Other applications of the FMA to our model can be found in Refs. 40, 42, 43, and 46.

The FMA is based on the principle that the Green's function can be expanded in a series of separated variables, for which only a few terms need to be retained when the source point $\boldsymbol{x}_l$ and the evaluation point $\boldsymbol{x}$ are far enough from one another. Thus, for a point $\boldsymbol{O}$ (origin of the expansion) close to $\boldsymbol{x}$ but far from $\boldsymbol{x}_l$, we have,

$$G(\boldsymbol{x}, \boldsymbol{x}_l) \approx \frac{1}{4\pi} \sum_{k=0}^{p} \sum_{m=-k}^{k} \rho^k Y_k^{-m}(\alpha, \beta) \frac{Y_k^m(\theta, \varphi)}{r^{k+1}} \tag{3.18}$$

where $\boldsymbol{x} - \boldsymbol{O} = (\rho, \alpha, \beta)$ and $\boldsymbol{x}_l - \boldsymbol{O} = (r, \theta, \varphi)$ in spherical coordinates. The functions $Y_k^{\pm m}$ are the spherical harmonics defined from Legendre polynomials. A hierarchical subdivision of the domain, with regular partitioning automatically verifying distance criteria, is defined to determine for which nodes this approximation applies. Thus, close interactions are evaluated by direct computation of the full Green's functions, whereas far interactions are approximated by successive local operations based on the subdivision into cells and the expansion of the Green's function into spherical harmonics. The underlying theory for this approximation is well established in the case of Laplace's equation. In particular, error and complexity analyses are given in the monograph by Greengard.[71]

In our case, Laplace's equation has been transformed into a BIE and a specific discretization has been used. Thus, the FMA must be adapted in order to be part of the surface wave model, but the series expansion (3.18) remains the same. Hence, with the FMA, Eq. (2.2) can be rewritten as

$$\alpha(\boldsymbol{x}_l)\, \phi(\boldsymbol{x}_l) \approx \frac{1}{4\pi} \sum_{k=0}^{p} \sum_{m=-k}^{k} M_k^m(\boldsymbol{O}) \frac{Y_k^m(\theta, \varphi)}{r^{k+1}} \tag{3.19}$$

where moments $M_k^m(\boldsymbol{O})$ are defined as

$$M_k^m(\boldsymbol{O}) = \int_\Gamma \left\{ \frac{\partial \phi}{\partial n}(\boldsymbol{x})\, \rho^k Y_k^{-m}(\alpha,\beta) - \phi(\boldsymbol{x})\, \frac{\partial}{\partial n}\left(\rho^k Y_k^{-m}(\alpha,\beta)\right)\right\} \mathrm{d}\Gamma\,. \tag{3.20}$$

Instead of considering mutual interactions between two points on the boundary, we now need to look at the contribution of an element of the discretization to a collocation point. The local computation of several elements, grouped together into a multipole, relies on a BEM analysis using the spherical harmonic functions instead of the Green's function. The integration of the normal derivative of the spherical harmonics is done by taking care of avoiding an apparent singularity, which could generate numerical errors. The BEM discretization only applies to the computation of the moments. Thus, the rest of the FMA is unchanged, especially regarding translation and conversion formulas, which allow to pass information through the hierarchical spatial subdivision, from the multipole contributions to the matrix evaluation for each collocation node. In the model, the implementation of the FMA thus only affected programs that involved the assembling and the solution of the algebraic system matrix. The storage of coefficients that are used several times for each time step, for instance, is now done inside the cells of the hierarchical subdivision. The rigid mode and multiple nodes techniques, which *a priori* modified the matrix before the computation of matrix-vector products, are now considered as terms correcting the result of such products, so that the linear system keeps the same properties.

The accelerated model benefits from the faster Laplace's equation solver at each time step. The FMA model performance was tested by comparing new results with results of the former model using GMRES, for a 3D application which requires great accuracy: the propagation of a solitary wave on a sloping bottom with a transverse modulation, leading to a plunging jet.[36] The consistency of the new solution was checked but, more importantly, the accuracy and stability of results and their convergence as a function of discretization size was verified. In fact, by adjusting the parameters of the FMA, i.e. the hierarchical spatial subdivision and the number of terms p in the multipole expansions, one can essentially obtain the same results as with the former model. In this validation application, for discretizations having more than 4,000 nodes, the computational time was observed to increase nearly linearly with the number of nodes.[49,70] Using this model for other applications,[40,43,46] similar properties of the solution were observed as a function of the discretization when applying the model to the generation of tsunamis by underwater landslides, of freak waves by directional focusing, and of waves by a moving surface disturbance, respectively.

3.1.5. *Image method*

Some applications may have a horizontal symmetry and/or a flat bottom in the computational domain. In such cases, an image method can be applied when computing the BIEs, with respect to the planes $z = -d$ and/or $y = 0$, to remove parts of the discretization. Doing so, the 3D free space Green's function is modified in the BIE, by adding contributions of each image source. Grilli and Brandini[38] show details and results of this approach in the case of the generation of extreme waves by a directional wavemaker over a flat bottom. The image method was also applied for a horizontal symmetry to compute cases of wave generation by underwater landslides,[42,43] and for a flat bottom and a horizontal symmetry, to the generation of waves by a moving disturbance on the free surface, over a flat bottom.[54]

Note, in the FMA, when the original source point is far from the collocation point, so are the images. Thus, image contributions are simply added to the multipole associated with the original point. In the usual application of the FMA, images should be accounted for at a coarser subdivision level than that of the original source points, since they are further away from the evaluation point.

3.2. *Time integration*

A second-order explicit scheme based on Taylor series expansions is used to update the position $\boldsymbol{R}$ and velocity potential ϕ on the free surface, as,

$$\boldsymbol{R}(t + \Delta t) = \boldsymbol{R} + \Delta t \frac{\mathrm{D}\boldsymbol{R}}{\mathrm{D}t} + \frac{\Delta t^2}{2} \frac{\mathrm{D}^2 \boldsymbol{R}}{\mathrm{D}t^2} + O\left(\Delta t^3\right) \tag{3.21}$$

$$\phi(t + \Delta t) = \phi + \Delta t \frac{\mathrm{D}\phi}{\mathrm{D}t} + \frac{\Delta t^2}{2} \frac{\mathrm{D}^2 \phi}{\mathrm{D}t^2} + O\left(\Delta t^3\right) \tag{3.22}$$

where Δt is the varying time step and all terms in the right-hand sides are evaluated at time t.

First-order coefficients in these Taylor series are given by Eqs. (2.4) and (2.5), which requires calculating ϕ, $\partial\phi/\partial n$ at time t on the free surface. Second-order coefficients are obtained from the Lagrangian time derivative of Eqs. (2.4) and (2.5), as

$$\frac{D^2\boldsymbol{R}}{Dt^2} = \frac{D\boldsymbol{u}}{Dt}$$
$$\frac{D^2\phi}{Dt^2} = -gw + \boldsymbol{u} \cdot \frac{D\boldsymbol{u}}{Dt} - \frac{1}{\rho}\frac{Dp_a}{Dt}\,. \tag{3.23}$$

Appendix A.1 gives the expression for the Lagrangian acceleration on the boundary, which requires calculating $(\frac{\partial\phi}{\partial t}, \frac{\partial^2\phi}{\partial t\partial n})$ at time t, as well as their tangential derivatives. As mentioned above, this is done by solving a second BIE similar to Eq. (2.2) for $\frac{\partial\phi}{\partial t}$. Since both BIEs correspond to the same boundary geometry, the resulting linear algebraic system Eq. (3.15) only needs to be discretized and assembled once. Boundary conditions for the second BIE are expressed based on the solution of the first BIE. On the free surface, Bernoulli equation yields the Dirichlet condition,

$$\overline{\frac{\partial\phi}{\partial t}} = -gz - \tfrac{1}{2}\,\boldsymbol{\nabla}\phi\cdot\boldsymbol{\nabla}\phi - \frac{p_a}{\rho} \qquad \text{on } \Gamma_f(t)\,. \tag{3.24}$$

For wave generation by a moving boundary, such as a wavemaker (active or absorbing) or an underwater landslide, of velocity $\boldsymbol{u}_p(\boldsymbol{x}_p, t)$ and acceleration $\dot{\boldsymbol{u}}_p(\boldsymbol{x}_p, t)$, using Eqs. (2.6) and (2.24), and after some derivations, we obtain the Neumann condition,

$$\overline{\frac{\partial^2\phi}{\partial t\partial n}} = \dot{\boldsymbol{u}}_p\cdot\boldsymbol{n} + \boldsymbol{u}_p\cdot\dot{\boldsymbol{n}} - \phi_n\phi_{nn} - \frac{1}{(1-\kappa^2)^2}\Big\{(\phi_s - \kappa\phi_m)$$
$$(\phi_{ns} - \kappa\phi_{nm}) + (\phi_m - \kappa\phi_s)(\phi_{nm} - \kappa\phi_{ns})\Big\} \tag{3.25}$$

where the expression for ϕ_{nn} is given in Appendix A.1 and $\dot{\boldsymbol{n}} = \boldsymbol{\Omega}\times\boldsymbol{n}$ for a rigid body motion with angular velocity $\boldsymbol{\Omega}$.

On stationary boundaries we simply specify,

$$\overline{\frac{\partial^2\phi}{\partial t\partial n}} = 0 \qquad \text{on } (\Gamma_{r2}),\ (\Gamma_b) \tag{3.26}$$

depending on the considered problem.

Time step Δt in Eqs. (3.21) and (3.22) is adaptively selected at each time to satisfy a mesh Courant condition,

$$\Delta t = C_0\frac{\Delta r^{\min}}{\sqrt{gh}} \tag{3.27}$$

where C_0 denotes the Courant number, $\Delta r^{\min}$ is the instantaneous minimum distance between two neighboring nodes on Γ_f and h is a characteristic depth. Grilli et al.[36] performed a sensitivity analysis of numerical errors in the model (see next section) to the spatial and temporal discretization expressed as a function of the mesh size and Courant number. They showed that errors were minimum for $C_0 = 0.45$ to 0.5. This is the value used in the present applications.

This second-order time stepping scheme is explicit and uses all spatial derivatives of the field variables at time t to calculate the position and potential of water particles at $t + \Delta t$. This was shown in earlier applications to provide a good stability of the computed solution, which typically can proceed for thousands of time steps without requiring any smoothing.

Note that for absorbing wavemaker boundaries, the piston position is updated for the next time step using Taylor series expansions similar to those used for the free surface.

3.3. *Numerical errors*

In a number of earlier applications of the model to cases with known analytical solution, it was confirmed that maximum local numerical errors in the BEM are of more than third-order in mesh size $\Delta r^{\min}$ (typically 3.5-4th-order[46]), which is expected considering the third-order boundary elements used in the discretization. For a given mesh size, numerical errors were also shown to be of third-order in time step Δt, when mesh size was such as to satisfy the Courant condition Eq. (3.27).

For general nonlinear wave generation and propagation problems, when no a priori solution is known, the mesh size is selected based on expected wavelength in order to have a sufficient resolution of individual waves. Time step is then adaptively adjusted based on Courant condition Eq. (3.27). An estimate of numerical errors can thus only be made in a global sense. For instance, the global accuracy of the BEM solution can be checked at each time step by calculating a nondimensional error on boundary flux continuity as,

$$\varepsilon_C = \frac{\Delta t}{V_o} \int_\Gamma \frac{\partial \phi}{\partial n} \, \mathrm{d}\Gamma \tag{3.28}$$

where V_o denotes the initial domain volume.

The combined global accuracy of the BEM solution and time stepping scheme can then be estimated as a function of time, by checking the conservation of volume,

$$V(t) = \int_\Gamma z n_z \, \mathrm{d}\Gamma \tag{3.29}$$

and total energy,

$$E(t) = \tfrac{1}{2}\,\rho \int_{\Gamma} \left(\phi \frac{\partial \phi}{\partial n} + g z^2 n_z \right) \mathrm{d}\Gamma \tag{3.30}$$

where the first and second terms represent the kinetic and potential energy contributions of the flow respectively, and n_z is the vertical component of the unit normal vector. Thus, a time varying nondimensional error on volume conservation can be expressed as $\varepsilon_V = (V - V_o)/V_o$ while for the energy, the nondimensional error on total energy reads, $\varepsilon_E = (E - E_o)/E_o$. However, to calculate the latter error, one must either start from an initial wavefield with total energy E_o or wait for the model to achieve a quasi-steady state, with nearly constant time average energy, expressing a balance between wave generation and dissipation/absorption. Such error calculations for complex propagating and shoaling wave fields, dissipating their energy in an AB, were made extensively for 2D computations.[23]

As mentioned above, it was found in applications that errors on volume and energy conservation reach a minimum for $C_0 \simeq 0.45$ to 0.5, independently of mesh size. This implies in particular that Δt could not be imposed too small, otherwise numerical errors would accumulate faster when solving the BIEs at intermediate times. It is emphasized that no smoothing/filtering was used to stabilize the solution in all cases we considered in past applications as well as those reported later in this paper.

3.4. *Mesh regridding and time updating*

Due to the MEL time updating of the free surface nodes (identical to water particles), during computations of wave propagation, nodes may accumulate in some areas of the free surface (e.g., with converging flow such as overturning breaker jets) or move away from some areas (such as a wavemaker boundary generating nonlinear waves with a non-zero mean mass flux).

Hence, although this is not required in most cases to stabilize computations, it is desirable to have a means in the model of both refining the mesh discretization in areas of formation of breaker jets, prior to their occurrence or to periodically regrid sections of the free surface on a regular mesh. In 2D studies, regridding was done by implementing a node regridding method in which a specified number of nodes were regridded at constant arclength value in between two nodes selected on the free surface.[22] In the initial implementation of the 3D model,[36] a two-dimensional regridding method was implemented based on this principle. For practical reasons, the 3D regridding method required the free surface $\eta(x, y)$ to be single-valued. In other more recent applications, a pseudo-Lagrangian time updating was implemented,[45,46] in which free surface nodes can move at a velocity different from the flow velocity. By adjusting this velocity, free surface nodes can be updated on a purely Eulerian manner (i.e., vertically) in some regions of the boundary or on a fully Lagrangian matter in other regions, and all the range in between. The reader is referred to the published references for more information on this matter.

4. Numerical Results

4.1. *Shoaling and breaking of solitary waves on a sloping ridge*

After implementing the model, Grilli et al.[36] first tested its accuracy and efficiency by computing the propagation of the simplest possible nonlinear wave with numerically exact shape and potential, i.e., a solitary wave. Tanaka[51] developed a method for calculating fully nonlinear solitary wave shape and kinematics up to their maximum height of about 80% of the local constant depth. Grilli and Svendsen[52] adapted this method to calculating initial solitary solutions in their 2D model, and Grilli et al.[5,6] then calculated physical features for the shoaling and overturning of such waves over a variety of slopes. Upon implementing the 3D model,[36] such exact solitary waves propagating over constant depth and plane slopes were also used to validate the model and estimate the required spatial and temporal discretization sizes; this led to the optimal value of the mesh Courant number discussed before. To produce well-controlled 3D overturning jets, they then calculated the shoaling of solitary waves over a sloping ridge causing 3D focusing. Fochesato et al.[50] later used similar test cases to validate their implementation of a modified surface representation in the model. Since shallow water wave breaking on sloping beaches exhibits many of the features of a succession of solitary waves, Guyenne and Grilli[53] performed a detailed numerical study of the physics of 3D overturning waves over arbitrary bottom, using the same idealized case of a solitary wave shoaling over a sloping ridge. They analyzed shoaling and breaking wave profiles and kinematics (both on the free surface and within the flow) and observed that the transverse modulation of the ridge topography induces significant 3D effects on the time evolution, shape and kinematics of breaking waves. These effects were found to be similar to those observed for periodic 3D

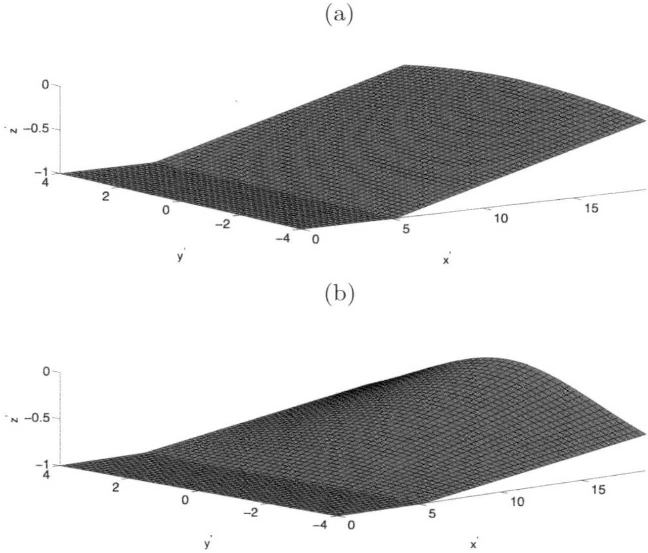

Fig. 4. Bottom topography and its initial discretization for the shoaling of a solitary wave over a sloping ridge with a $1:15$ slope in the x-direction and a lateral $\operatorname{sech}^2(ky')$ modulation. Configurations with $k =$ (a) 0.1; and (b) 0.25 are represented here.

overturning waves in deep water.[35] Comparing earlier 2D results[3,19] to 3D results in the middle cross-section of the ridge, Grilli and Guyenne showed remarkable similarities, especially for the shape and dynamics of the plunging jet, indicating that late in the overturning, wave breaking is quasi-2D and becomes almost independent from the background flow and boundary conditions (including bottom topography), that have induced breaking.

Solitary wave breaking on a sloping ridge is illustrated in the following. Figure 4 shows the typical sloping ridge geometry used.[53] [Primes hereafter indicate non-dimensional variables based on long wave theory, i.e. lengths are divided by h_0 and times divided by $\sqrt{h_0/g}$.] The ridge starts at $x' = 5.225$, with a $1:15$ slope in the middle cross-section and a transverse modulation of the form $\operatorname{sech}^2(ky')$. Here, we successively set $k = 0.1$ and 0.25, which correspond to different amplitudes for the transverse tails of the ridge. An incident 'Tanaka' solitary wave of height $H' = 0.6$ with the crest located at $x'_0 = 5.785$ for $t' = 0$ is used in computations.

The computational domain is of width 8 or $16h_0$, for each case respectively, in the y-direction and is truncated at $x' = 19$ in the x-direction, with minimum depth $z' = -0.082$ in the middle ($y' = 0$). The initial discretization for the bottom and free surface consists of 50×20 or 40 quadrilateral elements in the x- and y-directions, respectively ($\Delta x'_0 = 0.38$ and $\Delta y'_0 = 0.4$). The lateral boundaries have grid lines connecting the edge nodes of the bottom and free surface, with four elements specified in the vertical direction. Consequently, the total number of nodes is $N_\Gamma = 2862$ or 5102, and the initial time step is set to $\Delta t'_0 = 0.171$ for $C_0 = 0.45$. Computations are performed in this initial discretization as long as global errors on wave mass remain acceptable (i.e., less than 0.05% or so). A two-dimensional regridding to a finer resolution is then applied at a time when the wave profile is still single-valued. The discretization is increased to 60×40 or 70 quadrilateral elements, respectively, in the portion $8 \leq x' \leq 19$ of the bottom and free surface ($\Delta x' = 0.18$, $\Delta y' = 0.2$ and $N_\Gamma = 6,022$ or 9,982). For instance, regridding was applied at $t' = 4.900$ for $k = 0.25$, when errors on volume and energy conservation are 0.010% and 0.013% respectively.

Figure 5 shows results just before jet touch-down, at a time when global errors on mass and energy are still acceptable (typically less than 0.1% with respect to initial wave mass or energy). Figure 5a shows wave shapes for $k = 0.1$ at $t' = 8.958$, seen from two different angles, and Fig. 5b shows results for $k = 0.25$ at $t' = 9.268$. In the former case, the wave develops into a wide barely modulated breaker while in the latter case, for which the incident wave propagates over a laterally steeper ridge, the overturning wave develops into a narrow, well developed and more strongly plunging jet. These computations were performed in 2003-04, using GMRES as a solver, on a single processor Compaq Alpha GS160 computer. Without any particular optimization, the CPU time per time step for $N_\Gamma = 6,022$ was of $O(10)$ minutes (for a few hundred time steps). As we shall see, a much faster solution was achieved for more recent computations using the FMA method in combination with GMRES.

The model is general and various cases of breaking induced by complex nearshore bathymetry can be simulated. Figure 6, for instance, shows the breaking of a solitary wave over a sloping bar, a case that would be of interest for studying the sensitivity of wave-induced nearshore currents to breaker characteristics and bar shape designing submerged breakwaters used for coastal protection, or surfing reefs aimed at inducing certain breaker types and shape. For more results, details, and discussion of the physics of 3D overturning waves over shallow water topography, the interested reader is referred to the literature.[53]

(a)

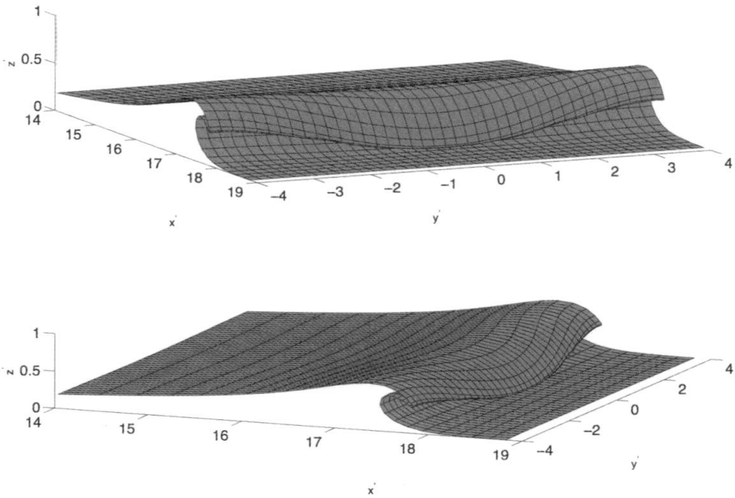

(b)

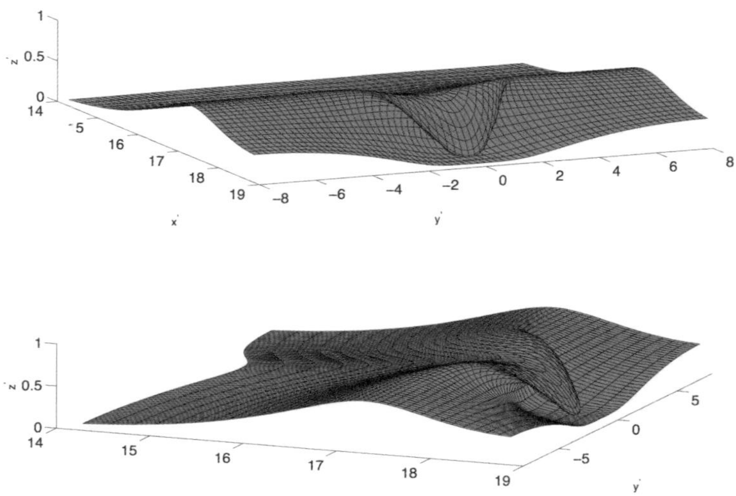

Fig. 5. Solitary wave profiles for (a) $k = 0.1$ ($t' = 8.958$), (b) $k = 0.25$ ($t' = 9.268$).

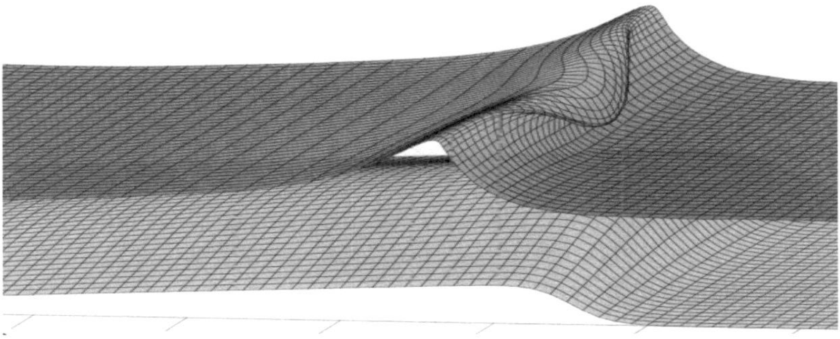

Fig. 6. Example of solitary wave breaking over a sloping bar.

4.2. *Generation of extreme waves by directional energy focusing*

Three-dimensional directional wave energy focusing is one of the mechanisms that contribute to the generation of extreme gravity waves in the ocean, also known as rogue waves. Grilli and Brandini[38,39] and Fochesato et al.[40] used the 3D-BEM FNPF model to simulate and analyze this phenomenon over constant depth, by specifying the motion of a snake wavemaker on one side of the model boundary and an open snake absorbing boundary on the other side, thus creating a 3D Numerical Wave Tank (NWT). The wavemaker law of motion was specified such as to linearly focus periodic waves in the middle of the tank.[72] Using the image method in the z direction to reduce the size of the discretization initial stages of 3D extreme breaking waves were simulated.[38,39] Computations however could not proceed further, in part because of the lack of resolution of the breaker jet and because the more accurate free surface representation developed later[50] was not used. By contrast, in more recent work,[40] all model improvements detailed in this paper were used, including the image method in both y and z directions, and the FMA combined with GMRES as a solver, whose efficiency made it possible to generate finely resolved 3D focused overturning waves and analyze their geometry and kinematics. In the following, we only present a typical simulation of an overturning rogue wave. A literature review, and further computations and sensitivity analyses of extreme wave geometry and kinematics to water depth and maximum angle of directional energy focusing can be found in Ref. 40. In particular, in the latter work, we show that an overturning rogue wave can have different properties depending on whether it is in the focusing or defocusing phase at breaking onset. The maximum focusing angle and the water depth largely control this situation, and therefore the main features of the rogue wave crest, such as its 3D shape and kinematics.

In the following application, we generate an overturning rogue wave by specifying the superposition of 30 wave components at the snake wavemaker. These wave components have identical frequency $\Omega = 0.8971$ rad/s (i.e., linear period of $T = 7$ s, linear wavelength $L = 2\pi/k = 72$ m, in depth $d = 20$ m) and amplitude $A = 0.19$ m, but directions varying between -45 and 45 degrees. With these parameters, linear wavemaker theory yields a stroke amplitude of $a = 0.2$ m for each individual wave. The amplitude at the linear focal point (specified here at the distance $x_f = 250$ m from the wavemaker) would thus theoretically be $A^* = 6.3$ m. This is clearly a large value, in accordance with our goal of generating a large overturning wave early in the generation process, before the wave reaches the far end of the tank where, despite the absorbing boundary condition, some reflection may occur that may perturb wave focusing. The NWT used in this computation has a 440 m length (or $22d$) and a 600 m width (or $30d$). For the selected focusing distance, this NWT length is such that, when overturning of the extreme wave occurs, only very few small waves will have reached the far end of the tank. Hence, the absorbing boundary condition will barely be activated in this computation. The half width of the NWT along y is divided into 70 elements ($\Delta y = 4.3$ m), and its depth into 4 elements. At the beginning of computations, the discretization has 90 elements in the $x-$longitudinal direction ($\Delta x = 4.9$ m), which corresponds to roughly 15 nodes per wavelength, which makes a total of $N_\Gamma = 7,626$ nodes (considering the image method eliminates the bottom and one lateral boundary). In order to better resolve the wave steepening towards breaking (defined as the occurrence of the first vertical tangent on the free surface), the x-resolution is later improved by using 120 elements with an irregular grid, refined around the breaking wave for $t > 43.39$ s ($= 6.20T$), bringing the total number of nodes to $N_\Gamma = 9,906$. The present simulations using the FMA required 2 min per time step (in scalar mode) on a biprocessor Xeon (3Ghz, 2Gb of RAM) and lasted for more than 350 time steps (i.e., a total of 700 min). Although processors clearly differ, such computational times per time step are clearly much faster than in the previous application that only used GMRES as a solver.

Figure 7 shows the free surface elevation at a time the focused wave has started to overturn. A detailed analysis of the earlier stages of this

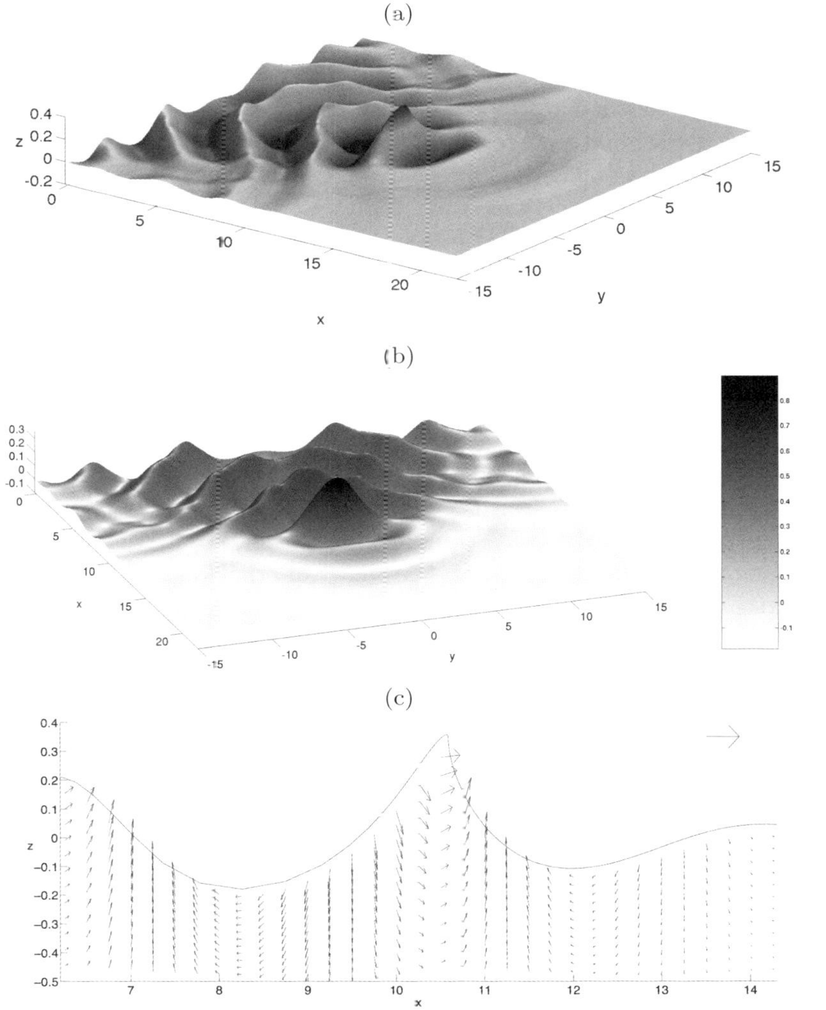

Fig. 7. Free surface evolution of focused wave shown at $t = 6.89T$, when overturning starts, with the wave crest located at $x = 211$ m (or 10.55 d): (a) wave elevation in arbitrary gray scale; (b) wave horizontal velocity $u/\sqrt{gd}$; (c) vertical section at $y = 0$, with the arrows showing the projected internal velocity vectors (the arrow in the upper-right corner represents the unit nondimensional velocity vector; the vertical axis scale is exaggerated by a factor of 9). [Note that contours shown in this figure are less smooth than the actual wave surface in the BEM, because of the plotting algorithm.]

computation would show that the initially flat NWT free surface starts oscillating near the wavemaker and, due to the ramp-up of the wavemaker motion specified over three wave periods, a first moderate amplitude wave is generated. This first wave then almost disappears at the plot scale while the wavemaker amplitude of oscillations further increases, to give rise in the middle of the tank to an even larger wave and, eventually, after the ramp-up is over to achieve complete focusing, with an even larger wave, that starts overturning around $x_c = 211$ m (or 10.55 d) as shown in Fig. 7a. This is closer to the wavemaker than the linearly estimated focal point ($12.5d$). Behind this breaker, we see on the figure that the phenomenon is starting to repeat itself, with a new curved crest line appearing and converging towards the center of the NWT. Figure 7b shows the horizontal velocity component $u/\sqrt{gd}$ values over the free surface; we see that, at overturning, very large values only occur in the upper third of the focused wave crest, approaching 0.9 at the wave crest tip, while the wave linear phase speed is $c/\sqrt{gd} = 0.73$. The focused wave crest thus tends to move forward faster than the phase velocity of its basic wave components, thus initiating overturning and breaking. The cross-section in Fig. 7c shows internal velocity fields, which illustrate the more intense kinematics at incipient breaking immediately below the wave crest. The full 3D fields would show that particle velocities are essentially upwards, with the upper part of the fields having nearly uniform values. Accelerations are negative, with greater values ($\simeq 2g$) nearest the crest.

The properties of this extreme focused wave agree qualitatively well with observed characteristics of rogue waves. Figure 7a shows a circular trough located just in front of the overturning wave (the so-called "hole in the sea" reported by rogue wave eyewitnesses). Behind the wave, an even deeper trough has formed (which is more clearly seen in Fig. 7c), separating the main wave from the curved crest line that follows it. This trough has more of a crescent shape, due to the directions of the incoming waves. Due to the significant directionality, the overturning part of the wave is quite narrow and also located in the middle of a curved front, hence illustrating strong 3D effects. The amplitude of the overturning wave is significantly larger than that of the following waves, which have not yet converged, and the wave has a strong back-to-front asymmetry (this is also more clearly seen in Figs. 7c and 8). In the vertical cross-section of Fig. 7c the wave profile appears similar to that observed in rogue wave measurements or observations (see for example the extreme wave measured under the Draupner platform in the North Sea on January 1st 1995), as well as in earlier 2D numerical

Fig. 8. Close-up of the overturning rogue wave for case of Fig. 7.

studies, for instance those related to modulational instabilities of a wave packet.[73] A large crest ($A_c = 7.16$ m or $0.358d$) is preceded and followed by two shallower troughs; the back trough is deeper than the front one ($A_{t1} = 3.60$ m and $A_{t2} = 2.14$ m, or $0.180d$ and $0.107d$, respectively). Wave height is $H_1 = A_c + A_{t1} = 10.76$ m or $0.538d$, which is less than the linearly predicted upper bound value $2A^* = 12.6$ m. This is because of the early breaking of the wave and the incomplete focusing. The wavelength of the nonlinear focused wave can be estimated to $\lambda \simeq 78.0$ m (or 3.90 d), by averaging the rear and front wavelengths, which is more than the linear value, due to amplitude dispersive effects. This yields a steepness $H/\lambda = 0.138$, which is greater than the limiting steepness predicted by Miche's law for this depth (about 0.132 for a symmetric maximum Stokes wave). Hence, the asymmetric and transient 3D extreme wave generated in the NWT in this application grows further than the theoretical limiting steepness, before it overturns. This may have important implications for structural design of offshore structures (e.g., Ref. 74). Finally, Fig. 8 shows a closeup of the development of the plunging jet in Fig. 7. We did not attempt to accurately follow the overturning jet beyond this stage, although as seen in the previous application, the model is clearly capable of doing so, given a proper discretization. Hence, here, we did not discuss wave breaking characteristics in detail, but limited our analyses to the initiation of breaking. Note nevertheless the similarity of this 3D breaking crest with some of the shallow water breaking waves shown in the previous section and in Ref. 53.

More computational cases and further discussions can be found in Ref. 40.

4.3. *Landslide tsunami generation*

While the scientific community has worked for decades on the modeling of tsunamis directly generated by the bottom motion caused by an earthquake (so-called co-seismic tsunamis), in the past ten years, the 1998 Papua New Guinea (PNG) tsunami has focused the interest of a sizeable part of the scientific community on the lesser studied landslide tsunamis. After years of field surveys and modeling work following the PNG event, an almost unanimous consensus was reached in the community that the large tsunami, whose coastal runup on the nearby shore reached 16 m at places, while associated with a moderately tsunamigenic 7.1 magnitude earthquake, had been triggered by an underwater mass failure, itself triggered (with a 15 min. delay) by the earthquake (e.g., ground acceleration and induced excess groundwater pore pressure). Field work further showed that the mass failure responsible for the triggering of the tsunami was a large rotational slide (i.e., a slump), of at least 6 km^3, initiated at an average depth of 1,500 m, about 60 km offshore of the main impacted area of PNG (see, e.g., Ref. 75 for a recent review of this event, including field work and modeling).

Thus, following PNG, a number of laboratory and modeling studies were conducted, aimed at better understanding the physics of landslide tsunami generation and relating initial tsunami parameters (i.e., the tsunami source), to geometrical, geological, and geotechnical parameters of the underwater slide. In most of the work so far, the more complex triggering phase was largely ignored to concentrate on tsunami generation by the moving underwater slide. While tsunamis, once generated, eventually behave in the far field as long gravity waves, which in deep water can even be well approximated by linear long wave theory, near their source, landslide tsunamis are generated by the complex 3D flow field induced by the moving slide. Moreover, if the slide is initiated in very shallow water (or is partly emerged as a so-called subaerial slide), initial waves may be strongly nonlinear or even breaking.

In view of these features of landslide tsunamis and to address the problem in the most general manner from the fluid point-of-view, Grilli and Watts[76] developed a 2D-FNPF model of landslide tsunami generation and initial propagation. The slide law of motion in this model was that of a solid body moving down a plane slope, under the action of gravity, buoyancy, basal friction, hydrodynamic drag and inertia (added mass) forces. Grilli and Watts[77] performed additional work in this direction, and applied an updated version of this initial model to a large set of parameters, in

Fig. 9. General view of experimental set-up for landslide tsunami experiments: slope, rail, landslide model, wave gauges/step motors, and supporting I-beams.

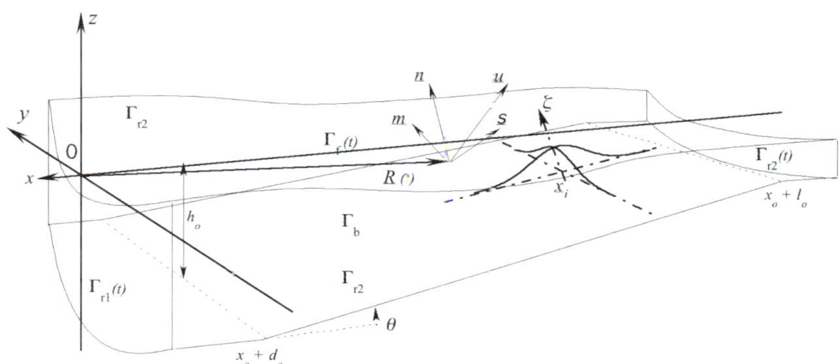

Fig. 10. BEM domain and variable set-up for 3D-FNPF computations of landslide tsunami generation.

combination with 2D laboratory experiments used for validation. In particular, these authors showed that most of the tsunami generation occurs for $t < t_0 = u_t/a_0$, a characteristic time, function of the slide terminal velocity u_t and initial acceleration a_o, both of these being themselves functions

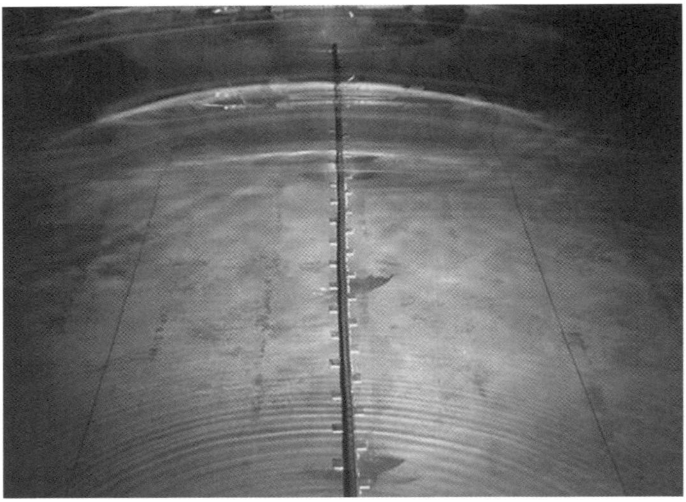

Fig. 11. Example of experimental surface elevations generated for $d = 0.12$ m at $t \simeq t_o$. Note, the submerged model slide is visible underwater at the top of the picture.

of the problem parameters. Based on this numerical work, Watts et al.[78] developed semi-empirical relationships expressing the main tsunami characteristics (such as initial depression) as a function of the slide parameters (for both translational landslides and slumps). Such relationships allowed to rapidly design a 3D landslide tsunami source and conduct case studies, using more standard long wave models (see Ref. 79 for details). While it was shown that landslide tsunamis are fairly directional and, hence, 3D effects in their sources are less prominent than for co-seismic tsunami sources, Grilli et al.[41] performed 3D-FNPF landslide tsunami simulations, by modifying the present FNPF model. Enet and Grilli[42,43,80] later performed large scale 3D laboratory experiments that validated both results of this model, as well as the empirical relationships developed earlier, based on 2D simulations. They also performed new simulations of their 3D experiments, using the more recent and optimized version of the 3D-FNPF model detailed in this paper.

The following illustrates some of these 3D landslide tsunami simulations and their experimental validation; more results and details can be found in the references. Figure 9 shows the experimental set-up for the earlier experiments.[42,80] A smooth bi-Gaussian-shaped aluminum body (length $b = 0.395$ m, width $w = 0.680$ m, and maximum thickness $T = 0.082$

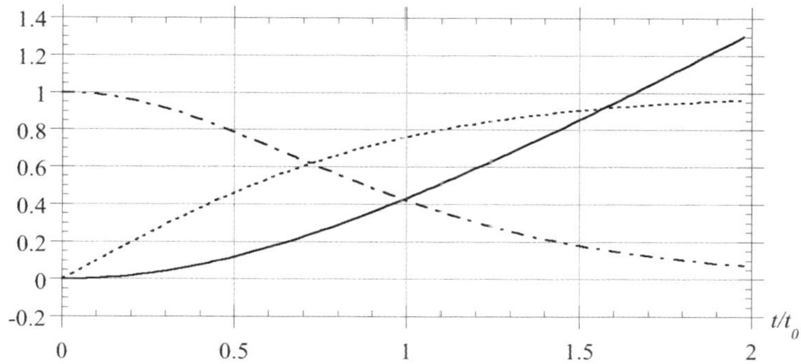

Fig. 12. Typical laws of motion for rigid underwater landslides, as a function of non-dimensional time t/t_o: dimensionless motion parallel to the slope s/s_0 (—), velocity u/u_t (- - - -), and acceleration a/a_0 (— - —), where (s_0, u_t, a_0) denote characteristic distance of motion, terminal velocity, and initial acceleration, respectively, all function of the slide and set-up physical parameters.

m) is used to represent a solid underwater slide moving down a 15° slope, from a series of initial submergence depths d. [This shape reduces flow separation and hence makes potential flow theory fully relevant to model such cases.] Experiments were performed in the 3.6 m wide, 1.8 m deep and 30 m long wavetank of the Department of Ocean Engineering, at the University of Rhode Island. Figure 10 shows the BEM model set-up used in numerical simulations, which is similar to the earlier work,[41] but uses two absorbing pistons, open boundaries, in the negative and positive x directions, representing the onshore and offshore directions, respectively. A shallow shelf of depth h_1 is modeled onshore of the slide, while the depth levels-up to h_o offshore of the slide. As mentioned, the model uses the image method to remove half of the unknowns for $y < 0$ and the FMA (combined with GMRES) is used for the BEM solution at each time step. For these simulations, which were performed on a PC-Pentium computer circa 2005, this led to about 4 min per time step for $N_\Gamma = 4146$ nodes (half domain).

Landslide motion, and the corresponding deformation of the bottom discretization, are specified as a boundary condition, based on laws of motion derived earlier[77] and adapted to the 3D model[80] (Fig. 12). Note that the landslide is represented in the BEM model as a space and time varying "wave" of elevation z, specified for the bottom elements (of fixed coordinates x and y) located on the underwater slope, between $x = x_o + l_o$ and

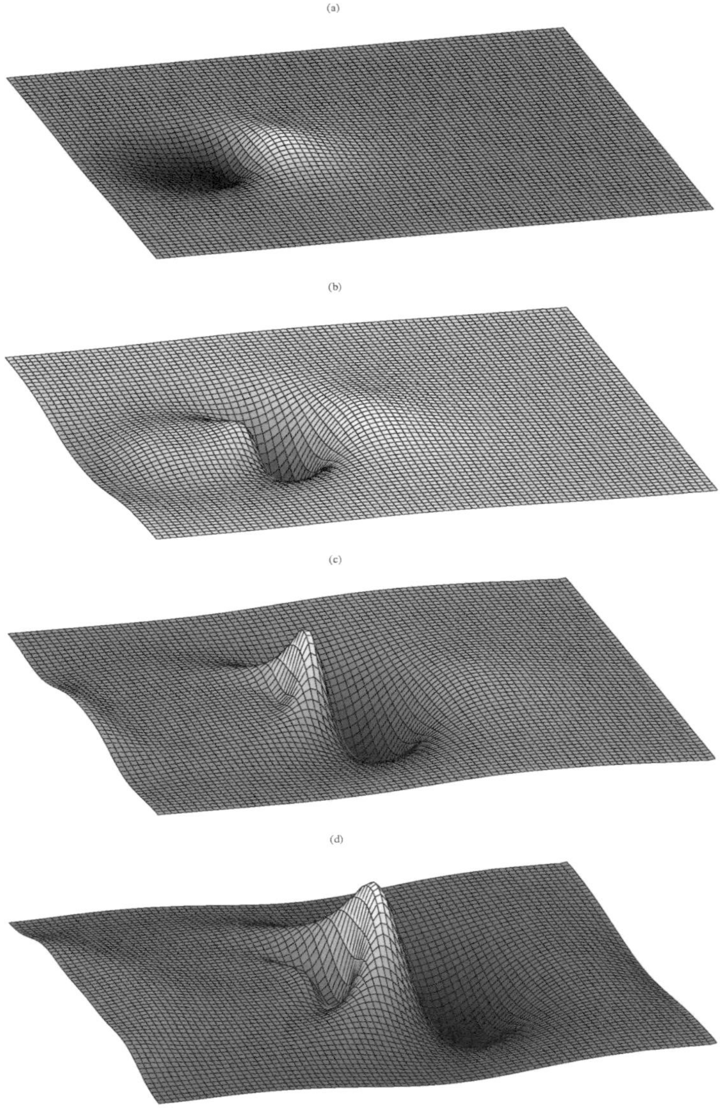

Fig. 13. Landslide tsunami shapes computed at $t = t_0/4$; (b) $t_0/2$; (c) $3t_0/4$; and (d) t_0, for initial slide submergence $d = 120$ mm, for which $t_0 = 1.74$ s at laboratory scale (vertical scale is exaggerated).

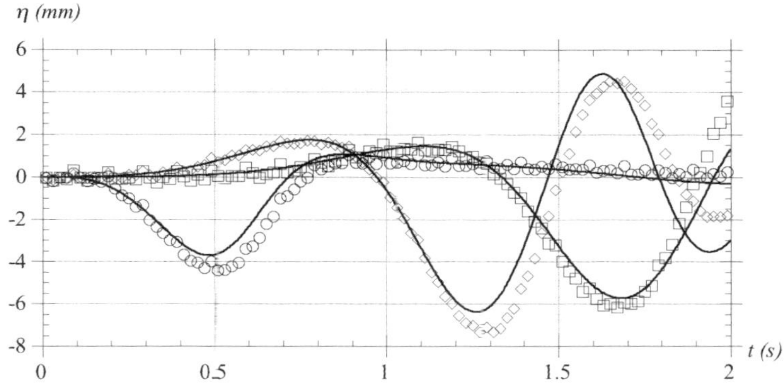

Fig. 14. Comparison of experiments (symbols) with model results (——) for $d = 0.140$ m at gauge: 1 ○; 2 ◇; and 4 □ (only 10% of experimental data points are shown).

$x_o + d_o$ (see Fig. 10, and[41] for detail). Figure 13 shows typical free surface elevations computed at various times $t \leq t_0$ for an initial slide submergence $d = 0.12$ m. Figure 11 shows a picture of the free surface observed in the tank at the time of Fig. 13d. Both observed and simulated surfaces exhibit a typical double crescent-shaped crests, in a $\simeq 30°$ angular sector centered in the direction x of slide motion. Figure 14 shows a comparison of numerical simulations with experiments for $d = 0.14$ m (for which $t_0 = 1.87$ s), for 3 wave gauges. [Note, in these simulations, in order to more closely model tank experiments, only one absorbing boundary was used on the offshore side of the NWT, while the length of the onshore shelf, of depth h_1 was adjusted to yield the same volume as for the sloping bottom.] The agreement between both of these is quite good, with absolute differences between computations and measurements on the order of 1 mm, which considering meniscus effects on the wave gauges, is within the experimental error. The first gauge (1) is located above the initial middle location of the slide at $x_i = 0.846$ and $y = 0$ (Fig. 10), and essentially records the initial tsunami depression induced by the slide motion (4 mm at model scale; also see Fig. 13b). As expected, this surface depression, which represents the initial tsunami generation (i.e., source), occurs for $t < 0.5t_0$. At later times, the depression 'rebounds' into large offshore propagating oscillatory waves, gradually spreading with a leading depression wave, and smaller onshore propagating waves (causing coastal runup), indicating strong dispersive effects (see Fig. 13). Gauge (2) is located further downslope and off-axis, at

$x = 1.469$ m and $y = 0.35$ m. Here, the tsunami has developed into a dispersive train (see Ref. 80 for longer time series of experimental measurements) whose first wave has a 11-12 mm height. Finally, at gauge (4), further offshore and off-axis at $x = 1.929$ m and $y = 0.50$ m, the tsunami also has an initial depression followed by a train of oscillatory waves. The interested reader will find more cases and details of experiments and simulations in Refs. 42,43 and 80.

4.4. *Waves generated by a surface disturbance*

In recent years there has been broad interest in high speed ships, not only for special purpose military crafts, but also for fast passenger ferries and commercial sealift. In this respect, one of the most promising concepts is the Surface Effect Ship (SES), which features an air cushion located, unlike hovercrafts, within a cavity built in a rigid hull. Numerical simulations were conducted with the 3D-FNPF model, as part of the design of a new type of SES with catamaran hull (the Harley SES), which also involved performing tow tank experiments and analyses with a 2.3 m long SES model. Specifically, among various resistance terms, experimentally measured wavemaking drag was compared to that calculated in an idealized numerical model of twin air cushions. Indeed for such very low draft planning ship hulls, wavemaking drag is the main hydrodynamic resistance component, which essentially corresponds to the integrated cushion pressure force acting on the sloping free surface $\eta(x, y)$ within the air cushions. For high speed sealift and hence high pressure in the cushions, free surface shape may become quite steep within the cushions and no longer be approximated by linear wave theory.

To simulate this problem, Sung and Grilli[44–46] modified the 3D-FNPF NWT detailed in this paper, by expressing time updating equations in a coordinate system (x', y, z) moving with the ship/cushion velocity $U(t)$, in the x direction. A side effect of this method is that the MEL updating will gradually transport free surface nodes downstream, at a mean velocity U. Hence, a variety of free-surface updating schemes were tested, that allowed for pseudo-Lagrangian updating of free surface nodes, allowing these to keep a fixed x location. Specifically, in this application, the potential is expressed as (more details can be found in the references),

$$\phi(\boldsymbol{x}, t) = Ux + \varphi(\boldsymbol{x}, t) \quad \text{with} \quad x' = x + \int_0^t U(\tau)\mathrm{d}\tau \qquad (4.1)$$

(a)

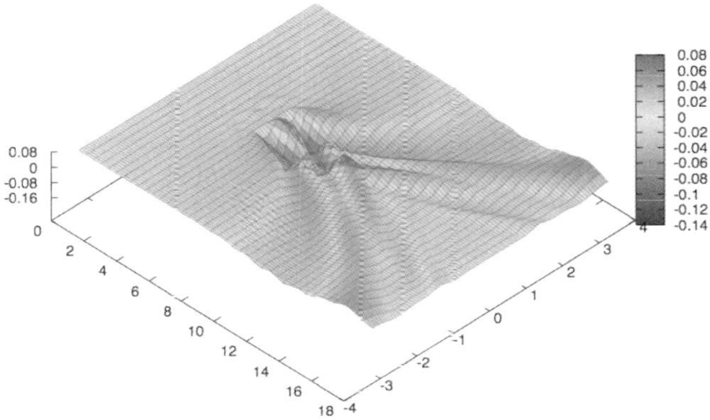

(b)

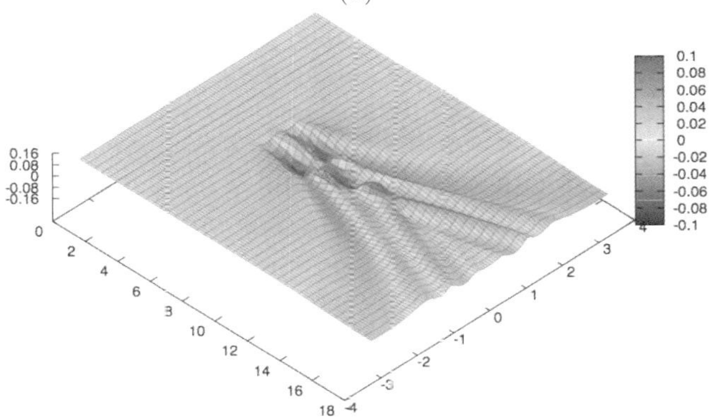

(c)

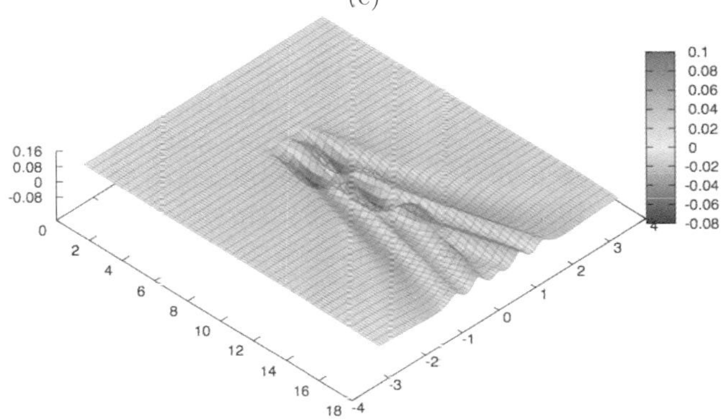

Fig. 15. Computed free surface elevations for waves generated by twin air cushions moving at speed $U =$ (a) 4, (b) 6, and (c) 8 m/s, with displacement $W = 445$ N. Quasi-steady state is established after 300 time steps (for $t = 18$, 12, and 9 s, respectively). [Note, dimensions are made nondimensional using cushion half-length $a = 0.745$.]

and the pseudo-Lagrangian operator is,

$$\frac{\tilde{\mathrm{D}}}{\mathrm{D}t} = \frac{\partial}{\partial t} + \frac{\partial \varphi}{\partial y}\frac{\partial}{\partial y} + \left\{ \frac{\partial \varphi}{\partial t} + \frac{\partial \eta}{\partial x}(U - \frac{\partial \varphi}{\partial x}) \right\} \frac{\partial}{\partial z} \, . \tag{4.2}$$

Another problem when using such relative axes is that, at the upstream edge of the NWT, a fast mean current flows under an essentially flat free surface, which typically causes the appearance of instabilities and high frequency oscillations in the model that must be damped using an absorbing beach. Here, this is done by specifying an artificial pressure over a narrow strip of free surface near the leading edge as, $p_a = -\nu\varphi$.

Although the 3D model is capable of simulating the full surface piercing ship hull, we only specify here the surface disturbance caused by a traveling air cushion. As mentioned, for a high speed SES, wavemaking drag is essentially due to cushion pressure effects. Following Harris and Grilli,[54] we perform simulations at model scale, for waves generated by a twin air cushion moving at speed U =4, 6, or 8 m/s in the x direction. A free surface pressure $p_a(x, y)$ is specified in the model (in Eq. (2.5)) over cushions of length $2a = 1.49$ m, width 0.23 m, and total surface area $S_c = 0.685$ m^2. The cushions are set one width apart and their initial x-location is 4.95 m (down from the tank leading edge), in a tank 13.4 m long, 2.3 m wide (10 cushion widths), and 7 m deep (corresponding to the actual experimental towing tank depth). The pressure distribution over the cushions is similar to that used earlier,[44,45] i.e., a double 'tanh' shape in x and y directions, with falloff parameters $\alpha = 5$ and $\beta = 10$. We assume a total displacement of $W = 445$ N for the modeled vehicle and hence an average pressure needed to support this displacement of $p_o = W/S_c = 649$ N/m^2. An absorbing boundary (snake wavemaker) condition is specified in the NWT downstream of the cushions, and no flow conditions on the side walls. We use 81, 45 and 25 nodes in the (x', y, z) directions, respectively. [Note, due to symmetry, only half the domain is represented in the y direction.] For a total of $N_\Gamma = 14,946$ nodes, the CPU time per time step, using the FMA-GMRES method is about 4.5 min on a 2 GHz x8664 single processor with 3Gb of RAM (i.e., a total of 1350 min; one of two processors on a Microways 4-bipro node cluster). In separate but similar applications, Sung and Grilli[46] reported a $\sim N_\Gamma^{1.32}$ growth of CPU time, using the FMA-GMRES method on a single processor computer. [Note, these computations included the set-up time of the BEM system matrices, which normally grows as N_Γ^2.]

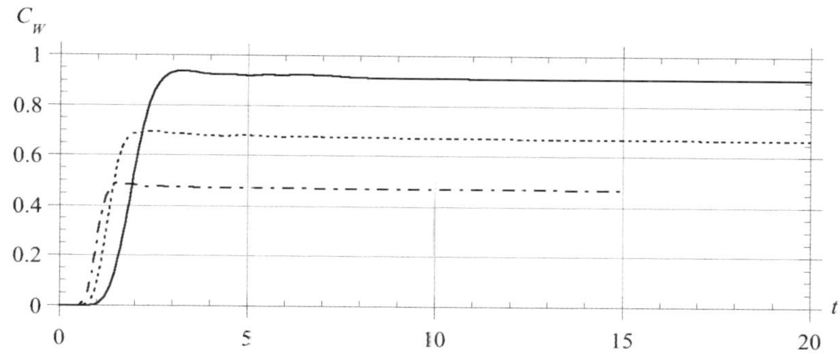

Fig. 16. Wave resistance coefficient as a function of time, for each test case in Fig. 15: (a) (—-); (b) (- - -); (c) (— - —).

After a smooth ramp-up of both cushion speed and pressure over 40 time steps, the model reaches a quasi-steady free surface elevation (Fig. 15) and resistance coefficient value (Fig. 16),

$$C_W = \frac{\rho g a}{W p_0} \int_{S_c} p \frac{\partial \eta}{\partial x} \mathrm{d}S \,. \tag{4.3}$$

Figure 15 shows free surfaces computed after 300 time steps, for the 3 cushion speeds, which all appear in a form similar to the classic Kelvin wave pattern. [Note, because of the finite depth, changes in patch velocity induce changes in the angle of the Kelvin wake.] Figure 16 indicates that the total wave drag coefficients reach stable values in the 3 test cases. Harris and Grilli[54] show that these values agree quite well with experimental measurements, particularly for the two highest speeds (for which the proportion of frictional drag in the experiments is very small).

5. Conclusions

We report on the development, validation, and application of an accurate and versatile model solving FNPF theory with a free surface, using both a higher-order BEM and MEL explicit free surface time updating, in a 3D domain of arbitrary geometry. Various boundary conditions can be specified to generate or absorb waves, such as complex snake wavemakers, actively absorbing boundaries, absorbing beaches, or moving underwater or free surface bodies. In particular, the model is well-suited to simulate wave interactions with surface-piercing moving bodies and the complex problems occurring in corners and edges, particularly with respect to well-posed boundary conditions and accurate numerical integrations. Although the model yields theoretically fully-populated system matrices, a sparse structure with improved performance can be achieved through using an efficient FMA, combined with the iterative solver GMRES. Recent computations on a single processor computer show a numerical complexity approaching a linear growth with mesh size, beyond some minimum number of nodes.

It should be stressed that our approach, is probably unique and differs in aim from that of more recent faster FNPF solvers that have been implemented to simulate complex sea states over large areas but are limited to non-breaking waves and/or use space periodic domains, and often approximate the FNPF problem to some extent, thus limiting the wave non-linearity that can be achieved (e.g., Refs. 30–33). Here, we sacrifice some potential gain in computational efficiency by implementing a very accurate method in both surface description (including higher-order inter-element continuity), MEL time updating, and costly numerical BEM integrations, that allow to accurately model overturning waves and their properties up to very close to the impact of breaking jets on the free surface, as shown in applications presented in this paper. Such results have been repeatedly shown to approach measurements in precision experiments, to within a surprising degree of accuracy. Another advantage of the present model is its easiness and versatility to simulate complex boundary condition and moving boundaries (both bottom and lateral), including those on surface piercing bodies. An additional advantage of our approach is the capability of validating simpler models and finding their range of validity.

Although the model is not capable of describing wave motion beyond the impact of the plunging jet on the free surface (which terminates computations), its results can be (and have been) used to accurately initialize wave kinematics and pressure close to the breaking point in numerical models solving full NS equations (e.g., using a volume of fluid (VOF) method for the interface reconstruction and tracking). Usually, NS models are computationally very costly (particularly in 3D) and suffer from numerical diffusion, leading to artificial loss of energy over long distances of wave propagation and hence limiting them to fairly small spatial extension. Provided they are properly initialized, such NS models can nevertheless realistically simulate the splash-up phenomenon, as reported in studies of 2D breaking waves using Reynolds averaged NS equations with a k–ϵ turbulent transport equation,[9] and using direct NS simulations,[10] or 3D breaking waves using a NS model with LES of the turbulent fields.[11,12] A promising development seems to be the actual coupling of BEM-FNPF and VOF-NS models, for pre- and post-breaking waves respectively. This was, e.g., done for 2D[13,14] and 3D[52] problems; the latter work simulated the breaking and post-breaking of solitary waves on a sloping ridge. Similarly, Corte and Grilli[74] initialized a 3D-VOF-NS model with an extreme focused wave, obtained in the 3D-FNPF model as in Ref. 40 or as shown in one of the applications reported in this paper, to calculate extreme wave loads on cylindrical piles of wind mills to be installed in 40-60 m of water in the North and Baltic Seas. The coupling of 3D-FNPF results to other methods such as those based on Smoothed Particle Hydrodynamics (e.g., Ref. 81) would also be an interesting line of work that some scientists have started investigating.

Some of the work reported here is still in progress, such as the FNPF-VOF coupling, the FNPF simulations of wave resistance and dynamic trim angle of various ships, with hulls of complex geometry (with or without an air cushion). In this respect, to be able to sufficiently resolve both the ship hull and the free surface, the image method is being fully integrated with the FMA (in both y and z directions) for the ship generated wave simulations, similar to the latest computations of rogue waves, and the FMA method is being parallelized, using a newly acquired library, to implement it on large computer clusters.

A.1. Detailed Expressions of Laplace's Equation and Accelerations on the Boundary

In the general non-orthogonal curvilinear system $(\boldsymbol{s}, \boldsymbol{m}, \boldsymbol{n})$ on the boundary, with $\kappa = \cos(\boldsymbol{s}, \boldsymbol{m})$, Laplace's equation for the potential reads,

$$\begin{aligned}\phi_{nn} = \frac{1}{1-\kappa^2}\Bigg\{ & 2\kappa\phi_{sm} - \phi_{ss} - \phi_{mm} + \frac{\phi_s}{1-\kappa^2}\{(\boldsymbol{x}_{ss} + \boldsymbol{x}_{mm} - 2\kappa\boldsymbol{x}_{sm})\cdot\boldsymbol{s} \\ & - \kappa(\boldsymbol{x}_{ss} + \boldsymbol{x}_{mm} - 2\kappa\boldsymbol{x}_{sm})\cdot\boldsymbol{m}\} - \frac{\phi_m}{1-\kappa^2}\{\kappa(\boldsymbol{x}_{mm} + \boldsymbol{x}_{ss} - 2\kappa\boldsymbol{x}_{sm})\cdot\boldsymbol{s} \\ & - (\boldsymbol{x}_{ss} + \boldsymbol{x}_{mm} - 2\kappa\boldsymbol{x}_{sm})\cdot\boldsymbol{m}\} + \phi_n\{\boldsymbol{x}_{ss} + \boldsymbol{x}_{mm} - 2\kappa\boldsymbol{x}_{sm}\}\cdot\boldsymbol{n}\Bigg\}.\end{aligned} \tag{A.1}$$

Then, substituting Equation (A.1) into the expression for the particle acceleration on the boundary yields,

$$\frac{\mathrm{D}\boldsymbol{u}}{\mathrm{D}t} = \left(\frac{\partial}{\partial t} + \boldsymbol{\nabla}\phi\cdot\boldsymbol{\nabla}\right)\boldsymbol{\nabla}\phi$$

$$= \frac{\boldsymbol{s}}{1-\kappa^2}\Bigg[\phi_{ts} - \kappa\phi_{tm} + \phi_n\phi_{ns} - \kappa\phi_n\phi_{nm} + \frac{1}{1-\kappa^2}\Big\{\phi_s\phi_{ss} - 2\kappa\phi_s\phi_{sm}$$

$$+ \kappa^2\phi_s\phi_{mm} - \kappa\phi_m\phi_{mm} + (1+\kappa^2)\phi_m\phi_{sm} - \kappa\phi_m\phi_{ss}\Big\} + \frac{\phi_s^2}{(1-\kappa^2)^2}$$

$$\Big\{(2\kappa\boldsymbol{x}_{sm} - \boldsymbol{x}_{ss} - \kappa^2\boldsymbol{x}_{mm})\cdot\boldsymbol{s} + \kappa(\boldsymbol{x}_{ss} - 2\kappa\boldsymbol{x}_{sm} + \kappa^2\boldsymbol{x}_{mm})\cdot\boldsymbol{m}\Big\}$$

$$+ \frac{\phi_s\phi_m}{(1-\kappa^2)^2}\Big\{(2\kappa\boldsymbol{x}_{ss} - (1+3\kappa^2)\boldsymbol{x}_{sm} + \kappa(1+\kappa^2)\boldsymbol{x}_{mm})\cdot\boldsymbol{s}$$

$$- ((1+\kappa^2)\boldsymbol{x}_{ss} - \kappa(3+\kappa^2)\boldsymbol{x}_{sm} + 2\kappa^2\boldsymbol{x}_{mm})\cdot\boldsymbol{m}\Big\} + \frac{\phi_m^2}{(1-\kappa^2)^2}\Big\{\kappa$$

$$((1+\kappa^2)\boldsymbol{x}_{sm} - \kappa\boldsymbol{x}_{ss} - \kappa\boldsymbol{x}_{mm})\cdot\boldsymbol{s} + (\kappa\boldsymbol{x}_{ss} - (1+\kappa^2)\boldsymbol{x}_{sm} + \kappa\boldsymbol{x}_{mm})$$

$$\cdot\boldsymbol{m}\Big\} + \frac{\phi_s\phi_n}{1-\kappa^2}\kappa\Big\{2\boldsymbol{x}_{sm} - \kappa\boldsymbol{x}_{ss} - \kappa\boldsymbol{x}_{mm}\Big\}\cdot\boldsymbol{n} + \frac{\phi_m\phi_n}{1-\kappa^2}\Big\{\kappa\boldsymbol{x}_{ss} -$$

$$(1+\kappa^2)\boldsymbol{x}_{sm} + \kappa^3\boldsymbol{x}_{mm}\Big\}\cdot\boldsymbol{n}\Bigg] + \frac{\boldsymbol{m}}{1-\kappa^2}\Bigg[\phi_{tm} - \kappa\phi_{ts} - \kappa\phi_n\phi_{ns} +$$

$$\phi_n\phi_{nm} + \frac{1}{1-\kappa^2}\Big\{(1+\kappa^2)\phi_s\phi_{sm} - \kappa\phi_s\phi_{ss} - \kappa\phi_s\phi_{mm} + \phi_m\phi_{mm} -$$

$$2\kappa\phi_m\phi_{sm} + \kappa^2\phi_m\phi_{ss}\Big\} + \frac{\phi_s^2}{(1-\kappa^2)^2}\Big\{(\kappa\boldsymbol{x}_{ss} - (1+\kappa^2)\boldsymbol{x}_{sm} + \kappa\boldsymbol{x}_{mm})$$

$$\cdot\boldsymbol{s} - \kappa(\kappa\boldsymbol{x}_{ss} - (1+\kappa^2)\boldsymbol{x}_{sm} + \kappa\boldsymbol{x}_{mm})\cdot\boldsymbol{m}\Big\} + \frac{\phi_s\phi_m}{(1-\kappa^2)^2}\Big\{(-2\kappa^2\boldsymbol{x}_{ss} +$$

$$\kappa(3+\kappa^2)\boldsymbol{x}_{sm} - (1+\kappa^2)\boldsymbol{x}_{mm})\cdot\boldsymbol{s} + (\kappa(1+\kappa^2)\boldsymbol{x}_{ss} - (1+3\kappa^2)\boldsymbol{x}_{sm} +$$

$$2\kappa\boldsymbol{x}_{mm})\cdot\boldsymbol{m}\Big\} + \frac{\phi_m^2}{(1-\kappa^2)^2}\Big\{\kappa(\kappa^2\boldsymbol{x}_{ss} - 2\kappa\boldsymbol{x}_{sm} + \boldsymbol{x}_{mm}\cdot\boldsymbol{s})\cdot\boldsymbol{s} +$$

$$(2\kappa\boldsymbol{x}_{sm} - \kappa^2\boldsymbol{x}_{ss} - \boldsymbol{x}_{mm})\cdot\boldsymbol{m}\Big\} + \frac{\phi_s\phi_n}{1-\kappa^2}\Big\{(\kappa^3\boldsymbol{x}_{ss} - (1+\kappa^2)\boldsymbol{x}_{sm} +$$

$$\kappa\boldsymbol{x}_{mm})\cdot\boldsymbol{n}\} + \frac{\phi_m\phi_n}{1-\kappa^2}\Big\{\kappa(2\boldsymbol{x}_{sm} - \kappa\boldsymbol{x}_{ss} - \kappa\boldsymbol{x}_{mm})\cdot\boldsymbol{n}\Big\}\Bigg] + \frac{\boldsymbol{n}}{1-\kappa^2}$$

$$\Bigg[(1-\kappa^2)\phi_{tn} + \phi_s\phi_{ns} - \kappa\phi_s\phi_{nm} - \kappa\phi_m\phi_{ns} + \phi_m\phi_{nm} - \phi_n\phi_{ss} +$$

$$2\kappa\phi_n\phi_{sm} - \phi_n\phi_{mm} + \frac{\phi_s^2}{1-\kappa^2}\Big\{(\boldsymbol{x}_{ss} - 2\kappa\boldsymbol{x}_{sm} + \kappa^2\boldsymbol{x}_{mm})\cdot\boldsymbol{n}\Big\} +$$

$$\frac{2\phi_s\phi_m}{1-\kappa^2}\left\{((1+\kappa^2)\boldsymbol{x}_{sm}-\kappa\boldsymbol{x}_{ss}-\kappa\boldsymbol{x}_{mm})\cdot\boldsymbol{n}\right\}+\frac{\phi_m^2}{1-\kappa^2}\left\{(\kappa^2\boldsymbol{x}_{ss}-\right.$$

$$\left.2\kappa\boldsymbol{x}_{sm}+\boldsymbol{x}_{mm})\cdot\boldsymbol{n}\right\}+\frac{\phi_s\phi_n}{1-\kappa^2}\left\{(\boldsymbol{x}_{ss}-2\kappa\boldsymbol{x}_{sm}+\boldsymbol{x}_{mm})\cdot\boldsymbol{s}+\kappa(2\kappa\boldsymbol{x}_{sm}\right.$$

$$\left.-\boldsymbol{x}_{ss}-\boldsymbol{x}_{mm})\cdot\boldsymbol{m}\right\}+\frac{\phi_m\phi_n}{1-\kappa^2}\left\{\kappa(2\kappa\boldsymbol{x}_{sm}-\boldsymbol{x}_{ss}-\boldsymbol{x}_{mm})\cdot\boldsymbol{s}+(\boldsymbol{x}_{ss}\right.$$

$$\left.\left.-2\kappa\boldsymbol{x}_{sm}+\boldsymbol{x}_{mm})\cdot\boldsymbol{m}\right\}+\phi_n^2\left\{(\boldsymbol{x}_{ss}-2\kappa\boldsymbol{x}_{sm}+\boldsymbol{x}_{mm})\cdot\boldsymbol{n}\right\}\right]. \quad \text{(A.2)}$$

For $\kappa = 0$, the orthogonal case, Eqs. (2.24), (A.1) and (A.2) simplify to the expressions given in[36] (Equations (60), (61) and (62)). Note that, when $\boldsymbol{s}\cdot\boldsymbol{m} = 0$, one has the identities $\boldsymbol{x}_{ss}\cdot\boldsymbol{m} = -\boldsymbol{x}_{sm}\cdot\boldsymbol{s}$ and $\boldsymbol{x}_{sm}\cdot\boldsymbol{m} = -\boldsymbol{x}_{mm}\cdot\boldsymbol{s}$.

References

1. Peregrine, D. H. (1983). Breaking waves on beaches. *Ann. Rev. Fluid Mech.*, **15**, 149–178.
2. Banner, M. L. and Peregrine, D. H. (1993). Breaking waves in deep water. *Ann. Rev. Fluid Mech.*, **25**, 373–397.
3. Dommermuth, D. G., Yue, D. K. P., Lin, W. M., Rapp, R. J., Chan, E. S. and Melville, W. K. (1988). Deep-water plunging breakers: a comparison between potential theory and experiments. *J. Fluid Mech.*, **189**, 423–442.
4. Skyner, D. J. (1996). A comparison of numerical predictions and experimental measurements of the internal kinematics of a deep-water plunging wave. *J. Fluid Mech.*, **315**, 51–64.
5. Grilli, S. T., Subramanya, R., Svendsen, I. A. and Veeramony, J. (1994). Shoaling of solitary waves on plane beaches. *J. Waterway, Port, Coastal, Ocean Engng*, **120**, 609–628.
6. Grilli, S. T., Svendsen, I. A. and Subramanya, R. (1997). Breaking criterion and characteristics for solitary waves on slopes. *J. Waterway, Port, Coastal, Ocean Engng*, **123**(3), 102–112.
7. Grilli, S. T., Svendsen, I. A. and Subramanya, R. (1998). Closure of: Breaking Criterion and Characteristics for Solitary Waves on Slopes. *J. Waterway, Port, Coastal, Ocean Engng*, **124**(6), 333–335.
8. Grilli, S. T., Gilbert, R., Lubin, P., Vincent, S., Legendre, D., Duvam, M., Kimmoun, O., Branger, H., Devrard, D., Fraunie, P., Abadie, S. (2004). Numerical modeling and experiments for solitary wave shoaling and breaking over a sloping beach. In *Proc. 14th Offshore and Polar Engng. Conf.* (ISOPE04, Toulon, France, May 2004), 306–312.
9. Lin, P. and Liu, P. L.-F. (1998). A numerical study of breaking waves in the surf zone. *J. Fluid Mech.*, **359**, 239–264.

10. Chen, G., Kharif, C., Zaleski, S. and Li, J. (1999). Two-dimensional Navier–Stokes simulation of breaking waves. *Phys. Fluids*, **11**, 121–133.
11. Christensen, E. D. and Deigaard, R. (2001). Large eddy simulation of breaking waves. *Coastal Engng.*, **42**, 53–86.
12. Lubin, P. (2004). *Large Eddy Simulations of Plunging Breaking Waves.* Ph.D Dissertation, University of Bordeaux I, France.
13. Guignard, S., Grilli, S. T., Marcer, R. and Rey, V. (1999). Computation of shoaling and breaking waves in nearshore areas by the coupling of BEM and VOF methods. In *Proc. 9th Intl Offshore and Polar Engng Conf.* (Brest, France), 304–309, ISOPE.
14. Lachaume, C., Biausser, B., Grilli, S. T., Fraunie, P. and Guignard, S. (2003). Modeling of breaking and post-breaking waves on slopes by coupling of BEM and VOF methods. In *Proc. 13th Intl Offshore and Polar Engng Conf.* (Honolulu, USA), 353–359, ISOPE.
15. Longuet-Higgins, M. S. and Cokelet, E. D. (1976). The deformation of steep surface waves on water. I. A numerical method of computation. *Proc. R. Soc. Lond.*, A **350**, 1–26.
16. Vinje, T. and Brevig, P. (1981). Numerical simulation of breaking waves. *Adv. Water Resour.*, **4**, 77–82.
17. Baker, G. R., Meiron, D. I. and Orszag, S. A. (1982). Generalized vortex methods for free-surface flow problems. *J. Fluid Mech.*, **123**, 477–501.
18. New, A. L. (1983). A class of elliptical free-surface flows. *J. Fluid Mech.*, **130**, 219–239.
19. New, A. L., McIver, P. and Peregrine, D. H. (1985). Computations of overturning waves. *J. Fluid Mech.*, **150**, 233–251.
20. Grilli, S. T., Skourup, J. and Svendsen, I. A. (1989). An Efficient Boundary Element Method for Nonlinear Water Waves. *Engng. Anal. Bound. Elem.*, **6**(2), 97–107.
21. Cointe, R. (1990). Numerical simulation of a wave channel. *Engng Anal. Bound. Elem.*, **7**, 167–177.
22. Grilli, S. T. and Subramanya, R. (1996). Numerical modelling of wave breaking induced by fixed or moving boundaries. *Comput. Mech.*, **17**, 374–391.
23. Grilli, S. T. and Horrillo, J. (1997). Numerical generation and absorption of fully nonlinear periodic waves. *J. Engng Mech.*, **123**, 1060–1069.
24. Romate, J. E. (1989). *The Numerical Simulation of Nonlinear Gravity Waves in Three Dimensions using a Higher Order Panel Method.* Ph.D. Dissertation. Department of Applied Mathematics, University of Twente, The Netherland, 1989.
25. Boo, S. Y., Kim, C. H. and Kim, M. H. (1994). A numerical wave tank for nonlinear irregular waves by three-dimensional higher-order boundary element method. *Intl J. Offshore and Polar Engng*, **4**(4), 265–272.
26. Broeze, J. (1993). Numerical modelling of nonlinear free surface waves with a three-dimensional panel method. PhD thesis, University of Twente, Enschede, The Netherlands.
27. Broeze, J., Van Daalen, E. F. G. and Zandbergen, P. J. (1993). A three-dimensional panel method for nonlinear free surface waves on vector computers. *Comput. Mech.*, **13**, 12–28.

28. Ferrant, P. (1996). Simulation of strongly nonlinear wave generation and wave–body interactions using a fully nonlinear MEL model. In *Proc. 21st Symp. on Naval Hydrodynamics*, Trondheim, Norway, pp. 93–108.
29. Celebi, M. S., Kim, M. H. and Beck, R. F. (1998). Fully nonlinear three-dimensional numerical wave tank simulations. *J. Ship. Res.*, **42**, 33–45.
30. Bonnefoy, F., Le Touze, D. and Ferrant, P. (2004). Generation of fully-nonlinear prescribed wave fields using a high-order spectral method, In *Proc. 14th Offshore and Polar Engng. Conf.* (ISOPE 2004, Toulon, France), 257–263, 2004.
31. Clamond, D., Fructus, D., Grue, J. and Kristiansen, O. (2005). An efficient model for three-dimensional surface wave simulations. Part II: Generation and absorption. *J. Comput. Phys.*, **205**, 686–705.
32. Fructus, D., Clamond, D., Grue, J. and Kristiansen, O. (2005). An efficient model for three-dimensional surface wave simulations. Part I: Free space problems. *J. Comput. Phys.*, **205**, 665–685.
33. Ducrozet, G., Bonnefoy, F., Le Touzé, D. and Ferrant, P. (2007). 3-D HOS simulations of extreme waves in open seas. *Nat. Hazards Earth Syst. Sci.*, **7**, 109–122.
34. Xü, H. and Yue, D. K. P. (1992). Computations of fully nonlinear three-dimensional water waves. In *Proc. 19th Symp. on Naval Hydrodynamics* (Seoul, Korea).
35. Xue, M., Xü, H., Liu, Y. and Yue, D. K. P. (2001). Computations of fully nonlinear three-dimensional wave–wave and wave–body interactions. Part 1. Dynamics of steep three-dimensional waves. *J. Fluid Mech.*, **438**, 11–39.
36. Grilli, S. T., Guyenne, P. and Dias, F. (2001). A fully nonlinear model for three-dimensional overturning waves over an arbitrary bottom. *Intl J. Numer. Meth. Fluids*, **35**, 829–867.
37. Guyenne, P., Grilli, S. T. and Dias, F. (2000). Numerical modelling of fully nonlinear three-dimensional overturning waves over arbitrary bottom. In *Proc. 27th Intl Conf. on Coastal Engng* (Sydney, Australia), 417–428, ASCE.
38. Brandini, C. and Grilli, S. T. (2001a). Modeling of freak wave generation in a three-dimensional NWT. In *Proc. 11th Intl Offshore and Polar Engng Conf.*, Stavanger, Norway, pp. 124–131, ISOPE.
39. Brandini, C. and Grilli, S. T. (2001b). Three-dimensional wave focusing in fully nonlinear wave models. In *Proc 4th Intl Symp. on Ocean Wave Measurement and Analysis* (San Francisco, USA), 1102–1111, ASCE.
40. Fochesato, C., Grilli, S. T. and Dias, F. (2007). Numerical modeling of extreme rogue waves generated by directional energy focusing. *Wave Motion*, **44**, 395–416.
41. Grilli, S. T., Vogelmann, S. and Watts, P. (2002). Development of a three-dimensional numerical wave tank for modelling tsunami generation by underwater landslides. *Engng Anal. Bound. Elem.*, **26**, 301–313.
42. Enet, F. (2006). *Tsunami generation by underwater landslides.* Ph.D Dissertation, Department of Ocean Engineering, University of Rhode Island.
43. Enet, F. and Grilli, S. T. (2005). Tsunami Landslide Generation: Modelling and Experiments. In *Proc. 5th Intl. n Ocean Wave Measurement and Analysis* (WAVES 2005, Madrid, Spain, July 2005), IAHR Pub., paper 88, 10 pp.

44. Sung, H. G. and Grilli, S. T., (2005). Numerical Modeling of Nonlinear Surface Waves caused by Surface Effect Ships. Dynamics and Kinematics. In *Proc. 15th Offshore and Polar Engng. Conf.* (ISOPE05, Seoul, South Korea, June 2005), 124–131.
45. Sung, H. G. and Grilli, S. T. (2006). Combined Eulerian-Lagrangian or Pseudo-Lagrangian, Descriptions of Waves Caused by an Advancing Free Surface Disturbance. In *Proc. 16th Offshore and Polar Engng. Conf.* (ISOPE06, San Francisco, California, June 2006), 487–494.
46. Sung, H. G. and Grilli, S. T. (2008). BEM Computations of 3D Fully Nonlinear Free Surface Flows Caused by Advancing Surface Disturbances. *Intl. J. Offshore and Polar Engng.*, **18**(4), 292–301.
47. Kervella, Y., Dutykh, D. and Dias, F. (2007). Comparison between three-dimensional linear and nonlinear tsunami generation models. *Theoretical and Computational Fluid Dynamics*, **21**, 245–269.
48. Tong, R. P. (1997) A new approach to modelling an unsteady free surface in boundary integral methods with application to bubble–structure interactions. *Math. Comput. Simul.*, **44**, 415–426.
49. Fochesato, C. and Dias, F. (2006). A fast method for nonlinear three-dimensional free-surface waves, *Proc. Roy. Soc. Lond. A*, **462**, 2715–2735.
50. Fochesato, C., Grilli, S. T. and Guyenne, P. (2005). Note on non-orthogonality of local curvilinear coordinates in a three-dimensional boundary element method. *Intl J. Numer. Meth. Fluids*, **48**, 305–324.
51. Tanaka, M. (1986). The stability of solitary waves. *Phys. Fluids*, **29**, 650–655.
52. Biausser, B., Grilli, S. T., Fraunie, P. and Marcer, R. (2004). Numerical analysis of the internal kinematics and dynamics of 3D breaking waves on slopes. *Intl J. Offshore and Polar Engng*, **14**, 247–256.
53. Guyenne, P. and Grilli, S. T. (2006). Numerical study of three-dimensional overturning waves in shallow water. *J. Fluid Mech.*, **547**, 361–388.
54. Harris, J. C., and Grilli, S. T. (2007). Computation of the wavemaking resistance of a Harley surface effect ship. In *Proc. 17th Offshore and Polar Engng. Conf.* (ISOPE07, Lisbon, Portugal, July 2007), 3732–3739.
55. Grilli, S. T., Voropayev, S., Testik, F. Y. and Fernando, H. J. S. (2003). Numerical Modeling and Experiments of Wave Shoaling over Buried Cylinders in Sandy Bottom. In *Proc. 13th Offshore and Polar Engng. Conf.* (ISOPE03, Honolulu, USA, May 2003), 405–412.
56. Dias, F., Dyachenko, A. and Zakharov, V. (2008) Theory of weakly damped free-surface flows: a new formulation based on potential flow solutions. *Physics Letters A* **372**, 1297–1302.
57. Dias, F. and Bridges, T. (2006). The numerical computation of freely propagating time-dependent irrotational water waves, *Fluid Dyn. Research*, **38**, 803–830.
58. Grilli, S. T. (1985). Experimental and Numerical Study of the Hydrodynamic Behavior of Large Self-propelled Floating Gates for Maritime Locks and Tidal-surge Barriers. Collection des Publications de la Faculté des Sciences Appliquées de l'Université de Liège, No. **99**, 447 pps, Liège, Belgium.

59. Grilli, S. T. and Svendsen, I. A. (1990). Corner Problems and Global Accuracy in the Boundary Element Solution of Nonlinear Wave Flows. *Engng. Analysis Boundary Elemt.*, **7**(4), 178–195.
60. Grilli, S. T. and Subramanya, R. (1994). Quasi-singular integrals in the modelling of nonlinear water waves in shallow water. *Engng Anal. Bound. Elem.*, **13**, 181–191.
61. Brebbia, C. A. (1978). *The Boundary Element Method for Engineers.* John Wiley and Sons.
62. Badmus, T., Cheng, A. H.-D. and Grilli, S. T. (1993). A three-dimensional Laplace transform BEM for poroelasticity. *Intl. Numer. Meth. Engng.*, **36**, 67–85.
63. Greengard, L. and Rokhlin, V. (1987). A fast algorithm for particle simulations. *J. Comput. Phys.*, **73**, 325–348.
64. Rokhlin, V. (1985). Rapid solution of integral equations of classical potential theory. *J. Comput. Phys.* **60**, 187–207.
65. Nishimura, N. (2002). Fast multipole accelerated boundary integral equation methods. *Appl. Mech. Rev.*, **55**, 299–324.
66. Korsmeyer, F. T., Yue, D. K. P. and Nabors, K. (1993). Multipole-Accelerated Preconditioned Iterative Methods for Three-Dimensional Potential Problems. Presented at *15th Intl. Conf. on Boundary Element*, Worcester, MA.
67. Scorpio, S. and Beck, F. (1996). A Multipole Accelerated Desingularized Method for Computing Nonlinear Wave Forces on Bodies. Presented at *15th Intl. Conf. Offshore Mech. Arctic Engng.* (Florence, Italy).
68. Graziani, G. and Landrini, M. (1999). Application of multipoles expansion technique to two-dimensional nonlinear free-surface flows. *J. Ship Research*, **43**, 1–12.
69. Srisupattarawanit, T., Niekamp, R. and Matthies, H. G. (2006). Simulation of nonlinear random finite depth waves coupled with an elastic structure. *Comput. Methods Appl. Mech. Engrg.*, **195**, 3072–3086.
70. Fochesato, C. (2004). *Modèles numériques pour les vagues et les ondes internes.* Ph.D Dissertation, Centre de Mathématiques et de Leurs Applications. Ecole Normale Supérieure de Cachan.
71. Greengard, L. (1988). *The Rapid Evaluation of Potential Fields in Particle Systems*, MIT Press, Cambridge, MA, 1988.
72. Dalrymple, R. A. (1989). Directional wavemaker theory with sidewall reflection, *J. Hydraulic Res.*, **27**(1), 23–34.
73. Kharif, C. and Pelinovsky, E. (2003). Physical mechanisms of the rogue wave phenomenon, *Eur. J. Mech. B/Fluids*, **22**, 603–634.
74. Corte, C. and Grilli, S. T. (2006). Numerical Modeling of Extreme Wave Slamming on Cylindrical Offshore Support Structures. In *Proc. 16th Offshore and Polar Engng. Conf.* (ISOPE06, San Francisco, CA, June 2006), 394–401.
75. Tappin, D. R., Watts, P. and Grilli, S. T. (2008). The Papua New Guinea tsunami of 1998: anatomy of a catastrophic event, *Nat. Hazards Earth Syst. Sci.*, **8**, 243–266, www.nat-hazards-earth-syst-sci.net/8/243/2008/.
76. Grilli, S. T. and Watts, P. (1999). Modeling of waves generated by a moving submerged body. Applications to underwater landslides. *Engng. Analysis Boundary Elemt.*, **23**, 645–656.

77. Grilli, S. T. and Watts, P. (2005). Tsunami generation by submarine mass failure Part I: Modeling, experimental validation, and sensitivity analysis. *J. Waterway Port Coastal and Ocean Engng.*, **131**(6), 283–297.
78. Watts, P., Grilli, S. T., Tappin, D. and Fryer, G. J. (2005). Tsunami generation by submarine mass failure Part II: Predictive Equations and case studies. *J. Waterway Port Coastal and Ocean Engng.*, **131**(6), 298–310.
79. Watts, P., Grilli, S. T., Kirby, J. T., Fryer, G. J. and Tappin, D. R. (2003). Landslide tsunami case studies using a Boussinesq model and a fully nonlinear tsunami generation model. *Natural Hazards and Earth Syst. Sc.*, **3**, 391–402.
80. Enet, F. and Grilli, S. T. (2007). Experimental Study of Tsunami Generation by Three-dimensional Rigid Underwater Landslides. *J. Waterway Port Coastal and Ocean Engng.*, **133**(6), 442–454.
81. Dalrymple, R. A. and Rogers, B. D. (2006). Numerical Modeling of Water Waves with the SPH Method. *Coastal Engineering*, **53**(2–3), 141–147.

CHAPTER 4

TIME DOMAIN SIMULATION OF NONLINEAR WATER WAVES USING SPECTRAL METHODS

Félicien Bonnefoy, Guillaume Ducrozet,
David Le Touzé and Pierre Ferrant*

Laboratoire de Mécanique des Fluides, Ecole Centrale Nantes
1 rue de la Noë, 44321 Nantes, France
**pierre.ferrant@ec-nantes.fr*

This chapter presents our progress on developing efficient nonlinear potential flow models for the simulation of 2D or 3D gravity wave propagation in wave tanks and oceanic domains. Spectral methods are used in combination with either second order perturbation series or fully-nonlinear High-Order Spectral method. Such techniques involve quick FFT resolution permitting accurate simulations of wave fields at low cost.

Both second order and fully nonlinear Numerical Wave Tanks are developed. Such models feature all the physical characteristics of a wavetank (snake wavemaker, experimentally calibrated absorbing zone, etc.). Several validation results are presented where numerical simulations are successfully compared to experiments on different 2D and 3D complex sea-states.

A periodic domain HOS modelling of open ocean wavefields is then applied to simulate long-time evolutions of 3D ocean in order to study the formation of freak waves. An investigation of the influence of directionality on these extreme wave events is shown through a study of the number and 3D shape of the detected freak waves.

1. Introduction

In open seas, ships and marine structures are periodically confronted to severe complex nonlinear directional sea-states, causing fatigue and damages. From times to times, these structures are also facing extreme waves, which constitute a major problem for both the structure integrity and the human safety. These extreme physical events are both three-dimensional and highly-nonlinear phenomena, making their numerical study uneasy.

More generally, the study of such 3D nonlinear sea-states and extreme events is challenging. To conduct experimental investigations, an increasing number of dedicated test facilities, called wave tanks, have been built. Nonetheless, besides the cost of the tests, the close reproduction of sea-states at model scale is not an easy task, if one considers that these wave patterns must be precisely generated and absorbed within this bounded structure.

The numerical simulation of the mentioned open-sea evolutions is as well very challenging. The temporal and spatial scales involved, and the shortness of the high-frequency waves included in the sea spectrum lead to huge requirements in terms of space and time grids (typically millions of spatial nodes and tens of thousands time steps). To model such long-time evolutions, potential methods are best suitable, due to their non-dissipative and numerical-efficiency characters. However, the Boundary Element Method, classically employed within the latter class of methods, remains too slow to reproduce square kilometers of ocean long-time evolution.

Another class of nonlinear potential methods has been developed during the last two decades, the spectral methods (a brief review can be found in Le Touzé[1]). With respect to classical time-domain models such as the Boundary Element Method, this spectral approach presents the two assets of its fast convergence and its resolution quickness (by means of FFTs), allowing to accurately simulate various wave evolutions with fine meshes. As it is the case for other potential methods, the resolution can be made through an expansion in perturbation series, as did Bonnefoy *et al.*[2,3] which addressed the problem of nonlinear wave generation and propagation in a wave tank by means of a complete formulation at second order. In a different way, the so-called Higher-Order Spectral (HOS) method, proposed in 1987 (West *et al.*,[4] Dommermuth & Yue[5]) and enhanced since (Bonnefoy[6]), permits the fully-nonlinear simulation of gravity waves within periodic unbounded 3D domains. This method is also related to the Dirichlet-Neumann Operator (Craig & Sulem[7]) which, in its accelerated version (Vijfvinkel[8]), is equivalent to HOS (*cf.* Schäffer[9]). Compared to approaches commonly used up to now to model ocean free-surface evolutions, essentially based on the Zakharov and Non-Linear Schrödinger equations, the fully-nonlinear character of the HOS method allows severe sea-state simulations, including the presence of extreme events. A number of studies have thus been carried out by using HOS; see *e.g.* Brandini[10] for the emergence of *natural* freak waves in a 2D spectrum, and the propagation of 2D regular waves with 3D modulational instabilities; Tanaka[11] for a study of the nonlinear evolution of a 3D spectrum; or Ducrozet *et al.*[12] for an investigation of 3D freak waves *natural* emergence within ocean evolutions. Another classical problem which has been studied using these spectral methods is the focusing of wave packets in 2D and 3D (see *e.g.* Bateman *et al.*,[13] Ducrozet *et al.*[12]). The book by Mei *et al.*[14] contains an interesting review of existing works done with HOS methods. These include the effect of forced pressure on the free surface, ambient current and bottom variation. The same kind of applications has also been investigated by using other methods based on spectral expansions. Fructus *et al.*[15,16] have derived another type of numerical iterative algorithm to compute the Dirichlet to Neumann operator It expands the operator as a sum of global convolution terms and local integrals with kernels that decay quickly in space. The global terms are computed very quickly via FFT while the local terms are evaluated by numerical integration. They were then able to investigate various problems such as the study of envelope solitons over long periods of time in Clamond *et al.*[17] and the study of crescent wave patterns in Fructus *et al.*[18]

The initially proposed HOS method is limited to unbounded domains, modelled with periodic conditions applied on the sides of the numerical domain. It therefore allows reproducing open-sea evolutions once an initial sea-state has been adequately defined; the definition of this initial state being not obvious (see Dommermuth[19]). Moreover, in this initial HOS formulation, no wave generation is possible, making difficult experimental validations. Indeed, sea evolutions reproduced in wave tanks present the additional difficulties of wall and beach reflections, generation of spurious free waves, and of starting from the rest. Nonetheless, recent developments have extended the method application range, allowing for instance the fully-nonlinear spectral modelling of the generation, propagation and absorption processes in a wave tank. Such a numerical wave tank (NWT) has been first derived at second-order (Bonnefoy *et al.*[2,3]), and then fully-nonlinear free-surface conditions have been taken into account (Ducrozet *et al.*[20]). Applications studied with these models include directional wave focusing, and reproducing of given sea-states and extreme events; such models constitute also a useful aid to experiments in physical wave tanks. Finally, other methods based on spectral expansions have also shown interesting results in bounded domains (see *e.g.* Chern *et al.*[21]).

The present chapter is divided as follows. The first part presents the general context of potential theory, in the framework of which are derived the models. The spectral basis is described for both cases of a wave tank and of an open periodic domain. The NWT involves to model the wavemaker that generates the wavefield in the domain. In the periodic approach, conversely, the initial wavefield is imposed and/or a pressure field is specified above the free surface. The implementation of the HOS model is described in appendix. A specific care has been paid to it, avoiding the convergence saturation of the original implementation (Dommermuth & Yue[5]) and resulting in a high level of accuracy even for high steepness.

The second part focuses on the free surface boundary conditions. Here again two approaches are described which either treat these conditions at second order in perturbation series or keep their fully nonlinear character. Time stepping method common to both methods is then presented as well as an absorbing zone.

The last two parts give examples of numerical results obtained with the previously described models. Irregular wave evolutions and focusing events are simulated in the developed NWT models, first at second order in perturbation series (model *SWEET*) and then using the fully-nonlinear HOS method (model *HOST*). Then studies of oceanic wave evolutions simulated by a HOS method in a periodic domain are presented in the last part of the chapter. Analysis of long-time 3D wavefield evolutions is given, including the detection of freak waves emerging in these evolutions, and the influence of the spectrum directionality is investigated.

2. Formulation and Spectral Expansion

We consider a rectangular fluid domain D of dimensions (L_x, L_y). We choose a cartesian coordinate system with the origin O located at one corner of the domain D. The O_z axis is vertical and oriented upwards, and the level $z = 0$ corresponds to the mean water level. The notation $\mathbf{x}$ stands for the (x, y) vector.

The fluid simulated is water, which is assumed to be incompressible and inviscid. The flow is also considered irrotational. With these assumptions one can apply potential flow theory, the velocity $\mathbf{V}$ derives from a potential $\mathbf{V}(\mathbf{x}, z, t) = \tilde{\nabla}\phi(\mathbf{x}, z, t)$, $\tilde{\nabla}$ representing the gradient and ϕ the velocity potential. Then, the continuity equation ($\tilde{\nabla}.\mathbf{V} = 0$) becomes the Laplace equation

$$\Delta\phi = 0 \quad \text{in } D\,. \tag{2.1}$$

The boundary conditions then come to close the system of equations. The slip conditions on the lateral boundaries and the bottom of the domain (if there is one) allow us to define a spectral basis on which the potential will be expanded. The condition on a flat bottom is written

$$\partial_z\phi(\mathbf{x}, z = -h, t) = 0 \tag{2.2}$$

h being the depth of the wave tank when modelling a wave tank, or infinite for oceanic simulations. ∂_z denotes the partial derivative with respect to z. On the free surface, the elevation is supposed to be single-valued and thus described by $z = \eta(\mathbf{x}, t)$. The slip condition and the continuity of the pressure give respectively the kinematic and dynamic free surface conditions, whose derivation and treatment are addressed in Sec. 3. With the spectral expansion of the velocity potential, the free surface conditions are used to advance the unknowns in time.

2.1. *Periodic domain*

The domain D is assumed periodic in both the x and y directions. This is expressed by the following two conditions

$$(\phi;\eta)(x=0,y,z,t) = (\phi;\eta)(x=L_x,y,z,t) \tag{2.3}$$
$$(\phi;\eta)(x,y=0,z,t) = (\phi;\eta)(x,y=L_y,z,t)\,. \tag{2.4}$$

The following basis functions ψ_{mn} individually satisfy the set of equations (2.1), (2.3) & (2.4)

$$\psi_{mn}(\mathbf{x},z) = \exp(ik_m x)\,\exp(ik_n y)\,\exp(k_{mn} z) \tag{2.5}$$

where $k_m = 2m\pi/L_x$, $k_n = 2n\pi/L_y$ and $k_{mn} = \sqrt{k_m^2 + k_n^2}$ are the wavenumbers associated with the mode $(m,n) \in \mathbb{Z}^2$. A similar basis without depth dependance is used for the free surface elevation η. The velocity potential ϕ and η can thus be expressed on this basis as

$$\phi(\mathbf{x},z,t) = \sum_{m=-\infty}^{+\infty}\sum_{n=-\infty}^{+\infty} A_{mn}(t)\,\psi_{mn}(\mathbf{x},z) \tag{2.6}$$

$$\eta(\mathbf{x},t) = \sum_{m=-\infty}^{+\infty}\sum_{n=-\infty}^{+\infty} B_{mn}(t)\,\exp(ik_m x)\,\exp(ik_n y) \tag{2.7}$$

$$\phi^s(\mathbf{x},t) = \sum_{m=-\infty}^{+\infty}\sum_{n=-\infty}^{+\infty} C_{mn}(t)\,\exp(ik_m x)\,\exp(ik_n y) \tag{2.8}$$

where A_{mn}, B_{mn} and C_{mn} are the modal amplitudes of ϕ, η and ϕ^s respectively. Here the free surface potential $\phi^s = \phi(\mathbf{x},\eta,t)$ is introduced (see Sec. 3.2). The spectral basis for the potential taken at $z=0$ and the one for surface quantities are adequate for use of Fourier transforms.

► **Initial condition** In most cases simulations in the periodic domain are started with imposing an initial free surface elevation and velocity potential. The wavefield is then let evolved. Some authors also generate the wave by means of a pressure acting on the free surface. Dommermuth & Yue[5] used a sine moving patch to study the amplification of regular waves by such a linear forcing. Touboul *et al.*[22] used a pressure patch from Jeffrey's theory to model the effect of sheltering. Their study was devoted to the influence of asymmetric wind forcing on the evolution of extreme events and to comparisons with experiments.

Without wind forcing, the initial condition is in general obtained from linear theory. The relation between the elevation in the discrete Fourier domain B_{mn} and the spectrum $\Phi(\omega, \theta)$ is

$$\frac{1}{2}\left|B_{mn}\right|^2 = \Phi(\omega, \theta)\Delta k_x \Delta k_y \tag{2.9}$$

with Δk_x and Δk_y the modal discretization in both directions. This equation allows to calculate the norm $|B_{mn}(t = 0)|$ for the desired directional spectrum $\Phi(\omega, \theta)$. Then, the phase of each component $B_{mn}(t = 0)$ is determined by a random number in $[0,\, 2\pi]$. The free surface potential is linearly evaluated from the elevation $C_{mn} = -i\omega_{mn}\, B_{mn}$ where ω_{mn} is given by the adequate dispersion relation, either $\omega_{mn}^2 = gk_{mn}$ or $\omega_{mn}^2 = gk_{mn} \tanh\left(k_{mn}h\right)$ for infinite or finite depth respectively. Finally, the initial wavefield is constructed with an inverse Fourier transform.

The latter initialization process corresponds to a superposition of linear components, each component having its own frequency and direction. It has been demonstrated by Dommermuth,[19] that the definition of linear initial wavefields for fully-nonlinear computations can lead to numerical instabilities. Indeed, he indicates that a transition period (lasting from 5 to 10 peak periods of propagation) exists with such an initialization. After this establishment delay, the initial wavefield becomes a *realistic* fully-nonlinear one. However, instabilities can appear in this specified sea-state after the transition period. To avoid this problem, the transition period is taken into account through the use of a relaxation scheme allowing the use of such linear initial conditions (Dommermuth[19]). The free surface boundary conditions can be written in the following form

$$\frac{\partial \phi^s}{\partial t} + g\eta = F, \qquad \frac{\partial \eta}{\partial t} - W^{(1)} = G \tag{2.10}$$

where $W^{(1)}$ is the linear vertical velocity while F and G stand for the nonlinear parts of the dynamic and kinematic free surface boundary condition respectively. So these nonlinear terms can be adjusted as follows

$$\frac{\partial \phi^s}{\partial t} + g\eta = F\left(1 - \exp\left[-\left(\frac{t}{T_a}\right)^n\right]\right) \tag{2.11}$$

$$\frac{\partial \eta}{\partial t} - W^{(1)} = G\left(1 - \exp\left[-\left(\frac{t}{T_a}\right)^n\right]\right) \tag{2.12}$$

and following conclusions of Dommermuth,[19] we choose to fix parameters $T_a = 10\, T_p$ and $n = 4$ (T_p is the peak period of propagation).

2.2. *Wave tank*

The domain D is now supposed to be a rectangular box with perfectly reflective side walls. In that case, the normal velocity vanishes on the walls. We further assume that the origin is located at a corner of the basin and the horizontal axis oriented towards the basin. The boundary conditions thus read

$$\partial_x \phi(x = 0, y, z, t) = \partial_x \phi(x = L_x, y, z, t) = 0 \tag{2.13}$$

$$\partial_y \phi(x, y = 0, z, t) = \partial_y \phi(x, y = L_y, z, t) = 0\,. \tag{2.14}$$

We can then build a basis of functions for the potential satisfying (2.1), (2.13) and (2.14)

$$\psi_{mn}(\mathbf{x}, z) = \cos(k_m x)\, \cos(k_n y)\, \exp(k_{mn} z) \tag{2.15}$$

where $k_m = m\pi/L_x$, $k_n = n\pi/L_y$ and $k_{mn} = \sqrt{k_m^2 + k_n^2}$ are the wavenumbers associated with the mode $(m, n) \in \mathbb{N}^2$. The velocity potential ϕ can thus be expressed on this basis as

$$\phi(\mathbf{x}, z, t) = \sum_{m=0}^{+\infty} \sum_{n=0}^{+\infty} A_{mn}(t)\, \psi_{mn}(\mathbf{x}, z) \tag{2.16}$$

$$\cos(k_n y)\, \exp(k_{mn} z) \tag{2.17}$$

$$\eta(\mathbf{x}, t) = \sum_{m=0}^{+\infty} \sum_{n=0}^{+\infty} B_{mn}(t)\, \cos(k_m x)\, \cos(k_n y)\,. \tag{2.18}$$

Once more, these basis are adequate for use of Fourier transforms.

Note that for both the periodic domain and wave tank formulations, the effect of finite depth can be simply included by replacing the vertical exponential $\exp(k_{mn} z)$ by $\cosh k_{mn}(z + h)/\cosh k_{mn} h$ in the velocity potential. In both cases the velocity potential and the elevation are fully determined by the knowledge of the time-dependent coefficients $A_{mn}(t)$ and $B_{mn}(t)$ respectively.

► **Wavemaker modelling** In the periodic model described in previous Sec. 2.1, the wave propagation is initiated with the specification of η and ϕ^s. Another possibility is to use the spectral formulation to model wave tanks like the ECN wave basin (50m×30m×5m). In that case, the physical wavemaker has to be modelled in order to get a NWT able to reproduce physical experiments carried out in wave basins.

However, we consider a homogeneous condition on the wall ($x = 0$) (Eq. (2.13)) which allows us the decomposition of our unknowns on the natural

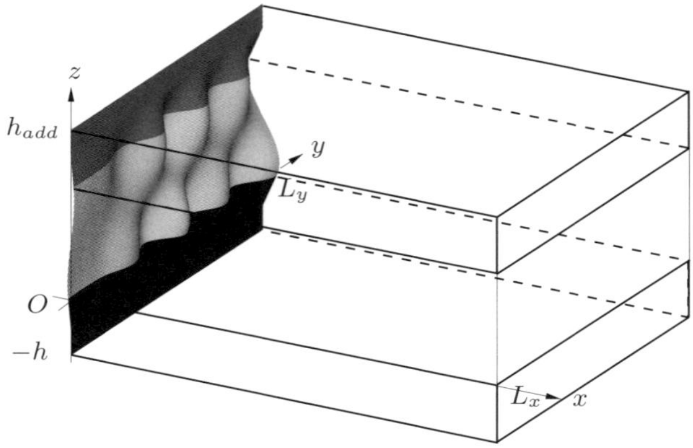

Fig. 1. Extended computational domain for calculating the additional potential.

modes of the wave basin. With a wavemaker on this wall ($x = 0$), such decomposition did not seem to be possible. Indeed, the wavemaker condition is expressed as a no-flux boundary condition

$$\partial_x \phi = \frac{\partial X}{\partial t} + (\boldsymbol{\nabla}_v X) \cdot (\boldsymbol{\nabla}_v \phi) \tag{2.19}$$

at the position of the wavemaker $x = X(y, z, t)$ with $\boldsymbol{\nabla}_v = (\partial_y, \partial_z)$.

Thus, an idea is, following Agnon and Bingham,[23] to introduce the concept of additional potential. The potential ϕ, solution of the whole problem (generation and propagation), is decomposed into two parts:

$$\phi = \phi_{spec} + \phi_{add} \tag{2.20}$$

where ϕ_{spec} is the previously described spectral potential in the fixed-geometry tank with its free surface, and ϕ_{add} is the additional potential accounting only for the wavemaker. Each of the three potentials ϕ, ϕ_{spec} and ϕ_{add} individually satisfies the Laplace equation. ϕ_{spec} accounts only for the propagation of waves in the fixed-geometry tank (that is to say, with the homogeneous condition on ($x = 0$)), while ϕ_{add} satisfies the wavemaker condition (Eq. (2.19)) without satisfying any condition on the free surface.

It is then possible, knowing the wavemaker motion imposed in experiments, to solve at each time-step the problem for the additional potential only. This problem is solved in a new extended domain of computation (see Fig. 1). This new domain has been chosen in order to be able to solve this additional problem again by means of a fully-spectral formulation, *i.e.* by using only Fourier transforms. It is to notice that the wavemaker condition (Eq. (2.19)) is thus exactly solved (up to the order of nonlinearity desired). This process takes into account the real geometry of the physical *snake* wavemaker (*i.e.* a series of independent flaps hinged side by side on the same axis) as well as the control law of the wavemaker motion (including, for instance, the time ramp at the beginning of experiments). Once the additional problem solved, we use its solution, the additional potential, in the whole problem formulated for the total potential. Thus, the latter problem reduces to a boundary value problem for ϕ_{spec} only (in the fixed-geometry tank, with the homogeneous condition on $x = 0$) and in which some forcing terms appear in free surface conditions due to ϕ_{add}. This second problem is still formulated in a fully-spectral manner, which allows again a resolution based on Fourier transforms. Please refer to Bonnefoy *et al.*[2] and Ducrozet *et al.*[20] for further details on this procedure.

3. Free Surface Conditions

3.1. *Perturbation series*

The kinematic and dynamic free surface conditions are respectively

$$\frac{\partial \eta}{\partial t} = \frac{\partial \phi}{\partial z} - \boldsymbol{\nabla}\phi \cdot \boldsymbol{\nabla}\eta \tag{3.1}$$

$$\frac{\partial \phi}{\partial t} = -g\eta + \frac{1}{2}|\boldsymbol{\nabla}\phi|^2 \tag{3.2}$$

on $z = \eta$. The nonlinear quantities are expanded in powers of the wave steepness. According to this so-called Stokes expansion, the nonlinear quantities $q = (\eta; \phi)$ are expressed as $q = \varepsilon q^{(1)} + \varepsilon^2 q^{(2)} + O(\varepsilon^3)$ for an expansion limited to second order. Moreover, since the deformation of the free surface is relatively small, one can expand in Taylor series around its level at rest the two conditions (3.1) and (3.2). At each time step, equations (3.1) and (3.2) provide successively the time derivatives of $\eta^{(1)}, \phi^{(1)}$ and $\eta^{(2)}, \phi^{(2)}$. Each evaluation of the elevations and of the potentials on $z = 0$ is made through Fourier transforms on the horizontal directions. Spatial derivatives are also obtained by Fourier transforms. This is a key point of the model since fast algorithms provide low computational cost transforms, namely Fast Fourier Transforms (FFTs).

Note that the wavemaker condition can also be developed in power series of the steepness so that we can build a consistent model for both generation and propagation of the waves. This is achieved by a Taylor expansion about the rest position $x = 0$ of the wavemaker. This approach is used either with the second order wave basin model or the fully nonlinear one.

3.2. *Fully nonlinear model*

Following Zakharov,[24] the fully-nonlinear free surface boundary conditions can be written in terms of surface quantities, namely η and the surface potential $\phi^s(\mathbf{x}, t) = \phi(\mathbf{x}, \eta, t)$

$$\frac{\partial \phi^s}{\partial t} = -g\eta - \frac{1}{2}|\boldsymbol{\nabla}\phi^s|^2 + \frac{1}{2}\left(1 + |\boldsymbol{\nabla}\eta|^2\right)\left(\frac{\partial \phi}{\partial z}\right)^2 \tag{3.3}$$

$$\frac{\partial \eta}{\partial t} = \left(1 + |\boldsymbol{\nabla}\eta|^2\right)\frac{\partial \phi}{\partial z} - \boldsymbol{\nabla}\phi^s \cdot \boldsymbol{\nabla}\eta \tag{3.4}$$

on $z = \eta(\mathbf{x}, t)$. This way, the only remaining non-surface quantity is the vertical velocity $W(\mathbf{x}, t) = \partial_z\phi(\mathbf{x}, \eta, t)$ which is evaluated thanks to the order-consistent High-Order Spectral (HOS) scheme of West *et al.*[4] (see appendix A.1). It must be underlined that within this HOS procedure, all the evaluations of the potentials $\phi^{(m)}$ are made on $z = 0$. Thus, all the quantities present in equations (3.4) and (3.4) can again be calculated thanks to FFTs. Validations of the evaluation of W for high steepness are presented in appendix A.3. In this process, it is crucial to perform a careful dealiasing to preserve the method convergence and accuracy for waves close to Stokes' limit (see appendix A.2).

3.3. *Time stepping*

The resolution of the problem in time is conducted the same way in the wave tank and periodic domain approaches. The unknowns are marched in time using an efficient 4th order Runge-Kutta Cash-Karp[25] scheme with adaptive step size and in which the linear part of the equations is integrated analytically. In both cases the free surface conditions provide the time derivative of the unknowns. The only difference between the two approaches is that in the NWT model, at each time-step the problem for the additional potential is solved first, providing additional forcing terms appearing in the free-surface conditions then marched in time. Further details on this procedure can be found in Bonnefoy *et al.*[2] and Ducrozet *et al.*[20]

In this time-stepping procedure, one has also to pay attention to the so-called aliasing, which can dramatically degrade the accuracy of the numerical solution. This phenomenon appears each time two spectral quantities are multiplied in the physical domain (see appendix A.2), as it is the case in the free-surface conditions (3.1) to (3.4), and also within the HOS procedure (see appendix A.1).

3.4. *Absorption*

An absorbing zone is included in the numerical model through a local modification of the free surface dynamic boundary condition. A pressure term is thus added on the free surface in order to absorb the outgoing waves near the end wall $x = L_x$. The chosen expression for the pressure is $p_a(x,t) = \rho\nu(x)\phi_n$ so that the corresponding power

$$\mathcal{P} = -\int_S p_a\,\phi_n dS = -\rho\int_{S_a} \nu(x)\,\phi_n^2\,dS \tag{3.5}$$

is always negative provided $\nu(x)$ is positive in the region S_a where absorption is required and zero elsewhere. Note that p_a reduces simply to $p_a = \rho\nu(x)\,\partial_t\eta$ from the kinematic boundary condition (see Clamond *et al.*[16] and reference therein for alternative techniques). The damping function ν is chosen simply as a polynomial $\nu(x) = \nu_0\,u^2(3-2u)$ with $u = 1 - x/x_a$. To correctly reproduce the physical beach, the reflection coefficient has first been experimentally evaluated with the method of Mansard and Funke[26] for a set of wavemaker frequencies. Then, the suitable length x_a and strength ν_0 of the numerical absorbing zone have been determined to suit with the experimental results (see Bonnefoy *et al.*[2,3] or Bonnefoy[6]).

4. Numerical Wave Tanks

4.1. *Second order basin*

After having given in the previous sections an outline of the NWT and periodic-domain spectral models we have developed, we briefly present here some of the results obtained by using our second order spectral model of wave basin (see Bonnefoy *et al.*[2,3] for more details).

4.1.1. *Moderate steepness events*

The adopted second-order description of the wavefield is accurate for modelling waves of moderate steepness in the basin. A verification of our model *SWEET* in that sense has been carried out on the case of 2D focused wave packets, by comparison with experimental results. Following the methodology adopted by Baldock *et al.*[27] (see Appendix A.4), one can investigate some nonlinear properties of the experimental wave packet which can then be compared to the numerical results. In particular, for wave groups of moderate steepness, the odd elevation $\eta_{\rm odd}$ extracted from the experimental signals is expected to correspond to the linear

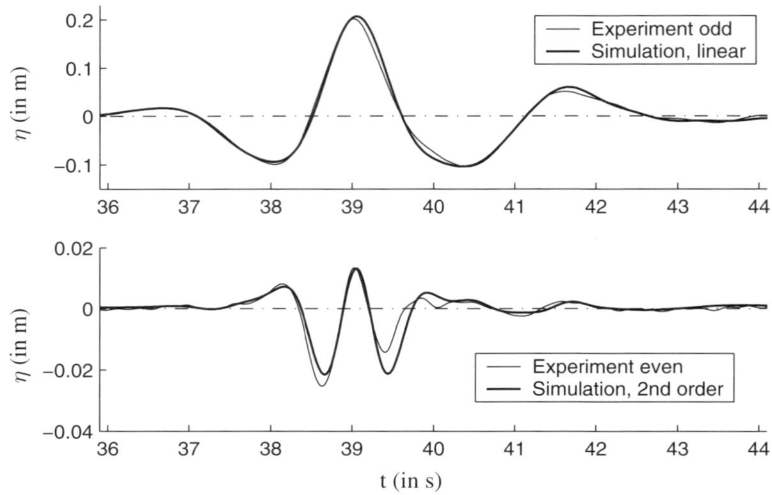

Fig. 2. Comparison between experimental and numerically-predicted elevations.

wavefield. The even elevation η_{even} is assumed to be an estimation of the second-order field. Comparisons can thus be drawn between the NWT and the adequately separated nonlinear experimental components.

The wave groups studied here are characterized by a wave height of $H = 0.3$ m for a peak wavelength $\lambda = 7.7$ m (*i.e.* a steepness of $H/\lambda = 3.9\%$). This steepness is chosen to be high enough to obtain an accurate measurement of the low-amplitude second-order elevation, but not too high to avoid sensible higher-order effects. In the top part of Fig. 2, one can observe that the odd elevation and the linear numerical result superimpose well. In the bottom part, the even elevation and the second-order result also show a good overall agreement. Some discrepancies can be noticed however, which are attributed to higher-order nonlinear effects. One well known effect of the third-order nonlinearities is indeed the modification of the phase velocity. In the odd elevation, the phases of the different wave components are thus modified at the focusing point, and the superimposition of these time-shifted components leads to a wave packet elevation different from the linear one. This third-order phase velocity modification is also present in the even elevation, as shown in Appendix A.4, since this elevation represents the bound waves at second order.

4.1.2. *Directional focused waves*

Another interesting kind of simulation is the generation of a directional irregular wavefield. The wavefield simulated here is created from a Bretschneider spectrum with zero-crossing period $T_Z = 1.4$ s and significant wave height $H_s = 2$ cm. The directional spreading used is $\cos^{2n}[(\theta - \theta_o)/2]$ with $n = 10$; 320 wave components are employed. The phases are adjusted so that all the components focus at the location ($x = 20$ m, $y = 15$ m) at $t = 80$ s. Following the conclusions drawn for the geometric focusing, the Dalrymple method is used to control the wavemaker. The mesh used in the simulation includes 256×128 nodes on the free surface and 128×32 nodes on the wavemaker. The whole simulation from which Fig. 3 is extracted covered 600 s real-time, for which 28800 time-steps were necessary, corresponding to about 17h CPU-time on a 3GHz-Xeon single-processor PC. Figure 3 shows two views of the free surface at $t = 80$ s when the wave packet is fully focused, on the left at first order and on the right at second order. It can be seen that the second-order component is much greater inside the focused wave packet than everywhere else in the random directional wavefield, as expected since the bound waves are related to the square of the linear ones.

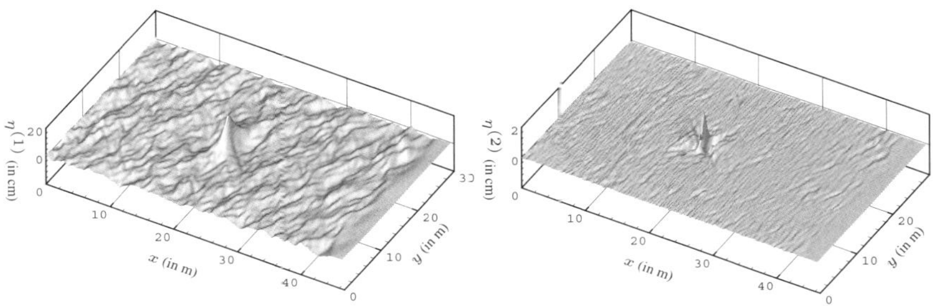

Fig. 3. Directional waves: first (left) and second (right) order elevations.

4.1.3. *Irregular waves*

To further illustrate the behaviour of the numerical model with irregular waves, a comparison has been performed in the case of another long-time generation, this time of 2D irregular waves.

Figure 4 shows the wave-elevation history of a high-amplitude wave group within the 2D irregular wavefield, after 755 s real-time of generation (about 540 waves) for two steepnesses ε_c (defined here as the ratio of the significant wave height H_s over the peak wavelength λ_p of the Bretschneider spectrum). The probe is located 20 m away from the wavemaker in a 50 m-long basin of five meter depth. The comparison at the lower steepness $\varepsilon_c = 1.5\%$ shows a good agreement

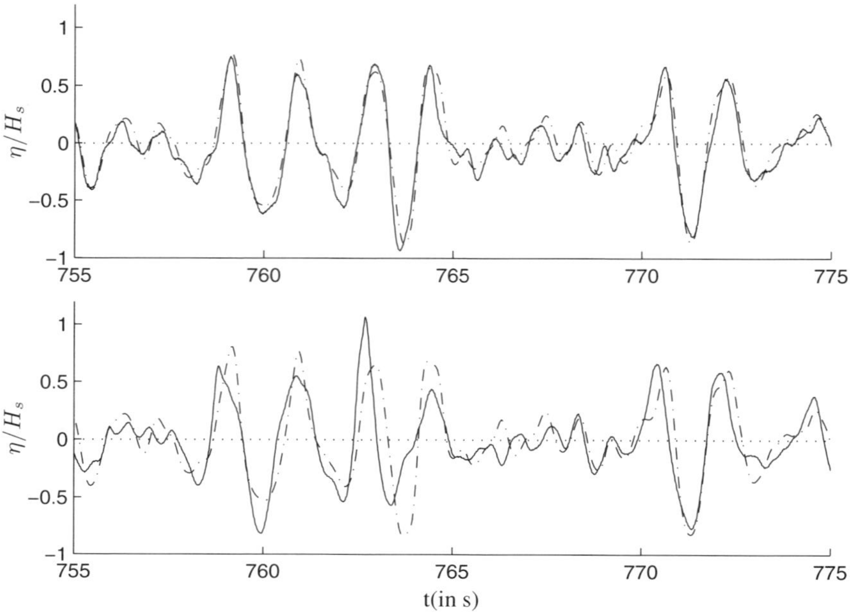

Fig. 4. 2D irregular-wave elevation history: comparison of experimental data (—) and second-order simulation (- · -). Top: $\varepsilon_c = 1.5\%$, bottom: $\varepsilon_c = 3\%$.

between the simulated wavefield and the experimental data, for the high-amplitude wave packet between $t = 760$ and 765 s as well as for the low-amplitude one in the range $765 < t < 770$ s. This emphasizes the ability of the model to simulate long-time evolutions in the wave tank, which confirms the quality of the wave generation process as well as the performance of the absorbing zone. The higher steepness studied $\varepsilon_c = 3\%$ illustrates, however, the limits of the model which fails to correctly reproduce the high-wave packets in this case. At this steepness, a model involving higher-order nonlinearities is required. This is why we have then developed a fully-nonlinear version of our NWT model. The results obtained with this model called *HOST* are presented in next section.

4.2. *HOS basin*

Most of the developments involved in the second order NWT have been used then to build a HOS NWT. In the latter model, the free surface conditions are kept fully nonlinear, thanks to the use of the HOS technique. Regarding the wave generation, it has first been included in a linear manner by using the first order additional potential previously described. The resulting model, called *HOST-wm1*, has been validated for long time generation of irregular wavefields and provides satisfactory improvements when compared to the second order NWT (see Sec. 4.2.1). Then a second order generation has been applied to enhance the wavemaker modelling for accurately simulating steep focused wave packets. This was achieved in a second version of HOS NWT, called *HOST-wm2*, by applying the same procedure as for the second order NWT to evaluate the first and second order additional potentials. The latter potentials then provide a nonlinear forcing in the fully nonlinear free surface equations (3.3) and (3.4), see Ducrozet *et al.*[20] for further details. Some of the results obtained with these two versions of HOS NWT are presented here.

4.2.1. *Irregular waves*

The HOS NWT is devoted to capture more accurately the higher than second order nonlinear effects. For instance, Fig. 5 shows the same irregular wavefield as in bottom of Fig. 4. One can clearly observe the improvements provided by the fully nonlinear approach. The steep wave packet located between $t = 760$ s and 765 s is this time correctly reproduced by the nonlinear NWT.

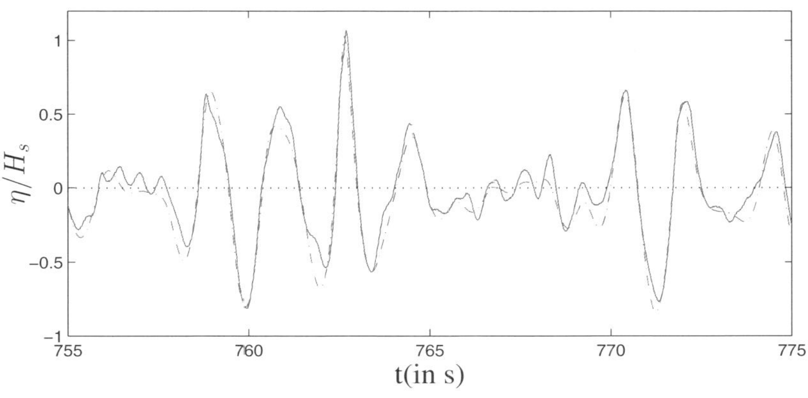

Fig. 5. 2D irregular-wave elevation history: comparison of experimental data (—) and HOS simulation (- · · -). Steepness $\varepsilon_c = 3\%$.

The nonlinearities of order greater than two in the propagation process are therefore more influent than the nonlinearities of the generation process.

4.2.2. *Focused waves*

Indeed, one has to keep in mind that the previous validation was obtained with the *HOST-wm1* formulation in which the generation is considered linearly.

However, with highly nonlinear phenomena, slight discrepancies appear with experiments. Therefore, the enhanced *HOST-wm2* model has been developed. Several experiments, with different ranges of nonlinearity, have been run for its validation. Comparisons between the *HOST-wm1* and *HOST-wm2* formulations and the experiments have thus been performed on different test-cases. The case of a 2D focused wave packet embedded in an irregular sea is presented here (efficiency and accuracy of the model on a complex 3D case is also shown in Ducrozet *et al.*[20]).

In this experimental test the wavemaker motion is highly nonlinear. Indeed, the generation of a focused wave packet implies a wide range of wavemaker motions with high nonlinearities. The comparisons between experiments and numerical simulations are made with the following characteristics

- ECN wave tank ($50\text{m}\times30\text{m}\times5\text{m}$)
- Unidirectional wavefield
- Superposition of two wavefields:
  - Irregular wavefield defined by a Bretschneider spectrum $H_s = 0.5$ m; $T_p = 3.13$ s
  - Focused wave packet defined by a Bretschneider spectrum $H_s = 0.075$ m; $T_p = 3.13$ s with phase adjusted to produce a focused wave at the position of a wave probe in the physical basin

The choice of these parameters has been made on the basis of experimental observations. It is the steepest wavefield we could generate without observing wave-breaking (this phenomenon can not be simulated with our numerical models). Wave-breaking occurs just after the focused wave reaches the experimental set up (different wave gauges located close to the position of the expected wave focusing). Thus, we are able to obtain a wave elevation record corresponding to a highly nonlinear phenomenon. It is to be noticed that, as expected, the numerical simulation breaks down when the wave-breaking occurs in the experiments.

These numerical simulations were performed with $N_x = 513$ modes on the free surface, $N_z = 65$ vertical modes, and a HOS order $M = 8$, for a total CPU time of around 5 minutes on a single processor, corresponding to 20 seconds real-time. In Fig. 6 is presented the comparison between first order (*HOST-wm1*) and second order (*HOST-wm2*) wave generation and experiment. This is made on the free surface elevation at one position of the basin, corresponding to the focusing location.

One can see in Fig. 6 that the agreement between experiment and numerical simulations is pretty good. However, the trough just before the focused wave seems to be better captured with the second order generation. A zoom on the more demonstrative part is made in Fig. 7. It is clear in this figure that the nonlinearities of the wavemaker motion have to be taken into account in certain cases. In fact, the agreement between experiment and *HOST* calculation with second order generation is almost perfect. The slight errors observed with a linear wave generation are corrected with the *HOST-wm2* scheme.

Next, in the perspective of the evaluation of the influence of the wavemaker motion nonlinearities, different experiments have been performed with a factor applied on the significant wave heights of the irregular wavefield and of the focused wave. Thus, we have now $H_s = \alpha \times 0.5$ m and $H_s = \alpha \times 0.075$ m for the irregular wavefield and the focused wave respectively, with the amplitude factor α varying in a range from 0.5 to 1. The results are presented in the next tables.

In Table 1, we first interest ourselves to the error on the focused wave height $\text{err}_H = H/H_{exp}$, where H is the focused wave height defined by $H = \eta_{\text{crest}} - \eta_{\text{trough}}$, the trough considered being the one preceding the crest. In this table, effects of wavemaker motion nonlinearities are clearly highlighted. Indeed, parameters of the simulations are exactly the same for both models, except the additional potential imposed to model the wavemaker. Thus, the generation appears as a nonlinear phenomenon, as the propagation is. With the first order wave generation (*HOST-wm1*), error on the focused wave height could be more than 10%, even if we have a very efficient fully-nonlinear model for the wave propagation. On the other hand, with a second order wave generation (*HOST-wm2*), agreement

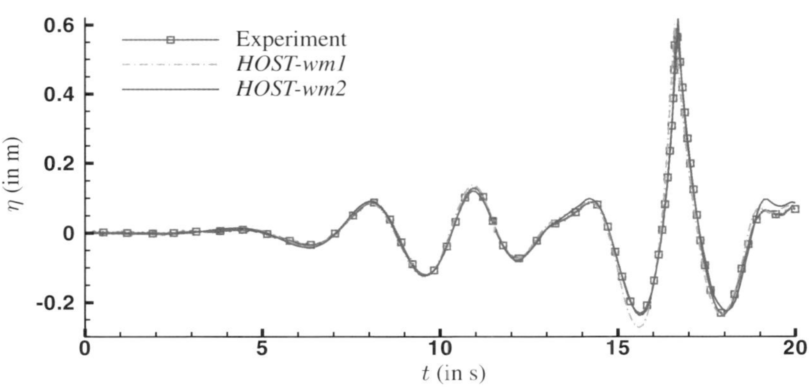

Fig. 6. Comparison between first order and second order wave generation and experiment.

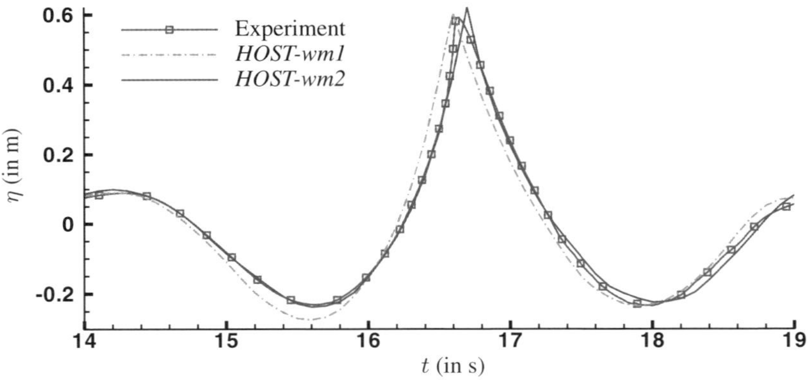

Fig. 7. Comparison between first order and second order wave generation and experiment.

Table 1. Error err_H on the focused wave height H for different amplitudes.

Amplitude factor α	HOST *wm1*	HOST *wm2*
0.50	6.3%	2.0%
0.60	7.7%	2.4%
0.70	10.0%	2.8%
0.80	11.0%	4.1%
0.90	8.8%	2.1%
1.00	10.7%	4.6%

between experiments and numerical simulations is greatly improved with an error less than 5% even for the more steep case.

We focus then, in Table 2, on the mean error err_η during the focused event $t \in [0, 20]$

$$\text{err}_\eta = \frac{\int_t |\eta_{probe}^{num} - \eta_{probe}^{exp}|}{\int_t |\eta_{probe}^{exp}|} \tag{4.1}$$

with η_{probe}^{exp} being the experimental signal of the wave probe and η_{probe}^{num} the results of numerical simulations for the free surface elevation located at the probe position.

Similar conclusions can be drawn as for the error on the wave height. Indeed, evolution of error is clearly rising with wave amplitude, that is to say, with nonlinearities. Thus, the second order wave generation is again more accurate than the linear one. Error is reduced by 30% to 50%.

Table 2. Mean error err_η on the wave probe record for different amplitudes.

Amplitude factor α	HOST *wm1*	HOST *wm2*
0.50	10.8%	7.4%
0.60	12.1%	8.1%
0.70	14%	9.1%
0.80	17.8%	11.9%
0.90	21.3%	11.8%
1.00	21.8%	11%

5. Oceanic Domain

After having exposed some results obtained with our NWT models, we present in this section different results of open-sea calculations. In the latter we simulate the evolution of sea-states, initially prescribed as described in paragraph 5.1. We put our interest on 3D long-time simulations of large domains (typically $\simeq 100$ km^2 of ocean are computed) with our fully nonlinear HOS model. A study of freak wave emergence in a directional wavefield is given here as an illustration, as well as an analysis of the influence of directionality on the statistics and chracteristics of such extreme events. One has to be cautious when drawing conclusions from such studies related to rarely occuring and strongly nonlinear events. Especially, validation issues are essential, as the simulation of the evolution of severe sea states over long durations requires an accurate and reliable model. In this respect, significant validation of our 2D and 3D periodic models for long simulations of steep sea-states may be found in Ducrozet *et al.*[12]

5.1. *Initial conditions*

The initial sea state is defined by a directional wave spectrum, here we use a JONSWAP spectrum. We classically define the directional spectrum $\Phi(\omega, \theta)$ as

$$\Phi(\omega, \theta) = \psi(\omega) \times G(\theta) \tag{5.1}$$

and the spectrum is typically

$$\psi(\omega) = \alpha g^2 \omega^{-5} \exp\left(-\frac{5}{4}\left(\frac{\omega}{\omega_p}\right)^{-4}\right) \gamma^{\exp\left[-\frac{(\omega-\omega_p)^2}{2\sigma^2\omega_p^2}\right]} \tag{5.2}$$

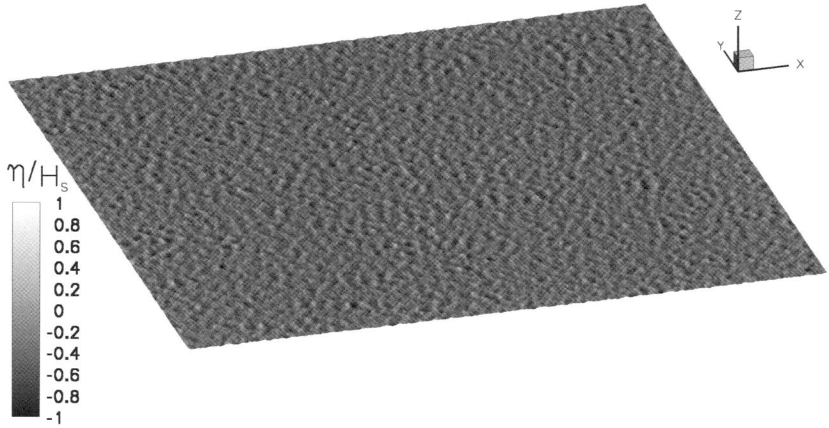

Fig. 8. Initial free surface elevation, $n = 2$.

with α being the Phillips constant and ω_p the angular frequency at the peak of the spectrum. The directionality function is defined by

$$G(\theta) = \begin{cases} A_n \cos^n \theta & |\theta| \leq \frac{\pi}{2} \\ 0 & |\theta| > \frac{\pi}{2} \end{cases} \tag{5.3}$$

with

$$A_n = \begin{cases} \dfrac{(2^p!)^2}{\pi(2p)!} & \text{if } n = 2p \\ \dfrac{(2p+1)!}{2(2^p p!)^2} & \text{if } n = 2p+1\,. \end{cases} \tag{5.4}$$

Space and time scales are defined respectively as $\widetilde{L} = 1/k_p$ where k_p is the peak wavenumber and $\widetilde{T} = \sqrt{L/g}$. Thus, $\Phi(\omega, \theta)$ is the JONSWAP spectrum when

$$\alpha = 3.279E, \qquad \gamma = 3.3, \qquad \sigma = \begin{cases} 0.07\ (\omega < 1) \\ 0.09\ (\omega \geq 1) \end{cases} \tag{5.5}$$

where E is the dimensionless energy density of the wavefield. The significant wave height can be estimated by $H_s \approx 4\sqrt{E}$ (*cf.* Tanaka[28]). Figure 8 represents a view of the 3D initial free surface elevation with directionality parameter $n = 2$.

5.2. *Long-time 3D simulations*

5.2.1. *Wavefield analysis*

In order to analyse the wavefield and determine its characteristics, we use the following analysis procedure. Its principle is to decompose the surface elevation into a set of indivual waves, so as to be able to exhibit different characteristics of the sea state such as the significant wave height H_s, the maxima of quantities (particularly η_{max} and H_{max}), the mean wavelength λ_0 ...

In two dimensions, the analysis is performed thanks to a classical zero up-and-down crossing technique. A wave is thus defined as the event separated by two zero up-crossings or two zero down-crossings. We can then isolate each individual wave and determine its height (crest to trough) H and its wavelength λ_x, which allow us to deduce characteristics of the whole sea state.

In three dimensions, we choose to define the height of a wave as its height in the direction of propagation. We first perform the same kind of analysis than in two dimensions (zero up-and-down crossing along x-axis which is the mean direction of propagation of our wavefield). Thus, we obtain the characteristics of each wave and can deduce those of the whole wavefield (H_s, η_{max}, H_{max}, λ_0, ...). A similar analysis is then performed in the transverse direction. We then deduce the transverse height H_y, the transverse wavelength λ_y, or the crest length in the transverse direction λ_c. Information on the transversal extent of waves is very interesting in terms of incidence of waves on a structure for example.

Further, this wavefield analysis is typically performed 5 times per wave-spectrum peak period, which gives access to the time dependence of the sea-state parameters.

5.2.2. *Freak waves modelling*

Freak wave events are obtained in our HOS periodic-domain model by using different kinds of configurations: either i) we impose an initial 3D directional spectrum with phases adjusted so as to form a *forced* focused event after a while, or ii) we let 2D and 3D wavefields defined by a directional wave spectrum evolve up to the *natural* emergence of freak waves. Here, the results presented concern the natural emergence of freak waves inside 3D wavefields; a wider range of applications can be found in Ducrozet *et al.*[12]

The wavefield and numerical conditions we consider are:

- Wavefield characterized by $E = 0.005$ *i.e.* $\alpha = 0.016$ and $H_s = 0.28$ (non-dimensional form),

- Domain length: $L_x \times L_y = 42\lambda_p \times 42\lambda_p$,
- Number of modes used: $N_x \times N_y = 1024 \times 512$, HOS order $M = 5$,
- If we fix for instance $T_p = 12.5$ s (typical in North Atlantic): $H_s = 10.8$ m and $\lambda_p = 244$ m, we obtain the following domain area: 10250 m $\times 10250$ m (*i.e.* $\simeq 105$ km^2).

The choice of energy $E = 0.005$ corresponds to a mean steepness of the wavefield $H_s/\lambda_p = 0.05$, that is among the steepest real sea-states (see Socquet-Juglard *et al.*[29]). A view of the initial wavefield used is exposed in Fig. 8. With these numerical conditions, the simulation of 250 peak periods takes around 10 days on a single 2.4GHz-Opteron processor. Note that considering the domain size and number of peak periods simulated, this CPU time has to be considered as very reasonable. This performance is the result of a thorough optimization of the numerical code.[a]

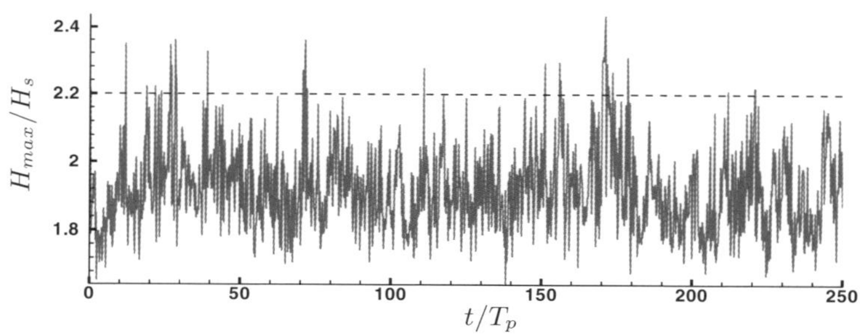

Fig. 9. Time evolution of the parameter H_{max}/H_s during the simulation, $n = 2$.

This wavefield is propagated during $250\,T_p$ with the surface elevation being analysed 5 times per peak period. The zero-crossing analysis described before is performed and permits us to study the evolution of the parameter H_{max}/H_s. This is the parameter typically used for detecting freak waves which are defined as the waves exceeding a certain ratio between H and H_s (see Kharif & Pelinovsky[30]). More precisely, the modern definition of a freak wave, accounting for the non-linearity of the process, states that it is a wave whose height exceeds $2.2H_s$. In the recent Rogue Waves Workshop (2004), several papers used this definition for freak waves, that we will also use in the following. Figure 9 represents the time evolution of H_{max}/H_s. Thus, we are able to detect the extreme events following the evolution of this parameter. At $t/T_p = 0$, a spurious extreme event is

[a]A parallelization of the model will then allow us to simulate long evolutions of thousands of square kilometers of ocean.

created by the linear initialization of the simulation. Then, one can notice that the threshold of 2.2 is overshot several times in this simulation. Two different kinds of extreme events are detected: i) isolated events corresponding to extreme events with very short life-time, *e.g.* at $t/T_p = 110$, and ii) events which stay longer in the wavefield. The latter case corresponds to a localized wavegroup of high amplitude which remains coherent for several periods in a row and produces successive extreme events, *e.g.* at $t/T_p = 170$ to 180.

As we carry out a three dimensional simulation, it is also interesting to look at the shape of the freak events. Figure 10 shows the free surface elevation at $t/T_p = 171$ when the strongest extreme event is observed. The small white square at the top left corner encloses the peak of which a closer view is given in Fig. 11. This freak event propagates in the domain, having a rather long life-time (around $20\,T_p$). The crest-to-trough height is $H = 2.4H_s$ for this case. In dimensional quantities, if we fix $T_p = 12.5$ s, *i.e.* $H_s = 10.8$ m, it represents a wave of $H = 25.9$ m and $\lambda_x = 244$ m. In the main direction of propagation, this extreme event has a $1.2\lambda_p$ wavelength. In the transverse direction, the wave group is rather short: the observed event is shaped like a single peak wave whose width is less than a peak wavelength. This *pyramidal* shape does not seem to correspond to the freak waves observed by seafarers. It has often been reported that these extreme events look like walls of water. But, one has to notice that we used for these first simulations a directionality parameter $n = 2$, which corresponds to a large

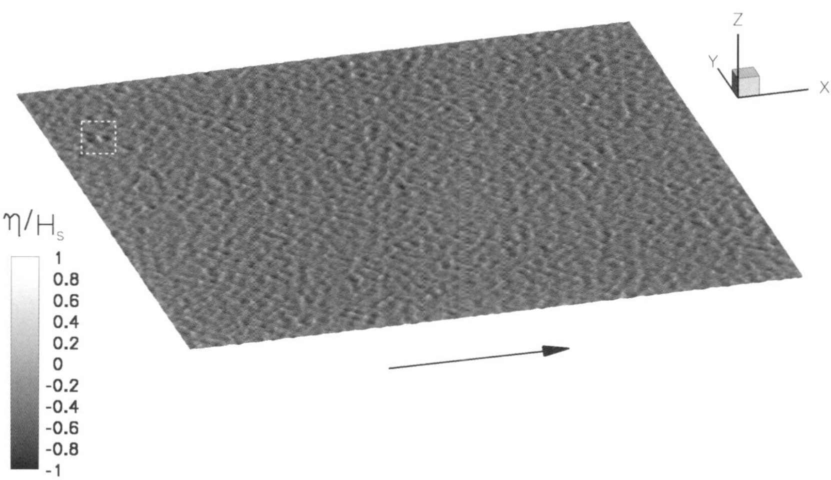

Fig. 10. Free surface elevation at the instant of a freak wave formation, $t = 171T_p$, $n = 2$.

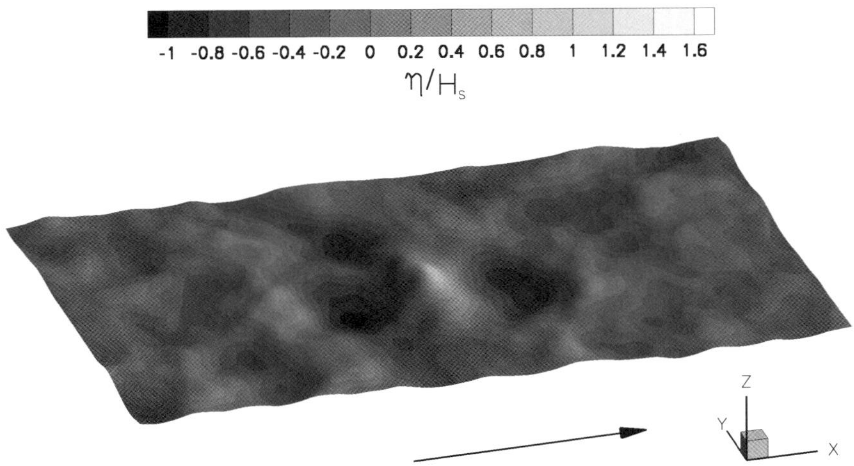

Fig. 11. Zoom of Fig. 10: freak wave at $t = 171T_p$, $n = 2$.

directional spreading of the wavefield. Usually observed sea-states tend to be more unidirectional (*i.e.* n is higher). The directionality is thus expected to play a significant role in the modification of the shape of the extreme events.

5.3. *Influence of directionality*

The choice $n = 2$ not being representative of real sea states, simulations have been performed with more realistic directionality parameters: $n = 15$, $n = 30$ and $n = 90$. Figure 12 represents the initial sea-state defined by $n = 90$. As expected, this less spread wavefield looks more realistic.

5.3.1. *Number of waves*

It is to notice that the number of waves inside the domain is highly dependent of the directionality. Indeed, following Krogstad *et al.*,[31] an approximation of this number of waves is given by the following equation

$$N_{space/time} \sim \sqrt{2\pi}\frac{N_x N_y \lambda_p^2 T}{6\lambda_0 \lambda_c T_z}\,. \tag{5.6}$$

For a JONSWAP spectrum we have

$$\frac{T_z}{T_p} = 0.6063 + 0.1164\gamma^{\frac{1}{2}} - 0.01224\gamma\,. \tag{5.7}$$

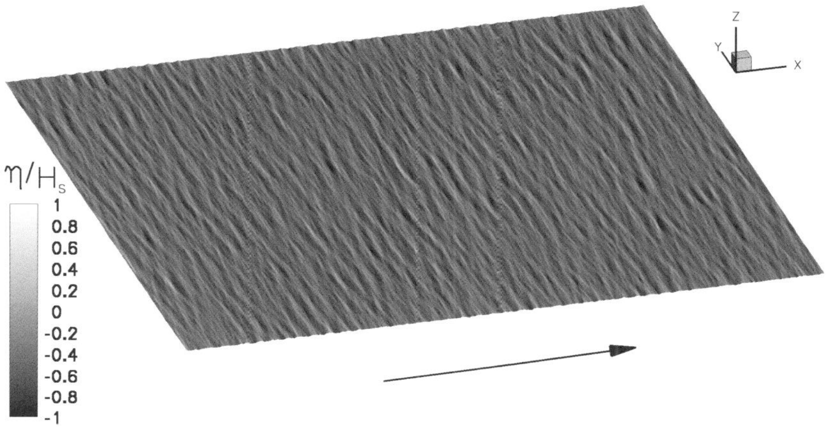

Fig. 12. Initial free surface elevation, $n = 90$.

Table 3. Dimensions λ_0, λ_c and number of waves $N_{space/time}$ in the domain, as functions of directionality.

	$n = 2$	$n = 15$	$n = 30$	$n = 90$
λ_0	$1.1\lambda_p$	$1.07\lambda_p$	$1.05\lambda_p$	$0.97\lambda_p$
λ_c	$0.69\lambda_p$	$0.91\lambda_p$	$1.10\lambda_p$	$3.34\lambda_p$
$N_{space/time}$	9.7×10^5	7.9×10^5	6.4×10^5	2.3×10^4

The lengths λ_0 and λ_c are determined by successive zero-crossing analyses. The following Table 3 is then obtained As expected, the number of waves decreases when the directional spreading is reduced. This is an important point to keep in mind when comparing simulations performed with different parameters for directionality.

5.3.2. *Shape of freak waves*

The zero-crossing analysis presented in paragraph 5.2.1 permits to study the shape of the detected extreme events. Figure 13 presents all the freak waves observed during the simulations for the four directionality cases. Each rectangle represents an extreme event (*i.e.* a wave with $H/H_s > 2.2$) with its own dimensions (*i.e.* its wavelength in the direction of propagation λ_x, and its wavelength in the transverse direction λ_y).

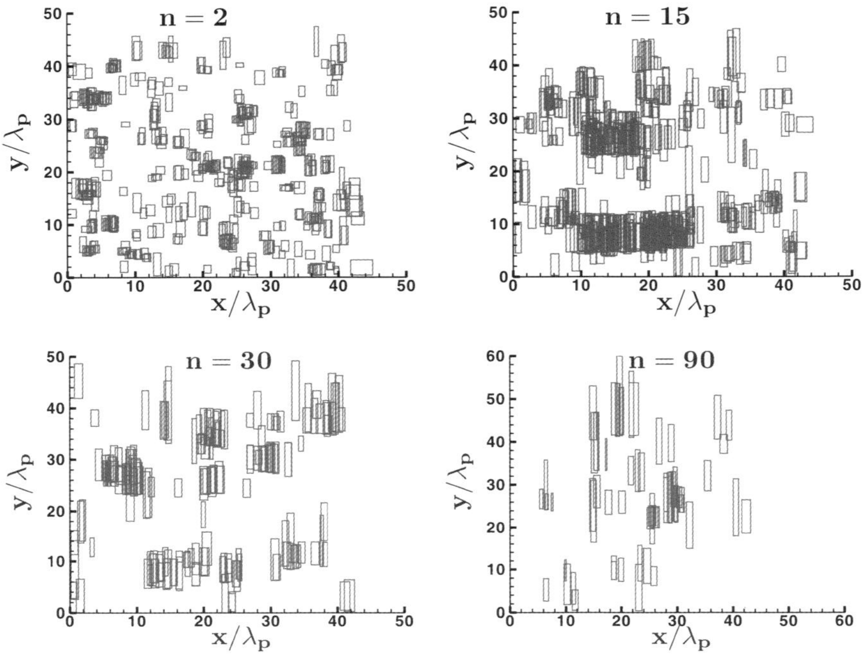

Fig. 13. Position and shape of the freak waves detected in the simulation domain for 4 different directionality parameters.

First of all, these figures can characterize the behaviour of the detected events. Indeed, two different kinds of freak waves exist, as suggested by the study of H_{max}/H_s (case $n = 2$ in Fig. 9): isolated extreme events, and freak waves appearing in groups. For instance, for the case $n = 2$ a group of large waves appears at $(x/\lambda_p, y/\lambda_p) = (6, 10)$ and remains coherent during several periods. Extreme wave groups can be similarly detected for each directional case: at $(x/\lambda_p, y/\lambda_p) = (11, 8)$ for $n = 15$, at $(x/\lambda_p, y/\lambda_p) = (6, 28)$ for $n = 30$, and at $(x/\lambda_p, y/\lambda_p) = (29, 28)$ for $n = 90$. These events have to be compared with isolated events of very brief life-time (*e.g.* at $(x/\lambda_p, y/\lambda_p) = (21, 1)$ for $n = 2$).

Then, one can observe that the number of waves strongly depends on the directional spreading, cases $n = 2$ and $n = 15$ presenting a large number of waves. The transversal extent is also, as expected, closely linked to the directional spreading: when the spreading rises (n decreases) λ_y is reduced. Thus, sea states with small directional spreading will exhibit long-crested freak events. For instance, for the case $n = 90$, the mean transverse crest length will be (if we fix $T_p = 12.5$ s)

$\lambda_c \simeq 1000$ m. We are here in presence of the observed walls of water. Conversely, wavefields with large directional spreading (short-crested) will have a pyramidal shape as seen in Fig. 11.

Table 4. Number of freak waves and their mean dimensions, as functions of the directionality parameter n.

	$n = 2$	$n = 15$	$n = 30$	$n = 90$
N_{freaks}	376	475	152	62
R	3.9×10^{-4}	6.0×10^{-4}	2.4×10^{-4}	2.7×10^{-3}
λ_x	$1.10\lambda_p$	$1.05\lambda_p$	$1.08\lambda_p$	$1.02\lambda_p$
λ_y	$1.59\lambda_p$	$3.98\lambda_p$	$4.77\lambda_p$	$8.75\lambda_p$

Main characteristics of the detected freak waves are reported in Table 4. It is interesting is to examine the ratio $R = N_{freaks}/N_{space/time}$, with N_{freaks} standing for the number of detected freak events. Indeed, it indicates approximately the probability of occurrence of the extreme events, showing again that the number of waves is highly dependent on the directionality. Actually, whereas the number of extreme events decreases in long-crested seas, their probability tends to increase because of the relatively low number of waves present in the domain. This conclusion that the occurrence of freak waves is higher in long-crested than in short-crested seas is conform to the simulations by Socquet-Juglard *et al.*,[29] based on an improved Non-Linear Schrödinger (NLS) model, and to the experimental observations made by Onorato *et al.*[32] Conversely, when the directionality parameter decreases the probabilities become quite close for the different simulations. It has to be noticed that with our model, the freak wave probabilities of occurrence can be investigated more in detail through the study of the probability distributions of surface elevations, crest-to-trough heights or crest heights (see *e.g.* Ducrozet[33]).

6. Conclusion

In this chapter have been presented some applications of nonlinear numerical models based on spectral potential methods, and some validations through comparisons with wavetank experiments. The pseudo-spectral solution method allows to model details of complex wavefields with both high accuracy and high rate of convergence, while the FFT-based resolution leads to $O(NlogN)$ CPU costs. In the presented models, special attention had been paid to the numerical implementation. In particular, a full dealiasing of nonlinear products in physical space is achieved, and an optimized time-marching procedure is used, which is based on a change of variable and a fourth-order Runge-Kutta scheme with adaptive step size control. The reliability of the model was assessed by careful validations including comparisons to both experiments and other numerical models. Comparisons between NWT simulations and experiments show that irregular wavefields can been reproduced satisfactorily up to steepness of $H_s/\lambda_p = 3\%$. Higher-steepness experiments show local wave breaking which is out of the scope of the present model. The stability of the HOS method, that was sometimes controverted, was assessed here by studying the accuracy of the estimation of W and by long-time simulation of regular wave. We confirm that the controversial results in Dommermuth & Yue[5] and Skandrani *et al.*[34] are due to the inconsistent approach in Dommermuth & Yue.[5] The model described in West *et al.*[4] gives however satisfactory results without any saturation. For steep waves, the long-time evolution we obtained is similar to the one obtained with other spectral methods, *e.g.* in Fructus *et al.*[18] or in Clamond *et al.*[16] (see Appendix A.3).

On the one hand spectral methods have been applied to develop Numerical Wave Tanks (NWTs) which reproduce all the characteristics of a physical wave tank (snake wavemaker, experimentally calibrated absorbing beach, etc.). Two approaches have been used to model the nonlinear wave propagation while an additional potential technique is used to model the generation. First, an expansion in perturbation series has permitted to build a second-order NWT (model *SWEET*). Satisfactory comparisons to experiments are presented for moderate-steepness focused and irregular wavefields. Limitations however arise for high steepness events. A second model has thus been developed then, in which fully nonlinear free surface conditions are taken into account through the use of the High-Order Spectral method, this in combination with first or second order additional potentials to model the generation. Validations have been achieved on the steep case of a two-dimensional focused wave embedded in an irregular sea. Results are really convincing with a non-negligible improvement when using the second order wave generation (model *HOST-wm2*) compared to the linear one (model *HOST-wm1*). This new powerful formulation is also able to model accurately complex 3D steep sea patterns.

On the other hand the same HOS method has been applied more classically to the simulation of wave evolutions (no generation) in periodic domains. A parametric study of directionality in long-time 3D wavefield evolutions has been especially carried out. The latter reveals that the number and shape of the detected freak wave events largely depends on the directional spreading. For widely spread seas, the number of extreme events is high and their width in the transverse direction tends to be limited. When the spreading decreases, the freak events number decreases and their transverse extent increases, leading to the well known *walls of water* for quasi uni-directional seas. Another remarkable feature shown by our approach is the groupiness of freak events, especially when the directional spreading is limited, as shown in Fig. 13 where several successive freak wave events appear in the same wave group.

A.1. HOS Method

Here is briefly described the HOS method first introduced by West *et al.*[4] and Dommermuth & Yue.[5] The evaluation process of the vertical velocity at the exact free surface position ($W = \frac{\partial\phi}{\partial z}|_{z=\eta(\mathbf{x},t)}$) is based on a series expansion of the velocity potential and the vertical velocity. This series expansion with respect to ε, being a measure of the wave steepness, is written

$$\phi(\mathbf{x},z,t) = \sum_{m=1}^{M} \phi^{(m)}(\mathbf{x},z,t) \quad \text{and} \quad W(\mathbf{x},t) = \sum_{m=1}^{M} W^{(m)}(\mathbf{x},t) \qquad \text{(A.1)}$$

where M is the order of nonlinearity and $\phi^{(m)}$ a quantity of order $O(\varepsilon^m)$. Reporting this series into the definition of the surface velocity potential ϕ^s, we perform a Taylor expansion of the potential ϕ around $z = 0$. Arranging according to order m we obtain the following scheme:

$$\phi^{(1)}(\mathbf{x},0,t) = \phi^s(\mathbf{x},t) \qquad \text{(A.2)}$$

$$\phi^{(m)}(\mathbf{x},0,t) = -\sum_{k=1}^{m-1} \frac{\eta^k}{k!} \frac{\partial^k \phi^{(m-k)}}{\partial z^k}(\mathbf{x},0,t) \quad \text{for} \quad m > 1\,. \qquad \text{(A.3)}$$

This way, the complicated Dirichlet problem for $\phi(\mathbf{x},z,t)$ on $z = \eta(\mathbf{x},t)$ is transformed into M simpler Dirichlet problems for $\phi^{(m)}(\mathbf{x},z,t)$ on $z = 0$. Afterwards, the simplified problems are solved successively by a spectral method using the spectral basis functions described in Sec. 2. The vertical velocity is then evaluated with the same Taylor development than previously described

$$W^{(m)}(\mathbf{x},t) = -\sum_{k=0}^{m-1} \frac{\eta^k}{k!} \frac{\partial^{k+1} \phi^{(m-k)}}{\partial z^{k+1}}(\mathbf{x},0,t)\,. \qquad \text{(A.4)}$$

The two formulations of West *et al.*[4] and Dommermuth & Yue[5] and are equivalent up to this point. However, we employ the one of West *et al.*[4] which propose a consistent treatment with respect to ε of the free surface boundary conditions. This means that we make sure every term is calculated up to the maximum order M, especially the terms $\left(1 + |\boldsymbol{\nabla}\eta|^2\right) W^a$ with $a = 1$ or 2. For instance, if we denote by W_M and W_{M-2} the truncation at order M and $M - 2$ respectively of the vertical velocity defined by expression (A.1), we write in equation (3.4)

$$\left(1 + |\nabla\eta|^2\right) W = W_M + |\nabla\eta|^2 W_{M-2}\,. \tag{A.5}$$

The same consistent approach is required for both the calculation of W^2 and $|\nabla\eta|^2\, W^2$. For more details, see *e.g.* West *et al.*,[4] Tanaka[11] or Le Touzé.[1]

A.2. Dealiasing

The nonlinear products of the free surface boundary conditions (3.3) & (3.4) are computed in the physical space instead of the modal space. This leads to the well-known aliasing phenomenon which has to be taken into account (see *e.g.* Canuto *et al.*[35]). We perform dealiased computations using spectra extended with zero padding.

The number of points in the physical space is then chosen in order to remove errors on multiple products (which are at the maximum M products). Applying the *half* rule, the number of point to use in the physical space (N_{x_d}, N_{y_d}) to get a full dealiasing is

$$N_{x_d} = \frac{M+1}{2} N_x, \qquad N_{y_d} = \frac{M+1}{2} N_y\,. \tag{A.6}$$

This is generally called the $(M+1)$-*half* rule, cf. Canuto *et al.*[35] This total dealiasing is the one used originally by West *et al.*[4] It provides very accurate results even for steep waves (see appendix A.3). However, for high values of M, the rising of computational effort and memory allocations due to this procedure could become prohibitive in 3D computations. Thus, we have chosen to perform this total dealiasing in 2D computations and a partial dealiasing for the 3D computations with $M > 3$ defined as follows.

We introduce the partial dealiasing as a dealiasing of the products of order $M' < M$. This is applied within the HOS scheme described in Sec. A.1 until order M is reached. For instance, to dealiase a product of order 4 (η^4) with $M' = 3$ we will rewrite ($\eta^4 = \eta^3 \times \eta$) and perform the dealiasing on η^3 with the *four-half* rule (here $M' + 1 = 4$). This allows to maintain reasonable CPU time and memory storage as well as a good accuracy for large 3D computations (see Ducrozet[33]).

A.3. Validation of the HOS Model

We briefly present some validation of the periodic HOS model. Some of the first papers to use HOS model (see *e.g.* Dommermuth & Yue[5] and Skandrani *et al.*[34]) showed that the approximation error made on the vertical velocities would not decrease past a given threshold: the model would saturate for steepness greater than $H/\lambda = 6.4\%$ (or equivalently $ka = 0.2$ with k the wavenumber and a the wave amplitude defined as half the wave height H). Based on that several authors claimed the HOS model to be unstable and non convergent at high steepness (*e.g.* Clamond & Grue[36] or Fructus *et al.*[15]). We report here our investigation on this saturation and found that consistency and dealiasing technique both play a key role for convergence at high steepness (see also[17]). Applying the total dealiasing presented in A.2 we were able to obtain better accuracy than with the inconsistent partial dealiasing scheme used by Dommermuth & Yue[5] and Skandrani *et al.*[34] Following the test made by Dommermuth & Yue (only the highest steepness $H/\lambda = 12.7\%$, *i.e.* $ka = 0.4$, is presented), we provide the reference η and ϕ_s from the model of Rienecker and Fenton[37] and evaluate the vertical velocity with the HOS scheme. Table A.1 gives three sets of the absolute error on W obtained with the HOS model and the grey cells points out out when the present implementation gives a smaller error than in Dommermuth & Yue.[5] The first set is the error obtained by Dommermuth & Yue[5] while the second one is obtained with the present model and a partial dealiasing at $M' = 2$. Note that such results should be identical to those from Dommermuth & Yue however we obtain a better accuracy. Note that we also recomputed the tests in Skandrani *et al.*[34] obtained with $M' = 2$: results with the present implementation does not reproduce the saturation and show indeed better accuracy. This is surely to be linked with the fact that we have chosen the consistent approach of West *et al.* for the evaluation of $|\nabla\eta|^2 W$ and $\left(1 + |\nabla\eta|^2\right) W^2$.

Table A.1. Maximum absolute erreur on W with $ka = 0.4$

N_x	M				
	6	8	10	12	14
32	2.8×10^{-3}	8.1×10^{-3}		/	
64	1.5×10^{-3}	3.5×10^{-4}	9.1×10^{-4}		
128	1.5×10^{-3}	3.0×10^{-4}	8.9×10^{-4}		
32	2.0×10^{-4}	2.0×10^{-3}	2.4×10^{-3}	1.7×10^{-3}	4.7×10^{-4}
64	1.5×10^{-3}	3.2×10^{-4}	2.5×10^{-4}	1.1×10^{-3}	4.0×10^{-3}
128	1.5×10^{-3}	3.0×10^{-4}	6.0×10^{-5}	1.3×10^{-5}	4.7×10^{-5}
32	1.5×10^{-3}	3.0×10^{-4}	6.5×10^{-5}	1.9×10^{-5}	1.0×10^{-5}
64	1.5×10^{-3}	2.9×10^{-4}	6.0×10^{-5}	1.2×10^{-5}	2.5×10^{-6}
128	1.5×10^{-3}	3.0×10^{-4}	6.0×10^{-5}	1.2×10^{-5}	2.5×10^{-6}

The third set of results in Table A.1 is obtained with the total dealiasing. In that case the error decreases all along the range for M so that no saturation is observed. This means that provided the West *et al.* approach is used for both total dealiasing and consistent evaluations in the boundary conditions, the HOS model gives accurate results without the saturation previously observed.

A second validation concerns the accuracy of the model for long time simulation. We follow the study of Fructus *et al.*[16,18] and let a steep nonlinear regular wave evolve for over 1000 periods. Here again the reference solution of Rienecker and Fenton[37] provides the initial wavefield at $ka = 0.3$ and $ka = 0.4$ (*i.e.* respectively $H/\lambda = 10\%$ and $H/\lambda = 13\%$). It is reminded here that the Stokes' limit, defining the steepest stable regular wave, is $H/\lambda = 14.1\%$ (see *e.g.* Schwartz[38]). Simulations over 1000 periods are run with $L_x = \lambda$ and $N_x = 15$ so that seven harmonics are taken into account, the HOS order being set accordingly to $M = 7$. A number of $N_{x_d} = 60$ modes are used for a fully dealiased result. We observe as in Fructus *et al.*[18] that the phase shift increases linearly in time provided that the tolerance in the Runge-Kutta scheme is sufficiently low (10^{-9} here). After 1000 periods, the numerical solution at $ka = 0.3$ is only 1.9 degrees out of phase with the reference solution. The relative error on the energy is about 10^{-4}. At $ka = 0.4$, the phase shift is 160 degrees after 1000 periods. These two results are the same order of magnitude as the one obtained by Fructus *et al.*[18] (18 degrees for $ka = 0.3$ after 1000 periods and 20 degrees for $ka = 0.4$ after 100 periods only, see the corresponding reference for more details).

Figure A.1 shows the reference elevation at $t = 0$ and the elevation from the HOS model after 1000 periods (T). One can observe the previously mentioned phase shift. We also simulate backwards from $t = 1000T$ to $t = 0$ by simply changing the sign of the time step at $t = 1000$. The final elevation at $t = 0$ is also plotted on Fig. A.1 ; it is superimposed with the initial reference solution which validate the reversibility of the numerical scheme and shows that accumulating errors are of negligible amount.

A.4. Separation of Nonlinear Components

Two focused wavefields are generated in the basin. The *crest focused* group is first obtained with a given set of wave components to build the wavemaker motion. The second wavefield, *trough focused*, is the inverse of the former, obtained by using the same set of components but with amplitudes of opposite sign. The wave elevations at the focusing point are recorded in both cases, respectively giving η and η^*. Then the odd and even elevations are built by combinations: $\eta_{\text{odd}} = (\eta - \eta^*)/2$ and $\eta_{\text{even}} = (\eta + \eta^*)/2$.

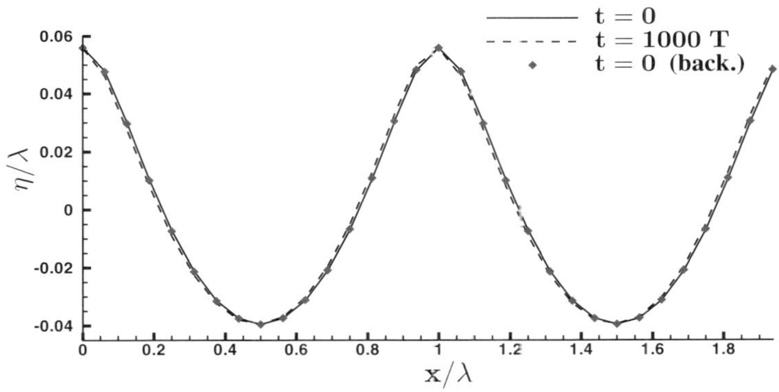

Fig. A.1. Forward and backward simulation of Stokes wave over 1000 periods ($ka = 0.3$).

The theoretical linear elevation can be written as the real part of

$$\eta^{(1)} = \sum_n \underline{a_n}\, \mathrm{e}^{\mathrm{i}(\omega_n t - k_n x)} . \tag{A.7}$$

Using these linear components, the odd and even elevations are given at third order by

$$\eta_{\mathrm{odd}} = \sum_n \underline{a'_n}\, \mathrm{e}^{\mathrm{i}(\omega_n t - k'_n x)} \tag{A.8}$$

$$+ \sum_{m,n,p} \underline{a_m}\, \underline{a_n^\pm}\, \underline{a_p^\pm}\, F_{mnp}^{\pm,\pm}\, \mathrm{e}^{\mathrm{i}[(\omega_m \pm \omega_n \pm \omega_p)t - (k'_m \pm k'_n \pm k'_p)x]} \tag{A.9}$$

$$\eta_{\mathrm{even}} = \sum_{m,n} \underline{a_m}\, \underline{a_n^\pm}\, G_{mn}^\pm\, \mathrm{e}^{\mathrm{i}[(\omega_m \pm \omega_n)t - (k'_m \pm k'_n)x]} \tag{A.10}$$

where $a'_n = a_n(1 + A_n)$ and $k'_n = k_n(1 + K_n)$ are the amplitudes and wavenumbers correct to the third order, and $G_{mn}^\pm$ and $F_{mnp}^{\pm,\pm}$ some appropriate coefficients which are independent of the linear amplitudes $\underline{a_n}$. The second order coefficients G_{mn} can be found in *e.g.* Dalzell.[39] The superscript $\pm$ denotes the sum and difference terms. The amplitudes $\underline{a_n^\pm}$ are respectively equal to $\underline{a_n}$ for the sum term (+), and to its conjugate for the difference term (-). Concerning the even elevation, one can notice that third-order nonlinearities are included in the term $k'_m \pm k'_n$ in (A.10), and thus modify the phase velocity of the sum and difference terms.

References

1. D. Le Touzé. *Méthodes spectrales pour la modélisation non-linéaire d'écoulements à surface libre instationnaires*. PhD thesis, École Centrale de Nantes, France, (2003).
2. F. Bonnefoy, D. Le Touzé, and P. Ferrant, A fully-spectral 3D time-domain model for second-order simulation of wavetank experiments. Part A: Formulation, implementation & numerical properties, *App. Ocean Res.* **28**, 33–43, (2006).
3. F. Bonnefoy, D. Le Touzé, and P. Ferrant, A fully-spectral 3D time-domain model for second-order simulation of wavetank experiments. Part B: Validation, calibration versus experiments & sample applications, *App. Ocean Res.* **28**, 121–132, (2006).
4. B. J. West, K. A. Brueckner, R. S. Janda, D. M. Milder, and R. L. Milton, A new numerical method for surface hydrodynamics, *J. Geophys. Res.* **92**(C11), 11,803–11,824 (Oct., 1987).
5. D. G. Dommermuth and D. K. Yue, A high-order spectral method for the study of nonlinear gravity waves, *J. Fluid Mech.* **184**, 267–288, (1987).
6. F. Bonnefoy. *Modélisation expérimentale et numérique des états de mer complexes*. PhD thesis, École Centrale de Nantes, France, (2005).
7. W. Craig and C. Sulem, Numerical simulation of gravity waves, *J. Comp. Phys.* **108**, 73–83, (1993).
8. E. Vijfvinkel. Focused wave groups on deep and shallow water. Master's thesis, University of Groningen, The Netherlands, (1996).
9. H. A. Schäffer. On the Dirichlet-Neumann Operator for nonlinear water waves. In *Proc.* 20th *Int. Workshop on Water Waves and Floating Bodies*, (2005).
10. C. Brandini. *Nonlinear interaction processes in extreme waves dynamics*. PhD thesis, Università di Firenze, Florence, Italy, (2001).
11. M. Tanaka, A method of studying nonlinear random field or surface gravity waves by direct numerical simulation, *Fluid Dyn. Res.* **28**, 41–60, (2001).
12. G. Ducrozet, F. Bonnefoy, D. Le Touzé, and P. Ferrant, 3-D HOS simulations of extreme waves in open seas, *Nat. Hazards Earth Syst. Sci.* **7**, 11–14, (2007).
13. W. J. D. Bateman, C. Swan, and P. H. Taylor, On the efficient numerical simulation of directionally spread surface water waves, *J. Comp. Phys.* **174**, 277–305, (2001).
14. C. C. Mei, M. Stiassnie, and D. K.-P. Yue, *Theory and applications of ocean surface waves. Part 2; Nonlinear aspects*. vol. 23, *Adv. series on ocean eng.*, (World Scientific Publishing Co. Pte. Ltd., 2005).
15. D. Fructus, D. Clamond, J. Grue, and . Kristiansen, An efficient model for three-dimensional surface wave simulations. Part I: Free space problems, *J. Comp. Phys.* **205**, 665–685, (2005).
16. D. Clamond, D. Fructus, J. Grue, and . Kristiansen, An efficient model for three-dimensional surface wave simulations. Part II: Generation and absorption, *J. Comp. Phys.* **205**, 686–705, (2005).
17. D. Clamond, M. Francius, J. Grue, and C. Kharif, Long time interaction of envelope solitons and freak wave formations, *Eur. J. Mech., B/Fluids*. **25**(5), 536–553, (2006).
18. D. Fructus, C. Kharif, M. Francius, . Kristiansen, D. Clamond, and J. Grue, Dynamics of crescent water wave patterns, *J. Fluid Mech.* **537**, 155–186, (2005).
19. D. Dommermuth, The initialization of nonlinear waves using an adjustment scheme, *Wave Motion*. **32**, 307–317, (2000).

20. G. Ducrozet, F. Bonnefoy, D. Le Touzé, and P. Ferrant, Implementation and validation of nonlinear wave maker models in a HOS Numerical Wave Tank, *Int. J. Offshore Polar Eng.* **16**(3), 161–167, (2006).
21. M. Chern, A. Borthwick, and R. Eatock Taylor, A pseudospectral σ-transformation model of 2-D nonlinear waves, *J. Fluids Struct.* **13**, 607–630, (1999).
22. J. Touboul, J. Giovanangeli, C. Kharif, and E. Pelinovsky, Freak waves under the action of wind: experiments and simulations, *Eur. J. Mech., B/Fluids.* **25**, 662–676, (2006).
23. Y. Agnon and H. B. Bingham, A non-periodic spectral method with applications to non linear water waves, *Eur. J. Mech., B/Fluids.* **18**, 527–534, (1999).
24. V. Zakharov, Stability of periodic waves of finite amplitude on the surface of a deep fluid, *J. App. Mech. Tech. Phys.* **9**, 86–94, (1968).
25. J. R. Cash and A. H. Karp, A variable order Runge-Kutta method for initial value problems with rapidly varying right-hand sides, *ACM Trans. Math. Softw.* **16**(3), 201–222, (1990).
26. E. Mansard and E. Funke. The measurement of incident and reflected spectra using a least squares method. In *Proc. of the* 17th *Int. Conf. on Coastal Eng.*, pp. 154–172, (1980).
27. T. E. Baldock, C. Swan, and P. H. Taylor, A laboratory study of nonlinear surface waves on water, *Phil. Trans. Royal Soc. London.* **A354**, 649–676, (1996).
28. M. Tanaka, Verification of Hasselmann's energy transfer among surface gravity waves by direct numerical simulations of primitive equations, *J. Fluid Mech.* **444**, 199–221, (2001).
29. H. Socquet-Juglard, K. Dysthe, K. Trulsen, H. E. Krogstad, and J. Liu, Probability distributions of surface gravity waves during spectral changes, *J. Fluid Mech.* **542**, 195–216, (2005).
30. C. Kharif and E. Pelinovsky, Physical mechanisms of the rogue wave phenomenon, *Eur. J. Mech., B/Fluids.* **22**, 603–634, (2003).
31. H. Krogstad, J. Liu, H. Socquet-Juglard, K. Dysthe, and K. Trulsen. Spatial extreme value analysis of nonlinear simulations of random surface waves. In *Proc.* 23rd *Int. Conf. on Offshore Mech. and Arctic Engng.*, (2004).
32. M. Onorato, A. R. Osborne, and M. Serio, Extreme wave events in directional, random oceanic sea states, *Phys. Fluids.* **14**, L25–L28, (2002).
33. G. Ducrozet. *Modélisation des processus non-linéaires de génération et de propagation d'états de mer par une approche spectrale.* PhD thesis, École Centrale de Nantes, France, (2007).
34. C. Skandrani, C. Kharif, and J. Poitevin, Nonlinear evolution of water surface waves: the frequency down-shift phenomenon, *Contemp. Math.* **200**, 157–171, (1996).
35. C. Canuto, M. Y. Hussaini, A. Quarteroni, and T. A. Zang, *Spectral methods in fluid dynamics*. Springer series in comp. phys., (Springer-Verlag, 1988). ISBN 3-540-52205-0.
36. D. Clamond and J. Grue, A fast method for fully nonlinear water-wave computations, *J. Fluid Mech.* **447**, 337–355, (2001).
37. M. M. Rienecker and J. D. Fenton, A Fourier approximation method for steady water waves, *J. Fluid Mech.* **104**, 119–137, (1981).

38. L. Schwartz, A computer extension and analytic continuation of Stokes' expansion for gravity waves, *J. Fluid Mech.* **62**, 553–578, (1974).
39. J. Dalzell, A note on finite depth second-order wave-wave interactions, *App. Ocean Res.* **21**, 105–111, (1999).

CHAPTER 5

QALE-FEM METHOD AND ITS APPLICATION TO THE SIMULATION OF FREE-RESPONSES OF FLOATING BODIES AND OVERTURNING WAVES

Q.W. Ma* and S. Yan†

School of Engineering and Mathematical Sciences, City University
Northampton Square, London, EC1V 0HB, UK
**q.ma@city.ac.uk*
†*s.yan@city.ac.uk*

The QALE-FEM (Quasi Arbitrary Lagrangian-Eulerian Finite Element Method) has been developed very recently. It is based on the fully nonlinear potential theory. As is known, the successes of conventional FEMs rely on a good mesh. The mesh may be generated with high quality but may become over-distorted during simulation since the fluid domain is continuously changing with oscillation of the free surface and/or motion of floating bodies. To overcome the over-distortion, they must be regenerated frequently or even every time step but regenerating mesh may take a major proportion of computational costs if an unstructured mesh is used. In the QALE-FEM, the complex mesh is generated only once at the beginning and is moved at all other time steps in order to conform to motions of the free surface and bodies by using a spring analogy method specially developed for problems concerning nonlinear waters waves and their interaction with fixed or floating bodies. Extensive numerical instigations have shown that the QALE-FEM can be much faster than other methods with similar capability. It has been applied to model the free-responses of 3D floating bodies with 6 degrees of freedom and 3D overturning waves. This chapter will review the development of this method and its applications.

1. Introduction

Waves and wave-body interactions have been a research focus for many years and studied by a large number of researchers. However, computation of very steep and/or overturning waves, their loads on offshore structures as well as fully nonlinear responses of floating bodies to such waves is still a very challenging and time-consuming task. Because of strong nonlinearity involved, solutions based on linear or other simplified theories may be insufficient and so fully nonlinear theory is necessary for this kind of problem. Two types of fully nonlinear model, i.e. NS model (governed by the Navier-Stokes and the continuity equations together with proper boundary conditions) and FNPT model (fully nonlinear potential theory model), may be used. The latter is relatively simpler and needs relatively less computational resource than the former with satisfactory accuracy if post-breaking of waves are not of interest and/or structures involved are large. We concentrate on the FNPT model in this chapter.

The problems formulated by the FNPT model are usually solved by a time marching procedure. In this procedure, the key task is to solve a boundary value problem by using a numerical method, such as a boundary element or desingularized boundary integral method (both are shortened as BEM here) and a finite element method (FEM). The BEM has been attempted by many researchers, such as Vinje et al. (1981)[1], Lin, et al. (1984)[2], Kashiwagi (1996)[3], Cao, et al. (1994)[4], Celebi, et al. (1998)[5], Grilli, et al. (2001)[6] and Guyenne, et al. (2006)[7]. Details about this method can be found in other chapters of this book, such as the one authored by Grilli, et al. (Ch 3).

This chapter will focus on development and applications related to the FEM. The FEM for fully nonlinear water waves described by potential theory was developed by Wu, et al. in 1994[8] and 1995[9] for 2-dimensional (2D) transient waves and radiation problems, respectively. It was followed by Westhuis, et al. (1998)[10], Clauss, et al. (1999)[11], Sriram, et al. (2006)[12] and Wang, et al. (2005)[13] in this direction. From 1995, the FEM method started to be used for dealing with various 3-dimensional (3D) wave problems. It was first extended to 3D problems in a circular tank in 1995 by Wu, et al.[14,15] Then it was further extended to deal with 3D problems in rectangular tank with waves generated by a wavemaker or motion of a tank by Wu, et al. (1996)[16], Ma, et al. (1997[17], 1998[18,19]). The work on interaction between waves and multiple bodies was published by Ma, et al. (2001)[20,21]. The work on waves generated by moving a vertical cylinder was given by Hu, et al. (2002)[22] and Wang, et al. (2006).[23] Wang, et al. (2007)[24] applied the FEM to modelling waves by oscillating a cylinder with flare and to estimating wave loads on such a structure, fixed in an incoming wave. Although significant progress has been made, none of the above papers have demonstrated that the FEM could model overturning waves and fully nonlinear interaction between steep waves and 3D floating bodies with free responses of 6 degrees of freedom (DoF).

Compared to the BEM, the FEM requires less memory and is therefore computationally more efficient for fully nonlinear waves, as pointed out by Wu, et al.[8] This has been further demonstrated by Ma, et al. (2008)[25] and Yan, et al. (2008)[26]. A disadvantage of the FEM, however, is that a complex unstructured mesh, which is necessary for complicated geometries to achieve accurate results, may need to be regenerated at every time step to follow the motion of waves and bodies. Repeatedly regenerating such a mesh may take a major part of CPU time and so makes the overall simulation very slow. In order to reduce the CPU time spent on meshing, simple structured meshes have been used in some publications[8,14-22]. These were improved by Wu, et al. (2004)[27], in which the unstructured meshes are generated only on the free surface while the mesh below it is obtained by drawing lines or curves from each node on the free surface, depending on the shape of the body. Turnbull, et al. (2003)[28] and Heinze (2003)[29] adopted a hybrid structured-unstructured mesh for 2D problems, which is unstructured near a body and structured in the outer region. However, those techniques are either still time-consuming or restricted to the cases for bodies with special shapes and/or undergoing only translational motions. The problem associated with meshing has become a bottleneck in the development of more efficient methods dealing with fully nonlinear waves and its interaction with bodies by the FEM. To overcome the problem, Yan, et al. (2005)[30] and Ma, et al. (2006)[31] have developed a new method called QALE-FEM (Quasi Arbitrary Lagrangian-Eulerian Finite Element Method). The main idea of this method is that the complex unstructured mesh is generated only once at the beginning of calculation and is moved at other time steps to conform to motions of boundaries by using a robust spring analogy method. This allows one to use an unstructured mesh with any degree of complexity without the need of regenerating it at every time step. Its promising features have been shown by simulating various 2D and 3D problems including free responses of 3D floating bodies with 6 DoFs and overturning waves[25,26,32,33]. In the cited papers by the authors, the QALE-FEM was compared with other methods in terms of computational efficiency and accuracy for various cases. It concluded that the QALE-FEM can be many times faster than other methods (some details to be given below). It is noted that Sudharsan, et al. (2004)[34] has also tried to test the idea for moving meshes but they did not suggest a special spring analogy method for the free surface problems.

This chapter will summarise the main developments of the QALE-FEM and its applications to nonlinear wave problems including 3D overturning waves and free responses of 3D floating structures in steep waves with 6 degrees of freedom.

2. Governing Equations

The computational domain is in a tank formed by four vertical walls and a bed, as illustrated in Fig. 1. The bed of the tank can be arbitrary. If there are floating bodies, they will be moored to the seabed or the walls. Two types of methods may be used to generate nonlinear waves. The first one is to utilise a piston or paddle wavemaker mounted at the left end of the tank. The second one is to specify the initial condition for the elevation of and the velocity potential on the free surface. In any case, the main direction of the wave propagation is assumed to be along the x-axis. To reduce the reflection, a damping zone with a Sommerfeld condition is applied at the right end of the tank. Details about this will not be given here but can be found in Ma, et al. (2001)[20,21]. A Cartesian coordinate system is adopted with the oxy plane on the mean free surface and with the z-axis being positive upwards. The origin of the system can be chosen to be anywhere for the sake of convenience, though it is illustrated to be at the centre of the tank in Fig. 1. If a floating body is involved, a body-fixed coordinate system ($O_b x_b y_b z_b$, which is not shown in Fig. 1 for clarity) may be added, in which all axes go through the centre of gravity of the body with x_b-axis parallel to the keel and z_b-axis initially parallel to z-axis.

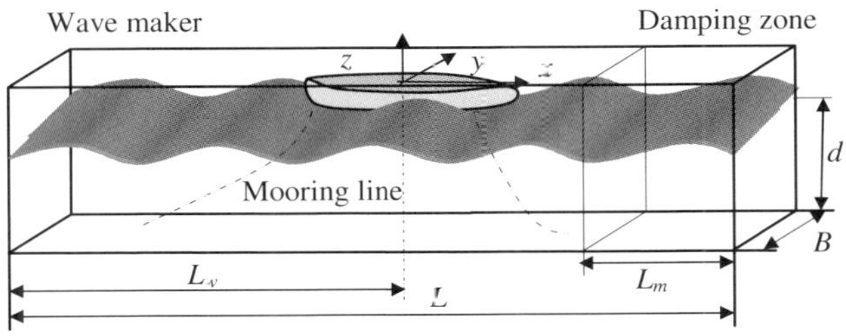

Fig. 1. Illustration of computational domain.

2.1. *Governing equations for fluid*

Fluid is assumed to be incompressible and its velocity ($\vec{u}$) can be described by $\vec{u} = \nabla\phi$, where ϕ is the velocity potential. The velocity potential satisfies Laplace's equation,

$$\nabla^2\phi = 0 \tag{1}$$

in the fluid domain. On the free surface $z = \zeta(x, y, t)$, it satisfies the kinematic and dynamic conditions in the following Lagrangian form,

$$\frac{Dx}{Dt} = \frac{\partial\phi}{\partial x}, \frac{Dy}{Dt} = \frac{\partial\phi}{\partial y}, \frac{Dz}{Dt} = \frac{\partial\phi}{\partial z} \tag{2}$$

$$\frac{D\phi}{Dt} = -gz + \frac{1}{2}|\nabla\varphi|^2 \tag{3}$$

where $\frac{D}{Dt}$ is the substantial (or total) time derivative following fluid particles and g is the gravitational acceleration. In Eq. (3), the atmospheric pressure has been taken as zero. On all rigid boundaries, such as the wavemaker and the floating body, the velocity potential satisfies

$$\frac{\partial\phi}{\partial n} = \vec{n} \cdot \vec{U}(t), \tag{4}$$

where $\vec{U}(t)$ and $\vec{n}$ are the velocity and the outward unit normal vector of the rigid boundaries, respectively.

2.2. *Governing equations for floating bodies*

If there is any floating body involved, its displacements, velocities and accelerations are governed (see, e.g., Refs. 25 and 32) by

$$[M]\dot{\vec{U}}_c = \vec{F}\,, \tag{5}$$

$$[I]\dot{\vec{\Omega}} + \vec{\Omega}\times[I]\vec{\Omega} = \vec{N}\,, \tag{6}$$

$$\frac{d\vec{S}}{dt} = \vec{U}_c\,, \tag{7}$$

$$[B]\frac{d\vec{\theta}}{dt} = \vec{\Omega}\,, \tag{8}$$

where $\vec{F}$ and $\vec{N}$ are the external forces and moments acting on the floating body, $\vec{U}_c$ and $\dot{\vec{U}}_c$ the translational velocity and acceleration of its gravitational centre, $\vec{\Omega}$ and $\dot{\vec{\Omega}}$ its angular velocity and acceleration, $\vec{\theta}(\alpha,\beta,\gamma)$ the Euler angles and $\vec{S}$ the translational displacements. In Eq. (5) and (6), $[M]$ and $[I]$ are mass and inertia-moment matrixes, respectively; and $[B]$ in Eq. (8) is the transformation matrix formed by Euler angles and defined as,

$$[B] = \begin{bmatrix} \cos\beta\cos\gamma & \sin\gamma & 0 \\ -\cos\beta\sin\gamma & \cos\gamma & 0 \\ \sin\beta & 0 & 1 \end{bmatrix}. \tag{9}$$

Once $\vec{U}_c$ and $\vec{\Omega}$ are known, the velocity at a point on the body surface is determined by

$$\vec{U} = \vec{U}_c + \vec{\Omega}\times\vec{r}_b\,, \tag{10}$$

where $\vec{r}_b$ is the position vector of a point on the body surface relative to the gravitational centre.

The external forces ($\vec{F}$) and moments ($\vec{N}$) acting on a body in Eqs. (5) and (6) can be evaluated by,

$$\vec{F} = -\rho\iint_{S_b}\left(\frac{\partial\phi}{\partial t} + \frac{1}{2}|\nabla\phi|^2 + gz\right)\vec{n}ds + \vec{f}_m\,, \tag{11}$$

$$\vec{N} = -\rho\iint_{S_b}\left(\frac{\partial\phi}{\partial t} + \frac{1}{2}|\nabla\phi|^2 + gz\right)\vec{r}_b\times\vec{n}ds + \vec{N}_m\,, \tag{12}$$

where the terms in brackets represent the pressure given by Bernoulli's equation, S_b denotes the wetted body surface, $\vec{f}_m$ and $\vec{N}_m$ are forces and moments due to mooring lines. In this work, the mooring lines are approximated by using nonlinear springs, i.e.,

$$\vec{f}_m = k_m \vec{S}_m \tag{13}$$

$$\vec{N}_m = \vec{r}_m \times \vec{f}_m \tag{14}$$

in which $\vec{S}_m$ is the displacement of a mooring point, k_m is the spring stiffness (possibly depending on $\vec{S}_m$) and $\vec{r}_m$ is the position vector of a mooring point relative to the gravitational centre.

As can be seen, the time derivative of the velocity potential ($\partial\phi/\partial t$) is required and is critical for accurately calculating forces and moments. Although it may be estimated by using a finite difference method, it is more stable to find the term by solving a boundary value problem (BVP) about it. The BVP about $\partial\phi/\partial t$ is defined by,

$$\nabla^2\left(\frac{\partial\phi}{\partial t}\right) = 0 \tag{15}$$

in the fluid domain. On the free surface $z = \zeta(x, y, t)$, it is given by

$$\frac{\partial\phi}{\partial t} = -g\zeta - \frac{1}{2}|\nabla\phi|^2 . \tag{16}$$

On all rigid boundaries, it satisfies

$$\frac{\partial}{\partial n}\left(\frac{\partial\phi}{\partial t}\right) = [\dot{\vec{U}}_c + \dot{\vec{\Omega}} \times \vec{r}_b] \cdot \vec{n} - \vec{U}_c \cdot \frac{\partial\nabla\phi}{\partial n} + \vec{\Omega} \cdot \frac{\partial}{\partial n}[\vec{r}_b \times (\vec{U}_c - \nabla\phi)] . \tag{17a}$$

The second order derivative in the normal direction in this equation is not easy to deal with and direct calculation of it can degrade the numerical accuracy. To avoid this, this equation is equivalently changed to:

$$\frac{\partial}{\partial n}\left(\frac{\partial\phi}{\partial t}\right) = [\dot{\vec{U}}_c + \dot{\vec{\Omega}} \times \vec{r}_b - \vec{\Omega} \times \vec{U}_c] \cdot \vec{n} - (\vec{\Omega} \times \vec{n}) \cdot \nabla\phi - \frac{\partial\nabla\phi}{\partial n} \cdot (\vec{U}_c + \vec{\Omega} \times \vec{r}_b) \tag{17b}$$

with

$$\frac{\partial\nabla\phi}{\partial n} = -\left[\frac{\partial}{\partial\tau_1}\left(\frac{\partial\phi}{\partial\tau_1}\right) + \frac{\partial}{\partial\tau_2}\left(\frac{\partial\phi}{\partial\tau_2}\right)\right]\vec{n} + \frac{\partial}{\partial\tau_1}\left(\frac{\partial\phi}{\partial n}\right)\vec{\tau}_1 + \frac{\partial}{\partial\tau_2}\left(\frac{\partial\phi}{\partial n}\right)\vec{\tau}_2 , \tag{17c}$$

where $\vec{\tau}_1$ and $\vec{\tau}_2$ are two tangential unit vectors at the same point on the body surface as $\vec{n}$. More details about derivation and numerical implementation of these equations are given by Yan (2006)[42].

There is a difficulty with solving Eqs. (15) to (17) together with Eqs. (1) to (12) if bodies freely respond to excitation of waves. As can be seen from Eq. (17), the velocity and acceleration of the bodies must be known prior to solving the BVP about $\partial\phi/\partial t$. However, in cases involving freely-responding floating bodies, they depend on the forces and moments in Eqs. (11) and (12) as shown in Eqs. (5) and (6). In turn, to find the forces and moments, one needs the term of $\partial\phi/\partial t$. The two sets of equations governing waves and floating bodies are fully coupled. The schemes to overcome this difficulty will be discussed below.

3. Numerical Procedure and Techniques

This section will summarise the numerical procedure and techniques to solve the various problems by using the QALE-FEM.

3.1. *ISITIFMB_M*

The problem described by Eqs. (1) to (17) is solved by using a time step marching procedure as mentioned before. At each time step, the position of the free surface and the potential values on it are updated by integrating Eqs. (2) and (3). Thus, the boundary condition for the potential on the free surface can be replaced by a Dirichlet condition:

$$\phi = \phi_f \text{ on free surface.} \tag{18}$$

Therefore, the unknown velocity potential in the fluid domain at the next time step can be found by solving a mixed boundary value problem which is defined by Eqs. (1), (4) and (18). The problem about $\partial\phi/\partial t$ is formulated in a similar way. Its Dirichlet condition on the free surface is given by Eq. (16) without any change. The whole procedure for QALE-FEM is illustrated in Fig. 2, which is called ISITIFMB_M (Iterative Semi Implicit Time Integration Method for Floating Bodies-Modified). Although this procedure is devised for solving problems associated with fully nonlinear interaction between waves and floating bodies, it can be applied to other problems, such as wave dynamics without bodies.

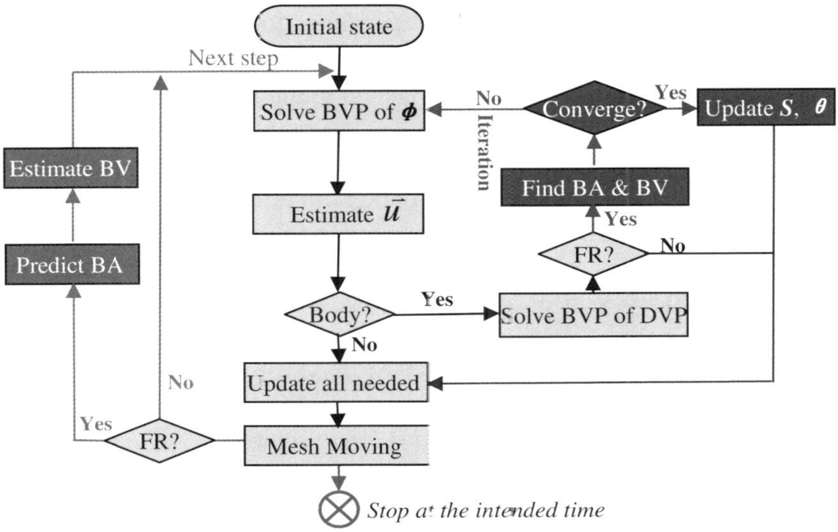

Fig. 2. Illustration of the ISITIFMB_M (DVP: time derivative of ϕ; BA: body acceleration; BV: body velocity; FR: free response of bodies).

To interpret the procedure, one should consider three situations: (1) waves without any body; (2) waves interacting with fixed bodies or bodies moving as prescribed and (3) waves interacting with floating bodies freely responding to wave excitation. For the first situation, the procedure starts, at each time step, with solving the BVP of the velocity potential (ϕ), then estimating the velocity on the free surface and updating all the parameters on the free surface, including its position and potential value on it. The main computing tasks in this situation at each time step include only these in the central part in Fig. 2. For the second situation, bodies are considered together with waves. The velocity and position of the bodies are prescribed and thus they are not part of solution. The additional work compared to that in the first situation is the computation of pressure and forces on these bodies. To do so, the BVP about $\partial\phi/\partial t$ needs to be solved in addition to seeking the solution for the velocity potential. The procedure for the two situations has been employed and described in many publications (see, e.g., Vinje, et al.,[1] Guyenne, et al.,[7] Wu, et al.,[9] Ma, et al.[21]).

When floating bodies freely responding to wave excitation are involved (the third situation), the problem becomes much more complicated than the previous two. That is mainly due to the fact that in such a situation, the attitude, position and velocity of floating bodies are all the part of the solution and fully coupled with the unknown variables of water waves, as pointed out in the paragraph after Eq. (17).

To overcome the difficulty associated with full coupling, four types of methods have been suggested by other researchers, which include the indirect method by Wu, et al.,[35,36] the mode-decomposition method by Vinje, et al.,[1] the Dalen & Tanizawa's method by Dalen[37] and Tanizawa[38] and the iterative method by Cao, et al.[39] In the indirect method, six auxiliary functions were introduced and thus the time derivative of the velocity potential is not explicitly required if the pressure on body is not of interest. (If the pressure is required, another BVP for $\partial\phi/\partial t$ may need to be solved after the body acceleration and velocity are obtained by using the auxiliary functions.) In the mode-decomposition method, the body acceleration is decomposed into several modes (7 modes in 3D cases). In both the methods, each of modes or auxiliary function is found by solving a BVP similar to that for the velocity potential but under different boundary conditions. The matrixes for all the BVPs are the same and so the CPU time is almost the same as that for solving one BVP of this kind if a direct solution scheme (such as Gauss Elimination) is employed to solve the linear algebraic system involved. However, the direct solution scheme is unlikely to be suitable for solving the linear algebraic system containing a very large number of unknowns, such as in cases involving 3D floating bodies or overturning waves. In such cases, an iterative solution scheme for solving the linear algebraic system is much more efficient in terms of CPU time and storage requirement. When using an iterative solution scheme, each of the BVPs needs almost the same CPU time, even they have the same matrix. That means that the indirect method and the mode decomposition method may require considerably more CPU time due to solving 7 extra BVPs. In the method proposed by Dalen and Tanizawa, the body accelerations are implicitly substituted by the Bernoulli's equations and thus the velocity potential and its time derivative are solved without the need to explicitly calculate accelerations of the floating bodies. However, this method results in a special matrix for

$\partial\phi/\partial t$ which is different from the one for ϕ and whose properties have not been sufficiently studied. This is likely to increase the difficulty for solving the algebraic equations associated with $\partial\phi/\partial t$ and also needs more CPU time for generating such a special matrix. This would be the main reason why this method has not been commonly used.

In the iterative method, suggested by Cao, et al.,[39] the acceleration and velocity of bodies are explicitly and iteratively calculated at each time step; in this way, the need to solve extra equations in the first two methods and the problem with the third method is eliminated. It is noted that for the purpose of time marching, a standard explicit 4th-order Runge-Kutta scheme is generally used in all these methods, which requires an extra three sub-step calculations at one time step forward. In each sub-step, the geometry of the computational domain may or may not be updated. If it is not updated, it is called a frozen coefficient method; if it is updated, it is called a fully-updated method. The CPU time spent on updating in the fully-updated method is roughly equal to 4 times as much as that in the frozen coefficient method. However, the frozen coefficient method may not lead to stable and reasonable results for problems with large motions of floating bodies, as indicated by Koo, et al.[40] It is also noted that the body velocity is estimated from the acceleration at previous time steps in all the above methods, i.e., the corresponding time marching procedure is explicit. The explicit procedure may be satisfactory if time steps and so changes in the velocity and acceleration in one step are sufficiently small; otherwise, it may degrade the accuracy and even lead to instability.

The authors of this chapter have developed an iterative method called ISITIMFB (Iterative Semi Implicit Time Integration Method for Floating Bodies).[32] The distinct feature of this method is that body velocities are estimated implicitly and iteratively by considering the acceleration at both previous and current time steps. This feature enables one to use relative larger time step without the need of sub-step calculations. As demonstrated by numerical tests in Yan, et al.,[32] the ISITIMFB is accurate, stable and efficient, and matches the QALE-FEM very well. In our recent paper[25], this method was modified to further improve its efficiency. The basic idea of the modification is to estimate more parameters implicitly and iteratively. As can be understood, for problems associated with freely floating bodies, update at a time step needs to be performed not only for body velocities but also for the positions of and the velocity potential on the free surface. Therefore, the modified ISITIMFB includes implicit and iterative estimation of the velocity potential on the free surface in addition to dealing with body velocities in such a way. The modified ISITIFMB is named as ISITIFMB_M and includes all these tasks shown in Fig. 2. It is pointed out that at the end of each time step, body velocities and acceleration are predicted by using extrapolation based on their values in previous time steps, as shown in the right part of Fig. 2. These predicted values are used as initial values of boundary velocity and acceleration for solution at next time step. With the initial values, solving the BVPs about ϕ and $\partial\phi/\partial t$ is similar to that in the first two situations. However, in this situation, iteration is performed, as shown in the right part of Fig. 2, in which the body velocity and acceleration as well as the fluid velocity and its potential on the free surface are refined. This iteration continues until the computed forces and moments on bodies converge within the specified accuracy. After this, the body position and attitude as well as the position of free surface are updated. It is noted that the prediction of body velocity and acceleration at the end of each time step can markedly reduce the number of iterations and play an important role in achieving a high efficiency of procedure. The equations associated with these tasks in Fig. 2 will not be given here due to space limitation. Readers are referred to Ma et al.[25] and Yan, et al.[26]

3.2. *FEM formulation*

As pointed above, one needs to solve the BVPs about ϕ and $\partial\phi/\partial t$ in each time step. For this purpose, we use finite element formulation. To do so, the fluid domain is discretised into a set of small tetrahedral elements and the velocity potential is expressed in terms of a linear shape function, $N_J(x, y, z)$:

$$\phi = \sum_J \phi_J N_J(x, y, z)\,, \tag{19}$$

where ϕ_J is the velocity potential at Node J. Using the Galerkin method, the Laplace equation and the boundary conditions are discretised as follows,

$$\iiint_{\forall} \nabla N_I \cdot \sum_{\substack{J \\ J \notin S_P}} \varphi_J \nabla N_J d\forall = \iint_{S_n} N_I f_n dS - \iiint_{\forall} \nabla N_I \cdot \sum_{\substack{J \\ J \in S_P}} (f_p)_J \nabla N_J d\forall \,, \tag{20}$$

where S_P represents the Dirichlet boundary on which the velocity potential f_p is known and S_n represents the Neumann boundary on which the normal derivative of the velocity potential f_n is given. Eq. (20) can further be written in the matrix form:

$$[A]\{\phi\} = \{B\} \,, \tag{21a}$$

where

$$\{\phi\} = [\phi_1, \phi_2, \phi_3, \ldots, \phi_I, \ldots]^T \qquad (I \notin S_P) \,, \tag{21b}$$

$$A_{IJ} = \iiint_{\forall} \nabla N_I \cdot \nabla N_J d\forall \qquad (I \notin S_P, J \notin S_P), \tag{21c}$$

$$B_I = \iint_{S_n} N_I f_n dS - \iiint_{\forall} \nabla N_I \cdot \sum_{J \in S_P} (f_p)_J \nabla N_J d\forall \qquad (I \notin S_P). \tag{21d}$$

The algebraic system in Eq. (21) is solved by using a conjugate gradient iterative scheme with SSOR pre-conditioner and optimised parameters as given in Ma (1998).[41] It is noted that in the FEM, one node is only affected by those which are connected with the node through elements. Therefore, in each row of Matrix [A], there are only a very small number of nonzero entries, such as 27 in the 3D case if using a structured mesh. To take the advantage of the properties of Matrix [A], Ma (1998)[41] suggested that only the nonzero entries are stored. The algebraic solver requires only the nonzero entries during the calculation. This technique greatly reduces requirements on computer resources and is one of reasons why our FEM is very efficient.

The problem about $\partial\phi/\partial t$ described in Eqs. (15) to (17) is also solved by using the above method with ϕ and the boundary conditions for it are replaced by $\partial\phi/\partial t$ and corresponding boundary conditions for $\partial\phi/\partial t$.

3.3. *Mesh moving scheme*

The QALE-FEM includes three key elements in comparison with the conventional FEM method, such as that described by Wu, et al.[9] and Ma, et al.[21]: (1) the scheme for moving mesh, (2) the method for estimating the velocity on the free surface and body surfaces and (3) the iterative method to deal with full coupling between floating bodies and water waves. The iterative method has been described in Section 3.1 and the velocity calculation will be discussed in Section 3.4. This section will summarise the mesh moving scheme developed specially for the problems about water waves and their interaction with fixed or floating bodies.

As pointed out above, the mesh is generated only once at the beginning of calculation and is moved at other time steps in the QALE-FEM. The initial mesh can be generated using any mesh generator with any degrees of complexity, either structured or unstructured, or even mixed. There is no limitation on the mesh structure for the QALE-FEM. In all our work so far[25,26,31,32], the initial mesh is generated using an in-house mesh generator based on the mixed Delaunay triangulation and the advancing front technique. To model a complex fluid domain, one may assign a different representative mesh size (*ds*) for different areas of the fluid domain to the mesh generator, which indicates the characteristic distance between two connected nodes. For example, near the free surface but far from the bodies, the representative mesh size would be equal to about one thirtieth of a wave length while it is further reduced near the body surface. Although this mesh size is not precisely equal to the real mesh size, it largely indicates how fine the mesh is.

Obviously, the technique for moving the mesh is more crucial than generating the initial mesh in this method to achieve high robustness and high efficiency. For this purpose, a novel methodology has been suggested and adopted, in which interior nodes and boundary nodes are considered separately; and the nodes on the free surface and on rigid boundaries are considered separately. Further more, the nodes on the free surface and on rigid boundaries are split into three groups: those on waterlines, those (inner-free-surface nodes) on the free surface but not on waterlines and those (inner-body-surface nodes) on rigid boundaries but excluding those on waterlines. Different methods are employed for moving different groups of nodes. The sequences and the methods are illustrated in Fig. 3. In the following subsection, more details are described.

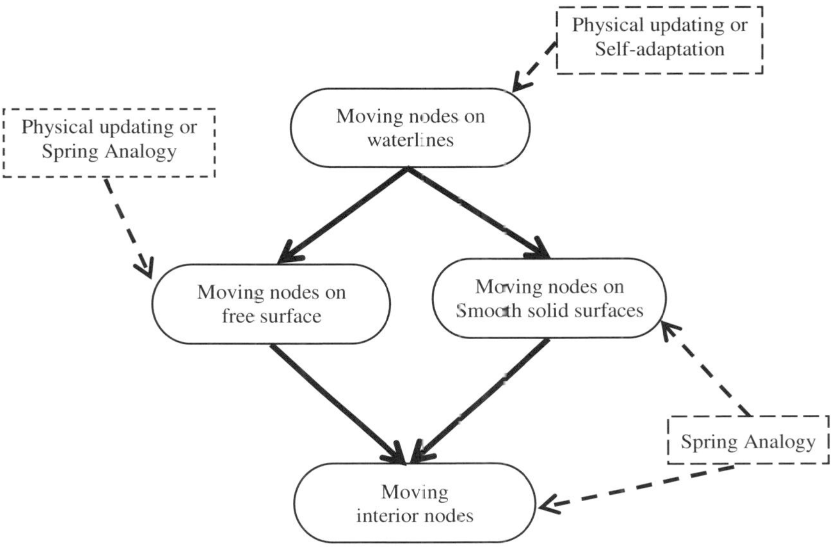

Fig. 3. Illustration of the methodology for moving mesh (⬭: sequence; [- -]: method).

3.3.1. *Moving interior nodes*

To move the interior nodes which do not lie on boundaries, a spring analogy method is used. In this method, nodes are considered to be connected by springs and the whole mesh is then deformed like a spring system. Specifically, the nodal displacement is determined by

$$\Delta\vec{r}_i = \sum_{j=1}^{N_i} k_{ij}\Delta\vec{r}_j \Big/ \sum_{j=1}^{N} k_{ij} , \tag{22}$$

where $\Delta\vec{r}_i$ is the displacement of Node i; k_{ij} is the spring stiffness and N_i is the number of nodes that are connected to Node i. As pointed out in our previous publication[31], the spring analogy method was originally adopted to cope with aerodynamic problems[43,44,45] without the free surface and without floating bodies on the free surface. In order to apply it to the problems associated with the large deformation of the free surface, the authors of this chapter have modified the method considerably by proposing a robust and distinctive method for computing the spring stiffness:

$$k_{ij} = k_{ij}^0 \Psi^{fs}\Psi^{bs} , \tag{23a}$$

with k_{ij}^0 , Ψ^{bs} and Ψ^{fs} determined, respectively, by

$$k_{ij}^0 = \frac{1}{l_{ij}^2} + \alpha \sum_{m=1}^{NE_{ij}} \frac{1}{\sin^2 \theta_m^{ij}} \tag{24b}$$

$$\Psi^{bs} = e^{\gamma_b (\hat{w}_i + \hat{w}_j / 2)} \Big/ q_{\min}^{ij} \tag{24c}$$

$$\Psi^{fs} = e^{\gamma_f \left[1 + \left(z_i + z_j\right)/2d\right]} (1 + \gamma_{jet} \delta_x \delta_y \delta_z) \tag{24d}$$

where l_{ij} is the distance between Nodes i and j, θ_m^{ij} is the facing angle in the mth element of NE_{ij} elements sharing spring i-j (Fig. 4), z_i and z_j are the vertical coordinates of the nodes (i,j); and γ_f is a coefficient that can be taken as $\gamma_f = 1.7$ based on numerical tests so far. $\hat{w}$ is a function that yields a stiffer spring near the body surface and given by

$$\hat{w} = \begin{cases} 0 & d_f > D_f \\ 1 - d_f / D_f & d_f \le D_f \end{cases}, \tag{25}$$

where d_f is the minimum distance from the concerned node to the body surface as shown in Fig. 5; D_f is the distance between the body surface and the boundary of a near-body-region and is estimated by,

$$D_f = \varepsilon \min(B_b / 2, L_b / 2, D_r)\,, \tag{26}$$

where L_b, B_b and D_r are the length, breadth and draft of the body, respectively. Numerical tests show that the coefficient $\varepsilon = 1.5$ is suitable. γ_b is related to the curvature of the body surface and evaluated by

$$r_b = \gamma_f \tilde{\omega}_{ij}\ , \tag{27a}$$

where $\tilde{\omega}_{ij}$ is determined by

$$\tilde{\omega}_{ij} = \begin{cases} 1 & \mathrm{K}_{\max} = \mathrm{K}_{\min} \\ (\mathrm{K}_{ij} - \mathrm{K}_{\min})/(\mathrm{K}_{\max} - \mathrm{K}_{\min}) & \mathrm{K}_{\max} \ne \mathrm{K}_{\min} \end{cases}. \tag{27b}$$

In the above equation, K is defined as $\mathrm{K} = \sqrt{\kappa_1^2 + \kappa_2^2}$, where κ_1 and κ_2 are the curvature of the intersecting curves between the body surface and each of two inter-perpendicular planes normal to the body surface at a node. $\mathrm{K}_{\max}$ and $\mathrm{K}_{\min}$ are, respectively, the maximum and minimum values of K at all nodes on the smooth part of the body surface. K_{ij} is the value of K at a body-surface node that is the nearest to the centre of Spring *i-j*. It should be noted that the value of K at a node may vary with time as the node is continuously moving on the body surface; however, it is not necessarily updated at every time step because one just needs its approximate value to move the mesh. In fact, it can be taken as the initial value calculated at the first time step, which works well for all the cases we have modelled so far. Nevertheless, it is envisaged that the value may need to be updated at every certain number of time steps for other cases considered in future and that there is no extra difficulty for doing so. In Eq. (24c), $q_{\min}^{ij}$ is the minimum value of the quality indexes of all the elements sharing Spring *i-j*. The quality index for an element *e* is defined as:

$$q_e = \frac{3R_i^e}{R_c^e}, \tag{28}$$

where R_i and R_c are the inradius and circumradius of the element, respectively. This quality index is based on the fact that the best tetrahedral element is a regular tetrahedron whose circumradius is three times its inradius. The range of the quality index is from 0 to 1. It equals to 1 for regular tetrahedrons and 0 for elements whose 4 points are located on a plane. The coefficient α in Eq. (24b) is determined by

$$\alpha = \begin{cases} 0 & q_{\min} > q_0 \\ 1 & q_{\min} \le q_0 \end{cases}, \tag{29}$$

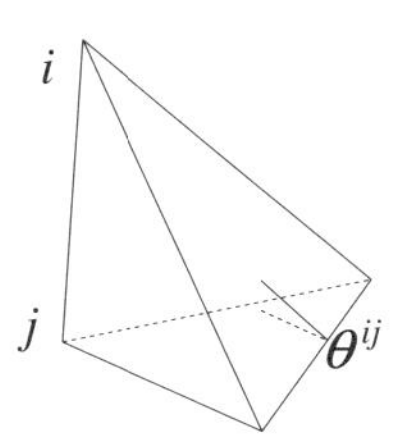

Fig. 4. Illustration of facing angle.

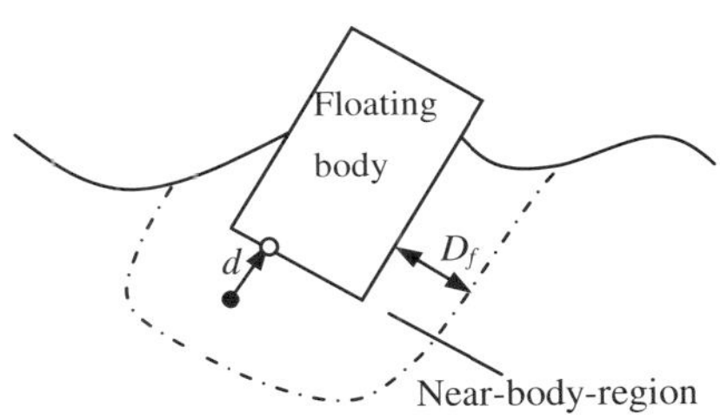

Fig. 5. Region near a body surface.

where q_0 is a control parameter chosen as equal to 0.02 based on numerical tests; $q_{\min}$ is the minimum value of the quality indexes of all elements in the whole computational domain.

When an overturning wave occurs, the mesh near its jet tends to be distorted significantly and quickly. To prevent this from taking place, a coefficient and three functions are introduced in Eq. (24d). γ_{jet} is a coefficient which is nonzero only if the free surface near the node concerned becomes vertical or overturning; δ_x, δ_y and δ_z are correction functions in x-, y- and z-direction, respectively. They are all in a similar form and one of them is given by

$$\delta_x = \begin{cases} 1 - d_x / D_x^{jet} & d_x < D_x^{jet} \\ 0 & d_x \geq D_x^{jet} \end{cases}, \tag{30}$$

where subscript x is replaced by y or z to give δ_y and δ_z, d_x (d_y or d_z) is the distance between the centre of Spring i-j and the nearest jet node (which is on the free surface and near or at the tip of an overturning jet; details about how to find such a node was given in Yan, et al.[26]) in x- (y- or z-) direction; D_x^{jet}, D_y^{jet} and D_z^{jet} indicate the maximum distance in different directions, within which the correction is applied. According to the numerical tests so far, γ_{jet} = 0.5, $D_x^{jet} = D_y^{jet} = 10ds$ and $D_z^{jet} = 0.5H$ are appropriate, where H is the wave height and ds is the representative mesh size.

3.3.2. *Moving nodes on free surface and body surfaces*

As has been pointed out, the nodes on free and body surfaces are split into three groups: waterline nodes, inner-free-surface nodes and inner-body-surface nodes. The positions of the inner-free-surface and waterline nodes are determined by physical boundary conditions, i.e., following the fluid particles, at most time steps. The nodes moved in this way may become too close to or too far from each other. To prevent this from happening, these nodes are relocated at every certain number of time steps, e.g., every 40 time steps. When doing so, the nodes on the waterlines are re-distributed by adopting a principle for self-adaptive mesh, requiring that the weighted length of the curved segments is the same. Specifically, the new position of the waterline nodes is determined by the following equation

$$\varpi_i \Delta s_i = C_s \,, \tag{31}$$

where ϖ_i is a weighted function and can be taken as 1, Δs_i the arc-segment length between two successive nodes and C_s a constant.

To redistribute the inner-free-surface nodes, two methods have been suggested, both based on a spring analogy method for a free surface. In the first method, the inner-free-surface nodes are first projected onto a horizontal plane and then they are moved in the plane. After that the elevations of the surface corresponding to the new coordinates are evaluated by an interpolating method. This method called MNPP (Moving Nodes in Projected Plane) method. For this method, the spring stiffness coefficients used for moving the nodes are different in x- and y-directions and given, respectively

$$k_{ij}^{(x)} = \frac{1}{l_{ij}^2}\sqrt{1+\left(\frac{\partial \zeta}{\partial x}\right)^2}\, e^{\gamma_b(\hat{w}_i+\hat{w}_j/2)} \tag{32a}$$

$$k_{ij}^{(y)} = \frac{1}{l_{ij}^2}\sqrt{1+\left(\frac{\partial \zeta}{\partial y}\right)^2}\, e^{\gamma_b(\hat{w}_i+\hat{w}_j/2)} \tag{32b}$$

where $\frac{\partial \zeta}{\partial x}$ and $\frac{\partial \zeta}{\partial y}$ are the local slopes of the free surface, and other parameters are the same as defined in Eq. (24). This method is not applicable for multi-valued free surfaces, such as overturning waves. For them the second method should be adopted. The second method is based on a local coordinate system formed by the local tangential lines and the normal line at the node concerned. When determining the new position of a inner-free-surface node, it is first moved in the tangential plane by

$$\Delta \vec{r}_{i\vec{\tau}_i} = \sum_{j=1}^{N_i} k_{ij} \Delta \vec{r}_{j\vec{\tau}_i} \Big/ \sum_{j=1}^{N} k_{ij} \tag{33a}$$

where $\Delta \vec{r}_{i\vec{\tau}_i}$ represents the displacement in the tangential plane at node i. After that, the new position of the nodes on the free surface is found by interpolation in the local coordinate system. This method is called MNLTP (Moving Nodes on Local Tangential Plane) method. For this method, the spring stiffness is assigned as

$$k_{ij} = \frac{1}{l_{ij}^{\;2}}(1 + \gamma_{jet}\delta_x\delta_y\delta_z)\,, \tag{33b}$$

where $\gamma_{jet}, \delta_x, \delta_y$ and δ_z are the same as those in Eq. (24).

Inner-body-surface nodes are moved at every time step. Depending on shapes of floating bodies, the inner-body-surface nodes may be further split into three groups: those on wetted corners (submerged, intersecting points of two or more edges), those on edges and those on smooth body surfaces. Nodes on wetted corners are not moved relative to bodies. Nodes on an edge are moved relative to the edge by using the same method for waterlines. The nodes on the smooth body surface may be moved relative to the surface by either the MNPP or MNLTP method when the body surface is single-valued in the vertical direction but the latter is preferred because it can deal with problems about floating bodies with large motions of 6 DoFs. When applying the MNLTP to move inter-body-surface nodes, γ_{jet} Eq. (33b) is taken as 0.

3.4. *Velocity calculation on body and free surfaces*

It is crucial to accurately compute the fluid velocity on body and free surfaces after the velocity potential is found. It is not easy because the mesh is unstructured and continuously moving. A robust technique has been developed for this purpose. Basically, the fluid velocity is split into the normal component and the tangential components, which are separately dealt with.

It is not necessary to compute the normal velocity component of fluid on body surfaces from the velocity potential; instead, it can be determined from the body velocity in the same direction by using the boundary condition on these surfaces (Eq. 4). The normal velocity component ($\vec{v}_n$) at Node *I* on the free surface is estimated by using a three point method given by

$$\vec{v}_n = \left[\frac{2}{3h_{I1}}\left(\frac{2h_{I1} + h_{I2}}{h_{I1} + h_{I2}} + \frac{1}{2}\right)\phi_I - \left(\frac{2}{3h_{I2}} + \frac{1}{h_{I1}}\right)\phi_{I1} + \frac{2}{3h_{I2}}\left(\frac{h_{I1}}{h_{I1} + h_{I2}}\right)\phi_{I2}\right]\vec{n}_s\,, \tag{34}$$

where $\vec{n}_s$ is the normal vector of the surface at the node, *I1* and *I2* represent the two points selected along the normal line (perhaps not coinciding with any nodes); h_{I1} and h_{I2} are the distances between *I* and *I1* and between *I1* and *I2,* respectively; and ϕ_I, ϕ_{I1} and ϕ_{I2} denote the velocity potentials at Node *I* and the two points (*I1* and *I2*); the latter two, ϕ_{I1} and ϕ_{I2}, are found by a moving least square method described in Ma (2005)[46]. After the normal component of the velocity is determined, the tangential components of the velocity (on body or free surfaces) is calculated using a least square method, in which each of equations is given by

$$\vec{v}_{\tau_1} \cdot \vec{l}_{IJ_k} + \vec{v}_{\tau_2} \cdot \vec{l}_{IJ_k} = \vec{l}_{IJ_k} \cdot \nabla\phi - \vec{v}_n \cdot \vec{l}_{IJ_k} \quad (k=1,2,3,\ldots\ldots,m), \quad (35)$$

where $\vec{l}_{IJ_k}$ is the unit vector from Node *I* to Nodes J_k ($k=1,2,3,\ldots\ldots,m$) that are the free-surface nodes connected to Node *I* on the free surface; $\vec{v}_{\tau_1}$ and $\vec{v}_{\tau_2}$ represent the velocity components in $\vec{\tau}_1$ and $\vec{\tau}_2$ directions, respectively. $\vec{\tau}_1$ and $\vec{\tau}_2$ can be any two orthogonal unit vectors in the tangential plane of the surface at node *I*. To be specific, they are determined by $\vec{\tau}_1 = \vec{e}_y \times \vec{n}_s$ and $\vec{\tau}_2 = \vec{n}_s \times \vec{\tau}_1$ if $|\vec{e}_y \times \vec{n}_s| \neq 0$; otherwise $\vec{\tau}_2 = \vec{n}_s \times \vec{e}_x$, $\vec{\tau}_1 = \vec{\tau}_2 \times \vec{n}_s$, where $\vec{e}_x$ and $\vec{e}_y$ are the unit vectors in the *x*- and *y*-directions, respectively.

The above method may not be suitable for nodes near overturning jets and near the waterlines of a body. In these areas, the two points along the normal vector in Eq. (34) may not have enough neighbour nodes or may be outside of the fluid domain, so degrading the accuracy or rendering the method to fail. To tackle the difficulty, it is suggested that the normal and tangential vectors in Eqs. (34) and (35) are replaced by other orthogonal vectors while these equations are still used to compute the fluid velocity in the areas. Specifically, near overturning jets, the normal vector $\vec{n}_s$ is replaced by another vector $\vec{n}_r$. The later is determined by the following. Assume that Node *I* lies on the free surface and $j_1, j_2, j_3, \ldots, j_M$ are the interior nodes connected with it. The angle α_{IK} between $\vec{n}_s$ and each vector $\vec{x}_I - \vec{x}_{j_k}$ ($k=1,2,\ldots M$) is given by $\cos\alpha_{Ik} = \vec{n}_s \cdot (\vec{x}_I - \vec{x}_{j_k}) / |(\vec{x}_I - \vec{x}_{j_k})|$. If $j_{k\min}$ is the node whose angle is $\alpha_{Ik\min} = \min\{\alpha_{I1}, \alpha_{I2}, \ldots, \alpha_{IM}\}$, $\vec{n}_r$ is then chosen to pass Node *I* and satisfy $\vec{n}_r = (\vec{x}_I - \vec{x}_{j_{k\min}}) / |\vec{x}_I - \vec{x}_{j_{k\min}}|$. After determining $\vec{n}_r$, the other two vectors ($\vec{\tau}_{r1}$ and $\vec{\tau}_{r2}$) replacing the tangential vectors ($\vec{\tau}_1$ and $\vec{\tau}_2$) are yielded using a similar method to that for $\vec{\tau}_1$ and $\vec{\tau}_2$, i.e., $\vec{\tau}_{r1} = \vec{e}_y \times \vec{n}_r$ and $\vec{\tau}_{r2} = \vec{n}_r \times \vec{\tau}_{r1}$ if $|\vec{e}_y \times \vec{n}_r| \neq 0$; otherwise $\vec{\tau}_{r2} = \vec{n}_r \times \vec{e}_y$, $\vec{\tau}_{r1} = \vec{\tau}_{r2} \times \vec{n}_r$. By using three new vectors, Eqs. (34) and (35) are still used to evaluate the velocities but with $\vec{n}_s$ replaced by $\vec{n}_r$ and ($\vec{\tau}_1$, $\vec{\tau}_2$) replaced by ($\vec{\tau}_{r1}$, $\vec{\tau}_{r2}$), respectively.

In the area near the waterline of a body, calculation of the velocities at nodes on the body surface and on the free surface must be distinguished. For nodes on the free surface and near the waterline, the method for estimating their velocities is almost the same as that for the nodes near overturning jets. The only difference is the selection of the vector $\vec{n}_r$. In this case, $\vec{n}_r$ is determined by rotating $\vec{n}_s$ in the plane normal to the body surface and passing $\vec{n}_s$ to the direction in parallel with the tangential plane of the body surface. The other two orthogonal vectors are found in the same way as above. The velocities at these nodes are again calculated by Eqs. (34) and (35) using these new vectors.

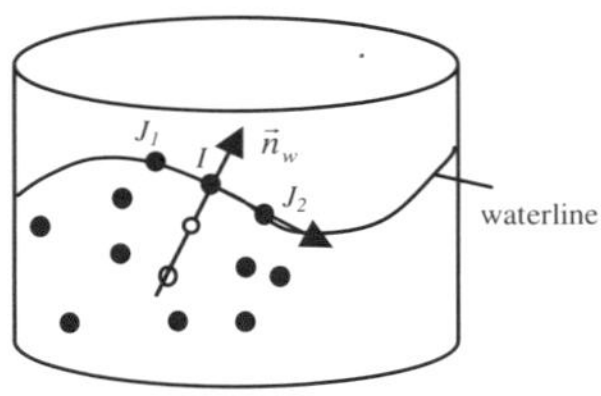

Fig. 6. Illustration of tangential and normal vectors of waterline.

For the nodes on the body surface and near the waterline, special treatment is required only for these nodes on the waterline, e.g. Node I shown in Fig. 6. At such a node, the normal velocity in the normal direction of the body surface is still determined from the body velocity based on the boundary condition (Eq. 4). The tangential velocity components are replaced by those along the other two orthogonal vectors: one is tangent to the waterline ($\vec{\tau}_w$) and another ($\vec{n}_w$) is perpendicular to it but, of course, both are located in the plane tangential to the body surface at the node. The velocity component $\vec{v}_{nw}$ in the direction of Vector $\vec{n}_w$ is estimated by using the same three-point method as in Eq. (34). The velocity component $\vec{v}_{\tau w}$ in the direction of $\vec{\tau}_w$ can be computed by using Eq. (35) with $\vec{v}_{\tau_1}$ and $\vec{v}_{\tau_2}$ being replaced by $\vec{v}_{nw}$ and $\vec{v}_{\tau w}$, respectively. Only two neighbour nodes (J_1 and J_2) of Node I on the waterline is considered for this purpose. Therefore, Eq. (35) actually becomes a central difference scheme for determining the component $\vec{v}_{\tau w}$ if the segments from J_1 to J_2 lie on the same line.

4. Validation

The QALE-FEM has been validated in various cases by comparing its numerical results with experimental data and results from other numerical methods available in the literature. Its convergent properties have also investigated by applying different mesh sizes and time steps. All these could be found from publications of the authors on this topic (e.g., Ma, et al.,[25,31] Yan, et al.[26,32] and so on). Only two cases are presented here due to space limitation in this chapter.

In the rest of the chapter, the parameters with a length scale are nondimensionalised by the water depth *d*; and other parameters by ρ, g and *d,* such as the time and frequency becoming

$$t \to \tau\sqrt{d/g} \text{ and } \omega \to \omega\sqrt{g/d},$$

unless mentioned otherwise. Where indicated, the wave length determined by using the linear wave dispersive relation $\lambda = 2\pi \tanh(2\pi/\lambda)/\omega^2$ is used as a characteristic length.

The first case concerns a solitary overturning wave over a slope seabed. The solitary wave with a height of 0.48 is generated by specifying the initial elevation and the velocity as well as the velocity potential on the free surface, which are calculated by using Tanaka's method (Tanaka[47]). The length and width of the tank are 32 and 2, respectively. From the left wall (x=0) to x=19.8, the seabed is flat. A sloping seabed with a slope of 0.0773 on the right starts from x=19.8 and truncated at x =32. Initially, the wave crest is located at x=9.5. This set-up is similar to that used by Guignard et al.,[48] who presented some experimental data and the numerical results obtained by the VOF method based on their NS Model. For modelling this case by using the QALE-FEM, the representative mesh size is specified as 0.06 and the time step as 0.02, which are sufficiently small for this case as discussed in Yan, et al.[26] Figure 7 shows the free surface profiles recorded at different instances together with the results of Guignard et al.[48] Figure 7(a) shows the free surface profile corresponding to the breaking point ($\tau = \tau_{ref}$). Figures 7(b) and (c) presents those at two other instances. Because

Guignard et al.[48] did not give the VOF results at the time step shown in Figs. 7(b) and (c), the results from their VOF method are not shown. At all the instances, our numerical results are very close to the experimental data. The relative error (estimated by using the same method as in Ma, et al.[31]) is about 1% in the region in which the experimental data are given.

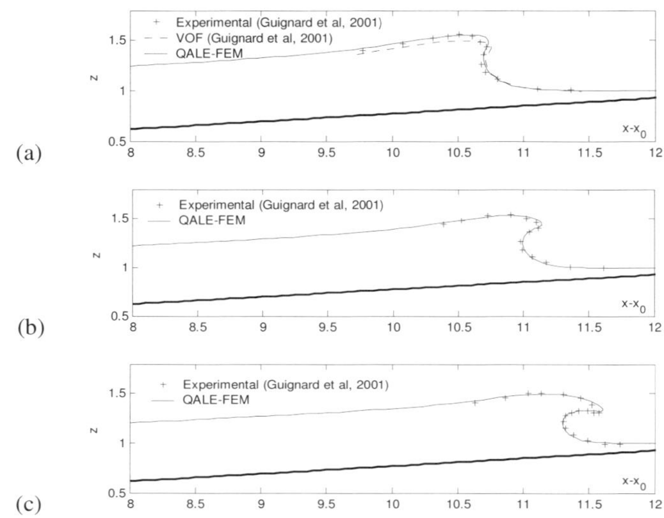

Fig. 7. Free surface profiles recorded at (a) $\tau = \tau_{ref}$; (b) $\tau = \tau_{ref} + 0.313$; (c) $\tau = \tau_{ref} + 0.595$.

The second case to show the accuracy of the QALE-FEM concerns a SPAR platform moored to the seabed. It is subjected to a bichromatic wave generated by a wavemaker. The bichromatic wave has dimensionless amplitudes and periods of (0.02, 0.02) and (1.943, 2.125), respectively. The diameter of the SPAR is 0.135; its initial draft is 0.6607. Its dimensionless mass and the radius of gyration (pitch and roll) are 0.005926 and 0.2078, respectively. The centre of gravity from keel is 0.308 and the fairlead of mooring lines from keel is 0.3087. The mooring line is considered as a nonlinear spring with its stiffness taken as 2.163×10^{-4} up to an offset of 0.04567 and 4.512×10^{-4} beyond it. The details of this case can be found in Weggel, et al.[49] The cases are run in a numerical tank with a length of L=8 and a width of B=2. The time step is specified as T/128, where T=1.943 is the smaller period of the two bichromatic wave components. The mesh is unstructured with the representative mesh size being specified by $ds \approx \lambda / 30$, where λ is the wave length corresponding to T=1.943. In Ma et al.,[25] smaller mesh size and time step were tested but did not lead to significant difference. Figure 8 shows the comparison of the pitch motions of the SPAR platform obtained by the QALE-FEM with experimental data (Weggel, et al.[49]). The agreement between them is reasonably good.

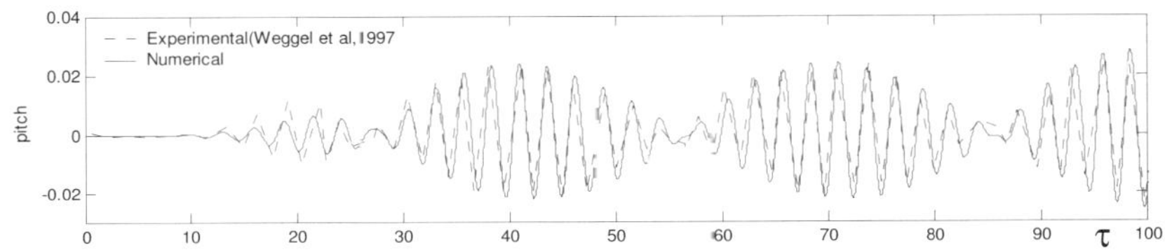

Fig. 8. Comparison between numerical and experimental results for pitch of the SPAR platform.

Both the cases demonstrate that the QALE-FEM can achieve satisfactory accuracy for problems with very different configurations.

5. Efficiency

Ma, et al.[31] pointed out that the QALE-FEM method might use only 15% of the CPU time required by the conventional FEM. Yan, et al.[26] also compared its efficiency with that of the BEM. Some results in the latter paper are discussed here.

Grilli, et al.[6] developed a high-order BEM model and applied it to modelling 3D overturning waves as pointed out above. They used a tank of 19 long and 8 wide. The seabed geometry is described by

$$z_{bed} = (x - x_0)\operatorname{sec}h^2(k_c y)s_c \tag{36}$$

where x_0 is the location where the sloping seabed starts, s_c is the slope in the central (y=0) longitudinal vertical plane of the tank and sech$^2(k_c y)$ is the transverse modulation of the water depth along y-direction with the coefficient k_c being 0.25. The solitary wave is generated by using the same method as that for Fig. 7. The wave height is 0.6 with the initial crest centred at x=5.7 (x=0 is the left end of the tank). To obtain the results up to $\tau \approx 8.57$, they used a coarser quadrilateral grid (50×20×4) for the first 70 time steps and then used a finer grid (60×30×4) for the remaining 120 time steps. The total CPU time spent on the two stages is about 52.8 hours on a supercomputer (CRAY-C90). Fochesato, et al.[50] developed a fast BEM method, which may be 6 times faster than the conventional BEM (Grilli, et al.[6]). Their calculations for the same case were also split into two stages. They used a coarser grid (40×10×4) with 1,422 boundary nodes for the first stage ($\tau < 6$, about 54 time steps) and then a finer grid (60×40×4) with 6,022 boundary nodes for the remainder of the calculation (200 time steps). Totally, they spent about 19 hours to achieve the results up to $\tau \approx 8.57$ by using a PC (2.2GHz processor, 1G RAM). We also applied the QALE-FEM to model the case with the same fluid domain for the same duration of calculation. Some wave profiles along the middle cross-section together with those from the references are shown in Figs. 9 and 10. Our simulation is performed on a PC with 2.53GHz processor and 1G RAM. The QALE-FEM takes only 0.9h (or 54 minutes) to achieve the result in the figure with acceptable accuracy in terms of mass and energy conversation (ε_m = 0.1% and ε_e = 0.26%, respectively). Even to achieve higher accuracy of ε_m = 0.09% and ε_e = 0.16% (which are smaller than those errors given by Fochesato, et al.[50]), the CPU time taken by the QALE-FEM is only 1.8h (or 108 minutes). More details about the comparison can be found in Yan, et al.[26] For this particular case, the QALE-FEM method can be at least 10 times faster than the fast BEM method.

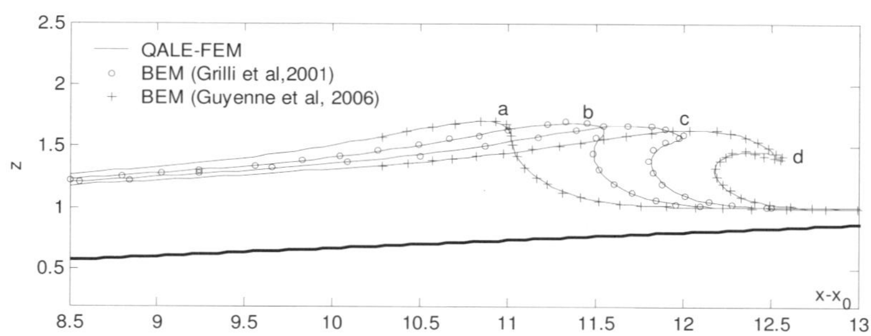

Fig. 9. Free surface profiles at y=0 (H=0.6; Curve a: $\tau \approx$ 7.89; b: $\tau \approx$ 8.25; c: $\tau \approx$ 8.57; d: $\tau \approx$ 8.827; thick solid line represents the seabed geometry).

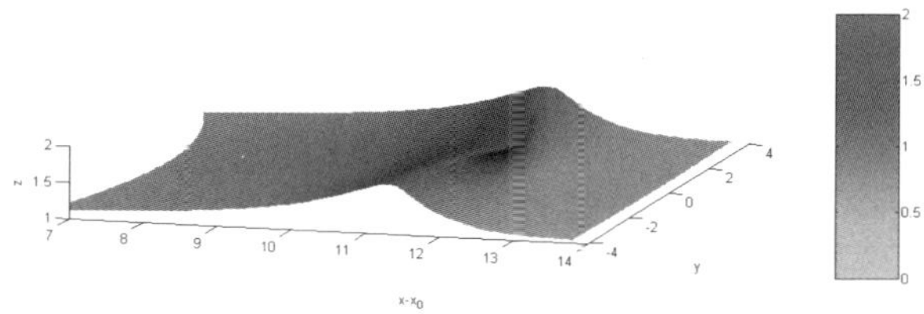

Fig. 10. A snapshot of a part of free surface at $\tau \approx 8.57$ for 3D overturning wave.

6. Mesh Quality

One of the distinctive features of the QALE-FEM is that the unstructured mesh is moved at every time step to conform to the motion of boundaries. It is understood that mesh quality is continuously changing during simulation. Two methods (Ma, et al.[25]) are adopted to look at how mesh quality changes quantitatively: (1) quality distribution and (2) aggregate quality. The quality distribution denotes how the quality index (q_e) of elements calculated by Eq. (28) is distributed at a certain instance. The aggregate quality of the mesh is estimated by

$$Q_s(t) = M_t / \sum_{e=1}^{M_t} \frac{1}{q_e}, \tag{37}$$

where M_t is the total number of the elements in the whole fluid domain. The aggregate quality $Q_s(t)$ has only one value at a time step. It is obvious that the quality distribution is suitable to show the mesh quality at a time step while the aggregate quality is convenient to illustrate how the mesh quality changes during the whole period of simulation.

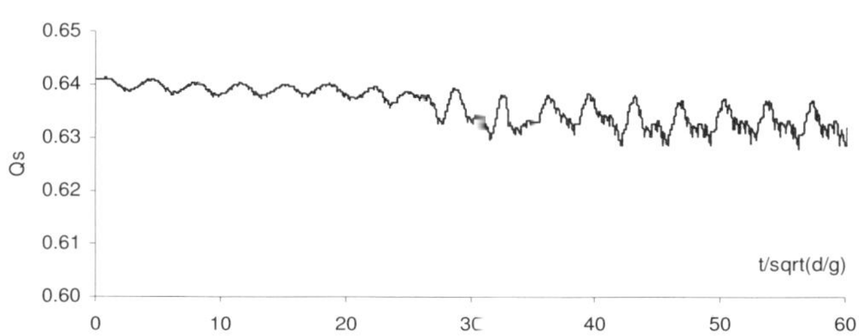

Fig. 11. Change of aggregate quality with time.

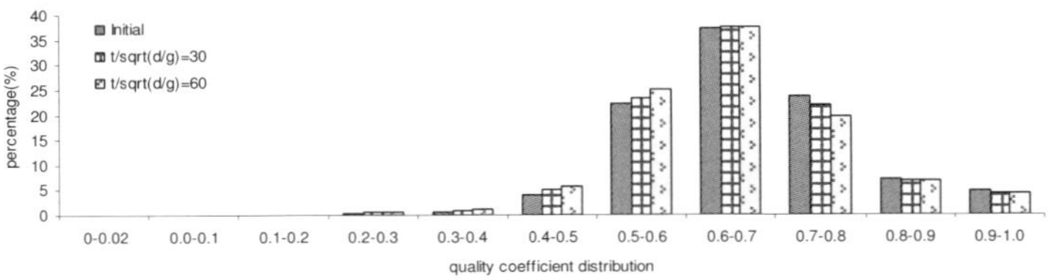

Fig. 12. Quality distributions at three time steps.

In all the simulations, the mesh quality is monitored by using the two methods. In order to show how the mesh quality changes during a simulation, one typical example is presented here. In this example, the two Wigley Hulls are put into a tank of 15 long and 6 wide and subjected to a wave in the direction of head sea, which generated by the wavemaker subjected to cosine motion with the amplitude and frequency being $a = 0.04$ and $\omega = 1.7691$ respectively. The shape of each Wigley Hull is expressed as

$$\eta = (1-\varsigma^2)(1-\xi^2)(1+0.2\xi^2) + \varsigma^2(1-\varsigma^8)(1-\xi^2)^3, \tag{38a}$$

where

$$\xi = 2x / L_b, \eta = 2y / B_b, \varsigma = z / D_r \tag{38b}$$

and L_b, B_b and D_r are the length, breadth and draft of the Wigley Hull, specified as $L_b = 1.0$, $B_b = 0.2$ and $D_r = 0.15$, respectively. The gravity centres of the Wigley hulls are initially located at $x = 0$, $y = -0.3$ (No. 1) and $x = 0$, $y = 0.3$ (No.2), respectively, where $x = 0$ denote the middle of the tank. The distance (d_{sb}) between the centres of these two hulls is 0.6 (3 times the width of the hull). Although they are all in the head sea, each of them will undergo motions of 6 DoFs with the sway, roll and yaw being solely caused by the interaction between the two hulls. Due to symmetry, only half the domain with a single body is considered by applying the symmetrical condition on the middle cross-section. The half fluid domain is discretised into about 847,254 elements and 154,342 nodes. The time step used for this case is $T/128$. The aggregate quality of the mesh during the long-time calculation is shown in Fig. 11. One can see that the aggregate quality is about 0.64 initially and it always remains larger than 0.62. Figure 12 depicts the quality distribution of all elements at two time steps apart from that at the initial step, where the horizontal axis denotes the ranges of quality index and the vertical axis gives the percentage of elements whose quality indexes fall in a range of quality index. For example, quality indices of about 40% elements fall in the range 0.6-0.7. The aggregate quality at the two time steps ($\tau = 30$ and $\tau = 60$) almost reaches the minimum as could be seen from Fig. 11. Even at such instances, the quality distributions are almost the same as that of the initial step - about 80% of elements having the quality index larger than 0.5, as seen in Fig. 12. More importantly, very bad elements (quality indexes less than 0.1) never appear during the calculation. Observation on the quality of mesh in other cases modelled so far by the QALE-FEM shows similar characteristics to that presented here. This fact implies that the good quality of the mesh is retained throughout simulation.

7. Typical Examples of Application

The QALE-FEM has been applied to model various cases including those for waves only and those for interaction between waves and floating bodies. We have not enough space in the chapter to present all these results. Only some typical examples are given in this section. More could be found in our other papers[25,26,31,32].

7.1. *3D Overturning of solitary waves over a non-symmetrical reef*

Numerical simulation of 3D overturning waves requires sophisticated techniques and is still very challenging. Grilli, et al.[6] applied their BEM model to model 3D solitary overturning waves over plane slopes and a sloping ridge. Guyenne, et al.[7] investigated kinematics and the effects of seabed geometry on overturning solitary waves. Similar attempts were also made by others[50] based on the BEM model. To our best knowledge, overturning waves over more complicated seabeds have not been reported by other researchers and the research groups using the FEM model have not produced results for any overturning waves so far.

The seabed geometry considered by Grilli, et al.[6] is symmetrical about $y = 0$, given by Eq. (36) with k_c being a constant along y-direction. Some results from the QALE-FEM for such a case have been discussed in § 5. In this section, we will present a case with a non-symmetrical reef and discuss its effects on the wave properties. The seabed geometry is still described by Eq. (36) but the values of k_c are different for $y > 0$ (k_c^+) and $y < 0$ (k_c^-), as illustrated in Fig. 13.

The length and width of the tank and the solitary wave are the same as the case shown in Fig. 10. The seabed reef starts from $x_0 = 5.225$ with $s_c = 1/15$, which are also the same as for that figure. In order to investigate the effect of non-symmetry, different combinations of k_c^+ and k_c^- are selected. For this purpose, we chose $k_c^+ = 0.25$ but assign different values of 0.1, 0.5 and 0.75 to k_c^-. The representative mesh size is chosen as 0.07 and time step as 0.023.

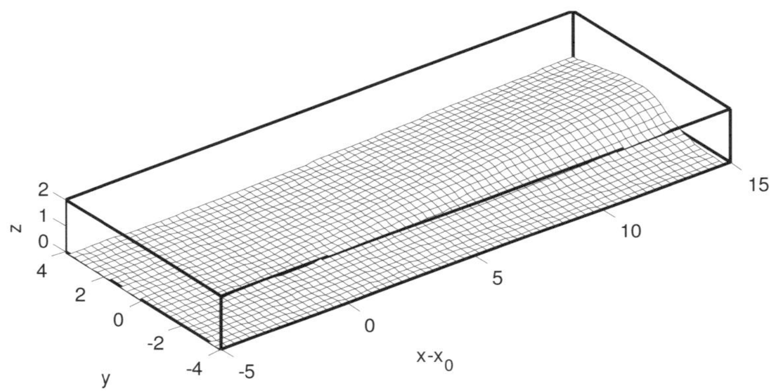

Fig. 13. The seabed geometry for the case with $s_c = 1/15$, $k_c = 0.25$ for y > 0 and 0.75 for y < 0.

Our results have suggested that the time when the overturning starts occurring increases as the increase of k_c^- (Yan, et al.[26]). In addition, the width of the overturning jet measured in y-direction decreases with the increase of k_c^-. These features are demonstrated in Fig. 14. This figure also shows that the centre of the jets tends to be nearer to the wall at $y = 4$ for the larger k_c^-. Specifically, it is at about $y = -1$ for $k_c^- = 0.1$, at about $y = 0$ for $k_c^- = 0.25$ but at about $y = 1$ for $k_c^- = 0.75$. From this fact, one may deduce that the overturning jets may be guided to occur in some areas by changing the seabed geometry in order to prevent them from happening at places where important structures sit.

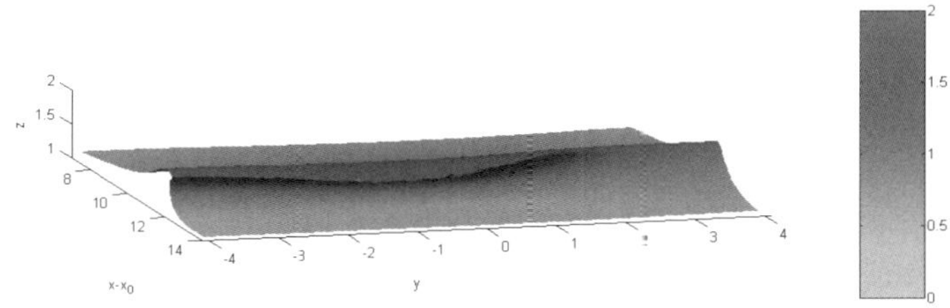

(a)

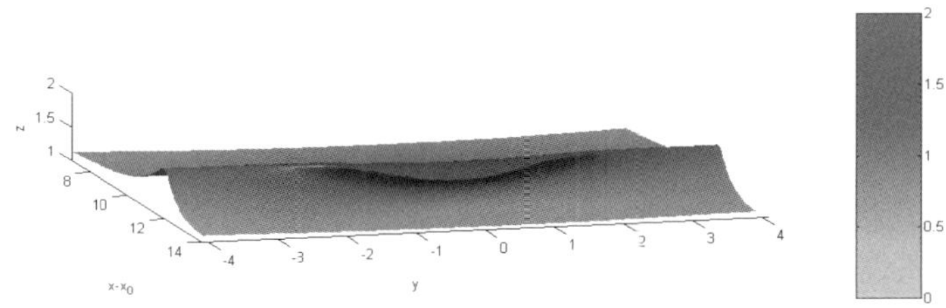

(b)

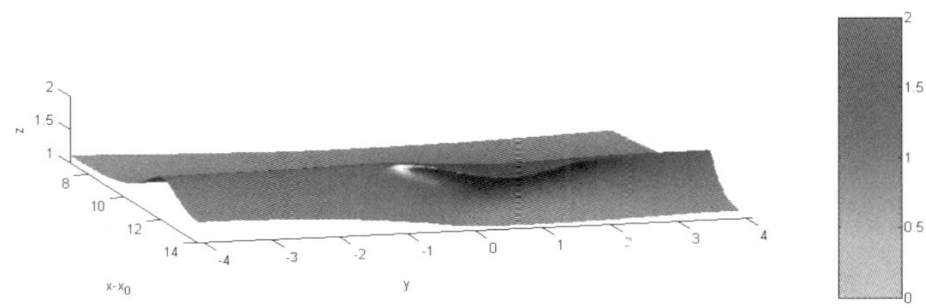

(c)

Fig. 14. Free surface profiles for (a): $k_c^- = 0.1$, $\tau \approx 8.70$; (b): $k_c^- = 0.25$, $\tau \approx 8.83$; (d): $k_c^- = 0.75$, $\tau \approx 9.10$; (the colour bar represents the value of total velocity ($|\nabla\phi|$) on the free surface).

7.2. *3D Overturning of transient waves propagating over multiple artificial reefs on a sloping seabed*

Overturning transient oscillating waves near beaches or over submerged reefs are more commonly observed than overturning solitary waves. The research on overturning of transient oscillating waves over 3D artificial reefs is very limited. Xu, et al.[51,52] modelled 3D overturning stokes waves in space-periodic numerical tank. In their BEM model, the waves are generated by specifying the pressure distribution on the free surface. This model has been extended by Xue, et al.[53] to simulate crescent waves, which are generated by specifying the initial wave elevation and the velocity potential on the free surface based on linear theory, in infinite depth of water, again in a space-periodic domain. Although they presented some very impressive results, including the kinematic property of the overturning waves, their calculations are made in a spatially periodic domain without considering the seabed effects, which could be very rarely observed in reality.

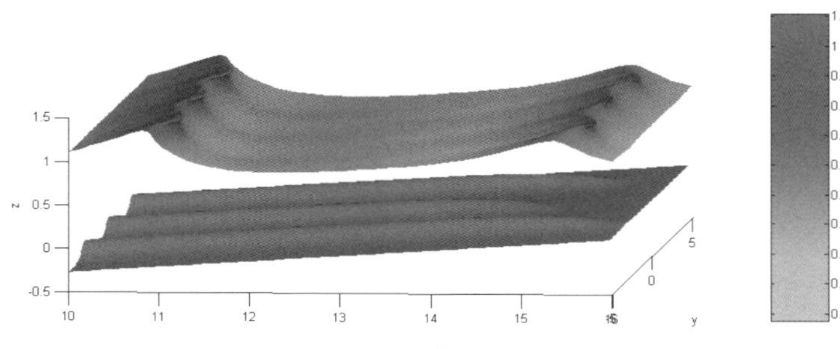

(a)

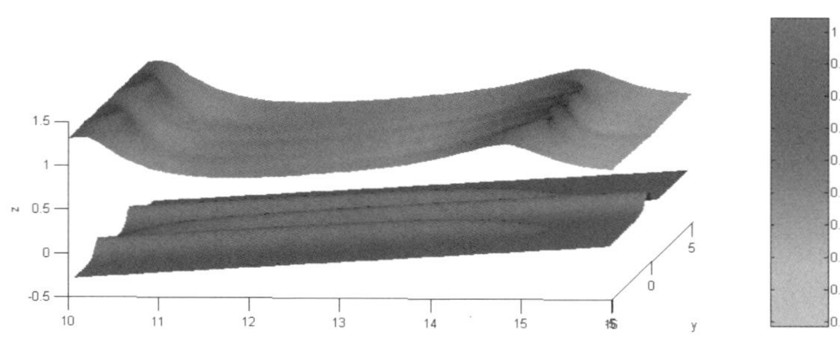

(b)

Fig. 15. Free surface snapshots in cases with three artificial reefs for (a): $\tau \approx 26.68$, three reefs centred at (11, 0), (11.5, 3) and (12, –3), respectively; (b): $\tau \approx 25.65$, three reefs centred at (14, 0), (11, 2) and (11, –2), respectively; (The profiles below the frees surface is the seabed geometry which is shifted by z-0.5; the colour bar represents the value of total velocity ($|\nabla\phi|$) on the free surface).

We have modelled various cases involving transient oscillating waves propagating over single or multiple reefs[26]. Only two cases are presented here. In these cases, there are three parallel artificial reefs with different configuration. Each of them is described by

$$z = z_{reef}\gamma_{reef} \sec h^2[k(y - y_c)], \tag{39}$$

where z_{reef} is the height of the artificial reef. Similar to Eq. (36), $\sec h^2[k(y - y_c)]$ is used to specify the variation in the y-direction. γ_{reef} gives the variation in the x-direction by

$$\gamma_{reef} = 1 - e^{-c_\alpha(|x - x_c| - c_\beta)}, \tag{40}$$

in which c_α and c_β are coefficients and (x_c, y_c) represents the centre of a reef. For the cases presented here $z_{reef} = 0.2$, $c_\alpha = 1.7$, $c_\beta = 4.0$ and $k = 2.0$; and the multiple reefs sit on a sloping plane seabed, whose slope is 1/15 in the x-direction starting from $x = 7.0$. The width and the length of the tank are taken as 10 and 20, respectively. The wave is generated by a piston wavemaker located at $x = 0$ with $a = 0.2$ and $\omega = 1.0$. Other parameters are given in the figure caption.

The corresponding free surface profiles at an instant are depicted in Fig. 15. It is striking to see that there are two groups of overturning jets at the same time in each case and the number of jets in each group depend on the location of the reefs. In Fig. 15a, there are 3 jets in each group, whose shapes are different. In Fig. 15b, there are two in the left group and three in the right group. Another interesting point is that in both cases, the overturning jets in the right groups seem to occur beyond the reefs.

7.3. *Multiple floating bodies*

As reviewed by Ma, et al.,[25] the work on 3D floating bodies with free responses of 6 DoFs in fully nonlinear waves are hardly found at present in the public domain. The QALE-FEM has been used to simulate various cases of this kind, which include SPAR platforms, barge-type floating bodies and one or two Wigley Hulls in head seas or in oblique waves. For some selected cases, the numerical results are compared with experimental data available in the literature and satisfactory agreements are achieved. One example will be presented here. More results could be found in Ma, et al.[25]

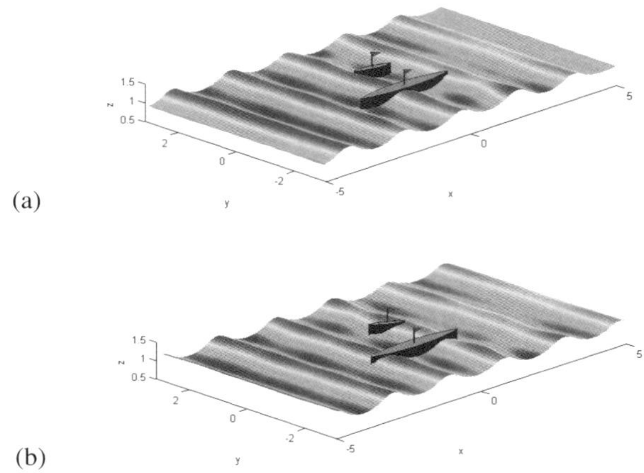

Fig. 16. Snapshot of the free surfaces for the cases with two different Wigley hulls at (a) $\tau \approx 42.3$ and (b) $\tau \approx 48.0$.

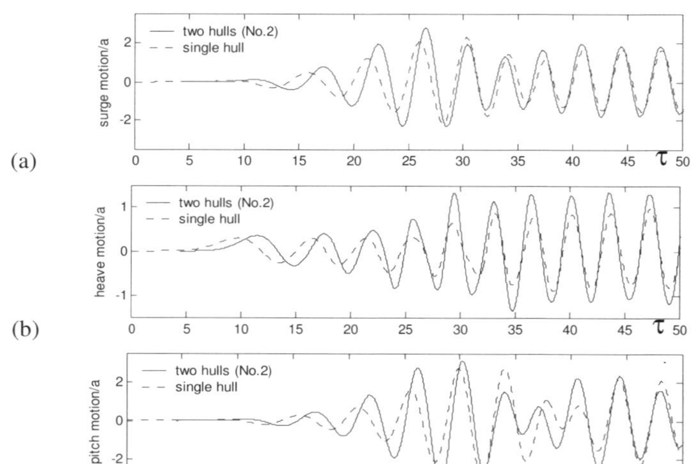

Fig. 17. Comparison on the (a) surge, (b) heave and (c) pitch motion of hull No. 2 with the corresponding results of a single hull in the same wave condition.

In this example, we consider two different Wigley Hulls. Both of them can be described by Eq. (38), but their sizes and wave incident angles (θ_i) are different. The sizes of the larger one are $L_b = 3.0$, $B_b = 0.4$, and $D_r = 0.3$ (No.1) while the smaller one (No.2) has $L_b = 1.0$, $B_b = 0.2$, and $D_r = 0.15$. The incident angle to the former is $\theta_i = 0^0$ and $\theta_i = 20^0$ to the latter. The distance d_{sb} between their centres is 1.5. The mass and moments of inertia for roll, pitch and yaw are 0.2017, 0.07326, 1.0237, 4.05 for No. 1 and 0.01677, 0.005470, 0.00875, 0.025 for No. 2, respectively. The amplitude and the frequency of the wavemaker motion are 0.03 and 1.7691, respectively. The simulation is carried out in a numerical tank with length of 15 and width of 6. Using this tank, the reflection from the side walls is negligible when $\tau \leq 50$, according to our numerical tests.

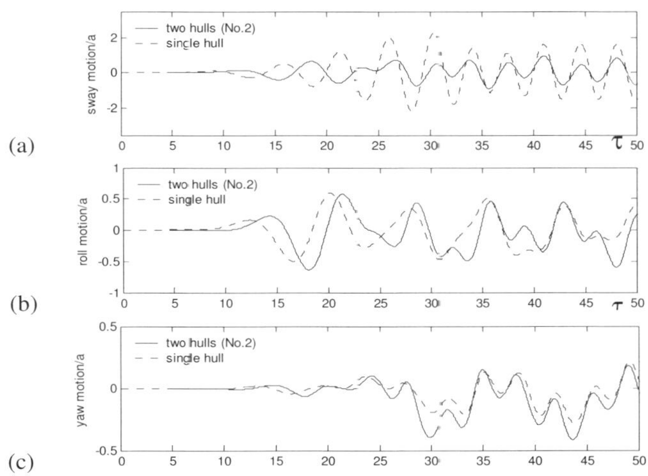

Fig. 18. Comparison on the (a) sway, (b) roll and (c) yaw motion of hull No. 2 with the corresponding results of a single hull in the same wave condition.

The snapshots at two instances for this case are shown in Fig. 16. The responses of the Wigley Hull No. 2 are plotted in Fig. 17 for the surge, heave & pitch motion and Fig. 18 for the sway, roll & yaw motion. The corresponding results for a single hull without the larger one under the same wave condition are also plotted together. From these two figures, it is found that the maximum value of surge (Fig. 17a), heave (Fig. 17b), pitch (Fig. 17c) and yaw (Fig. 18c) motions in the case with two hulls are larger than that in the case for a single hull. This implies that the effect from the other floating structures moored near it has to be considered when predicting its response to steep waves

8. Conclusions

In this chapter, we review the developments, features and applications of the QALE-FEM, which is based on fully nonlinear potential theory. This method has three distinct elements: (1) mesh moving schemes based on a robust spring analogy method specially developed for problems involving waves and their interaction with floating bodies, (2) techniques for computing the velocities on free and body surfaces which are suitable for moving unstructured meshes and (3) the semi-implicit iterative method for dealing with fully coupling between motions of waves and floating bodies.

According to comparison with results in the literature, this method is reasonably accurate and very efficient. It has been adopted to simulate many cases with or without floating bodies, though only a few cases are presented due to the limitation of the space here. Even from these cases, one may see its flexibility and powerfulness, from modelling complex 3D overturning waves with many jets to modelling fully nonlinear interaction between waves and multiple 3D floating bodies with 6-DoFs responses to steep waves. We have not found other methods based on the FNPT model could produce such results so far.

Acknowledgments

Our research work is supported by the Leverhulme Trust, UK and by EPSRC, UK. For them we are very grateful.

References

1. T. Vinje and P. Brevig, *Proc. 3rd Int. Conf. on Numerical Ship Hydrodynamics*, Paris, France, 257–268 (1981).
2. W.M. Lin, J.N. Newman and D.K. Yue, *Proc. 15th Symp. on Naval Hydrology*, Hamburg, Germany, 33–49 (1984).
3. M. Kashiwagi, *J. Soc. Nav. Archit. Japan.* Vol. 180, pp 373–381 (1996).
4. Y. Cao, R.F. Beck, and W.W. Schultz, *Proc. 9th Int. Workshop on Water Waves and Floating Bodies,* Kuju, Oita, Japan , 33–37 (1994).
5. M.S. Celebi, M.H. Kim and R.F. Beck, *J. Ship Res.*, Vol. 42, pp 33–45 (1998).
6. S.T. Grilli, P. Guyenne and F. Dias, F, *Intl J. Numer. Meth. Fluids*, Vol. 35, pp 829–867 (2001).
7. P. Guyenne and S.T. Grilli, *J. Fluid Mech.*, Vol. 547, pp 361–388 (2006).

8. G.X. Wu and R. Eatock Taylor, *Appl. Ocean Res.*, 16 363–372 (1994).
9. G.X. Wu and R. Eatock Taylor, *Ocean Eng.*, 22 785–798 (1995).
10. J.H. Westhuis and A.J. Andonowati, *Proc. 13th Int. Workshop on Water Waves and Floating Bodies*, Hermans, The Netherlands, 171–174 (1998).
11. G.F. Clauss and U. Steinhagen, *Proc. 9th International Offshore and Polar Engineering Conference*, Brest, France, 368–375 (1999).
12. V. Sriram, S.A. Sannasiraj and V.J. Sundar, *Fluids and Struct.*, 22(5) (2006).
13. C.Z. Wang and B.C. Khoo, *Ocean Engineering*, 32, 107–133 (2005).
14. G.X. Wu and Q.W. Ma, *Proceedings 14th Int. Conf. OMAE*, Copenhagen, Vol. 1, pp 329–340 (1995).
15. G.X. Wu, Q.W. Ma and R. Eatock Taylor, *Proceedings of 10th International Workshop for Water Waves and Floating Bodies*, Oxford, (1995).
16. G.X. Wu, Q.W. Ma and R. Eatock Taylor, *Proceedings of 21st Symposium on Naval Hydrodynamics*, 110–119 (1996).
17. Q.W. Ma, G.X. Wu and R. Eatock Taylor, *Proceedings of 12th International Workshop for Water Waves and Floating Bodies*, France (1997).
18. Q.W. Ma, G.X. Wu and R. Eatock Taylor, *Proceedings of 13th International Workshop for Water Waves and Floating Bodies*, The Netherlands (1998).
19. G.X. Wu, Q.W. Ma and R. Eatock Taylor, *Applied Ocean Research*, Vol. 20, pp. 337–355 (1998).
20. Q.W. Ma, G.X. Wu and R. Eatock Taylor, *Int. J Numer. Meth. Fluids*, Vol. 36, pp. 265–285 (2001).
21. Q.W. Ma, G.X. Wu and R. Eatock Taylor, 2001, *Int. J. Numer. Meth. Fluids*, Vol. 36, pp 265–285 (2001).
22. Peixin Hu, G.X. Wu and Q.W. Ma, *Ocean Engineering*, 29 (14): 1733–1750 (2002).
23. C.Z. Wang and G.X. Wu, *J. Fluid Struct.* 22, 441–461 (2006).
24. C.Z. Wang, G.X. Wu and K.R. Drake, *Ocean Engineering*, 34 (8), 1182–1196 (2007).
25. Q.W. Ma and S. Yan, "QALE-FEM for Numerical Modelling of Nonlinear Interaction between 3D Moored Floating Bodies and Steep Waves," accepted for publication by International Journal for Numerical Methods in Engineering.
26. S. Yan and Q.W. Ma. "Modelling 3D overturning waves over complex seabeds," submitted to a journal but some results are given at www.staff.city.ac.uk/q.ma. (2008).
27. G.X. Wu and Z.Z. Hu, *Proc. R. Soc. Lond. A*, 460, 3037–3058 (2004).
28. M.S. Turnbull, A.G.L. Borthwick and R. Eatock Taylor, *Appl. Ocean Res.*, 25, 63–77 (2003).
29. C. Heinze, Nonlinear hydrodynamic effects on fixed and oscillating structures in waves, PhD Thesis, Department of Engineering Science, Oxford University, 2003.
30. S. Yan and Q.W. Ma, *Proceedings of 20th International Workshop for Water Waves and Floating Bodies, Longyearbyen, Norway* (2005).
31. Q.W. Ma, and S. Yan, *J. Comput. Phys.*, Vol. 212, Issue 1, 52–72 (2006).

32. S. Yan and Q.W. Ma, *J. Comput. Phys.,* 221, 666–692 (2007).
33. S. Yan and Q.W. Ma, *Proceeding of 17th International Offshore and Polar Engineering Conference (ISOPE)*, Lisbon, Portugal, 2192–2199 (2007).
34. N.M. Sudharsan, R. Ajaykumar, K. Murali and K. Kurichi, *J. Mech. Engg. Sci., Proceedings of the Institution of Mechanical Engineers*, U.K. C, 218(3) (2004).
35. G.X. Wu and R. Eatock Taylor, *Proceedings of the 11th International Workshop on Water Waves and Floating Bodies*, Hamburg (1996).
36. G.X. Wu and R. Eatock Taylor, *Ocean Eng.,* 30, 387–400 (2003).
37. E.F.G. Dalen, Numerical and theoretical studies on water waves and floating bodies, PhD thesis, University of Twente, Enschede (1993).
38. K. Tanizawa, *10th International Workshop on Water Waves and Floating Bodies,* Oxford, UK (1995).
39. Y. Cao, R.F. Beck and W.W. Schultz, *Proceedings of 9th International Workshop on Water Waves and Floating Bodies*, Kuju, Oita, Japan, pp 33–37 (1994).
40. W. Koo and M. Kim, *Ocean Eng.,* 31, 2011–2046 (2004).
41. Q.W. Ma, *Numerical Simulation of Nonlinear Interaction between Structures and Steep Waves, PhD Thesis*, Department of Mechanical Engineering, University College London, UK (1998).
42. S. Yan, *Numerical Simulation of Nonlinear Response of Moored Floating Structures to Steep Waves, PhD Thesis*, School of Engineering and Mathematical Sciences, City University, London, UK (2006).
43. J.T. Batina, IAA Paper 89-0115, 27th AIAA Aerospace Sciences Meeting (1989).
44. C. Farhat, C. Degand, B.M. Koobus and M. Lesoinne, *Comput. Struct.,* 80, 305–316 (2002).
45. C.L. Bottasso, D. Detom and R. Serra, *Comp. Meth. Appl. Mech. and Engrg.*, 194, 4244–4264 (2005).
46. Q.W. Ma, *J. Comput. Phys.*, Vol. 205, 611–625 (2005).
47. M. Tanaka, *Phys. Fluid.* 29 (3), 650–655 (1986).
48. S. Guignard, R. Marcer, V. Rey, C. Kharif and P. Fraunie, *Eur. J. Mech. B Fluids* 20, 57–74 (2001).
49. D.C. Weggel, J.M. Roesset, R.L. Davies, A. Steen and F. Frimm, *Offshore Technology Conf.*, OTC 8382, 2, 237–251 (1997).
50. C. Fochesato and F. Dias, *Proc. R. Soc. Lond. A,* 462, 2715–2735 (2006).
51. H. Xu and D.K.P. Yue, *Proc. 7th Int. Workshop on Water Waves and Floating Bodies*, Cointe, France, 303–307 (1992).
52. H. Xu and D.K.P. Yue, *Civil Engineering in the Oceans V,* 81–98 (1992).
53. M. Xue, H. Xu, Y. Liu and D.K.P. Yue, *J. Fluid Mech.*, 438, 11–39 (2001).

CHAPTER 6

VELOCITY CALCULATION METHODS IN FINITE ELEMENT BASED MEL FORMULATION

V. Sriram*, S.A. Sannasiraj and V. Sundar

Department of Ocean Engineering, Indian Institute of Technology Madras
Chennai 600 036, India
**send2sriram@rediffmail.com*

The simulation of nonlinear waves can be carried out by using the conventional methods like Finite Element Method (FEM), Boundary Element Method (BEM) based on Mixed Eulerian and Lagrangian (MEL) formulation. The simulation based on FEM has the advantages of extending the code easily to viscous flow and to three-dimensional (3D) tank with complex geometry. While adopting FEM, the derivatives are usually found from differentiating the shape function, which is the direct differentiation of the velocity potential. The approximation of velocity field thus obtained is inferior than the approximation of the velocity potential. In time-dependent problems, this play an important role. Thus, researchers have been focusing on obtaining the derivatives through different methods such as Global Projection, Local Finite Difference (FD), mapped FD, least square method or by using cubic spline approximation. The present chapter shows a detailed review of these methods for calculating the derivatives including the advantages and disadvantages in the context of simulation of nonlinear free surface waves using structured/unstructured FEM.

1. Introduction

Till the recent past, the emphasis in understanding the behaviour of marine structures under wave loading has been mostly through physical model tests, requiring large hydrodynamic testing facilities with a controlled wave generation systems. Due to rapid progress in the field of computers and simulation techniques in the recent past, numerical simulation of hydrodynamic processes and development of the Numerical Wave Flume (NWF) have become more popular and handy. The NWF has the flexibility of reproducing several scenarios of the predefined wave characteristics and their interaction with structures within hours, which otherwise through physical model tests might take several days or even months.

In NWF, the free surface nonlinearity should be taken into account to replicate laboratory conditions. There are two different approaches that are being used for the simulation of nonlinear waves: 'the frequency domain analysis based on the perturbation method' and 'the time stepping simulation'. In deriving an analytical solution, it is difficult to fit in irregular boundaries and is tedious to obtain solutions for problems higher than second order, whereas, numerical modeling can easily handle the above problems. In numerical modeling, two approaches exist of which, one is based on inclusion of viscous effects while, the other is based on inviscid assumption. The former approach uses Navier-stokes equations, which is time consuming and for long time simulation it leads to significant energy dissipation. Hence, it is good to use this approach where the viscous effects are mandatory like the breaking event, in the vicinity of submerged structures in shallow water, as used by Guignard[38] by coupling with inviscid code. For the simulation of the time stepping problem using inviscid flow, Longuet-Higgins and Cokelet[1] proposed a Mixed Eulerian and Lagrangian (MEL) method. This methodology has been widely adopted by several researchers for the simulation of nonlinear waves among which: the higher order BEM (Grilli *et al.*[2]; Boo[3]), BEM (Sen and Maiti[4]; Ohyama[5]) and FEM (Wu and Eatock Taylor[6]; Westhuis[7]; Ma *et al.*[8]) are worth mentioning. Most of the conventional methods need smoothing or regridding even for a wave steepness of about 0.05, except the Spline-BIEM[4].

The BEM is a mixed method. Hence, the potential and the derivatives on the boundary are approximated independently, usually by the same set of shape functions. Since no differentiation of the shape function is required, the BEM results for the derivatives are of the same order of approximation as the potential. However, in FEM, the derivatives are usually found from differentiating the shape function, which is the direct differentiation of the velocity potential that induces further approximation in the velocity field than the approximation of potential. In time-dependent problems, this plays an important role. Hence, Wu and Eatock Taylor[6] solved a fully nonlinear wave problem that is based on the potential flow formulation for fluid in a container, considering the velocity potential as an unknown (FEM) or both velocity potential and velocity as unknowns [Mixed Finite element method (MFEM)]. The advantages and accuracy of both the methods were compared and it was suggested that the MFEM is less accurate. A five point smoothing technique[1] was applied at every time step for the simulation of the waves, in order to rectify the mesh instabilities. Westhuis[7] adopted a polynomial function for the calculation of the velocity in which, a correction vector to the final velocity was adopted in order to minimise the drawbacks in the calculations using the global projection method[6]. The global projection method corresponds to re-sampling the velocity at the Gauss – Lagrange integration points. A more accurate approximation of the velocity field can be obtained from this method compared to the direct differentiation of velocity potential representations. Westhuis[7] showed the inaccuracy of the global projection method by linear stability analysis. A number of techniques[8,9,10] address the main drawback due to inaccurate calculation of velocity for handling the simulation of nonlinear waves using FEM. Sriram *et al.*[12] adopted cubic spline approach for the estimation of velocity from the velocity potential in order to avoid smoothing/regridding. In the case of unstructured mesh, recently, Wang and Wu[14] used the global projection method for tackling non-wall-sided boundaries but the unstructured mesh was regenerated at every time step requiring a higher computational cost. Other possible approach is the mesh movement procedure which is being widely used in solid mechanics. In free surface simulation, this strategy has been adopted by Sudharsan *et al.*[15] and Ma and Yan[16]. Though, the approach remains the same, the latter named it as Quasi-ALE. Bai and Kim[17] applied the FEM to the nonlinear problems like the ship motions, wave resistance and lifting problems. Wu and Eatock Taylor[18] showed that the FEM is advantageous than BEM in the generation of fully nonlinear waves in terms of its computational efficiency and in the accuracy of the results. Kim *et al.*[20] reviewed the recent research and development in the simulation of nonlinear waves in regard to numerical implementations, methods of wave generation and absorption. But, it is to be mentioned that this review focused mainly on BEM and no detail review on the progress in FEM for the simulation of nonlinear free surface waves is available. The progress in the base formulation given by Wu and Eatock Taylor[6] has been quite drastic till date, the details of which are projected through a flow chart in Fig. 1. It should be noted that only the base research work of different authors are quoted.

In this chapter, the formulation of the FEM and the velocity calculation methods proposed by different authors are presented. The advantages and the disadvantages of the methods are discussed.

The formulation based on FEM and the various velocity calculation techniques are presented in the following section. The error estimation based on standing wave problem to understand the importance of the velocity calculation method is discussed under section 3. The simulation of nonlinear free surface waves and its comparison with the experimental measurements are presented in section 4, whereas, section 5 briefly deals with the implementations related to the unstructured mesh.

2. Formulation of the Problem

2.1. *General*

The two-dimensional fluid motion is defined with respect to the fixed Cartesian coordinate system, *Oxz*, with the *z*-axis positive upwards. The water depth h is assumed to be a constant. The fluid is assumed to be incompressible and the flow as irrotational. Forces due to viscosity are neglected. This simplifies the flow problem, that can then be defined with Laplace's equation involving a velocity potential $\Phi(x,z,t)$, i.e.,

$$\nabla^2\Phi=0. \tag{1}$$

A potential flow in a rectangular flume with a wave maker at one end and the nonlinear free surface boundary conditions are considered. The schematic representation of the computational domain and the prescribed Neumann and Dirichlet boundary conditions on the three boundaries (bottom, left and right) and at the free surface are shown in Fig. 2.

Considering the flume bottom as flat and impermeable

$$\frac{\partial \Phi}{\partial n}=0 \text{ at } z=-h \text{ on } \Gamma_B. \tag{2}$$

The far field is a fully reflecting wall, leading to

$$\frac{\partial \Phi}{\partial x}=0 \text{ at } x=l \text{ on } \Gamma_\infty. \tag{3}$$

Motion of the wave paddle at the left end can be enforced by,

$$\frac{\partial \Phi}{\partial x}=\dot{x}_p(t) \text{ at } x=x_p(t) \text{ on } \Gamma_p \tag{4}$$

where $x_p(t)$ is the time history of the piston type wave paddle motion.

The nonlinear dynamic free-surface condition to be satisfied at the air-water interface can be written as,

$$\frac{\partial \Phi}{\partial t}+\frac{1}{2}\nabla\Phi\nabla\Phi+g\eta=0 \text{ on } z=\eta(x,t) \tag{5a}$$

where, '*g*' is the acceleration due to gravity. The kinematic free surface boundary condition can be written as

$$\frac{\partial \Phi}{\partial z}=\frac{\partial \eta}{\partial t}+\frac{\partial \Phi \partial \eta}{\partial x \partial x} \text{ on } z=\eta(x,t). \tag{5b}$$

The above equations are written in Lagrangian form[1], after substituting in substantial derivatives as

$$\frac{Dx}{Dt}=\frac{\partial \Phi}{\partial x} \tag{6a}$$

$$\frac{Dz}{Dt}=\frac{\partial \Phi}{\partial z} \tag{6b}$$

$$\frac{D\Phi}{Dt}=\frac{1}{2}\nabla\Phi\nabla\Phi-g\eta. \tag{6c}$$

The Eulerian form of the free surface boundary condition restricts the movement of the nodes in the horizontal direction but allows only vertical motion, which is given by

$$\frac{\delta \Phi}{\delta t}+\frac{1}{2}\nabla\Phi\nabla\Phi+g\eta+\frac{\partial \Phi}{\partial z}\frac{\partial \eta}{\partial t}=0 \tag{7a}$$

$$\frac{\partial \eta}{\partial t}=\frac{\partial \Phi}{\partial z}-\nabla \Phi \nabla \eta. \tag{7b}$$

Upon expansion, these can be written as

$$\frac{\partial \eta}{\partial t}=\frac{\partial \Phi}{\partial z}-\frac{\partial \eta}{\partial x}\frac{\partial \Phi}{\partial x}, \tag{7c}$$

$$\frac{\delta \Phi}{\delta t}=-\frac{1}{2}\left[\left(\frac{\partial \Phi}{\partial x}\right)^2-\left(\frac{\partial \Phi}{\partial z}\right)^2\right]-g\eta-\frac{\partial \Phi}{\partial z}\frac{\partial \eta}{\partial x}\frac{\partial \Phi}{\partial x}. \tag{7d}$$

In the above equations, the kinematic free surface boundary condition is still in the Eulerian form, whereas, the dynamic boundary condition is strictly not in the Eulerian form. The derivative δ/δt is different from ∂/∂t (local derivative) and *D/Dt* (total derivative) in the sense that the node is allowed to move only in vertical direction with the rate of change of time. Thus, this form is also known as the Semi-Lagrangian approach, due to the restriction of nodes against motion in the horizontal direction. The advantage of this method compared to the Lagrangian form is that the process of regridding is not required due to the restriction of nodes in the horizontal direction. In the literature, it is stated that for floating bodies or for handling breaking waves, the Lagrangian approach is more suitable. For fixed structures, such as submerged obstacles or multiple cylinders under non-breaking waves, the Semi-Lagrangian approach will be more appropriate. In the present chapter, the Lagrangian approach is used unless otherwise stated.

The above dynamic and kinematic free surface condition are modified as follows to include the wave absorbing beach,

$$\frac{Dx}{Dt}=\frac{\partial \Phi}{\partial x} \tag{8a}$$

$$\frac{Dz}{Dt}=\frac{\partial \Phi}{\partial z}-v(x)z \tag{8b}$$

$$\frac{D\Phi}{Dt}=\frac{1}{2}\nabla \Phi \nabla \Phi-g\eta-v(x)\Phi \tag{8c}$$

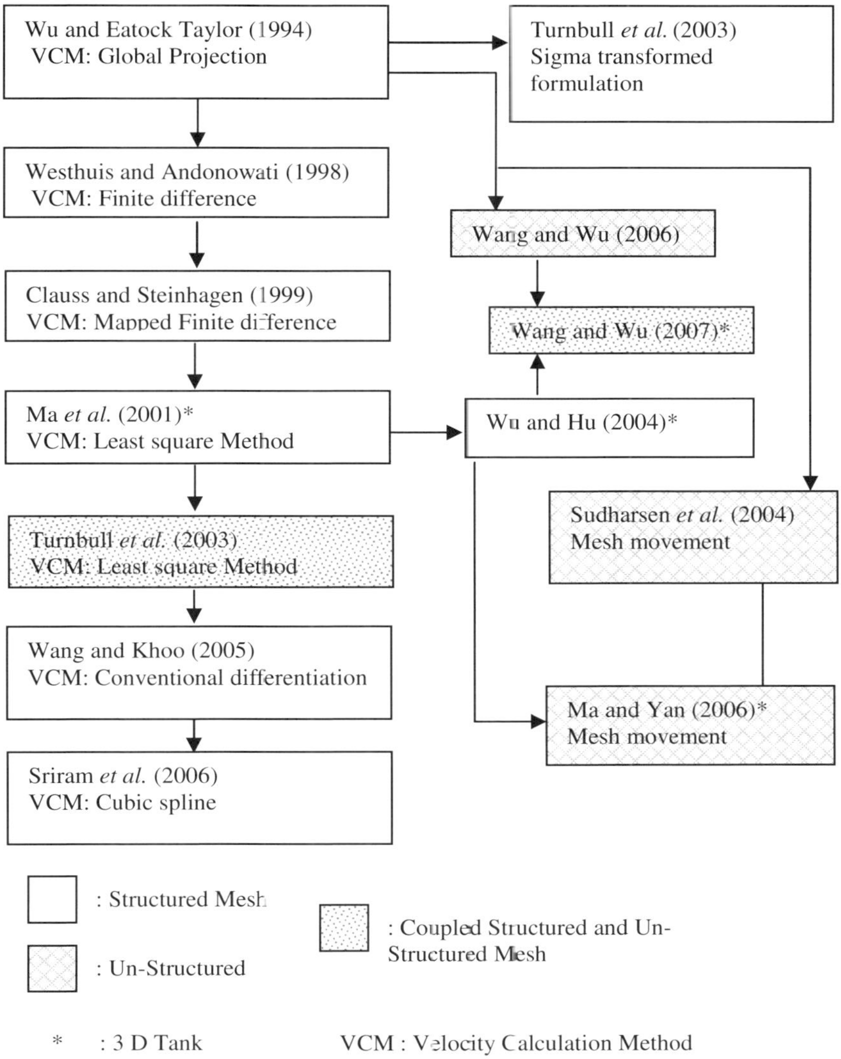

Fig. 1. Progress in the Finite Element method proposed by Wu and Eatock Taylor[6].

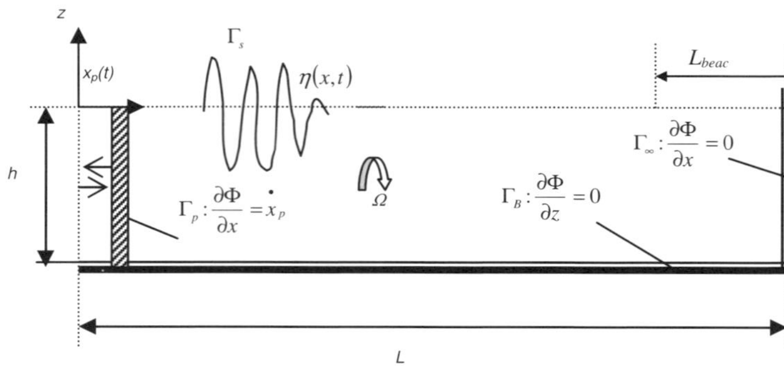

Fig. 2. Computational domain with prescribed boundary conditions.

where $v(x)$is a damping coefficient defined by

$$v(x) = \begin{cases} 0, & x < L - L_{beach} \\ \omega_{beach}\left(\dfrac{x-(L-L_{beach})}{L_{beach}}\right)^2, & L - L_{beach} \le x \le L. \end{cases} \tag{8d}$$

The damping frequency (ω_{beach}) is used to control the strength of the damping zone, while the parameter L_{beach} (beach length) is used to control the length of the damping zone. For effective wave absorption, the choice of damping coefficient is crucial.

For the initial condition (t = 0), the free surface elevation $\eta(x,0)$ and the velocity potential $\Phi(x,z,0)$ at the free surface are assumed to be zero for the wave generation problem in order to represent the free surface elevation at rest during the start of the simulation. For the case of standing wave problem, a known free surface elevation has been assumed keeping the velocity potential at the free surface as zero.

The solution for the above boundary value problem is sought using the finite element scheme. Formulating the governing Laplace's equation subject to the associated boundary conditions leads to the following finite element system of equations:

$$\int_{\Omega} \nabla N_i \sum_{j=1}^{m} \phi_j \nabla N_j \, d\Omega \Big|_{j,i \notin \Gamma_s} = - \int_{\Gamma_p} N_i \dot{x}_p(t) d\Gamma - \int_{\Omega} \nabla N_i \sum_{j=1}^{m} \phi_j \nabla N_j \, d\Omega \Big|_{j \in \Gamma_s, i \notin \Gamma_s}, \tag{9a}$$

where 'm' is the total number of nodes in the domain and the potential inside an element $\Phi(x,z)$ can be expressed in terms of its nodal potentials, ϕ_j, as

$$\Phi(x,z) = \sum_{j=1}^{n} \phi_j N_j(x,z). \tag{9b}$$

Here, N_j is the shape function and n is the number of nodes in an element. The above formulation is found to be effective in dealing with the singularity at the intersection point between the free surface and the wave maker[6]. Linear 3 noded triangular elements were used for the simulations. The structured mesh has been generated using a simple formula[6] having an exponential decay given by,

$$z_{i,j} = -(h+\eta_i) \frac{1-\exp\left(\alpha_z (h+\eta_i)(NZ+1-j)/NZ\right)}{1-\exp\left(\alpha_z (h+\eta_i)\right)} + \eta_i$$

$$\text{for } i = 1,2 \ldots NX+1 \text{ and } j = 1,2 \ldots NZ+1 \tag{10}$$

where α_z is the parameter controlling the mesh size along the vertical direction, taken as 2.0. Z_i are the vertical coordinates, $NX+1$ and $NZ+1$ are the number of nodes along the horizontal and vertical directions respectively.

2.2. *Velocity calculation*

In contrast with a linear formulation of the boundary value problem, the horizontal water particle velocity at the free surface needs to be evaluated in order to extract the free surface elevation at each time step. Once the velocity potential is obtained by solving Eq. (9a), the free surface, horizontal and vertical velocities can be evaluated. However, the need for smoothing or regridding arises due to the inaccuracies in the evaluation of the velocity from the velocity potential. The direct differentiation of the velocity potential results in the approximation of the velocity field at an order lower than the approximation of potential as

$$\nabla \Phi = \sum_{j=1}^{n} \phi_j \nabla N_j . \tag{11}$$

The above approximation has been used by Wang and Khoo[39], considering 8 noded isoparametric elements, which, in the case of linear element, is not accurate enough. The reason for choosing linear element by most of the authors is that the analytical evaluation of Eq. (9a) is easy to implement for complex geometries. However, in order to achieve a higher accuracy in velocity, several approaches are available such as the global projection method[6] and local finite differences[7,9,18,23]. The application of the global projection method (considering structured mesh) to the nonlinear free surface problem leads to unstable high frequency waves that will be discussed later. The local finite difference technique is more accurate compared to the global projection method, requiring local smoothing or local regridding. After obtaining the horizontal and vertical velocities, the new positions of the free surface and the velocity potential are evaluated using the dynamic and kinematic equations as discussed earlier. The time integration is carried out using the standard fourth order methods like Runge-Kutta method/Adam Bashforth method as both methods are stable in our investigation. The general procedure for the simulation is shown in Fig. 3. The smoothing or regridding at each time step has to be minimized for the successful simulation of nonlinear waves to avoid possible energy diffusion. Four methods of velocity calculation methods are discussed herein.

2.2.1. *Global projection method*[6]

The velocity vector $\mathrm{u} = u\vec{i} + v\vec{j}$ is written in terms of shape function similar to equation (9b). The Galerkin method is used to approximate $\nabla \Phi = \vec{u}$ in the form

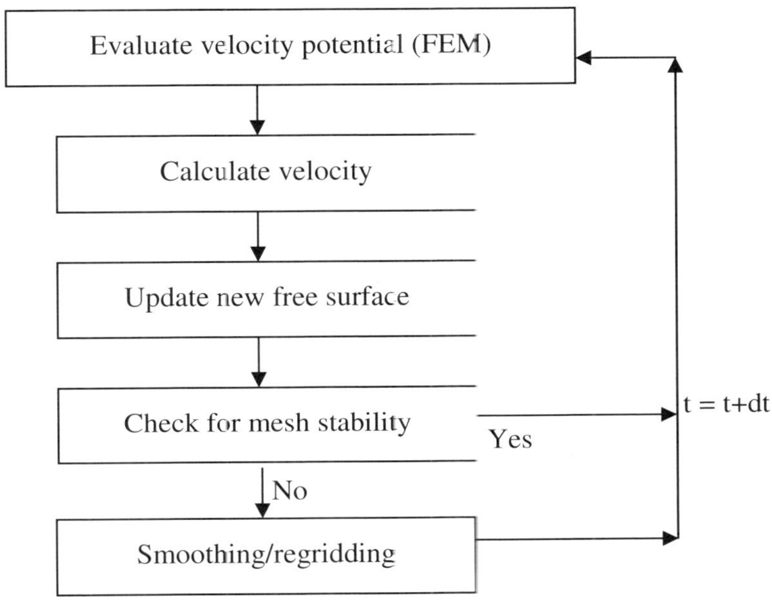

Fig. 3 General procedure for the simulation.

$$\int_{\Omega} N_i(\nabla \Phi - u)d\Omega = 0. \tag{12a}$$

This leads to the following equation in matrix form,

$$[C]\{u\} = [D1]\{\phi\} \text{ and } [C]\{v\} = [D2]\{\phi\} \tag{12b}$$

where,

$$C(i,j) = \int_{\Omega} N_i N_j d\Omega$$

$$D1(i,j) = \int_{\Omega} N_i \frac{\partial N_j}{\partial x} d\Omega \ , \quad D2(i,j) = \int_{\Omega} N_i \frac{\partial N_j}{\partial z} d\Omega \ , \tag{12c}$$

u and v correspond to horizontal and vertical velocities at node. The disadvantage of this method is that it requires a quality mesh generation at every time step and more computational time the details of which are brought out in the following sections.

2.2.2. *Least square method*

The least square method[10] to estimate the horizontal velocity is explained briefly as follows. Consider an arbitrary free surface node (i) connected to k neighboring nodes in the Finite Element mesh. Let $\mathbf{I}^{\mathbf{k}}$ denotes the position vector connected to the free surface nodes (i) to the nth (n=1,…,k) node under consideration. Then, the velocities are estimated by using the following least square approximation in the matrix form as,

$$\begin{pmatrix} \sum_{n=1}^{k} x,_l^n\, x,_l^n & \sum_{n=1}^{k} x,_l^n\, z,_l^n \\ \sum_{n=1}^{k} x,_l^n\, z,_l^n & \sum_{n=1}^{k} z,_l^n\, z,_l^n \end{pmatrix} \begin{Bmatrix} u \\ v \end{Bmatrix} = \begin{Bmatrix} \sum_{n=1}^{k} x,_l^n \phi,_l^n \\ \sum_{n=1}^{k} z,_l^n\, \phi,_l^n \end{Bmatrix} \tag{13a}$$

where, $x,_l^n = \dfrac{(x_i - x_n)}{\sqrt{(x_i - x_n)^2 + (z_i - z_n)^2}}$, $z,_l^n = \dfrac{(z_i - z_n)}{\sqrt{(x_i - x_n)^2 + (z_i - z_n)^2}}$

$$\phi,_l^n = \frac{(\phi_i - \phi_n)}{\sqrt{(x_i - x_n)^2 + (z_i - z_n)^2}}.$$

If one knows the vertical velocity at i^{th} node the above equation reduces to the following

$$u_i = \frac{\sum_{n=1}^{k} x,_l^n\, \phi,_l^n - \sum_{n=1}^{k} x,_l^n\, z,_l^n\, v_i}{\sum_{n=1}^{k} x,_l^n\, x,_l^n}. \tag{13b}$$

Similar kind of equations but without dividing by the distance has been implemented for the simulation of waves of predefined characteristics in a three dimensional tank[8]. On the other hand, the vertical velocity can be estimated based on the backward finite difference

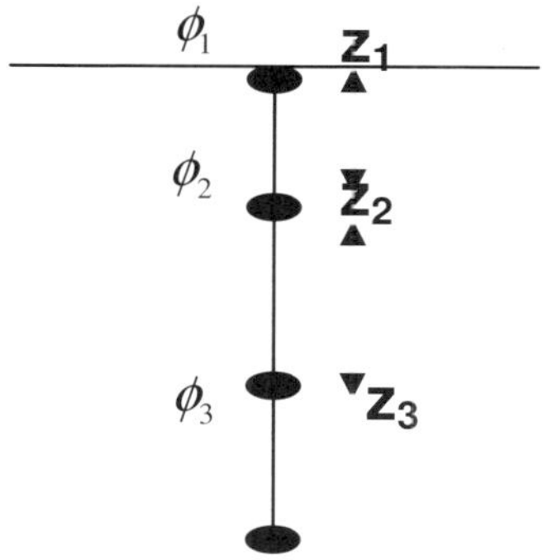

Fig. 4. The Node configuration.

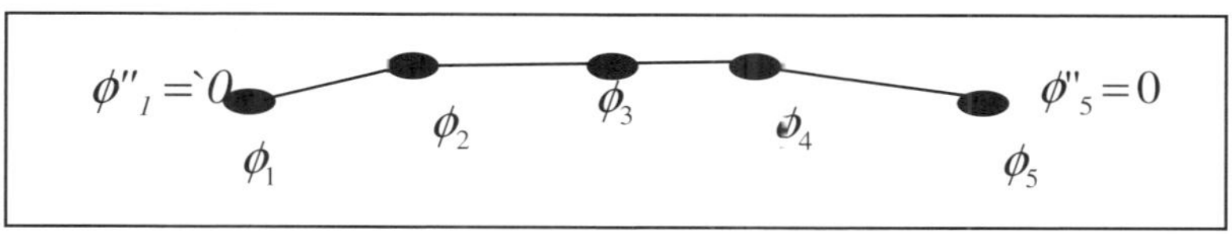

Fig. 5. Cubic Spline approximation using five nodes.

scheme taking advantage of distributing the nodes in a vertical line during mesh generation.

Consider ϕ_i as the velocity potential at the nodes corresponding to z_i, where i = 1, 2, 3 as shown in Fig. 4. The vertical velocity at the free surface node can then be obtained as

$$\frac{\partial \Phi}{\partial z}=\frac{\left(\alpha^2-1\right)\phi_1-\alpha^2\phi_2+\phi_3}{\alpha\left(\alpha-1\right)\left(z_1-z_2\right)}, \tag{14}$$

where, $\alpha=(z_1-z_3)/(z_1-z_2)$.

When the nodes are equidistant (i.e., $\alpha=2$), the above equation reduces to the standard backward Finite Difference (FD) scheme. The procedure for estimating horizontal velocity holds good irrespective of the mesh structure, for known vertical velocity. The estimation of horizontal and vertical velocity can be replaced by tangential and normal velocity component respectively. This modification has been implemented by Ma and Yan[16] in unstructured mesh discretization; the details are given in chapter 5 of this volume.

2.2.3. *Mapped finite difference*

The estimation of vertical and horizontal velocities are evaluated based on mapped finite difference scheme. The wavy surface domain is transformed to a rectangular domain, i.e., the mesh is transformed from the physical coordinate system (*x, z*) to a mapped coordinate system (ξ, ς) as follows,

$$\xi = x$$

$$\varsigma = \frac{h+z}{h+\eta(x)}. \tag{15a}$$

Then, the velocities are estimated based on the second order FD schemes which are based on the velocity potentials with respect to the new coordinate system (ξ, ς). After estimation of velocity, it is again transformed to the physical coordinate system using the following equations obtained through the chain rule of differentiation.

$$\frac{\partial}{\partial x} = \frac{\partial}{\partial \xi} - \frac{\partial}{\partial \varsigma}\frac{h+z}{(h+\eta(x))^2}\frac{\partial \eta(x)}{\partial x} \tag{15b}$$

$$\frac{\partial}{\partial z} = \frac{\partial}{\partial \varsigma}\frac{1}{h+\eta(x)}. \tag{15c}$$

This methodology is adopted by Steinhagen[27]. From the procedure, it is clear that it leads to accurate velocity estimation but the mapping has to be done at each time step. Limitations that arise by using this procedure are the impossibility of simulation of overturning waves and the difficulty in the implementation for unstructured mesh.

2.2.4. *Cubic spline*

In order to minimize the need for smoothing or regridding, splines are used as a velocity calculation method[12]. Splines provide a better approximation for the behaviour of functions that have abrupt local changes. Further, splines perform better than higher order polynomial approximations. The efficient implementation of cubic splines as numerical differentiation for the evaluation of the tangential velocity in the simulation of waves using the lower order BEM has been adopted by Sen *et al.*[24] Sriram *et al.*[12] evaluated the horizontal velocity by fitting a cubic spline to the 'x'-coordinates and $\phi(x, z)$ values. The end conditions are considered as the natural spline condition. To evaluate the smooth first derivative at the ith node, five nodes are considered (two nodes on either side of the ith node) in order to minimize the effect of boundary constraints (natural spline condition).

Let us consider that f_i, f'_i, f''_i, are continuous over a given interval. Based on the continuity condition, we have

$$\frac{\delta x_i}{6} f''_{i-1} + \frac{\delta x_i + \delta x_{i+1}}{3} f''_i + \frac{\delta x_{i+1}}{6} f''_{i+1} = \frac{1}{\delta x_{i+1}}(f_{i+1} - f_i) - \frac{1}{\delta x_i}(f_i - f_{i-1})$$

$$i = 2,3...k-1. \tag{16a}$$

The above equation leads to a set of (k-2) linear equations for the k unknown functional values, f_i. The horizontal spacing (δx) between two nodes is a known parameter. The above stated equation is solved by using the tridiagonal system of matrix assuming the second derivatives at the ends are zero i.e., the natural spline condition. In the present simulation, assuming $f_i = \phi_i$, the derivatives at a particular node (ϕ_i) are found out by considering two nodes on either side (k = 5) as can be seen in Fig. 5, with the second derivatives (ϕ_{i-2}, ϕ_{i+2}) at the end nodes being set to zero. Following the evaluation of the second derivatives, the first derivatives can be estimated through Eq. (16a) at the required node (ϕ_3), which are derived in the intermediate steps of the cubic spline interpolation[25].

$$2f''_i + f''_{i+1} = \frac{6}{\delta x_i}\left(\frac{f_{i+1} - f_i}{\delta x} - f'_i\right). \tag{16b}$$

It should be noted that the above formula is valid only for calculating at the intermediate nodes and not at the end nodes. At the wave paddle, the velocity is assumed to be the input velocity and the velocity at the second node is evaluated by interpolation between the wave paddle and the third node (which will be known by the above method). Similarly, at the end of tank, the velocity is assumed to be zero. For the evaluation of velocity in the vertical direction, second order FD scheme is used as described in section 2.2.2.

One of the main advantages of this cubic spline approach is its capability of estimating smooth first derivatives, which minimize the requirement of smoothing/regridding when adopting the Lagrangian approach and smoothing in the case of Semi-Lagrangian approach, which will be discussed in the following section. On the other hand Eq. (16a) does not hold good for the very steep wave fronts, due to the near by nodes falling on a vertical line for which the equation becomes singular. More over, the procedure holds good for two-dimensional tank only. A modification is certainly required for an extension for simulation of waves in a three-dimensional tank.

2.3. *Algorithm*

To summarize this section, the algorithm for the numerical procedure is briefly reported herein. Assume an initial velocity potential and surface elevation. Digitize the entire domain using the required number of nodes and establish the element connectivity. Apply FEM and obtain the velocity potential inside the fluid domain. The challenging task in the simulation is the velocity estimation. Recover the horizontal velocity using a suitable methodology on the free surface. Based on the velocities, update the free surface nodes using the dynamic and kinematic boundary condition based on Lagrangian/Semi-Lagrangian form. The time integration can be carried out using the fourth order Runge-Kutta method that requires repeated evaluation of velocity potential and velocity at the intermitted time steps to obtain the new position at next time step. The time integration using Adam-Bashforth method is half expensive than the Runge-kutta method. Hence, depending upon the problem in hand, choose the respective one. After the computation of the new free surface position and velocity potential, repeat the calculation as many times required based on the termination time. In the following sections, the applicability of various velocity calculation method has been brought out.

3. Simulation of Standing Waves

3.1. *Steep standing waves in a container*

Initially, the behavior of the global projection method is studied. The generation of standing waves in a container, for which analytical and numerical solutions of Wu and Eatock Taylor[6] is considered. Let $l = 2h$, where l is the length of the tank and h is the water depth. The initial water surface elevation is assumed as

$$\eta_i = \frac{H}{2}\cos\left(\frac{2\pi}{\lambda}x_i\right) \tag{17}$$

where H is the wave height, λ is the wave length and i is the free surface node index. From the given free surface profile, the wave propagation is initiated by no flow boundary conditions on the sidewalls of an impervious container and the propagation is governed by Eq. (1). Comparison of the simulated free surface profile with results based on cubic spline, the global projection method[6] and the second-order analytical solution for $H/\lambda = 0.05$ and 0.1 are shown in Figs. 6 and 7, respectively. The number of nodes used for cubic spline and global projection method simulations in the horizontal and vertical directions are 65 and 17, respectively. A time step of 0.06s lead to Courant number[26] of 0.44 is adopted. In these simulations, no smoothing was found necessary for applying the cubic spline approximation on the free surface. However, Wu and Eatock Taylor[6] stated the need for smoothing in the global projection method.

The CPU time required for the cubic spline simulation by evaluating only the free surface velocity is 0.8750s per time step, whereas, for evaluating velocity at all the grid nodes, it is 1.2188s per time step, and in the case of global projection method this was 1.8438s per time step. This simulation was carried out on a Pentium IV with 2.8GHz processor.

Thus, the cubic spline methodology is computationally inexpensive too.

3.2. *Error analysis*

To quantitatively examine the energy conservation, a relative error analysis has been carried out. The comparisons are made between the Cubic spline method, the results of Westhuis[7], the global projection method[6] and the analytical approach.

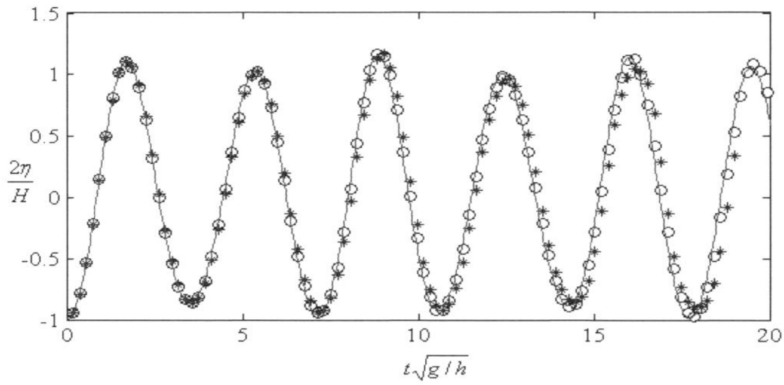

Fig. 6. Time history of the free surface profile at the center of the container for steepness H/λ = 0.05 (——— , analytical (up to 2nd order); ***, Global Projection Method; ooo, cubic spline).

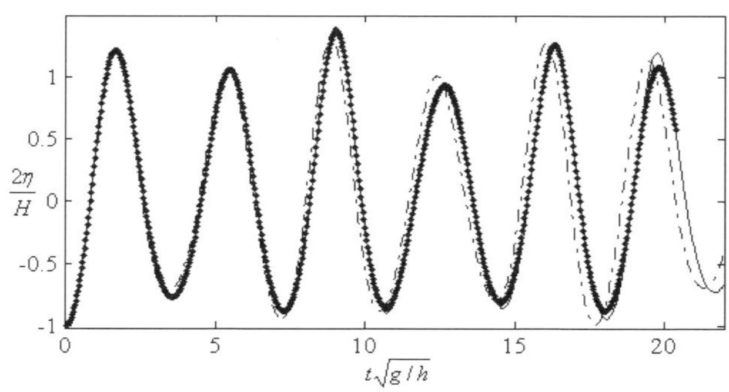

Fig. 7. Time history of the free surface profile at the center of the container for steepness H/λ = 0.1 (-.-, analytical (up to 2nd order); ———, cubic spline; ***, Global Projection Method).

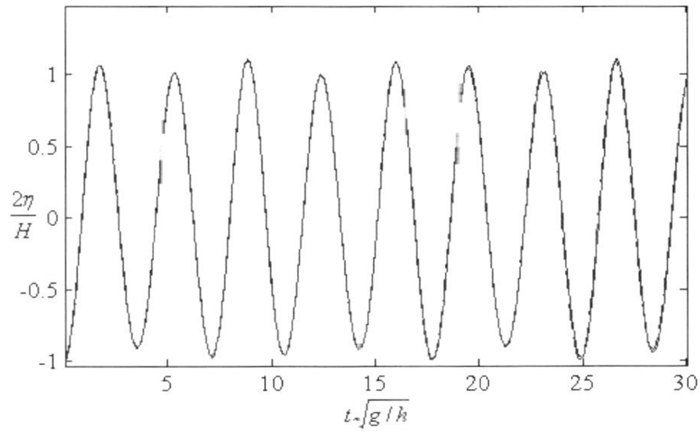

Fig. 8. Time history of free surface wave profile at the center of the container for the steepness $H/\lambda = 0.033$ (———, analytical (up to 2nd order); ----, Cubic spline).

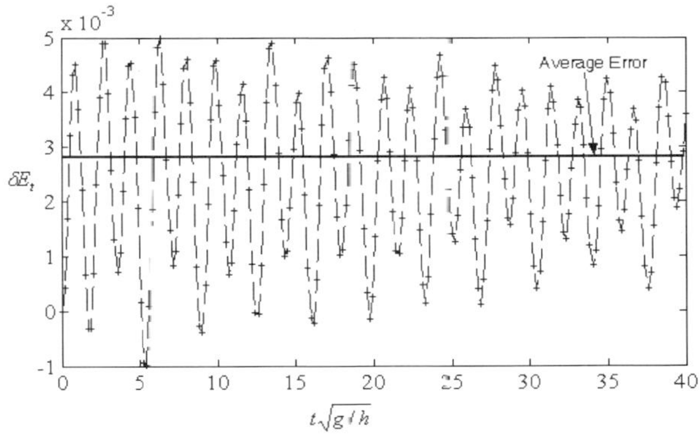

Fig. 9. Relative energy error (δE_t) while using global projection method.

The simulation has been performed using the initial condition defined by Eq. (17) for a steepness of 0.033 with the number of nodes in the x and z directions being 31 and 11, respectively.

The total energy in the system is estimated from,

$$E(t) = \int_0^l \int_{-h}^{\eta} \frac{1}{2} \left\| \nabla \Phi \right\|^2 dz dx + \int_0^l \frac{1}{2} (h + \eta)^2 \, dx. \tag{18}$$

The relative energy error (δE_t) for this simulation has been calculated using

$$\delta E_t = \frac{E(t) - E(0)}{E(0) - e_0}, \tag{19}$$

where $E(t)$ is the total discrete energy at any time t, $E(0)$ is the initial discrete energy in the container. The first term in Eq. (18) is the absolute of convective inertia term and e_0 is the total potential energy in the system when $\eta_i = 0$.

The second-order analytical solution of wave time history for the standing wave problem[6], has been derived for the more general case that

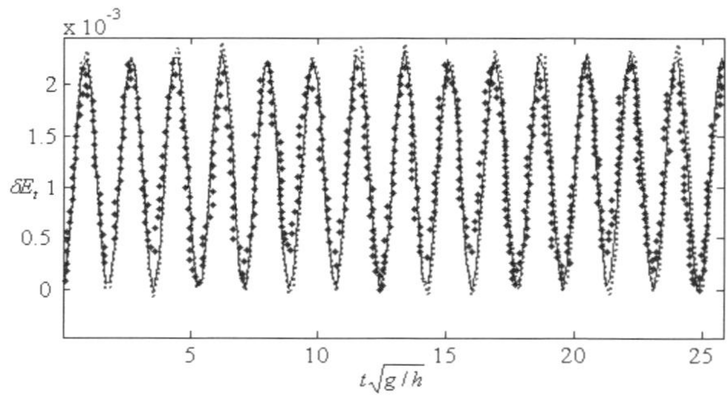

Fig. 10. Comparison of relative energy error (δE_t) (-----, Cubic spline; ***, Westhuis[7]; and, ———, analytical solution (up to 2nd order)).

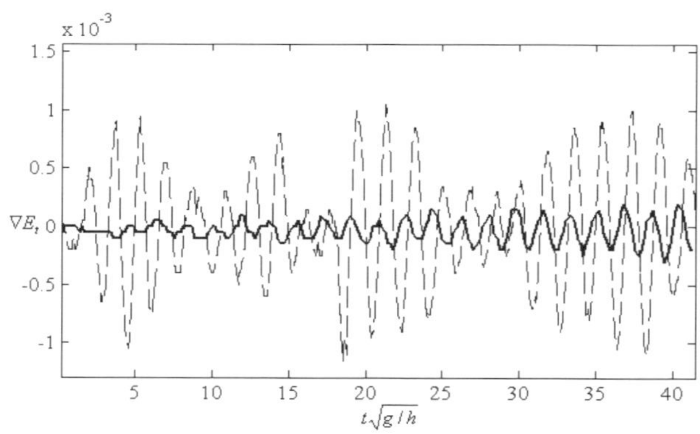

Fig. 11a. Relative Energy loss (∇E_t) with respect to second order analytical solution (- - -, Westhuis[7]; and, ———, Cubic spline).

leads to the first and second order potential and the surface elevation at each time step in the entire domain. The total energy is evaluated using Eq. (18) and the integration is carried out numerically. A comparison between the cubic spline simulation and the analytical solution for the wave profile at the center of the container is shown in Fig. 8. The relative energy error (δE_t) for the simulation using the global projection method is presented in Fig. 9. The average relative energy error is of the order of 2.8×10^{-3}. The comparison of relative energy error using the cubic spline simulation with the results of Westhuis[7] and second order analytical solution are depicted in Fig. 10. It is clearly seen that the relative error is of the same order as that of analytical results. Thus, it is inferred that the global projection method[6] leads to relatively higher energy loss due to relatively inaccurate calculation of velocity (leading to some high frequency waves) and mesh instability. The cubic spline method has an average relative energy error of an order of 1×10^{-3}.

Subsequently, the relative energy loss [∇E_t] with respect to the energy calculated from the second order analytical solution has been derived using

$$\nabla E_t = \frac{E(t) - E_2(t)}{E_2(t)} \tag{20}$$

where $E_2(t)$ is the second order energy at any time t.

Typical comparison of the relative energy loss obtained from the method of Westhuis[7] and global projection method with the cubic spline method are shown in Figs. 11a and 11b. It should be mentioned that the digitized result of Westhuis[7] shown in the earlier figure has been used for evaluating $E(t)$ to estimate the relative energy loss ∇E_t. From the results, the energy loss in the cubic spline method is found to be less than the other methods. The error is found to accumulate with an increase in simulation time in all the methods. It should be mentioned here that in all the simulations, mesh is regenerated at every time step based on the simple mesh generation technique suggested by Wu and Eatock Taylor[6], without checking the mesh properties like skewness, aspect ratio. Recently, Wang and Wu[14] successfully used the global projection method for the non-wall sided boundaries by generating the mesh at every time step using an independent mesh generator, where all these considerations are taken care. But the results given by Sudharsan *et al.*[15] based on global projection method using mesh moving method is only for the short simulation time. The behavior of the global projection method when the mesh structure has been taken care, is discussed in the subsequent section.

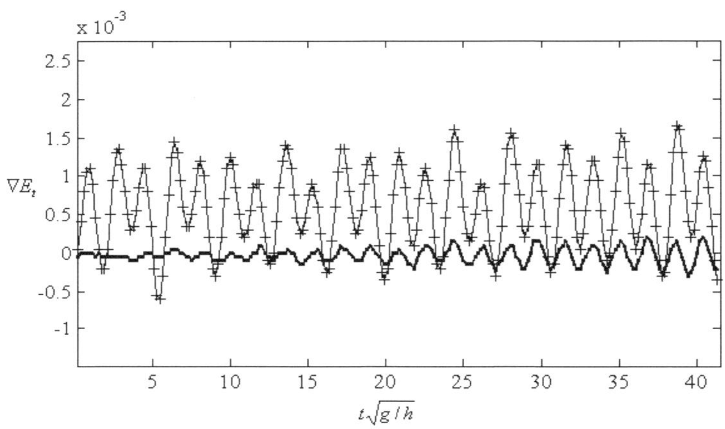

Fig. 11b. Relative Energy loss (∇E_t) with respect to second order analytical solution (--+--, Global projection method; and, ——— , Cubic spline).

4. Simulation Using Wave Paddle – Structured Mesh

4.1. *General*

The advantage of using present Finite Element method compared to the Boundary Element method is that incorporation of even sudden startup of the wave paddle in the model is possible. For the simulation of regular waves, one end of the tank is considered to be a 'piston' type wave maker. The paddle displacement $x_p(t)$ is given by,

$$x_p(t) = -\frac{S}{2}\cos(\omega t). \tag{21}$$

The velocity of the paddle is,

$$\dot{x}_p(t) = \omega \frac{s}{2} \sin(\omega t), \tag{22}$$

where, '*s*' is the maximum stroke of the wave paddle and ω is the circular wave frequency.

4.2. *Simulation of medium steep waves*

Four different velocity calculation methods as mentioned above are used to simulate a regular wave of steepness (H/λ) 0.03 in a tank of length 40m. The water depth considered is 1m. The number of nodes used in the horizontal and vertical directions is 416 (corresponding to 30 nodes per wavelength) and 13, respectively. The stroke of the wave maker is 0.025m and the circular wave frequency is $1.45\sqrt{(g/h)}$. No smoothing/regridding strategy has been used for the simulation carried

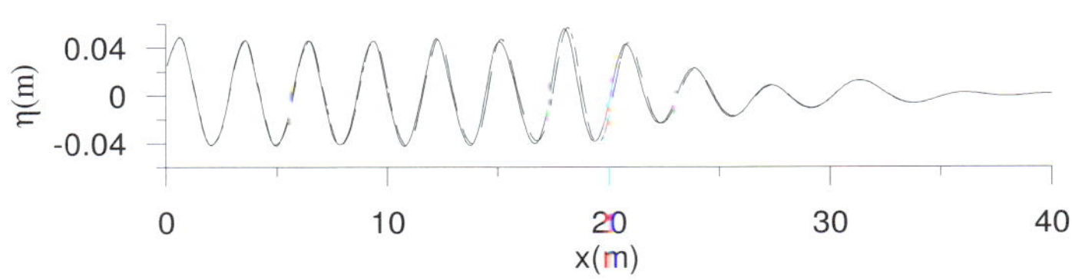

Fig. 12. The free surface profile at 20s along the length of the tank. (------ Least square method, _____ Cubic spline method, -·-·-·-·- Mapped FD).

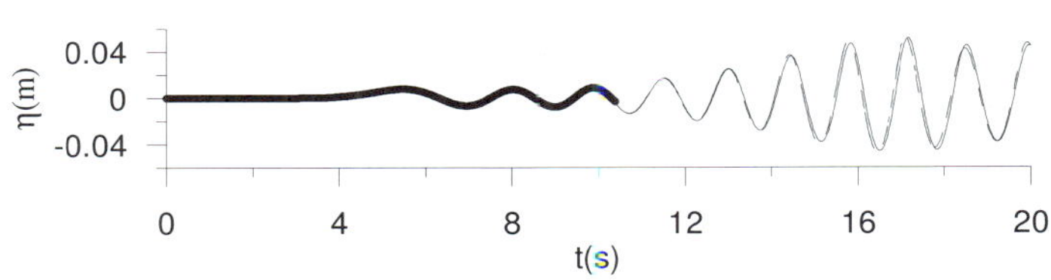

Fig. 13. The time history at 12m from the wave paddle. (------ Least square method, _____ Cubic spline method, -·-·-·-·- Mapped FD, ——— Global projection method).

out using the Cubic spline, least square and Mapped FD methodologies. The need for smoothing or regridding in the context of simulation of nonlinear waves is discussed in the next section. The free surface profiles at 20s along the length of the tank using the three different methodologies are shown in Fig. 12. It was observed that the simulation breaks down while adopting the global projection method at 3.16s.

Hence, smoothing and regridding were done after 10s by increasing the number of nodes in horizontal direction to 1101 and in vertical direction to 21. For smoothing, a five point Chebynev polynomial with unequal spacing was adopted, whereas for regridding, the method based on cubic spline arc length fitting following the suggestion of Dommermuth and Yue[26] was applied. Even after adopting smoothing and regridding, the simulation was observed to break down at 10.4s. Due to frequent smoothing and regridding, a small damping in amplitude has been noticed. The time series at 12m from the wave paddle is depicted in Fig. 13. The dark black line shows the break down of the simulation using Global projection method. In the same figure, an excellent agreement between the other three methods namely, Cubic spline, Least square method and Mapped FD schemes is noticed. The time taken to run the entire computation using Cubic spline was 4.939 min, whereas, Least square method took 5.0859 min and Mapped FD took 5.0760 min. The simulations were carried out in Pentium(R) CPU 3 GHz, with 1 GB of RAM.

4.3. *Simulation of steep waves*

In this section, simulation of steep waves is discussed. The stroke of the wave maker alone was increased to 0.0415m, so as to produce a steepness of 0.05 with the remaining parameters being same as given in the previous section. The free surface elevation near the wave paddle is shown in Fig. 14. It is clearly seen that the simulations carried out using Least square method and Mapped FD show a higher amplitude than the given input near 17 to 20s, whereas Cubic spline does not exhibit this property. The reason is the x-coordinate of the first node at the free surface is set to the position of wave paddle, the horizontal gap to the

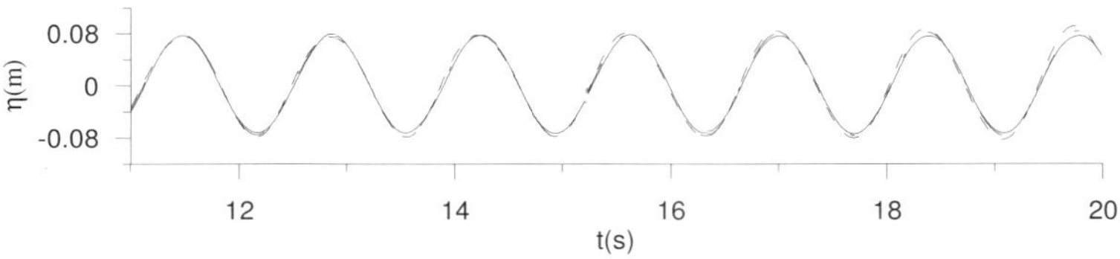

Fig. 14. The free surface elevation near the wave paddle, showing instability after 17s. (------ Least square method, ____ Cubic spline method, -—-—-— Mapped FD).

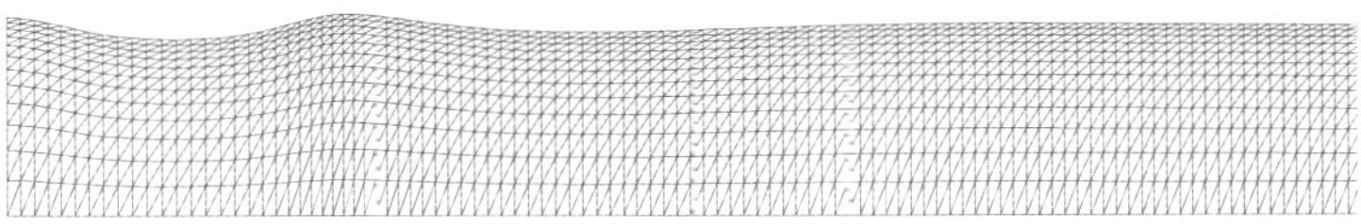

Fig. 15a. The mesh structure near the wave paddle at 2s.

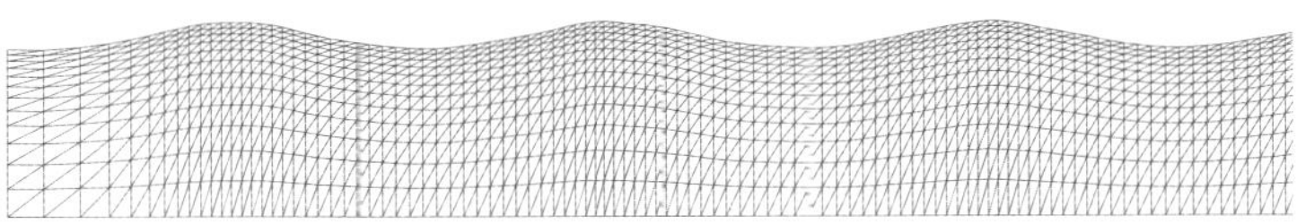

Fig. 15b. The mesh structure near the wave paddle, showing node movement around 18s.

nearby node increases with an increase in time due to the Lagrangian motion characteristics. Hence, in the case of nonlinear waves, as the steepness increases, due to mass transport, the node movement quickly leads to instability of the mesh. This phenomenon is shown by considering a snap shot of the node behavior nearby the wave paddle initially (Fig. 15a) and after certain time steps (Fig. 15b). The figure shows an increase in nodal spacing near the wave paddle. This is a common phenomena encounter in the generation of high nonlinear waves near to the paddle. Hence, the calculation of velocity (by least square/ Mapped FD) leads to inaccurate free surface estimation, whereas the cubic spline method fits a curve through these nodes and estimates the velocity from the curve, which has overcome the difficulties that are faced by the other methods. To avoid the instability posed by other methods, the number of nodes has to be increased in the entire domain.

Another possibility, as suggested by Steinhagen[27] is a dynamic inclusion of the new nodes whenever, the spacing is greater than twice the initial grid spacing. The values of velocity potential and velocities at new nodes are evaluated using linear interpolation based on the neighboring node details. Physically this is correct in the linear approximation but it leads to instability near the wave paddle and, hence, a local smoothing scheme was suggested. Instead of this technique, the authors attempted to increase the initial number of nodes. If the number of nodes is more, collision of nodes takes place, hence the mesh has to be regridded/ smoothed after certain time steps. Consequently, the energy loss accumulates. This also should be taken care in the increased number of nodes in horizontal direction. Thus, the simulations were repeated considering the number of nodes in the horizontal direction as 825 (corresponding to 60 nodes per wavelength) when using Least square and Mapped FD. In usual practice, the regridding is not highly sensitive and depends on the frequency of the waves. For the present case, it is not required. In general, for the simulation of steep nonlinear waves, regridding can be adopted every 20 to 40 time steps depending upon the frequency. After this modification, the result shows that there is no change in amplitude, as depicted in Fig. 16. The simulations were later carried out by increasing the stroke of the wave maker to 0.0572m such that the generated wave has a steepness of about 0.07. For this case, the mesh was regridded at every 50 time steps for all the methodologies.

The time history at 12m from the wave paddle is shown in Fig. 17 for three velocity calculation methods. The Least square and Mapped FD yielded exactly same results, whereas, cubic spline shows a phase shift, It is due to the fact that as the wave steepness increases, the cubic spline approach based on the neighboring x-coordinates along the free surface predicts a very smoothened first derivatives, leading to loss of information (filtered high frequency components in steep nonlinear waves). The global projection method for the above two cases, breaks down after 10s. Considering the computational aspect of the simulation for this case, the Mapped FD took 26.7318 min., whereas, Least square method and Cubic spline method took 26.6659 min. and 26.1508 min., respectively. This clearly shows that all the three methods exhibit similar computational performance.

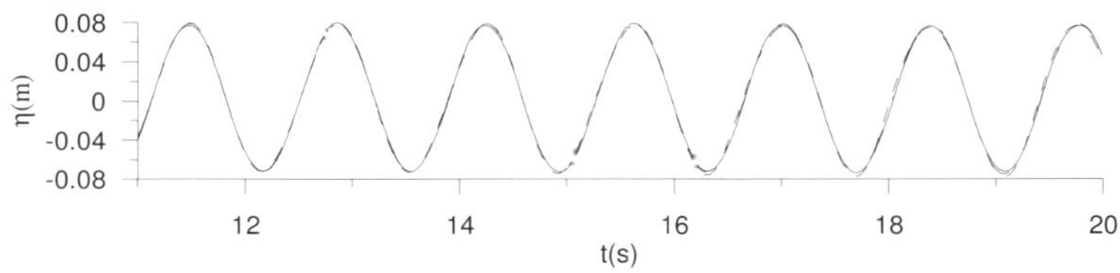

Fig. 16. The free surface elevation near the wave paddle after increasing the number of nodes to 825. (------ Least square method, ____ Cubic spline method, -–-–-–-– Mapped FD).

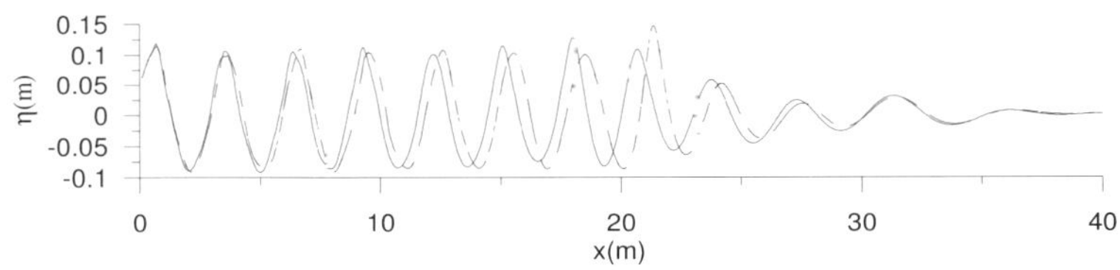

Fig. 17. The free surface elevation along the length of the tank for steepness 0.7. (------ Least square method, ____ Cubic spline method, -–-–-–-– Mapped FD).

4.4. *Propagation of steep waves over submerged bar*

While a wave propagates over a submerged bar, the transmitted wave disintegrates into shorter waves. The spatial evolution of this kind of wave form has been dealt experimentally by Beji and Battjes[28] and numerically by Beji and Battjes[29]. This particular setup has been reproduced in Sriram *et al.*[12] for waves of medium steepness using the cubic spline methodology. The accurate description of the wave nonlinearity and wave dispersive characteristics is essential since the dispersive character of free higher harmonics is a major factor i.e., the amount of energy transferred between the harmonics.

When the wave form propagates over submerged obstacles, the lee side of which is in deeper water region, nonlinearity would tend to become weak leading to the non-existence of bound waves (nonlinear distortion of the long waves). A drastic change in the wave form takes place due to the higher harmonics that travel with different phases. Thus, predicting the above said phenomena proves to be a classical case study to test the applicability of the numerical model for wave – propagation models. For this purpose, the experiment results of Ohayama *et al.*[30] were considered. Their experimental setup consisting of a tank of length 65m and water depth of 0.5m with a submerged bar is shown in Fig. 18. The bar is placed at 28.3m from the wave paddle. The locations of the wave gauges are also shown in the above figure. The same setup is reproduced numerically by using the structured FEM mesh with the number of nodes in the horizontal and vertical direction as 1700 and 15, respectively. This numerical setup is kept constant and the simulation is carried out for all the test cases. Semi-Lagrangian approach is being used for the present simulations. The characteristics of the regular waves adopted for the study are shown in Table 1. The free surface profile over the length of the tank along with the moving mesh configuration is shown in Fig. 19. The spatial evolution showing drastic change in the wave form is clearly visible compared to the long wave profile before the bar. While, Case C and Case D represent the higher wave heights, Case A and Case B correspond to the short and smaller waves, respectively.

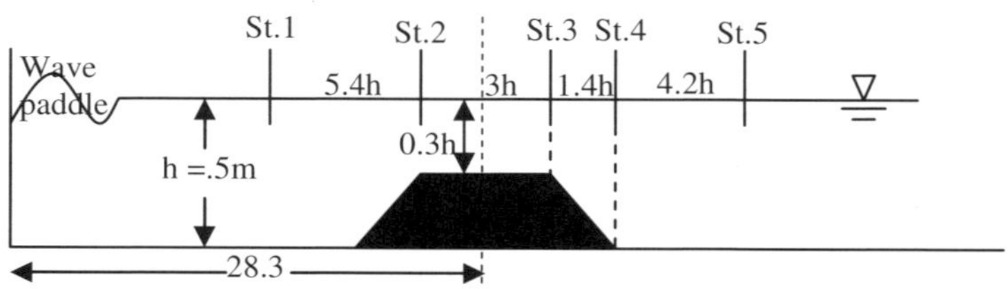

Fig. 18. Experimental setup of ohyama *et al.* (1995).

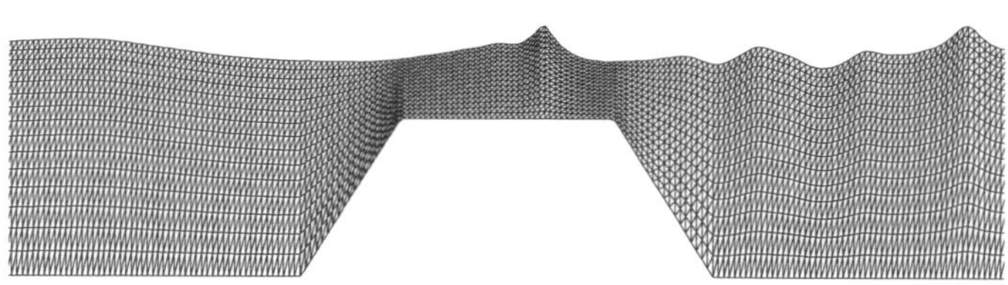

Fig. 19. Free surface profile and the mesh configuration, when the wave reaches the deeper water region from the shallow water region.

Table 1. Simulated input parameters on a water depth of 0.5m.

Case	Wave Period (T in sec)	Wave Height (H in m)	Ursell parameter over the shelf
A	1.34	0.025	21.6
B	1.34	0.05	43.3
C	2.01	0.05	108.7
D	2.68	0.05	201.5

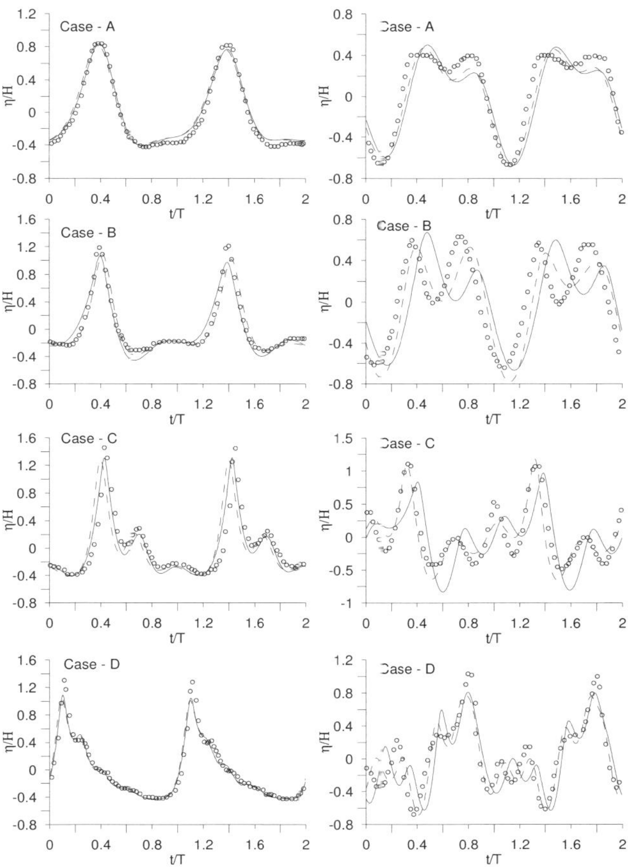

Fig. 20. Comparison with the experimental and numerical simulation, Left side figures are at station 3 and Right side figures are at station 5. (Dotted – Experiments[30], Line – Cubic spline, Dashed line – Least square method).

The comparison between the experimental and numerical computations using cubic spline and least square approaches are shown in Fig. 20. The left side of the figure shows the comparison at station 3 and the right side of the figure shows the comparison at station 5. Considering Case A and Case B at station 3, the comparison is in good agreement with the experimental measurements. Behind the submerged bar, in the region of dispersive wave field, the agreement is quite satisfactory but there is a slight phase shift for both Cubic spline and Least square approach. Considering the steep wave cases i.e., cases C and D, at station 3, a wiggling tail is seen both in experiments and numerical simulations. This is similar to the splitting of solitary waves as studied in detail using the cubic spline method of Sriram *et al.*[13] The numerical simulations are in good agreement but there is a slight reduction in wave heights over the bar. However, behind the bar, slight discrepancy between the numerical results based on cubic spline and that of the experimental measurements of Ohyama *et al.*[30] is more when compared to Least square approach. The discrepancy is due to the disadvantage in using Cubic spline as a velocity calculation technique that is it fits a curve to the neighboring nodes on the free surface, hence, loss of resolution in the higher gradients (removes the high frequency components).

However, minor changes in the computed wave form using Least square method may be due to the viscous effect that is predominant during the wave interaction with the bar in the experiments. The mapped FD behaves in a similar manner as that of Least square method and hence not reported due to clarity of figures.

The simulation using the VOF method (Shen *et al.*[31]) with the inclusion of turbulence effect could not predict well at station 5 and hence had claimed room for further improvement in viscous code. This also justifies the well known fact that the potential flow assumption is a better approximation for the nonlinear wave simulation problems.

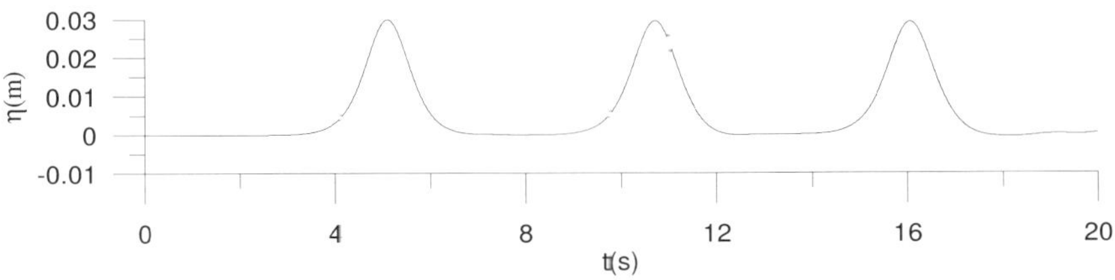

Fig. 21. Time history at the center of the tank. (------ Least square method, ____ Cubic spline method, -·-·-·- Mapped FD).

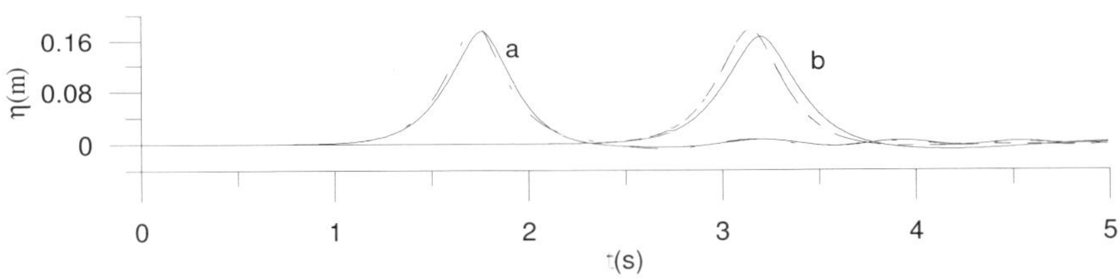

Fig. 22. Time history at a) 2m and b) 5m. (------ Least square method, ____ Cubic spline method, -·-·-·- Mapped FD).

4.5. *Simulation of solitary waves*

The simulation of solitary waves and its interaction with coastal structures have been a topic of research over the past two decades that has gained momentum after the recent Indian Ocean Tsunami. In numerical experiments, tsunami can be well modeled as solitary wave for understanding its propagation over an uneven bottom topography. Hence, the comparison of the different velocity calculation techniques for the simulation of solitary waves is carried out. The solitary wave is generated by using the wave paddle motion given by Goring[32], with the suitable modification as made by Grilli and Svedsen[33] to truncate the initial motion of the infinitely long solitary waves. First, the simulation is carried out for the waves of less steepness of 0.1. The steepness in the case of shallow water waves is H/h. The length of the tank is 10m having a water depth of 0.3m. The number of nodes considered is

301 and 13 along the horizontal and vertical directions, respectively. No smoothing/regridding has been applied. The simulation has been carried out over 20s, which means that the wave reflected from the vertical wall, once again get re-reflected from the wave paddle. This is clearly depicted in Fig. 21 which shows three profiles recorded at the centre of the tank (5m). All the three methodologies are in exact agreement. The simulation was then carried out for a solitary wave of large steepness of 0.6 maintaining all other parameters remain the same. The simulation was carried out only for 5s. While the crest of the wave reaches the wall, the simulation breaks down due to the crossing of the nodal points. This particular effect can be explained by the fact that the pressure becomes zero and eventually the water falls from the free surface suddenly, i.e., the wave breaking takes place. As the fluid is inviscid and in addition the mesh is structured, breaking of waves could not be modelled. The comparisons are illustrated in Fig. 22, which shows that Least square method and Mapped FD are in exact agreement, whereas, a phase shift and a reduction in amplitude are clearly visible in cubic spline method. Thus, cubic spline leads to inaccurate estimation in the case of steep waves

It is worth mentioning to note that the results reported in Sriram *et al.*[13] for the case of run up due to very steep solitary waves on the vertical wall using the cubic spline approach shows a marked decrease in amplitude compared to experimental results. The reason to the discrepancies with the experimental measurements quoted in that paper can be linked to the above said observation.

4.6. *Long time wave simulation*

In order to understand the damping mechanism in all the velocity calculation methodologies, the simulation has been carried out over a long time. For validation of such simulation, experimental measurements form the basis for comparison. Various methodologies are compared with the experimental results[9] for a long time wave simulation. The length of the tank was 200m and water depth was 4m in the

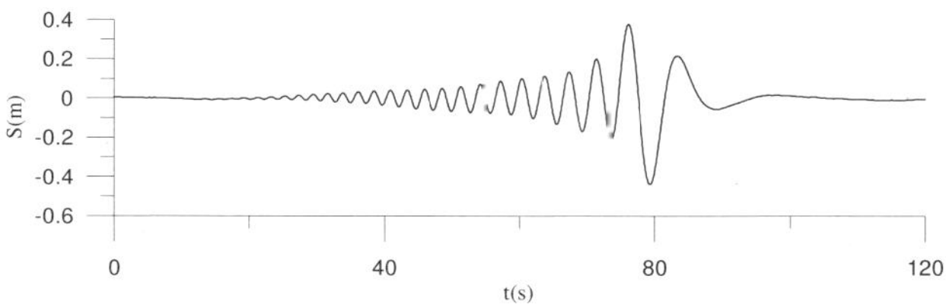

Fig. 23a. Paddle displacement for the simulation of transient wavepackets[9].

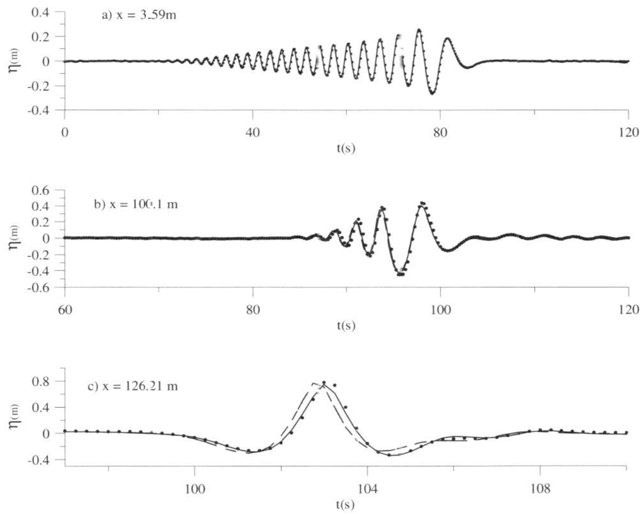

Fig. 23b. Time history comparisons at various locations. (°°°° Experiments[9], ------ Least square method, _____ Cubic spline method, -—-—-— Mapped FD).

experimental setup. The wave paddle motion with a sampling interval of 0.05s is given in Fig. 23(a). The duration of the simulation is 120s. For the numerical modeling, the number of nodes in the horizontal and vertical directions is taken as 501 and 21, respectively. The time step adopted is 0.05s. The corresponding free surface elevation at various locations along the tank is shown in Fig. 23(b). The figure shows a reasonable agreement between all the methodologies with the experimental measurements. The focusing point of the transient wave at a very long distance (126.21m) is also well predicted by cubic spline, whereas, Least square and mapped FD show a phase delay of 0.2s.

The long time simulation of shallow water waves such as Cnoidal wave has been attempted using the different velocity calculation methodologies. The experimental setup of Jeong[34] has been considered for validating the numerical simulation. The length of the tank was 300m and water depth was 4m in the experimental setup. The wave paddle motion with a sampling interval of 0.05s is given in Fig. 24a. The duration of the simulation was 100s. For the numerical modeling, the number of nodes in the horizontal and vertical directions is taken as 801 and 21, respectively. The time step adopted is 0.01s. The corresponding

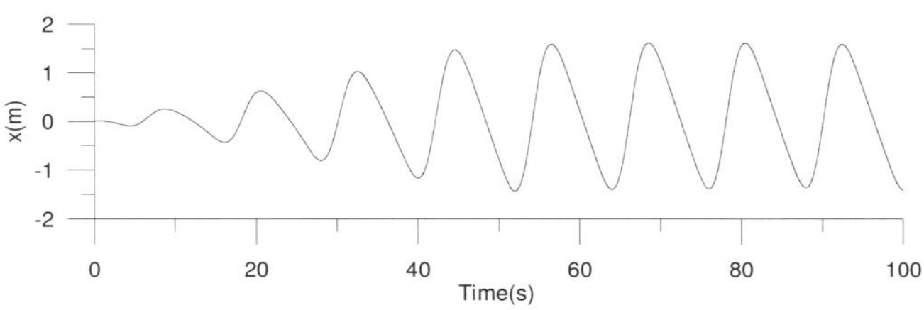

Fig. 24a. Paddle displacement for the simulation of cnoidal waves[34].

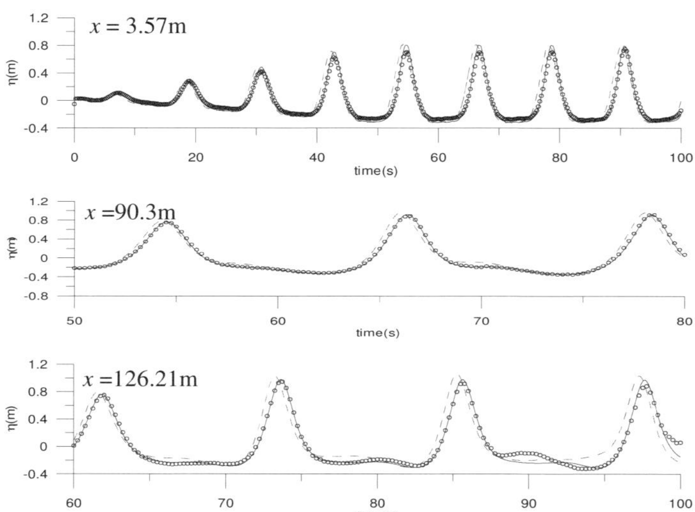

Fig. 24b. Time history comparisons at various locations. (°°°° Experiments[34], ------ Jeong[34], ____ Cubic spline method).

free surface elevation at different locations along the tank is shown in Fig. 24b. Since, all the three methodologies produced identical results; we have plotted only the cubic spline approach. The figure shows an excellent agreement with the experimental measurements even after a long distance. The numerical results of Jeong[34] using the finite volume method are superposed in the above figure and these exhibit a phase difference compared with the present numerical results. The velocity vector plot at a particular time step along the length of the tank shown in Fig. 25 demonstrates a return flow under the trough. These velocity vector plots are obtained by solving the Eq. (12b) from the known velocities at the free surface (cubic spline), side walls and the bottom boundary condition at a particular time step.

5. Simulation Using Wave Paddle – Unstructured Mesh

5.1. *General*

The simulation of nonlinear waves dealing with structured mesh described in the previous sections show that Least square method and mapped FD are providing good results whereas, cubic spline method holds good only for small steep waves, while global projection method usually breaks down for all the cases considered. For modeling complex geometry or for the simulation in the presence of floating bodies, one needs to go for unstructured mesh simulation. In this section, we have described briefly the implementation issues related to the simulation of nonlinear free surface waves in the context of unstructured FEM. For successful implementation, suitable velocity calculation methodology should be adopted. The mapped FD is difficult to implement when the mesh is generated using unstructured mesh where the node numbers are not in regular orientation, whereas, cubic spline holds good for small steep waves. Hence, Least square, Cubic spline and global projection methods are considered in this section.

5.2. *Implementation issues*

5.2.1. *Mesh considerations*

In the case of structured mesh, the element connectivity remains the same, whereas, the node positions at every time step are evaluated based on the new free surface nodes with the vertical elevation calculated by using a simple formula (Eqn. (10)). The computational cost is not expensive in that case, thus regeneration of mesh nodes has been done with ease. In the case of unstructured mesh, one has to resort to the external mesh generation code (First approach) or commercial CFD mesh generators (second approach). While, using the first approach, regeneration of mesh is possible at every time step by simply calling the external code from the source code, whereas, in the second approach, it is not possible to get at every time step automatically. Wang and Wu[14] used the first approach of regenerating the mesh at every time step using the public domain code called BAMG. In the second approach, one can use the commercial CFD mesh generators like GAMBIT, ICEM-CFD and create the initial mesh. Then at every time step, a mesh moving techniques[15,16,35] like Laplacian smoothing/Torsional spring/spring analogy method has been invoked to find the new nodal position. The second approach is more popular in the field of aerodynamics. The method is similar to Arbitary Lagrangian and Eulerian method. Hence, the researchers called them as Semi-ALE (SALE) or Quasi-ALE (QALE) but the basic principle remains the same.

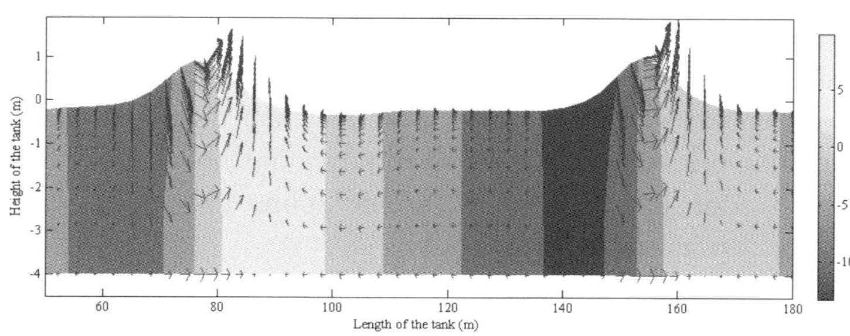

Fig. 25. Velocity vector plot along the length of the tank at 80s.

Sudharsan *et al.*[15] compared three different mesh movement schemes namely Laplacian smoothing, Torsional spring and Spring analogy for the nonlinear free surface problem. It was concluded that the spring analogy is a good choice for handling complex geometries. There are two different methods usually used in spring analogy; one is vertex method and the other is segment method. The vertex spring analogy was originally used for smoothing a mesh after mesh generation or refinement. The segment spring analogy[36] is used for the deformation of the mesh in a moving boundary problem. Ma and Yan[16] preferred improved segment spring approach for the simulation, whereas, Sudharsan *et al.*[15] showed that improved vertex method is more suitable than the segment method. But it should be noted that the stiffness of the springs play a major role. Blom[37] has reported a detailed analysis on the application of both segment and vertex methods.

It is reported in the literature that using normal vertex and spring methodologies led to mesh skewness near the boundary. So, the method has been improved by using modified stiffness and hence, the name 'improved vertex/segment spring methodologies'. Readers can read Blom[37] to know the analytical background by modifying the stiffness. Blom[37] has suggested that the stiffness should be increased in the boundary layers when compared to in the interior layers. Considering the applicability in the field of simulation of nonlinear waves, Sudharsan *et al.*[15] assumed the stiffness in the boundary layer alone and concluded that vertex method is superior than all the available methods to treat wave-structure interaction problems. However, Ma and Yan[16] used segment method and showed promising results, by adopting the stiffness in such a way that the adjacent layers (i.e. along the entire water depth) were also stiffened.

Sriram *et al.*[40] have proved that both the method work well if the spring stiffness is properly defined. The discussion related to principles underlying the methodologies and comparative studies based on numerical coefficients for obtaining the stiffness are beyond the scope of this chapter, hence, the final results are only shown.

5.2.2. *Velocity calculation*

The implementation of the above formulation based on FEM to evaluate velocity potential and global projection method (to calculate velocity) to extract the velocity is straight forward irrespective of the mesh whether it is structured or unstructured. The estimation of the horizontal velocity using Least square method also still holds good, but in the previous section, investigators in the past have taken the advantage of the vertical distribution of mesh and evaluated the vertical velocity accurately using FD schemes. However, in the case of unstructured mesh, it does not hold good. Ma and Yan[16] proposed a technique of effectively using this strategy by drawing the normal line with respect to free surface and used an effective Moving Least square method for the interpolation of the two points over the normal line. But the computational cost to carry out the interpolation is very expensive since it deals with a group of old nodes. After the estimation of normal velocity the Least square method is used to evaluate the tangential velocity [suitable modification of Eq. (13b)]. Thus, cubic spline can also hold good for 2D unstructured mesh.

5.3. *Global projection method and its improvement*

From the previous sections, it has been revealed that the global projection method exhibits poor results for the simulation of nonlinear

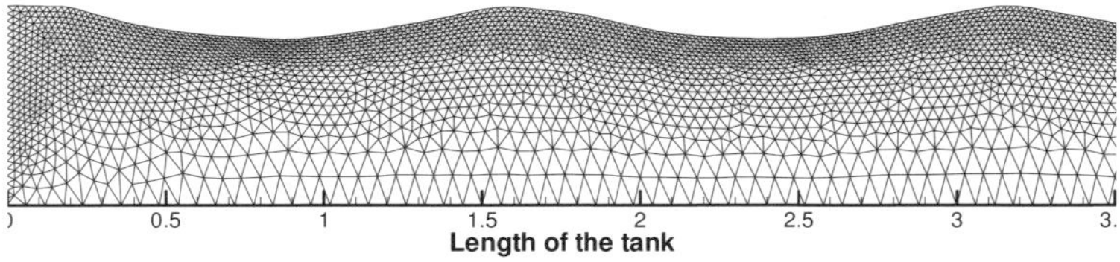

Fig. 26. The mesh configuration at 10.4s.

waves. If the mesh structure has been taken care, the results will be improved. The details of which are discussed below. The mesh moving strategy of vertex method has been adopted[15]. The length of the tank (L) is 9m and water depth (h) is 0.6m. The stroke of the wave maker (S) is $0.05h$ and the wave frequency (ω) is $1.5539\sqrt{g/h}$. The number of elements in the structured mesh corresponds to 16000, while, in the unstructured mesh it is 13598. The initial mesh was generated using GAMBIT commercial software. Time step of 0.005s was adopted that correspond to 200 time step per wave period (T) in both the simulations. The beach length (L_{beach}) and the frequency to be damped in the beach (ω_{beach}) are assumed to be equal to wavelength and the frequency of the of the input wave respectively. Smoothing and regridding have been adopted after every 20 time steps. The simulation was carried out over a long time. The mesh configuration at 10.4s is shown in Fig. 26. The time histories near the wave paddle and at a distance of 5m from the wave paddle are shown in Figs. 27a and 27b respectively. The free surface

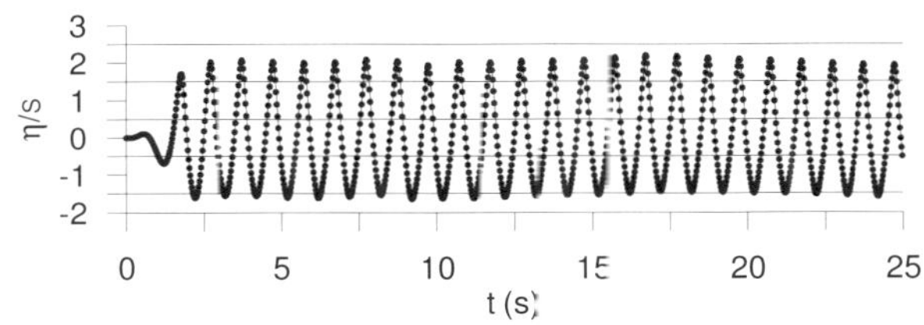

a) Time history of free surface profile near the wave paddle.

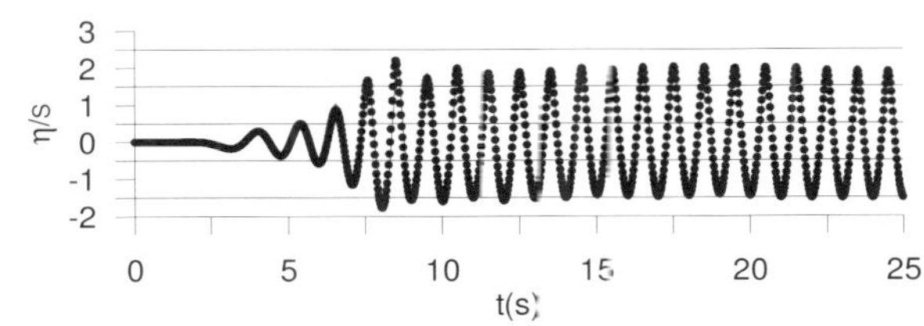

b) Time history of free surface profile at 5m from the wave paddle.

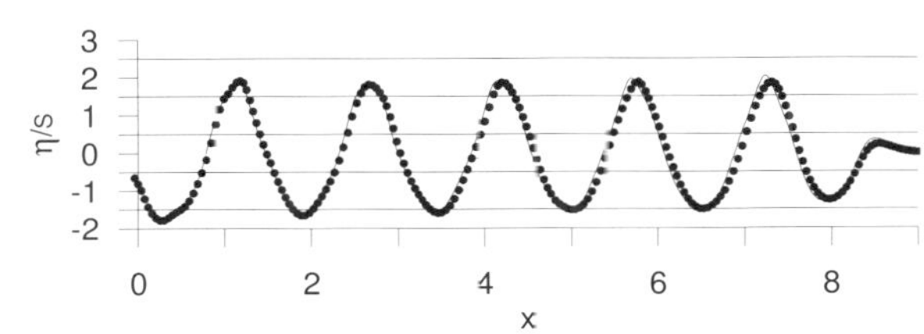

c) Free surface profile along the length of the tank at 25s.

Fig. 27. Comparison between structured and unstructured code for Stroke of the wave maker is 0.05h (*** unstructured, ___ structured[2]).

profile over the entire length of the tank at 25s is shown in Fig. 27c. The comparison between the two methods is found to be good. Thus, it is clear that the global projection can also be used as a velocity calculation technique as adopted by Wang and Wu[14] but the quality of the mesh has to be considered with utmost care.

6. Summary and Outlook

This chapter has focused exclusively on the basic approaches that are used for the simulation of nonlinear free surface waves by applying the Finite element method considering the potential flow theory. The computational methods for the calculation of velocity such as global projection, mapped FD, cubic spline and Least square methodologies are discussed in brief. The following are the salient conclusions drawn from the study.

The main reason for the requirement of regridding/smoothing has been attributed to the fact of lagrangian motion characteristics of the nodes near the moving wave making boundary for medium steep waves. But, in most of the cases the regridding/smoothing technique is not used unless and otherwise needed. This highly depends on the frequency and the problem in hand. The computational time of all the methods are also reported.

The global projection method can be used efficiently when the mesh structure has been taken care at every time steps whereas, while, adopting regeneration of mesh using a simple formula, the simulation breaks down most of the time, certainly needs smoothing and regridding after every 20 time steps.

The Cubic spline hold good for both structured and unstructured meshes in two dimensional tanks for medium steep waves, whereas, for very high steepness, the phase lag was noticed. The method certainly needs an improvement even though it requires very less number of nodes with less smoothing/regridding strategies required.

The Mapped FD and Least square method are promising approaches for the simulation of nonlinear free surface waves using FEM. The former method holds good for the structured mesh, since transformation to the computational grid using unstructured grid is difficult to implement at every time step. The least square method show good results with the flexibility of adapting to all kinds of mesh structures. It can also be implemented for three-dimensional tanks with ease.

References

1. Longuet-Higgins M.S., and Cokelet E.D., *Proc. R. Soc. Lond.* **A.**, 350 (1976).
2. Grilli S.T., Skourup J.. and Svendsen I.A., *Engg. Anal. with Bound. Elemt.,* 6(2) (1989).
3. Boo S.Y., *Ocean Engg.*, 29 (2002).
4. Sen D., and Maiti S., International Conference on Ocean Engineering, IIT, Chennai, India, p. 165-170, (1996).
5. Ohyama T., *Fluid Dyn. Res.,* 8 (1991).
6. Wu G.X., and Eatock Taylor R., *Appl. Ocean Res.,* 16 (1994).
7. Westhuis J.H., *Ph.D. Thesis, Universiteit Twente,* The Netherlands (2001).
8. Ma Q.W., Wu G.X., and Eatock Taylor R., *Int. J. Numer. Meth. Fluids.,* 36 (2001).
9. Clauss G.F., and Steinhagen U., In Proc. of the 9th International Offshore and Polar Engineering Conference, Brest, France, 368-375 (1999).
10. Turnbull M.S., Borthwick A.G.L., and Eatock Taylor R., *Appl. Ocean Res.,* 25 (2003).
11. Turnbull M.S., Borthwick A.G.L., and Eatock Taylor R., *Int. J. Numer. Meth. Fluids.,* 42 (2003).
12. Sriram V., Sannasiraj S.A., and Sundar V.*J. Fluids and Struct.,* 22(5) (2006).
13. Sriram V., Sannasiraj S.A., and Sundar V., *Marine Geodesy.,* Special Issue on Tsunamis, 29(1) (2006).
14. Wang C.Z., and Wu G.X., *J. Fluids and Struct.,* 22 (2006).
15. Sudharsan, N.M., Ajaykumar, R., Murali, K., and Kurichi, K., *J. Mech. Engg. Sci.*, Proceedings of the Institution of Mechanical Engineers U.K. **C**, 218(3) (2004).
16. Ma Q.W., and Yan S., *J. Comput. Phys.,* 212(1) (2006).
17. Bai K.J., and Kim. J.W. Eds., Chakrabarti S., Numerical Models in Fluid-Structure Interaction (WIT Press, 2005).
18. Wu G.X., and Eatock Taylor R., *Ocean Engg.,* 22 (2005).
19. Westhuis J.H., and Andonowati A.J., *In 13th Int. Workshop on Water Waves and Floating Bodies,* Alphen aan denRijin, The Netherlands, Eds. Hermans A.J., p. 171-174, (1998).
20. Kim C.H., Clement A.H., and Tanizawa K., *Int. J. Offshore and Polar Engg.,* 9(4) (1999).
21. Wu G.X. and Hu Z.Z., *Proc. R. Soc. Lond.* **A.**, 460 (2004).
22. Wang C.Z., and Wu G.X., *J Fluids and Struct.,* 23(4) (2007).
23. Cai X., Langtangen H.P., Nielsen B.F., and Tveito A., *J. Comput. Phys.,* 143 (1998).

24. Sen D., Pawlowski J.S., Lever J., and Hinchey M.J., Proceedings of the 5th International Conference on Numerical Modeling of Ship Hydrodynamics, Hiroshima, p. 351-373, (1989).
25. Jain M.K., Iyengar S.R.K., and Jain R.K., (New Age International Publishers, New Delhi, 2003).
26. Dommermuth D.G., and Yue D.K.P., *J. Fluid Mech.,* 178 (1987).
27. Steinhagen U., Ph.D. Thesis, Technische Universitaet Berlin (1999).
28. Beji, S. and Battjes, J.A. *Coastal Engg.,* 19 (1993).
29. Beji, S. and Battjes, J.A. *Coastal Engg.,* 23 (1994).
30. Ohyama, T., Kioka, W. and Tada, A. *Coastal Engg.,* 24 (1995).
31. Shen, Y.M., Ng, C.O. and Zheng, Y.H. *Ocean Engg.,* 31(1) (2004).
32. Goring D.G., PhD thesis of the California Institute of Technology, Pasadena, California, (1979).
33. Grilli, S., and Svendsen, I.A., *Water Wave Kinematics,* (Proc. NATO-ARW, Molde, Norway, May 89) (ed. A. Torum and O.T. Gudmestad) NATO ASI Series E: Applied Sciences Vol. 178, p. 387-412. Kluwer Academic Publishers (1990).
34. Jeong, S.J., Ph.D. Thesis, Technische Universitaet Berlin, D83 (2003).
35. Sriram V., Sannasiraj S.A., and Sundar V., *J. Coastal Res., Special Issue, 50* (2007).
36. Batina J.T., *AIAA Journal.*, 28(8) (1990).
37. Blom F.J., *Int. J. Numer. Meth. Fluids.*, 32 (2000).
38. Guignard, S., Grilli, S.T., Marcer, R. and Rey, V. In *Proc. 9th Offshore and Polar Engng. Conf.* (ISOPE99, Brest, France, May 1999), Vol. 3, p. 304-309 (1999).
39. Wang C.Z., Khoo B.C., *Ocean Engg.,* 32 (2005).
40. Sriram V., Sannasiraj S.A. and Sundar V., 7th International Conference on Coastal and Port Engineering in Developing Countries, *PIANC-COPEDEC VII*, Dubai., (2008).

CHAPTER 7

HIGH-ORDER BOUSSINESQ-TYPE MODELLING OF NONLINEAR WAVE PHENOMENA IN DEEP AND SHALLOW WATER

Per A. Madsen* and David R. Fuhrman†

Technical University of Denmark, Department of Mechanical Engineering
Nils Koppels Allé, DTU - Building 403, DK-2800 Kgs. Lyngby, Denmark
**prm@mek.dtu.dk*
†drf@mek.dtu.dk

In this work, we start with a review of the development of Boussinesq theory for water waves covering the period from 1872 to date. Previous reviews have been given by Dingemans,[1] Kirby,[2,3] and Madsen & Schäffer.[4] Next, we present our most recent high-order Boussinesq-type formulation valid for fully nonlinear and highly dispersive waves traveling over a rapidly varying bathymetry. Finally, we cover applications of this Boussinesq model, and we study a number of nonlinear wave phenomena in deep and shallow water. These include (1) Kinematics in highly nonlinear progressive deep-water waves; (2) Kinematics in progressive solitary waves; (3) Reflection of solitary waves from a vertical wall; (4) Reflection and diffraction around a vertical plate; (5) Quartet and quintet interactions and class I and II instabilities; (6) Extreme events from focused directionally spread wavefields; (7) Bragg scattering from an undular sea bed; (8) Run-up of non-breaking solitary waves on a beach; and (9) Tsunami generation from submerged landslides.

1. Milestones in the Development of Boussinesq Theory

Classical low-order Boussinesq wave theory represents a shallow-water approximation to the fully dispersive and nonlinear water wave problem. It is a first correction to the nonlinear shallow water equations allowing for a non-hydrostatic pressure due to vertical accelerations, and horizontal velocities, which are non-uniform over depth. In general the vertical variation of the horizontal velocity is expressed by a polynomial expansion in the vertical coordinate and at lowest order it takes the form of a parabolic variation. The classical equations are typically expressed in terms of the depth-averaged velocity or in terms of the horizontal velocity variable at the still water level or at the sea bottom. There are two important parameters associated with Boussinesq theory: The nonlinearity represented by the ratio of amplitude to depth ($\varepsilon = a_0/h_0$), and the dispersion represented by wave number times water depth ($\mu = k_0 h_0$). The linear limit is represented by $\varepsilon \to 0$, while $\mu \to 0$ represents the non-dispersive limit.

The original formulation by Boussinesq (1872)[5] was expressed in terms of the depth-averaged velocity. It was derived on a constant depth, and was restricted to weak dispersion ($\mu \ll 1$) as well as weak nonlinearity ($\varepsilon \ll 1$), assuming $\varepsilon = O(\mu^2)$ and retaining terms of order $O(\mu^2)$. Similar assumptions were applied by Korteweg & de Vries (1895)[6] in their derivation of the classical KdV equation for wave motion in a single direction only. Later, Serre (1953)[7] and Su & Gardner (1969)[8] derived a so-called fully nonlinear version of the Boussinesq equations by assuming $\varepsilon = O(1)$ and retaining terms of order $O(\varepsilon^3\mu^2)$.

Mei & Le Méhauté (1966)[9] and Peregrine (1967)[10] extended Boussinesq's original equations to an uneven bottom and the equations were expressed in alternative velocity variables such as the horizontal velocity at the sea bed and at the still water level. This work also initiated the first computer models simulating e.g. the transformation of solitary waves on beaches (see Peregrine[10] and Madsen & Mei[11]). Dingemans (1973)[12] was the first to derive higher-order Boussinesq equations (μ^4) by retaining more terms in the polynomial velocity expansion. These formulations, however, were later shown by Madsen & Schäffer[13] to incorporate singularities in linear or non-linear properties, which explains why they were never used for implementation. Benjamin et al. (1972)[14] made an important contribution by discussing the linear dispersion properties in the framework of KdV equations, and they pointed out how leading order terms could be used to manipulate the third-derivative dispersive term from e.g. u_{xxx} to u_{xxt}, by which numerical stability and dispersion were improved. Following up on this idea, Mei[15] provided four different versions of the dispersive KdV terms.

In the late 1970s, the first commercial model systems based on classical low-order Boussinesq formulations were developed (see e.g. Abbott et al.[16–18]), and these models quickly became popular in coastal engineering. This shifted the focus from long waves such as solitary and cnoidal waves to nonlinear irregular waves and started a series of systematic studies in order to establish the practical range of application of the governing equations (see e.g. McCowan,[19–22] Madsen & Warren,[23] Schaper & Zielke,[24] Rygg,[25]

Abbott & Madsen[26] and Prüser & Zielke[27]). As a consequence, the inherent shallow water limitations in dispersion and nonlinearity became of concern and efforts were initiated to improve the applicability of Boussinesq formulations.

The work by Witting (1984)[28] on a so-called unified model for nonlinear waves contained many brilliant and novel ideas. First, Witting used the exact dynamic surface condition expressed in terms of a velocity variable defined directly on the free surface. Second, he used the exact depth-integrated continuity equation expressed in terms of the depth-averaged velocity. Both of these velocity variables were then expressed in terms of a utility variable using power series expressions with arbitrary coefficients. Finally, these coefficients were chosen, so that the resulting linear dispersion relation would incorporate Padé approximants in powers of kh. This resulted in extraordinarily good linear characteristics. The formulation, however, was relatively complicated and fundamentally different from conventional Boussinesq formulations. Furthermore, it was restricted to a constant depth and a single horizontal dimension, and not straight-forward to extend. To our knowledge, Witting never extended his work to uneven bottom or two horizontal dimensions. Nevertheless, Witting's pioneering work became highly important for the development of modern Boussinesq formulations, and his introduction of the concept of Padé approximants within differential equations kick-started a chain reaction, which is still on-going.

Madsen et al. (1991)[29] utilized the ideas of Witting[28] and Benjamin et al.[14] in the framework of conventional Boussinesq equations. First, Madsen et al.[29] demonstrated that the Boussinesq dispersion relation can be modified in two ways: (1) By choosing different velocity expansion variables such as e.g. the horizontal velocity at the bottom, the still water level or the depth-averaged velocity; (2) By using the leading order manipulation technique proposed by Benjamin et al.[14] It all boils down to different mixtures of spatial and temporal derivatives in the governing equations. Madsen et al.[29] refined the ideas of Benjamin et al.[14] to introduce a linear operator enhancement technique, which could introduce Padé (2,2) dispersion properties in any low-order Boussinesq formulation e.g. in terms of the depth-integrated velocity. Madsen & Sørensen[30] extended this work to an uneven bottom and optimized the linear shoaling gradient. Soon after, Nwogu (1993)[31] presented an alternative way of achieving Padé (2,2) dispersion characteristics, as he derived a low-order formulation in terms of the horizontal velocity variable at an arbitrary vertical location (close to mid-depth). In practice these works significantly extended the linear application range of low-order Boussinesq formulations from wave number times water depth $kh \simeq 1$ to $kh \simeq 3$. Later, Schröter et al.[32] and Schäffer & Madsen[33] applied the linear operator enhancement technique on Nwogu's equations to obtain low-order equations incorporating Padé (4,4) dispersion characteristics accurate up to $kh \simeq 6$.

To evaluate the nonlinear characteristics of Boussinesq equations, Madsen & Sørensen (1993)[34] suggested to apply a Stokes-type perturbation analysis, while using as a reference Stokes third-order theory for regular waves, and second order theory for bichromatic waves (including transfer functions for sum- and difference-interactions). At first sight, this procedure appeared to be somewhat dubious due to the well-known fact that both theories diverge for finite amplitude waves in shallow water. What is important, however, is that the *transfer functions* originating from the fully dispersive problem and from the approximate Boussinesq equations converge to the same solution in the shallow water limit, while they will inevitably deviate for larger values of kh (see e.g. Whitham,[35] Chap. 13.13). Furthermore, it is a convenient way of separating linear and nonlinear effects e.g. in the amplitude dispersion. Based on this analysis it could be concluded that the linear enhancements of Madsen et al.,[29] Nwogu[31] and Schäffer & Madsen[33] all left a lot to be desired with respect to nonlinearity, where a 10% error was exceeded for $kh \simeq 0.5$.

In order to improve these properties Wei & Kirby (1995)[36] and Wei et al.[37] adopted the idea of Serre[7] to allow for full nonlinearity i.e. $\varepsilon = O(1)$ during the derivation. Hence, they extended the formulation of Nwogu[31] to include all nonlinear terms up to the retained order of dispersion $O(\mu^2)$ and included terms of order $O(\varepsilon^3\mu^2)$ in their formulation. In fact, this extension turned out to be much more successful than the original work by Serre[7] and Su & Gardner:[8] As demonstrated by Kirby,[2] the Serre formulation produces solitary wave crests which are too broad relative to the exact solutions, while the Wei & Kirby[36] formulation leads to quite accurate solitary wave profiles. The reason is most likely that although both formulations formally include $O(\varepsilon^3\mu^2)$ terms, the Wei & Kirby[36] formulation *effectively* also includes $O(\mu^4)$ terms because of the Padé (2,2) dispersion characteristics. A Stokes-type analysis of the Wei & Kirby[36] equations showed that a significant improvement of nonlinearity was achieved for $kh < 1.25$ relative to the Nwogu[31] formulation (see the analysis by Madsen & Schäffer[4,13]). On the other hand, both formulations gave poor nonlinearity for $kh > 1.5$.

Madsen & Schäffer (1998)[13] derived and analyzed a number of dispersion-enhanced low-order and high-order Boussinesq formulations in terms of the depth-averaged velocity U as well as the arbitrary-level velocity $\hat{u}$, while using $\varepsilon = O(\mu^2)$ as well as $\varepsilon = O(1)$ assumptions. Their low order equations included either $O(\mu^2, \varepsilon)$ terms or $O(\mu^2, \varepsilon^3\mu^2)$ terms, while their high-order equations included either $O(\mu^4, \varepsilon^3\mu^2)$ terms or $O(\mu^4, \varepsilon^5\mu^4)$ terms. It was concluded that higher-order equations are generally useless unless they are Padé enhanced. This is due to the inherent singularities, which show up in either the linear dispersion characteristics or in the nonlinear transfer functions. Typically, these problems vanish when Padé expansions replace Taylor expansions. Despite the achieved improvements of applicability, the overall outcome of Madsen & Schäffer's work was disappointing with respect to nonlinear characteristics, and even for the best forms the nonlinear transfer functions became inaccurate for $kh > 1.7$.

Some of the formulations by Madsen & Schäffer[13] re-appeared in Zou.[38,39] First, Zou[38] presented a Padé (2,2) enhanced $O(\mu^2, \varepsilon^3\mu^2)$ formulation in terms of the depth-averaged velocity U. This formulation happens to be linearly and nonlinearly identical to that of Madsen & Schäffer,[13] their Chapter 4 (example 3, figure 4; see also their figure 17*b* for an analysis of sum- and difference-interactions). While the embedded difference-interactions are reasonably accurate, the sum-interactions are severely underestimated for $kh > 1.2$. Zou tried to repair this problem by introducing a free parameter on one of the four $O(\varepsilon\mu^2)$-terms. By changing this coefficient from 1/10 to 1/2, Zou claimed to improve the overall error of self-self interaction in the interval $0 < kh < 3$, but in fact he destroyed the Taylor expansion of the second harmonic and made it only leading order accurate. Furthermore, he increased the error significantly in the interval $0 \leq kh \leq 1$, from a few percent to a maximum of 20%. We cannot support this procedure. Soon after, Zou[39] presented a Padé (4,4) enhanced $O(\mu^2, \varepsilon^3\mu^2)$-formulation in terms of a pseudo velocity $\hat{u}$. This formulation turns out to be linearly (but not non-linearly) identical to Madsen & Schäffer,[13] their Chapter 6 (example 1, figure 12). Again the nonlinear optimization technique was applied, this time with somewhat more success, yet fundamentally we disagree with this procedure (reasons are given in what follows).

A similar but more consistent enhancement procedure was suggested by Kennedy et al. (2001),[40] who extended the Wei & Kirby[36] $(\mu^2, \varepsilon^3\mu^2)$-formulation to allow for an expansion level proportional to not only the still water depth but also to the time-varying surface elevation i.e.

$\hat{z} = \alpha h(x,y) + \beta\eta(x,y,t)$. This provided an extra free parameter, which was used to match the second order Stokes amplitude up to $(kh)^4$. As a result of this effort, the limit of 5% errors in second order self-self interactions was improved from $kh \simeq 1.0$ (Wei & Kirby[36]) to $kh \simeq 2.5$. Other second order sum-interactions were also improved, while difference-interactions showed little difference from Wei & Kirby's results, reaching a 5% error already at $kh \simeq 1.0$. Kennedy et al.[40] concluded that an asymptotic improvement of all nonlinear measures would likely require a higher order model. Consequently, they tried to apply their nonlinear enhancement procedure also on the $O(\mu^4, \varepsilon^5\mu^4)$-formulation by Madsen & Schäffer,[13] but they reported that this attempt failed to improve the nonlinear characteristics, which over the range of $1 < kh < 3$ actually were found to be less accurate than before. This example supports our general concern about the concept of nonlinear optimization: Linear dispersion (and linear shoaling for that matter) is represented by a single term at a given order and hence it can be matched with a few free parameters. In contrast, nonlinear interactions are indefinite and nonlinear terms are numerous at each order and therefore very problematic to curve-fit even with a relatively high number of free parameters. While the second-order self-self interaction and some of the nearby sum-interactions may be improved, the technique generally does not capture difference-interactions or higher-order sum-interactions.

A breakthrough with respect to the nonlinear characteristics was finally made by Agnon, Madsen & Schäffer (1999).[41] First of all, they expressed and time-stepped the exact nonlinear dynamic and kinematic surface boundary conditions in terms of velocity variables defined directly on the free surface. Secondly, they satisfied the linear Laplace problem and the kinematic bottom condition by an infinite power series expansion of the velocity field from the still water level. With this expression at hand, the corresponding velocity components at the free surface could be determined and a closure achieved. In principle this formulation was fully nonlinear and fully dispersive. Some fundamental properties were derived on the basis of this infinite series expansion e.g. the exact linear dispersion relation and the exact linear shoaling gradient. For practical solutions, Agnon et al.[41] truncated the infinite series expansions to achieve Padé (4,4) linear characteristics. No further truncations were made e.g. when squaring a variable containing high-derivatives. As a result, they were the first to achieve a complete carry over in the quality of linear dispersion characteristics to the nonlinear transfer functions. In many ways, this work was inspired by Witting's[28] formulation, and as a penalty the system of equations was just as complicated, although straight-forward to formulate on an uneven bottom in two horizontal dimensions. Witting, however, failed to utilize the full potential of his formulation, partly because he actually truncated all equations at $O(\mu^6)$ and partly because he mixed a Padé expansion of linear terms with a Taylor expansion of nonlinear terms. A complete analysis of Witting[28] versus Agnon et al.[41] can be found e.g. in the review by Madsen & Schäffer,[4] chapter 5.1.

At the same time, another line of development was presented by Wu (1998–2001),[42–44] who also provided fully nonlinear and fully dispersive formulations. He also utilized the exact surface conditions expressed in terms of surface velocity variables and combined them with infinite series expansions in terms of the bottom velocity, the surface velocity, the arbitrary depth velocity and the depth-averaged velocity, respectively. Wu acknowledged that the inverted series expansions had only finite convergence radii (a subject which was later discussed in detail by Madsen & Agnon[45]), but he did not determine the radius for the different formulations. More importantly he did not invoke Padé techniques to obtain accurate truncated versions of his infinite formulations.

With significantly improved linear and nonlinear characteristics at hand, research now started to focus on the quality of the interior velocity profile. In this respect, the high-order Agnon et al.[41] formulation based on the still-water velocities was relatively disappointing, leading to 2% errors already at $kh \simeq 1.7$. Other low-order formulations such as Boussinesq,[5] Peregrine,[10] Whitham,[35] Mei,[15] Madsen et al.[29] and Dingemans[1] using the depth-averaged or depth-integrated velocity were applicable up to $kh \simeq 1$, while the low-order formulation by Nwogu[31] and Wei & Kirby[36] using the mid-depth velocity was accurate up to $kh \simeq 1.5$. It also turned out that adding higher order terms as e.g. in Madsen & Schäffer[13] did not have much effect on the quality of the velocity profile.

An interesting approach was presented by Gobbi et al. (2000)[46] using an enhanced high order formulation expressed in terms of the weighted average of the velocity potential at two arbitrary reference levels and including terms of order $(\mu^4, \varepsilon^5\mu^4)$. In many respects, this was a special case of the procedure suggested by Schäffer & Madsen (1995),[47] who proposed introducing generalized velocity variables including e.g. integrals and moments of physical velocities to achieve accurate characteristics in Boussinesq formulations (however, their method was never implemented and never pursued beyond the initial formulation). Gobbi et al.[46] achieved Padé (4,4) linear characteristics, but the nonlinear characteristics were not as impressive:

Typically, sum-interactions were accurate up to $kh \simeq 3$, while difference-interactions were only accurate up to $kh \simeq 1.5$. This performance was similar to Kennedy et al.,[40] and certainly an improvement compared to Madsen & Schäffer,[13] but compared to Agnon et al.[41] it left a lot to be desired. It was, however, the quality of the interior velocity field, which really made the formulation by Gobbi et al.[46] interesting, as linear velocity profiles were shown to be accurate for wave numbers as high as $kh \simeq 5$. This was a huge improvement compared to any other formulation at the time.

Kennedy et al. (2002)[48] and Madsen & Agnon (2003)[45] systematically analyzed a number of Boussinesq formulations with special emphasis on the linear velocity characteristics. While Kennedy et al.[48] concentrated on enhanced low-order formulations, Madsen & Agnon[45] considered formulations of arbitrarily high order. Generally, it was shown that formulations in terms of the depth-averaged velocity U are limited by a convergence radius $kh_{Limit} = \pi$, while formulations given in terms of $\hat{u}$ are limited by $kh_{Limit} = \pi/(2 + 2\sigma)$, where $\sigma = \hat{z}/h$ defines the reference level for expansion (i.e. $0 \geq \sigma \geq -1$). This explains why formulations in terms of e.g. the mid-depth velocity have the potential of being more accurate than formulations in terms of the still-water velocity. The radius of convergence is governed by the distance to the nearest complex pole of the target function, and a power series representation of this function cannot exceed this radius no matter how many terms are included. If, however, the power series is combined with or replaced by fractions of power series, the nearest singularities may be captured, and hence the convergence radius extended to the next singularity. Madsen & Agnon[45] showed that this could be achieved by keeping the vertical velocity $\hat{w}$ as an unknown instead of introducing the explicit determination and elimination of $\hat{w}$, which is common practice in nearly all other Boussinesq formulations (except for Agnon et al.[41] and Madsen et al.[49,50]). As an important additional step, Madsen et al.[49] showed that Padé efficiency could be introduced in the velocity profile by expanding the physical velocity variables $(\hat{u}, \hat{w})$ in terms of pseudo-velocity variables $(\hat{u}^*, \hat{w}^*)$ using linear operator techniques. On this basis, they derived a velocity formulation involving fifth-order spatial derivatives and accurate up to $kh \simeq 12$. An even more accurate profile was derived by Madsen & Agnon:[45] With genuine Padé properties at every z-level, this was shown to be accurate up to $kh \simeq 19$, but because of its complexity it has yet only been implemented for the case of a horizontal sea bottom (see Toledo & Agnon[51]).

Madsen et al. (2002, 2003)[49,50] utilized their new highly accurate velocity formulation satisfying the Laplace equation and the bottom boundary condition and combined it with the splitting procedure by Agnon et al.[41] i.e. the exact nonlinear surface boundary conditions were time-stepped directly in terms of the free surface velocities $(\widetilde{u}, \widetilde{w})$, while the connection between these velocities and the pseudo-velocities $(\hat{u}^*, \hat{w}^*)$ at the expansion level $z = \hat{z}(x, y)$ was established and inverted numerically. The outcome of this procedure was a major improvement of linear and nonlinear properties: (1) With fifth-derivative operators and a choice of $\hat{z} = -0.2h$, linear and nonlinear properties could accurately be represented up to $kh \simeq 40$, while the interior velocity field was relatively inaccurate; (2) With fifth-derivative operators and a choice of $\hat{z} = -0.5h$, linear and nonlinear properties could accurately be represented up to $kh \simeq 25$, while the interior velocity field was accurate up to $kh \simeq 12$, even for highly nonlinear waves. For this reason, we generally recommend using the latter alternative. While the original formulation by Madsen et al.[49,50] was restricted to mildly sloping bathymetries, this formulation was recently extended to deal with rapidly varying bathymetries by Madsen, Fuhrman & Wang (2006)[52] and they studied e.g. wave transformation over steep trenches and undular sea-beds (Bragg-scattering). Finally, it should be mentioned that Madsen et al.[50] also derived a simplified set of their equations truncated at third-order spatial derivatives and this set has recently been implemented in terms of a velocity potential on a constant depth by Jamois et al.[53,54]

Motivated by a concern about the complexity involved in handling the fifth-derivative operators present in the formulations by Gobbi et al.[46] and Madsen et al.,[49] Lynett & Liu (2004)[55] presented a two-layer approach to Boussinesq theory. In this approach two separate quadratic velocity polynomials were used and matched at the interfaces between the layers. Furthermore, they utilized the optimization ideas of Kennedy et al.[40] and expressed the expansion levels of the two velocity profiles as well as the interface between the layers in terms of a parameter times the still water depth and another parameter times the free surface elevation. This introduced six free parameters for linear and nonlinear optimization. The optimization was made on the basis of a global minimization of errors over a span of wave numbers: (1) Linear optimization was based on a sum of errors in phase celerity, group velocity and shoaling gradient (starting from $kh = 0.1$); (2) Nonlinear optimization was based on a sum of errors in second order wave amplitude, second order sum-interactions and second order difference-interactions (for unclear reasons starting from $kh = 1.0$). As an outcome of this effort, the linear properties were accurate up to $kh \simeq 6-8$, with linear velocity profiles of the same quality as Gobbi et al.[46] With respect to nonlinearity, the optimized two-layer model approximately doubled the range of applicability (in terms of kh) compared to the one-layer model by Kennedy et al.:[40] Again the second-order self-self interactions could easily be improved by optimization and they now showed a 5% error at $kh \simeq 6$. Again the difference-interactions were unaffected by optimization and showed a 100% error at $kh \simeq 5$ and a 5% error at $kh \simeq 2.0$ (see their figure 5*b*). This again demonstrates short-comings inherent within the nonlinear optimization procedure.

Lynett & Liu (2004)[56] generalized the two-layer approach to a multi-layer approach considering three-, four- and five-layer formulations. They provided a linear analysis of these formulations and showed that typically one can obtain linear Padé ($2N$, $2N$) characteristics with a N-layer approach. Rather than pursuing this option, they preferred to optimize celerity, group velocity and shoaling gradients and with their optimized sets they obtained the following linear applicability: Two-layer ($kh \simeq 6$), three-layer ($kh \simeq 17$), four-layer ($kh \simeq 30$). Unfortunately, they did not provide any analysis of nonlinear transfer functions, and it is presently not clear to which extent the added number of layers might improve nonlinearity with or without nonlinear optimization. Numerical simulations presented by Hsiao et al.[57] using two- and three-layer models indicated some nonlinear improvement. On the other hand, this paper concluded that for modulated (unstable) waves, the applicable water depth for the multi-layer model was much shallower than what its linear characteristics showed, and that further nonlinear analysis and optimization was required. As discussed earlier, we still have a profound concern about the concept of nonlinear optimization.

Recently, a novel approach to modeling nonlinear fully dispersive water waves over gently varying bathymetry has been initiated by Schäffer (2004–2005).[58–60] In this approach a convolution method is developed for the explicit determination of the vertical SWL-velocity (still water level) in terms of the horizontal SWL-velocity. Along with a time-stepping of the two free surface boundary conditions, the convolution integral closes the system without the need for inversion of systems of algebraic equations that usually dominates the computational effort in e.g. Boussinesq-type models. The convolution integrals involve impulse response functions that decay exponentially with the distance from the point of interest and in practice the infinite limits of the integrals can be reduced to horizontal distances of a few water depths. Nonlinearity and the connection between free surface velocities and still water velocities are incorporated by the perturbation approach of West et al.[61]

Finally, it should be mentioned that a number of special features have been incorporated in the framework of Boussinesq models during the last 20 years: (1) Wave-current interactions have been studied by e.g. Yoon & Liu,[62] Chen et al.,[63,64] Madsen & Schäffer,[4,13] and Shen;[65] (2) Wave-ship interaction has been studied by e.g. Madsen & Sørensen,[66] Jiang[67] and Nwogu & Demirbilek;[68] (3) Surf-zone dynamics including breaking waves and wave-induced currents have been studied by e.g. Tao,[69] Zelt,[70] Karambas & Koutitas,[71] Schäffer et al.,[72] Madsen et al.,[73,74] Sørensen et al.,[75] Chen et al.,[76–78] Kennedy et al.,[79] and Veeramony & Svendsen.[80] We shall not discuss these features in any detail in this paper, and the interested reader is referred to the original papers.

2. Formulation of the High-Order Boussinesq Model

2.1. *Introduction*

In this section, we provide the governing equations for the high-order Boussinesq formulation for nonlinear waves over a rapidly varying bathymetry as originally presented by Madsen, Fuhrman & Wang.[52] We consider the irrotational flow of an incompressible inviscid fluid with a free surface. The coordinate system is Cartesian with the x-axis and y-axis located on the still water plane and with the z-axis pointing vertically upwards. The fluid domain is bounded by the sea bed at $z = -h(x, y)$ and the free surface at $z = \eta(x, y, t)$. The problem consists of solving the Laplace equation in the interior plus boundary conditions at the sea bottom and at the free surface.

2.2. *Exact solutions to the Laplace equation*

As shown by Rayleigh[81] and Agnon et al.[41] an exact solution to the Laplace equation can be expressed as

$$\mathbf{u}(x, y, z, t) = \cos(z\nabla)\,\mathbf{u}_0 + \sin(z\nabla)\,w_0, \tag{1}$$

$$w(x, y, z, t) = \cos(z\nabla)\,w_0 - \sin(z\nabla)\,\mathbf{u}_0, \tag{2}$$

where cos- and sin-operators are defined by

$$\cos(z\nabla) \equiv \sum_{n=0}^{\infty}(-1)^n \frac{z^{2n}}{(2n)!}\nabla^{2n}, \qquad \sin(z\nabla) \equiv \sum_{n=0}^{\infty}(-1)^n \frac{z^{2n+1}}{(2n+1)!}\nabla^{2n+1}, \tag{3}$$

while $\mathbf{u}_0$ is the horizontal velocity vector and w_0 the vertical velocity component, both defined at $z = 0$. Furthermore, ∇ is the horizontal gradient operator defined by $\nabla = (\partial/\partial x, \partial/\partial y)$, and the interpretation of powers of ∇ depends on whether this operator is acting on a scalar or a vector. In this context the following set of rules apply

$$\nabla^{2n}\mathbf{u}_0 \equiv \nabla\left(\nabla^{2n-2}(\nabla\cdot\mathbf{u}_0)\right), \qquad \nabla^{2n+1}\mathbf{u}_0 \equiv \nabla^{2n}(\nabla\cdot\mathbf{u}_0),$$
$$\nabla^{2n}w_0 \equiv \nabla^{2n}w_0, \qquad \nabla^{2n+1}w_0 \equiv \nabla\left(\nabla^{2n}w_0\right).$$

Unfortunately, truncated versions of (1)–(3) are very inaccurate and as shown by Madsen & Agnon,[45] their Table 4, a 2% velocity error is reached for $kh \simeq 2$.

To improve the accuracy of the velocity profile, Madsen et al.[49,50] generalized (1) and (2) to

$$\mathbf{u}(x,y,z,t) = L_I\hat{\mathbf{u}} + L_{II}\hat{w}, \qquad w(x,y,z,t) = L_I\hat{w} - L_{II}\hat{\mathbf{u}}, \tag{4}$$

where $\hat{\mathbf{u}}$ is the horizontal velocity vector and $\hat{w}$ the vertical velocity component, both defined at an arbitrary level $z = \hat{z}$. On a constant depth, the operators in (4) were determined to be

$$L_I\hat{\mathbf{u}} = \cos((z-\hat{z})\nabla)\,\hat{\mathbf{u}}, \qquad L_{II}\hat{\mathbf{u}} = \sin((z-\hat{z})\nabla)\,\hat{\mathbf{u}}. \tag{5}$$

This was further extended by Madsen et al.[52] to allow for a rapid spatial variation of $\hat{z}(x,y)$, in which case the operators become

$$L_I\hat{\mathbf{u}} = \hat{\mathbf{u}} + \sum_{m=1}^{\infty}\sum_{n=0}^{2m} \frac{(-1)^{m+n}}{n!(2m-n)!} z^n \nabla^{2m-1}\left(\hat{z}^{2m-n}\nabla\hat{\mathbf{u}}\right), \tag{6}$$

$$L_{II}\hat{\mathbf{u}} = \sum_{m=0}^{\infty}\sum_{n=0}^{2m+1} \frac{(-1)^{m+n+1}}{n!(2m-n+1)!} z^n \nabla^{2m}\left(\hat{z}^{2m-n+1}\nabla\hat{\mathbf{u}}\right). \tag{7}$$

Truncated versions of (4)–(5) including up to fifth-order spatial derivatives, provide a 2% velocity error at $kh \simeq 6$ for an optimal choice of $\hat{z}/h = -0.53$ (see Madsen & Agnon[45]).

Madsen et al.[49,50] showed that further improvement of the accuracy of the velocity profile can be achieved by expanding the physical velocities

$(\hat{\mathbf{u}}, \hat{w})$ in terms of utility velocities using

$$\hat{\mathbf{u}} = L(\hat{z}\nabla)\,\hat{\mathbf{u}}^*, \qquad \hat{w} = L(\hat{z}\nabla)\,\hat{w}^*, \qquad L(\hat{z}\nabla) \equiv \sum_{n=0}^{2N} \lambda_{2n}\hat{z}^{2n}\nabla^{2n}, \tag{8}$$

where L is a linear operator with coefficients chosen so that Padé properties can be introduced in the velocity profile. By combining (8) with the constant depth expression (5), they found that a 2% velocity error would be reached at $kh \simeq 12$ for an optimal choice of $\hat{z}/h = -0.5$ (see also Madsen & Agnon[45]).

Madsen et al.[52] tried to combine (8) with (6) and (7), and obtained a general recipe for deriving high-order velocity expressions valid for a rapid variation of $\hat{z}$. Unfortunately, these expressions turned out to be quite complicated, containing triple summations, and they decided to follow a more practical approach, which is summarized in the following section.

2.3. *The truncated velocity formulation satisfying the Laplace equation*

As a basic assumption we now assume that the expansion level $\hat{z}(x, y)$ varies slowly in space, while the sea bottom $h(x, y)$ is allowed to be rapidly varying. By inserting (8) into (6)–(7), while assuming a mild variation of $\hat{z}(x, y)$ we obtain the following expressions for the vertical distribution of the velocity field in terms of the utility-velocity variables $\hat{\mathbf{u}}^*$ and $\hat{w}^*$ taken at $\hat{z}$:

$$\mathbf{u}(x, y, z, t) = J_I\hat{\mathbf{u}}^* + J_{II}\hat{w}^*, \qquad w(x, y, z, t) = J_I\hat{w}^* - J_{II}\hat{\mathbf{u}}^*, \tag{9}$$

where

$$J_I = J_{01} + \nabla\hat{z}J_{11}, \qquad J_{II} = J_{02} + \nabla\hat{z}J_{12}, \tag{10}$$

and

$$J_{01} \equiv 1 + \left(-\frac{\psi^2}{2} + \frac{\hat{z}^2}{18}\right)\nabla^2 + \left(\frac{\psi^4}{24} - \frac{\hat{z}^2\psi^2}{36} + \frac{\hat{z}^4}{504}\right)\nabla^4,$$

$$J_{02} \equiv \psi\nabla + \left(-\frac{\psi^3}{6} + \frac{\hat{z}^2\psi}{18}\right)\nabla^3 + \left(\frac{\psi^5}{120} - \frac{\hat{z}^2\psi^3}{108} + \frac{\hat{z}^4\psi}{504}\right)\nabla^5,$$

$$J_{11} \equiv \psi\nabla + \left(-\frac{\psi^3}{2} - \frac{\hat{z}\psi^2}{9} + \frac{\hat{z}^2\psi}{18}\right)\nabla^3 + \left(\frac{\psi^5}{24} + \frac{\hat{z}\psi^4}{54} - \frac{\hat{z}^2\psi^3}{36} - \frac{\hat{z}^3\psi^2}{126} + \frac{\hat{z}^4\psi}{504}\right)\nabla^5,$$

$$J_{12} \equiv \left(\psi^2 + \frac{\hat{z}\psi}{9}\right)\nabla^2 + \left(-\frac{\psi^4}{6} - \frac{\hat{z}\psi^3}{18} + \frac{\hat{z}^2\psi^2}{18} + \frac{\hat{z}^3\psi}{126}\right)\nabla^4,$$

with $\psi \equiv z - \hat{z}$.

2.4. *The kinematic bottom condition*

The kinematic bottom condition is defined by $w_b + \nabla h \cdot \mathbf{u}_b = 0$, where $\mathbf{u}_b$ is the horizontal velocity vector and w_b the vertical velocity component, both taken at $z = -h$. By utilizing the velocity formulation (9)-(10) we get

$$\begin{aligned}
&\left(1 + \left(\frac{\hat{z}^2}{18} - \frac{\psi_b^2}{2}\right)\nabla^2 + \left(\frac{\hat{z}^4}{504} - \frac{\hat{z}^2\psi_b^2}{36} + \frac{\psi_b^4}{24}\right)\nabla^4\right)\hat{w}^* - \\
&\left(\psi_b\nabla + \left(\frac{\hat{z}^2\psi_b}{18} - \frac{\psi_b^3}{6}\right)\nabla^3 + \left(\frac{\hat{z}^4\psi_b}{504} - \frac{\hat{z}^2\psi_b^3}{108} + \frac{\psi_b^5}{120}\right)\nabla^5\right)\hat{\mathbf{u}}^* + \\
\nabla h\cdot\ &\left[\left(1 + \left(\frac{\hat{z}^2}{18} - \frac{\psi_b^2}{2}\right)\nabla^2 + \left(\frac{\hat{z}^4}{504} - \frac{\hat{z}^2\psi_b^2}{36} + \frac{\psi_b^4}{24}\right)\nabla^4\right)\hat{\mathbf{u}}^* + \right.\\
&\left.\left(\psi_b\nabla + \left(\frac{\hat{z}^2\psi_b}{18} - \frac{\psi_b^3}{6}\right)\nabla^3 + \left(\frac{\hat{z}^4\psi_b}{504} - \frac{\hat{z}^2\psi_b^3}{108} + \frac{\psi_b^5}{120}\right)\nabla^5\right)\hat{w}^*\right] + \\
\nabla\hat{z}\cdot\ &\left[\left(\psi_b\nabla + \beta_{13}\left(\frac{\hat{z}^2\psi_b}{18} - \frac{\hat{z}\psi_b^2}{9} - \frac{\psi_b^3}{2}\right)\nabla^3 + \right.\right.\\
&\left.\beta_{15}\left(\frac{\hat{z}^4\psi_b}{504} - \frac{\hat{z}^3\psi_b^2}{126} - \frac{\hat{z}^2\psi_b^3}{36} + \frac{\hat{z}\psi_b^4}{54} + \frac{\psi_b^5}{24}\right)\nabla^5\right)\hat{w}^* - \\
&\left.\left(\beta_{12}\left(\frac{\hat{z}\psi_b}{9} + \psi_b^2\right)\nabla^2 + \beta_{14}\left(\frac{\hat{z}^3\psi_b}{126} + \frac{\hat{z}^2\psi_b^2}{18} - \frac{\hat{z}\psi_b^3}{18} - \frac{\psi_b^4}{6}\right)\nabla^4\right)\hat{\mathbf{u}}^*\right] \\
&= 0, \qquad (11)
\end{aligned}$$

where $\psi_b = -(h + \hat{z})$. Notice that the default values of $(\beta_{12}, \beta_{13}, \beta_{14}, \beta_{15})$ are unity, while optimized linear shoaling properties can be obtained for the set

$$\beta_{12} = 0.95583, \quad \beta_{13} = 0.72885, \quad \beta_{14} = 0.51637, \quad \beta_{15} = 0.28478,$$

which provide excellent linear shoaling properties for $kh < 30$.

In contrast to Madsen et al.,[49,50] we have left $\hat{z}$ as an independent function, which is not necessarily a constant fraction of the still water depth h. This provides the flexibility to allow for a rapidly varying bottom combined with a slowly varying $\hat{z}$. The mild-slope formulation by Madsen et al.[49,50] is obtained by neglecting the $O(\nabla\hat{z})$ terms and using $\hat{z} = -0.5h$. We note that Madsen et al.[52] also included terms of order $(\nabla\hat{z})^2$ and $\nabla^2\hat{z}$. As we rarely apply these terms, they are omitted here.

2.5. *The exact nonlinear time-stepping equations*

The remaining step is to express the nonlinear kinematic and dynamic surface conditions in terms of velocity variables defined directly on the free surface and we obtain

$$\frac{\partial \eta}{\partial t} - \widetilde{w} + \nabla \eta \cdot \widetilde{\mathbf{u}} = 0, \tag{12}$$

$$\frac{\partial \widetilde{\mathbf{V}}}{\partial t} + g\nabla \eta + \frac{1}{2}\nabla \left(\widetilde{\mathbf{V}} \cdot \widetilde{\mathbf{V}} - \widetilde{w}^2 \left(1 + \nabla \eta \cdot \nabla \eta\right)\right) = 0, \tag{13}$$

where

$$\widetilde{\mathbf{V}} \equiv \widetilde{\mathbf{u}} - \widetilde{w}\nabla \eta. \tag{14}$$

Here $\widetilde{\mathbf{u}}$ is the horizontal velocity vector and $\widetilde{w}$ the vertical velocity component, both determined at the free surface defined by $z = \eta(x, y, t)$, while g is the gravitational acceleration (9 81 m/s^2). This form of the surface conditions was also utilized by Agnon et al.,[41] and Madsen et al.[49,50,52]

The connection between $(\widetilde{\mathbf{u}}, \widetilde{w})$ and $(\hat{\mathbf{u}}^*, \hat{w}^*)$ is determined by setting $z = \eta$ in (9)–(10), leading to the implicit relation

$$\begin{pmatrix} \widetilde{\mathbf{u}} \\ \widetilde{w} \end{pmatrix} = \begin{pmatrix} J_I & J_{II} \\ -J_{II} & J_I \end{pmatrix} \begin{pmatrix} \hat{\mathbf{u}}^* \\ \hat{w}^* \end{pmatrix}.$$

This expression is inverted numerically for each time step.

2.6. *The numerical solution procedure*

Our standard numerical solution procedure is based on finite difference discretizations on an equidistant grid, and an explicit four-stage fourth-order Runge-Kutta scheme is used for the time integration. In one horizontal dimension a stencil size of seven points is used in order to apply up to fifth-derivative operators and all derivatives make use of the entire stencil. A detailed description of the scheme can be found in Madsen et al.[49] for one horizontal dimension, and in Fuhrman & Bingham[82] for two horizontal dimensions. As shown by Fuhrman et al.,[83] the discretization satisfies standard linear stability criteria valid on a constant depth. However, for nonlinear waves and on an uneven bottom, it becomes necessary to apply minor amounts of smoothing in order to maintain numerical stability. We generally apply high-order Savitzky & Golay[84]-type smoothing filters at various intervals, which has minimal impact on modes of physical interest (see e.g. the Fourier analysis provided in the Appendix of Fuhrman et al.[85]).

The finite difference code has been applied for a number of applications, some of which will be discussed in the next section.

Recently, we have also developed a code based on the discontinuous Galerkin finite-element method (DG-FEM). The continuous equations are discretized using nodal polynomial basis functions of arbitrary order in space on each element of an unstructured computational domain. A detailed description of this scheme can be found in Engsig-Karup et al.[86]

3. Nonlinear Wave Phenomena in Deep and Shallow Water

Numerical models based on the high-order formulations by Madsen et al.[49,52] have proven useful for the study of numerous wave phenomena in deep and shallow water. These include the kinematics of progressive highly nonlinear solitary waves (Madsen et al.[49]), pressure forces on a vertical wall exposed to highly nonlinear solitary waves (Madsen et al.[49]), nonlinear shoaling on a mildly sloping beach (Madsen et al.[49]), non-linear refraction and diffraction (Fuhrman & Bingham[82]), side-band instabilities (Class I) of unidirectional and short-crested nonlinear wave trains (Madsen et al.,[49] Fuhrman, Madsen & Bingham[85]), crescent wave instabilities (Class II) of nonlinear wave trains (Fuhrman, Madsen & Bingham[87]), monochromatic short-crested waves (Fuhrman & Madsen[88]), extreme waves from focused directionally spread wave fields (Fuhrman & Madsen[89]), nonlinear wave-structure interactions (Fuhrman, Bingham & Madsen,[90] Jamois et al.[54]), Bragg scattering (Class I, II and III) over an undular sea bottom (Madsen et al.[52]); reflection/transmission of waves over steep trenches (Madsen et al.[52]), and run-up on beaches of tsunamis and long waves (Madsen & Fuhrman,[91,92] Fuhrman & Madsen[93,94]).

In this section, we will present highlights from a number of the papers mentioned above. The cases have been chosen to demonstrate the versatility and accuracy of this approach on a wide range of physically challenging problems. In what follows, a brief description of the various problems considered is provided, however, for full details the interested reader should consult the original references.

3.1. *Kinematics in highly nonlinear progressive deep-water waves*

As a first example, we present a spectral solution to the Boussinesq formulation for highly nonlinear progressive waves in deep water, and compare it with the stream function solution for the exact problem by Fenton.[95] For this purpose, we first expand the surface elevation and the utility velocity variables in terms of the Fourier series

$$\eta(x) = \sum_{j=1}^{M} A_j \cos(jkx),$$

$$\hat{u}^*(x) = \overline{u} + \sum_{j=1}^{M} B_j \cos(jkx), \qquad \hat{w}^*(x) = \sum_{j=1}^{M} C_j \sin(jkx),$$

insert them in the Boussinesq velocity expansion and determine the corresponding expressions for the surface velocities $\widetilde{u}$ and $\widetilde{w}$. We note that the kinematic bottom boundary condition leads to the constraint

$$C_j = \lambda_j \left(\frac{1 + \frac{1}{9}\lambda_j^2 + \frac{1}{945}\lambda_j^4}{1 + \frac{4}{9}\lambda_j^2 + \frac{1}{63}\lambda_j^4} \right) B_j, \qquad \lambda_j \equiv jk(h + \hat{z}).$$

Next, the exact dynamic surface condition is applied at $M+1$ equally spaced points from the wave trough to the wave crest, while the exact kinematic surface condition is applied at M staggered points (midway between the others). Given the four inputs H (wave height), h (water depth), L (wavelength), and $\overline{u_E}$ (the mean Eulerian velocity), there are three additional kinematic constraints:

$$H = \eta(0) - \eta(L/2), \qquad c = \frac{L}{T}, \qquad c + \overline{u} - \overline{u_E} = 0.$$

These add up to $2M + 4$ nonlinear equations for the unknowns T (wave period), c (wave celerity), $\overline{u}$ (time-averaged velocity), R (Bernoulli constant), and the coefficients A_j and B_j. The system is solved by Newton's method with linear theory as the initial conditions.

Figure 1 shows the results for the case of highly nonlinear deep-water waves with $H/L = 0.135$ and $kh = 10$. The spatial variation of the horizontal and vertical surface velocities normalized by the wave celerity is shown in figure 1*a*. The results are not distinguishable from the exact stream-function solution. Figure 1*b* shows the vertical distribution of the horizontal velocity under the wave crest. From the free surface down to approximately $z/h = -0.1$ the agreement with the stream-function theory is excellent. Below this level and to the sea bed deviations of the Boussinesq profile given by (9) and (10) start to show up for this very extreme case. Typically deviations will appear as oscillations about the true solution in contrast to lower-order formulations which deviate by a one-sided offset. Further details can be found in Madsen et al.[50]

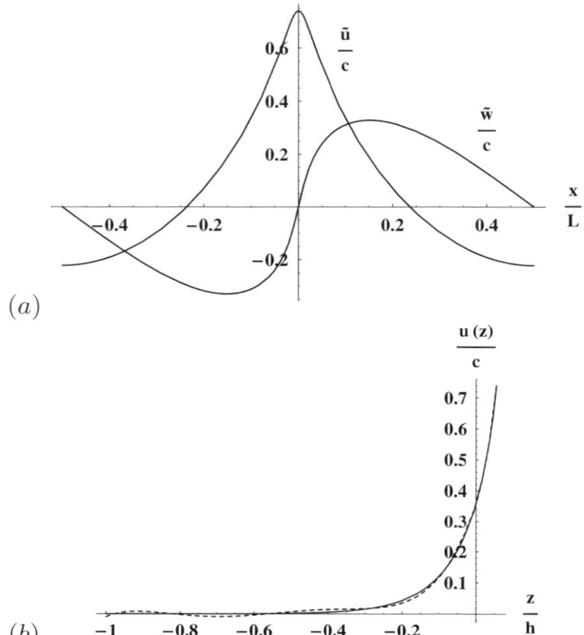

Fig. 1. Steady deep water wave with $kh = 10$ and $H/L = 0.135$. (*a*) The spatial variation of the horizontal and vertical surface velocities. (*b*) The vertical distribution of the horizontal velocity under the wave crest. Full line: exact stream-function solution; dashed line: Boussinesq formulation with $\hat{z} = -0.5h$.

3.2. *Kinematics in progressive solitary waves*

Our second example deals with the kinematics of nonlinear solitary waves propagating on a fluid at rest over a horizontal bottom. We consider a constant water depth $h = 1.0$ m, a grid size of 0.05 m and a time step of 0.025 s. The boundary conditions at the end of the model domain are determined by computing the surface elevation and the surface velocity directly from the solution by Tanaka.[96] Any discrepancies between the exact input solution and the numerical Boussinesq solution will show up as a dispersive tail which is left behind. At the present resolution, the initial conditions by Tanaka,[96] turn out to be steady up to a steepness of $H/h = 0.75$, after which some tail begins to appear. Figure 2 shows the vertical distribution of the horizontal velocity for the case of $H/h = 0.65$, which is seen to be in almost perfect agreement with the profile of Williams.[97] This result is superior to what was achieved by Gobbi et al.[46] Figure 3 shows

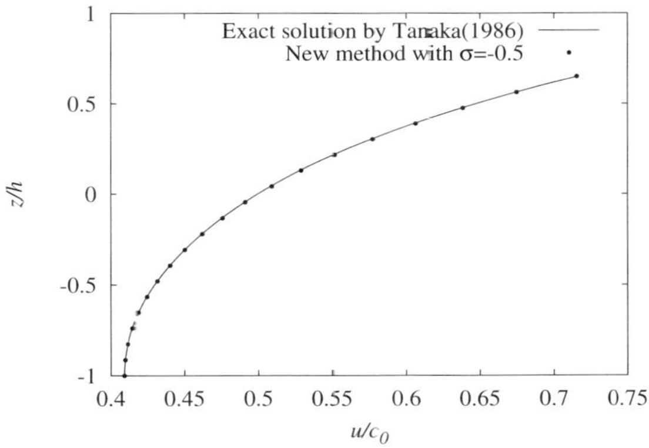

Fig. 2. The vertical profile of the horizontal velocity under the crest of a solitary wave. Computed with $\hat{z} = -0.5h$. Specifications: $h = 1.0$ m, $H/h = 0.65$, $c/c_0 = 2.653$, $\Delta x = 0.05$ m, $\Delta t = 0.025$ s.

the wave celerity and crest velocity for a number of solitary waves, varying the relative amplitudes H/h from 0.1 to 0.83. Again an excellent agreement with Williams' solution is obtained except for the highest waves.

Inaccuracies do occur for relative amplitudes H/h higher than 0.78 and the problem is most likely connected to the sharpening of the wave crest, which for the highest solitary waves rapidly approaches the limiting angle of 120°. This makes it difficult to resolve the finest details with a finite-difference formulation and it is expected that numerical inaccuracies are introduced by the successive use of fifth-order spatial derivatives, a problem which was also recognized by Gobbi et al.[46] Further details can be found in Madsen et al.[49]

3.3. *Reflection of solitary waves from a vertical wall*

Our third example deals with nonlinear reflection of high-amplitude solitary waves from a vertical wall. This problem has previously been studied by e.g. Grilli & Svendsen,[98] Svendsen & Grilli[99] and Cooker et al.[100] Figure 4 shows a number of snapshots of the computed surface elevation for a solitary wave with $H/h = 0.7$ moving from right to left. We notice that the maximum run-up is significantly larger than twice the incoming amplitude. Figure 5 shows the instantaneous wall force determined by integrating the pressure distribution over the wall from the sea bed to the free surface. The pressure distribution is obtained from integrating the vertical Euler

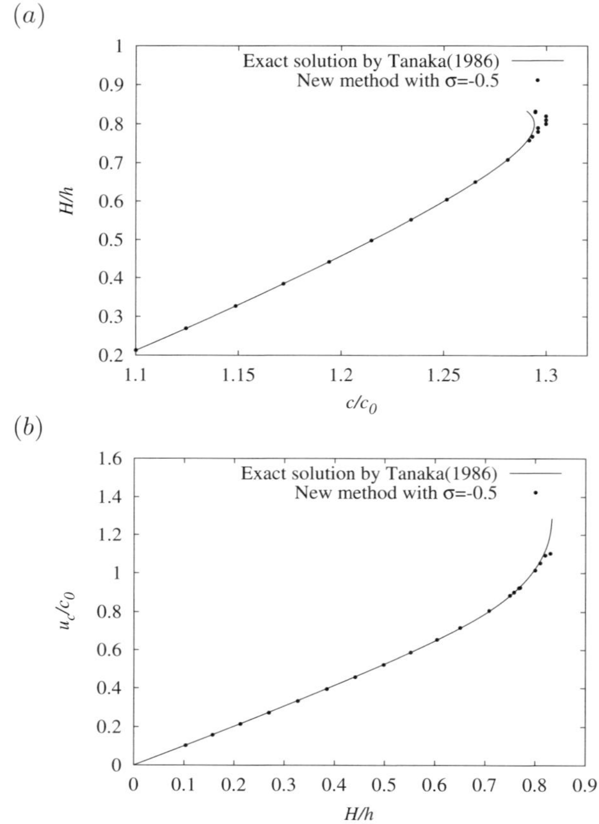

Fig. 3. The (*a*) wave celerity and (*b*) crest velocity versus wave amplitude for solitary waves. Computed with $\hat{z} = -0.5h$.

equation from the free surface to an arbitrary z-datum. The computed force is shown as a function of time relative to t_0 (time of maximum run-up) for selected incoming values of H/h. Except for the highest waves, our computations agree very well with the boundary integral method of Cooker et al.,[100] shown as dots. The overall trend is that for H/h up to 0.3 the wall force is single peaked, while it becomes double peaked for higher waves. The reason is that the force is dominated by hydrostatic pressure for smaller waves, while vertical accelerations come into play for larger waves. This increases the pressure above the hydrostatic value and a local maximum occurs before the maximum run-up. During the phase of maximum run-up, the acceleration forces counteract the hydrostatic forces leading to a local minimum. After the maximum run-up, the water slides back down the wall and the deceleration again increases the pressure above the hydrostatic value. More details can be found in Madsen et al.[49]

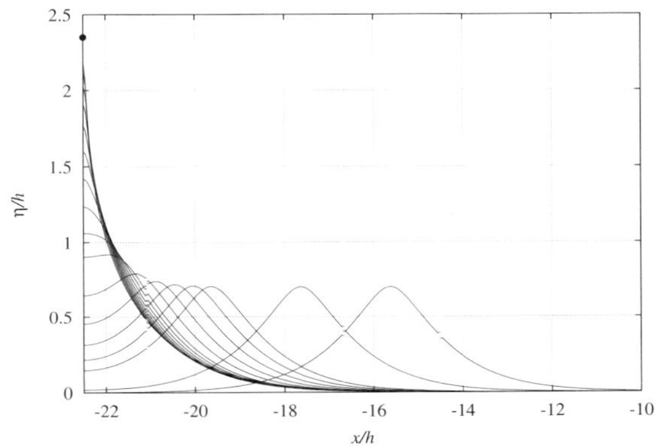

Fig. 4. Reflection of a solitary wave from a vertical wall. Snapshots of the surface elevation moving from right to left. Computed with $\hat{z} = -0.5h$. Dot: maximum run-up as computed by Cooker et al.[100] Specifications: $h = 1.0m$, $H/h = 0.70$, $\Delta x = 0.05$ m, $\Delta t = 0.025$ s.

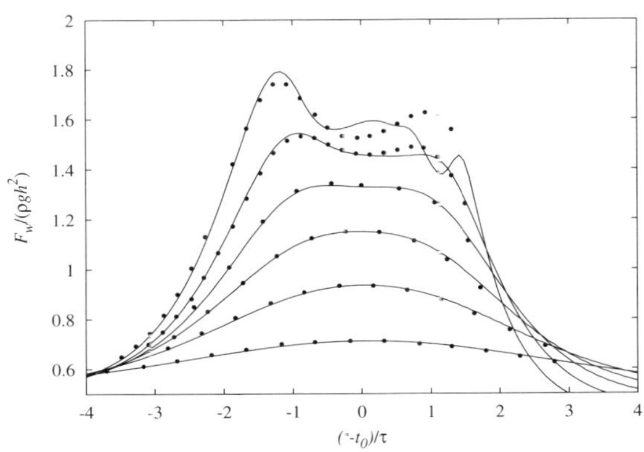

Fig. 5. Reflection of a solitary wave from a vertical wall. Time variation of depth-integrated wall force computed for a range of different incoming wave amplitudes ($H/h = 0.1, 0.2, 0.3, 0.4, 0.5, 0.6$). Full line: Boussinesq results with $\hat{z} = -0.5h$; symbols: computed by Cooker et al.[100]

3.4. *Nonlinear wave-structure interactions*

The present model (in two horizontal dimensions) also allows for the inclusion of piecewise-rectangular, bottom-mounted, surface-piercing structures, as described by Fuhrman et al.[90] As a demonstration, we will here consider a simulation involving highly-nonlinear wave run-up on a vertical bottom-mounted plate, observed experimentally by Molin et al.[101] Analysis therein also explains that a refraction-type phenomenon occurs due to celerity modification from third-order interactions between the incident and reflected wavefields, resulting in focused wave elevations in front of the plate significantly exceeding those expected from linear theory. The model setup involves a 1.2 m plate extended perpendicularly from the bottom (in plan) sidewall at a distance of 19.3 m from the wavemaker. We use plane incident waves with period $T = 0.88$ s and waveheight $H = 0.058$ m on a flat depth $h = 0.6$ m. A stream function solution[95] then gives the wavelength $L = 1.223$ m ($H/L = 0.0474$, $kh \approx \pi$ i.e. deep water).

A computed free surface in the vicinity of the plate is shown in figure 6*a*, while a comparison between computed and measured free surface envelopes in front of the plate is provided in figure 6*b*. As can be seen in figure 6*a*, the simulation results in extremely steep nearly standing waves in front of the plate. While the incident steepness is moderate $H/L = 0.0474$, a distance of roughly half a wavelength in front of the plate the local wave steepness in fact reaches $H/L = 0.203$, which is remarkably close to the theoretical standing wave limit $H/L = 0.204$ (taking the lower estimate of Schwartz & Whitney[102]). In terms of the maximum surface elevation this corresponds to an amplification factor of 5.26, whereas an otherwise identical linear simulation yields a factor of just 2.37, clearly demonstrating the importance of nonlinear effects in this problem. Even in these rather extreme physical circumstances, the model can be seen to maintain excellent quantitative accuracy, as demonstrated in figure 6*b*. For additional simulations of this phenomena see Fuhrman et al.,[90] as well as Jamois et al.[54]

3.5. *Deep water class I and II instabilities*

We shall now consider wave-wave interactions and instabilities occurring in deep water. Specifically, we will consider numerical simulations of both class I (i.e. those involving quartet interactions) and II (quintet interactions) wave instabilities.

The high-order Boussinesq model has been used to simulate class I (Benjamin-Feir type) instabilities by Madsen et al.[49] and Fuhrman et al.,[85]

(*a*)

(*b*)

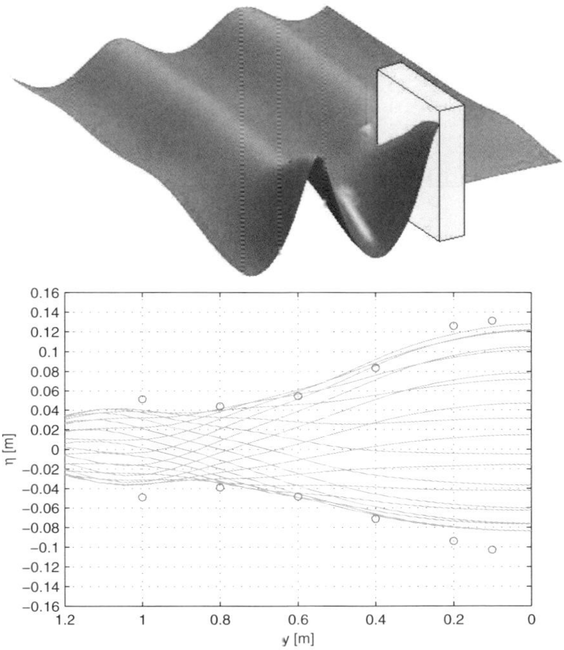

Fig. 6. (*a*) Computed free surface around the plate and (*b*) computed (lines) and measured (circles) free surface envelopes in front of the plate.

the latter study focusing on the numerical simulation of three-dimensional short-crested wave instabilities. Figure 7 shows a comparison of the evolution of the primary and side-band frequency amplitudes within a computed (two-dimensional) class I plane wave instability and a (three-dimensional) class Ia[103] short-crested wave instability, where the carrier short-crested waves are formed by two progressive waves at a (nearly grazing) incident angle $\theta = 80°$. Both cases have an incident carrier wave steepness of $ak = 0.10$, depth $kh = 2\pi$, and use a relative perturbation strength for the unstable modes of $\epsilon = 0.02$. Figure 8 additionally shows free surface segments from the computed short-crested simulation at the beginning of the evolution, as well as at a highly-modulated state. As can be seen, the class I simulation (figure 7*a*) results in a well-known recurrence phenomenon, with the side-bands initially growing exponentially, before shrinking back to a much smaller level. Interestingly, however, in the three-dimensional short-crested wave simulation (figure 7*b*), after the initial growth the lower

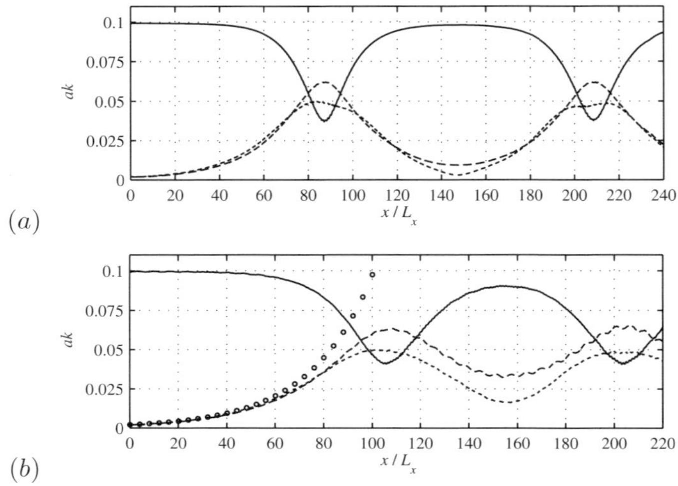

Fig. 7. Computed evolution of the primary frequency (full line) and lower/upper side-bands (long/short dashed lines) from simulations (with $ak = 0.10$) of (*a*) class I plane wave instability and (*b*) a three-dimensional class Ia short-crested wave instability. The circles in (*b*) show the theoretical exponential growth of the side-bands.

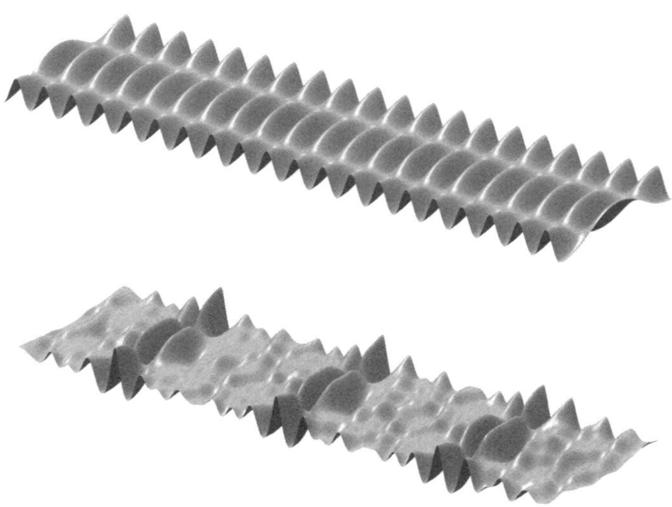

Fig. 8. Segments of the computed free surface from the simulation of the class Ia short-crested wave instability with $ak = 0.10$ and $\theta = 80^\circ$ covering (top) $0 \leq x \leq 20L_x$ and (bottom) $102L_x \leq x \leq 122L_x$. The vertical scale is exaggerated 30 times.

side-band maintains an amplitude significantly above that of the upper, demonstrating qualitative differences between the two- and three-dimensional scenarios. This tendency can become even more pronounced when the nonlinearity is further increased, as demonstrated by Fuhrman et al.,[85] and for full details the interested reader is referred to this work.

We also here note that the model has recently been utilized to shed light on the physical generation of steady monochromatic short-crested waves in deep water by Fuhrman & Madsen,[88] explaining a number of features observed experimentally by Hammack et al.,[104] as confirmed by the further experiments of Henderson et al.[105]

We now turn our attention to the simulation of spectacular crescent (or horseshoe) wave patterns, arising from the class II (three-dimensional) plane wave instability of McLean.[106] The class II plane wave instability dominates the class I instability for waves with steepness $H/L > 0.10$, thus models capable of simulating very steep waves (essentially up to the breaking point) are necessary for the study of this phenomena at physically relevant conditions. For full details of the present study see Fuhrman et al.[87] Crescent waves are generated in the model by superimposing three dimensional perturbations over initially steep stream function solutions for plane progressive waves, exciting a quintet resonant interaction. An example of a computed phase-locked crescent wave pattern is shown in figure 9, where the incident waves have $H/L = 0.105$ and $kh = 2\pi$. The free surface shown in figure 9 shows the final $3L$ from the computational domain covering a total of roughly 12 incident wavelengths. Just after the instant shown, the simulation breaks down, presumably due to wave breaking. From this figure many of the distinguishable features described in the experiments of Sue et al.[107,108] can clearly be seen: The waves have noticeable front-back asymmetry, the crests are shifted by one-half the width of the crescents on successive rows (i.e. the classical so-called L_2 pattern), deep troughs appear in front of the crescent face, and flattened troughs are evident directly behind the crests. Table 1 also shows computed and measured[108] characteristic ratios along the crescent center line, as defined in figure 10. The computed values compare well both with measurements, as well as with similar comparisons by fully nonlinear models in Xue et al.[109] and Fructus et al.,[110] generally confirming the quantitative accuracy of the Boussinesq model in these highly-nonlinear circumstances.

We also note that Collard & Caulliez[111] have observed non phase-locked (so-called oscillating) crescent wave patterns in physical experiments. Such patterns have also been simulated with the Boussinesq model in Fuhrman

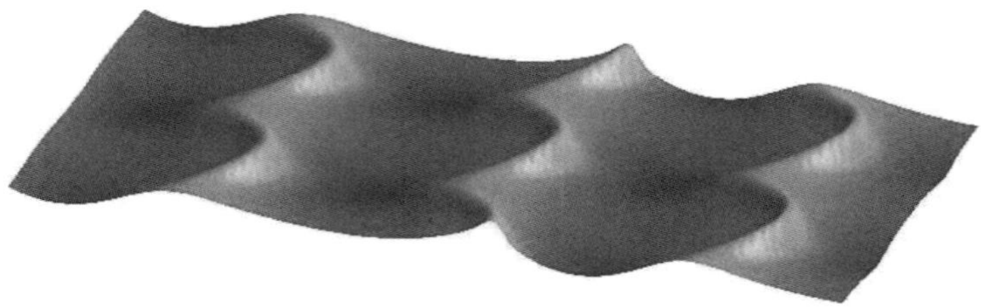

Fig. 9. Computed free surface (to scale) for phase-locked L_2 crescent waves with $H/L = 0.105$ and $\epsilon = 0.05$.

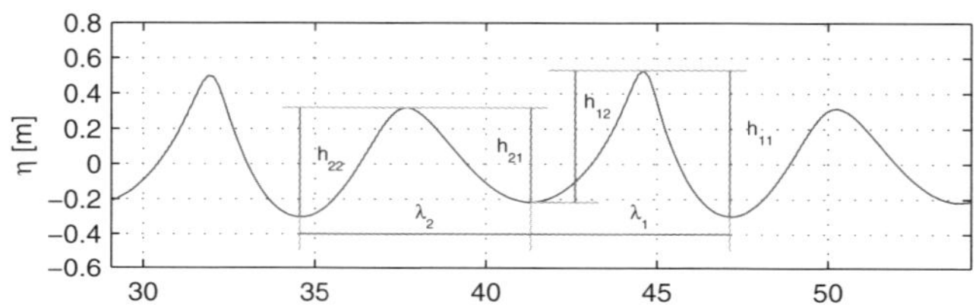

Fig. 10. Computed free surface elevations at $t = 10T$ with $H/L = 0.105$ and $\epsilon = 0.05$ along the crescent centerline $y = L_y/2$. Here characteristic lengths are also defined.

Table 1. Characteristic ratios for measured[108] and computed phase-locked L_2 crescent wave patterns with $H/L = 0.105$ (at full-period intervals), for two values of the relative initial perturbation strength ϵ. Here s_{max} refers to the maximum surface slope.

		$\epsilon = 0.05$				$\epsilon = 0.16$			
	Su[108]	$t/T = 8$	9	10	11	4	5	6	7
λ_2/λ_1	1.28	1.16	1.16	1.22	1.29	1.22	1.26	1.16	1.22
h_{11}/h_{12}	1.10	1.11	1.11	1.18	1.19	1.13	1.23	1.15	1.30
h_{21}/h_{22}	0.88	0.89	0.86	0.78	0.74	0.86	0.77	0.81	0.67
h_{11}/h_{21}	1.66	1.36	1.55	1.80	2.05	1.44	1.62	1.79	2.11
s_{max}	0.65, 1.02	0.51	0.66	0.79	1.42	0.62	0.59	0.83	0.92

et al.,[87] and further details on their possible selection in experimental wave-tanks can be found in Fuhrman & Madsen.[112]

3.6. *Extreme events from focused directionally spread wavefields*

As another demonstration of the fully-nonlinear deep-water capabilities of the present high-order Boussinesq approach, we will consider the numerical simulation of extreme wave events created from the focusing of

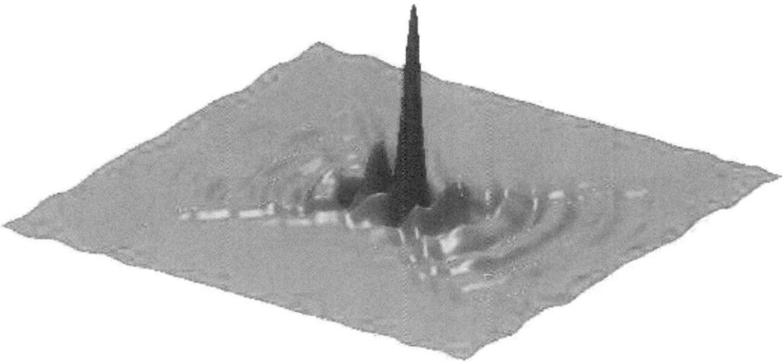

Fig. 11. Computed free surface near the time of focusing for the extreme wave simulation (case D93s4).

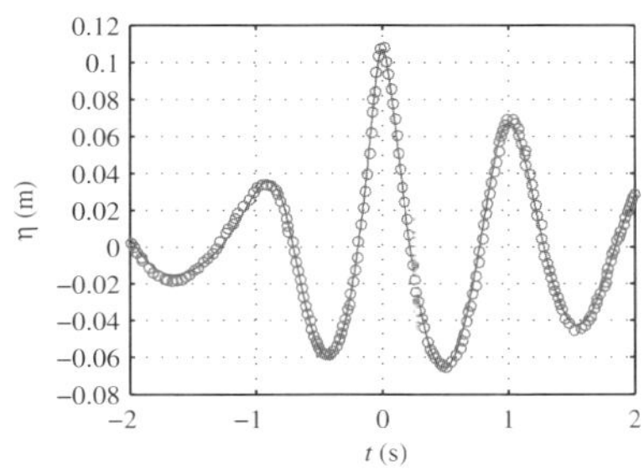

(a)

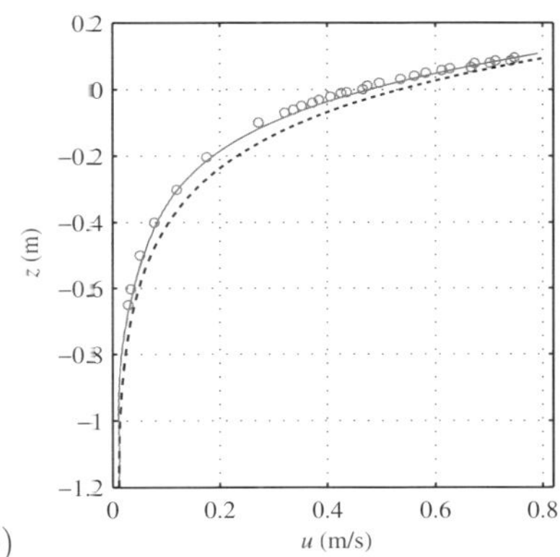

(b)

Fig. 12. Comparison of computed (full line), measured (circles), and linear (dotted line) (a) free surface time series and (b) horizontal velocity distribution at the time of focusing for the extreme wave simulation.

directionally spread wavefields. Such events have been studied experimentally by Johannessen & Swan,[113] who conducted a series of 84 experiments in a three-dimensional wave basin, where the frequency-amplitude spectrum, the directional-amplitude spectrum, and the input amplitude sum were all systematically varied. They considered three different frequency spectra, where the amplitude a was varied with the frequency f in each case according to $a \propto f^{-2}$, scaled such that their linear amplitude sum A is given by $A = \sum_{n=1}^{N} a_n$, with N being the total number of frequency components.

As an example, we will here consider a single case from their narrow-banded experiments, covering the frequency range $53/64 \leq f \leq 80/64$ Hz. The experiments were performed on a depth $h = 1.2$ m, which (from the linear dispersion relation) means that these cases respectively cover the range of dimensionless depths $3.32 \leq kh \leq 7.55$. In particular, we will consider their case D93s4, having linear amplitude sum $A = 93$ mm, and directional spreading parameter $s = 4$. This represents the most three-dimensional case considered, while also containing the largest amplitude waves. For the simulation we specify an initial condition corresponding to the previous experimental description, where the relative phase of each wave component is determined from linear theory, such that all the wave crests will superimpose at a particular point in space and time. The free surface near the time of nonlinear focusing is presented in figure 11, clearly showing an extreme focusing event. To validate the accuracy of the simulation, a comparison of the computed free surface time series and velocity kinematics underneath the focusing event are presented in figure 12, generally showing excellent accuracy. Fuhrman & Madsen[89] have made further comparisons with the present model against both their additional narrow-banded and as well as their broad-banded experiments (containing waves up to non-dimensional depth $kh = 13.25$), demonstrating similar accuracy as shown here. The results are comparable to those achieved previously using a pseudo-spectral approach by Bateman et al.,[114,115] generally confirming the claimed kinematic accuracy of the present Boussinesq method for highly-nonlinear waves in very deep water. Other recent numerical solutions to this problem can be found in Fochesato et al.,[116] who used a fully nonlinear potential flow solver.

3.7. *Bragg scattering from an undular sea bed*

As an example involving finite water depths, we now consider the Bragg scattering of waves from an undular (sinusoidal) sea bed. We have previously considered the simulation of so-called class I, II, and III Bragg scattering, and for full details the reader should consult Madsen et al.[52] For brevity, we will here only present class I and III results.

Class I Bragg resonance defines the second-order triad interaction involving a single bottom wavenumber $\mathbf{K}$ (with angular frequency $\Omega = 0$)

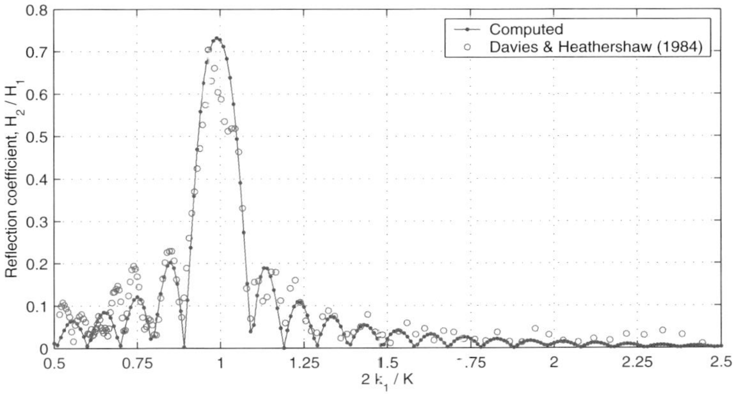

Fig. 13. Comparison of the computed Boussinesq results for class I Bragg scattering against the experimental measurements of Davies & Heathershaw.[117]

and two surface components having wavenumbers $\mathbf{k_1}$ (the incoming wave) and $\mathbf{k_2}$ (the reflected wave). A class I Bragg resonance occurs when

$$\mathbf{k}_1 \pm \mathbf{k}_2 \pm \mathbf{K} = 0, \qquad \omega_1 \pm \omega_2 = 0, \tag{15}$$

where ω_i is the angular frequency corresponding to surface wavenumber $\mathbf{k}_i$. The simplest case involves normal incidence (i.e. a single horizontal dimension), where $k_1 = -k_2 = K/2$ and $\omega_1 = \omega_2$, i.e. the reflected wave (subscript 2) has the same wavenumber and frequency as the incoming wave (subscript 1). To simulate this phenomenon we consider the most demanding case from the experiments of Davies & Heathershaw,[117] where surface waves interact with a 10-ripple bottom patch with $Kd = 0.31$ and $d/h = 0.16$, where d is the ripple amplitude. We determine the reflection coefficients for a range of wave periods corresponding to the interval $0.5 \leq 2k_1/K \leq 2.5$. The results are shown in figure 13, where a very good agreement between our computations and their measurements is observed. The peak reflection occurs near the theoretical point of resonance at $2k_1/K = 1$.

Class III Bragg resonance occurs when nonlinear surface waves interact with an undular sea bottom, as first discussed by Liu & Yue.[118] This class defines a third-order quartet wave-ripple interaction involving one bottom wavenumber and three surface wavenumbers. In this case the resonance condition reads

$$\mathbf{k}_1 \pm \mathbf{k}_2 \pm \mathbf{k}_3 \pm \mathbf{K} = 0, \qquad \omega_1 \pm \omega_2 \pm \omega_3 = 0. \tag{16}$$

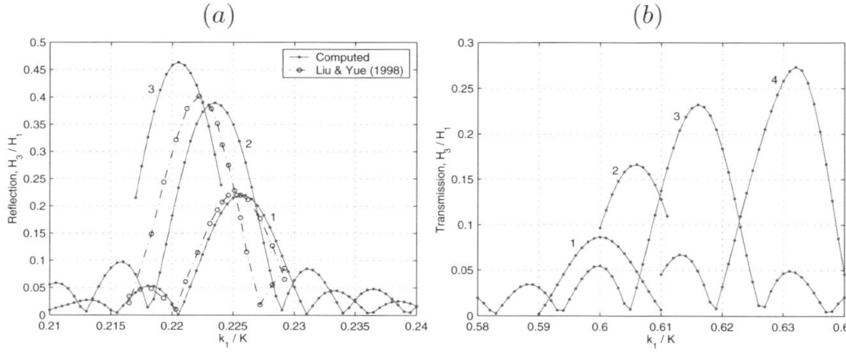

Fig. 14. Computed results for class III Bragg scattering. Figure (a) shows the subharmonic reflection with 1–3: $k_1H_1 = 0.0609$, 0.1272, 0.18. Figure (b) shows the superharmonic transmission with 1–4: $k_1H_1 = 0.06$, 0.12, 0.18, 0.24. The computed results of Liu & Yue[118] are also shown in figure (a) for comparison.

This condition can be most simply satisfied (in a single horizontal dimension) by accounting for the incident wave twice i.e. with $k_1 = k_2$, in which case (16) leads to a reflected subharmonic $k_3 = 2k_1 - K$ or a transmitted superharmonic $k_3 = 2k_1 + K$, both having frequency $\omega_3 = 2\omega_1$. As an example we consider the scenario of Liu & Yue,[118] who used a patch of 36 sinusoidal bottom ripples with $Kd = 0.25$ on an otherwise flat bed, where $Kh = 2.642$. Using the linear dispersion relation as a first approximation for the free waves (subscript 1 and 3), this can be shown to predict a subharmonic reflection at $k_3/K = 0.546$ for an incident wavenumber $k_1/K = 0.227$, and a superharmonic transmission at $k_3/K = 2.195$ for an incident wavenumber $k_1/K = 0.598$. Computed results in the vicinity of the predicted resonance locations, for various incident wave steepness, are shown in figure 14 for both the reflected and transmitted components. For low steepness the location of the peak reflection/transmission compares well with that predicted from linear theory. For larger steepness, however, we observe an interesting downshift/upshift for the reflection/transmission case. These trends have been quantitatively explained by Madsen et al.[52] and Madsen & Fuhrman[119] by considering nonlinear amplitude dispersion effects between the interacting wavefields.

3.8. *Run-up of non-breaking solitary waves on a beach*

Recently, the high-order Boussinesq model has also been extended to allow moving wet-dry boundaries by Fuhrman & Madsen,[93] e.g. for the

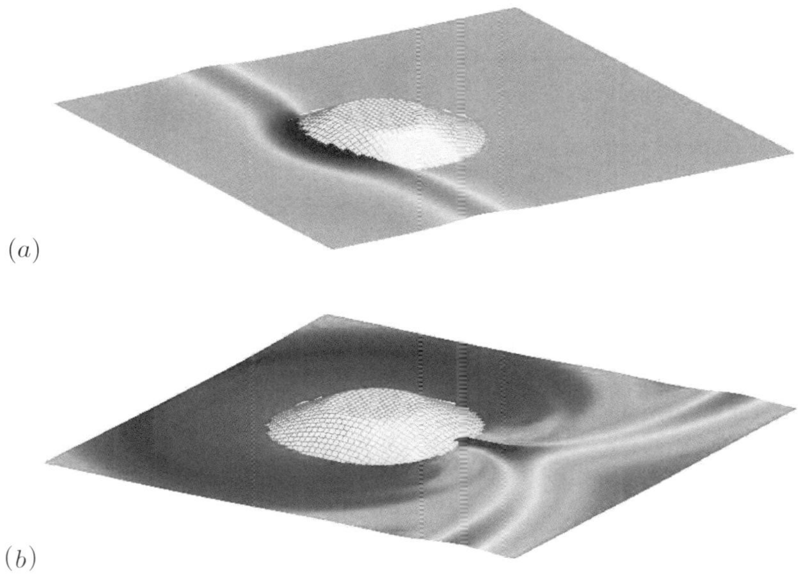

Fig. 15. Computed free surfaces near the time of maximum run-up at the (*a*) front ($t = 10.08$ s) and (*b*) back ($t = 13.68$ s) of the conical island, with $H/h = 0.18$. Both surfaces show the same spatial section in the vicinity of the island, but at different viewing angles. The vertical scale is exaggerated by a factor of 2.

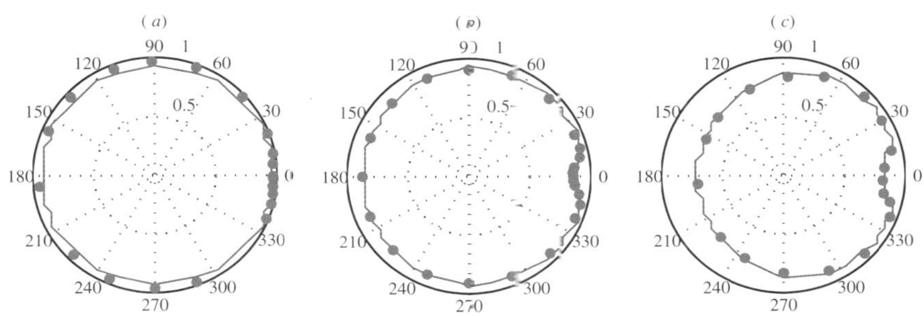

Fig. 16. Computed (line) and measured (dots) maximum run-up around the conical island with (*a*) $H/h = 0.045$, (*b*) 0.09, and (*c*) 0.18.

simulation of wave run-up on beaches. The implemented run-up algorithm is based on the extrapolating boundary approach as utilized previously by Lynett et al.[120] As a demonstration of this capability in two horizontal dimensions we present results from three cases involving solitary wave run-up on a circular conical island, which has been studied experimentally by Briggs et al.[121] and Liu et al.[122] We consider solitary waves with initial waveheight to depth ratio $H/h = 0.045$, 0.09, and 0.18 impacting a conical island (with base radius 3.60 m and side slope $s = 1/4$), where the depth away from the island is $h = 0.32$ m. Examples of computed free surfaces from the steepest case are shown in figure 15, demonstrating the initial impact on the island front face, as well as the eventual collision of edge waves circling the island on the back side. We note that dissipative wave breaking is neglected in this simulation, though weak breaking was experimentally reported in this case. A comparison of the maximum computed inundation around the island for all three cases is shown in figure 16, demonstrating excellent agreement with the experimental measurements, confirming the accuracy of the implemented run-up algorithm. A number of additional run-up simulations in both one and two horizontal dimensions can be found in Fuhrman & Madsen[93] and Madsen & Fuhrman,[91,92] which further confirm the accuracy of both free surface elevations, as well as horizontal velocities, within the swash zone.

3.9. *Tsunami generation from submerged landslides*

The high-order Boussinesq model has also recently been extended to allow wave generation from user specified bottom motions, with obvious applications in the simulation of earthquake- and landslide-induced tsunamis. This extension involves replacing the zero on the right-hand-side of (11) with a $-\partial h/\partial t$ term. As an example of this feature, we consider a numerical simulation based on the recent experimental work of Enet[123] and Enet & Grilli,[124] where a submerged mass with thickness 0.082 m, length 0.395 m, and width 0.680 m slides down an incline of 15° until reaching a constant offshore depth $h = 1.5$ m. The experiments were performed in a 3.7 m wide, 30 m long tank, and we here consider their case where the initially at-rest sliding mass has minimum submergence depth $d = 61$ mm, with initial acceleration 1.12 m/s^2, and terminal velocity 1.70 m/s. The resulting rigid landslide position is then prescribed according to its theoretical motion. At the nearshore beach the moving boundary algorithm of Fuhrman & Madsen[89] is again applied, allowing the generated waves to run up and

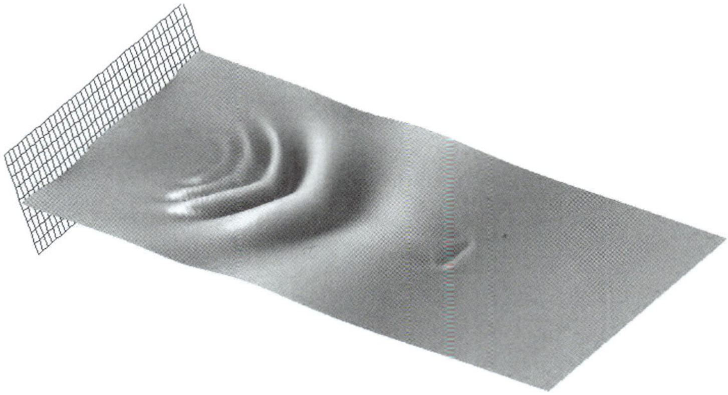

Fig. 17. Computed free surface for the three-dimensional submerged landslide case at $t = 3.5$ s. The vertical scale is exaggerated 10 times.

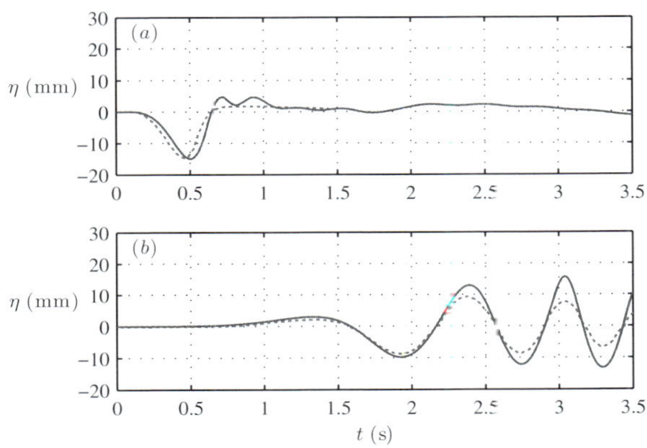

Fig. 18. Computed (full lines) and measured (dashed lines) results for the simulation of a submerged landslide at (a) gauge 1, $(x, y) = (x_0, 0)$, where $x_0 = 0.551$ m is the initial x-location of the sliding mass; and (b) gauge 4, $(x, y) = (1.929, 0.4988)$ m. Note that the origin here represents the still water shoreline, with the x axis pointing offshore.

down the beach. At the other three boundaries fully reflecting numerical walls are imposed. As an indication of computational efficiency, this case (discretized on a 401×41 grid, simulated for 801 time steps) requires 5.5 hr on a single 3.2 GHz Pentium 4 processor.

A computed free surface near the end of the simulation is shown in figure 17, demonstrating the resulting three-dimensional wavefield.

Additionally, comparisons between our computed free surface values and the measurements at two locations (one at the initial mass location, and one further offshore) are provided in figure 18, generally showing good agreement, though the waveheights offshore are somewhat over-predicted. Similarly good agreement with other cases from these experiments, as well as numerous other cases involving waves generated by moving bottoms can be found in Fuhrman & Madsen.[94]

Acknowledgments

We thank the Danish Center for Scientific Computing for providing the invaluable supercomputing time used in the numerical simulations. This work was financially supported by the Danish Technical Research Council (STVF grant no. 9801635). This support is greatly appreciated.

References

1. M. Dingemans, *Water wave propagation over uneven bottoms, Parts 1 & 2.* (World Scientific, Singapore, 1997).
2. J. T. Kirby, Nonlinear, dispersive long waves in water of variable depth. In ed. J. N. Hunt, *Gravity Waves in Water of Finite Depth*, pp. 55–125. Computational Mechanics Publications, (1997).
3. J. T. Kirby, Boussinesq models and applications to nearshore wave propagation, surfzone processes and wave-induced currents. In ed. V. C. Lakhan, *Advances in Coastal Modelling*, pp. 1–41. Elsevier, (2003).
4. P. A. Madsen and H. A. Schäffer, A review of Boussinesq-type equations for gravity waves. In ed. P. L.-F. Liu, *Advances in Coastal and Ocean Engineering*, vol. 5, pp. 1–95. World Scientific, (1999).
5. J. V. Boussinesq, Théorie des ondes et des remous qui se propagent le long d'un canal rectangulaire horizontal, en communiquant au liquide contenu dans ce canal des vitesses sensiblement pareilles de la surface au fond, *J. Math. Pures Appl.* **17**, 55–108, (1872).
6. D. J. Korteweg and F. de Vries, On the change of form of long waves advancing in a rectangular canal, and on a new type of long stationary waves, *Philos. Mag.* **39**, 422–443, (1895).
7. P. F. Serre, Contribution à l'étude des écoulements permanents et variables dans les canaux, *La Houille Blanche.* pp. 374–388; 830–872, (1953).
8. C. H. Su and C. S. Gardner, Korteweg-de Vries equation and generalizations. III. Derivation of the Korteweg-de Vries equation and Burgers equation, *J. Math. Phys.* **10**, 536–539, (1969).
9. C. C. Mei and B. Le Méhauté, Note on the equations of long waves over an uneven bottom, *J. Geophys. Res.* **71**, 393–400, (1966).
10. D. H. Peregrine, Long waves on a beach, *J. Fluid Mech.* **27**, 815–827, (1967).

11. O. S. Madsen and C. C. Mei, The transformation of a solitary wave over an uneven bottom, *J. Fluid Mech.* **39**, 781–791, (1969).
12. M. Dingemans, Water waves over an uneven bottom; a discussion of long-wave equations. Technical Report R729, part 2, Delft Hydraulics, Delft, The Netherlands, (1973).
13. P. A. Madsen and H. A. Schäffer, Higher order Boussinesq-type equations for surface gravity waves–derivation and analysis, *Phil. Trans. R. Soc. Lond. A.* **356**, 3123–3181, (1998).
14. T. B. Benjamin, J. L. Bona, and J. J. Mahony, Model equations for long waves in non-linear dispersive systems, *Phil. Trans. R. Soc. Lond. A.* **272**, 47–78, (1972).
15. C. C. Mei, *The Applied Dynamics of Ocean Surface Waves.* (World Scientific, Singapore, 1989).
16. M. B. Abbott, H. M. Petersen, and O. Skovgaard, On the numerical modelling of short waves in shallow water, *J. Hydr. Res.* **16**, 173–203, (1978).
17. M. B. Abbott, A. McCowan, and I. R. Warren, Numerical modelling of free-surface flows that are two-dimensional in plan. In *Transport models for Inland and Coastal Water Symp. on Predictive Ability.* Academic Press, New York, (1981).
18. M. B. Abbott, A. McCowan, and I. R. Warren, Accuracy of short-wave numerical models, *J. Hydr. Eng. ASCE.* **110**, 1287–1301, (1984).
19. A. D. McCowan, Numerical simulation of shallow water waves. In *Proc. 4th Aust. Conf. on Coastal and Ocean Engineering*, Adelaide, (1978).
20. A. D. McCowan, Developments in numerical short wave modelling. In *Proc. 5th Aust. Conf. on Coastal and Ocean Engineering*, Perth, (1981).
21. A. D. McCowan, Equation systems for modelling dispersive flow in shallow water. In *Proc. 21st IAHR Congress*, Melbourne, (1985).
22. A. D. McCowan, The range of application of Boussinesq type numerical short wave models. In *Proc. 22nd IAHR Congress*, Lausanne, (1987).
23. P. A. Madsen and I. R. Warren, Performance of a numerical short-wave model, *Coast. Eng.* **8**, 73–93, (1984).
24. H. Schaper and W. Zielke, A numerical solution of Boussinesq type wave equations. In *Proc. 19th Int. Conf. on Coastal Engineering*, pp. 1057–1071, (1984).
25. O. B. Rygg, Nonlinear refraction-diffraction of surface waves in intermediate and shallow water, *Coast. Eng.* **12**, 191–211, (1988).
26. M. B. Abbott and P. A. Madsen, Modelling of wave agitation in harbours. In *The Sea*, vol. 9, pp. 1067–1103. Wiley, New York, (1990).
27. H. H. Prüser and W. Zielke, Irregular waves on a current. In *Proc. 22nd Int. Conf. Coastal Engineering*, pp. 1088–1101, Delft, (1990).
28. J. M. Witting, A unified model for the evolution of nonlinear water waves, *J. Comput. Phys.* **56**, 203–236, (1984).
29. P. A. Madsen, R. Murray, and O. R. Sørensen, A new form of the Boussinesq equations with improved linear dispersion characteristics. Part 1., *Coast. Eng.* **15**, 371–388, (1991).

30. P. A. Madsen and O. R. Sørensen, A new form of the Boussinesq equations with improved linear dispersion characteristics. Part 2. A slowly-varying bathymetry, *Coast. Eng.* **18**, 183–204, (1992).
31. O. Nwogu, Alternative form of Boussinesq equations for nearshore wave propagation, *J. Waterw. Port C-ASCE.* **119**, 618–638, (1993).
32. A. Schröter, R. Mayerle, and W. Zielke, Optimized dispersion characteristics of the Boussinesq wave equations. In *Proc. Waves - Physical and Numerical Modelling*, pp. 416–425, Vancouver, B.C., Canada, (1995).
33. H. A. Schäffer and P. A. Madsen, Further enhancements of Boussinesq-type equations, *Coast. Eng.* **26**, 1–15, (1995).
34. P. A. Madsen and O. R. Sørensen, Bound waves and triad interactions in shallow water, *Ocean Eng.* **20**, 359–388, (1993).
35. G. B. Whitham, *Linear and Nonlinear Waves.* (Wiley-Interscience, 1974).
36. G. Wei and J. T. Kirby, Time dependent numerical code for extended Boussinesq equations, *J. Waterw. Port C-ASCE.* **121**, 251–261, (1995).
37. G. Wei, J. T. Kirby, S. T. Grilli, and R. Subramanya, A fully nonlinear Boussinesq model for surface waves. Part 1. Highly nonlinear unsteady waves, *J. Fluid Mech.* **294**, 71–92, (1995).
38. Z. L. Zou, Higher order Boussinesq equations, *Ocean Eng.* **26**, 767–792, (1999).
39. Z. L. Zou, A new form of higher order Boussinesq equations, *Ocean Eng.* **27**, 557–575, (2000).
40. A. B. Kennedy, J. T. Kirby, Q. Chen, and R. A. Dalrymple, Boussinesq-type equations with improved nonlinear performance, *Wave Motion.* **33**, 225–243, (2001).
41. Y. Agnon, P. A. Madsen, and H. A. Schäffer, A new approach to high order Boussinesq models, *J. Fluid Mech.* **399**, 319–333, (1999).
42. T. Y. Wu, Nonlinear waves and solitons in water, *Physica D.* **123**, 48–63, (1998).
43. T. Y. Wu, Modelling nonlinear dispersive water waves, *J. Eng. Mech.* **125**, 747–755, (1999).
44. T. Y. Wu, A unified theory for modelling water waves, *Adv. Appl. Mech.* **37**, 1–88, (2001).
45. P. A. Madsen and Y. Agnon, Accuracy and convergence of velocity formulations for water waves in the framework of Boussinesq theory, *J. Fluid Mech.* **477**, 285–319, (2003).
46. M. F. Gobbi, J. T. Kirby, and G. Wei, A fully non-linear Boussinesq model for surface waves. Part 2. Extension to O(kh^4), *J. Fluid Mech.* **405**, 181–210, (2000).
47. H. A. Schäffer and P. A. Madsen, A new formulation of higher order Boussinesq equations. In *Proc. COASTAL '95*, pp. 159–166, Cancun, Mexico, (1995).
48. A. B. Kennedy, J. T. Kirby, and M. F. Gobbi, Simplified higher order Boussinesq equations. 1: Linear considerations, *Coast. Eng.* **44**, 205–229, (2002).
49. P. A. Madsen, H. B. Bingham, and H. Liu, A new Boussinesq method for fully nonlinear waves from shallow to deep water, *J. Fluid Mech.* **462**, 1–30, (2002).

50. P. A. Madsen, H. B. Bingham, and H. A. Schäffer, Boussinesq-type formulations for fully nonlinear and extremely dispersive water waves: derivation and analysis, *Proc. R. Soc. Lond. A.* **459**, 1075–1104, (2003).
51. Y. Toledo and Y. Agnon, On uniformly accurate high-order Boussinesq difference equations for water waves, *Int. J. Numer. Meth. Fluids.* **50**, 925–945, (2006).
52. P. A. Madsen, D. R. Fuhrman, and B. Wang, A Boussinesq-type method for fully nonlinear waves interacting with a rapidly varying bathymetry, *Coast. Eng.* **53**, 487–504, (2006).
53. E. Jamois, B. Molin, F. Remy, and O. Kimmoun, Nonlinear wave amplification in front of reflective structures, *Eur. J Mech. B/Fluids.* **25**, 565–573, (2006).
54. E. Jamois, D. R. Fuhrman, H. B. Bingham, and B. Molin, A numerical study of nonlinear wave run-up on a vertical plate, *Coast. Eng.* **53**, 929–945, (2006).
55. P. Lynett and P. L.-F. Liu, A two-layer approach to wave modeling, *Proc. R. Soc. Lond. A.* **460**, 2637–69, (2004).
56. P. Lynett and P. L.-F. Liu, Linear analysis of the multi-layer model, *Coast. Eng.* **51**, 439–454, (2004).
57. S.-C. Hsiao, P. Lynett, H.-H. Hwung, and P. L.-F. Liu, Numerical simulations of nonlinear short waves using a multilayer model, *J. Eng. Mech.* **131**, 231–243, (2005).
58. H. A. Schäffer, Accurate determination of internal kinematics from numerical wave model results, *Coast. Eng.* **50**, 199–211, (2004).
59. H. A. Schäffer, A convolution method for 1-D nonlinear dispersive wave transformation over a mild-slope bottom. In *Waves 2005: Proc. 5th Int. Symp. on Ocean Wave Measurement and Analysis*, Madrid, (2005).
60. H. A. Schäffer, Progress on a convolution method for nonlinear dispersive wave transformation over gently varying bathymetry. In *Proc. 30th Int. Conf. on Coastal Engineering*, pp. 149–156, (1984).
61. B. J. West, K. A. Brueckner, R. S. Janda, D. M. Milder, and R. Milton, A new numerical method for surface hydrodynamics, *J. Geophys. Res.* **92 (C11)**, 11803–11824, (1987).
62. S. B. Yoon and P. L.-F. Liu, Interactions of currents and weakly nonlinear water waves in shallow water, *J. Fluid Mech.* **205**, 397–419, (1989).
63. Q. Chen, P. A. Madsen, H. A. Schäffer, and D. R. Basco, Wave-current interaction based on an enhanced Boussinesq-type approach, *Coast. Eng.* **33**, 11–40, (1998).
64. Q. Chen, P. A. Madsen, and D. R. Basco, Current effects on interactions of shallow water waves, *J. Waterw. Port C-ASCE.* **125**, 176–187, (1999).
65. C. Y. Shen, Constituent Boussinesq equations for waves and currents, *J. Phys. Oceanogr.* **31**, 850–859, (2001).
66. P. A. Madsen and O. R. Sørensen, Extension of the Boussinesq equations to include wave propagation in deeper water and wave-ship interaction in shallow water. In *Proc. 22nd Int. Conf. Coastal Engineering*, pp. 3112–3125, Delft, (1990).

67. T. Jiang, Simulation of shallow water waves generated by ships using Boussinesq equations solved by a flux-difference-splitting method. In *Proc. 11th Int. Workshop on Water Waves and Floating Bodies*, Hamburg, (1996).
68. O. Nwogu and Z. Demirbilek, Numerical modeling of ship-induced currents in confined waterways. In *Proc. 29th Int. Conf. on Coastal Engineering*, pp. 256–268, (2004).
69. J. Tao, Numerical modelling of wave runup and breaking on a beach, *Acta Oceanologica Sinica*. **6**, 692–700, (1984). (in Chinese).
70. J. A. Zelt, The run-up of nonbreaking and breaking solitary waves, *Coast. Eng.* **15**, 205–246, (1991).
71. T. K. Karambas and C. Koutitas, A breaking wave propagation model based on the Boussinesq equations, *Coast. Eng.* **18**, 1–19, (1992).
72. H. A. Schäffer, P. A. Madsen, and R. A. Deigaard, A Boussinesq model for waves breaking in shallow water, *Coast. Eng.* **20**, 185–202, (1993).
73. P. A. Madsen, O. R. Sørensen, and H. A. Schäffer, Surf zone dynamics simulated by a Boussinesq type model. Part 1: Model description and cross-shore motion of regular waves, *Coast. Eng.* **32**, 255–283, (1997).
74. P. A. Madsen, O. R. Sørensen, and H. A. Schäffer, Surf zone dynamics simulated by a Boussinesq type model. Part 2: Surf beat and swash oscillations for wave groups and irregular waves, *Coast. Eng.* **32**, 289–320, (1997).
75. O. R. Sørensen, H. A. Schäffer, and P. A. Madsen, Surfzone dynamics simulated by a Boussinesq model. Part III: Wave induced horizontal nearshore circulations, *Coast. Eng.* **33**, 155–176, (1998).
76. Q. Chen, R. A. Dalrymple, J. T. Kirby, A. Kennedy, and M. C. Haller, Boussinesq modelling of a rip current system, *J. Geophys. Res.* **104**, 20.617–20.637, (1999).
77. Q. Chen, J. T. Kirby, R. A. Dalrymple, A. B. Kennedy, and A. Chawla, Boussinesq modeling of wave transformation, breaking, and runup. II: 2D, *J. Waterw. Port C-ASCE.* **126**, 48–56, (2000).
78. Q. Chen, J. T. Kirby, R. A. Dalrymple, F. Shi, and E. B. Thornton, Boussinesq modelling of longshore currents, *J. Geophys. Res.* **108 (C11)**, 3362, (2003).
79. A. B. Kennedy, Q. Chen, J. T. Kirby, and R. A. Dalrymple, Boussinesq modeling of wave transformation, breaking, and runup. Part I: 1D, *J. Waterw. Port C-ASCE.* **126**, 39–47, (2000).
80. J. Veeramony and I. A. Svendsen, The flow in surf-zone waves, *Coast. Eng.* **39**, 93–122, (2000).
81. Lord Rayleigh, On waves, *Philos. Mag.* **1**, 257–279, (1876).
82. D. R. Fuhrman and H. B. Bingham, Numerical solutions of fully non-linear and highly dispersive Boussinesq equations in two horizontal dimensions, *Int. J. Numer. Meth. Fluids.* **44**, 231–255, (2004).
83. D. R. Fuhrman, H. B. Bingham, P. A. Madsen, and P. G. Thomsen, Linear and non-linear stability analysis for finite difference discretizations of high-order Boussinesq equations, *Int. J. Numer. Meth. Fluids.* **45**, 751–773, (2004).

84. A. Savitzky and M. J. E. Golay, Smoothing and differentiation of data by simplified least squares procedures, *Anal. Chem.* **36**, 1627–1639, (1964).
85. D. R. Fuhrman, P. A. Madsen, and H. B. Bingham, Numerical simulation of lowest-order short-crested wave instabilities, *J. Fluid Mech.* **563**, 415–441, (2006).
86. A. P. Ensig-Karup, J. S. Hesthaven, H. B. Bingham, and P. A. Madsen, Nodal DG-FEM solution of high-order Boussinesq-type equations, *J. Eng. Math.* **56**, 351–370, (2006).
87. D. R. Fuhrman, P. A. Madsen, and H. B. Bingham, A numerical study of crescent waves, *J. Fluid Mech.* **513**, 309–341, (2004).
88. D. R. Fuhrman and P. A. Madsen, Short-crested waves in deep water: a numerical investigation of recent laboratory experiments, *J. Fluid Mech.* **559**, 391–411, (2006).
89. D. R. Fuhrman and P. A. Madsen, Numerical simulation of extreme events from focused directionally spread wavefields. In ed. J. M. Smith, *Proc. 30th Int. Conf. on Coastal Engineering*, vol. 1, pp. 772–781, San Diego, (2006).
90. D. R. Fuhrman, H. B. Bingham, and P. A. Madsen, Nonlinear wave-structure interactions with a high-order Boussinesq model, *Coast. Eng.* **52**, 655–672, (2005).
91. P. A. Madsen and D. R. Fuhrman, Run-up of tsunamis and long waves in terms of surf-similarity, *Coast. Eng.* **55**, 209–223, (2008).
92. P. A. Madsen and D. R. Fuhrman, Analytical and numerical models for tsunami run-up. In ed. A. Kundu, *Tsunami and nonlinear waves*, pp. 209–236. Springer-Verlag, Berlin, (2007). ISBN 978-3-540-71255-8.
93. D. R. Fuhrman and P. A. Madsen, Simulation of nonlinear wave run-up with a high-order Boussinesq model, *Coast. Eng.* **55**, 139–154, (2008).
94. D. R. Fuhrman and P. A. Madsen, Tsunami generation, propagation, and run-up with a high-order Boussinesq model. (2008). (submitted to Coastal Engineering).
95. J. D. Fenton, The numerical solution of steady water wave problems, *Comput. Geosci.* **14**, 357–368, (1988).
96. M. Tanaka, The stability of solitary waves, *Phys. Fluids.* **29**, 650–655, (1986).
97. J. M. Williams, *Tables of Progressive Gravity Waves*. (Pitman, 1985).
98. S. T. Grilli and I. A. Svendsen, The propagation and runup of solitary waves on steep slopes. Center for Applied Coastal Research, University of Delaware, Research Report No. 91-4, (1990).
99. I. A. Svendsen and S. T. Grilli, Nonlinear waves on steep slopes, *J. Coast. Res.* **SI 7**, 185–202, (1990).
100. M. J. Cooker, P. D. Weidman, and D. S. Bale, Reflection of a high-amplitude solitary wave at a vertical wall, *J. Fluid Mech.* **342**, 141–158, (1997).
101. B. Molin, F. Remy, O. Kimmoun, and E. Jamois, The role of tertiary wave interactions in wave-body problems, *J. Fluid Mech.* **528**, 323–354, (2005).
102. L. W. Schwartz and A. K. Whitney, A semi-analytic solution for nonlinear standing waves in deep water, *J. Fluid Mech.* **107**, 147–171, (1981).

103. M. Ioualalen and C. Kharif, On the subharmonic instability of steady three-dimensional deep water waves, *J. Fluid Mech.* **262**, 265–291, (1994).
104. J. Hammack, D. M. Henderson, and H. Segur, Progressive waves with persistent two-dimensional surface patterns in deep water, *J. Fluid Mech.* **532**, 1–52, (2005).
105. D. M. Henderson, M. S. Patterson, and H. Segur, On the laboratory generation of two-dimensional, progressive, surface waves of nearly permanent form on deep water, *J. Fluid Mech.* **559**, 413–427, (2006).
106. J. W. McLean, Instabilities of finite-amplitude water waves, *J. Fluid Mech.* **114**, 315–330, (1982).
107. M. Y. Su, M. Bergin, P. Marler, and R. Myrick, Experiments on nonlinear instabilities and evolution of steep gravity-wave trains, *J. Fluid Mech.* **124**, 45–72, (1982).
108. M. Y. Su, Three-dimensional deep-water waves. Part 1. Experimental measurement of skew and symmetric wave patterns, *J. Fluid Mech.* **124**, 73–108, (1982).
109. M. Xue, H. Xü, Y. Liu, and D. K. P. Yue, Computations of fully nonlinear three-dimensional wave-wave and wave-body interactions. Part 1. Dynamics of steep three-dimensional waves, *J. Fluid Mech.* **438**, 11–39, (2001).
110. D. Fructus, C. Kharif, M. Francius, O. Kristiansen, D. Clamond, and J. Grue, Dynamics of crescent water wave patterns, *J. Fluid Mech.* **537**, 155–186, (2005).
111. F. Collard and G. Caulliez, Oscillating crescent-shaped water wave patterns, *Phys. Fluids.* **11**, 3195–3197, (1999).
112. D. R. Fuhrman and P. A. Madsen, Potential dominance of oscillating crescent waves in finite width tanks, *Phys. Fluids.* **17**, 038102/1–4, (2005).
113. T. B. Johannessen and C. Swan, A laboratory study of the focusing of transient and directionally spread surface water waves, *Proc. R. Soc. Lond. A.* **457**, 971–1006, (2001).
114. W. J. D. Bateman, C. Swan, and P. H. Taylor, On the efficient numerical simulation of directionally spread surface water waves, *J. Comput. Phys.* **174**, 277–305, (2001).
115. W. J. D. Bateman, C. Swan, and P. H. Taylor, On the calculation of the water particle kinematics arising in a directionally spread wavefield, *J. Comput. Phys.* **186**, 70–92, (2003).
116. C. Fochesato, S. T. Grilli, and F. Dias, Numerical modeling of extreme rogue waves generated by directional energy focusing, *Wave Motion.* **44**, 395–416, (2007).
117. A. G. Davies and A. D. Heathershaw, Surface-wave propagation over sinusoidally varying topography, *J. Fluid Mech.* **144**, 419–443, (1984).
118. Y. Liu and D. K. P. Yue, On generalized Bragg scattering of surface waves by bottom ripples, *J. Fluid Mech.* **356**, 297–326, (1998).
119. P. A. Madsen and D. R. Fuhrman, Third-order theory for bichromatic bi-directional water waves, *J. Fluid Mech.* **557**, 369–397, (2006).
120. P. J. Lynett, T.-R. Wu, and P. L.-F. Liu, Modeling wave runup with depth-integrated equations, *Coast. Eng.* **46**, 89–107, (2002).

121. M. J. Briggs, C. E. Synolakis, G. S. Harkins, and D. R. Green, Laboratory experiments of tsunami runup on a circular island, *Pure Appl. Geophys.* **144**, 569–593, (1995).
122. P. L.-F. Liu, Y.-S. Cho, M. J. Briggs, U. Kanoglu, and C. E. Synolakis, Runup of solitary waves on a circular island, *J. Fluid Mech.* **302**, 259–285, (1995).
123. F. Enet, *Tsunami generation from underwater landslides.* PhD thesis, University of Rhode Island, (2006).
124. F. Enet and S. T. Grilli, Experimental study of tsunami generation by three-dimensional rigid underwater landslides, *J. Waterw. Port C-ASCE.* **133**, 442–454, (2007).

CHAPTER 8

INTER-COMPARISONS OF DIFFERENT FORMS OF HIGHER-ORDER BOUSSINESQ EQUATIONS

Z.L. Zou*, K.Z. Fang and Z.B. Liu

The State Key Laboratory of Coastal and Offshore Engineering
Dalian University of Technology, Dalian 116024, P. R. China
**zlzou@dlut.edu.cn*

Seven sets of higher order Boussinesq equations are inter-compared in order to verify the effects of different approximation levels of nonlinearity and the mild slope assumption on numerical results. The models have linear dispersion accurate to a Padé [2,2] or Padé [4,4] expansion, respectively and different nonlinearities with $\varepsilon=O(\mu^2)$, $\varepsilon=O(\mu^{2/3})$ or $\varepsilon=O(1)$ (ε is the ratio of wave height to water depth and μ is the ratio of water depth to wave length). Four sets of them are formulated with the mild slope assumption, and the other three sets are not. One of the models is developed by using a newly developed enhancement for nonlinear characteristics embodied in equations. The effects of different approximation levels of nonlinearity are examined by simulations of the wave propagation over a plane beach, submerged breakwaters and by simulations of the nonlinear evolution of wave groups. The applicability of the mild slope assumption is examined by comparing the numerical results of the models with and without the mild slope assumption.

1. Introduction

A variety of Boussinesq-type equations have been developed in order to make the models more applicable to most water wave motions in coastal and offshore water. Earlier works concentrated on the improvement of the embedded dispersion properties of the classical Boussinesq equations, and the improvement methods include (1) adding higher order terms to the equations (e.g. Madsen et al.[2,3]); (2) taking the velocity at an arbitrary level as the velocity variable (e.g. Nwogu[4]; Wei et al.[5]; Gobbi et al.[6,7]); (3) introducing a pseudo velocity (e.g. Witting[8]; Zou[9]). Weak nonlinearity in the classical Boussinesq equations is another research focus for the improvement of the Boussinesq equations, and different levels of higher order nonlinearity have been incorporated into the Boussinesq models. This work goes back to the models of Serre[10], Su and Gardner[11], Green and Naghdi[12], Mei[13]. All of them are fully nonlinear up to $O(\mu^2)$. Wei et al.[5] derived a set of higher order Boussinesq equations with the same nonlinearity, but a linear dispersion to Padé [4,4] expansion is achieved by the formulation of the equations in terms of the weighted velocity from the velocities at two arbitrary levels. Stronger nonlinearity which is fully nonlinear up to $O(\mu^4)$ can be included (e.g. Gobbi et al.[6,7]; Madsen and Schäffer[14]; Zou and Fang[15]). Zou and Fang[1] shows that like the dispersion the nonlinear characteristics embodied in the equations can also be improved by introducing a nonlinear velocity expansion for the determination of a computation velocity. An approach which breaks through the framework of Boussinesq-type equations is also developed by Agnon et al.,[16] Madsen et al.[17,18] The model consists of four partial differential equations resulting from the dynamic and kinematic surface and bottom boundary conditions with truncated solutions of Laplace equation, and can incorporate much higher order nonlinearity and dispersion. However, a sophisticated numerical scheme is also needed for this model, especially for a horizontal 2-D case (Fuhrman[19]).

The present study investigates the applicability of different forms of higher order Boussinesq equations. Seven sets of higher order Boussinesq models with different nonlinearities and with or without the mild slope assumption are inter-compared, including a newly developed model which can improve the nonlinear characteristics for large relative water depth (Zou and Fang[1]). Inter-comparisons are carried out for the wave propagations over a plane slope, submerged breakwaters with different slopes, and for the nonlinear evolution of wave groups in a wave flume.

2. Mathematical Models

In this section, we first present six sets of the Boussinesq-type equations which will be investigated in the present study. These sets of Boussinesq equations can be divided into two groups: the first three sets adopt the mild slope assumption either $\nabla h = O(\mu^2)$ or $\nabla h = O(\mu^4)$ (h is water depth, ∇ is horizontal gradient operator) for the higher order dispersion and nonlinear terms, so they have relatively simple forms; the latter three sets do not, and have relatively complex forms. Therefore, the two groups of models present two different formulations of improved Boussinesq models. Following these discussions, another set of Boussinesq equations is presented, which uses a newly developed enhancement for nonlinear characteristics embodied in equations (Zou and Fang[1]).

The first three sets of equations are obtained by extending the two models given by Zou[9,20], which are of linear dispersion accurate to the Padé [2,2] and Padé [4,4] expansion, respectively and have the second order ($O(\varepsilon\mu^2)$) nonlinearity. This is done through replacing the assumption $\varepsilon = O(\mu^2)$ for the above two models by the new assumption of $\varepsilon = O(\mu^{2/3})$ or $\varepsilon = O(1)$. The assumption $\varepsilon = O(\mu^{2/3})$ will lead to a new model which includes all the nonlinear terms at $O(\mu^2)$, i.e., being fully nonlinearity up to $O(\mu^2)$; the assumption $\varepsilon = O(1)$ will lead to a new model which includes all the nonlinear terms at $O(\mu^4)$, i.e., being fully nonlinearity up to $O(\mu^4)$. These new assumptions mean that the nonlinearities of water waves considered here are stronger than that considered by the original models. The higher order Boussinesq equations given by Zou[9,20] read, respectively,

$$\eta_t + \nabla \cdot [(h+\eta)\bar{\boldsymbol{u}}] = 0, \tag{2.1}$$

$$\bar{\boldsymbol{u}}_t + (\bar{\boldsymbol{u}} \cdot \nabla)\bar{\boldsymbol{u}} + g\nabla\eta + \boldsymbol{G} = \frac{1}{2} h\nabla[\nabla \cdot (h\bar{\boldsymbol{u}}_t)] - \frac{1}{6} h^2 \nabla(\nabla \cdot \bar{\boldsymbol{u}}_t)$$
$$+ B_1 h^2 \nabla[\nabla \cdot (\bar{\boldsymbol{u}}_t + g\nabla\eta)] + B_2 \nabla[\nabla \cdot (h^2 \bar{\boldsymbol{u}}_t + g h^2 \nabla\eta)], \tag{2.2}$$

with
$$\boldsymbol{G} = \frac{h^2}{3} \nabla[(\nabla \cdot \bar{\boldsymbol{u}})^2 - (\bar{\boldsymbol{u}} \cdot \nabla)\nabla \cdot \bar{\boldsymbol{u}} - \frac{B_3}{10} \nabla(\nabla \cdot (\bar{\boldsymbol{u}} \cdot \bar{\boldsymbol{u}}))]$$
$$- h\nabla\eta\nabla \cdot \bar{\boldsymbol{u}}_t - \frac{2}{3}\eta h \nabla(\nabla \cdot \bar{\boldsymbol{u}}_t), \tag{2.3}$$

and

$$\eta_t + \nabla \cdot (d\tilde{\boldsymbol{u}}) = a_1 \nabla \cdot [dh^2 \nabla(\nabla \cdot \tilde{\boldsymbol{u}})] + a_2 \nabla \cdot [dh \nabla(\nabla \cdot (h\tilde{\boldsymbol{u}}))]$$
$$+ \frac{b_1}{a} \nabla \cdot \{h^2 \nabla[\eta_t + \nabla \cdot (d\tilde{\boldsymbol{u}})]\} + \frac{b_2}{a} \nabla \cdot \{\nabla[h^2 \eta_t + h^2 \nabla \cdot (d\tilde{\boldsymbol{u}})]\}, \quad (2.4)$$

$$\tilde{\boldsymbol{u}}_t + (\tilde{\boldsymbol{u}} \cdot \nabla)\tilde{\boldsymbol{u}} + g\nabla\eta + \boldsymbol{G} = (a_1 - \frac{1}{6}) h^2 \nabla(\nabla \cdot \tilde{\boldsymbol{u}}_t)$$
$$+ (a_2 + \frac{1}{2}) h \nabla[\nabla \cdot (h\tilde{\boldsymbol{u}}_t)] + c_1 h^2 \nabla[\nabla \cdot (\tilde{\boldsymbol{u}}_t + g\nabla\eta)]$$
$$+ c_2 \nabla[\nabla \cdot (h^2 \tilde{\boldsymbol{u}}_t + gh^2 \nabla\eta)] \quad (2.5)$$

with
$$\boldsymbol{G} = \frac{h^2}{3} \nabla[(\nabla \cdot \tilde{\boldsymbol{u}})^2 - (1+3a)\tilde{\boldsymbol{u}} \cdot \nabla^2 \tilde{\boldsymbol{u}} - \frac{3}{2} c \nabla^2 (\tilde{\boldsymbol{u}} \cdot \tilde{\boldsymbol{u}})]$$
$$- h\nabla\eta\nabla \cdot \tilde{\boldsymbol{u}}_t - \frac{2}{3} h\eta\nabla(\nabla \cdot \tilde{\boldsymbol{u}}_t), \quad (2.6)$$

where $d = h + \eta$, $a = a_1 + a_2$, $b = b_1 + b_2$, $c = c_1 + c_2$. Equations (2.4) and (2.5) were derived using the following velocity expression to replace the depth averaged velocity $\bar{\boldsymbol{u}}$ in (2.1) and (2.2) by the computation velocity $\tilde{\boldsymbol{u}}$.

$$\bar{\boldsymbol{u}} = \tilde{\boldsymbol{u}} - a_1 h^2 \nabla^2 \tilde{\boldsymbol{u}} - a_2 h \nabla^2 (h\tilde{\boldsymbol{u}}) - b_1 h^4 \nabla^2 \nabla^2 \tilde{\boldsymbol{u}} - b_2 h \nabla^2 \nabla^2 (h^3 \tilde{\boldsymbol{u}}) \quad (2.7)$$

here $\nabla^2 \equiv \nabla\nabla \cdot$. The three parameters ($B_1$, B_2, B_3) in Eq. (2.2) satisfy the following three relations: $B_1+B_2=1/15$ for optimization of dispersion accurate to the Padé [2,2] expansion of the Stokes linear dispersion relation, $B_2=2/59$ for optimization of shoaling property, $B_3=4.5$ for the optimization of nonlinear characteristics.

The six parameters (a_1, a_2, b_1, b_2, c_1, c_2) in Eqs. (2.4) and (2.5) satisfy the relation: a=-0.029, b/a=0.039, c=0.101 for optimization of dispersion accurate to a Padé [4,4] expansion, and have the following values for optimization of the model's shoaling and nonlinear properties.

$$(a_1, a_2, b_1, b_2, c_1, c_2) =$$
$$(-0.0067, -0.0220, -0.0017, -0.0005, 0.0393, 0.0613). \quad (2.8)$$

The $\boldsymbol{G}$ terms given by (2.3) and (2.6) for the two models are accurate to $O(\varepsilon\mu^2)$, this was from the consideration of keeping the derived equations accurate to second order with the assumption $\varepsilon = O(\mu^2)$. If we want to improve third order and higher nonlinear characteristics, this limitation can be relaxed by using the new nonlinearity scale $\varepsilon = O(\mu^{2/3})$ or $\varepsilon = O(1)$ to replace the original assumption $\varepsilon = O(\mu^2)$. When using $\varepsilon = O(\mu^{2/3})$, all the nonlinear terms at order $O(\mu^2)$ can be retained and balance the $O(\mu^4)$ dispersion terms. Then, the **G** term will have the following form (Zou[22]):

$$\boldsymbol{G} = \nabla[\frac{d^2}{2}(\nabla\cdot\bar{\boldsymbol{u}})^2 - \frac{d^2}{3}(\bar{\boldsymbol{u}}\cdot\nabla)\nabla\cdot\bar{\boldsymbol{u}} - B_3\frac{h^2}{10}\nabla(\nabla\cdot(\bar{\boldsymbol{u}}\cdot\bar{\boldsymbol{u}}))]$$

$$+\frac{1}{3}d\eta_t\nabla(\nabla\cdot\bar{\boldsymbol{u}}) - \nabla\eta d\nabla\cdot\bar{\boldsymbol{u}}_t - \frac{1}{3}\eta(2h+\eta)\nabla(\nabla\cdot\bar{\boldsymbol{u}}_t) \qquad (2.9)$$

for Eqs. (2.1) and (2.2), and

$$\boldsymbol{G} = \nabla[\frac{d^2}{2}(\nabla\cdot\tilde{\boldsymbol{u}})^2 - (\frac{d^2}{3} + ah^2)(\tilde{\boldsymbol{u}}\cdot\nabla)\nabla\cdot\tilde{\boldsymbol{u}} - \frac{h^2}{2}c\nabla(\nabla\cdot(\tilde{\boldsymbol{u}}\cdot\tilde{\boldsymbol{u}}))]$$

$$+\frac{1}{3}d\eta_t\nabla(\nabla\cdot\tilde{\boldsymbol{u}}) - d\nabla\eta d\nabla\cdot\tilde{\boldsymbol{u}}_t - \frac{1}{3}(2h+\eta)\eta\nabla(\nabla\cdot\tilde{\boldsymbol{u}}_t) \qquad (2.10)$$

for Eqs. (2.4) and (2.5). Hereafter, Eqs. (2.1) and (2.2) with **G** given by (2.3) or (2.9) are referred to as the three-parameter equations, and Eqs. (2.4) and (2.5) with **G** given by (2.6) or (2.10) are referred to as the six-parameter equations, since they contain the three parameters (B_1, B_2, B_3) and the six-parameters (a_1, a_2, b_1, b_2, c_1, c_2), respectively. The six-parameter equations with **G** given by (2.10) are similar in dispersion and nonlinearity to the enhanced Boussinesq model in terms of velocity at an arbitrary z level given by Madsen and Schäffer[14] (Eqs. (6.3a) and (6.3b) in their paper), which is obtained by enhancing the $O(\mu^2, \varepsilon^3\mu^2)$ Boussinesq equations of Wei et al.,[5] and is fully nonlinear up to $O(\mu^2)$ and has a Padé [4,4] dispersion (obtained by applying the enhancement technique of Schäffer and Madsen[21]).

The above models only retain the linear part of $O(\mu^4)$ terms (the $O(\mu^4)$ dispersion terms) on the basis of the nonlinearity scale $\varepsilon = O(\mu^{2/3})$. If we want to retain all the nonlinear terms at $O(\mu^4)$, we should use another nonlinearity scale $\varepsilon = O(1)$; then, similar to the derivation of Eqs. (2.4) and (2.5) with **G** given by (2.10), the following form of Boussinesq equations, fully nonlinear up to $O(\mu^4)$, can be obtained by extending the formulation to $O(\mu^4)$.

$$\eta_t + \nabla\cdot(d\tilde{\boldsymbol{u}}) = a_1\nabla\cdot[dh^2\nabla(\nabla\cdot\tilde{\boldsymbol{u}})] + a_2\nabla\cdot[dh\nabla(\nabla\cdot(h\tilde{\boldsymbol{u}}))]$$

$$+\frac{b_1}{a}\nabla\cdot\{h^2\nabla[\eta_t + \nabla\cdot(d\tilde{\boldsymbol{u}})]\} + \frac{b_2}{a}\nabla\cdot\{\nabla[h^2\eta_t + h^2\nabla\cdot(d\tilde{\boldsymbol{u}})]\}, \qquad (2.11)$$

$$\tilde{\boldsymbol{u}}_t + (\tilde{\boldsymbol{u}}\cdot\nabla)\tilde{\boldsymbol{u}} + g\nabla\eta + \boldsymbol{G}_1 + \boldsymbol{G}_2 =$$

$$(a_1 - \frac{1}{6})h^2\nabla(\nabla\cdot\tilde{\boldsymbol{u}}_t) + (a_2 + \frac{1}{2})h\nabla[\nabla\cdot(h\tilde{\boldsymbol{u}}_t)]$$

$$+c_1h^2\nabla[\nabla\cdot(\tilde{\boldsymbol{u}}_t + g\nabla\eta)] + c_2\nabla[\nabla\cdot(h^2\tilde{\boldsymbol{u}}_t + gh^2\nabla\eta)] \qquad (2.12)$$

with $$\boldsymbol{G}_1 = \nabla[\frac{d^2}{2}(\nabla\cdot\tilde{\boldsymbol{u}})^2 - (\frac{d^2}{3} + ah^2)(\tilde{\boldsymbol{u}}\cdot\nabla)\nabla\cdot\tilde{\boldsymbol{u}} - \frac{h^2}{2}c\nabla^2(\tilde{\boldsymbol{u}}\cdot\tilde{\boldsymbol{u}})]$$

$$+\frac{1}{3}d\eta_t\nabla^2\tilde{\boldsymbol{u}}) - d\nabla\eta d\nabla\cdot\tilde{\boldsymbol{u}}_t - \frac{1}{3}(2h+\eta)\eta\nabla^2\tilde{\boldsymbol{u}}_t), \qquad (2.13)$$

where $$a = a_1 + a_2, b = b_1 + b_2, c = c_1 + c_2,\ \nabla^2 \equiv \nabla\nabla\cdot, \qquad (2.14a,b)$$

$$\tilde{d}^2 = d^2 - 6ah^2,\quad \tilde{d}^4 = d^4 + 20ad^2h^2 + 120bh^4. \qquad (2.14c,d)$$

The expression for $\boldsymbol{G}_2$ is given in Appendix. The values of the parameters (a_1, a_2, b_1, b_2, c_1, c_2) are still given by (2.8). The $\boldsymbol{G}_1$ term is the $O(\mu^2)$ nonlinear term, it is the same as the $\boldsymbol{G}$ term in the previous model, see (2.10). The $\boldsymbol{G}_2$ term is the nonlinear term at $O(\mu^4)$, which includes all the nonlinear terms at $O(\mu^4)$. The above equation can be seen to be obtained by adding all the nonlinear terms at $O(\mu^4)$ (the $\boldsymbol{G}_2$ term) to the momentum of the six-parameter equations (2.4) and (2.5), and this make the resulting equations (2.11) and (2.12) be fully nonlinear up to $O(\mu^4)$, which is the direct result of using the assumption $\varepsilon = O(1)$. Hereafter we refer to this set of equations as the $O(\mu^4)$ six-parameter equations.

In the expressions for $\boldsymbol{G}_1$ and $\boldsymbol{G}_2$, all the terms related to bottom variation have been ignored (the derivation is similar to Zou[9,20]), so the expressions for $\boldsymbol{G}_1$ and $\boldsymbol{G}_2$ are actually derived for a horizontal bottom case. This is achieved by using the mild slope assumption of $\nabla h = O(\mu^4)$. This mild slope assumption is also applied for the higher order dispersion terms (the terms containing the parameters (a_1, a_2, b_1, b_2, c_1, c_2)) of this model. For the three-parameter equations and the six-parameter equations mentioned above, the mild slope assumption $\nabla h = O(\mu^2)$ is applied (Zou[9,20]). This assumption is replaced by $\nabla h = O(\mu^4)$ in the $O(\mu^4)$ six-parameter equations, since the nonlinearity scale is now changed from $\varepsilon = O(\mu^{2/3})$ to $\varepsilon = O(1)$ and with this new nonlinearity scale, the bottom-gradient terms of order $O(\mu^2)$ in the $\boldsymbol{G}_1$ term can not be neglected any more for the equations accurate to $O(\mu^4)$. So the higher order mild slope assumption, $\nabla h = O(\mu^4)$, has to be adopted in order to still use a simple expression (2.13) for the $\boldsymbol{G}_1$ term in Eqs. (2.11) and (2.12). The applicability of the above mild slope assumptions will be validated in numerical simulations of Section 6 by comparing the models with these mild slope assumptions to the models without these assumptions. The latter models form the second group of models to be discussed below.

In contrast to the first group of models mentioned above, the formulations of the second group of models will not involve the mild slope assumption, and bottom variation is considered exactly by means of σ-transformation (Zou[23], Zou and Fang[15]). Here we describe this group of models staring from the following equations.

$$\eta_t + \nabla \cdot [(h+\eta)\bar{\boldsymbol{u}}] = 0, \tag{2.15}$$

$$\bar{\boldsymbol{u}}_t + (\bar{\boldsymbol{u}} \cdot \nabla)\bar{\boldsymbol{u}} + g\nabla\eta = \boldsymbol{P}^{(2)} + \boldsymbol{P}^{(4)} \tag{2.16a}$$

with

$$\boldsymbol{P}^{(2)} = -\frac{1}{2d}\nabla\{d^2 \mathrm{D}[\frac{2}{3}d\nabla\cdot\bar{\boldsymbol{u}} + \nabla h\cdot\bar{\boldsymbol{u}}]\} + \nabla h \mathrm{D}(\frac{1}{2}d\nabla\cdot\bar{\boldsymbol{u}} + \nabla h\cdot\bar{\boldsymbol{u}}), \tag{2.16b}$$

$$\boldsymbol{P}^{(4)} = -\frac{1}{24}h^3\nabla\nabla\cdot\nabla\nabla\cdot(h\bar{\boldsymbol{u}}_t) + \frac{1}{12}h^2\nabla\nabla\cdot[h\nabla\nabla\cdot(h\bar{\boldsymbol{u}}_t)]$$

$$+\frac{1}{120}h^4\nabla\nabla\cdot\nabla\nabla\cdot\bar{\boldsymbol{u}}_t - \frac{1}{36}h^2\nabla\nabla\cdot[h^2\nabla\nabla\cdot\bar{\boldsymbol{u}}_t] \tag{2.16c}$$

with $\mathrm{D}(\cdot)/\mathrm{D}t = \partial(\cdot)/\partial t + (\bar{\boldsymbol{u}}\cdot\nabla)$. This model is also based on the assumption of $\varepsilon = O(\mu^{2/3})$ as the three-parameter equation with $\boldsymbol{G}$ given by (2.9) and all the nonlinear terms at $O(\mu^2)$ (expressed by $\boldsymbol{P}^{(2)}$) are retained, which balance the $O(\mu^4)$ dispersion term (the $\boldsymbol{P}^{(4)}$ term), while the mild slope assumption is not used for this model and the bottom variation is considered exactly by applying σ-transformation in the derivation. For the details of derivation, see Zou[15,23]. But from the view point of practical application, this model is actually useless, since the dispersion relation of the model is the Padé [0,4] expansion in terms of the non-dimensional wave number kh (k is the wave number) and has a singularity at $kh \approx 4.2$. This weak point can be removed by the following two ways, and this leads to two sets of equations for the second group of models discussed in the present study. One way is to use the lower order momentum equation

$$\bar{\boldsymbol{u}}_t + (\bar{\boldsymbol{u}} \cdot \nabla)\bar{\boldsymbol{u}} + g\nabla\zeta = \frac{1}{2}h\nabla[\nabla \cdot (h\bar{\boldsymbol{u}}_t)] - \frac{1}{6}h^2\nabla\nabla \cdot \bar{\boldsymbol{u}}_t \tag{2.17}$$

in the expression (2.16c) to transform $\boldsymbol{P}^{(4)}$ into the following form

$$\begin{aligned}\hat{\boldsymbol{P}}^{(4)} &= \frac{1}{6}h^2\nabla\nabla \cdot \boldsymbol{C} - \frac{1}{8}h^3\nabla\nabla \cdot (\frac{\boldsymbol{C}}{h}) + \frac{1}{40}h^4\nabla\nabla \cdot (\frac{\boldsymbol{C}}{h^2}) \\ &+ \frac{1}{48}h^3\nabla\nabla \cdot [\nabla(\nabla h \cdot \bar{\boldsymbol{u}}_t) + \nabla h\nabla \cdot \bar{\boldsymbol{u}}_t] \\ &- \frac{1}{80}h^4\nabla\nabla \cdot \{\frac{1}{h}[\nabla(\nabla h \cdot \bar{\boldsymbol{u}}_t) + \nabla h\nabla \cdot \bar{\boldsymbol{u}}_t]\}\end{aligned} \tag{2.18}$$

with $\boldsymbol{C} = \bar{\boldsymbol{u}}_t + (\bar{\boldsymbol{u}} \cdot \nabla)\bar{\boldsymbol{u}} + g\nabla\eta$. Equations (2.15) and (2.16a,b) with $\boldsymbol{P}^{(4)}$ given by (2.18) form a working model of higher order Boussinesq equations, which reads (Zou[23])

$$\eta_t + \nabla \cdot [(h+\eta)\bar{\boldsymbol{u}}] = 0, \tag{2.19}$$

$$\bar{\boldsymbol{u}}_t + (\bar{\boldsymbol{u}} \cdot \nabla)\bar{\boldsymbol{u}} + g\nabla\eta = \boldsymbol{P}^{(2)} + \hat{\boldsymbol{P}}^{(4)}. \tag{2.20}$$

The dispersion relation of this set of equations will be the Padé [2,2] dispersion, and has no singularity due to the replacement of $\boldsymbol{P}^{(4)}$ by $\hat{\boldsymbol{P}}^{(4)}$. Hereafter, we refer to this set of equations as BouN2D2 (N2 denotes being fully nonlinear up to $O(\mu^2)$ and D2 denotes a Padé [2,2] dispersion). Since this model does not employ the mild slope assumption and has the same dispersion and nonlinearity as those of the three-parameter equations with $\boldsymbol{G}$ given by (2.9) and B_3=1.0 (a mild slope assumption is used for this model), the two models will be inter-compared in Section 6 to investigate the effect of the mild slope assumption on the numerical results. If the $O(\mu^4)$ dispersion term (the $\boldsymbol{P}^{(4)}$ term) is omitted, the model BouN2D2 will reduce to Green and Naghdi's equations[12] or Serre's equations[10] for horizontal bottom. The reduced model is exact up to $O(\mu^2)$, and this means that the limitation $\varepsilon = O(\mu^{2/3})$ for the nonlinearity parameter ε can be relaxed, i.e., ε can be as large as $O(1)$. In other words, this reduced model is fully nonlinear up to $O(\mu^2)$ (including all nonlinear terms up to $O(\mu^2)$).

Another way to improve the equations (2.15) and (2.16) is to add the following four higher order terms (in dimensionless form) (Zou and Fang[15])

$$(\alpha_2 - \alpha_1)\mu^2 h^2 \nabla\nabla \cdot \{\bar{\boldsymbol{u}}_t + \nabla\eta + \varepsilon(\bar{\boldsymbol{u}}\cdot\nabla)\bar{\boldsymbol{u}} - \mu^2 \boldsymbol{\Gamma}\} = O(\varepsilon\mu^4, \mu^6), \tag{2.21}$$

$$-\alpha_2 \mu^2 h^2 \nabla\nabla \cdot \{h[\bar{\boldsymbol{u}}_t + \nabla\eta + \varepsilon(\bar{\boldsymbol{u}}\cdot\nabla)\bar{\boldsymbol{u}} - \mu^2 \boldsymbol{\Gamma}]\} = O(\varepsilon\mu^4, \mu^6), \tag{2.22}$$

$$(\beta_2 - \beta_1)\mu^4 h^4 \nabla\nabla^2\nabla \cdot \{\bar{\boldsymbol{u}}_t + \nabla\eta\} = O(\varepsilon\mu^4, \mu^6), \tag{2.23}$$

$$-\beta_2 \mu^4 h^3 \nabla\nabla^2\nabla \cdot \{h(\bar{\boldsymbol{u}}_t + \nabla\eta)\} = O(\varepsilon\mu^4, \mu^6), \tag{2.24}$$

with
$$\boldsymbol{\Gamma} = \frac{1}{2} h\nabla[\nabla\cdot(h\bar{\boldsymbol{u}}_t)] - \frac{1}{6} h^2 \nabla(\nabla\cdot\bar{\boldsymbol{u}}_t) \tag{2.25}$$

to (2.16a) (rewritten in non-dimensional form) to form a new set of higher order Boussinesq equations with the Padé [4,4] dispersion. The resulting new equations read (in dimensional form)

$$\eta_t + \nabla\cdot[(h+\eta)\bar{\boldsymbol{u}}] = 0, \tag{2.26}$$

$$\bar{\boldsymbol{u}}_t + (\bar{\boldsymbol{u}}\cdot\nabla)\bar{\boldsymbol{u}} + g\nabla\eta = \tilde{\boldsymbol{P}}^{(2)} + \tilde{\boldsymbol{P}}^{(4)} \tag{2.27a}$$

with
$$\tilde{\boldsymbol{P}}^{(2)} = \boldsymbol{P}^{(2)} + \boldsymbol{L}_1\boldsymbol{C}, \ \tilde{\boldsymbol{P}}^{(4)} = \boldsymbol{P}^{(4)} - \boldsymbol{L}_1\boldsymbol{\Gamma} + \boldsymbol{L}_2(\bar{\boldsymbol{u}}_t + g\nabla\eta), \tag{2.27b,c}$$

$$\boldsymbol{L}_1 \equiv (\alpha_1 - \alpha_2)h^2\nabla\nabla\cdot + \alpha_2 h\nabla\nabla\cdot h, \tag{2.28}$$

$$\boldsymbol{L}_2 \equiv (\beta_1 - \beta_2)h^4\nabla\nabla^2\nabla\cdot + \beta_2 h^3\nabla\nabla^2\nabla\cdot h. \tag{2.29}$$

Hereafter, we referred to this model as BouN2D4 (N2 denotes being fully nonlinear up to $O(\mu^2)$ and D4 denotes a Padé [4,4] dispersion). The values of the parameters (α_1, α_2, β_1, β_2) are determined by matching the linear dispersion and shoaling property of the equations with theoretical solutions, the first two parameters are for the improvement of the linear dispersion, which have the values $\alpha_1 = 1/9$, $\beta_1 = -1/945$, and the last two for the improvement of the shoaling properties of the resulting equations, which have the values $\alpha_2 = 0.146488$ and $\beta_2 = -0.0019959$.

Similarly to the formulation of the $O(\mu^4)$ six-parameter equations, the further increase of the nonlinearity of the above Boussinesq models is to retain all nonlinear terms at $O(\mu^4)$. This means to extend the $\boldsymbol{P}^{(4)}$ term to include all nonlinear terms at $O(\mu^4)$, and to be consistent, to extend the momentum terms to the order $O(\mu^4)$ as well. We do this first for Eqs. (2.15) and (2.16) and have equations of the form (in dimensionless form):

$$\eta_t + \nabla\cdot(d\bar{\boldsymbol{u}}) = 0, \tag{2.30}$$

$$\bar{\boldsymbol{u}}_t + \varepsilon(\bar{\boldsymbol{u}}\cdot\nabla)\bar{\boldsymbol{u}} + \nabla\eta + \varepsilon\mu^4\boldsymbol{M}^{(4)} = \mu^2\boldsymbol{P}^{(2)} + \mu^4\boldsymbol{P}^{(4)} + O(\mu^6) \tag{2.31a}$$

with

$$\boldsymbol{M}^{(4)} = \frac{1}{12}(\boldsymbol{A}\cdot\nabla)\boldsymbol{A} - \frac{1}{24}[(\boldsymbol{A}\cdot\nabla)\boldsymbol{B} + (\boldsymbol{B}\cdot\nabla)\boldsymbol{A}] + \frac{1}{45}(\boldsymbol{B}\cdot\nabla)\boldsymbol{B}$$
$$+\frac{1}{12d}\boldsymbol{A}\nabla\cdot(d\boldsymbol{A}) - \frac{1}{24d}[\boldsymbol{A}\nabla\cdot(d\boldsymbol{B}) + \boldsymbol{B}\nabla\cdot(d\boldsymbol{A})] + \frac{1}{45}\boldsymbol{B}\nabla\cdot(d\boldsymbol{B}), \tag{2.31b}$$

$$\boldsymbol{P}^{(2)} = \frac{1}{2d}\nabla\{d^2\mathrm{D}[\frac{2}{3}d\nabla\cdot\bar{\boldsymbol{u}} + \nabla h\cdot\bar{\boldsymbol{u}}]\} - \nabla h\mathrm{D}(\frac{1}{2}d\nabla\cdot\bar{\boldsymbol{u}} + \nabla h\cdot\bar{\boldsymbol{u}}), \tag{2.31c}$$

$$\boldsymbol{P}^{(4)} = \frac{1}{d}\nabla\{d^2\mathrm{D}[\frac{1}{12}d\nabla\cdot\boldsymbol{A} - \frac{1}{24}d^2\nabla\cdot(\frac{\boldsymbol{A}}{d}) + \frac{1}{12}\varepsilon\nabla\eta\cdot\boldsymbol{A}$$
$$-\frac{1}{36}d\nabla\cdot\boldsymbol{B} + \frac{1}{120}d^3\nabla\cdot(\frac{\boldsymbol{B}}{d^2}) - \frac{1}{24}\varepsilon\nabla\eta\cdot\boldsymbol{B}]\}$$
$$-\nabla h\mathrm{D}[\frac{1}{4}d\nabla\cdot\boldsymbol{A} - \frac{1}{6}d^2\nabla\cdot(\frac{\boldsymbol{A}}{d}) - \frac{1}{12}d\nabla\cdot\boldsymbol{B} + \frac{1}{24}d^3\nabla\cdot(\frac{\boldsymbol{B}}{d^2})]$$
$$-\varepsilon\frac{1}{12d}\nabla\{d^2[(\boldsymbol{A} - \frac{1}{2}\boldsymbol{B})\cdot(\frac{\boldsymbol{A}}{d} + \varepsilon\nabla\eta\nabla\cdot\bar{\boldsymbol{u}}) + \frac{1}{30}\boldsymbol{B}\cdot(\frac{\boldsymbol{B}}{d} + \nabla d\nabla\cdot\bar{\boldsymbol{u}})]\}$$
$$+\varepsilon\frac{1}{12}\nabla h(\boldsymbol{A} - \frac{1}{2}\boldsymbol{B})\cdot(\frac{\boldsymbol{B}}{d} + \nabla d\nabla\cdot\bar{\boldsymbol{u}})$$
$$-\varepsilon\frac{1}{12d}\nabla\{d^2[\frac{1}{2}\nabla\cdot(d\boldsymbol{A}) - \frac{7}{30}\nabla\cdot(d\boldsymbol{B})]\nabla\cdot\bar{\boldsymbol{u}}\}$$
$$+\varepsilon\frac{1}{12}\nabla h[\nabla\cdot(d\boldsymbol{A}) - \frac{1}{2}\nabla\cdot(d\boldsymbol{B})]\nabla\cdot\bar{\boldsymbol{u}} \tag{2.31d}$$

where $\boldsymbol{A} = d[\nabla\nabla\cdot(h\bar{\boldsymbol{u}}) + \varepsilon\eta\nabla\nabla\cdot\bar{\boldsymbol{u}}]$, $\boldsymbol{B} = d^2\nabla\nabla\cdot\bar{\boldsymbol{u}}$, $d = h + \varepsilon\eta$.

Equations (3.30) and (3.31) form the fourth-order Boussinesq equations, which are fully nonlinear up to $O(\mu^4)$ (including all nonlinear terms up to $O(\mu^4)$) and have the embedded linear dispersion accurate to $O(\mu^4)$ as well. Madsen and Schäffer[14] derived a set of higher order Boussinesq equations accurate to $O(\mu^4)$ in terms of the depth-averaged velocity (Eqs. (3.2) and (3.7) in their paper), but only the $O(\varepsilon\mu^4)$ nonlinear terms at $O(\mu^4)$ are considered. The formulation (3.31d) for $\boldsymbol{P}^{(4)}$ includes the total nonlinear terms at $O(\mu^4)$ and is written in compact form.

Similarly to Eqs. (2.26) and (2.27), the dispersion accuracy of the above model can be improved by adding the following terms of order $O(\mu^6)$ to the momentum equation (2.31a).

$$(\alpha_2-\alpha_1)\mu^2h^2\nabla\nabla\cdot\{\bar{\boldsymbol{u}}_t+\varepsilon(\bar{\boldsymbol{u}}\cdot\nabla)\bar{\boldsymbol{u}}+\nabla\eta-\mu^2\boldsymbol{P}^{(2)}\}=O(\mu^6)\,, \quad (2.32)$$

$$-\alpha_2\mu^2h\nabla\nabla\cdot\{h[\bar{\boldsymbol{u}}_t+\varepsilon(\bar{\boldsymbol{u}}\cdot\nabla)\bar{\boldsymbol{u}}+\nabla\eta-\mu^2\boldsymbol{P}^{(2)}]\}=O(\mu^6)\,, \quad (2.33)$$

$$(\beta_2-\beta_1)\mu^4h^4\nabla\nabla^2\nabla\cdot\{\bar{\boldsymbol{u}}_t+\varepsilon(\bar{\boldsymbol{u}}\cdot\nabla)\bar{\boldsymbol{u}}+\nabla\eta\}=O(\varepsilon\mu^4,\mu^6)\,, \quad (2.34)$$

$$-\beta_2\mu^4h^4\nabla\nabla^2\nabla\cdot\{h(\bar{\boldsymbol{u}}_t+\varepsilon(\bar{\boldsymbol{u}}\cdot\nabla)\bar{\boldsymbol{u}}+\nabla\eta)\}=O(\varepsilon\mu^4,\mu^6)\,. \quad (2.35)$$

Then, the third set of the second group of models is obtained in a dimensional form (Zou and Fang[15]):

$$\eta_t+\nabla\cdot(d\bar{\boldsymbol{u}})=0\,, \quad (2.36)$$

$$\bar{\boldsymbol{u}}_t+(\bar{\boldsymbol{u}}\cdot\nabla)\bar{\boldsymbol{u}}+g\nabla\eta+\boldsymbol{M}^{(4)}=\boldsymbol{P}'^{(2)}+\boldsymbol{P}'^{(4)} \quad (2.37a)$$

with $$\boldsymbol{P}'^{(2)}=\boldsymbol{P}^{(2)}+\boldsymbol{L}_1\boldsymbol{C}\,,\ \boldsymbol{P}'^{(4)}=\boldsymbol{P}^{(4)}-\boldsymbol{L}_1\boldsymbol{P}^{(2)}+\boldsymbol{L}_2\boldsymbol{C}\,, \quad (2.37b,c)$$

$$\boldsymbol{L}_1\equiv(\alpha_1-\alpha_2)h^2\nabla\nabla\cdot+\alpha_2h\nabla\nabla\cdot h\,, \quad (2.38)$$

$$\boldsymbol{L}_2\equiv(\beta_1-\beta_2)h^4\nabla\nabla^2\nabla\cdot+\beta_2h^3\nabla\nabla^2\nabla\cdot h\,. \quad (2.39)$$

The values of the parameters (α_1, α_2, β_1, β_2) are the same as those for Eqs. (2.26) and (2.27). The above model is fully nonlinear up to $O(\mu^4)$ and has a Padé [4,4] dispersion. A model with the same nonlinearity and dispersion was also derived by Gobbi et al.,[7] but is in terms of a weighted velocity. Hereafter, we refer to equation (2.36) and (2.37) as BouN4D4 (N4 denotes being fully nonlinear up to $O(\mu^4)$ and D4 denotes Padé [4,4] dispersion). This model can reduce to the enhanced equations in terms of the depth-averaged velocity presented by Madsen and Schäffer[14] (Eqs. (3.2) and (4.2) in their paper), which are referred to as MS98-1 hereafter, when the nonlinear terms higher than $O(\varepsilon\mu^4)$ are all neglected. As seen in Section 7, these neglected terms have significant contribution to amplitude dispersion effects (see Figs. 11 and 13).

Although the nonlinearities of the $O(\mu^4)$ six-parameter equations and BouN4D4 have been increased to be full nonlinear up to $O(\mu^4)$ with $\varepsilon = O(1)$, the third order nonlinear characteristics of these models, especially the amplitude dispersion, are still not good for larger relative water depth, kh (see Fig. 11). So here we present anther way to further improve the nonlinear characteristics of Boussinesq models. This approach is to replace the original velocity expression (2.7) for the formulation of Eqs. (2.4) and (2.5) by the computation velocity $\tilde{\boldsymbol{u}}$ determined by following new velocity expression.

$$\bar{\boldsymbol{u}} = \tilde{\boldsymbol{u}} - a_1 h^2 \nabla^2 \tilde{\boldsymbol{u}} - a_2 h \nabla^2 (h\tilde{\boldsymbol{u}}) - b_1 h^4 \nabla^2 \nabla^2 \tilde{\boldsymbol{u}} - b_2 h \nabla^2 \nabla^2 (h^3 \tilde{\boldsymbol{u}})$$
$$- d_1 h \eta \nabla^2 \tilde{\boldsymbol{u}} - d_2 \eta^2 \nabla^2 \tilde{\boldsymbol{u}} \tag{2.40}$$

Compared to (2.7), this expression includes two extra nonlinear terms (the last two terms). The coefficients of these terms, free parameters d_1 and d_2, can be determined by optimizing the third nonlinear characteristics of the resulting equations (see Fig. 11), and have the values of d_1=0.13 and d_2=0.495 (Zou and Fang[1]). Similarly to the derivation of Eqs. (2.4) and (2.5) (see also the formulation of (3.10) and (3.11)), substituting (2.40) into (2.1), (2.2) and (2.9) leads to the following set of equations:

$$\eta_t + \nabla \cdot (d\tilde{\boldsymbol{u}}) = a_1 \nabla \cdot [d h^2 \nabla(\nabla \cdot \tilde{\boldsymbol{u}})] + a_2 \nabla \cdot [d h \nabla(\nabla \cdot (h\tilde{\boldsymbol{u}}))]$$
$$+ \frac{b_1}{a} \nabla \cdot \{h^2 \nabla[\eta_t + \nabla \cdot (d\tilde{\boldsymbol{u}})]\} + \frac{b_2}{a} \nabla \cdot \{\nabla[h^2 \eta_t + h^2 \nabla \cdot (d\tilde{\boldsymbol{u}})]\}$$
$$+ \nabla \cdot [d\eta(d_1 h + d_2 \eta) \nabla(\nabla \cdot \tilde{\boldsymbol{u}})], \tag{2.41}$$

$$\tilde{\boldsymbol{u}}_t + (\tilde{\boldsymbol{u}} \cdot \nabla)\tilde{\boldsymbol{u}} + g\nabla\eta + \boldsymbol{G} =$$
$$(a_1 - \frac{1}{6}) h^2 \nabla(\nabla \cdot \tilde{\boldsymbol{u}}_t) + (a_2 + \frac{1}{2}) h \nabla[\nabla \cdot (h\tilde{\boldsymbol{u}}_t)]$$
$$+ c_1 h^2 \nabla[\nabla \cdot (\tilde{\boldsymbol{u}}_t + g\nabla\eta)] + c_2 \nabla[\nabla \cdot (h^2 \tilde{\boldsymbol{u}}_t + g h^2 \nabla\eta)] \tag{2.42}$$

with $$\boldsymbol{G} = \nabla[\frac{d^2}{2}(\nabla\cdot\tilde{\boldsymbol{u}})^2 - (\frac{d^2}{3} + ah^2 + d_1 h\eta + d_2\eta^2)(\tilde{\boldsymbol{u}}\cdot\nabla)\nabla\cdot\tilde{\boldsymbol{u}}$$
$$-\frac{h^2}{2}c\nabla(\nabla\cdot(\tilde{\boldsymbol{u}}\cdot\tilde{\boldsymbol{u}}))] - [(\frac{2}{3}+d_1)h\eta + (\frac{1}{3}+d_2)\eta^2]\nabla(\nabla\cdot\tilde{\boldsymbol{u}}_t)$$
$$-d\nabla\eta\nabla\cdot\tilde{\boldsymbol{u}}_t - (d_1 h + 2d_2\eta - \frac{1}{3}d)\eta_t\nabla\nabla\cdot\tilde{\boldsymbol{u}}\ . \qquad (2.43)$$

Compared to the original Eqs. (2.4) and (2.5), the above equations contain new nonlinear terms related to the parameters d_1 and d_2. Since in the derivation for Eqs. (2.41) and (2.42), the starting Eqs. (2.1) and (2.2) with $\boldsymbol{G}$ given by (2.9) are already fully nonlinear up to $O(\mu^2)$, the above new nonlinear terms will cancel each other at $O(\mu^2)$, i.e., they will not affect the nonlinearity at $O(\mu^2)$, while they will contribute to the nonlinear characteristics of the equations at higher order in term of μ^2 (see Fig. 11). For the detailed analysis, see Zou and Fang[1]. Hereafter, we refer Eqs. (2.41) and (2.42) as the eight-parameter equations, since they contain eight parameters (a_1, a_2, b_1, b_2, c_1, c_2, d_1, d_2).

The physical implication of the expression (2.40) for the computation velocity $\tilde{\boldsymbol{u}}$ can be seen from the following theoretical relation between the depth-averaged velocity $\bar{\boldsymbol{u}}$ and the velocity at an arbitrary z location $\hat{\boldsymbol{u}}$ for horizontal bottom case[14],

$$\bar{\boldsymbol{u}} = \hat{\boldsymbol{u}} + (\frac{1}{3}h^2 + z_\alpha(h+\frac{1}{2}z_\alpha) - \frac{1}{3}h\eta - \frac{1}{6}\eta^2)\nabla\nabla\cdot\hat{\boldsymbol{u}}$$
$$+(\frac{2}{15}h^4 + \frac{1}{4}z_\alpha h^3 + \frac{13}{24}z_\alpha^2 h^2 + \frac{7}{12}z_\alpha^3 h + \frac{5}{24}z_\alpha^4)\nabla\nabla\cdot\nabla\nabla\cdot\hat{\boldsymbol{u}} \qquad (2.44)$$

where z_α is the vertical coordinate of the velocity $\hat{\boldsymbol{u}}$. Comparing (2.44) with the following expression of (2.40) for horizontal bottom case,

$$\bar{\boldsymbol{u}} = \tilde{\boldsymbol{u}} - (a_1 + a_2)h^2\nabla\nabla\cdot\tilde{\boldsymbol{u}} - (b_1 + b_2)h^4\nabla\nabla\cdot\nabla\nabla\cdot\tilde{\boldsymbol{u}}$$
$$-d_1 h\eta\nabla\nabla\cdot\tilde{\boldsymbol{u}} - d_2\eta^2\nabla\nabla\cdot\tilde{\boldsymbol{u}}\ , \qquad (2.45)$$

we see that if we take

$$a = a_1 + a_2 = -(\frac{1}{3} + \alpha(1+\frac{1}{2}\alpha)),\ a_1 = \frac{1}{3},\ d_2 = \frac{1}{6} \qquad (2.46a)$$

$$b = b_1 + b_2 = -\left(\frac{2}{15} + \frac{1}{4}\alpha + \frac{13}{24}\alpha^2 + \frac{7}{12}\alpha^3 + \frac{5}{24}\alpha^4\right) \quad (2.46b)$$

where $\alpha = z_\alpha / h$, then (2.40) recovers the exact relation between $\bar{u}$ and $\hat{u}$, that is, $\tilde{u}$ is just the velocity at an arbitrary z-level for horizontal bottom. For other values of the parameters, the velocity $\tilde{u}$ given by (2.40) is not a physical velocity, but a pseudo velocity. The depth-averaged velocity $\bar{u}$ can be obtained by this pseudo velocity with the application of (2.40), and from $\bar{u}$ we can obtain the velocity $\hat{u}$ at an arbitrary z location from a relation between $\bar{u}$ and $\hat{u}$, such as that shown by (2.44). The detail is shown in Zou and Fang[1].

3. Discussion about the Free Parameters Needed for the Improvement of Equations

We have seen from the previous section that there are a number of free parameters for improved Boussinesq models. Here we give some explanations about the free parameters. There are three purposes for introduction of free parameters, improvements of dispersion, shoaling property and nonlinear characteristics of the Boussinesq equations. The introductions of the parameters for improvements of equation's nonlinear characteristics are explained in Zou[20] for the parameter B_3 and in the previous section of the present paper for the parameters (d_1, d_2). Further discussions are given in Section 7. The introduction of the parameters B_2 for Eqs. (2.1) and (2.2), a_2 and b_2 in Eqs. (2.4) and (2.5), and the parameters α_2 and β_2 in BouN2D4 and BouN4D4 for the improvement of shoaling property is relatively simple: just by including the higher order terms related to bottom gradient during the process of the improvement of dispersion. The introduction of free parameters for the improvement of dispersion is more of a subtle manipulation, and there are many ways to do this. The development of the approach is more an art than science. For example, to achieve a Padé [2,2] dispersion for the $O(\mu^2)$ Boussinesq equations, there are several ways for the introduction of the parameter, one way, given by Witting[8], is replacing the coefficients of Taylor expansion of velocity by arbitrary constants, which are used as free parameters of the derived equations. Although two parameters are contained in Witting's model, only one is actually needed to obtain a

Padé [2,2] dispersion (see Madsen and Schäffer[2]). Another way, given by Nwogu[4], is taking the velocity at an arbitrary vertical level as the velocity variable of derived equations, and the vertical location of the velocity is the free parameter for this purpose. The third way, given by Madsen et al.,[2] is adding higher order terms to the classical Boussinesq equations, which contain a free parameter for the optimization of model's dispersion. The fourth way, given by Zou[20], is expressing the second order velocity by its homogeneous solution, the coefficient of the solution can be adjustable to achieve a Padé [2,2] dispersion. For all these approaches, only one parameter is needed. If we go further to achieve the Padé [4,4] dispersion, more parameters are needed, and the number of free parameters depends on the equations to be enhanced. For the equations accurate to $O(\mu^4)$, like (2.15) and (2.16), (2.30) and (2.31), two free parameters (α_1, β_1) are needed, as shown by (2.21)-(2.24), (2.32)-(2.35). For the equations accurate to $O(\mu^2)$, like the equations of Nwogu[4] or the $O(\mu^2, \varepsilon^3\mu^2)$ equations of Wei et al.,[5] three parameters are needed: the vertical location (z_α) of an arbitrary velocity variable is taken as another parameter apart from α_1 and β_1 (see Wei et al.,[5] Schäffer and Madsen[21], Madsen and Schäffer[14]). Another way for this enhancement is shown by Zou[9]. In this approach free parameters are introduced by using the velocity expression (2.7), the resulting equations are obtained by substituting this velocity expression into the three-parameter equations (2.1) and (2.2) (with B_3=1.0) with a Padé [2,2] dispersion, and this leads to the six-parameter equations (2.4) and (2.5) with a Padé [4,4] dispersion. Three free parameters, (a, b/a, c), are needed by this model for the optimization of dispersion. Actually, the number of free parameter can also be reduced to be two, if the resulting equations are allowed to contain fourth and fifth spatial derivatives. For this case, we only need to retain the parameters (a, b) with the third parameter c dropped. Now we show this for the following reduced version of (2.30) and (2.31) for a horizontal-bottom.

$$\eta_t + \nabla\cdot[(h+\eta)\bar{\boldsymbol{u}}] = 0, \tag{3.1}$$

$$\bar{\boldsymbol{u}}_t + (\bar{\boldsymbol{u}}\cdot\nabla)\bar{\boldsymbol{u}} + g\nabla\eta + \boldsymbol{G}_1 + \boldsymbol{G}_2 = \frac{1}{3}h^2\nabla^2\bar{\boldsymbol{u}}_t + \frac{1}{45}h^4\nabla^2\nabla^2\bar{\boldsymbol{u}}_t \tag{3.2}$$

with $$\boldsymbol{G}_1 = \nabla[d^2(\frac{1}{2}(\nabla\cdot\overline{\boldsymbol{u}})^2 - \frac{1}{3}\overline{\boldsymbol{u}}\cdot\nabla^2\overline{\boldsymbol{u}})] + \frac{1}{3}d\eta_t\nabla^2\overline{\boldsymbol{u}}$$

$$-d\nabla\eta\nabla\cdot\overline{\boldsymbol{u}}_t \; -\frac{1}{3}(d^2-h^2)\nabla^2\overline{\boldsymbol{u}}_t\,. \tag{3.3}$$

The expression for $\boldsymbol{G}_2$ is given in Appendix. Substituting the velocity expression (2.7) into the above equations and taking the linearized form of the resulting equations yields

$$\eta_t + \nabla\cdot(h\tilde{\boldsymbol{u}}) = ah^3\nabla\cdot[\nabla\nabla\cdot\tilde{\boldsymbol{u}})] + bh^5\nabla\cdot[\nabla\nabla\cdot\nabla\nabla\cdot\tilde{\boldsymbol{u}}]\,, \tag{3.4}$$

$$\tilde{\boldsymbol{u}}_t + g\nabla\eta = (\frac{1}{3}+a)h^2\nabla\nabla\cdot\tilde{\boldsymbol{u}}_t + (\frac{1}{45}+b-\frac{1}{3}a)h^4\nabla\nabla\cdot\nabla\nabla\cdot\tilde{\boldsymbol{u}}_t\,. \tag{3.5}$$

This set of equations has the dispersion relation of the form

$$\frac{\omega^2}{k^2} = gh\frac{1+ak^2h^2-bk^4h^4}{1+(\frac{1}{3}+a)k^2h^2-(\frac{1}{45}+b-\frac{1}{3}a)k^4h^4}\,. \tag{3.6}$$

Matching it with the exact solution of dispersion relation

$$\frac{\omega^2}{k^2} = gh\frac{\tanh kh}{kh} = gh\frac{1+k^2h^2/9+k^4h^4/945}{1+4k^2h^2/9+k^4h^4/63} + O(k^{10}h^{10}) \tag{3.7}$$

yields $a=1/9$ and $b=-1/945$. The above dispersion Eq. (3.6) is the same as that of Eqs. (2.36) and (2.37) (or (2.26) and (2.27)) if we take $a=\alpha_1$ and $b=\beta_1$. This result shows that only two parameters are needed to obtain a Padé [4,4] dispersion for the improvement of equations accurate to $O(\mu^4)$, such as (2.30) and (2.31), while as a cost, the resulting Eqs., (3.4) and (3.5) (similarly (2.26) and (2.27), (2.36) and (2.37)) will contain the fourth and fifth spatial derivatives, which are not easier to deal with in numerical solutions. One way to get rid of this problem is to transform these higher order terms into second order spatial derivatives by applying the following lower order form of (3.4) and (3.5):

$$\eta_t + \nabla\cdot(h\tilde{\boldsymbol{u}}) = ah^3\nabla\cdot[\nabla\nabla\cdot\tilde{\boldsymbol{u}})] + O(\mu^4)\,, \tag{3.8}$$

$$\tilde{\boldsymbol{u}}_t + g\nabla\eta = (\frac{1}{3}+a)h^2\nabla\nabla\cdot\tilde{\boldsymbol{u}}_t + O(\mu^4) \tag{3.9}$$

to the fifth spatial derivative in (3.4) and fourth spatial derivative in (3.5), respectively. Then, we have

$$\eta_t + \nabla \cdot (h\tilde{\boldsymbol{u}}) = ah^3 \nabla \cdot [\nabla\nabla \cdot \tilde{\boldsymbol{u}})] + (\frac{b}{a}) h^2 \nabla^2 [\eta_t + \nabla \cdot (h\tilde{\boldsymbol{u}})] , \quad (3.10)$$

$$\tilde{\boldsymbol{u}}_t + g\nabla\eta = (\frac{1}{3} + a) h^2 \nabla\nabla \cdot \tilde{\boldsymbol{u}}_t + \frac{(1/45 + b - a/3)}{1/3 + a} h^2 \nabla\nabla \cdot (\tilde{\boldsymbol{u}}_t + g\nabla\eta) . \quad (3.11)$$

But it can be verified that the dispersion relation of the above equations is not capable of matching the Padé [4,4] dispersion (3.7). Actually, an extra parameter is needed to achieve this goal, this is shown by the formulation of the six-parameter equations (2.4) and (2.5), in which the transformations like (3.8) and (3.9) are also applied and in the meantime another free parameter $B(=B_1+B_2)$ is introduced into Eqs. (2.1) and (2.2). These results show that the number of free parameters is dependent on the accuracy of equations to be enhanced and also on the final form of the enhanced equations as well.

The corresponding nonlinear version of (3.4) and (3.5), which corresponds to Eqs. (2.11) and (2.12) but have four parameters (a_1, a_2, b_1, b_2) instead of six parameters, can be written out. This is straightforward and not presented here. By optimizing the shoaling property of the model, we can obtain the values of the model's parameters:

$$a_1 = 0.1759,\ a_2 = -0.0648,\ b_1 = -0\ 00206,\ b_2 = 0.001 . \quad (3.12)$$

4. Numerical Solution Method

For different forms of the Boussinesq models presented in the last section, similar numerical solving procedures can be adopted for the numerical solutions. Here we only take the 1-D six-parameter equations as an example to describe the numerical scheme briefly. (i) Rewrite Eqs. (2.4) and (2.5) with $\boldsymbol{G}$ given by (2.10) in the following forms

$$\zeta_t = E , \quad U_t = F + F^t \quad (4.1a,b)$$

where
$$\zeta = \eta - \frac{b_1}{a}(h^2\eta_x)_x - \frac{b_2}{a}(h^2\eta)_{xx} , \quad (4.2)$$

$$U = \tilde{u} - (a_1 + c_1 - \frac{1}{6})h^2\tilde{u}_{xx} - (a_2 + \frac{1}{2})h(h\tilde{u})_{xx} + c_2(h^2\tilde{u})_{xx}, \tag{4.3}$$

$$E = -(\mathrm{du})_{\mathrm{x}} + a_1(dh^2\tilde{u}_{xx})_x + a_2[dh(h\tilde{u})_{xx}]_x + \frac{b_1}{a}[h^2(d\tilde{u})_{xx}]_x + \frac{b_2}{a}[h^2(d\tilde{u})_x]_{xx}, \tag{4.4}$$

$$F = -\tilde{u}\tilde{u}_x - g\eta_x + c_1 gh^2\eta_{xxx} + c_2 g(h^2\eta_x)_{xx} - [\frac{d^2}{2}(\tilde{u}_x)^2 - (\frac{d^2}{3} + ah^2)\tilde{u}\tilde{u}_{xx} - \frac{1}{2}ch^2(\tilde{u}^2)_{xx}]_x \tag{4.5}$$

$$F^t = \eta_x d\tilde{u}_{xt} + (\frac{2}{3}h + \frac{1}{3}\eta)\eta\tilde{u}_{xxt} + \frac{d}{3}\eta_t\tilde{u}_{xx}. \tag{4.6}$$

(ii) Discretize the above equations in the spatial and temporal coordinates. A five-diagonal system is then obtained after using five-point difference formula for the terms at left hand side of equations and evaluating the terms at the right hand side of equations. (iii) Integrate equations using Adams-Bashforth-Moulton scheme and iterate the corrector step until the desired solutions are found. (iv) Update the variables and go to the step (i). To this end, the numerical scheme follows Funwave[24] or Gobbi and Kirby[6], but some illuminations are made as follows. First, the truncation error of difference formula adopted to evaluate the spatial derivatives are of higher order than dispersion accuracy embodied in the equations to avoid overwhelming the model's dispersion. Second, Shapiro's[25] 4th order filter is used every N_f time steps ($40<N_f<300$). This is necessary to remove extremely high frequency noise caused by bathymetry sharp variations, nonlinearity of wave motions or discretization errors. When the number of grids per wavelength is taken to be greater than 40, the use of this filter will not affect the solutions in any significant way while removing the Nyquist frequency oscillation.

All the models presented in Section 2 have been solved in the present study by using the above numerical scheme for both 1-D and 2-D simulations. In comparisons, the models with the mild slope assumption (three-parameter equations, six-parameter equations, eight-parameter equations) are easier for coding and more efficient than those without this assumption (BouN2D2, BouN2D4, BouN4D4). For this reason, the former type of models is recommended for practical applications. For this purpose, we need to make sure the effect of the mild slope assumption on numerical results, and this is shown in Section 6. Another concern for the practical applications of these models is the effect of different approximation levels of nonlinearity of these models on the numerical results, and this is investigated in Sections 5 and 7. Only the results for 1-D simulation are given due to the limited scope of this chapter.

5. Inter-Comparisons for Nonlinearity

The Boussinesq models shown in Section 2 have different nonlinearities, i.e., the nonlinearity $\varepsilon = O(\mu^2)$ for the three parameter equations ((2.1) and (2.2)) with $\boldsymbol{G}$ given by (2.3), the six-parameter equations ((2.4) and (2.5)) with $\boldsymbol{G}$ given by (2.6); the nonlinearity $\varepsilon = O(\mu^{2/3})$ for the three parameter equations with $\boldsymbol{G}$ given by (2.9), the six-parameter equations with $\boldsymbol{G}$ given by (2.10), BouN2D2 and BouN2D4; the nonlinearity $\varepsilon = O(1)$ for the $O(\mu^4)$ six-parameter equations ((2.11) and (2.12)) and BouN4D4. The effects of these different approximation levels of nonlinearity on the simulations of nonlinear wave propagation are first examined in this section for the solitary wave propagation on a plane beach. Figure 1 gives the numerical results of the six-parameter equations with $\boldsymbol{G}$ given by (2.6) and (2.10), respectively. The numerical solution of a three dimensional fully nonlinear potential theory by a boundary element method[26] (referred to as FNPF in the figure) is also given as an exact reference result. It is seen that the six-parameter equations with $\boldsymbol{G}$ given by (2.10) give the results which are closer to the exact reference result. This shows that the increase of nonlinearity from $\varepsilon = O(\mu^2)$ to $\varepsilon = O(\mu^{2/3})$ is necessary for this stronger nonlinear wave motions.

Another examination for the validation of the model's nonlinear property is for the wave propagation over a submerged breakwater. The experimental results of Luth et al.[27] are used here. Figure 2 gives the experiment setup and Table 1 gives the wave conditions. The six-parameter equations with $\boldsymbol{G}$ given by (2.10), the eight-parameter equations ((2.41) and (2.42)) and the $O(\mu^4)$ six-parameter equations ((2.11) and (2.12)) are inter-compared for this case. The third model has the same dispersion as the six-parameter equations but different nonlinearity, so the difference between the first and third models' results can show

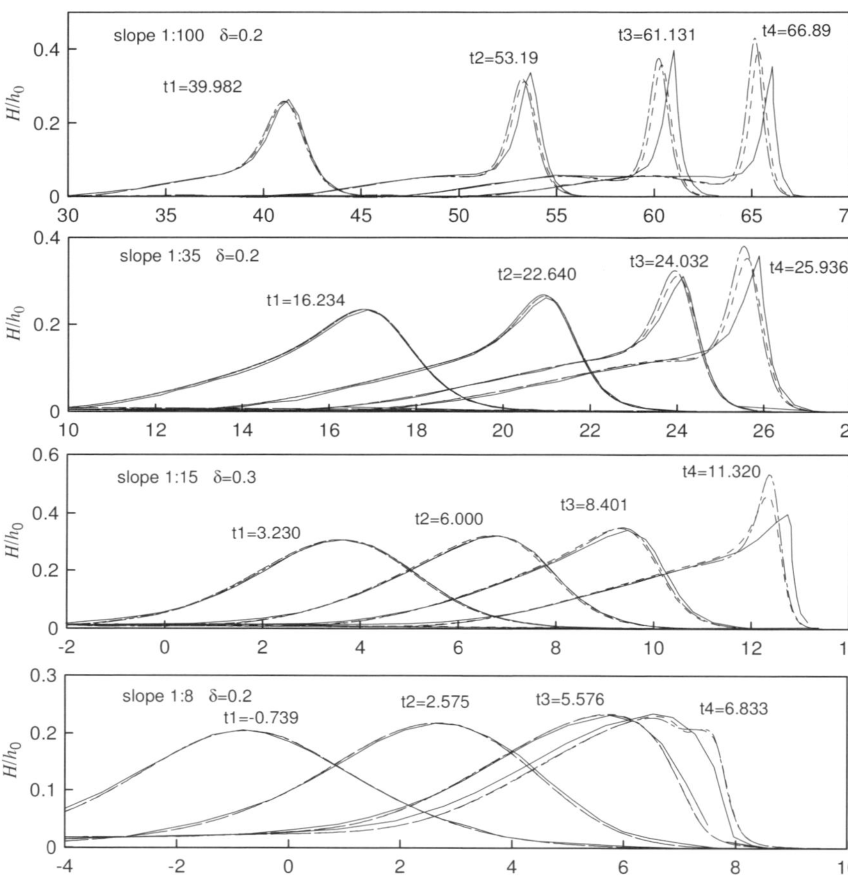

Fig. 1. Results for solitary wave profile over four plane beaches at four different moments. Dash-dot: six-parameter equations with $\boldsymbol{G}$ given by (2.6); Dash: six-parameter equations with $\boldsymbol{G}$ given by (2.10). Solid: FNPF. (δ is the ratio of initial wave height (H_0) to constant water depth at the toe of slope (h_0). $x=0$ is located at the toe of slope and the time when solitary wave crest reaches the toe is set as $t=0$).

Table 1. Wave conditions for Luth et al.[27] and Ohyama et al.[28]

	Luth et al.'s experiment		Ohyama et al.'s experiment		
Case	Case (a)	Case (c)	Case(2)	Case(4)	Case(6)
Wave height(m)	0.02	0.041	0.05	0.05	0.05
Wave period(s)	2.02	1.01	1.341	2.012	2.683

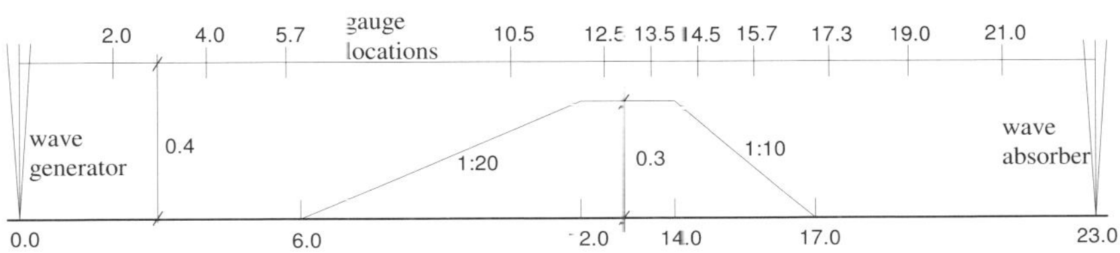

Fig. 2. Setup of Luth's[27] experiment. All dimensions in (m).

the contributions of the $O(\mu^4)$ nonlinear terms. The contributions of the new nonlinear terms related to parameters d_1 and d_2 in the eight-parameter equations, which are introduced by the velocity expression (2.40), can also be seen from the inter-comparisons, since the first model corresponds to $d_1=d_2=0$, and the difference between the results of this model and the eight-parameter equations can be attributed to the nonlinear terms related to the parameters d_1 and d_2. Figure 3 presents the time series of the wave free surface profiles and Fig. 4 presents the corresponding amplitudes of Fourier components of the time series. It is seen in Fig. 3 that for the longer wave case (case (a)), the three models give similar results and agree quite well with the measurement. The six-parameter equations give almost the same results as the $O(\mu^4)$ six-parameter equations, even though the former is without the $O(\mu^4)$ nonlinear terms. This shows that the $O(\mu^4)$ nonlinear effects are less important for the longer wave case. For the shorter period case (case (c)), there are some differences. This is due to the higher nonlinearity for this case (the nonlinearity of water waves increases with the decrease of wave period when the wave height is unchanged). This shows that the nonlinearity of $\varepsilon = O(\mu^{2/3})$ is not very appropriate for the stronger nonlinear waves. The eight-parameter equations give a little better result than the six-parameter equations. In Fig. 4, the eight-parameter equations also give the best results, even for the shorter wave case (case (c)). For

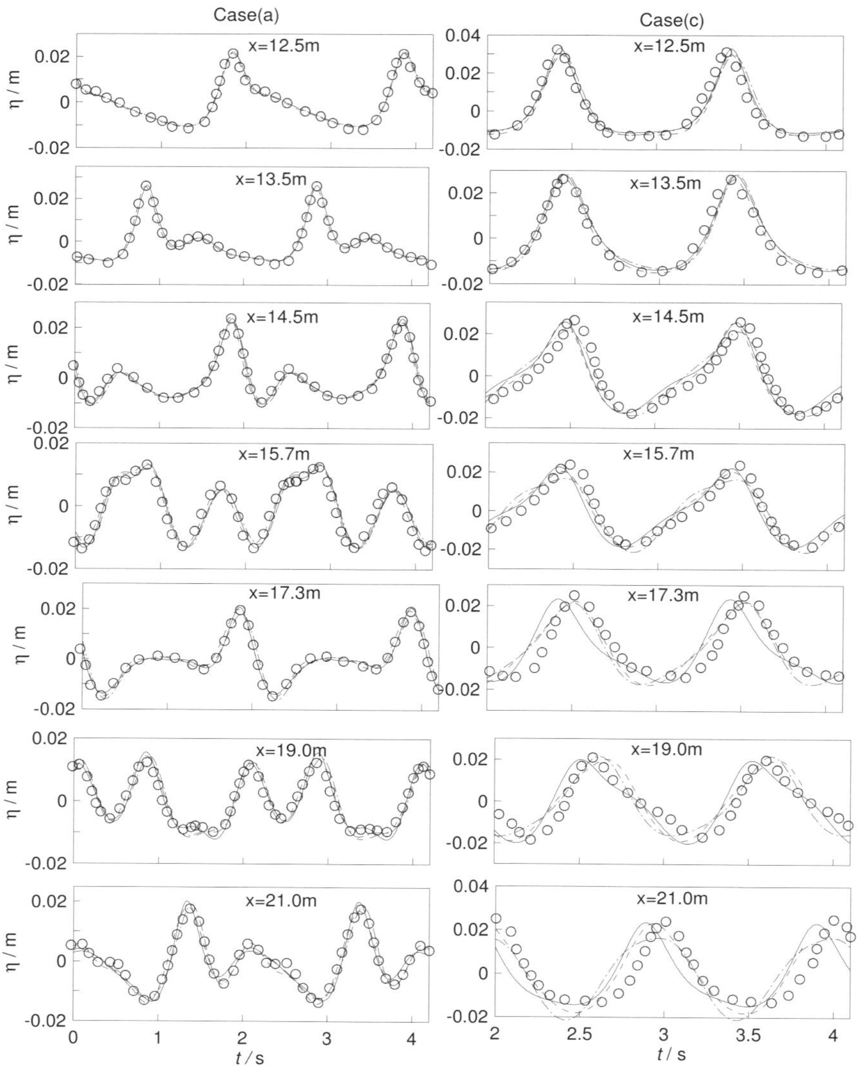

Fig. 3. Time series of free surface elevations for Luth's experiment. Left: case (a); Right: case (b). Symbol: data; Dash-dot: six-parameter equations with ***G*** given by (2.10). Dash: $O(\mu^4)$ six-parameter equations; Solid line: eight-parameter equations.

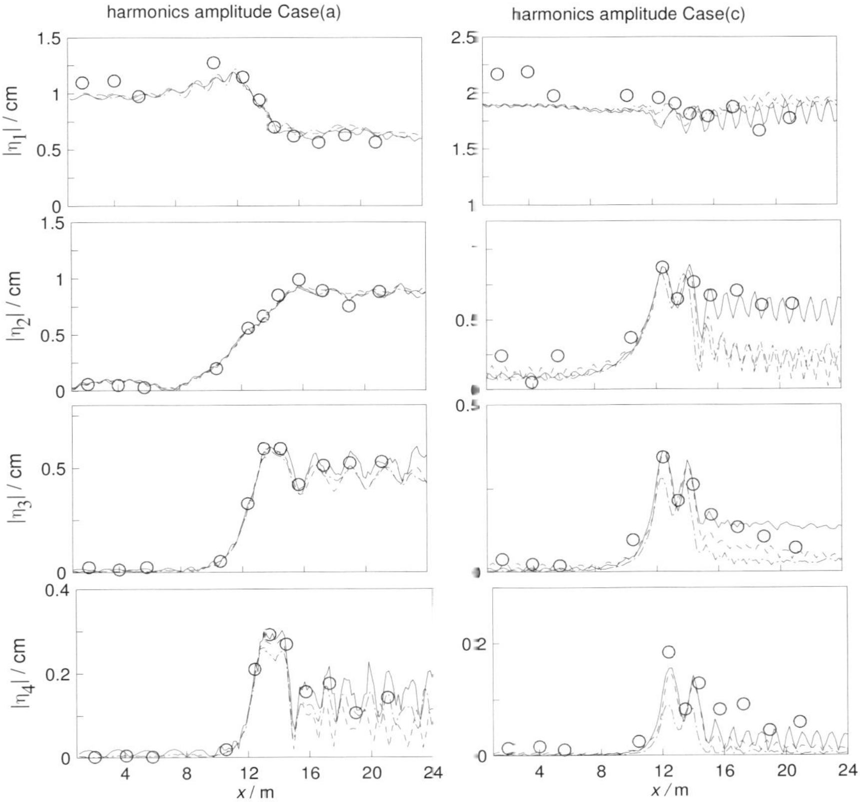

Fig. 4. Fourier components of wave free surface elevations shown in Fig. 3. Left: case (a); Right: case (b). Symbol: data; Dash-dot: six-parameter equations with ***G*** given by (2.10). Dash: $O(\mu^4)$ six-parameter equations; Solid line: eight-parameter equations.

this case, the other two models show bigger differences with measurement for higher order harmonics. The same trend also exists for case (a), but differences are very small.

The inter-comparisons between the six-parameter equations with ***G*** given by (2.6) and (2.10), respectively are also made for wave propagation over a breakwater of Luth et al.'s experiment, but it is shown that results of the two models are very similar, so the results are not given in Figs. 3 and 4 in order to save the figure space.

6. Inter-Comparisons for the Mild Slope Assumption

In order to validate the mild slope assumption used by the first group of models, which is defined by $\nabla h = O(\mu^2)$ for the three-parameter equations (2.1) and (2.2), the six-parameter equations (2.4) and (2.5) or $\nabla h = O(\mu^4)$ for the $O(\mu^4)$ six-parameter equations (2.11) and (2.12), the experimental results for wave propagations over breakwaters with steeper slopes are chosen here for the validation, they are from the two experiments and shown in Fig. 5. One experiment was conducted by Ohyama et al.,[28] the wave condition of this experiment includes both shorter and longer waves and is shown in Table 1. The front and back slopes of the breakwater in the experiment are both 1:2. Another experiment was by the present study and was done in the State Key Laboratory of Coastal and Offshore Engineering, Dalian University of Technology, P. R. China. Two breakwaters of the experiment are considered here. One is with 1:2 front slope and 1:10 back slope, and another is with 1:10 front slope and 1:2 back slope. These breakwaters can be seen as the extension of the breakwater of the Ohyama's experiment by changing one of breakwater slopes (the front slope or the back slope) of Ohyama et al.'s experiment from 1:2 to 1:10, see Fig. 5. Only the result for shorter wave condition with wave period 1.3s (the

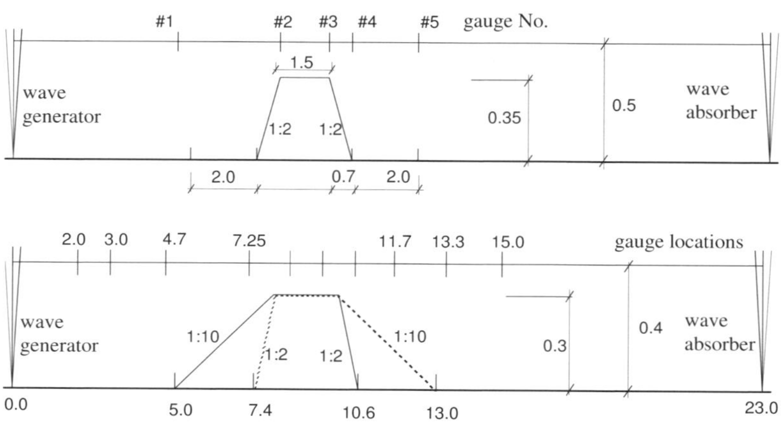

Fig. 5. Setups of Ohyama et al.[28] (up) and the present study experiment (down). All dimensions in (m).

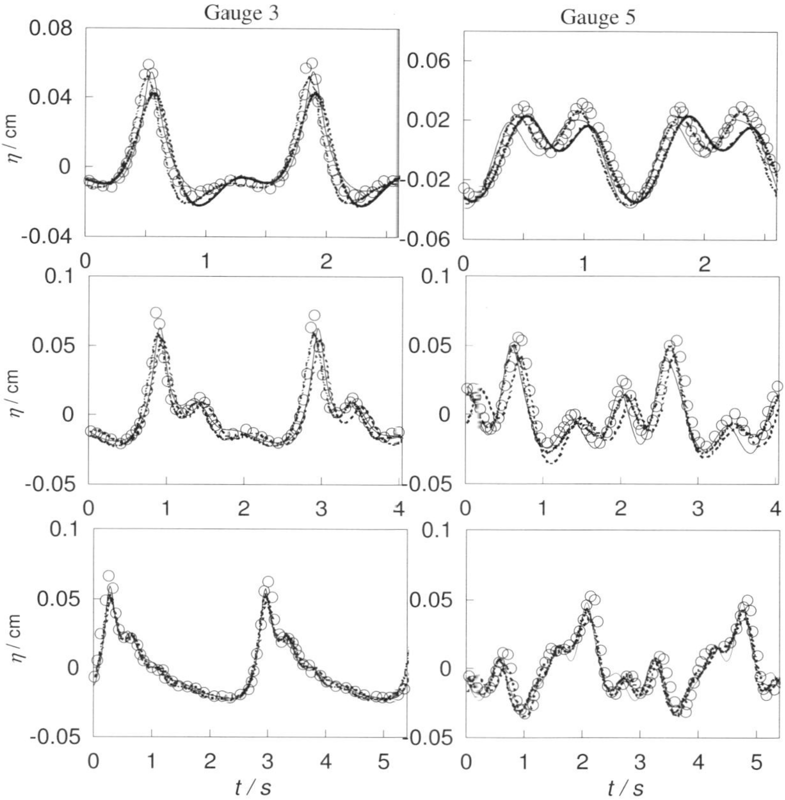

Fig. 6. Time series of free surface elevations for Ohyama's experiment (see Table 1). Top: case (2); Middle: case (4); Bottom: case (6). Symbol: data; Dash line: six-parameter equations ***G*** with by (2.10); Dash-dot: BouN2D4; Solid line: eight-parameter equations.

wave height is 0.02m) is considered here, since this is the case with a larger *kh* (or μ) and steep bottom slopes and can serve as a severe check of the mild slope assumption.

Figure 6 shows the results of the six-parameter equations with ***G*** given by (2.10) and BouN2D4 for the three cases of Ohyama et al.'s experiment, which have the same incident wave height but different wave periods. Since the two models have the same accuracy in dispersion and nonlinearity but with and without the mild slope assumption, the comparison of the two results can show the effect of the

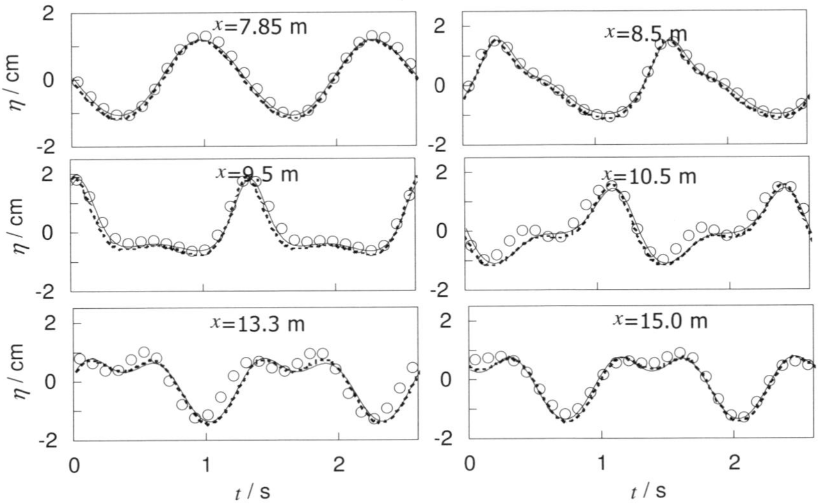

Fig. 7. Time series of free surface elevations over a breakwater with 1:2 front slope and 1:10 back slope. Symbol: data; Solid line: six-parameter equations ***G*** with by (2.10); Dash: BouN2D4.

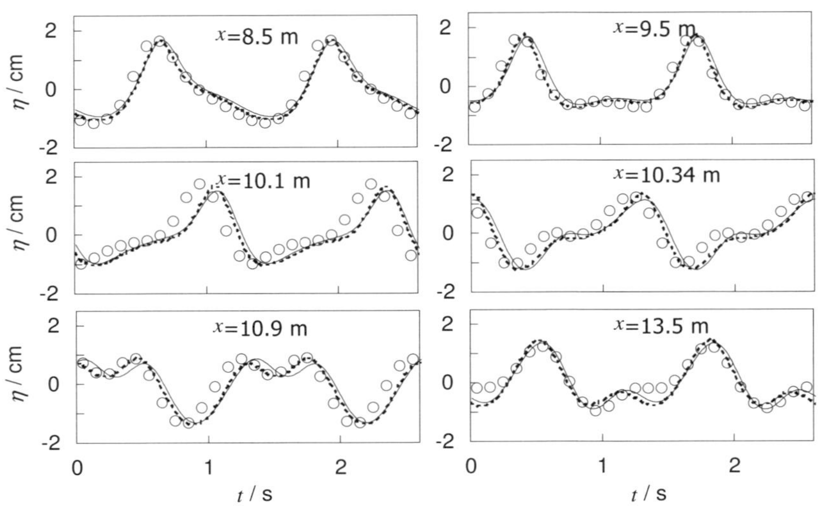

Fig. 8. Time series of free surface elevations over a breakwater with 1:10 front slope and 1:2 back slope. Symbol: data; Solid line: six-parameter equations ***G*** with by (2.10); Dash: BouN2D4.

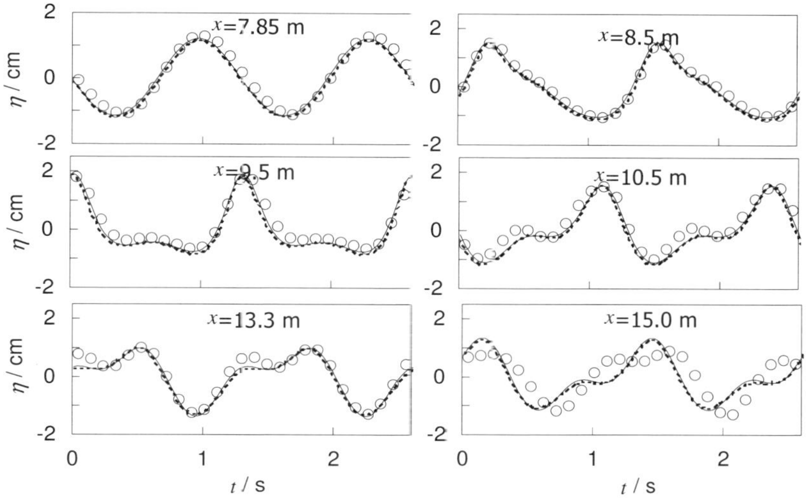

Fig. 9. Time series of free surface elevations over a breakwater with 1:2 front slope and 1:10 back slope. Symbol: data; Solid line: three-parameter equations ***G*** with by (2.9); Dash: BouN2D2.

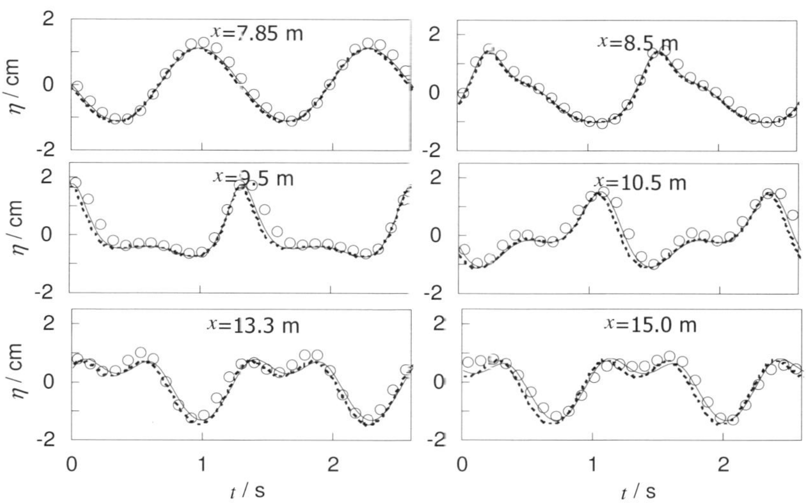

Fig. 10. Time series of free surface elevations over a breakwater with 1:2 front slope and 1:10 back slope. Symbol: data; Solid line: $O(\mu^4)$ six-parameter equations; Dash: BouN4D4.

mild slope assumption. It is seen from the figure that for the long wave case (case (6)) the difference between two model's results is small, while for the other two wave cases (cases (2) and (4)), it gets bigger. This shows that the mild slope assumption can lead to some error for shorter waves or a larger *kh* case. But it is found that for the case of the second experiment mentioned above (the breakwaters of this experiment have the same front or back slope (1:2) as that in the first experiment but another slope being 1:10 is different), the errors caused by the mild slope assumption are quite small. This can be seen from Figs. 7-10. Figures 7 and 8 show the numerical results of the six-parameter equations with $\boldsymbol{G}$ given by (2.10) and BouN2D4 for the two breakwaters of this experiment, respectively. Figure 9 shows the inter-comparison between three-parameter equations with $\boldsymbol{G}$ given by (2.9) and BouN2D2 in order to validate the mild slope assumption for different approximation levels of dispersion (the dispersions for Figs. 7 and 8 are the Padé [4,4] expansion while the dispersion for Fig. 9 is the Padé [2,2] expansion). Figure 10 shows the inter-comparison between the $O(\mu^4)$ six-parameter equations and BouN4D4 to validate the mild slope assumption for the different order of nonlinearity (the nonlinearity for Figs. 7 and 8 is $\varepsilon = O(\mu^{2/3})$ while the nonlinearity for Fig. 10 is $\varepsilon = O(1)$). The inter-comparisons in Figs. 9 and 10 are only done for the breakwater of 1:2 front slope and 1:10 back slope, since it is known from the results shown by Figs. 7 and 8 that the comparison results are almost the same for the two breakwaters. As in Figs. 7 and 8, Figs. 9 and 10 show that the difference between the results of the compared two models is also very small. This shows that the mild slope assumption can be applicable for these cases. This is contrary to the comparison result for the shorter wave case (case (2) and (4)) in Fig. 6. This may be because the breakwater of the present study experiment has one steep slope (1:2) with the other being mild slope (1:10) while the breakwater of Ohyama et al.'s experiment has two steep slopes (1:2).

7. Inter-Comparisons for Amplitude Dispersion Effects

In order to examine the effect of the accuracy of amplitude dispersion of different forms of higher order Boussinesq equations, we here first

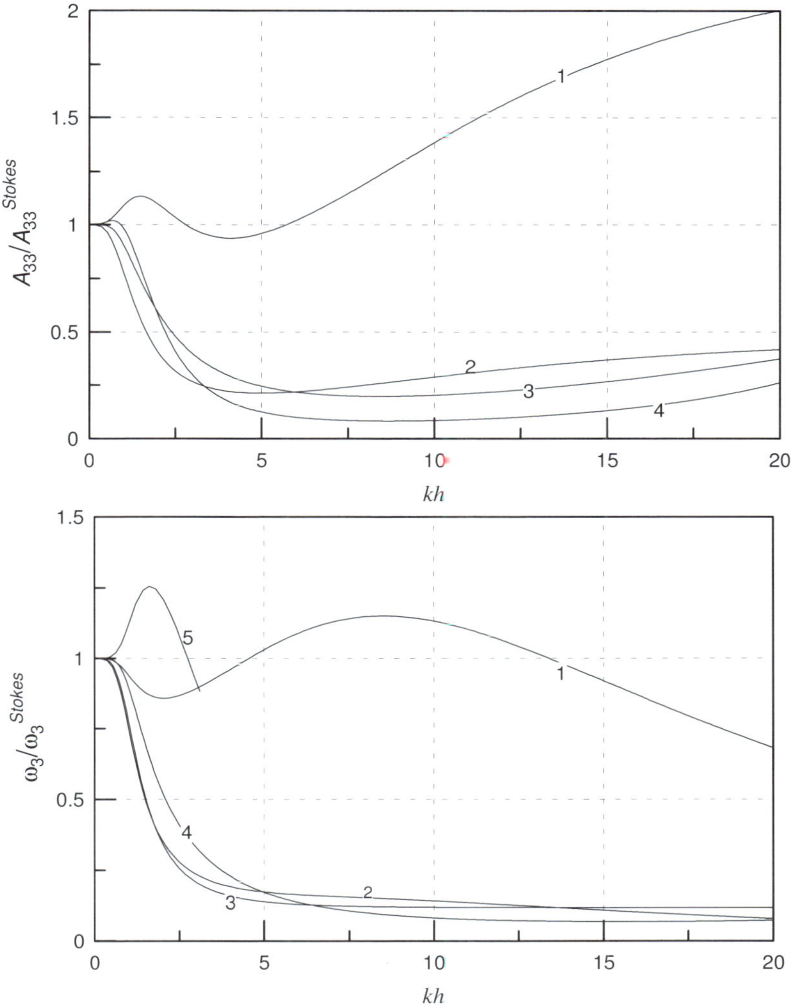

Fig. 11. Ratios of third harmonic A_{33} / A_{33}^{Stokes} and amplitude dispersion $\omega_3 / \omega_3^{Stokes}$. (1) eight-parameter equations; (2) six-parameter equations with $\boldsymbol{G}$ given by (2.10); (3) MS98-1 (4) enhanced (μ^2, $\varepsilon^3\mu^2$) equations in terms of the velocity $\hat{\boldsymbol{u}}$ at an arbitrary z location in Madsen and Schäffer[14] (the Eqs. (6.3a), (6.3b) in their paper). (5) equations of Gobbi, et al.[7]

compare the third harmonic amplitudes and amplitude dispersions of these models defined by the following Fourier expansions

$$\eta = A_1 \cos\vartheta + \varepsilon(A_{20} + A_{22})\cos 2\vartheta + \varepsilon^2 (A_{31}\cos\vartheta + A_{33}\cos 3\vartheta), \quad (7.1)$$

$$\tilde{u} = U_1 \cos\vartheta + \varepsilon U_{22}\cos 2\vartheta + \varepsilon^2 (U_{31}\cos\vartheta + U_{33}\cos 3\vartheta), \quad (7.2)$$

$$\omega = \omega_0 + \varepsilon^2 \omega_3 \quad (7.3)$$

where $\vartheta = kx - \omega t$, k and ω are the wave number and frequency. Then, we use these models to simulate the experiment of Shemer et al.[29] for the nonlinear evolution of wave groups in a flume. Figure 11 shows the ratios of the third harmonic amplitude A_{33} / A_{33}^{Stokes} and amplitude dispersion $\omega_3 / \omega_3^{Stokes}$ from different models (A_{33}^{Stokes} and ω_3^{Stokes} are the third harmonic amplitude and amplitude dispersion of the Stokes wave theory). It is seen that the eight-parameter equations have largely extended the model's application range compared to the original equations (2.4) and (2.5) (the six-parameter equations) with $\boldsymbol{G}$ given by (2.10) and to other forms of higher order Boussinesq equations shown in the figure as well.

The Shemer's experiment is a good check for the amplitude dispersion property of the Boussinesq model, since during the propagation of wave groups, wave components with higher amplitude travels faster than those with lower amplitude due to amplitude dispersion and thus wave envelope will show skewness. New wave components will be generated during the process. Here we consider the case of the Shemer et al.'s experiment in which wave maker is driven by the following signal

$$s(t) = s_0 \left|\cos(\Omega t)\right| \cos(\varpi_0 t) \qquad (\Omega = \varpi_0 / 20) \quad (7.4)$$

where, s_0 is the forcing amplitude, ϖ_0 is the carrier frequency, and Ω is the modulation frequency. The spectrum of the signal has a maximum at the carrier frequency ϖ_0 plus several smaller peaks at discrete frequencies spaced by $\varpi_0 / 10$, in which three dominant models exist (see Fig. 13). The period of carrier waves for the driving signal is 0.9s and the water depth is 0.6m, the corresponding value of *kh* is 3.0, so the wave motions are within deep water regime.

The six-parameter equations with $\boldsymbol{G}$ given by (2.10), BouD2D4, the eight-parameter equations, MS98-1 and BouN4D4 are inter-compared.

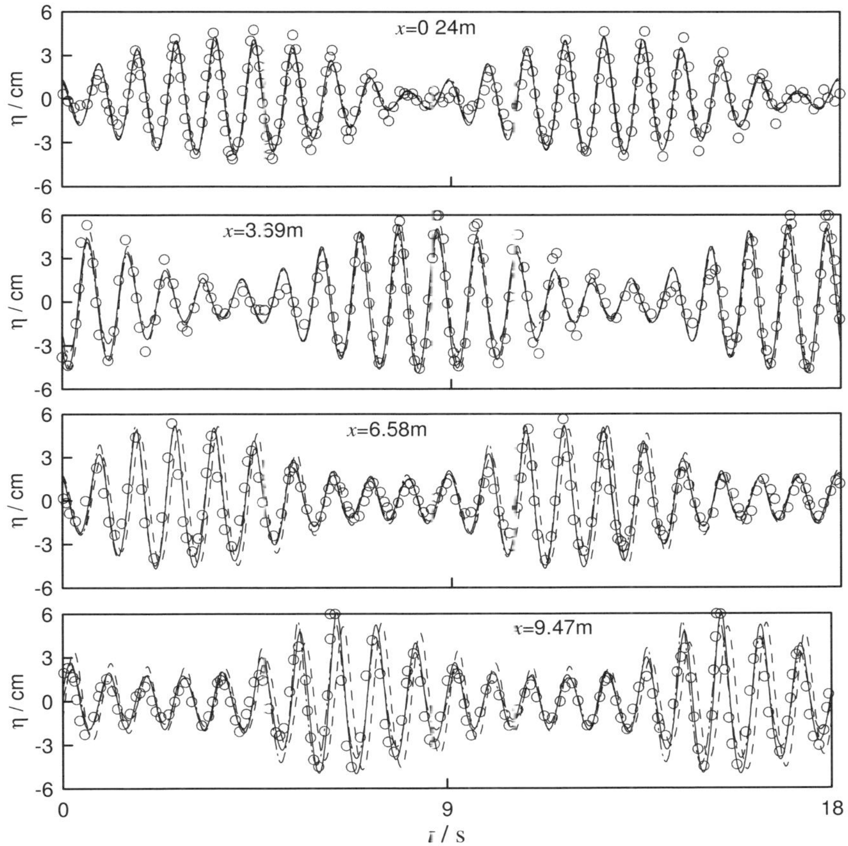

Fig. 12. Wave surface elevations of wave groups. Circle: data; Solid line: BouN4D4; Dash: BouN2D4; Dash-dot: eight-parameter equations.

The former three models have the nonlinearity of $\varepsilon = O(\mu^{2/3})$ (retaining all the nonlinear terms at $O(\mu^2)$) but are different in form, especially the third model, which contains the additional nonlinear terms related to the parameters d_1 and d_2; MS98-1 has the nonlinearity of $\varepsilon = O(\mu)$ (omitting all the nonlinear terms higher than $O(\varepsilon\mu^2)$); the last model has the nonlinearity of $\varepsilon = O(1)$ (being fully nonlinear up to $O(\mu^4)$). So the inter-comparisons between these models can show the effects of different nonlinearities and the enhancement for nonlinearity in the eight-parameter equations. The inter-comparison between the six-parameter equations with $\boldsymbol{G}$ given by (2.10) and BouD2D4 can check the accuracy of numerical schemes used for the two models, since the equations of the two models are of the same accuracy for the horizontal bottom case and should give the same results theoretically, although the forms of their dispersion terms are different. In the simulations, spatial and temporal steps are 0.02m and 0.01s, respectively.

Figure 12 presents the comparisons of wave surface elevations between the experiment data and the numerical results of BouN4D4, BouN2D4 and the eight-parameter equations. The wave maker is at position x=0m in a wave flume in the experiment. We see from this figure that near wave maker (x=0.24m), the numerical results of three models are almost same and all agree to the data, but as wave propagation along the wave flume, three models give different forms of wave envelopes. BouN4D4 and the eight-parameter predict elevations quite well, and give left-right asymmetric (skewed) wave envelopes which is similar to the data, the latter gives better result than the former, while BouN2D4 gives the basically left-right symmetric wave envelopes. Since change of wave envelope is due to amplitude dispersion, the above difference of the prediction results can be attributed to the different amplitude dispersions embodied in the models.

Figure 13 shows another phenomenon of nonlinear wave interaction, i.e. the generation of the new component waves due to resonant wave-wave interaction during wave evolution along the flume. This is a third order ($O(\varepsilon^3)$) nonlinear effect which causes the energy transfer from one frequency to another in a wave spectrum and is referred to as the four-wave resonant interaction[30]. For the present cases, although there are only three free-wave modes for the signal (7.4) near the wave maker (x=0.24), the four-wave resonant interaction will inspire other new free-wave modes to grow as waves propagate along the tank. This can be seen in Fig. 13 for the amplitude spectra at x=0.24m, 9.47m of the experiment. The spectra of the experiment are different at these two locations: compared to the spectrum at x=0.24m, the spectrum at x=9.47m has bigger wave amplitudes at the high frequency band (the right side of the spectrum). This asymmetry of the spectrum shape shows the generations of new wave components of higher frequency during wave groups propagate along the flume. Shemer et al.[29] used Zakharov equations to simulate this process in which 10 free-wave modes were chosen to calculate the evolution of wave spectrum along the tank. The cubic Schrödinger equation was also used by them for this simulation, but it is shown that the appearance of asymmetric wave groups and the

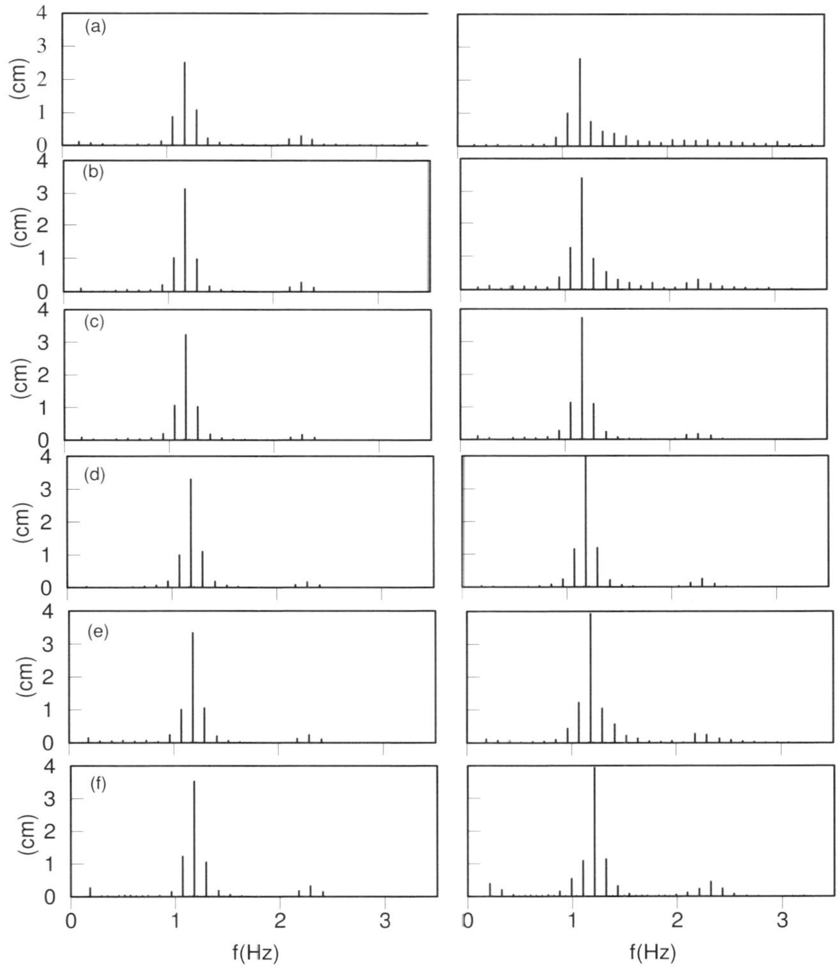

Fig. 13. Amplitude spectrum at x=0.24m (left), 9.47m (right) for wave group evolutions. (a) data; (b) eight-parameter equations; (c) six-parameter equations with G given by (2.10); (d) BouN2D4; (e) BouN4D4; (f) MS98-1.

generation of free wave modes could not be adequately described by this equation due to its symmetric properties and the lack of nonlinear resonant interactions of different wave components in the wave spectrum.

Among the results shown in Fig. 13, it is seen that the best is the eight-parameter equations, and the second is BouN4D4, but the six-parameter equations with ***G*** given by (2.10), BouD2D4, and MS98-1 give spectra which are almost the same for the two locations x=0.24m, 9.47m. This shows that the wave components given by the three models remain almost unchanged along the flume and they can hardly predict the development and generation of new wave components.

Besides the free-wave modes in the wave spectrum, there are also bound wave components, such as the waves with the sum and difference frequencies, but these components do not grow significantly along the tank as seen from the corresponding energy at these frequencies in the spectra of Fig. 13.

8. Conclusions

Inter-comparisons for seven sets of higher order Boussinesq equations are made in the present study in order to investigate the effects of different approximation levels of nonlinearity and the mild slope assumption on numerical results. The main conclusions are as follows.

(1) The inter-comparisons of numerical results of the six-parameter equations with ***G*** given by (2.6) and by (2.10) for solitary wave propagation on a beach show that the increase of nonlinearity from $\varepsilon = O(\mu^2)$ to $\varepsilon = O(\mu^{2/3})$ can increase the prediction accuracy. The inter-comparisons of the six-parameter equations with ***G*** given by (2.10), the six-parameter equations and the $O(\mu^4)$ six- parameter equations for wave propagation over a breakwater show that the eight-parameter equations give better results than the six-parameter equations with ***G*** given by (2.10). This shows that the contribution of the new nonlinear terms introduced by the nonlinear velocity expression (2.40) is positive for this case of wave motion. The $O(\mu^4)$ six- parameter equations also give better results than the six-parameter with ***G*** given by (2.10), and this shows that the increase of nonlinearity from $\varepsilon = O(\mu^{2/3})$ to $\varepsilon = O(1)$ can also increase the prediction accuracy for this weakly nonlinear wave motion.

(2) The mild slope assumption used for the higher order terms in the first group of models is checked by the inter-comparisons of the models with and without this assumption for the wave propagation over steep slope breakwaters (the Ohyama et al.'s experiment and the experiment of the present study). The similar results of the two groups of models (the three-parameter equations with $\boldsymbol{G}$ given by (2.9) vs. BouN2D2, the six-parameter equations with $\boldsymbol{G}$ given by (2.10) vs. BouN2D4, the $O(\mu^4)$ six-parameter equations vs. BouN4D4) for the experiment of the present study show that the mild slope assumption adopted by the first group of models is applicable for the case studied. But the numerical results for the breakwater of Ohyama et al.'s experiment with two steep slopes (1:2) have some errors for shorter wave cases, this indicates that for such a severe bottom variation (both slopes of the breakwater are 1:2) and shorter waves, the mild slope assumption may cause some errors.

(3) The effects of different amplitude dispersion accuracy of Boussinesq models are investigated by simulating the nonlinear evolution of wave groups in a flume and the results are compared with the corresponding measurement of Shemer et al.[29] It is shown that the nonlinearity of the order $O(\varepsilon^3\mu^2)$ possessed by the six-parameter equations with $\boldsymbol{G}$ given by (2.10) and BouN2D4 and the nonlinearity of the order $O(\varepsilon\mu^4)$ possessed by MS98-1 are not accurate enough for the simulation of amplitude dispersion effect on wave group evolution. The nonlinearity of the order of $O(\varepsilon^4\mu^4)$ possessed by BouN4D4 can lead to better results but not better than those given by the eight-parameter equations, the latter equations include the extra nonlinear terms which can improve the accuracy of amplitude dispersion for higher values of *kh*. Because four-wave resonant interaction is also involved in this process, the eight-parameter equations have also improved the prediction of this nonlinear behavior.

(4) In order to show the link between different forms of Boussinesq models presented in the present study, discussions are made about the number and types of the free parameters introduced for the improvements to the Boussinesq models. It is shown that the number of free parameters is dependent on the accuracy of equations to be enhanced and also on the final form of the enhanced equations, e.g., to obtain a Padé [4,4] dispersion, three parameters are needed for the $O(\mu^2)$ Boussinesq equations to be enhanced, while two parameters are needed

for the $O(\mu^4)$ Boussinesq equations to be enhanced. But if the transformations (3.8) and (3.9) are used for the latter equations, three parameters instead of two will be needed.

Acknowledgment

The authors would like to thank the financial support by the National Natural Science Foundation of China, under Grant Nos. 50479053, 10672034 and the support by Program for Changjiang Scholars and Innovative Research Team in University.

Appendix Expressions for G_2

$$\begin{aligned}
\boldsymbol{G}_2 = &\frac{1}{2}\nabla\{(\frac{1}{4}d^4+\frac{1}{36}\tilde{d}^2\tilde{d}^2)(\nabla^2\tilde{\boldsymbol{u}})^2-\frac{1}{3}d^4(\nabla\cdot\tilde{\boldsymbol{u}})\nabla^2(\nabla\cdot\tilde{\boldsymbol{u}})\\
&+(\frac{1}{12}d^4-\frac{1}{60}\tilde{d}^4)\tilde{\boldsymbol{u}}\cdot\nabla^2\nabla^2\tilde{\boldsymbol{u}}+\frac{1}{18}d^2\tilde{\boldsymbol{u}}\cdot\nabla^2(d^2\nabla^2\tilde{\boldsymbol{u}})\}+\frac{1}{6}d^3\nabla\eta\nabla^2(\nabla\cdot\tilde{\boldsymbol{u}}_t)\\
&+[\frac{1}{30}(d^4-h^4)+\frac{1}{3}(a-\frac{1}{6})(d^2-h^2)h^2]\nabla^2\nabla^2\tilde{\boldsymbol{u}}_t\\
&-(\frac{1}{30}d^3+\frac{1}{3}adh^2)\eta_t\nabla^2\nabla^2\tilde{\boldsymbol{u}}-\frac{1}{18}d^2\nabla^2((d^2-h^2)\nabla^2\tilde{\boldsymbol{u}})_t\\
&+\frac{1}{6}\nabla\{d^2(\nabla\cdot\tilde{\boldsymbol{u}})\nabla\cdot(\tilde{d}^2\nabla^2\tilde{\boldsymbol{u}})-\frac{1}{2}d^2[(\tilde{\boldsymbol{u}}\cdot\nabla)\nabla\cdot(\tilde{d}^2\nabla^2\tilde{\boldsymbol{u}})+(\tilde{d}^2\nabla^2\tilde{\boldsymbol{u}})\cdot\nabla)\nabla\cdot\tilde{\boldsymbol{u}}]\}\\
&-\frac{1}{6}d\nabla\eta\nabla\cdot(\tilde{d}^2\nabla^2\tilde{\boldsymbol{u}})_t+\frac{1}{18}d\eta_t\nabla^2(d^2\nabla^2\tilde{\boldsymbol{u}})
\end{aligned}\tag{A.1}$$

for Eqs. (2.11) and (2.12), and

$$\begin{aligned}
\boldsymbol{G}_2 = &\frac{1}{2}\nabla[d^4(\frac{5}{18}(\nabla^2\overline{\boldsymbol{u}})^2-\frac{1}{3}(\nabla\cdot\overline{\boldsymbol{u}})\nabla^2\nabla\cdot\overline{\boldsymbol{u}}+\frac{1}{15}\overline{\boldsymbol{u}}\cdot\nabla^2\nabla^2\overline{\boldsymbol{u}})+\frac{1}{18}d^2\overline{\boldsymbol{u}}\cdot\nabla^2(d^2\nabla^2\overline{\boldsymbol{u}})]\\
&+[\frac{1}{30}(d^4-h^4)-\frac{1}{18}(d^2-h^2)h^2]\nabla^2\nabla^2\cdot\overline{\boldsymbol{u}}_t+\frac{1}{6}d^3\nabla\eta\nabla^2(\nabla\cdot\overline{\boldsymbol{u}}_t)-\frac{1}{30}d^3\eta_t\nabla^2\nabla^2\overline{\boldsymbol{u}}\\
&+\frac{1}{18}d\eta_t\nabla^2(d^2\nabla^2\overline{\boldsymbol{u}})-\frac{1}{18}d^2\nabla^2((d^2-h^2)\nabla^2\overline{\boldsymbol{u}})_t-\frac{1}{6}d\nabla\eta\nabla\cdot(d^2\nabla^2\overline{\boldsymbol{u}})_t\\
&+\frac{1}{6}\nabla\{d^2(\nabla\cdot\overline{\boldsymbol{u}})\nabla\cdot(d^2\nabla^2\overline{\boldsymbol{u}})-\frac{d^2}{2}[(\overline{\boldsymbol{u}}\cdot\nabla)\nabla\cdot(d^2\nabla^2\overline{\boldsymbol{u}})+(d^2\nabla^2\overline{\boldsymbol{u}})\cdot\nabla)\nabla\cdot\overline{\boldsymbol{u}}]\}
\end{aligned}\tag{A.2}$$

for Eqs. (3.1) and (3.2).

References

1. Zou, Z.L., Fang, K.Z., submitted to Wave Motion (2008a)
2. Madsen, P.A. and Schäffer, H.A., Coastal Engineering, 15, 371-388 (1991)
3. Madsen, P.A., Sørensen, O.R., Coastal Engineering, 18, 183-204 (1992)
4. Nwogu, O., J. Waterway. Port. Coast. and Ocean Eng., 119 (6), 618-638 (1993)
5. Wei G., Kirby J.T., Grilli S.T. and Subramanya R., J. Fluid Mech., 294, 71-92. (1995)
6. Gobbi, M.F., Kirby, J.T., Coastal Engineering, 37, 57-96 (1999)
7. Gobbi, M.F., Kirby, J.T. and Wei, G., J. Fluid Mech., 405, 181-210 (2000)
8. Witting, J.M., J. Comp. Phys., 56, 203-236 (1984)
9. Zou, Z.L., Ocean Engineering, 27, 557-575 (2000)
10. Serre, P.F., La Houille Blanche 8, 374-388 (1953)
11. Su, C.H., Gardner, C.S., J. Math. Phys., 10, 536-539 (1969)
12. Green, A.E., Naghdi, P.E., J. Fluid Mech., 78, 237-246 (1976)
13. Mei, C.C., New York: John Wiley (1983)
14. Madsen, P.A. and Schäffer, H.A., Phil. Trans. R. Soc. Lond. A, 356: 3123-3184 (1998)
15. Zou, Z.L., Fang, K.Z., Coastal Engineering, 55, 506-521 (2008b)
16. Agnon, Y., Madsen, P.A. and Schäffer, H.A., J. Fluid Mech., 399, 319-333 (1999)
17. Madsen, P. A., Bingham, H.B. and Liu, H., J. Fluid Mech., 462, 1-30 (2002)
18. Madsen, P.A., Bingham, H.B. and Schäffer, H.A., Phil. Trans. R. Soc. Lond. A, 459: 1075-1104 (2003)
19. Fuhrman, D. R., Ph.D thesis, Technical University of Denmark, Denmark (2004)
20. Zou, Z.L., Ocean Engineering, 26, 767-792 (1999)
21. Schäffer, H.A., Madsen, P.A., Coastal Engineering, 26, 1-14 (1995)
22. Zou, Z.L., Science Press, Beijing, China (in Chinese) (2005)
23. Zou, Z.L., Oceanologica Sinica, 23(1): 109-119 (in Chinese) (2001)
24. Kirby, J.T., Wei, G., Chen, Q., Kennedy, A.B., and Dalrymple, R.A., Research report NO. CACR-98-06, University of Delaware, Newark, DE(1998)
25. Shapiro, R., Review of geophysics and space physics, 8(2), 359-386 (1970)
26. Grilli, S.T., Skourup, J. and Svendsen, I.A., Engineering Analysis with Boundary Elements, 6(2), 97-107 (1989)
27. Luth, H.R., Klopman, G., Kitou, N., Report H-1573, Delft hydraulics, 40 (1994)
28. Ohyama, T., Kitoa, W., Tada, A., Coastal Engineering, 24, 213-279 (1994)
29. Shemer, L., Jiao, H.Y., Kit, E., Agnon, Y., J. Fluid Mech., 427,107-129 (2001)
30. Phillips, O.M., J. Fluid Mech., 9, 193-217 (1960).

CHAPTER 9

METHOD OF FUNDAMENTAL SOLUTIONS FOR FULLY NONLINEAR WATER WAVES

Der-Liang Young*, Nan-Jing Wu and Ting-Kuei Tsay

Department of Civil Engineering, National Taiwan University
Sec. 4, Roosevelt Rd., Taipei, 106, Taiwan
**dlyoung@ntu.edu.tw*

A meshless numerical model for fully nonlinear water waves by the method of fundamental solutions is presented in this chapter. An approach for handling the moving free surface boundary is proposed. Using the method of fundamental solutions of the Laplace equation as the radial basis functions and locating the source points outside the computational domain, we are able to solve the problem by collocating only a few boundary points since governing equation is satisfied automatically. Very good agreement is observed for four examples for (1) the heights of a wavemaker; (2) the generation of periodic finite – amplitude steep waves; (3) modulation of monochromatic waves over a submerged obstacle and (4) 3D wave generation by a submerged moving spheroid in an infinite water depth as comparing with analytical, experimental and other numerical works.

1. Introduction

Nonlinear water waves are of paramount importance to the coastal, offshore and ocean engineering but are very difficult to deal with due to the nature of moving free surfaces and involved nonlinear phenomena. The study of fully nonlinear water wave problems in general uses either the fully nonlinear potential theory or Euler or Navier-Stokes equations.

They are usually handicapped or prohibited by the nonlinear features of the kinematic and dynamic free surface boundary conditions plus the quasi-linear advection effects for the Euler or Navier-Stokes equations. To overcome the difficulty of the moving and nonlinear free surfaces and governing equations, extensive works to simulate the propagation of surface gravity water waves have been carried out analytically or numerically or experimentally over the decades. Analytical and numerical solutions are sought in the fully nonlinear potential theory, while most Euler or Navier-Stokes models have to rely on the numerical schemes or experimental studies. Non-fixed (deforming) meshes are needed for traditional numerical methods; such as finite element method (FEM) or finite difference method (FDM) or finite volume method (FVM) for examples see Glauss,[1] Lo and Young.[2] If one tries to solve the problem by conventionally numerical methods such as mixed Eulerian-Lagrangian (MEL) or arbitrary Lagrangian-Eulerian methods (ALE), an excessive number of nodes (moving and fixed) and a huge size of matrix usually accompany with the formulation. Besides, these mesh-based methods will encounter the over-distorted mesh and re-meshing problems during the simulation of water wave evolutions.

Another popular numerical scheme to simulate the nonlinear water waves is the boundary element method (BEM), or named as boundary integral equation method (BIEM). Time-domain boundary element method has been employed for its applicability to compute steep water wave propagation since nonlinear free surface conditions were fully incorporated (Longuet-Higgins and Cokelet;[3] Issacson;[4] Dommermuth and Yue;[5] Grilli et al.;[6] Cooker et al.;[7] Ohyama and Nadaoka;[8] Grilli et al.[9]). Particle trajectories and the nonlinear boundary conditions on the free surface can be predicted by the MEL or ALE time marching scheme through the numerical integration in the time domain. The velocity potential of linear boundary value problem can then be solved directly at the further time step. The review paper of Tsai and Yue[10] explained the MEL approach in detail. Lo and Young[2] concentrated on the work of ALE method. Dual BEM has also been introduced to simulate the degenerated linear water wave problems in wave-structure interaction by Chen et al.[11,12] As BEM is based on solving integral equations, numerical integration along the boundaries is always inevitable. Evaluating these integrals is a tedious and non-economic task since there are singularities on the boundaries, either fixed or moving ones. Furthermore, BEM produces full dense matrices, which is time consuming when solving the linear algebraic system, although BEM is a mesh-reduction and dimension-reduction mesh-based method. Though the dense matrices can be efficiently compressed to some sparse ones and the computation may be speeded up by some means of approximation such as fast multipole method: Rokhlin,[13] panel clustering: Hackbusch and Nowak[14] or wavelet compression: Beylkin et al.,[15] the need of computing tedious singular integrals still remains in BEM with some difficulties for implementing free surface water wave problems. Beck, Schultz and their coworkers: Cao et al.;[16,17] Celebi and Beck;[18] Scorpio and Beck,[19] and Young et al.[20] alternatively proposed an approach by moving the singularities away from the boundaries and outside the computational domain to desingularize the integral equations. This approach is named as the desingularized boundary integral equation method (DBIE) or desingularized boundary element method (DBEM). There are two versions of DBIE, direct and indirect. The direct version of DBIE still requires the boundary elements/panels (and meshes) and integration over the panels, but the integration are easier as a result of desingularization. In the indirect version of DBIE, the integration can be replaced by summation of singularities (or their derivatives) outside the computational domain and no integration is needed. The approach has been successfully applied[16-19] to simulate both steady and time-dependent water wave problems.

An innovative class of meshless numerical methods has appeared in the past decade, which attempts to overcome the requirement of computational grids or meshes of mesh-based methods. These schemes have several names, such as meshless, mesh-free, grid-free, element-free, mesh-reduction, particle or finite point methods, including the Galerkin and collocation methods. All of them share the common characteristics of no need for mesh and associated explicit connectivity information, which is built by the method in a process designated by the users. The connectivity includes node-to-node distances, azimuths and the sequential relations among these nodes. The basic idea of meshless method is the construction of radial basis functions (RBFs) that only states the relationship between the two-point distances which are very easily extended to the multi-dimensional problems. The RBFs were first developed for interpolating scattered data: Hardy;[21] Franke,[22] and later became a kind of artificial neural network kernels: Moody and Darken.[23] The applications of the concepts of RBFs to solve partial differential equations (PDEs) have been introduced for years: Kansa.[24] Many meshless methods have been reported in the recent literature, such as element free Galerkin method: Belytschko et al.,[25] diffusion element method: Nayroles et al.,[26] reproducing kernel particle method: Liu et al.,[27] smoothed particle hydrodynamics (SPH) method: Monaghan,[28] particle finite element method: Onate et al.,[29] meshless local Petrov-Galerkin (MLPG) method: Atluri;[30] Ma,[31] method of fundamental solutions (MFS): Kupradze and Aleksidze;[32] Fairweather and Karageorghis;[33] Wu et al.[34] and so on. Among them, the SPH, MLPG and MFS have been widely used to simulate water wave problems.

Generally speaking, there are two kinds of meshless methods, namely the domain-type and the boundary-type methods. In the domain-type meshless method, any single RBF does not satisfy the governing equations, and there must be a large number of collocation points in both the computational domain as well as the boundary to obtain a better solution. Thus neither the number of nodes nor the size of matrix is reduced too much, and the advantages of domain meshless approach are just for easy programming and avoiding the chore of mesh generation. There are many works in the literature on solving fluid dynamics problems with domain-type meshless methods such as: Du;[35,36] Zhou et al.;[37] Young et al.;[38] Ata and Soulaimani;[39] Ma.[31] On the other hand, in the boundary meshless method in some PDEs (such as Laplace, Helmholtz, etc. for example), one can choose the fundamental solutions of the linear operators to be the RBFs: Golberg and Chen;[40] Young et al.,[41] which will automatically satisfy the governing equations except at the centers of RBFs. If all the source points (centers of RBFs) are set outside the computational domain, there will be no singularity in the computational domain at all and only collocation on the boundary points is needed to solve the problems. One kind of these meshless methods is named as the method of fundamental solutions (MFS). The idea of MFS is similar to the indirect DBIE, but MFS only employs collocation about the boundary points without needing any mesh generation and numerical integration. Therefore the MFS is simple and efficient to model nonlinear water wave problems. It should be noted that the integrals in the indirect DBIE in which the integration is carried out outside the domain can be reduced to the solution form of MFS by using the one-point quadrature integrating scheme.

For the meshless modeling on the linear and nonlinear water wave problems, Young et al.[42] solved the simple linear water wave problem with a semi-infinite domain of normal incident water wave past a submerged breakwater using the hypersingular meshless method (HMM) which is a further improvement of the MFS in selecting the source positions. Wu et al.[34,43] used the MFS to solve the two-dimensional fully nonlinear water wave problems governed by the fully nonlinear potential theory. Ma[31,44,45] employed the meshless local Petrov-Galerkin (MLPG) methods to simulate the two-dimensional nonlinear water wave problems for an incompressible and inviscid fluid based on the Euler equations. Hon et al.[46] first used the multiquadrics (MQ) method to study the two-dimensional shallow water wave equations and applied to the Hong Kong harbor. They used the partial derivatives of RBFs and scattered collocation points to solve the shallow water equations problems in irregular topographic water bodies. However we are going to limit ourselves in this chapter to review the investigation of fully nonlinear water wave problems governed by the fully nonlinear potential theory using the meshless MFS methods only, since this topic is still in its infancy as far as meshless numerical modeling of linear and nonlinear water waves is concerned.

2. Governing Equations and Boundary Conditions

The problem of a free surface water wave can be considered as a hydrodynamics problem for the inviscid, irrotational and incompressible fluids with free surface boundary. Thus, there exists a velocity potential satisfying the Laplace equation

$$\nabla^2 \phi = 0 \tag{1}$$

where $\phi(\vec{x}, t)$ is the velocity potential, in which $\vec{x} = x\vec{i} + y\vec{j} + z\vec{k}$ denotes the position vector in three dimensional space, and the velocity vector is defined as $\vec{v} = \nabla \phi$.

At the free surface, the boundary conditions are specified by the following equations

$$\left.\frac{\partial \phi}{\partial z}\right|_{z=\eta} = \frac{\partial \eta}{\partial t} + (\nabla_h \phi)_{z=\eta} \cdot \nabla_h \eta \tag{2}$$

$$\left.\frac{\partial \phi}{\partial t}\right|_{z=\eta} = -g\eta - \frac{1}{2}(\nabla \phi \cdot \nabla \phi)_{z=\eta} \tag{3}$$

where $\eta(x, y, t)$ is the free surface displacement, $\nabla_h = \frac{\partial}{\partial x}\vec{i} + \frac{\partial}{\partial y}\vec{j}$ is a horizontal (two-dimensional) operator.

The above two equations are called the kinematic and dynamic free surface boundary conditions (KFSBC and DFSBC), respectively.

Beneath the free surface, no-flux boundary condition is applied on the body surface (and at the bottom in finite depth problems). That is

$$\nabla \phi\big|_{\vec{x}=\vec{x}_b} \cdot \vec{n} = \vec{v}_b \cdot \vec{n} \tag{4}$$

where $\vec{n}$ is the normal vector outward the body surface and $\vec{v}_b$ is the velocity of the body surface while $\vec{x}_b$ denotes the body surface. The right hand side of this condition is usually zero because most of the wave problems are with fixed bottom in finite depth.

3. Numerical Implementation — MFS

The numerical solution at each time step is assumed as the linear combination of N radial basis functions. That is

$$\phi(\vec{x}, t) = \sum_{i=1}^{N} \alpha_i(t) q_i(\vec{x}) \tag{5}$$

where $q_i(\vec{x})$ is the radial basis function (RBF) whose center (also named source point) is positioned at $\vec{x}_i$ with $\alpha_i(t)$ its weight at the instant of t. The type of RBF chosen for ϕ in water wave problems can be the fundamental solution of a 3-D Laplacian operator. That is

$$q_i(\vec{x}) = 1 / r_i(\vec{x}) \tag{6}$$

where $r_i(\vec{x}) = |\vec{x} - \vec{x}_i|$ is the distance from any position in the computational domain to that specific source point (for 2D, $q_i(\vec{x}) = 1/\ln r_i(\vec{x})$). The solution form satisfies the governing equation automatically throughout the computational domain if all the RBF centers are chosen outside the computational domain. From this point, we can discretize the time domain by using the symbol (n) to denote the n-th time step. The solution can be obtained by solving $\alpha_i^{(n)}$ from the matrix collocated from the boundary conditions only. At each time step, there are N unknowns ($\alpha_i^{(n)}$, $i=1,\cdots,N$) to be solved. Thus, N boundary points are needed for the method of collocation. Based on the solution form in Eq. 5, the partial derivatives of the velocity potential are shown as

$$(\nabla\phi)^{(n)}=\sum_{i=1}^{N}\alpha_i^{(n)}\nabla q_i^{(n)}. \tag{7}$$

Then the collocation of surface boundary points is

$$\sum_{i=1}^{N}\alpha_i^{(n)}q_i^{(n)}=\phi\Big|_{z=\eta}^{(n)} \tag{8}$$

where $\phi^{(n)}$ at $z=\eta$ is obtained from the KFSBC and DFSBC. The time marching schemes for the KFSBC of Eq. 2 and DFSBC of Eq. 3 can be obtained from the second-order finite difference in time. That is:

$$\eta^{(n)}=\eta^{(n-2)}+2\Delta t\left(\frac{\partial\phi}{\partial z}\bigg|_{z=\eta}-\nabla_h\phi\big|_{z=\eta}\cdot\nabla_h\eta\right)^{(n-1)} \tag{9}$$

$$\phi\Big|_{z=\eta^{(n-1)}}^{(n)}=\phi_{z=\eta^{(n-1)}}^{(n-2)}-2\Delta t\left(g\eta+\frac{1}{2}(\nabla\phi\cdot\nabla\phi)\big|_{z=\eta}\right)^{(n-1)}. \tag{10}$$

The collocation boundary points beneath the free surface is

$$\sum_{i=1}^{N}\alpha_i^{(n)}(\nabla p_i\cdot\vec{n})_{\vec{x}=\vec{x}_b}^{(n)}=(\vec{v}_b\cdot\vec{n})_{\vec{x}=\vec{x}_b}^{(n)}. \tag{11}$$

Apart from the above description, the gradient of free surface in Eq. 9 is needed for solving the problem. Wu et al.[34] proposed the second-order finite difference in space to approximate the gradient of free surface for two dimensional potential free surface flow problems It was not really meshless, for boundary points at the free surface had to be coded in sequence and arranged in equal spacing. It would be all right for two dimensional potential free surface flow problems. However, it is not good when applying it to three dimensional problems, because one needs to build up the connectivity among these free surface nodes, including node-to-node distances, azimuths and the sequential relations. Also, the requirement of equal nodal spacing for finite difference limits its applicability to those problems in which higher density of free surface boundary points are required in some regions. For example, in problems of variable water depth, higher density of free surface boundary points is required in the region of deep water region than in shallow water region because of the differences of the wave lengths. To establish a really meshless model, Wu et al.[43] adopt another interpolating method of RBF type artificial neural network and choose the Gaussian function to be the RBF to fit the free surface. That is

$$\eta(x, y, t) = \sum_{i=1}^{N_s} \beta_i(t) p_i(x, y) \tag{12}$$

$$p_i(x, y) = \exp\left(\frac{-((x-(x_s)_i)^2 + (y-(y_s)_i)^2)}{\sigma_i^2}\right) \tag{13}$$

where $p_i(x, y)$ is the RBF whose center is at $(x_s)_i, (y_s)_i$, and $\beta_i(t)$ is its weight. Following the above-mentioned discretization in the time domain, $\beta_i(t)$ can be written as $\beta_i^{(n)}$ at the n-th time step. The values of σ_i are the shape parameters denoting the range of influence of these RBFs. The reason for choosing the Gaussian RBFs to fit the free surface displacements is that when solving the weights of these RBFs in Eq. 12 the matrix will approach sparse and banded and can be solved rapidly by iterating solvers, if the values of the σ_i are small. However, as σ_i approach to zero, the Gaussian RBFs perform like the delta functions and the value of $\eta(x, y, t)$ is always zero except when (x, y) is at one of the RBF centers. The value of σ_i should be chosen properly to avoid this kind of singular condition and to solve the problem quickly. Wu et al.[43] suggested that the shape parameter should be greater than half of nodal spacing to keep off the singularity, but we found choosing the shape parameter greater than or equal to the nodal spacing would be better. As the free surface at the n-th time step $\eta^{(n)}$ is known, the weights $\beta_i^{(n)}$ can be solved from the linear algebraic system, and then the gradient of free surface can be obtained as

$$(\nabla_h \eta)^{(n)} = \sum_{i=1}^{N_s} \beta_i^{(n)} \nabla_h p_i^{(n)}(x, y). \tag{14}$$

The procedures of numerical implementation are illustrated as follows:

Step 1: The velocity potential in the entire domain ϕ and the particle displacement at the free surface η are all set equal to zero initially (at $n=-1$ and $n=0$).

Step 2: The positions of collocation points at the wave paddle for the $(n)^{th}$ step are updated.

Step 3: The free surface displacement and the velocity potential at the free surface are calculated respectively by Eqs. 9 and 10. The positions of collocation points at the free surface for the $(n)^{th}$ step are then updated.

Step 4: After calculating the partial derivatives of the RBFs at all collocation points, the body boundary condition beneath the free surface can then be applied. And the solution of the velocity potential in the entire domain at $(n)^{th}$ time step can thus be obtained.

Step 5: The number of time steps proceeded is checked. If the requirement of time steps is not achieved, step 2 for the next time step is then processed.

4. Model Applications

4.1. *Wave heights of a piston-type wavemaker*

Ursell et al.[47] observed and discussed the waves generated by a finite-amplitude harmonically oscillating paddle of a piston-type wavemaker. The wave heights were less than predicted by the small amplitude wave theory because of the finite-amplitude effects. Ursell et al.[47] carried out 20 test runs for small amplitude waves, and 4 test runs for high steep waves. The experimental works were to test the validation of the small-amplitude wave theory and also to find the finite-amplitude effects on the wave characteristics. The wave conditions included the wave period T, water depth h, wave height H, and the wavemaker stroke S. Table 1 lists the wave conditions for high wave-steepness cases. Since this is a two dimensional problem, the governing equation (Eq. 1) can be reduced to

$$\frac{\partial^2\phi}{\partial x^2}+\frac{\partial^2\phi}{\partial z^2}=0 \qquad (15)$$

while the free surface boundary conditions(Eqs. 2 and 3) can be reduced to the following equations.

$$\left.\frac{\partial \phi}{\partial z}\right|_{z=\eta}=\frac{\partial \eta}{\partial t}+\left.\frac{\partial \phi}{\partial x}\right|_{z=\eta}\frac{\partial \eta}{\partial x} \tag{16}$$

$$\left.\frac{\partial \phi}{\partial t}\right|_{z=\eta}=-g\eta-\frac{1}{2}\left.((\frac{\partial \phi}{\partial x})^2+(\frac{\partial \phi}{\partial z})^2)\right|_{z=\eta}. \tag{17}$$

At the bottom, the no flux boundary condition (Eq. 4) can be expressed as

$$\left.\frac{\partial \phi}{\partial z}\right|_{z=-h}=-\left.\frac{\partial \phi}{\partial x}\right|_{z=-h}\frac{dh}{dx} \tag{18}$$

where $h(x)$ is the water depth. The no flux boundary condition at the wave generator (also Eq. 4) can be expressed as

$$\left.\frac{\partial \phi}{\partial x}\right|_{x=\xi}=\frac{\partial \xi}{\partial t} \tag{19}$$

where $\xi(z,t)$ is the displacement of the wave paddle. For piston type wave makers, the displacement of the wave paddle can be express as

$$\xi(t)=-\xi_0 \cos \omega t \tag{20}$$

where ω is the radian frequency and $\xi_0(=S/2)$ is the amplitude of the harmonically oscillating paddle.

The boundary condition at the end of the flume is treated as the radiation boundary condition which means waves are always outgoing. It is shown as:

$$\left.\frac{\partial \phi}{\partial x}\right|_{x=x_r}=-\left.\left(\frac{1}{C}\frac{\partial \phi}{\partial t}\right)\right|_{x=x_r} \tag{21}$$

where C is the wave speed.

The arrangement of source points and boundary points is shown in Fig. 1. Horizontal spacing of adjacent boundary points along the free surface is 0.1 m (about 1/10 ~ 1/13 wave length). The interval of numerical time step is with the order of 1/50 wave period. Wave evolution nearby the wave paddle for the case of T = 0.79sec, S = 2.54cm, h = 2.00ft is displayed in Fig. 2, from which one can be sure that in the range of $x \le 3m$, the quasi-steady state wave profile nearby the wave paddle is achieved when the simulation time is more than 10 wave periods, since the free surface profile at t = 10T almost coincides with those of t = 15T and t = 20T.

When quasi-steady solution nearby the wave paddle is obtained, the wave height at each specific position can be calculated by subtracting the lowest water level from the highest during the time interval of one wave period. The envelopes of the wave crest and the wave trough in the range of $x \leq 3m$ is depicted in Fig. 3. From the method mentioned above, the wave height distributions near the wave paddle for these cases are shown in Fig. 4. For short wave cases one can find the evanescent waves are significant nearby the wave paddle. Since the actual position of wave gage was not mentioned by Ursell et al.,[47] we calculate the average wave height in the range of $x = 1.0m \sim 3.0m$ in order to compare with experimental data and the small amplitude wave theory. This distance is far enough to avoid taking the evanescent waves into account in these cases. The comparisons are illustrated in Fig. 5. For the case of deep water wave ($h/L > 0.5$), the numerical result matches to the experimental data quite well. While for the other cases, the wave heights generated by present model is slightly higher than experimental data, but still much lower than predicted by the small amplitude wave theory. This demonstrates that the wave heights by the small amplitude wave theory are generally over predicted. As far as wave heights are concerned, the nonlinearity plays a vital role for the finite-amplitude waves.

Table 1. List of Ursell et al.'s[47] finite amplitude wave generation conditions.

run	T(sec)	S(cm)	h(ft)	$H_{measured}$ (cm)
21	0.79	2.54	2.00	4.77
22	0.85	3.15	1.50	5.25
23	0.95	4.50	1.00	5.47
24	0.96	5.73	0.66	5.14

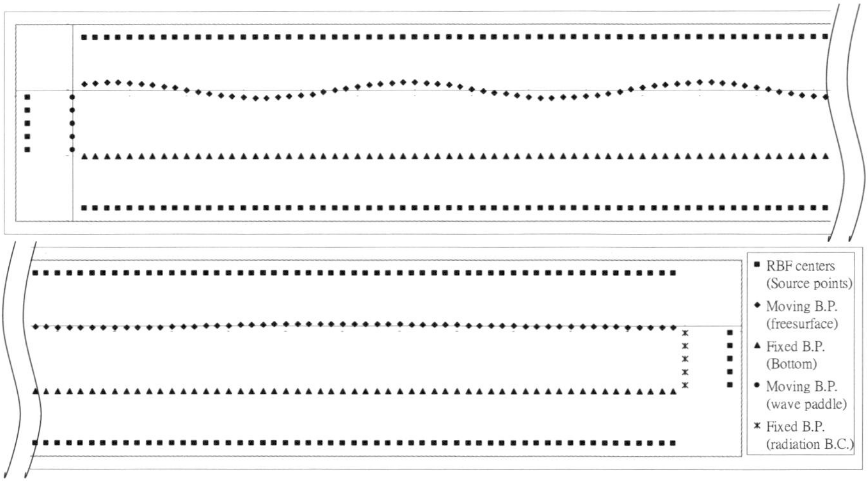

Fig. 1. The arrangement of source points and boundary points.

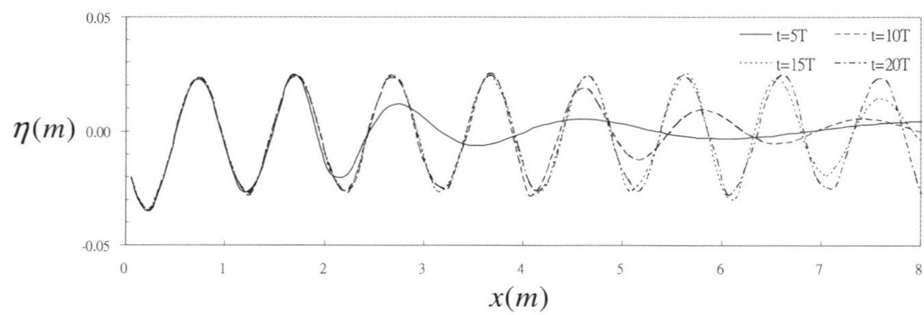

Fig. 2. Wave evolution nearby the wave paddle for the case of T = 0.79sec, S = 2.54cm, h = 2.00ft.

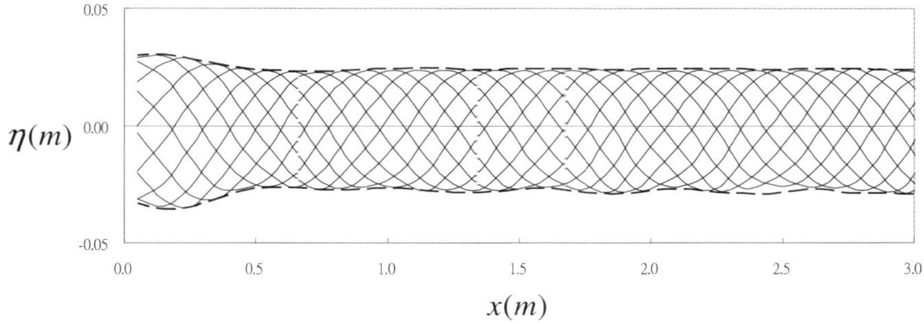

Fig. 3. Envelopes of the wave crest and the wave trough nearby the wave paddle for the case of T = 0.79sec, S = 2.54cm, h = 2.00ft.

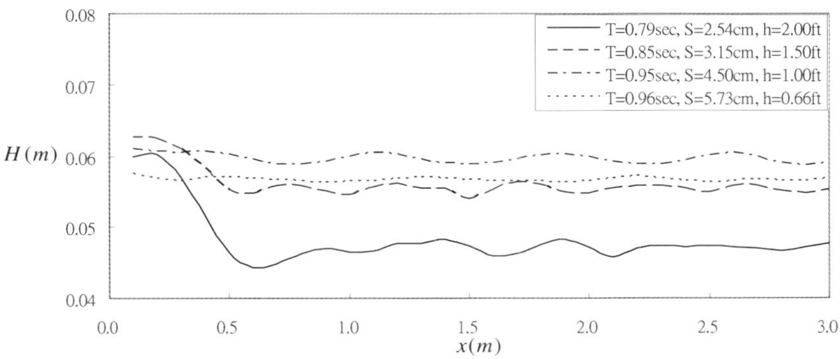

Fig. 4. Wave height distributions near the wave paddle.

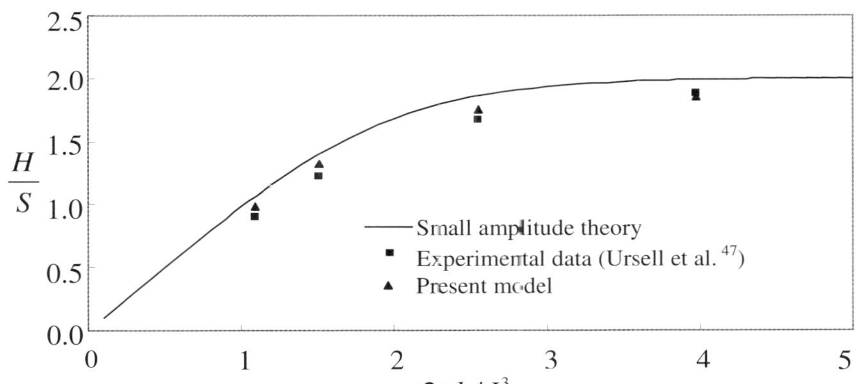

Fig. 5. Comparison with the experimental data about the deviation from wave maker theory due to finite wave steepness.

4.2. *Second harmonic wave components generated by a high-stroke wavemaker*

As the stroke of the wave paddle of piston-type wavemaker gets higher, the nonlinearity becomes more dominant. While solving the second-order equations for the wave making problem, Madsen[48] found that if the motion of the wave paddle was prescribed as Eq. 21, the generated surface profile would have this form

$$\eta = -a\cos(k_p x - \omega t) - a_{2p}\cos 2(k_p x - \omega t) + a_{2f}\cos(k_f x - 2\omega t) \tag{22}$$

where

$$a = \frac{\xi_0 \tanh k_p h}{n_1} \tag{23}$$

$$n_1 = \frac{1}{2}\left(1 + \frac{2k_p h}{\sinh 2k_p h}\right) \tag{24}$$

$$a_{2p} = \frac{k_p a^2}{4}\frac{(2+\cosh 2k_p h)\cosh k_p h}{\sinh^3 k_p h} \tag{25}$$

$$a_{2f} = \frac{a^2 \cosh k_p h}{2h}\left(\frac{3}{4\sinh^2 k_p h} - \frac{n_1}{2}\right)\frac{\tanh k_f h}{n_2} \tag{26}$$

$$n_2 = \frac{1}{2}\left(1 + \frac{2k_f h}{\sinh 2k_f h}\right). \tag{27}$$

In Eqs. 22-27 k_p and k_f denote the wave number of the fundamental wave and the free wave, respectively. The wave numbers can be obtained from the following dispersion relations

$$\omega^2 = gk_p \tanh k_p h \tag{28}$$

$$4\omega^2 = gk_f \tanh k_f h\,. \tag{29}$$

The first term on the right hand side of Eq. 22 is the first-order wave component. The second term represents the Stokes second-order waves traveling at the same speed as the first-order wave. The third term is the second-order free wave, which travels at its own speed.

Madsen[48] performed some experiments to verify the above theories. One of the experiments was conducted in a wave channel with a water depth $h = 38$ cm and a period of the first harmonic $T = 2.75$ sec. The motion of the piston-type wavemaker was taken to be sinusoidal with an amplitude ξ_0 = 6.1 cm. This case was numerically simulated by Dong and Huang[49] using a finite-analytic model directly solving the continuity equation and the Navier-Stokes equations. The length of our numerical wave flume is 40 m. A moderate horizontal nodal spacing for simulating nonlinear free surface waves is empirically taken less than 1/20 local wave length. We choose the horizontal spacing of adjacent boundary points on the free surface to be 0.25 m as a constant, about 1/21 local wave length. For convenience, we also set the horizontal spacing of adjacent boundary points at the flume bottom to be a constant of 0.25 m. The horizontal spacing of source points above the free surface and beneath the bottom boundary is also 0.25 m. There are 324 source points outside the computational domain. We tested our model and found that numerical instability may occur if the offset distance to the sources from the boundary is less than the nodal spacing. The interval of numerical time step is taken to be $T/44$. The wave evolution is shown in Fig. 6. From Fig. 6, one can see that after about four wave periods, the quasi-steady steady is achieved in the region of $x \leq 10m$. The history of free surface elevation at two stations, x = 4.9 m and 8.7 m, are illustrated respectively in Fig. 7. The numerical results of free surface displacement at these two stations are illustrated in Figs. 8a and 8b, respectively. The comparisons show that the numerical results coincide with the experimental data and the analytical results as well as the results of the other numerical models. The numerical results of free surface profile at t = 21.8 sec along the wave flume are shown in Fig. 9. Our potential results compared well with those of Dong and Huang[49] although they used the viscous and rotational Navier-Stokes model. From the shape of the wave profile, one can find that the sinusoidally moving wave paddle does generate second-order free waves, which disrupt the permanent wave form. Observations from Figs. 8 and 9 indicate that the inertia force dominates the wave motion and ignoring the viscous effect is a reasonable assertion when solving such kind of water wave problems.

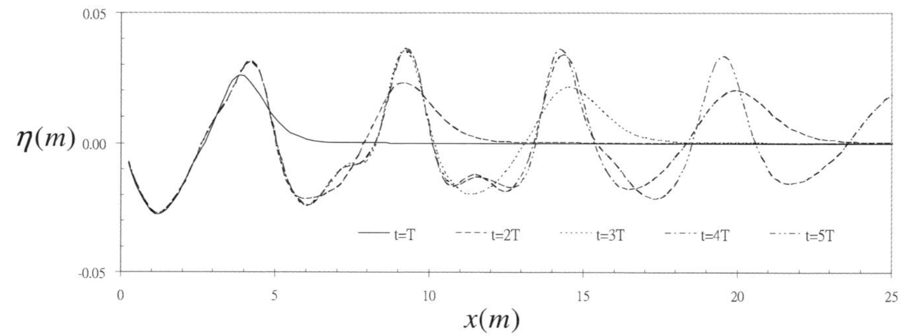

Fig. 6. Wave evolution nearby the wave paddle for the case of T = 2.75sec, S = 12.2cm, h = 38cm.

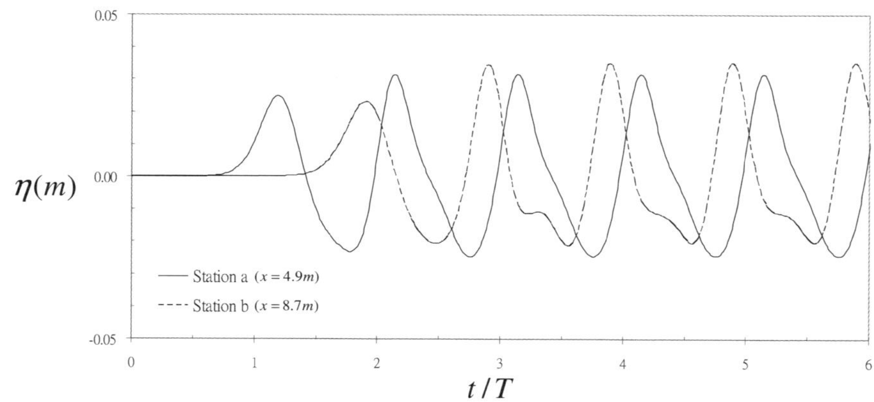

Fig. 7. History of free surface elevation at two stations, x = 4.9m and 8.7m, for the case of T = 2.75sec, S = 12.2cm, h = 38cm.

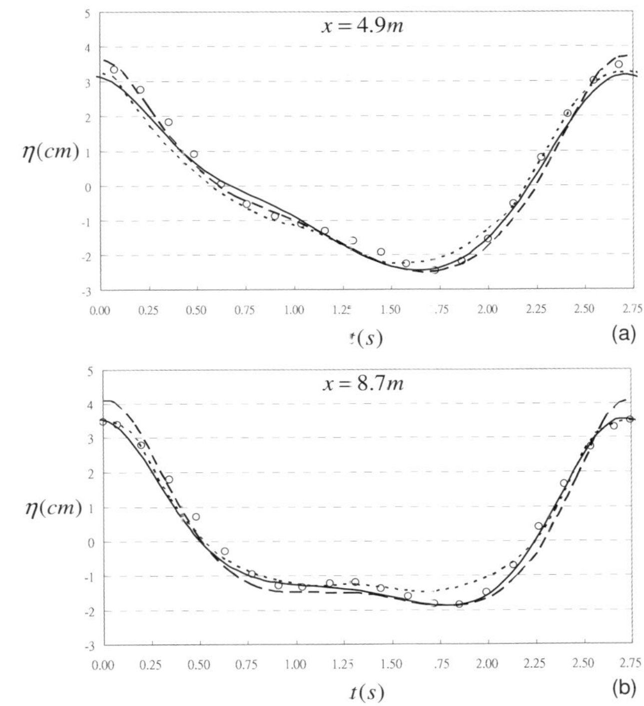

Fig. 8. Comparison between observed, predicted, and numerical surface displacements generated by sinusoidally moving wavemaker: (- - - - -) experimental data (Madsen[48]); (○) analytical solution from Eq. (22); (— — —) numerical results of Dong and Huang[49]; (———) results from present numerical model.

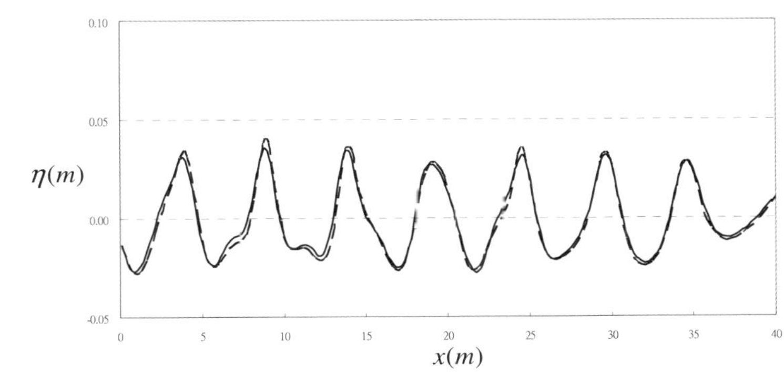

Fig. 9. Numerical results of free surface profile at t = 21.8 s along the wave flume: (— — —) numerical results of Dong and Huang[49]; (———) results from present numerical model.

4.3. *Monochromatic waves passing over a submerged obstacle*

Ohyama and Nadaoka[8] developed a BEM model to solve the Laplace equation for the irrotational flow of an inviscid liquid. In order to verify their numerical model and to analyze the modulation of the frequency spectrum for waves traveling over a submerged structure, Ohyama et al.[50] carried out near breaking wave generation experiments in a physical wave flume, both monochromatic waves and random ones were generated. The layout of the physical wave flume is shown in Fig. 10. Present model is applied to two monochromatic wave tests, with two different frequencies (0.5 Hz and 0.8 Hz, referred to as the "long" and "short" waves respectively). The wave heights for the two experiments are respectively 2 cm and 2.5 cm. Data for comparison were digitized from the paper of Ohyama et al.[50] We applied the radiation boundary condition on the slope at the position of 0.15 m water depth in the rear part of the flume. For the long wave test, horizontal spacing of adjacent boundary points in front of the shelf varies from 0.1 m to 0.2 m, about 1/20 local wave length. We set the horizontal spacing above the shelf and in the lee region as a constant of 0.1 m, to obtain fine resolution for strong nonlinear waves. There are 398 source points outside the computational domain. For the short wave test, the boundary points are arranged in a similar regulation, while the horizontal spacing of adjacent boundary points is in the range from 0.08 m to 0.125 m, and the number of source points is 552. We also simulate these cases by arranging the free surface nodes in constant space, with 508 and 634 source points for long and short wave cases respectively. Wu et al.[34] tested the convergence of the horizontal-spacing of adjacent boundary points along the free surface with the model that needs constant free surface nodal spacing for the case of wave height 2.0 cm and wave period 2.02 sec, in which the wave flume is arranged the same as present case while the wavemaker is a hinge-type one (Fig. 11). It is observed that the nodal spacing of 0.1m, which is about 1/20 local wave length above the submerged obstacle, is sufficient.

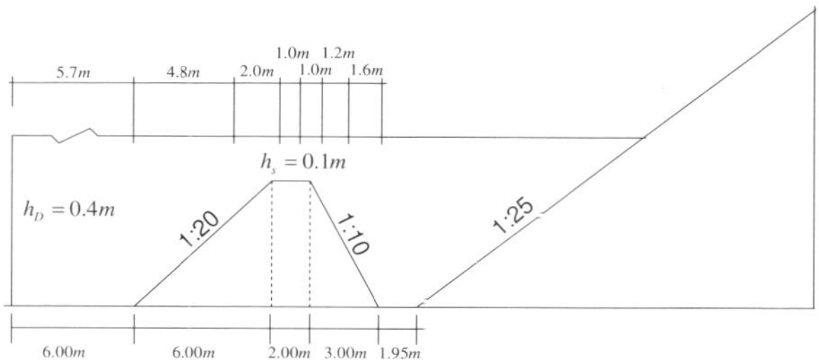

Fig. 10. Wave flume and locations of wave gages (Ohyama et al.[50]).

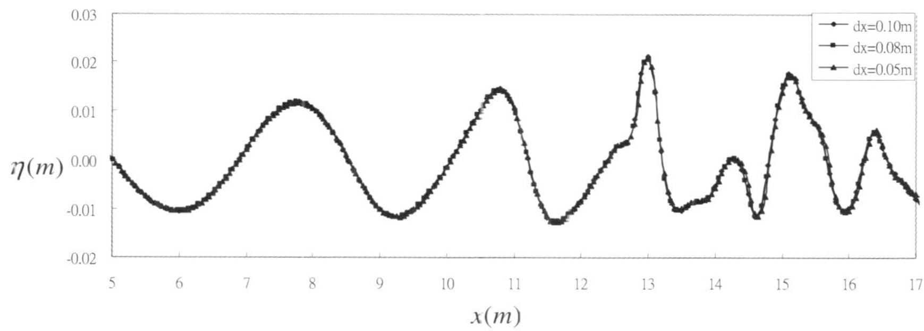

Fig. 11. Results of different horizontal-spacing of adjacent boundary points along the free surface for the case of wave height 2.0 cm and wave period 2.02 sec (Wu et al.[34]).

The numerical results of the free surface displacements for long and short monochromatic waves are shown in Figs. 12 and 13, respectively. The comparison shows that present model performs very well as the other experimental or numerical results.

It is worthwhile to observe how the initially sinusoidal wave form evolves as it travels over the obstacle. In the upslope region, saw-toothed shape waves with the slight forward pitch can be observed. On the top of the shelf, both short and long waves are reminiscent of the soliton formation behind a solitary wave propagating into a shallow depth. In the deepening region, because the higher harmonic free components travel at their own speeds, no permanent wave shape can be formed. One can discover that the evolution of the short waves is less appreciable. This is because the short waves are less obstructed by the seabed than the long waves.

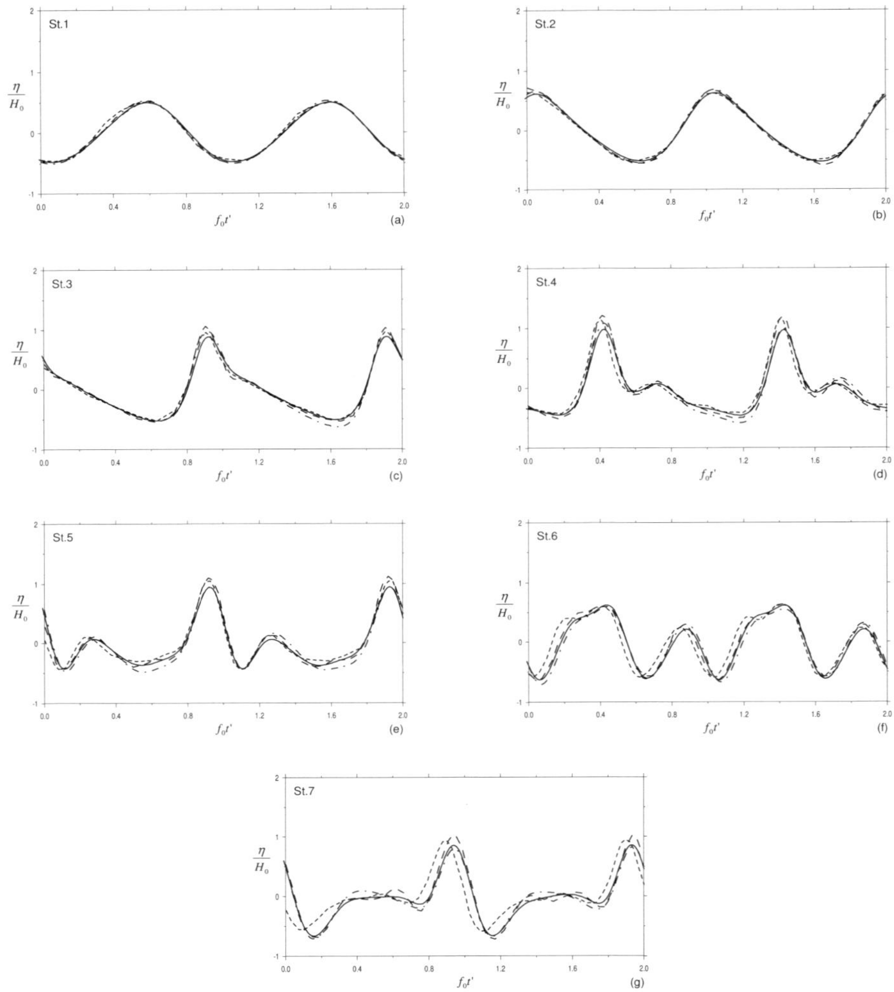

Fig. 12. Free surface displacements for long monochromatic waves at stations 1-7: (- - - -) experimental data (Ohyama et al.[50]); (— — —) numerical results of Ohyama et al.;[50] (— · — ·) present results with constant nodal spacing; (———) present results with varying nodal spacing.

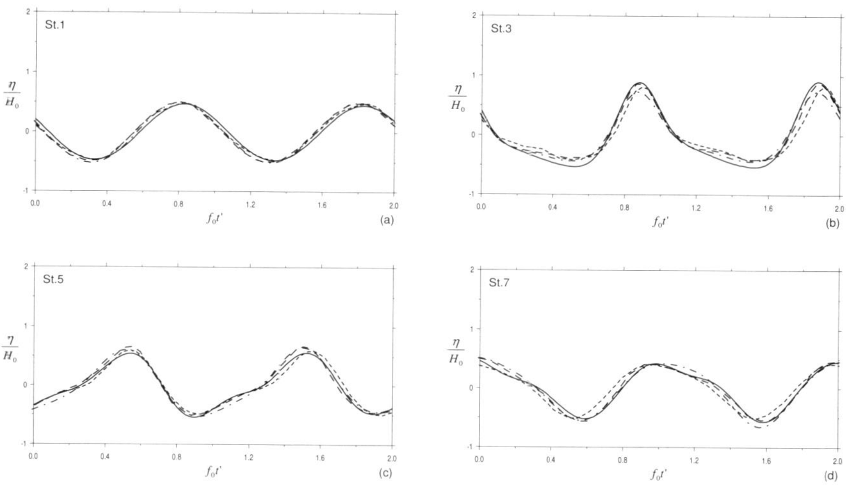

Fig. 13. Free surface displacements for short monochromatic waves at stations 1, 3, 5, and 7: (- - - - -) experimental data (Ohyama et al.;[50]) (— — —) numerical results of Ohyama et al.;[50] (— · —) present results with constant nodal spacing; (———) present results with varying nodal spacing.

4.4. *Three dimensional application — Wave generation by a submerged moving spheroid in infinite water depth*

Different from those models of depth-integrated wave models, each with a fixed bottom in finite depth treated as a function of the plane coordinate, present numerical method is applicable to problems with infinite depth or with deforming bottom. Also, present method is applicable to wave problems with objects moving under water. To show the advantage of present approach, we choose the problem of wave generation by a submerged moving spheroid in infinite water depth for demonstration. The problem of wave generation by a submerged immobile spheroid in a steady stream of infinite depth, which equates to the problem of wave generation in still water by a submerged moving spheroid with constant speed, has been investigated previously by others (Havelock,[51]; Farell,[52]; Doctors and Beck,[53]; Bertram et al.,[54]; Cao,[55]; Scullen,[56]; Tuck and Scullen,[57]), and so provides a useful benchmark for the comparison of present numerical results. As this case is a steady state problem while present model is a time-domain numerical one, the spheroid must be accelerated from still condition to a specific speed and then travels beneath the free surface with this speed. The spheroid is considered moving in the x direction with the speed of $V_b(t)$. A sieve technique is applied to allow the body boundary to be still initially and accelerated to the speed of U_b gradually over a short period. The speed of the submerged spheroid as a function of time can be adjusted as

$$V_b(t) = U_b S_v(t) \tag{30}$$

where the sieve function $S_v(t)$ is chosen as a smooth function

$$S_v(t) = \begin{cases} 3(t/t_c)^2 - 2(t/t_c)^3, & as\ 0 \le t \le t_c \\ 1 & , \quad as\ t > t_c \end{cases} \tag{31}$$

and t_c is the end time of the sieve treatment, which, in this work, is set at $t_c = 0.3L/U_b$ where L is the length of the spheroid. The spheroid is set to travel with a submergence depth h. The spheroid is allowed to move with its center from $x = 0$ to $x = 8L$ within the computation time. The computational domain is in the range of $x = -2.5L \sim 9.5L$, $y = -6L \sim 6L$, $z = \eta \sim -\infty$. Because treatment of the outgoing wave is a very big problem, we did not apply any radiation condition on the truncated boundary. The condition is the same as for other free surface points. The range of computational domain is set as large as we can, so that the outgoing waves are still far away from the truncation boundary when the simulation ends. The range in horizontal plane is far enough for the outgoing waves to travel before the spheroid reaches the terminal point in computation.

Since present method is named as "meshless", the boundary points on the free surface are unnecessarily arranged in regularity. One can reduce the nodal density in some regions according to his own judgment. We anticipate that wave decays when traveling away from the path of the spheroid. In the process of decaying, the wave lengthens. So, in the region of $|y| \ge 3L$, we set the nodal spacing to be 1/5 spheroid length, while in the regions of $3L > |y| \ge 2L$ and $2L > |y| \ge L$, we increase the nodal density by setting the nodal spacing to be 1/6 and 1/7 spheroid length. Near the path of the spheroid, we set the nodal spacing to be 1/8 spheroid length. The body surface of the spheroid is defined as follows.

$$\frac{(x-x_c)^2}{a^2}+\frac{(y-y_c)^2+(z-z_c)^2}{b^2}=1 \tag{32}$$

where x_c, y_c and z_c are the coordinates of the moving spheroid center, and a and b are the semi-lengths of the major and minor axes ($a=L/2$ and $b=d/2$ where d is the diameter at the center of the spheroid). The sketch of the spheroid and its motion is shown in Fig. 14. Schematic diagram of arrangement of the boundary and source points is plotted in Fig. 15. From prow to stern, 16 longitudinal lines are set up on the body surface. Along each longitudinal line, the nodal spacing is properly chosen so that it nearly equals to the nodal spacing in the lateral direction. It should be noted that the nodal spacing on the body surface is not constant and is much smaller than that on the free surface.

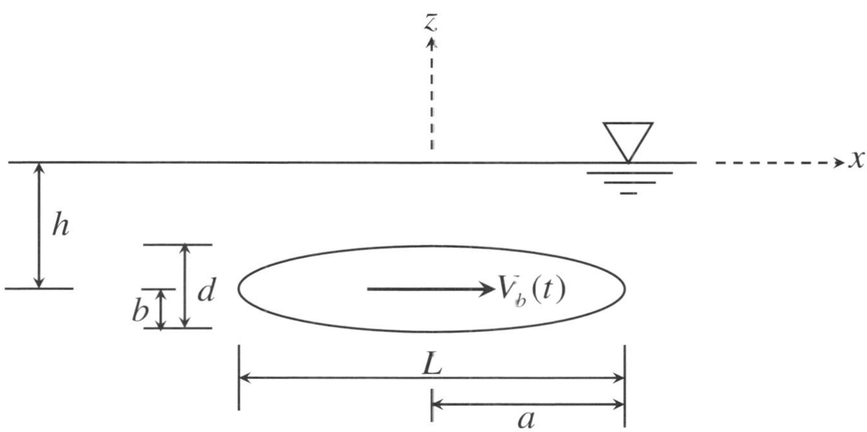

Fig. 14. Sketch of the spheroid and its motion.

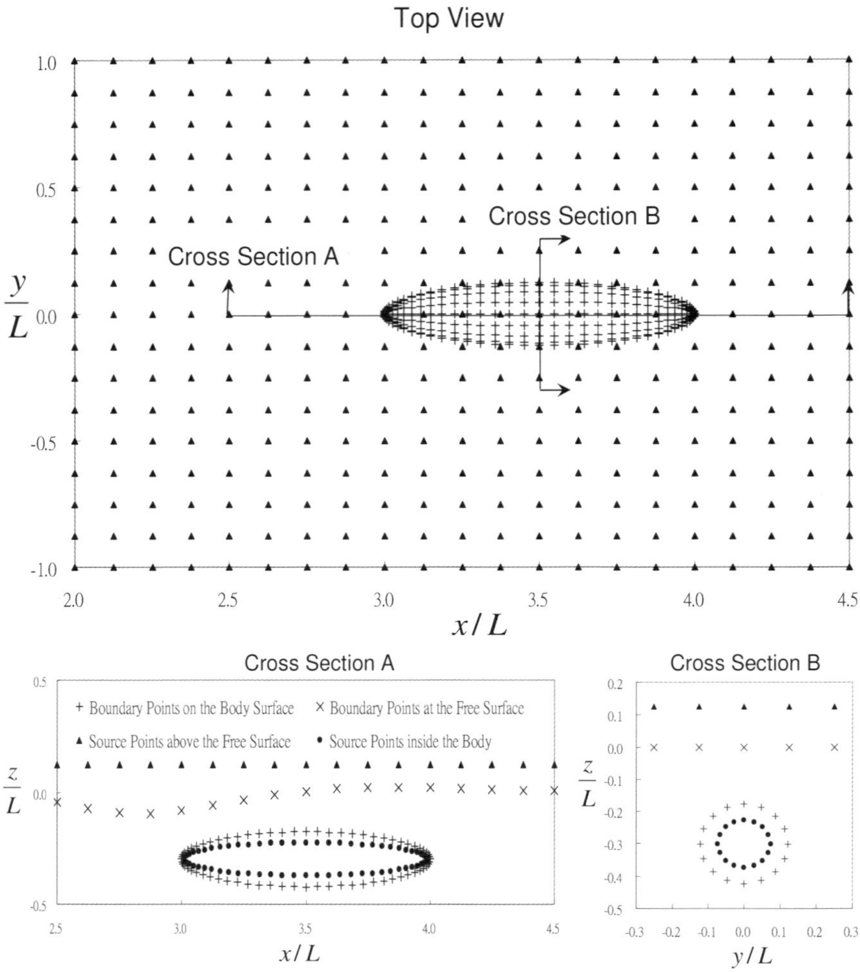

Fig. 15. Schematic diagram of boundary and source points arrangement.

For comparison with other numerical results, the spheroid of diameter-to-length ratio 0.1 is chosen. The diameter-to-submergence ratio is 1.0. Flow velocity is chosen for the case of Froude number ($F_L = U_b / \sqrt{gL}$) 0.5. Computed evolutions of both the free surface elevation contours and the central free surface profile are demonstrated in Figs. 16 and 17 respectively, while the three dimensional view of the wave pattern at stage of $x_c = 8L$ is shown in Fig. 18. It should be noted that the x and z axes in Fig. 17 are not drawn to scale.

When the spheroid has traveled for one hull length ($x_c = L$), the free surface humps up above the first half of the hull and sinks down to the lowest just right at the stern. As the spheroid goes further on for about two more hull lengths ($x_c = 3L$), the hollow aggravates and another hump arises behind the hollow while the first hump keeps its shape and the generated waves spread out to both sides. As the spheroid moves farther, the wave shape at the vicinity of the spheroid approaches to the steady state gradually and the generated waves keep on spreading out. From the evolution of the free surface wave, it is convinced that the wave pattern within the range of $3L$ from the center of the spheroid achieves the steady at stage of $x_c = 8L$.

Centerline surface elevations computed from other models, as a benchmark for comparison of present numerical results, are digitized from the work of Tuck and Scullen,[57] in which modification of the fully nonlinear model of Scullen,[56] were developed by approximating the nonlinear free surface and body boundary conditions with linearized and simplified ones. The comparison of the computed central free surface profile at steady state with those reported in the work of Tuck and Scullen,[57] is shown in Fig. 19. It should be noted that the results of Tuck and Scullen,[57] in which the spheroid rested firmly in a steady current flowing forward to the positive x-direction and the central free surface profile was demonstrated in dimensionless scale of $k_0 x$ and $k_0 z$, where $k_0 = g / U_b^2$, have been converted for comparison with present results, in dimensionless scale of $(x - x_c)/L$ and η / h. Comparing with the other three results of those linear models, better agreement between present result and the result of the nonlinear model developed by Scullen,[56] which was named as Neumann-Stokes model, can be observed.

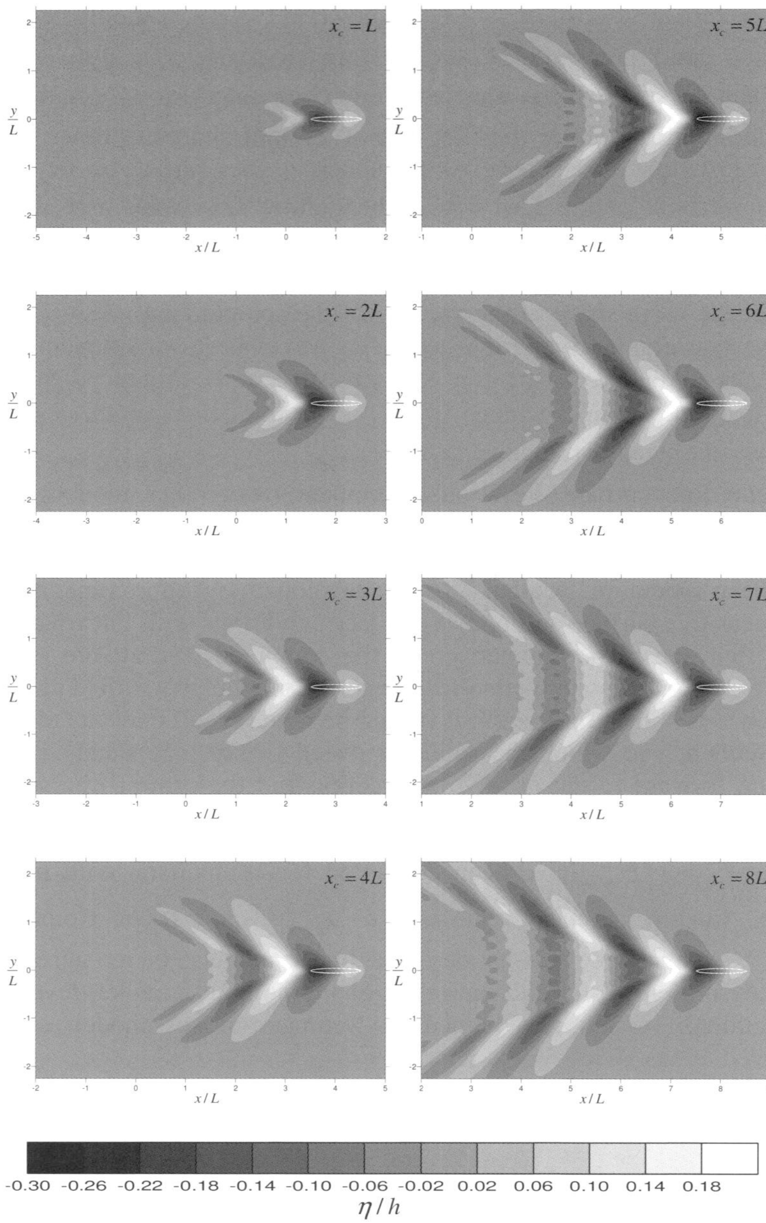

Fig. 16. Evolution of free surface elevation contours for the case of 1:10 spheroid with the diameter-to-submergence ratio of 1.0 and Froude number of 0.5.

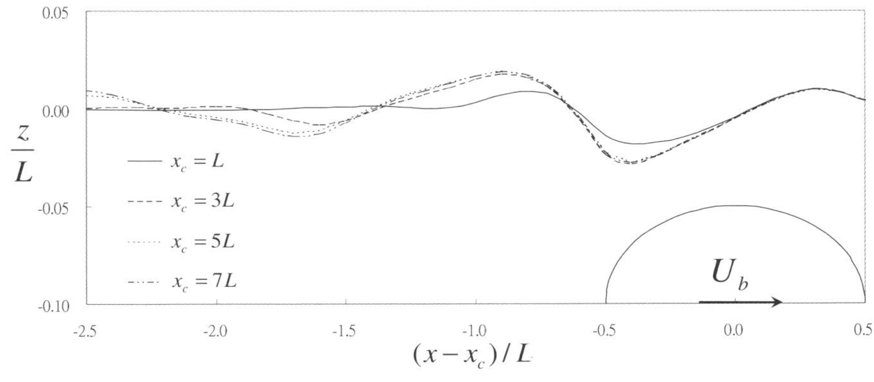

Fig. 17. Evolution of the central free surface profile for the case of 1:10 spheroid with the diameter-to-submergence ratio of 1.0 and Froude number of 0.5.

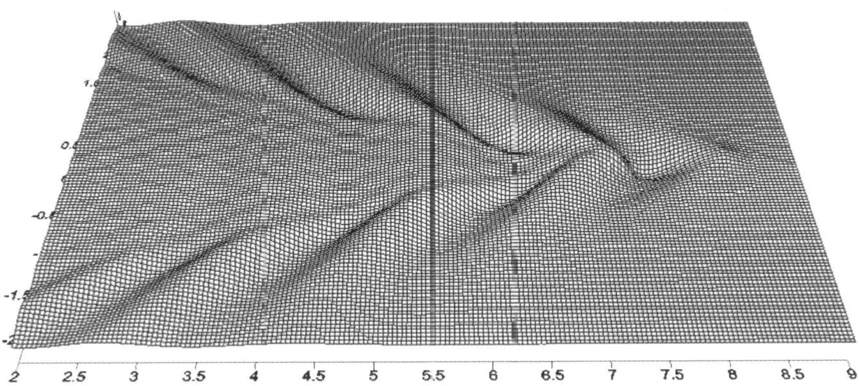

Fig. 18. Three dimensional view of the wave pattern at stage of $x_c = 8L$ for the case of 1:10 spheroid with the diameter-to-submergence ratio of 1.0 and Froude number of 0.5.

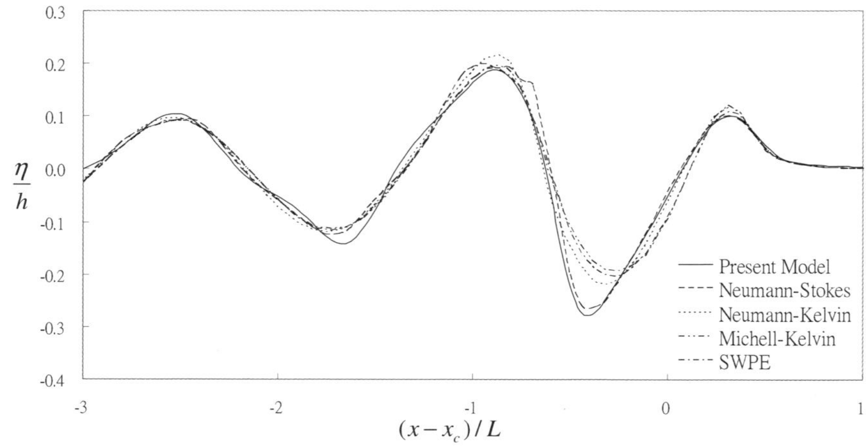

Fig. 19. Computed central free surface profile at steady state compared with other numerical results, for the case of 1:10 spheroid with the diameter-to-submergence ratio of 1.0 and Froude number of 0.5 (adapted from Tuck and Scullen[57]).

5. Conclusions

This boundary-type meshless method requiring neither domain nor surface meshing is suitable for solving problems with deforming and/or moving boundaries. In free surface water wave problems, fundamental solution of the Laplace operator is chosen to be the form of the radial basis function so that only collocation points along the boundary are needed. By fitting the free surface displacement with Gaussian radial basis function, a truly meshless model is established. The MFS is an alternative method with neither prolix scheme formulation nor tediously iterating procedures for simulating fully nonlinear free surface gravity water wave problems.

The MFS has been successfully applied to calculate the following three two dimensional nonlinear water wave problems: 1. Wave heights of a piston-type wavemaker, 2. Second harmonic wave components generated by a high stroke wavemaker, and 3. Monochromatic waves passing over a submerged obstacle. Very good agreement of these three case studies is observed as comparing with analytical solutions, experimental data and other numerical results.

Additionally, the problem of wave generation in infinite water depth by a submerged moving spheroid, which is a three dimensional nonlinear water wave problem, is also carried out. The numerical simulations are verified by comparison with other numerical results and show good agreement.

This study demonstrates that the MFS is a simple and powerful meshless numerical method to simulate the fully nonlinear water wave problems. However due to the global nature of the MFS solution, we may encounter the ill-condition and high condition number for large matrix system and also the sensitivity of locating source points for complex domains which obviously deserve further investigation.

Acknowledgment

The work presented in this chapter was supported by the National Science Council, Taiwan under the grant number 95-2221-E-002-406. It is greatly appreciated.

References

1. G. Glauss, *Appl. Ocean Res.*, **24**, 147 (2002).
2. D. C. Lo, and D. L.Young, *J. Comput. Phys.*, **195**, 175 (2004).
3. H. S. Longuet-Higgins, and E. D. Cokelet, *Proc. R. Soc. London*, **A350**, 1 (1976).
4. M. Issacson, *J. Fluid Mech.*, **120**, 267 (1982).
5. D. G. Dommermuth, and D. K. P. Yue, *J. Fluid Mech.*, **178**, 195 (1987).
6. S. T. Grilli, J. Skourup, and I. A. Svendsen, *Eng. Anal. Bound. Elem.*, **6**, 97 (1989).
7. M. J. Cooker, D. H. Peregrine, and O.Skovggard, *J. Fluid Mech.*, **215**, 1 (1990).
8. T. Ohyama, and K. Nadaoka, *Fluid Dyn. Res.*, **8**, 231 (1991).
9. S. T. Grilli, P. Guyenne, and F. Dias, *Int. J Numer. Methods Fluids*, **35**, 829 (2001).
10. W. T. Tsai, and D. K. P. Yue, *Annu. Rev. Fluid Mech.*, **28**, 249 (1996).
11. K. H. Chen, J. T. Chen, C. Y. Chou, and C. Y. Yueh, *Eng. Anal. Bound. Elem.*, **20**, 917 (2002).
12. K. H. Chen, J. T. Chen, S. Y. Lin, and Y. T. Lee, *J. Waterw. Port Coast. Ocean Eng.-ASCE*, **130**, 179 (2004).
13. V. Rokhlin, *J. Comput. Phys.*, **60**, 187 (1985).
14. W. Hackbusch, and Z. P. Nowak, *Numer. Math.*, **54**, 463 (1989).
15. G. Beylkin, R. Coifman, and V. Rokhlin, *Commun. Pure Appl. Math.*, **44**, 141 (1991).

16. Y. S. Cao, W. W. Schultz, and R. F. Beck, *Int. J. Numer. Methods Fluids*, **12**, 785 (1991).
17. Y. S. Cao, R. F. Beck, and W.W. Schultz, *Int. J. Numer. Methods Fluids*, **17**, 905 (1993).
18. M. S. Celebi, and R. F. Beck, *J. Ship Res.*, **41**, 17 (1997).
19. S. M. Scorpio, and R. F. Beck, *J. Offshore Mech. Arct. Eng. Trans. ASME*, **120**, 71 (1998).
20. D. L. Young, W. S. Hwang, C. C. Tsai, and H. L. Lu, *Eng. Anal. Bound. Elem.*, **29**, 224 (2005).
21. R. L. Hardy, *J. Geophys. Res.*, **76**, 1905 (1971).
22. C. Franke, *Math. Comput.,* **38**, 181 (1982).
23. J. Moody, and C. Darken, *Neural Comput.*, **1**, 281 (1989).
24. J. E. Kansa, *Comput. Math. Appl.*, **19**, 127 (1990).
25. T. Belytschko, Y. Y. Lu, and L. Gu, *Int. J. Numer. Methods Eng.*, **37**, 229 (1994).
26. B. Nayroles, G. Touzot, and P. Villon, *Comput. Mech.*, **10**, 307 (1992).
27. W. K. Liu, Y. Chen, S. Jun, J. S. Chen, T. Belytschko, C. Pan, R. A. Uras, and C. T. Chang, *Arch. Comput. Method Eng.*, **3**, 3 (1996).
28. J. J. Monaghan, *J. Comput. Phys.*, **110**, 399 (1994).
29. E. Onate, S. R. Idelsohn, F. Del. Pin, and R. Aubry, *Int. J. Comput. Methods*, **1**, 267 (2004).
30. S. N. Atluri, The meshless method (MLPG) for domain & bie discretization, Tech Science Press, Forsyth, GA, USA (2004).
31. Q. Ma, *J. Comput. Phys.*, **205**, 611 (2005).
32. V. D. Kupradze, and M. A. Aleksidze, *Comp. Maths. Math. Phys.*, **4**, 633 (1964).
33. G. Fairweather, and A. Karageorghis, *Adv. Comput. Math.*, **9**, 69 (1998).
34. N. J. Wu, T. K. Tsay, and D. L. Young, *Int. J. Numer. Methods Fluids*, **50**, 219 (2006).
35. C. J. Du, *J. Hydraul. Eng.-ASCE*, **125**, 621 (1999).
36. C. J. Du, Comput. Meth. Appl. Mech. Eng., **182**, 89 (2000).
37. X. Zhou, Y. C. Hon, and K. F. Cheung, *Eng. Anal. Bound. Elem.*, **28**, 967 (2004).
38. D. L. Young, S. J. Jane, C. Y. Lin, C. L. Chiu, and K. C. Chen, *Eng. Anal. Bound. Elem.*, **28**, 1233 (2004).
39. R. Ata, and A. Soulaimani, *Int. J. Numer. Methods Fluids*, **47**, 139 (2005).
40. M. A. Golberg, and C. S. Chen, *Comput. Mech. publ.*, 103 (1998).
41. D. L. Young, K. H. Chen, and C. W. Lee, *J. Comput. Phys.*, **209**, 290 (2005).
42. D. L. Young, K. H. Chen, J. T. Chen, and J. H. Kao, *CMES-Comp. Model. Eng. Sci.*, **19**, 197 (2007).
43. N. J. Wu, T. K. Tsay, and D.L Young, *J. Waterw. Port Coast. Ocean Eng.-ASCE*, **134**, 98 (2008).
44. Q. Ma, CMES-Comp. *Model. Eng. Sci.*, **9**,103 (2005).
45. Q. Ma, CMES-Comp. *Model. Eng. Sci.*, **18**, 223 (2007).

46. Y. C. Hon, K. F. Cheung, X. Z. Mao, and E. J. Kansa, *J. Hydraul. Eng.-ASCE*, **125**, 524 (1999).
47. F. Ursell, R. G. Dean, and Y. S. Yu, *J. Fluid Mech.*, **7**, 33 (1960).
48. S. Madsen, *J. Geophys. Res.*, **76**, 8672 (1971).
49. C. M. Dong, and C. J. Huang, *J. Waterw. Port Coast. Ocean Eng.-ASCE*, **130**, 143 (2004).
50. T. Ohyama, S. Beji, K. Nadaoka, and J. A. Battjes, *J. Waterw. Port Coast. Ocean Eng.-ASCE*, **120**, 637 (1994).
51. T. H. Havelock, *Proc. R. Soc. London*, **A131**, 275 (1931).
52. C. Farell, *J. Ship Res.*, **17**, 1 (1973).
53. L. J. Doctors, and R. F. Beck, *J. Ship Res.*, **31**, 227 (1987).
54. V. Bertram, W. W. Schultz, Y. Cao, and R. F. Beck, *Schiffstechnik*, **38**, 3 (1991).
55. Y. S. Cao, *Computations of nonlinear gravity waves by a desingularised boundary integral method* (Ph.D. dissertation, Department of Naval Architecture and Marine Engineering, The University of Michigan, 1991).
56. D.C. Scullen, *Accurate computation of steady nonlinear free-surface flows* (Ph.D. dissertation, Department of Applied Mathematics, The University of Adelaide. (downloadable from http://internal.maths.adelaide.edu.au/people/dscullen/Web_pages/publications.html), 1998).
57. E. O. Tuck, and D. C. Scullen, *J. Eng. Math.*, **42**, 255 (2002).

CHAPTER 10

APPLICATION OF THE FINITE VOLUME METHOD TO THE SIMULATION OF NONLINEAR WATER WAVES

Deborah Greaves*

School of Engineering, University of Plymouth, Drakes Circus
Plymouth, PL4 8AA, UK
deborah.greaves@plymouth.ac.uk

The finite volume method is discussed in its application to the simulation of highly nonlinear free surface waves in a viscous fluid. Solution of the Navier-Stokes equations is described for two-fluid flow calculations, in which both air and water are modelled. Different approaches to handling the pressure-velocity coupling are described and turbulence models used for free surface simulation are reviewed. Interface capturing approaches for dealing with the fluid interface are explained and numerical computations presented to compare selected schemes. An adaptive finite volume method based on quadtree grids with Cartesian cut cells and high resolution interface capturing for the free surface is described and used to simulate illustrative example calculations of complex free surface flow.

1. Introduction

The study of breaking waves is important both in oceanography and in ocean engineering. Wave breaking is an important mechanism for transfer of energy between the atmosphere and the ocean, and breaking waves impart much greater loading than non-breaking waves. This is because of the greater velocity and acceleration of fluid particles within them. It is important to understand the breaking process in order to

*Formerly at the *Department of Architecture and Civil Engineering, University of Bath Claverton Down, Bath, BA2 7AY, UK.*

predict the loading and run-up caused by breaking wave interaction with offshore and coastal structures. Traditional methods for free surface waves use potential flow theory for situations where the fluid viscosity may be ignored, and fully nonlinear free surface boundary conditions if the linearity assumption cannot be made. Fully non-linear waves and wave-body interactions have been solved by potential-based methods with moving boundary meshes[1–5] but generally these methods are limited to non-breaking waves. Simulations of a solitary wave breaking over a submerged reef calculated using a variety of different numerical methods, including compressible, incompressible, inviscid and viscous models, are compared by Helluy *et al.*[6]

In highly nonlinear wave loading situations, involving wave breaking and greenwater, traditional simplifications of the equations of fluid motion no longer apply. Beyond wave breaking, air entrainment is important, also fluid viscosity, turbulence and surface tension may be significant and so a viscous fluid solution method is necessary, in which both the air and water are simulated. In these cases, the full Navier-Stokes (NS) equations or Reynolds Averaged Navier-Stokes (RANS) equations, if the flow is turbulent, must be solved in a two-fluid system. The accuracy of the simulation depends on calculating the correct position of the air/water interface throughout the wave motion, and this becomes especially difficult when the wave overturns and merges with the water surface or when the interface breaks up into spray.

One aspect of the calculation is solution of the NS (or RANS) equations and another is prediction of the evolution of the free surface. The numerical discretisation may be by finite difference, finite element, finite volume or other discrete method, and here we focus on the finite volume method. The NS system for incompressible fluid has an inherent difficulty due to the lack of an equation for pressure. This leads to a need for coupling the momentum and continuity equations in some way. The governing NS equations are introduced in Section 2 and three broad methods: pressure correction, fractional step and artificial compressibility are discussed. Some recent advances for turbulence modelling in free surface simulations are also described. Prediction of the air-water interface for wave breaking cases needs an interface capturing approach. The two main approaches are the Level Set (LS)[7] and Volume of Fluid

(VoF)[8] method discussed in Section 3. Within the VoF method there are a variety of alternative methods for calculating the interface advection in time and its reconstruction, some of which are described in detail in Section 3. In Section 4, an adaptive quadtree finite volume method for viscous wave simulation is described and results presented for complex free surface flow.

2. Solution of the Governing Equations

The general form of the Navier-Stokes equations is given by

$$\rho\left(\frac{\partial \mathbf{u}}{\partial t}\right)+\rho(\mathbf{u}\cdot\nabla\mathbf{u})=\nabla\cdot\mathbf{T}+\mathbf{f}\,, \tag{1}$$

where $\mathbf{u}$ is the velocity vector with Cartesian components, u, v and w, ρ is the fluid density, $\mathbf{f}$ is the body force with Cartesian components $\{0 \quad -g \quad 0\}$ and $\mathbf{T}$ is the stress tensor,

$$\mathbf{T}=-\left(p+\frac{2}{3}\mu\nabla\cdot\mathbf{u}\right)\mathbf{I}+\mu\left(\nabla\otimes\mathbf{u}+(\nabla\otimes\mathbf{u})^T\right). \tag{2}$$

Here, p is the pressure, μ is dynamic viscosity and $\mathbf{I}$ is the unit tensor. Substituting the stress tensor into the Navier-Stokes equation gives

$$\rho\left(\frac{\partial \mathbf{u}}{\partial t}\right)+\rho(\mathbf{u}\cdot\nabla\mathbf{u})=-\nabla p+\mu\nabla^2\mathbf{u}+\mathbf{f}+(\nabla\otimes\mathbf{u})\cdot(\nabla\mu), \tag{3}$$

and the mass conservation equation is given by

$$\frac{\partial\rho}{\partial t}+\nabla\cdot(\rho\mathbf{u})=0\,. \tag{4}$$

For a Newtonian incompressible fluid,

$$\nabla\cdot\mathbf{u}=0\,, \tag{5}$$

and written in conservation form, the Navier-Stokes equation becomes

$$\frac{\partial\rho\mathbf{u}}{\partial t}+\nabla\cdot(\rho\mathbf{u}\otimes\mathbf{u})=-\nabla p+\nabla\cdot(\mu\nabla\otimes\mathbf{u})+\mathbf{f}+(\nabla\otimes\mathbf{u})\cdot(\nabla\mu)\,. \tag{6}$$

The last term on the right hand side is zero if the fluid viscosity is constant throughout the domain, though not in the two-fluid cases considered here.

In the finite volume discretisation, the partial differential equations are integrated over a control volume, of volume V, and time step, δt, thus the integral form is

$$\begin{aligned}
&\int_t^{t+\delta t}\left(\int_V\left(\frac{\partial \rho\mathbf{u}}{\partial t}\right)dV\right)dt+\int_t^{t+\delta t}\left(\int_V(\nabla\cdot(\rho\mathbf{u}\otimes\mathbf{u}))dV\right)dt\\
&-\int_t^{t+\delta t}\left(\int_V(\nabla\cdot(\mu\nabla\otimes\mathbf{u}))dV\right)dt=-\int_t^{t+\delta t}\left(\int_V(\nabla p)dV\right)dt \qquad (7)\\
&+\int_t^{t+\delta t}\left(\int_V(\rho\mathbf{g})dV\right)dt+\int_t^{t+\delta t}\left(\int_V((\nabla\otimes\mathbf{u})\cdot(\nabla\mu))dV\right)dt.
\end{aligned}$$

The control volume is bounded by a series of flat faces and so, using the divergence theorem, the integral can be written as a sum of integrals over the faces. The divergence theorem for an arbitrary function, φ is,

$$\int_V \nabla\cdot\varphi dV=\oint_{\partial V} d\mathbf{S}\cdot\varphi=\sum_{f=1}^{n}\left(\int_f d\mathbf{S}\cdot\varphi\right)\approx\sum_{f=1}^{n}\mathbf{A}_f\cdot\varphi_f\,, \qquad (8)$$

where $d\mathbf{S}$ is the surface area vector and ∂V the surface area of the control volume. $\mathbf{A}_f$ is the cell face area vector, n is the number of faces, f is the centre of the cell face and the midpoint rule is used to approximate the surface integrals.

Thus, the Navier-Stokes equation can be rewritten in partially discretised form for central cell P with n neighbours (and corresponding faces) as:

$$\begin{aligned}
&\int_t^{t+\delta t}\left(\left(\frac{\partial \rho\mathbf{u}}{\partial t}\right)_P V_P+\sum_{f=1}^{n}\rho F_f\mathbf{u}_f-\sum_{f=1}^{n}\mu_f\mathbf{A}_f\cdot(\nabla\otimes\mathbf{u})_f\right)dt\\
&=-\int_t^{t+\delta t}(\{-(\nabla p)_P+\mathbf{g}\rho+(\nabla\otimes\mathbf{u})_P\cdot(\nabla\mu)_P\}V_P)dt\,.
\end{aligned} \qquad (9)$$

Here, the convection term,

$$\int_V(\nabla\cdot(\rho\mathbf{u}\otimes\mathbf{u}))dV\approx\sum_{f=1}^{n}\rho F_f\mathbf{u}_f\,, \qquad (10)$$

where F_f, the convective flux through the cell face,

$$F_f = \mathbf{A}_f \cdot \mathbf{u}_f, \tag{11}$$

and the diffusion term,

$$\int_V \nabla \cdot (\mu \nabla \otimes \mathbf{u}) dV \approx \sum_{f=1}^{n} \mu_f \mathbf{A}_f \cdot (\nabla \otimes \mathbf{u})_f. \tag{12}$$

Using Euler implicit time stepping, the Navier-Stokes momentum equation is

$$\frac{\rho \mathbf{u}_P^{t+\delta t}}{\delta t} + \frac{1}{V_P} \sum_{f=1}^{n} \rho F_f \mathbf{u}_f^{t+\delta t} - \frac{1}{V_P} \sum_{f=1}^{n} \mu_f \mathbf{A}_f \cdot (\nabla \otimes \mathbf{u})_f^{t+\delta t} = \mathbf{S}_{\mathbf{u}_P} - (\nabla p)_P, \tag{13}$$

where

$$\mathbf{S}_{\mathbf{u}_P} = \frac{(\rho \mathbf{u})_P^t}{\delta t} + \mathbf{g}\rho + (\nabla \otimes \mathbf{u})_P \cdot (\nabla \mu)_P. \tag{14}$$

The discretised convection and diffusion terms contain contributions from neighbouring cells and the discrete equation may be written in compact form as

$$a_P \mathbf{u}_P^{t+\delta t} = \sum_{nb=1}^{n} a_{nb} \mathbf{u}_{nb}^{t+\delta t} + \mathbf{S}_{\mathbf{u}_P} - (\nabla p)_P, \tag{15}$$

in which a_i are coefficients of the central cell subscript P, and its neighbours, subscript nb. Rearranging,

$$\mathbf{u}_P^{t+\delta t} = \frac{H(\mathbf{u})_P}{a_P} - \frac{1}{a_P} (\nabla p)_P, \tag{16}$$

and

$$H(\mathbf{u})_P = \sum_{nb=1}^{n} a_{nb} \mathbf{u}_{nb}^{t+\delta t} + \mathbf{S}_{\mathbf{u}_P}. \tag{17}$$

The finite volume discretisation of the continuity equation is

$$\sum_{f=1}^{n} \mathbf{A}_f \cdot \mathbf{u}_f = 0. \tag{18}$$

In order to solve for the velocity and pressure field, we need to couple the momentum and continuity equations together. Various approaches to this are used in the literature and fall into three main categories: implicit pressure-correction, artificial compressibility and fractional step methods. In most cases, the resulting algebraic equations are solved by an iterative method, although a fully coupled approach may also be taken[9].

2.1. *The fractional step method*

The fractional step method[10,11] involves splitting the update of the velocity field into a series of steps in which a different operator is individually applied to partly update the variable. For example if the Navier-Stokes equations above can be written in symbolic form as

$$u_P^{n+1} = u_P^n + (C_P + D_P + P_P)\Delta t , \tag{19}$$

where C_P, D_P and P_P are the convective term, diffusion term and pressure term respectively. The update of the velocity may be split into a three step method

$$u_P^{*} = u_P^n + (C_P)\Delta t , \tag{20}$$

$$u_P^{**} = u_P^{*} + (D_P)\Delta t , \tag{21}$$

$$u_P^{n+1} = u_P^{**} + (P_P)\Delta t . \tag{22}$$

In the third step, P_P is the gradient of the pressure, or other pressure-like variable, that obeys a Poisson equation and must be chosen to satisfy continuity. In some methods the source term in the Poisson equation differs from that in the Poisson equation for pressure and so the variable is called the pseudo-pressure or pressure-like variable.

2.2. *The artificial compressibility method*

Artificial compressibility methods use the addition of a time derivative for pressure to the continuity equation to give the incompressible Navier-Stokes equations hyperbolic character. This means they can be solved using one of a variety of techniques developed for compressible Navier-Stokes equations. However, the addition of a time derivative for pressure means that the true incompressible equations are no longer being solved, and are only satisfied at convergence when the time derivative is zero[10]. The continuity equation is modified by the addition of a time derivative for pressure,

$$\frac{1}{\beta}\frac{\partial p}{\partial t} + \Delta \cdot \mathbf{u} = 0 , \tag{23}$$

where β is an artificial compressibility parameter. The larger the value of β the more close the equations are to incompressible.

2.3. *Implicit pressure correction*

Implicit pressure correction methods involve deriving a Poisson equation for the pressure or pressure correction by substituting discrete expressions for cell face velocities into the discretised continuity equation. Two well known methods are SIMPLE[12], in which a pressure correction equation is derived, and PISO[13], in which a pressure equation is derived.

The cell face values from the discretised momentum equation (16) can be written as

$$\mathbf{u}_f^{t+\delta t} = \left(\frac{H(\mathbf{u})}{a_P}\right)_f - \left(\frac{1}{a_P}\right)_f (\nabla p)_f . \tag{24}$$

The expression for the velocity at each face is substituted into the discretised continuity equation (18), which, after rearrangement, results in the discretised pressure equation

$$a2_P\, p_P = \sum_{nb=1}^{n} a2_{nb}\, p_{nb} + b . \tag{25}$$

Here, $a2_i$ are coefficients for node i and the source term is

$$b = \sum_{f=1}^{n} \mathbf{A}_f \cdot \left(\frac{H(\mathbf{u})}{a_P}\right)_f . \tag{26}$$

The algorithm for solution of the equations is

1. solve momentum equations (16) using a guessed pressure field
2. calculate $H(\mathbf{u})$ terms (17)
3. solve pressure equation (25)
4. calculate volumetric fluxes for use in momentum equation coefficients
5. correct velocity using (16) and calculated $H(\mathbf{u})$ terms.

The loop from (2) to (5) is repeated iteratively until a prescribed tolerance is achieved before proceeding to the next time step (1).

2.4. *Numerical instability*

Two well known areas of instability for solving the Navier-Stokes equations are discretisation of the convective term, equation (10), and pressure-velocity decoupling associated with use of a computational grid in which variables are collocated. Central differencing discretisation of the convective term will result in instability if the grid Peclet number, which is the ratio of convection to diffusion in a cell and equivalent to a cell Reynolds number, exceeds a limiting value. Pure upwinding eliminates the instability, but is considerably less accurate than central differencing and so blended schemes that contain both upwind and higher order differencing components are popular, such as the QUICK[14] scheme, blended differencing[15] and Jasak's[16] scheme.

Pressure-velocity decoupling, in which more than one independent solution to the pressure field coexists, may occur if variables are collocated in the grid. One solution is to use a staggered grid, in which the nodes for storing pressure and velocity are offset from one another. In the finite element method this is achieved by using shape functions of different order for the pressure and velocity, effectively a different number of nodes for pressure than for velocity.

However, a staggered grid can lead to more complicated data handling and is not practical for an unstructured mesh. An alternative treatment for collocated grids is to use a special interpolation for face velocity values in calculation of the convective flux, F_f. This is often referred to as Rhie–Chow treatment thanks to their early work in this area[17]. Face velocities in the face flux term are calculated in the following way; first the face velocity expression (24) is substituted into (11),

$$F_f = \mathbf{A}_f \cdot \mathbf{u}_f = \mathbf{A}_f \cdot \overline{\left(\frac{H(\mathbf{u})}{a_P}\right)_f} - \overline{\left(\frac{1}{a_P}\right)_f} (\nabla p)_f, \tag{27}$$

where the overbar represents linear interpolation,

$$\left(\frac{H(\mathbf{u})}{a_P}\right)_f = \ell_P \left(\frac{H(\mathbf{u})}{a_P}\right)_P + (1-\ell_P)\left(\frac{H(\mathbf{u})}{a_P}\right)_{nb}, \tag{28}$$

$$\left(\frac{1}{a_P}\right)_f = \ell_P \left(\frac{1}{a_P}\right)_P + (1-\ell_P)\left(\frac{1}{a_P}\right)_{nb}, \tag{29}$$

and the interpolation vector $\ell_P = \dfrac{|fnb|}{|fP|}$ is the ratio of distances between the cell face and adjacent nodes. Rather than using linear interpolation for the pressure gradient at the face, it is calculated using the adjacent nodal pressure values, thus the face velocity,

$$\mathbf{u}_f = \overline{\left(\frac{H(\mathbf{u})}{a_P}\right)_f} - \overline{\left(\frac{1}{a_P}\right)_f} \left(\frac{p_{nb} - p_P}{\mathbf{d}_f}\right)_f, \tag{30}$$

where $\mathbf{d}_f$ is the vector distance between node P and its neighbour nb. Equation (30) may be interpreted as modifying the interpolated cell face velocity by the difference between the interpolated pressure gradient and the gradient calculated at the cell face.

2.5. *Turbulence*

It is apparent from simple observation of breaking waves that turbulence is likely to be significant, but very little is known about the structure of turbulence near the air-water interface. Direct numerical simulation (DNS) of the NS equations will resolve the scales of turbulence if sufficient grid and temporal resolution is provided. However, this is extremely expensive and restricts DNS to low Reynolds number flows. The alternative for high Reynolds numbers is to use either Reynolds Averaged Navier Stokes, RANS, modelling, in which the NS equations are averaged over time and a turbulence model applied to relate the resulting Reynolds stress terms to the mean flow variables and thus close the equation system[10]. Another alternative is to use Large Eddy Simulation, LES, in which the NS equations are spatially filtered. The filtered NS equations contain sub-grid scale (SGS) components, which are modelled using a SGS model, such as Smagorinsky's[18] eddy viscosity model.

LES is generally considered to be superior to the RANS approach for unsteady complex flow simulation, although LES is usually more expensive[19]. Turbulent free surface flow simulation is highly complex because of the difficult free surface boundary condition and the highly anisotropic flow field near the free surface. Careful DNS studies[20] of free surface turbulence have helped to improve understanding of the physics involved and have been used as a basis from which to compare LES calculations[21] and developments in SGS modelling for free surface turbulence. Shen and Yue[21] use a linearised free surface boundary condition to model turbulent shear flow under a free surface and develop specialised SGS models for this problem.

A common simplification of the free surface boundary condition is to use a rigid free slip free surface. This approach is taken by Nagaosa[22] in DNS calculations of free surface channel flow and by Salvetti *et al.*[23], who investigate the use of dynamic SGS models with LES and compare with DNS results[24]. Li and Wang[25] allow movement of the free surface in their LES simulation of channel flow; the variation of the free surface is calculated from the hydrostatic pressure equation and surface shear stresses are set to zero. They discuss the use of appropriate inflow boundary conditions, compare results with experimental data[26] and propose a parabolic mixing length model for SGS terms.

RANS calculations have been used frequently for steady flow simulation in ship hydrodynamics[27] and more recently for unsteady calculations[28]. Rhee and Stern[28] simulate free surface wave interaction with a Wigley hull using RANS with a Baldwin Lomax turbulence model[29]. Grids are adjusted to conform to the moving free surface through the kinematic boundary condition. Hodges and Street[30] use a similar moving mesh approach to the free surface boundary condition, allowing free surface motion up to breaking. They use LES with a two-parameter dynamic model for the SGS stresses.

The free surface turbulence studies mentioned so far are limited to linear or nonlinear wave motion up to but not including breaking. This limitation is due to the free surface boundary condition, which is either linearised or applied on single valued boundary-conforming grids. In order to model wave breaking a more robust approach to the free surface boundary condition must be taken, in which interface merging and break up is allowed. Yue *et al.*[31] simulated turbulent free surface open channel flow using LES within a Level Set, LS, framework (discussed in Section 3). They found the dynamic Smagorinsky and two-parameter dynamic SGS models to give very similar performance in comparison with experiments[32]. Lin and Liu[33] used the RANS equations with a nonlinear Reynolds Stress model for turbulence to simulate breaking waves. Severe distortion of the free surface including wave overturning are handled using a volume of fluid (VoF) approach (discussed in Section 3). An impressive three-dimensional LES simulation of a plunging breaker is presented by Lubin *et al.*,[34] although they also mention the high computational cost of the 3-D LES. Labourasse et al.[35] investigate LES for two phase flow calculations. Interfacial turbulence is studied through the interaction of turbulence with a bubble and phase inversion involving complex deformation of the interface. Liovic and Lakehal[36] describe a 3-D LES formulation for deformable gas-liquid interfaces using VoF with piecewise planar interface reconstruction, which is applied to simulation of a plunging breaker.

3. Interface Modelling

Techniques for predicting the position of a moving free surface may be classified as: interface tracking methods, which include moving mesh, front tracking[37,38] and particle tracking schemes[39]; and interface capturing methods, which include volume of fluid (VoF) and level set techniques[40]. Lin and Lui[41] review surface tracking methods for applications in wave hydrodynamics. Front tracking methods use a piecewise polynomial fit to the interface, which is advected with the flowfield. Moving mesh methods have been applied successfully in simulation of drop formation[42], roll-coating flows[43] and non-breaking free surface waves[44]. However, the disadvantage of moving mesh and front tracking methods for simulating steep waves is that they can only calculate waves up to the limit of breaking as the free surface has to be single valued. Particle tracking methods tend to be expensive, although so-called meshless methods, such as smooth particle hydrodynamics, SPH[39] and the MLPG[45] (Meshless Local Petrov-Galerkin) method have become more popular recently. These have the advantage of eliminating the need for complicated mesh boundary tracking, as they rely on integrations over a distribution of points rather than a closed mesh.

On the other hand, front capturing methods can be used for modelling large-scale deformations of the interface including wave break up and merging. Front capturing methods differ from front tracking in that the solution is calculated in the combined air and water fluid domains, with the fluid properties changing at the interface. The interface is then located from the zero contour of a distance function in the case of level set[46,47], from the phase-field equation[48], from the discontinuity in the density field[49,50] or from the volume fraction field in the VoF method[51]. An early example of interface capturing is the marker and cell or MAC scheme of Harlow and Welch[52], which follows massless particles at the free surface.

3.1. *The Level Set, LS, method*

In the LS method, the fluid density and viscosity are defined as smooth functions of the distance from the interface. The interface is given a fixed thickness which is constant in time and across which the fluid properties vary smoothly. A scalar function that represents the signed distance from the interface is advected in time and the position of the interface at any time is captured as the zero contour of this function. The distance function needs to be re-initialised, usually at each time step, in order to keep the thickness of the transition zone constant.

The signed distance function, d, is computed for each grid point at $t = 0$ from the known position of the free surface, assuming that $d > 0$ in water and $d < 0$ in air. It is advected in time using,

$$\frac{\partial d}{\partial t} + \mathbf{u} \cdot \nabla d = 0, \tag{31}$$

and the position of the interface can be identified with the level $d = 0$.

The fluid properties, density and viscosity, are determined from the distance function at each time. The jump of fluid properties at the interface is smoothed by spreading the jump over a transition thickness, which is related to the cell size and kept constant in time. The fluid properties are calculated from

$$f(d)=\begin{cases} f_w & \text{if } d>\delta \\ f_a & \text{if } d<-\delta \\ (f_a+f_w)/2+(f_w-f_a)/2\sin(\pi d/2\delta) & \text{otherwise} \end{cases} \quad (32)$$

where δ is half the width of the region along which the jump is spread and $f(d)$ is a function representing the variation of fluid property (density and viscosity) in which the subscript a refers to air and w to water. The thickness δ is chosen so that the interface is spread over at least three computational cells in the direction normal to the interface. Equation (31) does not necessarily keep the interface thickness constant throughout the calculation and so, to avoid spreading or concentration of the interface, the distance function needs to be re-initialised at each time step[53,54]. Iafrati *et al.*[7,47] describe a level set method used with finite difference discretisation of the Navier-Stokes Equations to simulate unsteady free surface flows including breaking waves.

3.2. *The Volume of Fluid, VoF, method*

The volume of fluid method is similar to the Level Set method in that the fluid properties are defined by an additional scalar. In the case of the LS this is the distance function and in the VoF this is the volume fraction, α. The volume fraction expresses the proportion of water in a particular cell and is equal to 1 if the cell fully contains water and 0 if it fully contains air. The interface lies in cells with volume fraction values lying between 0 and 1. The fluid properties are calculated from

$$f(\alpha)=\alpha f_w+(1-\alpha)f_a \,. \quad (33)$$

The volume fraction satisfies an advection equation as the interface evolves and the solution advances in time, the interface is captured by the advection of the volume fraction and the fluid equations may be solved without explicitly calculating the interface. The interface shape may then be reconstructed from the volume fraction field. Alternatively, the interface may be reconstructed and advected at each time step in a piecewise manner[55].

Thus, the VoF method consists of two parts: an interface reconstruction algorithm for approximating the interface from the set of volume fractions, and a VoF transport algorithm for determining the volume fractions at the new time level from the velocity field and the reconstructed front. The basic method is robust and flexible and VoF schemes are widely used[41,56–58]. The traditional drawbacks of this method are its tendency to smear the interface and the high CPU cost due to the need for fine grids and small time steps, although recent advances in the development of new high resolution interface advection schemes[59–62] have alleviated these drawbacks somewhat. The volume fraction advection equation can be expressed as

$$\frac{\partial \alpha}{\partial t} + \nabla.(\mathbf{u}\alpha) = 0\,. \tag{34}$$

The original VoF scheme[51] uses a combination of upwinding and downwinding; when the interface is perpendicular to the flow, the downwind scheme is used and when the interface is parallel to the flow, upwinding is used. The advantage of the upwind scheme is that it is stable, but it is diffusive and may spread the interface over many cells. The downwind scheme is unstable, but sharpens the interface. Various VoF fluxing methods have been developed, most of which aim for a balance between the stability advantages of the upwind scheme and the front sharpening advantages of the downwind scheme.

3.2.1. *SURFER*

Lafaurie *et al.*[63] describe a fluxing scheme, implemented in the program SURFER, in which either upwind or modified downwinding is used depending on the orientation of the interface relative to the fluxing direction. A directional split algorithm is used that calculates fluxes in the *x*-direction for the entire grid and then in the *y*-direction. The downwind scheme is modified to constrain the volume fraction face value to prevent more fluid being outfluxed than the cell contains and to ensure that the Courant condition is satisfied.

The volume fraction equation (34) in discrete form is given by

$$\alpha^{n+1} = \alpha^n + \sum_{k=1}^{k=K} f_k \, , \tag{35}$$

where k is the cell face orientation and K is the number of cell faces. For a regular rectangular grid in two dimensions, $K = 4$ and $k = e, w, n$ or s. f_k represents the flux of α across the k direction cell boundary. Thus,

$$\alpha^{n+1} = \alpha^n + f_e - f_w + f_n - f_s \, , \tag{36}$$

and the cell face fluxes are

$$f_k = u_k \frac{\delta t}{\delta x} \alpha_k \, . \tag{37}$$

where δt and δx are the time step and mesh size and u_k is the velocity at face k. The upwind or modified downwind schemes for u_k are chosen depending on the angle between the interface normal and the fluxing direction.

3.2.2. *PLIC VoF*

The VoF scheme outlined above describes the interface implicitly since the volume fraction data must be inverted to find the approximate interface position. The interface may be reconstructed by SLIC[64] (Simple Line Interface Calculation) or by various PLIC[55,65,66] (Piecewise Linear Interface Calculation) methods. However, rather than solving (34), the interface may be propagated by Lagrangian advection from one time step to the next.

In the PLIC method, a straight line is defined in each interface cell that divides the cell into two parts, each of which contains the correct volume of one of the two fluids. This interface segment is propagated by the flow and the resulting volume of each fluid in the neighbouring cells is calculated. It is necessary both to calculate the volume of a fluid from the equation of the interface segment in the cell and inversely the equation of the interface from the volume of fluid in the cell.

The time stepping is performed separately in each spatial direction. The x-component of the velocity at each point on the line is a simple linear interpolation between the values on the cell faces, U_w and U_e,

$$u(x) = U_w\left(1 - \frac{x}{\delta x}\right) + U_e \frac{x}{\delta x}. \tag{38}$$

During the x-sweep, the x-coordinate of the line changes to the new value

$$x^* = x^n + u(x^n)\delta t = \left[1 + \left(\frac{U_e - U_w}{\delta x}\right)\delta t\right]x^n + U_w \delta t. \tag{39}$$

The superscript n denotes the value at time n and the asterisk is used to denote a fractional step, the x-sweep, to be followed by a similar step in y. Aulisa *et al.*[67] describe an extension of the PLIC algorithm in which a combination of the surface marker and volume fraction field approach is applied to calculation of free boundary flows.

3.2.3. *CICSAM VoF*

Ubbink[59] derived a compressive differencing scheme for discretisation of the volume fraction equation (34). The scheme is named CICSAM (Compressive Interface Capturing Scheme for Arbitrary Meshes) and is based on the Normalised Variable Diagram (NVD)[68]. CICSAM combines the Convection Boundedness Criteria (CBC) given by

$$\tilde{\alpha}_{fCBC} = \begin{cases} min\left\{1, \dfrac{\tilde{\alpha}_D}{c_D}\right\} & \text{if } 0 \le \tilde{\alpha}_D \le 1 \\ \tilde{\alpha}_D & \text{if } \tilde{\alpha}_D < 0, \tilde{\alpha}_D > 1 \end{cases}, \tag{40}$$

with the Ultimate Quickest (UQ) differencing scheme, which is a version of QUICK[14],

$$\tilde{\alpha}_{fUQ} = \begin{cases} \min\left\{\dfrac{8c_D\tilde{\alpha}_D + (1 - c_D)(6\tilde{\alpha}_D + 3)}{8}, \tilde{\alpha}_{fCBC}\right\} & \text{if } 0 \le \tilde{\alpha}_D \le 1 \\ \tilde{\alpha}_D & \text{if } \tilde{\alpha}_D < 0,\ \tilde{\alpha}_D > 1. \end{cases} \tag{41}$$

Here c_D is the Courant number and $\tilde{\alpha}_D$ is the normalised variable for the donor cell, calculated from

$$\tilde{\alpha}_D = \frac{\alpha_D - \alpha_U}{\alpha_A - \alpha_U}, \tag{42}$$

where subscript U indicates the upwind cell, A the acceptor and D the donor cell. These are determined depending on the velocity at a given face.

The normalised face value for the CICSAM differencing scheme is calculated by combining these two schemes through a weighting factor derived from the orientation of the interface to the direction of motion. The method is derived for arbitrary meshes and so a direction split approach is not necessary. Instead an implicit multi-dimensional implementation is utilised[59] that requires only one sweep through the mesh.

3.2.4. *CIP method for advection*

The constrained interpolation profile (CIP) algorithm described by Yabe *et al.* [69] is applied to advection computations and can be interpreted as a compact upwind scheme with sub-cell resolution. Their idea is that when considering advection of a variable, both the variable and its gradient are advected, where the interpolation approximation for the gradient of the variable is the gradient of the interpolation approximation of the variable. A cubic polynomial is derived from the continuity condition applied to both the variable and its gradient and then used to construct the interpolation function for the variable.

The CIP scheme may be used for advection of the velocity in the momentum equations as well as for advection of the volume fraction in the volume fraction equation[69–71]. The CIP method for interface tracking[72] is improved by combination with a tangent function transformation. The colour function, α, which identifies the material in a multi-phase calculation, is transformed using a tangent function,

$$F(\alpha) = \tan[(1-\varepsilon)\pi(\alpha - 1/2)], \tag{43}$$

where

$$\alpha = \arctan F(\alpha)/[(1-\varepsilon)\pi] + 1/2, \tag{44}$$

and ε is a small positive constant. The CIP algorithm is then applied to the transformed variable. This transformation improves the spatial resolution locally as most of the values of $F(\alpha)$ are concentrated near $\alpha = 0$ and $\alpha = 1$ and although α undergoes a rapid change from 0 to 1 at the interface, the transformed function $F(\alpha)$ shows regular behaviour.

3.3. *Adaptive quadtree-grid scheme*

Here, a method is demonstrated in which the high resolution CICSAM[59] is implemented on adapting quadtree grids[73,74]. An advantage of quadtree grids is that they can be readily adapted by the addition and removal of cells throughout a time dependent simulation. Grid refinement is used to follow the movement of the interface: remeshing of the grid operates by dividing a cell into four if it lies on the interface, i.e. if the value of α lies between 0 and 1; derefinement takes place by removing four sibling cells and replacing them with their parent. This only occurs where each of the four sibling cells lies away from the interface and the volume fraction value of each is equal to 0 or 1.

Interpolation and extrapolation of the volume fraction, α, during grid adaptation as cells are added and removed, is achieved using PLIC reconstruction. PLIC reconstruction of the interface in the divided cell is transformed to the coordinates of each newly created cell and the volume fraction determined from the equation of the interface line in each new cell[74].

3.4. *Comparison of interface tracking schemes*

A simple test to assess the CFD methodology for capturing interfaces between two immiscible fluids involves translation of a square interface through a uniform velocity field. The test case is calculated using the VoF schemes SURFER, PLIC, CICSAM and the accuracy of each method is compared. A regular grid size of 128 x 128 in a unit square domain is used and the initial contours for the square interface, size 0.125m x 0.125m, positioned at the centre of the grid, are shown in Fig. 1(a). The square is translated in a constant velocity field, u = 1m/s, v = -1m/s and the time step is determined such that the Courant number is equal to 0.125. With an accurate interface advection scheme the interface should be translated intact towards the bottom right hand corner of the domain. Figures 1(b) to (d) show the interface at time t = 0.375s, calculated using SURFER, PLIC and CICSAM. Four volume fraction contours, α = 0.05, α = 0.4, α = 0.6 and α = 0.95 are plotted. The interface translated using SURFER interface tracking is clearly distorted; the results obtained with PLIC and CICSAM are much closer to the initial square and similar to one another, although the original interface is better maintained with CICSAM.

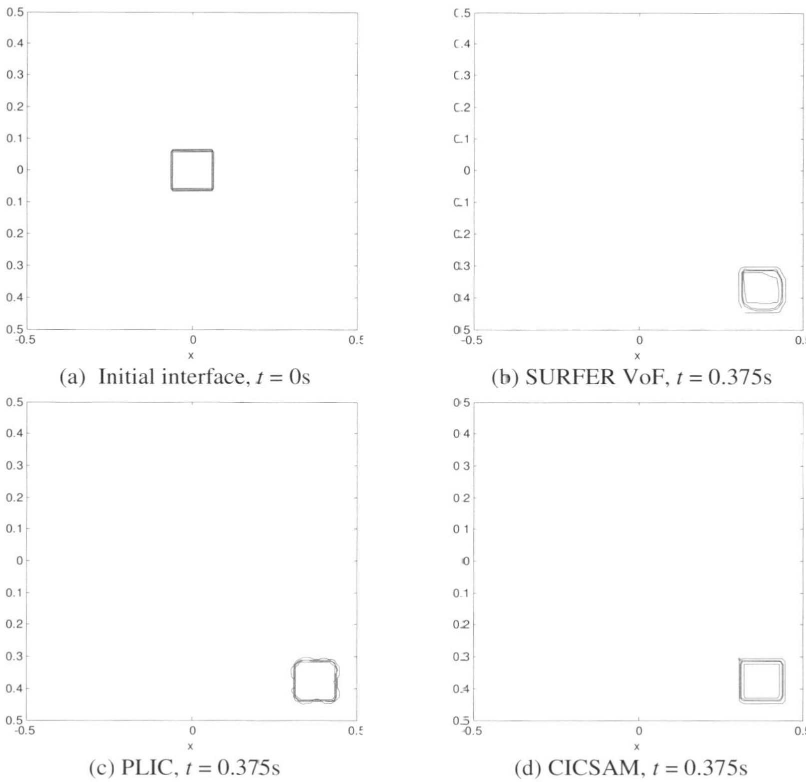

(a) Initial interface, t = 0s

(b) SURFER VoF, t = 0.375s

(c) PLIC, t = 0.375s

(d) CICSAM, t = 0.375s

Fig. 1. Advection of square interface, contours of α plotted at α = 0.05, α = 0.4, α = 0.6 and α = 0.95.

A more stringent test for the advection scheme is to operate in a shearing flow field. In this case[59,75], the velocity is prescribed to define a rotational vortex, in which the velocity components, $u = u_{max}\sin\theta\cos\theta$, $v = -v_{max}\cos\theta\sin\theta$, where u_{max} = 1.0m/s and v_{max} = -1.0m/s. An adapting quadtree grid is used, which adapts to follow the deformation of the interface. A circular interface of diameter 0.36m is initially positioned at x = 0.0m, y = -0.2m. The interface is first advected forward for a given length of time and then the velocities are reversed for the same length of time in order to return the interface to its initial position. A perfect advection scheme should result in the initial volume fraction field being recovered.

Figures 2 (a) and (b) show the interface and adapted grid at t = 5s and after the velocity has been reversed and the interface advected for 5 seconds in the opposite direction. Figures 2 (c) and (d) show similar results for advection up to and back from 10 seconds. After reversing from t = 5.0s, the resulting interface is very close to the initial circle, despite being highly distorted at t = 5.0s. At t = 10.0s, the interface has completed nearly two full revolutions, is distorted into a spiral and beginning to break up at the tail end. Even so, when tracked back for a further 10 seconds with velocities reversed, the resulting interface is close in appearance to the original circle.

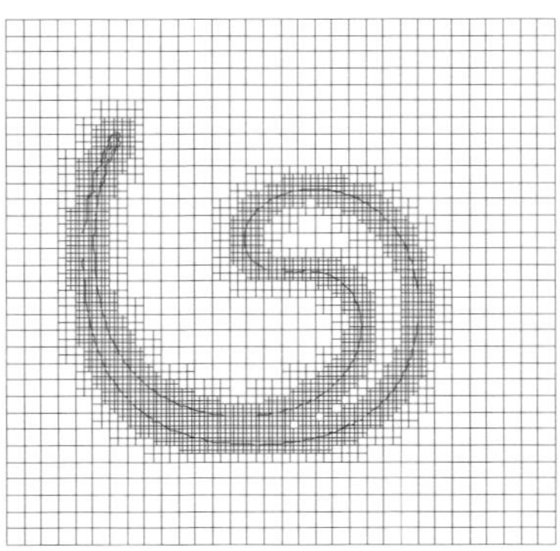

(a) Interface and adapted grid at t = 5.0s.

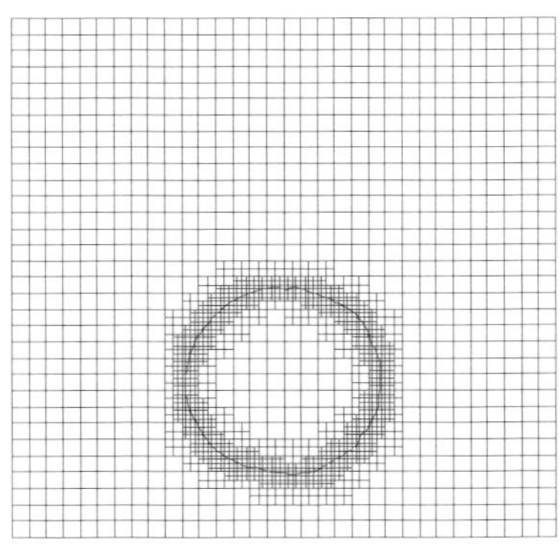

(b) Interface and adapted grid after reversing from t = 5.0s to t = 0.0s.

Fig. 2. Circular interface in shear flow.

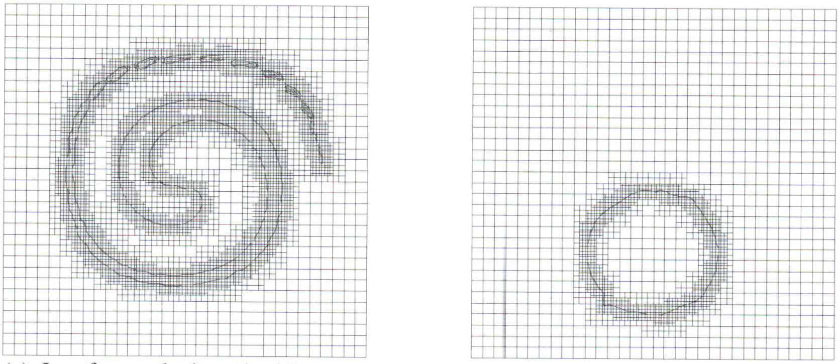

(c) Interface and adapted grid at t = 10.0s.

(d) Interface and adapted grid after reversing from t = 10.0s to t = 0.0s.

Fig. 2. *Continued.*

Further test cases for translating and rotating interfaces are presented by Greaves[74] to demonstrate the adaptive quadtree VoF scheme and compare it with other popular methods. Both PLIC and CICSAM schemes are combined with adaptive quadtree grids and the results compared with those from uniform grid calculations. Quadtree adaptation is shown to provide significant savings; typically the grid size is approximately ten times less than the equivalent uniform grid and the computation time approximately five times less.

4. Adaptive Finite Volume Method for Free Surface Flow Simulation Using Quadtree Grids and Cartesian Cut Cells

Having established in the previous Section that the adaptive quadtree-based approach combined with CICSAM interface advection achieves high accuracy with sharp definition of the interface even for highly sheared flows, it is natural to apply this scheme to calculation of fluid flow[76]. The Navier–Stokes equations are solved on adapting quadtree grids using a version of PISO[13] together with collocated variables. Quadtree grid generation, grid reference numbering schemes, grid data retrieval and neighbour finding routines are described by Greaves and Borthwick[73] and Yiu *et al.*[77] Special interpolations are needed at the interface between panels of different size, i.e. at hanging nodes, and during quadtree grid adaptation[74,76].

The adapting quadtree grids are combined with Cartesian cut cells[61,78] in order to model curved or moving boundaries smoothly. Moving boundaries are easily accommodated by the Cartesian cut cell method, simply by recalculating the cuts between the boundary contour and background grid for as long as the motion continues. Seeding points used in generation of the quadtree grid[73] may also be used to identify cut cells, to define cut faces and to assign a type to each cut cell[79]. Once the cut cell type has been assigned, parameters of the cut cell such as fluid volume, centroid coordinates (at which variables are stored) and cell face coordinates may be readily determined.

The cut cell process usually generates some very small cut cells around the boundary and these may require very small time steps for stability. This problem may be overcome by merging cut cells that have fluid volume less than a specified minimum[50,78]. Tseng and Ferziger[80] suggest an alternative to merging, known as the Ghost-Cell Immersed Boundary Method (GCIBM). In this method, the velocity and pressure fields are extrapolated from neighbouring fluid cells to ghost cells, lying out of the fluid domain, in order to enforce boundary conditions at the boundary.

4.1. *Collapse of a water column*

The collapse of a water column has been investigated by various researchers, both numerically[59–61,76,81,82] and by physical experiments[83,84]. A unit square tank contains a column of water 0.25m wide and 0.5m high held in place at $t = 0$s. The restraint is then removed instantaneously and the resulting motion of the water column as it collapses under gravity is simulated. Initially the velocity everywhere is zero; no-slip boundary conditions are applied on all walls; a free boundary condition for velocity is applied at the top of the tank and pressure at the top of the tank is fixed at zero.

Reconstructed interface and adapted quadtree grids are plotted together in Fig. 3, in which time is non-dimensionalised as $\tau = t\sqrt{g/a}$ and a is the width of the water column. The results agree well with other numerical solutions[59] and the video images taken by Koshizuka[83]. The time histories of the fronts are plotted in Fig. 4 together with the results calculated on the equivalent uniform grid and the experimental data[84]. The predicted height in Fig. 4(a) is the same for all three grids and agrees very well with the experimental data[84], although there is a discrepancy between the leading edge experimental and numerical results in Fig. 4(b), as also observed by others[58,59]. This is possibly due to the difficulty in determining the exact location of the leading edge, which shoots along the base of the tank in a thin layer, similar to a jet.

The time histories for the quadtree calculation are identical to those calculated on the equivalent uniform grid, thus showing that the same accuracy can be achieved on the quadtree grid as on its corresponding equivalent uniform grid. Further details are given by Greaves[76], who shows that use of adaptive quadtrees significantly reduces the size of the calculation grid and the CPU per time step without loss of accuracy.

4.2. *Collapse with an obstacle*

In a more extreme case[58,59,83] an obstacle is placed in the way of the wave front. The set-up is similar to that described above with the addition of a rectangular obstacle, 0.04m x 0.08m placed on the bottom of a tank with its leading edge at the centre of the tank. Results showing the reconstructed interface are given in Fig. 5. The free surface motion is predicted well and agrees with the experiments[83] for the first few time steps considered, but is in less agreement at later times when the surface is highly contorted with considerable spray, break up and overturning. This simulation demonstrates the capability of the method to deal with large-scale wave breaking, although it is likely that use of a finer grid is necessary to resolve better the finer details of the flow.

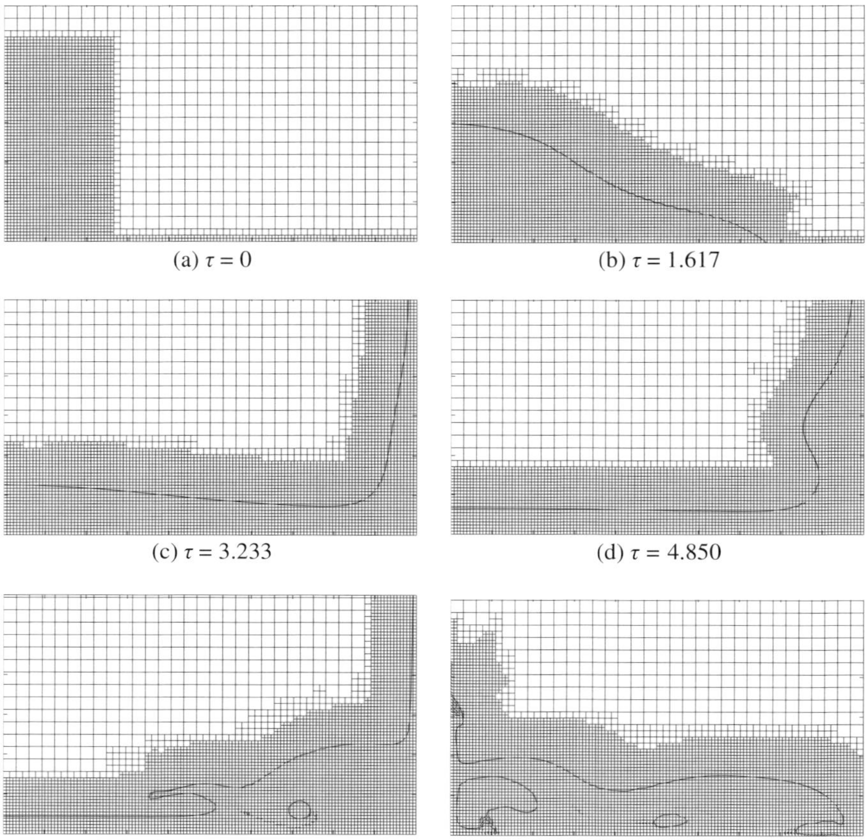

(a) $\tau = 0$ (b) $\tau = 1.617$

(c) $\tau = 3.233$ (d) $\tau = 4.850$

(e) $\tau = 6.466$ (f) $\tau = 8.029$

Fig. 3. Interface and adapted quadtree grid for dam collapse.

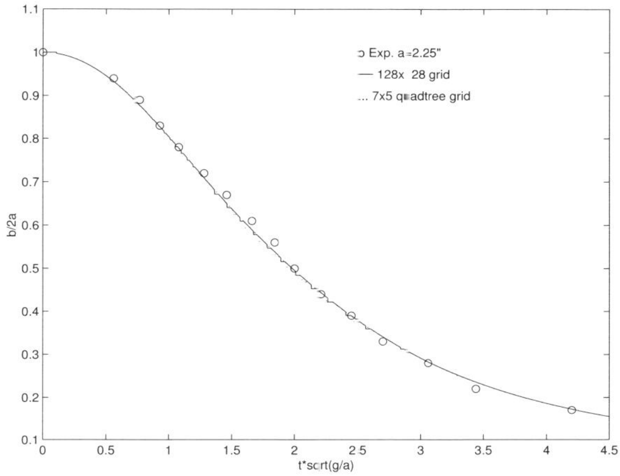

(a) Height of water column

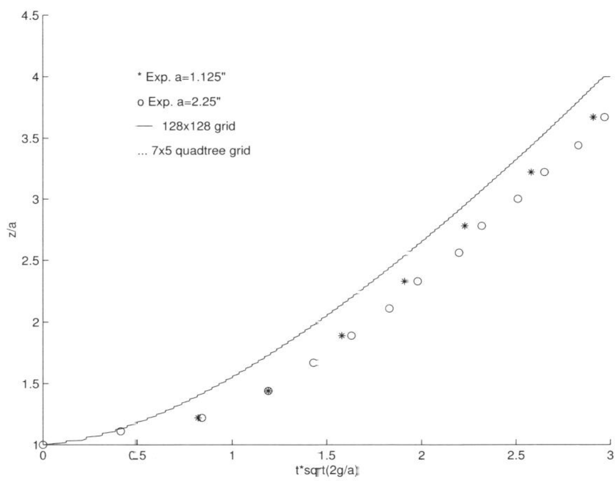

(b) Position of leading edge

Fig. 4. Interface comparison of adapted grid results with equivalent uniform grid.

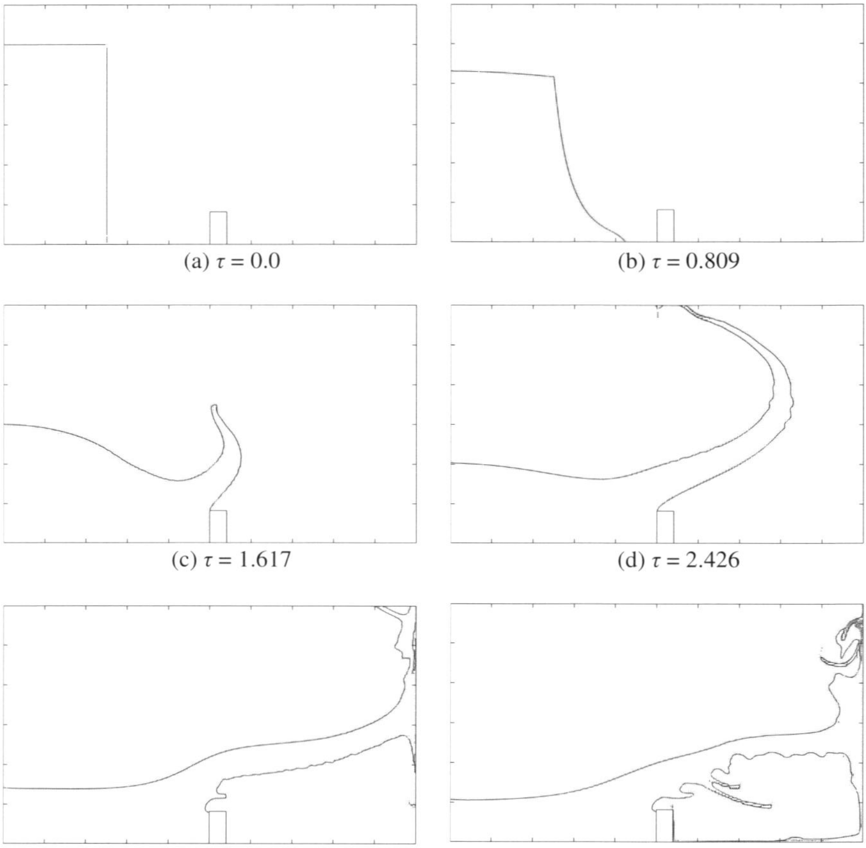

(a) $\tau = 0.0$ (b) $\tau = 0.809$

(c) $\tau = 1.617$ (d) $\tau = 2.426$

(e) $\tau = 3.234$ (f) $\tau = 4.043$

Fig. 5. Interface calculated using adapting quadtree grid for dam collapse with obstacle.

4.3. *Waves in viscous fluid over a submerged cylinder*

Details of viscous wave simulations using quadtree based VoF are given by Wang *et al.*[85] and Greaves[78,86]. Results for viscous wave interactions with a submerged cylinder in the tank are presented here. A submerged cylinder of diameter $d = 0.2h$ (where h is the undisturbed depth of water in the tank) is positioned at the horizontal centre of the unit square tank at depth $0.5h$ below the mean water level. The fluid Reynolds number for the wave, $Re = h\sqrt{gh}/\nu = 200$, where ν is the kinematic viscosity of the lower fluid, and the initial wave amplitude $a = 0.02h$. The initial and adapted grids at the first peak are shown in Fig. 6, in which $\tau = t\sqrt{g/h}$.

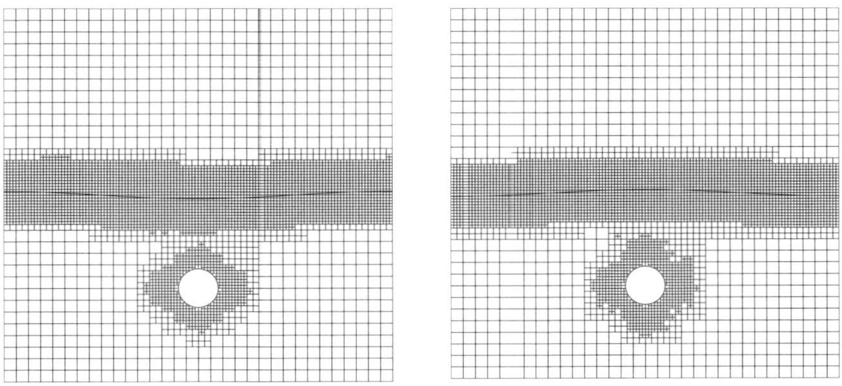

(a) $\tau = 0.0$ (b) $\tau = 1.754$

Fig. 6. Interface free surface and adapted quadtree grid.

The time history of the wave recorded at the centre of the tank is plotted in Fig. 7 together with the wave-only case without a cylinder for comparison. The results show that the presence of the submerged cylinder in the tank acts to damp out the wave motion at the free surface. A slight time lag in the wave history is also evident in the cylinder case when compared with the wave only case.

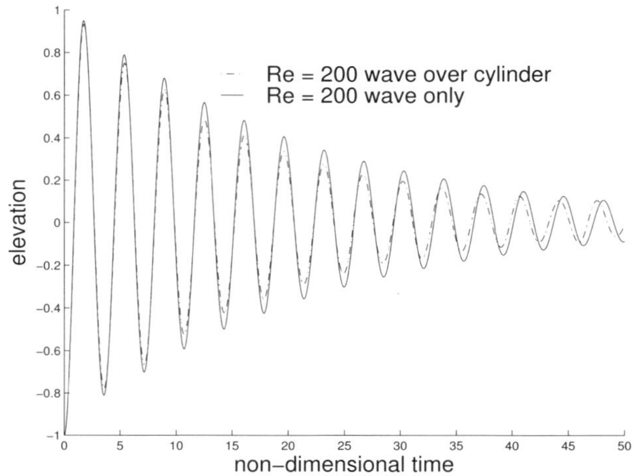

Fig. 7. Wave elevation time history at the centre of the tank, $Re = 200$.

The force on the cylinder is calculated by integrating the pressure and viscous shear stress around the cylinder and non-dimensionalised,

$$c_F = \frac{F}{\frac{1}{2}\rho U_m{}^2}, \tag{45}$$

where U_m is the maximum velocity at the cylinder position when no cylinder is present. The hydrostatic pressure component is subtracted and the horizontal and vertical force coefficients are plotted in Fig. 8. The horizontal force is approximately zero; the vertical force is much larger and oscillates with the same frequency as the wave motion, but in phase with the vertical fluid acceleration, indicating that it is dominated by inertia. Further investigations of wave loading on a submerged cylinder are given by Greaves[72].

4.4. *Water entry of a cylinder*

Preliminary results are presented here for the case of a cylinder moving through the initially still free surface. This demonstrates the capability of the Cartesian cut cells to model moving boundaries. A unit square tank half filled with water is simulated, into which a cylinder of radius,

a = 0.05m, descends with constant vertical velocity -0.2m/s, passing through the free surface until it is completely submerged. The Reynolds number, $Re = \rho_w h\sqrt{gh}/\mu_w$, and the non-dimensional time, $\tau = t\sqrt{g/a}$. Results showing the free surface and the velocity vectors in air and water for Re = 200 are presented in Fig. 9. In the highly viscous fluid, a bubble of air is dragged behind the cylinder, which then rises quickly to the surface. This can be seen more clearly in Fig. 10, in which the reconstructed interface is plotted without velocity vectors. Free surface waves can be seen radiating out from the cylinder and from the air bubble as it reaches the surface. In the less viscous fluids tested (not shown here), waves generated by the cylinder spread more quickly through the tank and the free surface settles down more quickly once the cylinder is submerged.

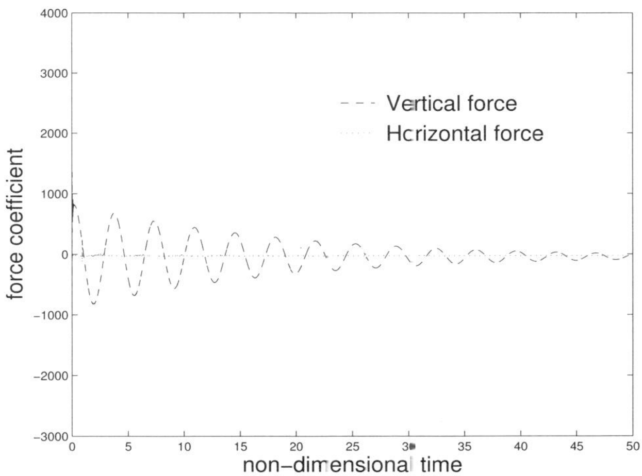

Fig. 8. Time history of forces on submerged cylinder, Re = 200.

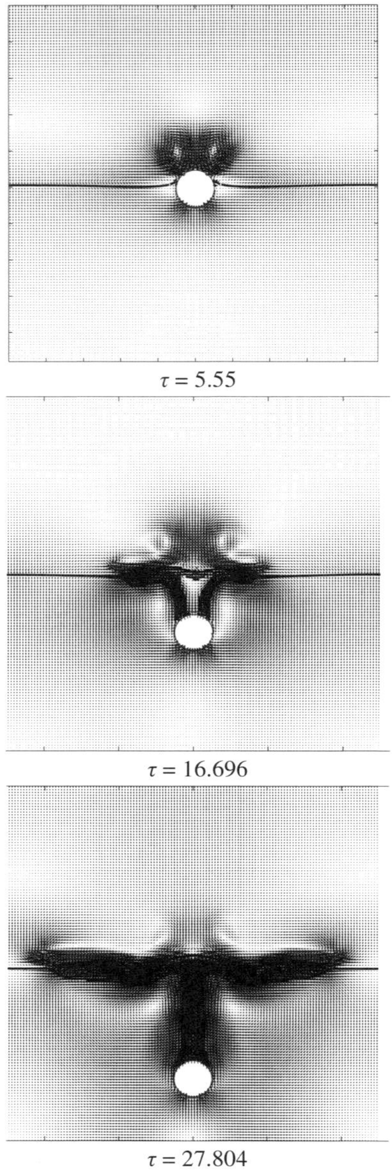

Fig. 9. Cylinder entering water at constant speed, $Re = 200$.

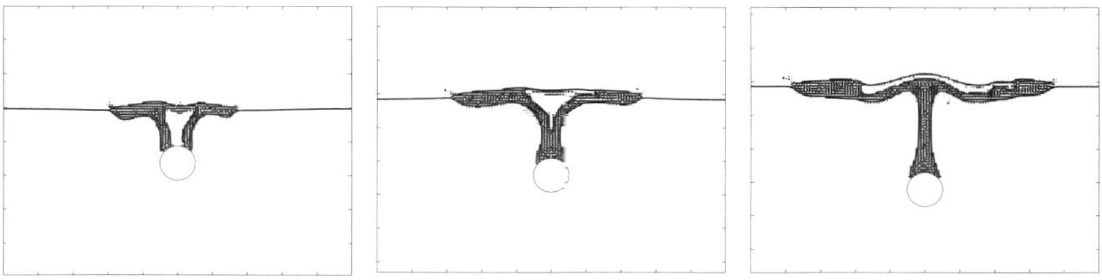

a) $\tau = 16.696$ b) $\tau = 22.173$ c) $\tau = 27.804$

Fig. 10. Cylinder entering water at constant speed, $Re = 200$: detail at cylinder.

4.5. *Wave breaking in a periodic domain*

A steep gravity water wave is simulated in a domain with periodic boundary conditions[86–89]. The width of the domain is one wavelength, $b = \lambda$, and the water depth is $h = b/2$. Following Chen *et al.*,[89] the ratio of density in air and water is 0.01 and the ratio of dynamic viscosities is 0.4. The initial condition for the wave elevation, η is

$$\eta(x,0) = a\cos(kx) + \frac{1}{2}a^2 k\cos(2kx) + \frac{3}{8}a^3 k^2 \cos(3kx), \quad (46)$$

where $k = 2\pi/\lambda$, a is the wave amplitude and the initial wave slope $\varepsilon = ak = 0.55$. The initial velocity field in the water is

$$u(x,y,0) = \frac{kga}{\omega} e^{ky} \cos(kx) \quad (47)$$

and

$$v(x,y,0) = \frac{kga}{\omega} ae^{ky} \sin(kx), \quad (48)$$

where $\omega = \sqrt{gk(1-k^2a^2)}$ and the air is initially at rest. The liquid Reynolds number, $Re = \left(g^{1/2}\lambda^{3/2}/\upsilon\right)_l = 3132$ and the simulation is calculated on a regular 256 x 256 grid.

The time sequence of free surface profiles is illustrated in Fig. 11, the profiles are separated vertically and a second wavelength is plotted on the right hand side of each to clarify the figure. The steepening of the wave and development of a plunging jet is clearly seen. The jet makes contact with the free surface, entraining air, and then splashes up and makes contact again with the free surface entraining a second air pocket. Further splashing up of the jet to begin entraining a third air pocket may be seen beyond this. Similar effects were observed in numerical calculations[89,90] of plunging breakers and in photographs of experiments[92].

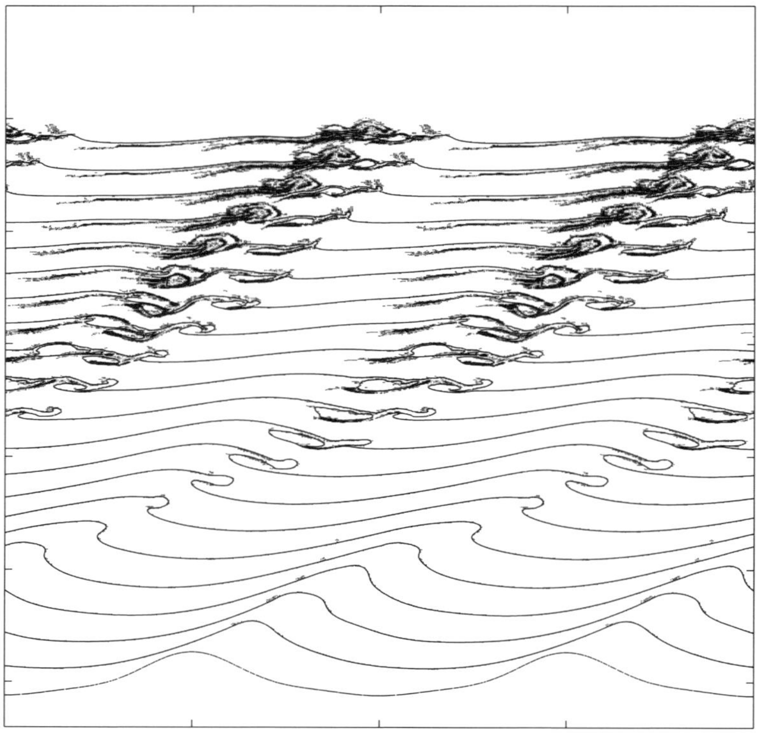

Fig. 11. Time sequence of wave profiles.

Individual free surface profiles are given in Fig. 12 at non-dimensional times, $\tau = t/\sqrt{\lambda/g} =$ 0.56 and 1.76, together with the velocity vectors plotted every 16 cells. The velocity fields are visually similar to those predicted by Chen *et al.*[89] (although the water depth is different in this case) and measured by Perlin *et al.*[93] with salient features evident, such as the high velocities in the prominent jet. The results are also similar to plunging breakers simulated by Song and Sirviente[94] using a finite difference scheme on a staggered grid (as described by Harlow and Welch[52]) with a hybrid of interface capturing and interface tracking. They investigated the influence of surface tension, initial wave steepness, initial wave profile (2nd, 3rd and 5th order are compared), Reynolds number, density ratio and viscosity ratio on the evolution of breaking waves.

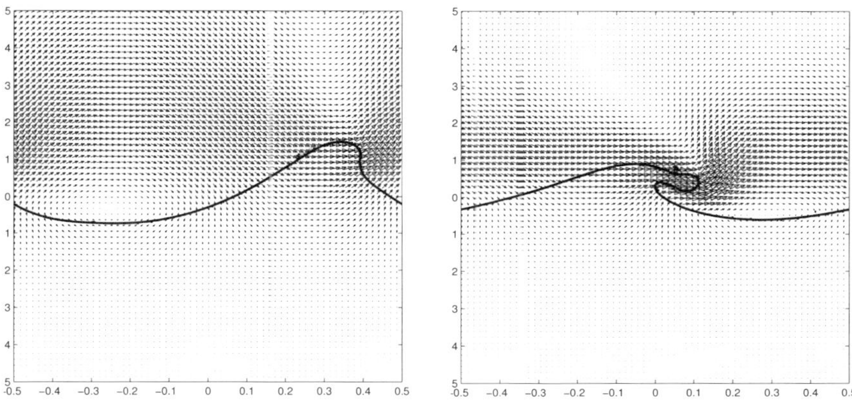

a) $\tau = 0.56$ b) $\tau = 1.76$

Fig. 12. Interface and velocity vectors for plunging breaker.

5. Conclusions

The finite volume method for discretisation of the Navier Stokes equations is discussed in the context of violent free surface flow calculations. The discussion focuses on applications where viscosity is important, so the NS or RANS must be solved, and where free surface motions are extreme, so the method must handle break up and merging of the interface. In these cases, common simplifications of the NS cannot be used and a two-fluid approach is usually necessary with interface capturing to resolve the free surface motion in time. Finite volume discretisation of the NS equations is described and different approaches to solving the equations are discussed. Approaches to handling the free surface boundary are also introduced and discussed and examples given of test problems.

An adaptive quadtree VoF method is described and applied to simulation of complex free surface flows. CICSAM differencing is used for advection of the interface together with a PISO-type NS solution method for adapting quadtree grids. Curved and moving boundaries are modelled using Cartesian cut cells. The adaptive quadtree results are in good agreement with analytical and other numerical data and use of adapting quadtree grids is found to save both grid size and CPU. The method can predict complex deformation and break up of the free surface as well as detailed flow kinematics and fluid loading on structures, and shows excellent potential for simulating violent interactions between waves and offshore structures.

Acknowledgment

The author gratefully acknowledges the support of the Royal Society in supporting this work through a Royal Society University Research Fellowship.

References

1. Wu, G.X. and Eatock Taylor, R. Finite element analysis of two-dimensional non-linear transient water waves. *Applied Ocean Research*, **16**, 363-372, (1994).
2. Wu, G.X., Ma, Q.W. and Eatock Taylor, R. Numerical simulation of sloshing waves in a 3D tank based on a finite element method. *Appl. Ocean Res.*, **20**, 337-355, (2001).
3. Dommermuth, D.G., Yue, D.K.P. Numerical simulations of non-linear axisymmetric flows with a free surface. *Journal of Fluid Mechanics*, **178**, 195-219, (1987).
4. Ma, Q.W., Wu, G.X., Eatock Taylor, R. Finite element simulation of fully nonlinear interaction between vertical cylinders and steep waves – Part 1: methodology and numerical procedure. *International Journal for Numerical Methods in Fluids*, **36**, 265-285, (2001).
5. Ma, Q.W., Wu, G.X., Eatock Taylor, R. Finite element simulation of fully nonlinear interaction between vertical cylinders and steep waves – Part 2: numerical results and validation. *International Journal for Numerical Methods in Fluids*, **36**, 287-308, (2001).

6. Helluy, P., Golay, F., Caltagirone, J-P, Lubin, P., Vincent, S., Drevard, D, Marcer, R, Fraunié, P., Seguin, N., Grilli, S., Lesage, A-C., Dervieux, A. and Allain, O. Numerical simulations of wave breaking, *Mathematical Modelling and Numerical Analysis*, **39** (3), 591-614, (2005).
7. Iafrati, A., Di Mascio, A. and Campana, E.F. A level set technique applied to unsteady free surface flows, *International Journal for Numerical Methods in Fluids*, **35**, 281-297, (2001).
8. Muzaferija, S. and Perić, M. Computation of free-surface flows using interface tracking and interface-capturing methods, *Advances in Fluid Mechanics*, **24**: 59-100, (1999).
9. P. Zwart. Numerical Modelling of Free Surface and Cavitating Flows. VKI Lecture Series: Industrial Two-Phase Flow CFD, (2005).
10. Ferziger, J.H., Perić, M. *Computational Methods for Fluid Dynamics*. Springer ISBN 3-540-59434-5, (1996).
11. Zang, Y., Street, R.L. and Koseff, R. A non-staggered fractional step method for time-dependent incompressible Navier-Stokes equations in curvilinear coordinates, *Journal of Computational Physics*, **114**, 18-33, (1994).
12. Patanker, S.V. *Numerical Heat Transfer and Fluid Flow*, Taylor and Francis, Hemishpere Publishing Corporation, USA, (1980).
13. Issa, R.I. Solution of the implicitly discretised fluid flow equations by operator-splitting. *Journal of Computational Physics*, **62**(1), 40-65, (1986).
14. Leonard, B.P. A stable and accurate convective modelling procedure based on quadratic upstream interpolation, *Computer Methods in Applied Mechanics and Engineering*, **19**, 59-98, (1979).
15. Perić, M. A finite volume method for the prediction of three-dimensional flow in complex ducts. PhD thesis, University of London, (1985).
16. Jasak, H. Weller, H.C., Issa, R.I. and Gosman A.D. High resolution NVD differencing scheme for arbitrarily unstructured meshes, *International Journal for Numerical Methods in Fluids*, **31**(2), 431-449, (1999).
17. Rhie, C.M. and Chow, W.L. A numerical study of the turbulent flow past an isolated airfoil with trailing edge separation. *AIAA Journal*, **21**, 1525-1532, (1983).
18. Smagorinsky, J. General circulation experiments with the primitive equations. Part 1: the basic experiment. *Monthly Weather Review*, **91**, 99-152, (1963).
19. Rodi, W. Large-eddy simulation and statistical turbulence models: complementary approaches. In New Tools in Turbulence Modelling, Metais, O. and Ferziger, J.H. (Eds.). Springer: Berlin, (1996).
20. Shen, L., Triantafyllou, G.S. and Yue, D.K.P. Turbulent diffusion near a free surface. *Journal of Fluid Mechanics*, **407**, 145-166, (2000).
21. Shen, L. and Yue, D.K.P. Large-eddy simulation of free-surface turbulence, *Journal of Fluid Mechanics*, **440**, 75-116, (2001).

22. Nagaosa, R. Direct numerical simulation of vortex structures and turbulent scalar transfer across a free surface in a fully developed turbulence, *Physics of Fluids*, **11**(6), 1581-1595, (1999).
23. Salvetti, M.V., Zang, Y., Street, R.L. and Banerjee, S. Large-eddy simulation of free-surface decaying turbulence with dynamic subgrid-scale models, *Physics of Fluids*, **9**(8), 2045-2419, (1997).
24. Pan, Y. and Banerjee, S. A numerical study of free-surface turbulence in channel flow, *Physics of Fluids*, **7**(7), 1649-1664, (1995).
25. Li, C.W. and Wang, J.H. Large eddy simulation of free surface shallow-water flow, *International Journal for Numerical Methods in Fluids*, **34**, 31-46, (2000).
26. Nezu, I. and Nakagawa, H. *Turbulence in Open-channel Flows*. A.A Balkema: Rotterdam, (1993).
27. Larsson, L., Stern, F. and Bertram, V. (Eds.) 2000 Proceedings of Gothenburg 2000 – A Workshop on Numerical Ship Hydrodynamics. Chalmers University of Technology: Gothenburg, Sweden, (2000).
28. Rhee, S.H. and Stern, F. Unsteady RANS method for surface ship boundary layer and wake and wave field, *International Journal for Numerical Methods in Fluids*, **37**, 445-478, (2001).
29. Wilcox, D. *Turbulence Modelling for CFD*, DCW Industries, La Canada, CA, (1998).
30. Hodges, B.R. and Street, R.L. On simulation of turbulent nonlinear free-surface flows, *Journal of Computational Physics*, **151**, 425-457, (1999).
31. Yue, W., Lin, C-L. and Patel, V.C. Large eddy simulation of turbulent open-channel flow with free surface simulated by level set method, *Physics of Fluids*, **17**, 025108, (2005).
32. Balachandar, R., Polatel, C., Hyun, B.-S., Yu, K., Lin, C.-L., Yue, W. and Patel, V.C., LDV, PIV and LES investigation of flow over a fixed dune, Proceedings of the Symposium held in Monte Verità: Sedimentation and Sediment Transport (Kluwer Academic, Dordrecht, 2002), pp. 171-178, (2002).
33. Lin, P and Liu, P. L-F. A numerical study of breaking waves in the surf zone, *Journal of Fluid Mechanics*, **359**, 239-264, (1998).
34. Lubin, P., Vincent, S., Caltagirone, J-P and Abadie, S. Fully three-dimensional large eddy simulation of vortices induced by plunging breaking waves, *Proc. ISOPE 2004 14th Intl. Offshore and Polar Engineering Conference and Exhibition*, **3**, 306-312, (2004).
35. Labourasse, E., Lacanette, D., Toutant, A., Lubin, P., Vincent, S., Lebaigue, O., Caltagirone, J-P, and Sagaut, P., Towards large eddy simulation of isothermal two-phase flows: Governing equations and *a priori* tests, *Int. Journal of Multiphase Flow*, **33**, 1-39, (2007).
36. Liovic, P. and Lakehal, D. Multi-physics treatment in the vicinity of srbitrarily deformable gas-liquid interfaces, *Journal of Computational Physics*, **222**, 504-535 (2007).

37. Hyman, J.M. Numerical Methods for Tracking Interfaces. *Physica*, **12D**, 396-407, (1984).
38. Tryggvason, G., Bunner, B., Esmaeeli, A., Juric, D., Al-Rawahi, N., Tauber, W., Han, J., Nas, S. and Jan, Y.-J. A Front-tracking method for the computations of multiphase flow, *Journal of Computational Physics*, **169**, 708-759, (2001).
39. Monaghan, J.J. Simulating free surface flows with SPH. *Journal of Computational Physics*, **65**, 179-214, (1994).
40. Losasso, F., Fedkiw, R., Osher, S. Spatially adaptive techniques for level set methods and incompressible flow. *Computers and Fluids*, **35**, 995-1010, (2005).
41. Lin, P and Liu P-F Free surface tracking methods and their applications to wave hydrodynamics, Advances in Coastal and Ocean Engineering, 5 (World Scientific) 213, (1999).
42. Wilkes, E.D., Phillips, S.D. and Basaran, O.A. Computational and experimental analysis of dynamics of drop formation, *Physics of Fluids*, **11**(12), 3577-3598, (1999).
43. Carvalho, M.S. and Scriven, L.E. Three-dimensional stability analysis of free surface flows: Application to forward deformable roll coating, *Journal of Computational Physics*, **151**, 534-562, (1999).
44. Greaves, D.M., Borthwick, A.G.L., Wu, G.X and Eatock Taylor, R. A moving boundary finite element method for fully non-linear wave simulations, *Journal of Ship Research*, **41**, 181-194 (1997).
45. Lin, H., Atluri, S.N. The Meshless Local Petrov-Galerkin (MLPG) Method for solving incompressible Navier-Stokes equations. *Computer Modelling in Engineering and Sciences*, **2**(2), 117-142, (2001).
46. Causon, D.M. An Efficient Front Tracking Algorithm for Multi-Component Fluid Calculations with Biomedical Applications. *Zeitscrift fur Angewandte Mathematik und Mechanik*, **76**(S1), 371-372, (1996).
47. Iafrati, A. and Campana, E.F. Free-surface fluctuations behind microbreakers: space-time behaviour and subsurface flow field, *Journal of Fluid Mechanics*, **529**, 311-347, (2005).
48. Sun, Y. and Beckermann, C. Sharp interface tracking using the phase-field equation, *Journal of Computational Physics*, **220**, 626-653, (2007).
49. Kelecy, F.J. and Pletcher, R.H. The development of a free surface capturing approach for multidimensional free surface flows in closed containers, *Journal of Computational Physics*, **138**, 939-980, (1997).
50. Qian, L. Causon, D.M., Mingham, C.G., Ingram, D.M. A free-surface capturing method for two fluid flows with moving bodies. *Proceedings A of Royal Society* **462**, 2065, 21-42, (2005).
51. Hirt, C.W., Nichols, B.D. Volume of Fluid (VOF) Method for the Dynamics of Free Boundaries. *Journal of Computational Physics*, **39**, 201-225, (1981).
52. Harlow, F.H. and Welch, J.E. Numerical calculation of time-dependent viscous incompressible flow of fluid with free surface, *Physics of Fluids*, **8**, 2182, (1965).

53. Adalsteinsson, D. and Sethian, J.A. The fast construction of extension velocities in level set methods, *Journal of Computational Physics*, **148**, 2-22, (1999).
54. Beux, F. and Banerjee, S. Numerical simulation of three-dimensional two-phase flows by means of a level set method. Proceedings of the 3rd ECCOMAS Conference, Paris, France. Wiley, pp. 563-569, (1996).
55. Gueyffier, D., Li, J., Nadim, A., Scardovelli, R. and Zaleski, S. Volume-of-Fluid Interface Tracking with Smoothed Surface Stress Methods for three-Dimensional Flows, *Journal of Computational Physics*, **152**, 423-456 (1999).
56. Harvie, D.J.E. and Fletcher, D.F. A new volume of fluid advection algorithm: the defined donating region scheme, *International Journal for Numerical Methods in Fluids*, **35**, 151-172, (2001).
57. Tomiyama, A. Sou, A. Minagawa, H., Sakaguchi, T. Numerical Analysis of a Single Bubble by VoF Method. *JSME Int. J*, Series B **36**, 51-56, (1993).
58. Andrillon, Y., Alessandrini, B. A 2D+T VOF fully coupled formulation for the calculation of breaking free-surface flow. *Journal of Marine Science Technology*, **8**, 159-168, (2004).
59. Ubbink, O. Numerical prediction of two fluid systems with sharp interfaces, PhD Thesis, Imperial College of Science, Technology and Medicine, London, (1997).
60. Ubbink, O., Issa., R.L. A method for capturing sharp fluid interfaces on arbitrary meshes. *Journal of Computational Physics*, **153**, 26-50, (1999).
61. Qian, L., Causon, D.M., Ingram, D.M., Mingham, C.G. An efficient two-fluid solver for hydraulic flow problems. *Journal of Hydraulic Engineering*, **129**(9), 688-696, (2003).
62. Darwish, M. and Moukalled, F. Convective Schemes for capturing interfaces of free-surface flows on unstructured grids, *Numerical Heat Transfer*, Part B, **49**, 19 – 42, (2006).
63. Lafaurie, B., Nardone, C., Scardovelli, R., Zaleski, S. and Zanetti, G. Modelling Merging and Fragmentation in Multiphase Flows with SURFER, *Journal of Computational Physics*. **113**, 134-147, (1994).
64. Noh, W.F. and Woodward, P. SLIC (Simple Line Interface Calculations). *Lecture Notes in Physics*, **59**, 330-340, (1976).
65. Ashgriz, N. and Poo, J.Y. FLAIR: Flux Line-Segment Model for Advection and Interface Reconstruction. *Journal of Computational Physics*, **93**, 449-468, (1991).
66. Puckett, E.G. Almgren, A.S., Bell, J.B., Marcus, D.L. and Rider, W.J. A high-order projection method for tracking fluid interfaces in variable density incompressible flows, *J Computational Physics*, **130**, 269, (1997).
67. Aulisa, E., Manservisi, S. and Scardovelli, R. A mixed markers and volume-of-fluid method for the reconstruction and advection of interfaces in two-phase and free-boundary flows, *Journal of Computational Physics*, **188,** 611 – 639, (2003).
68. Leonard, B.P. The ULTIMATE conservative difference scheme applied to steady one-dimensional advection, *Computer Methods in Applied Mechanics and Engineering*, **88**, 17-74 (1991).

69. Yabe, T, Xiao, F and Utsumi, T. The Constrained Interpolation Profile Method for Multiphase Analysis, *Journal of Computational Physics*, **169**(2), 556-593, (2001).
70. Hu, C. and Kashiwagi, M. A CIP-based method for numerical simulations of violent free-surface flows, *Journal of Marine Science Technology*, **9**, 143-157, (2004).
71. Xiao, F. Multi-fluid simulations using CIP finite volume method VSIAM3, Proceedings ECCOMAS 2004, Jyväskylä, 24-28 July 2004.
72. Yabe, T, and Xiao, F, Description of complex and sharp interface with fixed grids in incompressible and compressible fluid, *Computer Math. Applications*, **29**, 15, (1995).
73. Greaves, D.M., Borthwick, A.G.L. On the use of adaptive hierarchical meshes for numerical simulation of separated flows. *International Journal for Numerical Methods in Fluids*, **26**, 303-322, (1998).
74. Greaves, D.M. A quadtree adaptive method for simulating fluid flows with moving interfaces", *Journal of Computational Physics*, **194**/1, pp. 35-56, ISSN 0021-9991, (2004).
75. Rudman, M. Volume-tracking methods for interfacial flow calculations, *International Journal for Numerical Methods in Fluids*, **24**, 671-691, (1997).
76. Greaves, D.M. Simulation of viscous water column collapse using adapting hierarchical grids. *International Journal for Numerical Methods in Fluids*, **50**, 693-711, (2005).
77. Yiu, K.F.C., Greaves, D.M., Cruz, S., Saaleh., A., Borthwick, A.G.L. Quadtree grid generation: information handling, boundary fitting and CFD applications. *International Journal of Computers and Fluids*, **25**(8), 759-769, (1995).
78. Greaves, D.M. Viscous waves and wave-structure interaction in a tank using adapting quadtree grids, *Journal of Fluids and Structures*, **23**(8), 1149-1167, (2007).
79. Yang, G., Causon, D.M., Ingram, D.M., Saunders, R., Batten, P. A Cartesian cut cell method for compressible flows Part A: static body problems. *The Aeronautical Journal*, Paper 2119, (1997).
80. Tseng, Y-H, Ferziger, J.H. A Ghost-cell Immersed Boundary Method for flow in complex geometry. *Journal of Computational Physics*, **192**, 593-623 (2003).
81. Jeong, J.H. and Yang, D.Y. Finite element analysis of transient fluid flow with free surface using VOF (volume of fluid) method and adaptive grid, *International Journal for Numerical Methods in Fluids*, **26**, 1127-1154, (1998).
82. Jeong, J.H. and D.Y. Yang, D.Y. Three-dimensional finite element analysis of transient fluid flow with free surface using marker surface method and adaptive grid refinement, *International Journal for Numerical Methods in Fluids*, **29**, 657-684, (1999).
83. Koshizuka, S., Tamako, H. and Oka,Y. A particle method for incompressible viscous flow with fluid fragmentation, *Computational Fluid Mechanics Journal*, **113**, 134-147, (1995).

84. Martin, J.C. and Moyce, W.J. An Experimental Study of the Collapse of Liquid Columns on a Rigid Horizontal Plane. *Phil Trans Royal Society of London*, **A244**, 312-324, (1952).
85. Wang, J.P., Borthwick, A.G.L., Eatock Taylor, R. Finite-volume type VOF method on dynamically adaptive quadtree grids. *International Journal for Numerical Methods in Fluids*, **45**, 485-508 (2004).
86. Greaves, D.M. Viscous wave interaction with structures using adapting quadtree grids and Cartesian cut cells, European Conference on Computational Fluid Dynamics, ECCOMAS CFD 2006, Delft, The Netherlands 5-8 September 2006.
87. Cokelet, E.D. Breakins waves, the plunging jet and interior flow-field, Mechanics of Wave-Induced Forces on Cylinders, Ed. T.L. Shaw, 287-301.
88. Vinje, T. and Brevig, P. Numerical simulation of breaking waves, *J. Advances in Water Resources*, **4**, 77-82.
89. Chen, G. Kharif, C., Zaleski, S. and Li, J. Two-dimensional Navier-Stokes simulation of breaking waves, *Physics of Fluids*, **11**, 121-133 (1999).
90. Iafrati, A. Effect of the wave breaking mechanism on the momentum transfer, 21st International Workshop on water wave and floating bodies, Loughborough, 2-5 April 2006.
91. Iafrati, A. and Campana, E.F. A domain decomposition approach to compute wave breaking (wave-breaking flows), *International Journal for Numerical Methods in Fluids*, **41**, 419-445, (2003).
92. Rapp, R.J. and Melville, W.K. Laboratory measurements of deep-water breaking waves, *Phil. Trans. Royal Society London, Series A*, **331**:1622, pp. 735-800 (1990).
93. Perlin, M., He, J. and Bernal, L.P. An experimental study of deep water plunging breakers, *Physics of Fluids*, **8**, 2365-2374, (1996).
94. Song, C. and Sirviente, A.I. A numerical study of breaking waves, *Physics of Fluids*, **16**, 2649-2667, (2004).

CHAPTER 11

DEVELOPMENTS IN MULTI-FLUID FINITE VOLUME FREE SURFACE CAPTURING METHODS

D.M. Causon, C.G. Mingham and L. Qian

Centre for Mathematical Modelling and Flow Analysis
The Manchester Metropolitan University
Manchester M1 5GD, UK

Recent work at Manchester Metropolitan University has led to the development of AMAZON-SC, a multi-fluid free surface capturing solver for the Euler or Navier-Stokes equations. The free surface is captured automatically as a discontinuity in the density field in a manner similar to shock capturing in compressible flow. No surface tracking or reconstruction procedure is needed. This approach can simulate the breakup and recombination of the free surface and thus can model wave breaking. An artificial compressibility method is used to formulate the incompressible Navier-Stokes equations as a hyperbolic system which is solved using high-resolution Riemann-based upwind methods. Second-order accuracy is obtained by using linear interpolation of the stored cell centre data to cell interfaces and slope limiters to eliminate non-physical over/under-shoots in the interpolated data. A time-accurate solution is achieved using an implicit dual-time iteration technique in which the flow solution at each real physical time step is obtained by solving a steady-state problem in a pseudo-time domain. AMAZON-SC has been validated over a range of classical bench-mark problems for free surface flows including a collapsing water column inundating an obstruction, water sloshing in a tank, waves in a flume, slamming of a rigid wedge from air into water and to applications including wave overtopping at sea walls, wave energy devices and floating bodies. AMAZON-SC is implemented on a Cartesian cut cell mesh which automatically and efficiently produces a boundary conforming mesh simply by cutting out solid boundary regions from a fixed background Cartesian mesh. The cut cell technique accommodates moving internal solid boundaries very simply without the necessity to re-mesh globally or even locally. The method extends naturally to 3-D, can handle multi-fluid cases in which there are more than two fluid components as well as, in principle, compliant moving bodies.

1. Introduction

Simulation of free surface flows which may involve wave breaking and the presence of moving solid bodies is very challenging but of great benefit to the coastal engineering community. In the past there have been two main techniques for modelling free surface flows: surface-fitting methods and surface-tracking. Surface-fitting methods[1,2] treat the free surface as a moving upper boundary and solve the flow equations only in the liquid region. This method can be computationally efficient and accurate for small deformations of the free surface. However large motions of the free surface may degrade mesh quality and lead to geometrically induced inaccuracies in the solution. Furthermore no suitable boundary condition can be explicitly imposed at the free surface for wave breaking and fluid entrainment so the surface-fitting approach has limited applicability to real problems. Surface-tracking methods use a fixed mesh and, in contrast to surface-fitting methods, compute in each fluid region. The position of the free surface is described by a suitable marker function which satisfies an equation to be solved at each time step. In the widely used volume-of-fluid (VoF) method[3,4] the marker function is the volume fraction which satisfies a transport equation. At each time step the shape of the free surface is reconstructed from the volume fraction function. This method is robust and can define sharp interfaces in two dimensions but in three dimensions tracking and reconstructing the free surfaces is difficult.[5]

Recently, another approach, belonging to the family of free surface capturing methods, has been proposed initially by Kelecy and Pletcher[6] and Pan and Lomax,[7] based on the artificial compressibility method and using high resolution Riemann solvers. Here the free surface is treated as a contact discontinuity in the density field in a manner similar to shock-capturing in compressible flow. The free surface is regarded as the contour of average density between the two fluids which is captured automatically as part of the numerical solution. Complicated motions of the free surface which prove difficult or impossible to model using the methods outlined previously can be simulated in a straightforward manner. Furthermore this surface capturing method extends naturally and easily to three dimensional problems.

The method has been implemented on a Cartesian cut cell mesh[8,9] and developed into MMU's AMAZON-SC code for hydraulic and coastal engineering flow problems.[10,11] The Cartesian cut cell method provides a fully automated process for generating a boundary-fitted mesh. Solid regions, defined by line segments, are simply cut out of a stationary background

Cartesian mesh. Cut cells are subject to special treatment by the solver. Moving boundaries or internal moving bodies are accommodated easily by re-cutting the mesh at each time step. This is an efficient process as only a small percentage of cells are cut. There is no requirement to re-mesh globally or even locally as with other methods. The developed Cartesian cut cell free surface capturing solver thus has the capacity to tackle both free surfaces and stationary/moving bodies with wide applications in coastal and offshore engineering.

This chapter will summarise the essential features of the AMAZON-SC code and its applications to coastal engineering flow problems involving water waves and their nonlinear interaction with structures. This is accomplished by means of a brief introduction to the mathematical formulation of the method, followed by a detailed description of the numerical implementation of the interface capturing scheme on a Cartesian cut cell mesh with stationary or moving bodies. A number of test cases are presented illustrating the validation of the method including low amplitude wave sloshing, wave generation in a wave tank by a moving paddle, as well as water entry (slamming) of 2-D wedges. The applications of the code to the prediction of wave overtopping at coastal sea walls, wave interaction with floating bodies as well as the simulation of a novel wave energy device will also be included along with some initial results arising from the extension of the method to 3D cases.

2. Numerical Method

2.1. *Governing equations and boundary conditions*

The integral form of the 2-D incompressible Navier-Stokes equations for a fluid system with variable density field can be written as

$$\frac{\partial}{\partial t}\int\int_{\Omega} \mathbf{Q}\, d\Omega + \oint_{S} \mathbf{F}\cdot\mathbf{n}\, ds = \int\int_{\Omega} \mathbf{B}\, d\Omega \tag{2.1}$$

where Ω is the domain of interest or control volume, S is the boundary surrounding Ω, $\mathbf{n}$ is the unit normal to S in the outward direction, $\mathbf{Q}$ is the vector of conserved variables, $\mathbf{F}$ is the vector of flux function through S and $\mathbf{B}$ is the source term for body forces. For 2-D problems, this equation set consists of four scalar equations which, in turn, are the mass conservation (density) equation, x-direction momentum equation, y-direction momentum equation and an incompressibility constraint. It can be shown that the density equation is in fact the basis for deriving the volume fraction equation in the well known VoF method.[4,5] By using the artificial compressibility method (i.e. by introducing a fictitious time derivative of pressure into the constraint equation for incompressibility) and assuming the only body force is gravity, $\mathbf{Q}$, $\mathbf{F}$ and $\mathbf{B}$ are given as $\mathbf{Q} = [\rho, \rho u, \rho v, p/\beta]^T$, $\mathbf{F} = (f^I - f^V)\mathbf{n}_x + (g^I - g^V)\mathbf{n}_y$ and $\mathbf{B} = [0, 0, -\rho g, 0]$, where inviscid flux $f^I = [\rho(u - u_b), \rho u(u - u_b) + p, \rho v(u - u_b), u]^T$ and $g^I = [\rho(v - v_b), \rho u(v - v_b), \rho v(v - v_b) + p, v]^T$, viscous flux $f^V = [0, 2\mu\frac{\partial u}{\partial x}, \mu(\frac{\partial u}{\partial y} + \frac{\partial v}{\partial x}), 0]$ and $g^V = [0, \mu(\frac{\partial u}{\partial y} + \frac{\partial v}{\partial x}), 2\mu\frac{\partial v}{\partial y}, 0]$, $\mathbf{n}_x$ and $\mathbf{n}_y$ are the unit vectors along x- and y-directions respectively, u and v are the flow velocity components, ρ is the density, p is the pressure, β is the coefficient of artificial compressibility and g is the gravitational acceleration. In f^I and g^I, u_b and v_b are the velocities of the boundary S, which are zero when the boundary is stationary.

Introducing a fictitious time derivative of pressure into the continuity equation produces a system of hyperbolic equations which can then be solved by any of the recently developed upwind finite volume techniques such as the characteristics-based Godunov-type schemes.[12] Clearly, from the above formulation, any meaningful solution can only be achieved when a divergence-free velocity field is recovered, i.e. $\partial p/\partial t \to 0$. For unsteady flow problems, to achieve a fully time accurate solution, a divergence-free velocity must be attained at every physical time step. This can be achieved by using a dual-time stepping technique, subiterating the equations in a pseudotime domain to achieve a steady-state solution at each physical time step.

The boundary conditions implemented in the present study can be classified as

(1) solid stationary or moving boundary: At this boundary the no-penetration condition for inviscid flow or no-slip condition for viscous flow can be applied to the velocity and the density is assumed to have a zero normal gradient.
(2) outlet or open boundary: The pressure at this boundary is fixed and a zero gradient condition is applied to the velocity and density.

2.2. *The Cartesian cut cell mesh*

The Cartesian cut cell mesh is a simple, efficient and robust alternative to the traditional structured or unstructured body-fitted mesh. Conventional mesh generation is replaced by simple calculations to find the intersection points of the background Cartesian mesh lines with the lines delimiting internal or external solid boundaries. These intersection points define cut cells which have one of their sides coincident with a boundary segment so that the resulting cut cell mesh is boundary-fitted. If the flow domain is multiply-connected i.e. there is more than one solid region, these regions are specified in a similar manner with additional sets of data points. The majority of the flow domain contains purely Cartesian (uncut) cells. Moving bodies are boundary-fitted by recomputing local cell-boundary intersections at every time step. Unlike conventional mesh generation methods where local or even global remeshing would be required only those cells cut by the moving boundary are affected. Furthermore, the amplitude of the boundary motion is unrestricted and solid objects may traverse the flow domain.

In cases where a cut cell becomes very small it can be merged with a neighbouring cell to preserve solver accuracy and efficiency.

The cut cell method can be used with mesh adaptation stategies like adaptive mesh refinement (AMR) or quadtree/octree in two/three dimensions in order to better resolve flow features or geometries.

A detailed description of the principles of the cut cell method including the numerical procedures applied at static or moving solid boundaries have been given previously by the authors[8,9,13,14] including extensions of the cut cell algorithms to three dimensions[15] and the use of mesh adaptation.

2.3. *Numerical solution*

2.3.1. *Finite volume discretisation*

The flow equations are discretized using a finite volume approach on Cartesian cut cell meshes, see figure 1. For each cell (i, j), equation(2.1) can be written as

$$\frac{\partial \mathbf{Q}_{i,j} A_{i,j}}{\partial t} = -\oint_{\partial C_{i,j}} \mathbf{F} \cdot \mathbf{n}\, ds + \mathbf{B} A_{i,j} = -R(\mathbf{Q}_{i,j}) \tag{2.2}$$

where $\mathbf{Q}_{i,j}$ are the average quantities at cell (i, j) stored at the cell centre, and $\partial C_{i,j}$ and $A_{i,j}$ denote the boundary of the cell and area of cell i, j, respectively. The integral on the right side of equation(2.2) is evaluated by summing the flux vectors over each cell and the discrete form of the integral is

$$\oint_{\partial C_{i,j}} \mathbf{F} \cdot \mathbf{n}\, ds = \sum_{k=1}^{m} F_k \Delta l_k \tag{2.3}$$

where m is the number of sides of cell (i, j), F_k is the numerical flux through side k of cell (i, j), and Δl_k is the length of the side. In the following numerical experiments each fluid cell has four sides ($m = 4$), and each cut cell has three, four or five sides.

Inviscid numerical fluxes F_k^I are evaluated at each cell interface by the approximate solution of a local Riemann problem using Roe's method as follows:

$$F_k^I = \frac{1}{2}[F^I(\mathbf{Q}_k^+) + F^I(\mathbf{Q}_k^-) - |A|(\mathbf{Q}_k^+ - \mathbf{Q}_k^-)], |A| = R|\Lambda|L, \tag{2.4}$$

where $\mathbf{Q}_k^+$ and $\mathbf{Q}_k^-$ are the reconstructed right and left states at side k of cell (i, j) and A is the flux Jacobian evaluated by Roe's average state. The quantities R, L and Λ are the right and left eigenvectors of A and the eigenvalues of A. The flux Jacobian A is

$$A = \frac{\partial(\mathbf{F}\cdot\mathbf{n})}{\partial Q} = \begin{bmatrix} 0 & n_x & n_y & 0 \\ -u^2n_x - uvn_y & 2un_x + vn_y & un_y & \beta n_x \\ -uvn_x - v^2n_y & vn_x & un_x + 2vn_y & \beta n_y \\ -\frac{un_x}{\rho} - \frac{vn_y}{\rho} & \frac{n_x}{\rho} & \frac{n_y}{\rho} & 0 \end{bmatrix}, \tag{2.5}$$

and its eigenvalues are given by

$$\lambda_{1,2} = un_x + vn_y, \lambda_{3,4} = \tfrac{1}{2}(un_x + vn_y \pm c), \tag{2.6}$$

where $c = \sqrt{(un_x + vn_y)^2 + 4\beta/\rho}$.

To calculate the viscous fluxes F_k^V at a given cell face, a simple central difference approximation can be used to calculate the first order derivatives. The viscocity value required at the cell face is computed by a simple average of the values at the centres of the two neighbouring cells. It should be noted that since the viscosity is assumed to be uniform within each fluid, the numerical value of μ at a given point can be derived from a knowledge of the density distribution as follows. First, define a parameter α as

$$\alpha = \frac{\rho - \rho_2}{\rho_1 - \rho_2}. \tag{2.7}$$

It can be seen that the value of α is one in the first fluid and zero in the second fluid. Accordingly, the viscosity can be computed from

$$\mu = \alpha\mu_1 + (1 - \alpha)\mu_2 \tag{2.8}$$

where μ_1 and μ_2 are the dynamic viscosity coefficients for the first and second fluids respectively.

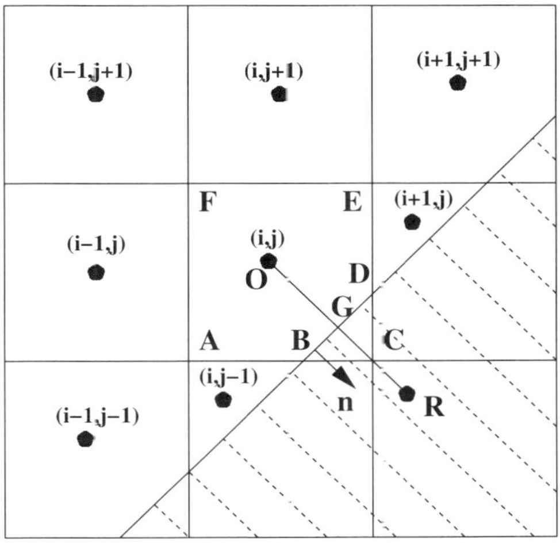

Fig. 1. Calculation of gradients at a cut cell.

2.3.2. *Calculation of gradients and reconstruction technique on moving boundaries*

To achieve second-order accuracy in smooth regions, a piecewise linear representation of the cell variables must first be reconstructed from the solution data before the Riemann states at either side of each cell interface are computed. This reconstruction is applied to the primitive variables, $\mathbf{U} = [\rho, u, v, p]^T$, which permits different limiters to be applied to different variables. A highly compressive limiter is applied to the density to help minimise numerical diffusion at the interface, while less compressive limiters or no limiting can be applied separately to the velocity components and pressure. In general for a cell with centre (i, j), for example, this requires the construction of the cell variables in the form

$$\mathbf{U}(x, y) = \mathbf{U}_{i,j} + \nabla \mathbf{U}_{i,j} \cdot \mathbf{r} \tag{2.9}$$

where $\mathbf{r}$ is the position vector from the cell centre to any point (x, y) within cell (i, j), $\mathbf{U}_{i,j}$ is the stored cell centre data, and $\Delta \mathbf{U}_{i,j}$ is the gradient of solution data at cell (i, j). Calculation of gradients depends on cell type. For a flow cell away from a solid boundary a central difference gradient of conserved variables is estimated using the known values at neighbouring cell centres. Then, a slope limiter function is applied to prevent spurious over- or under-shoots and/or to maintain a sharp interface. For cells near a solid boundary, however, a different gradient calculation is needed[9] to take the boundary condition for a moving body into account. If reflection boundary conditions are used on a solid boundary, the variables in the fictional cell R can be obtained by (see figure 1)

$$\rho_R = \rho_{i,j}, \tag{2.10}$$

$$\mathbf{v}_R = \mathbf{v}_{i,j} - 2(\mathbf{v}_{i,j} \cdot \mathbf{n})\mathbf{n} + 2(\mathbf{v_b}_{i,j} \cdot \mathbf{n})\mathbf{n}, \tag{2.11}$$

$$p_R = p_{i,j} - \rho_{i,j} g n_y |RO| \tag{2.12}$$

where $\mathbf{v} = u\mathbf{n}_x + v\mathbf{n}_y$, $\mathbf{v_b} = u_b\mathbf{n}_x + v_b\mathbf{n}_y$ and $|RO|$ is the distance between the centre of cell R and cell (i,j). The gradient on cut cell (i,j) may be of two types: fluid and solid. We calculate the fluid gradients and solid gradients separately, i.e.

$$\mathbf{U}_x^f = G\left(\frac{\mathbf{U}_{i+1,j} - \mathbf{U}_{i,j}}{\Delta x_{i+1/2,j}}, \frac{\mathbf{U}_{i,j} - \mathbf{U}_{i-1,j}}{\Delta x_{i-1/2,j}}\right), \tag{2.13}$$

$$\mathbf{U}_y^f = G\left(\frac{\mathbf{U}_{i,j+1} - \mathbf{U}_{i,j}}{\Delta y_{i,j+1/2}}, \frac{\mathbf{U}_{i,j} - \mathbf{U}_{i,j-1}}{\Delta y_{i,j-1/2}}\right), \tag{2.14}$$

where $\Delta x_{i+1/2,j} = x_{i+1,j} - x_{i,j}, \Delta y_{i,j+1/2} = y_{i,j+1} - y_{i,j}$ and

$$\mathbf{U}_x^s = G\left(\frac{\mathbf{U}_R - \mathbf{U}_{i,j}}{\Delta x_{i,R}}, \frac{\mathbf{U}_{i,j} - \mathbf{U}_{i-1,j}}{\Delta x_{i-1/2,j}}\right), \tag{2.15}$$

$$\mathbf{U}_y^s = G\left(\frac{\mathbf{U}_{i,j+1} - \mathbf{U}_{i,j}}{\Delta y_{i,j+1/2}}, \frac{\mathbf{U}_{i,j} - \mathbf{U}_R}{\Delta y_{j,R}}\right), \tag{2.16}$$

where $\Delta x_{i,R} = x_R - x_{i,j}, \Delta y_{j,R} = y_{i,j} - y_R$ and G is is a slope limiter function that is used to prevent over- or under-shoots and to maintain a sharp interface. The limiter function, among others, may take one of the following forms:
The van Leer limiter

$$G(a,b) = \frac{a|b| + |a|b}{|a| + |b|}. \tag{2.17}$$

The k limiter

$$G(a,b) = s \cdot max[0, min(k|b|, s \cdot a), min(|b|, ks \cdot a)] \tag{2.18}$$

where $s = sign(b)$ and $1 \le k \le 2$. In the present study, unless otherwise stated, the k limiter has been used for the calculations and it has been found that using $k = 2$ (which corresponds to the superbee limiter) for density and $k = 1$ (which corresponds to the minmod limiter) for velocity and pressure gives good performance in terms of solution stability and interface sharpness.

Unique gradients in cut cells are found by using a weighted averaging of values in neighbouring fluid and solid cells via:

$$\mathbf{U}_x = \frac{\Delta y_s \mathbf{U}_x^s + \Delta y_f \mathbf{U}_x^f}{\Delta y} \tag{2.19}$$

$$\mathbf{U}_y = \frac{\Delta x_s \mathbf{U}_y^s + \Delta x_f \mathbf{U}_y^f}{\Delta x} \tag{2.20}$$

where $\Delta x_f = |AB|$, $\Delta x_s = |BC|$, $\Delta y_s = |CD|$ and $\Delta y_f = |DE|$. Δx and Δy are the uncut cell side lengths in the x and y directions, respectively. Finally we have $\nabla \mathbf{U}_{i,j} = \mathbf{U}_x \mathbf{n}_x + \mathbf{U}_y \mathbf{n}_y$.

2.3.3. *Implicit time stepping and linearization of the discretised equations*

Equation(2.2) is discretised using the Euler implicit difference scheme:

$$\frac{(\mathbf{Q}A)^{n+1} - (\mathbf{Q}A)^n}{\Delta t} = -R(\mathbf{Q}^{n+1}) \tag{2.21}$$

where A is the cell area. To achieve a time-accurate solution at each physical time step in unsteady flow problems, equation(2.21) must be further modified to obtain a divergence free velocity field. This is accomplished by introducing a pseudotime derivative into the system of equations, as

$$\frac{(\mathbf{Q}A)^{n+1,m+1} - (\mathbf{Q}A)^{n+1,m}}{\Delta \tau} + I_{ta}\frac{(\mathbf{Q}A)^{n+1,m+1} - (\mathbf{Q}A)^n}{\Delta t} = -R(\mathbf{Q}^{n+1,m+1}), \tag{2.22}$$

where τ is the pseudotime and $I_{ta} = diag[1, 1, 1, 0]$. The right side of equation(2.22) is linearized using Newton's method at the $m+1$ pseudotime level to give

$$\left[I_m A + \frac{\partial R(\mathbf{Q}^{n+1,m})}{\partial \mathbf{Q}}\right](\mathbf{Q}^{n+1,m+1} - \mathbf{Q}^{n+1,m}) = -\left[I_{ta}\frac{(\mathbf{Q}^{n+1,m} - \mathbf{Q}^n)A}{\Delta t} + R(\mathbf{Q}^{n+1,m})\right],$$

where $I_m = diag[1/\Delta\tau + 1/\Delta t, 1/\Delta\tau + 1/\Delta t, 1/\Delta\tau + 1/\Delta t, 1/\Delta\tau]$. When $\Delta(\mathbf{Q}^{n+1})^m = \mathbf{Q}^{n+1,m+1} - \mathbf{Q}^{n+1,m}$ is iterated to zero, the density and momentum equations are satisfied, and the divergence of the velocity at time level $n+1$ is zero. The system of equations can be written in matrix form as

$$(D + L + U)\Delta\mathbf{Q}^s = RHS, \tag{2.23}$$

where D is a block diagonal matrix, L is a block lower triangular matrix, and U is a block upper triangular matrix. Each of the elements in D, L and U is a 4×4 matrix. An approximate LU factorization(ALU) scheme as proposed by Pan and Lomax[7] can be adopted to obtain the inverse of equation(2.23) in the form

$$(D + L)D^{-1}(D + U)\Delta\mathbf{Q}^s = RHS. \tag{2.24}$$

Within each time step of the implicit integration process, the subiteration is terminated when the L_2 norm of the change in successive sub-iterates

$$L_2 = \{[\sum_i^N (\mathbf{Q}^{s+1} - \mathbf{Q}^s)^2]/N\}^{1/2}, \tag{2.25}$$

is less than a specified limit ϵ. In the present study $\epsilon = 10^{-4}$.

3. Results

The AMAZON-SC code has been validated against a number of test cases involving free surface flows with both stationary and moving boundaries. The validated code has also been applied to study a number of real coastal engineering flow problems such as the evaluation of the hydrodynamic performance of several wave energy devices[16,17] and the prediction of wave overtopping events at sea defences.[18] The time step Δt used for advancing the solution was within the range of $5 \times 10^{-5}s$ to $5 \times 10^{-4}s$ and the pseudo-time step was normally of the order of 100 times the real physical time step in order to accelerate convergence. It was found that a β value between 400 and 2000 produced good results for all the test cases. The density ratio of water to air is taken throughout as 1000:1 and viscous effects have been ignored in all calculations. The value of the gravitational acceleration was taken as $9.8ms^{-2}$. The location of the free surface (air/water interface) was determined as the density contour with the average value of the two fluid densities. All the calculations presented here were performed on a 600MHz DEC-α computer or a NEC SX-6i vector machine. For a typical case using 100×100 mesh points and 20,000 time steps, the computational time is around 4 hours on the DEC-α machine.

3.1. *Low amplitude sloshing tank*

The sloshing of a liquid wave with finite amplitude under the influence of gravity is a classical test case for free surface flow problems and more recently for evaluating interface tracking methodology.[5]

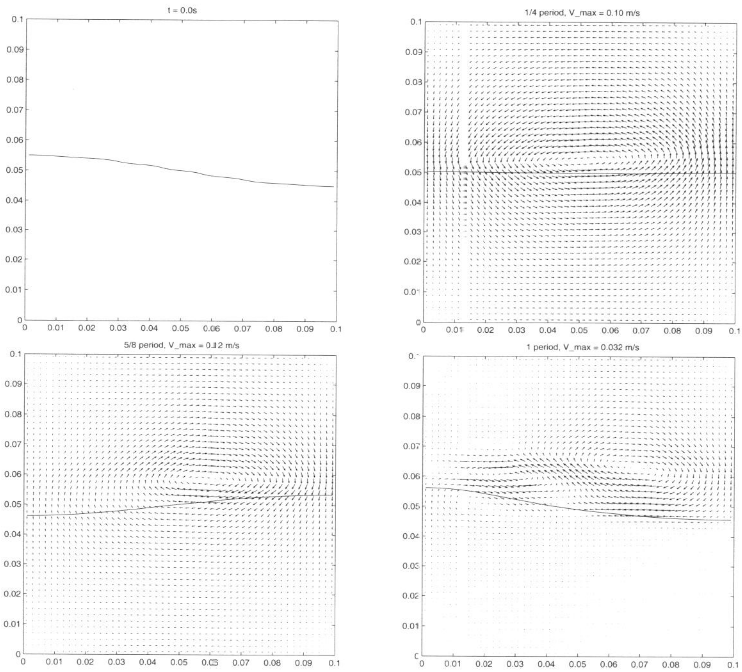

Fig. 2. Plots of the wave position and velocity vectors for the first period of the sloshing of an inviscid liquid under the influence of gravity.[10]

The initial conditions for the flow problem are shown in figure 2 for $t = 0.0$, where the water has an average depth of $0.05m$, and its surface is defined by one half of a cosine wave with an amplitude of $0.005m$. The computational domain used has a base length and a height of $0.1m$ and solid wall boundary conditions are applied on all sides of the container. For the results presented, the domain is discretized with 50×50 cells. Several snapshots of the results for the first period are also shown in figure 2, where both the location of the water surface and the velocity vectors are plotted. The maximum values of the velocity ($V_{\_max}$) on each figure have been calculated to show the scale of the velocity. After a quarter of a period the potential energy of the system has been transferred to kinetic energy and the velocities reach their maxima. After a half period all the kinetic energy has been transferred back into potential energy with the velocity almost back to zero.

Figure 3 shows plots of the position of the interface at the left boundary against time for the first five periods where the theoretical solution for the first mode of the problem with period of sloshing $P = 0.3739s$ is also shown. As found by Tadjbakhsh and Keller,[19] the second mode (which has been reflected in the numerical result) with half the above period is also important for this problem which accounts for slight differences between the predicted and the analytical values. In figure 3, the predicted interface position from a volume fraction based method[5] is also plotted and good agreement has been achieved between the results from the two different methods, especially for the peak values of the interface position.

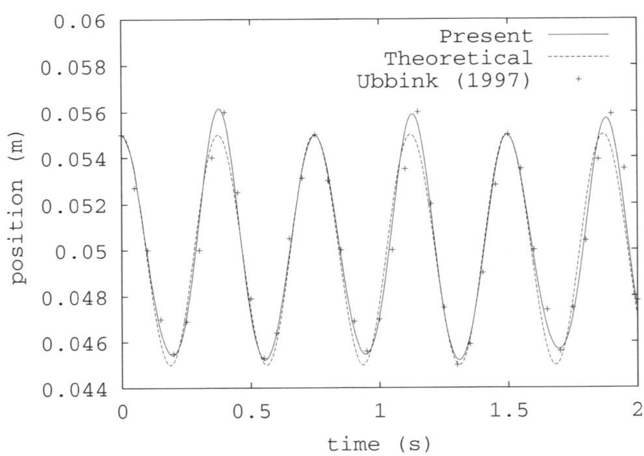

Fig. 3. Position of the interface at the left boundary plotted against time.[10]

To illustrate mesh convergence for the test case, two other set of meshes using 25×25 and 100×100 cells were used to calculate the same test case. In figure 4, the results for the position of the interface at the left boundary against time for the first three periods are compared for the three different mesh densities, from which it is clearly shown that the results from the meshes with 50×50 and 100×100 cells are nearly identical and thus an

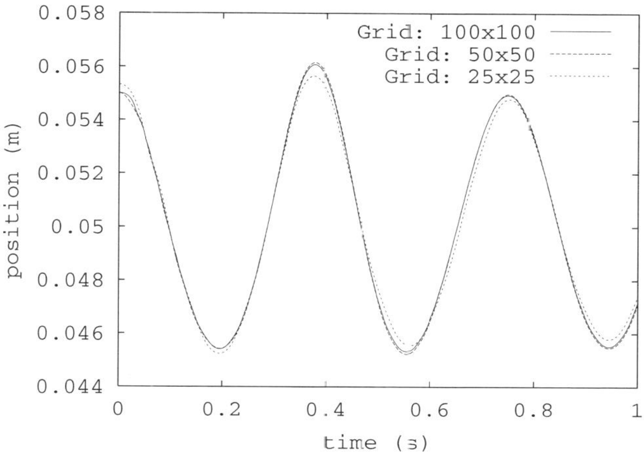

Fig. 4. Mesh convergence study for the sloshing tank test case.[10]

essentially converged solution can be obtained by using a mesh with 50×50 cells.

3.2. *Wave generation in a wave tank*

In wave tank tests, piston paddles oscillating horizontally are predominantly used for generating waves (wavemaker) in shallow water flow regimes. As this application involves a moving body and motion of a free surface, it is a good case to use for testing the current flow solver.

In our calculations, waves are numerically generated using a moving paddle situated at the seaward boundary, much like one used in a wave tank. The velocity of the moving paddle can be simply sinusoidal or specified as the actual velocity from the wave tank physical tests. In the first example, regular waves are generated in a $6m$ long numerical wave tank with a still water depth of $0.3m$ by prescribing the velocity of the paddle as $U_{in} = -0.2sin(2\pi t)$ (period of the wave $T = 1s$). A mesh with 120×40 cells was used for this simulation, the computational domain was extended $0.3m$ above the still water surface into the air region and the time step used was $\Delta t = 0.0001s$. The pseudo time step $\Delta\tau$ and β value were $0.005s$ and 500 respectively. Figure 5 shows wave patterns at times $t = 5.0s, 5.2s, 5.4s, 5.6s$ and $5.8s$ during one period of development. In this figure, the moving paddle is clearly seen at the left edge of each frame and moves transversely across three mesh cells.

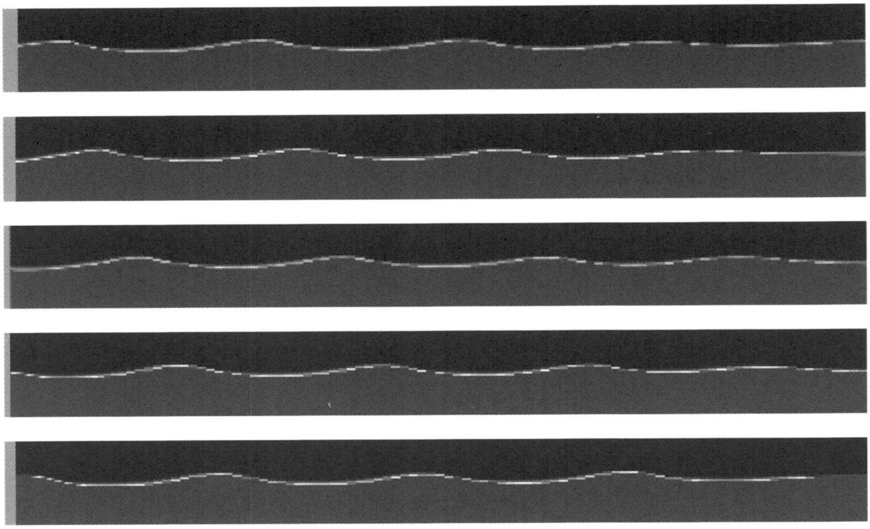

Fig. 5. Plots of the wave pattern during one period development at $t = 5.0, 5.2, 5.4, 5.6, 5.8s$.[11]

In the second example, the experiments conducted by Gao[20] are reproduced numerically. In this test, regular waves are generated in a $8.85m$ long wave tank with a still water depth of $0.28m$ by prescribing the velocity of the paddle to be the same as that used in the experiments as shown in figure 6 (period of the wave $T = 1s$). The values of the time step and β were the same as the previous test case. A total of $10s$ was simulated and figure 7 shows several snapshots of the location of the free surface and velocity vectors (for both water and air) for one period of development (from $t = 8.5s$ to $t = 9.5s$) using mesh size $DX = DY = 0.02m$. The free surface motion is one of the sources of vortex creation and this can be seen in the results around the free surface contour. It can be seen from the first and last pictures that the flow pattern almost repeats itself after one period. In figure 8, the free surface elevation at the location ($x = 3.55m$ from the initial position of the wave paddle) is compared with the corresponding experimental results.[20] To show mesh convergence in the calculations, three sets of meshes were used. The size of each mesh was $DX = DY = 0.04m$, $DX = DY = 0.02m$ and $DX = DY = 0.014m$ respectively. Unsurprisingly, while the coarsest mesh either under-predicted or over-predicted the peak values of the surface elevation, the values predicted on the two finer

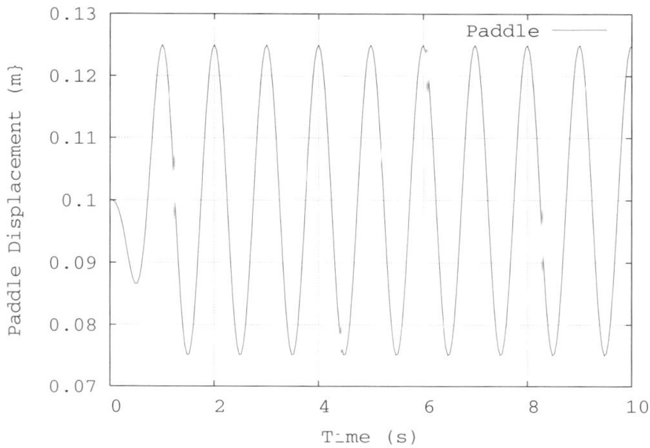

Fig. 6. Paddle displacement vs time used in wave tank test.[20]

meshes lie close to each other and are in satisfactory agreement with the experimental data which indicates that an accurate mesh independent solution can be achieved with the present numerical method.

3.3. *Water entry of a 2D rigid wedge*

The initial stages of water impact (slamming) of a 2-D rigid wedge has been a subject of extensive studies both theoretically and experimentally,[21,22,24,25] owing to its practical importance in wave impacts on fixed and floating offshore structures. As a body of data is available from theoretical studies, especially for the case of water entry of a rigid 2-D wedge with constant velocity, this flow problem was used as a test case for the validation of AMAZON-SC.

Three deadrise angles (the angle between the wedge slope and the horizontal or undisturbed water surface), namely $\alpha = 30°, 45°$ and $81°$ were selected for the calculations in a square container of size $2m \times 2m$ half occupied by water and half by air. Results are shown only for the case of $\alpha = 30°$. Further details can be found in Qian et al.[11]

The body was caused to move vertically downwards to an initially calm water surface at a prescribed unit velocity $V = 1$. In a theoretical study based on potential theory,[25] it was concluded that if $V \cdot T$ is small (T is the duration of the impact and thus $V \cdot T$ represents the penetration depth

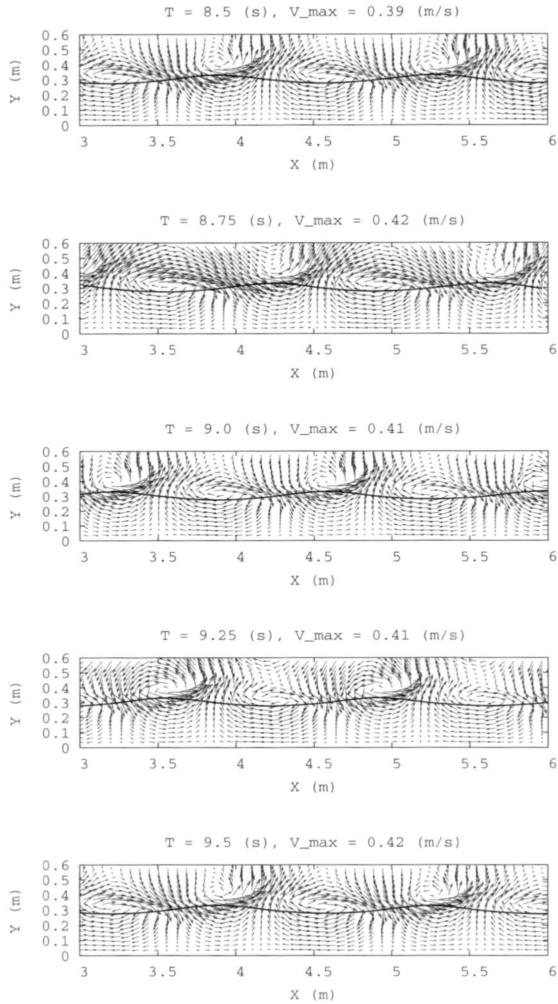

Fig. 7. Plots of the wave pattern and velocity vectors during one period development at $t = 8.5, 8.75, 9.0, 9.25, 9.5s$.[11]

(Z) of the apex of the wedge into the water), a similarity solution exists for this problem. Thus, in the early stages of water entry, the solutions are expected to be self-similar for different T if the variables are suitably non-dimensionalised. The computational domain was discretised using a

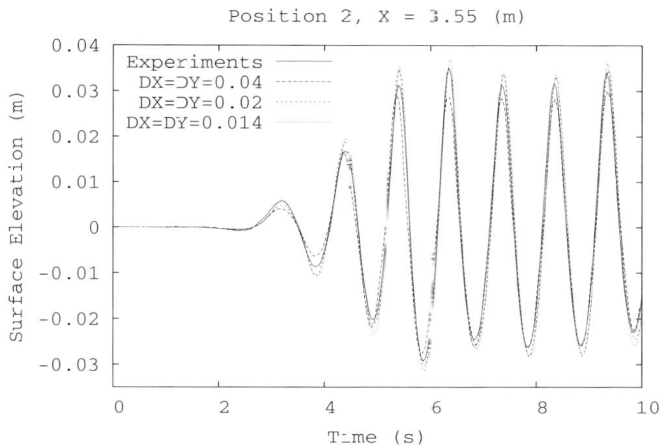

Fig. 8. Comparison with experimental results for free surface elevation.[11]

cut cell mesh with 160 × 160 cells ($DX = DY = 0.0125m$). This mesh size was chosen after conducting mesh convergence tests for the $\alpha = 30°$ case using three different meshes, as shown on the top of figure 9, where the data plotted corresponds to the time $T = 125ms$. The time step Δt and the pseudo time step $\Delta\tau$ were $0.00005s$ and $0.005s$ respectively. The β value was 500.

On the bottom of figure 9, the distribution of pressure coefficient $C_p = p/(\frac{1}{2}\rho V^2)$ along the base of the wedge is given at three different times after impact. For this deadrise angle, the numerical solutions showed good correlation (self similarity). A similarity solution and analytic solution[25] are also given in each plot on this figure, which are in close agreement with the present numerical results. The free surface profiles predicted by AMAZON-SC were also compared with the flow visualisation carried out by Greenhow and Lin[21] in figure 10 which confirms the accuracy of the code.

The tests were then extended to the complex process of a 2-D wedge-shaped rigid body ($\alpha = 45°$) entering initially calm water with a constant prescribed velocity until totally immersed. The geometry, initial position of the body and the calm water elevation are shown in the first frame of figure 11. The Froude number based on the downward velocity and the length of the base of the body is 0.71. A computational domain of $6m \times 6m$ was discretised using a uniform cut cell mesh with 120 × 120 cells and the

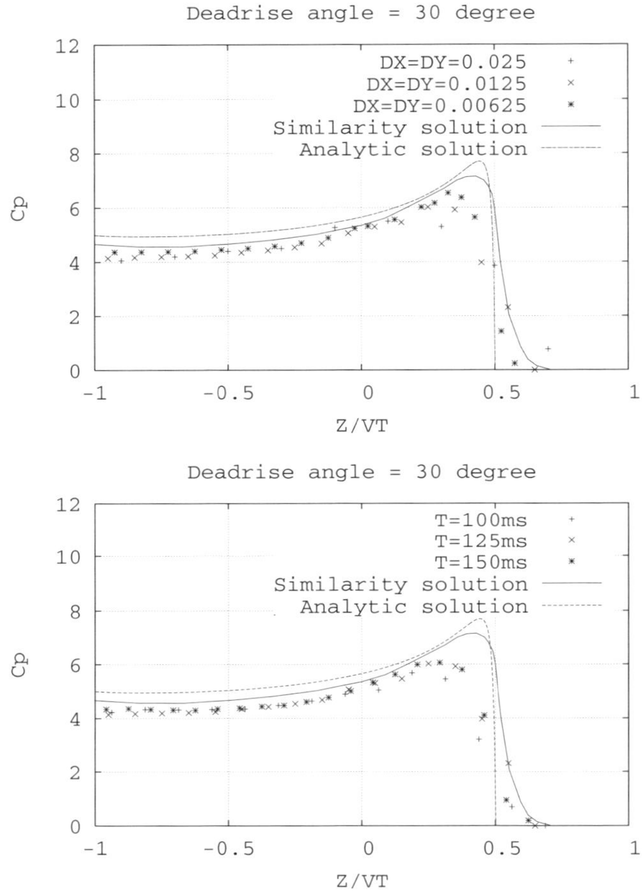

Fig. 9. Pressure coefficient distribution along the bottom of the wedge: grid convergence test (top) and self similarity test for the solution (bottom).[11]

time step used was $\Delta t = 0.0001s$. The pseudo time step $\Delta\tau$ and the β value were $0.05s$ and 500 respectively.

Figure 11 shows several snapshots of the flow development during the initial water entry and its subsequent total immersion in water. As the body pierces the surface, displacement waves are formed on each side of the body originating at the vertex of the wedge as the displaced water rises up the sides of the body. At the same time, the pocket of air previously trapped between the body and the water surface is forced around the sides of the

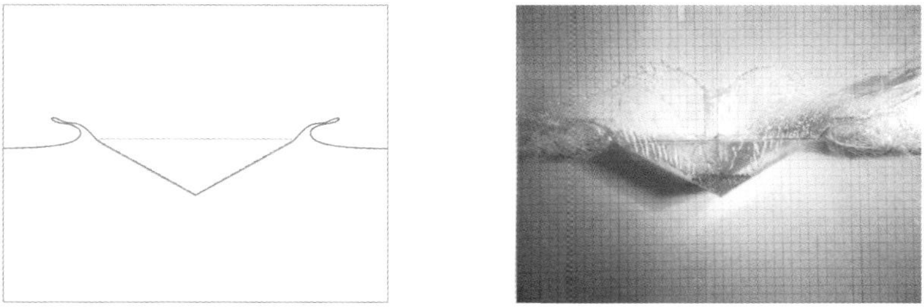

Fig. 10. Comparison of free surface profiles with experiments: wedge deadrise angle $= 30°$; number of cells: 320×320; mesh: $DX = DY = 0.00625m$.[11]

wedge and separates at the corner to form a pair of symmetric vortices. A second pair of vortices is created at the rear points of detachment forming a distinct recirculating base flow region behind the body in air. As the body travels further into the water, the displacement waves created on the free surface as the body penetrates the water surface rise and subsequently overturn towards each other, closing the gap between them. Finally, they are reconnected ($T = 2.23s$), leaving air bubbles trapped behind the body and creating a jet of air directed upwards above the reconnection point on the free surface.

3.4. *Simulation of a wave energy converter*

The Oscillating Wave Surge Converter (OWSC)[23] is a novel wave energy device. The device has one or more vanes with an axle spanning a recess in the shoreline or a caisson in the near-shore zone and responds to the predominant and amplified horizontal fluid motion in shallow and intermediate depth waves.

To demonstrate the capability of the code to simulate the complex flow around the OWSC device, the interaction between a wave driven rotating vane and a shoreline (a sloping backplane) has been simulated (bottom hinged case) The angle of the sloping backplane $\theta = 47°$ and the still water depth is $H = 0.20m$. A wave with a period of $1.524s$ is generated by a moving paddle. The vane can rotate freely around a centre located at $[3.0m, 0.0m]$. The density of the vane is $260\ kg/m^3$, so there will be a buoyancy force exerted on it. The angular velocity of the vane is derived from the motion of the waves, i.e. by calculating the torque from the pressure exerted on its surface. The displacement of the vane in terms of its relative rotation angle to the neutral position during the time of the

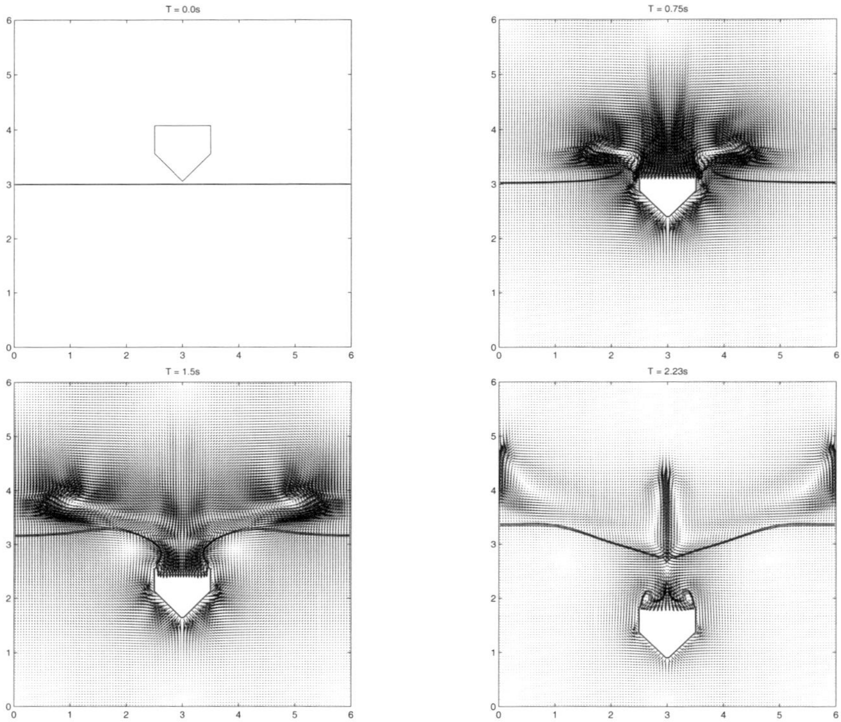

Fig. 11. Plots of the interface position and velocity vectors for a 2-D rigid body entering the water surface.[11]

simulation is shown in figure 12. Several snapshots for the wave profiles and velocity vectors around the device are also given in figure 13. This test case clearly demonstrates the potential of the numerical method to deal with complex wave/paddle interactions and real fluid flow problems such as the OWSC device.

3.5. *Wave/floating body interaction*

To validate the current numerical code for wave/floating body interaction problems, the experimental test as described by Hu and Kashiwagi[26] has been used. The 2-D experiment, as sketched in figure 14, was performed in a wave channel (10m long, 0.3m wide and 0.4 m deep). The floating body model used for the experiment has a length of 0.5m, a depth of 0.1m and a freeboard of 0.1m. A box type of super-structure is installed on the

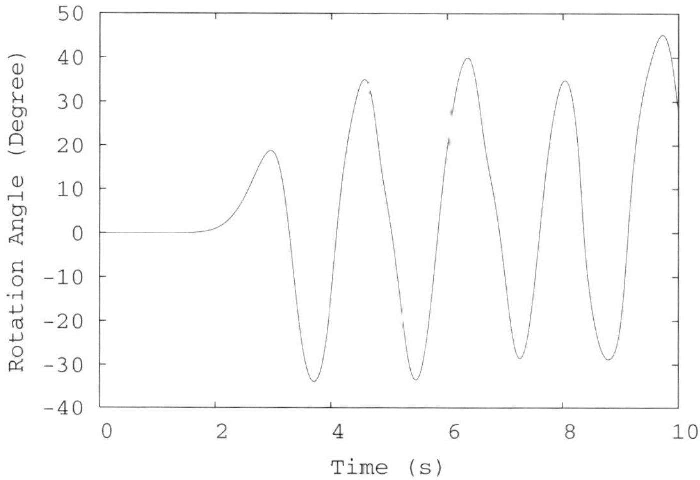

Fig. 12. OWSC test: bottom hinged case.

deck. The body can undergo heave and roll motions but the sway motion is restricted. A wedge type wave-maker is installed on the left of the tank which undergoes oscillating vertical movements. The exact same type of wave-maker has been used in the numerical simulation. The amplitude and period of the wave-maker displacement are $H_w = 0.025m$ and $T_w = 1s$ respectively. In the simulation, a total of 250×40 mesh cells have been used and the mesh size for each cell $DX = DY = 0.02m$. The time step Δt used for advancing the solution is $0.0001s$.

A total of $18s$ has been simulated and in figure 15, the free surface profiles and body positions at six instances during one period of wave development have been given, which clearly show the heave and roll motions of the body. In figure 16, the details of the roll motion of the body are compared with the experimental results, from which it can be seen that the rotation angle of the floating body is in a close agreement with the wave tank test data, an indication of the code's capability for simulating complex flow problems involving a free surface and freely floating bodies.

3.6. *Wave run up and over-topping at a smooth sea dike*

In this section, the test case (Troch et al[27]) of wave run-up and overtopping at a smooth impermeable sea dike was simulated by applying the AMAZON-SC code. The $0.8m$ tall dike with a $0.3m$ wide crest stands in

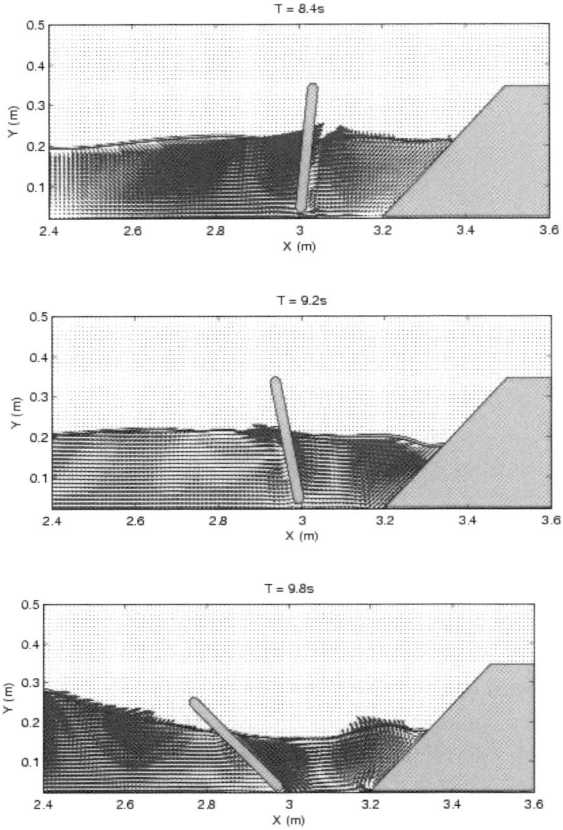

Fig. 13. Plots of free surface profiles and velocity vectors around the OWSC device at $t = 8.4, 9.2$ and $9.8s$.

$0.7m$ deep water, the seaward slope is 1:6, see figure 17. The computations were performed in a computational domain of $6.3m \times 1.6m$ which was discretized using a uniform grid of 200×80 cells. Numerical wave gauges were positioned at $0.01m$, $1.00m$ (co-located with the toe of the structure) and $3.81m$ from the seaward boundary. At the seaward boundary a series of random waves was introduced by specifying the local velocity and density distributions. These are calculated using linear wave theory, while the pressure is extrapolated from the interior of the flow domain. In the present study, a modified JONSWAP spectrum[28] was chosen as the random wave spectrum at the seaward boundary.

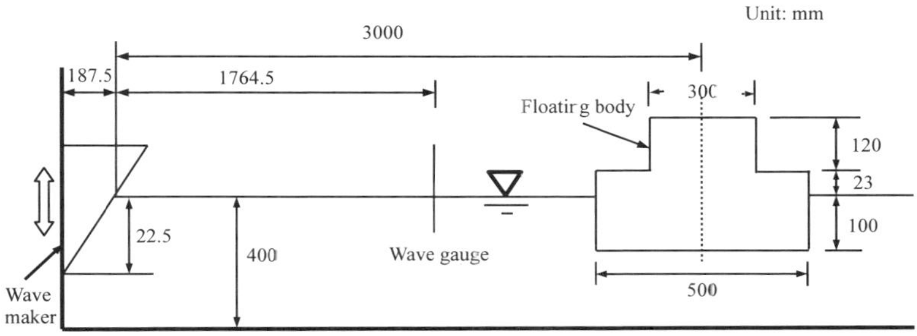

Fig. 14. Schematic view of the setup for the 2-D wave tank tests.

Figure 18 shows the computed density profiles between 3.5s and 7.0s, a period which corresponds to approximately two wave periods. As the waves start to run-up the dike shoaling occurs, causing the wave front to steepen and the wave height to increase. Close to the location of the third gauge the wave steepness may exceed the critical wave breaking criterion and the wave begins to break, forming a plunging breaker ($t = 5.5s$). Lower amplitude waves, for example those shown in figure 18 between 4.0s and 4.5s, still overtop the dike but under these conditions green water overtopping occurs without wave breaking.

3.7. *3D extensions*

The AMAZON-SC code has recently been extended to 3-D cases and partially vectorised. For a typical test case, a speed-up of at least a factor 12 was obtained for runs on the Departmental NEC SX6 vector computer with peak performance of 8 GFLOPS compared to runs on a high specification multicore desktop PC. To validate the numerical method, a simple experimental test case[29] involving a collapsing water column and its interaction with a small 3-D box has been chosen and simulated. A water column with a height 0.55 m and a width of 1.22 m was initially confined to the left in a container of size 3.22m x 1.0m x 1.0m (as shown in figure 19) which was discretized with 161 x 50 x 50 cells. The confinement was then suddenly withdrawn at time $t = 0.0s$. Under the effects of gravity, the collapsing water front reached the solid box at around $t = 0.4s$ and as it struck the wall of the box, a thin jet of water was formed at the front face of the box. At the same time, two streams of flow passed smoothly around both sides of the box and after some time, the water jet eventually collapsed onto the top of the box. During the simulation two water height gauges

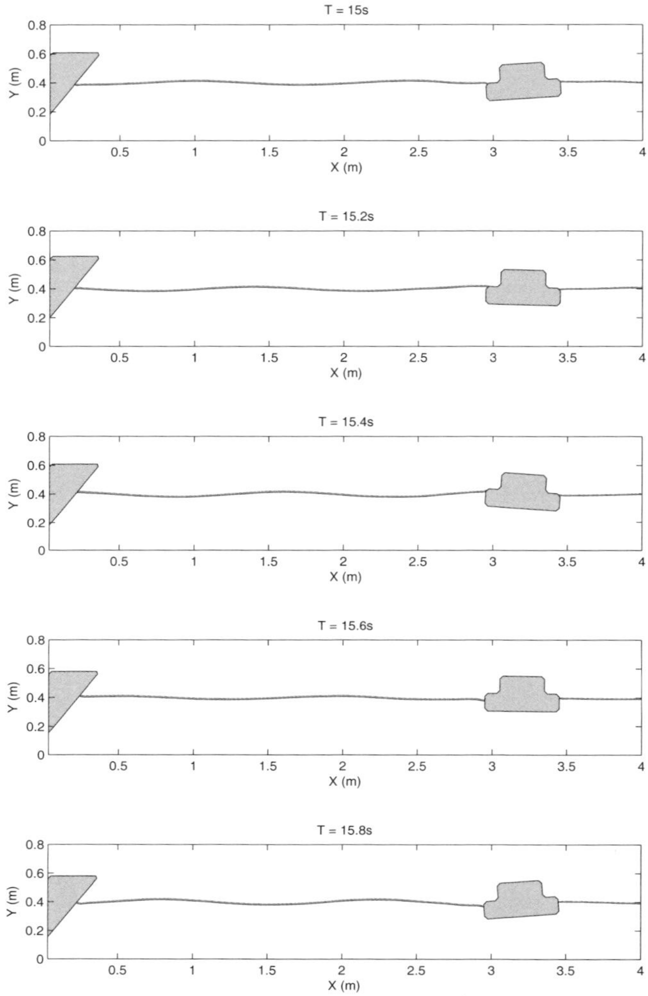

Fig. 15. Plots of the wave pattern and body positions during one period of development at $t = 15.0, 15.2, 15.4, 15.6, 15.8s$.

have been used: one in the water column and the other just in front of the small box. In figure 20, the time history of water height at the two locations is compared with the experimental data for a total of six seconds. For the water height in the reservoir, the numerical results agree with the

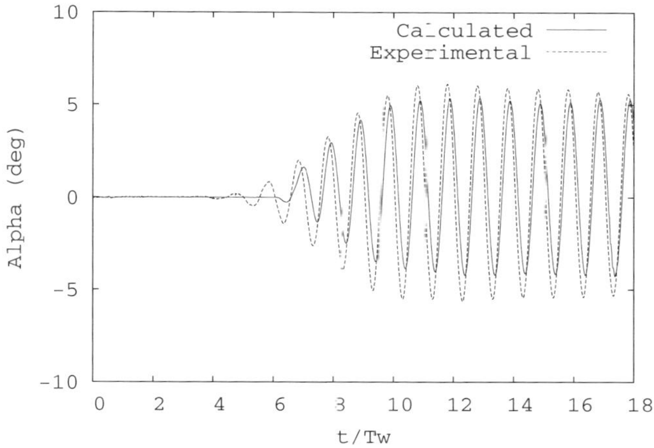

Fig. 16. Comparison of time histories with experimental results for roll motion of the floating body.

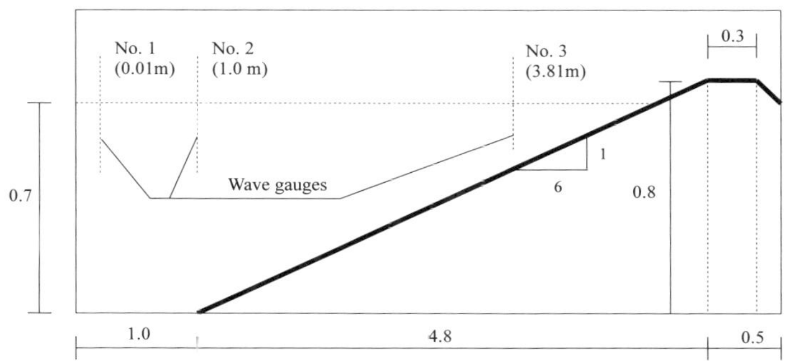

Fig. 17. The computational domain for the sea dike problem.

experimental data very well until the water front flows back to the reservoir and for the water height probe just in front of the box, the numerical code has predicted the arrival of the water front accurately and the global trend also agrees well with the test data. As viscosity has been ignored in the current simulations, the peak value of the water height in the second probe was over-predicted. These differences between the numerical predictions and the experimental data will be further investigated by including viscous effects and also via mesh refinement tests.

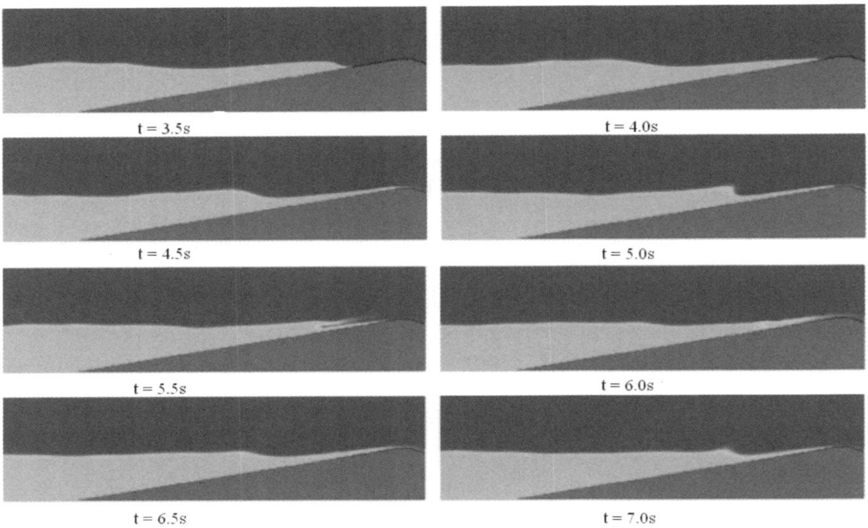

Fig. 18. Free surface motion for wave run-up and over-topping on a smooth sea dike.[18]

To demonstrate the applicability of the 3-D code to more general cases, a test case involving green water overtopping a more complex ship-like object has also be simulated. Similar to the previous test case, the incoming waves were generated by a collapsing water column of size 1.22m x 0.65m x 1.0m as shown in figure 21. A container ship-like object which remains fixed during the simulation is initially standing on a pool of still water with a depth of 0.15m. Transmissive boundary conditions were applied to the vertical outlet boundary on the right of the computational domain, so any arriving waves at this boundary pass freely through without reflecting back into the domain. A total of 2.5s has been simulated and in figure 21 several snapshots of the free surface positions have been shown: at time t = 0.4s, the wave front has hit the bow and two streams of water from both sides of the bow start to inundate the deck of the ship and at time t = 0.9s, the water has flooded the entire deck and also impacted upon the vertical wall of the upper structure; subsequently the water on the deck subsides away gradually until all the water is gone at the end of the simulation.

4. Conclusions

The development of AMAZON-SC, a two fluid free surface capturing code based on Cartesian cut cell mesh generation, has been reviewed in this

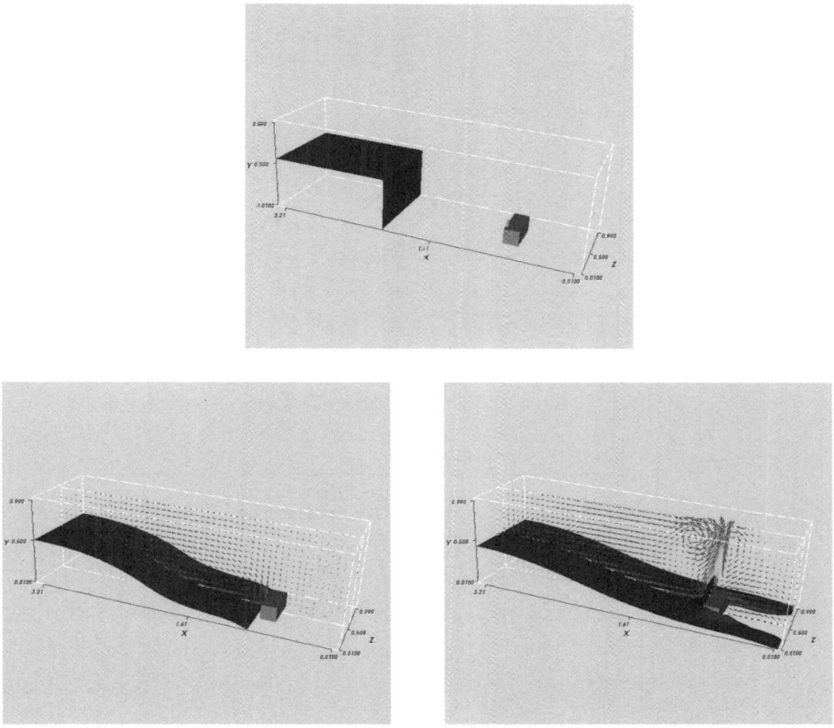

Fig. 19. Simulation of water from a dam break hitting a fixed obstacle in an enclosed box at $t = 0.0$, 0.4 and 0.6 seconds.

chapter. A number of test cases including a low amplitude wave sloshing in a tank, wave generation in a wave tank using a moving paddle, water entry and subsequent total immersion of a wedge-shaped rigid body as well as wave/floating body interaction problems have been used to validate the code and to show the code's capabilities for simulating a variety of free surface problems. The case of water penetration and immersion of a wedge-shaped rigid body shows the future promise of the method as a tool for studying free surface flow problems with interface break-up and recombination and entrapment of one fluid by another in the presence of moving bodies. Applications of the code to simulate a novel wave energy device and to predict wave overtopping events at sea defences were also presented. The method uses an efficient dual time-stepping artificial compressibility algorithm and modern Riemann-based upwind solvers of the

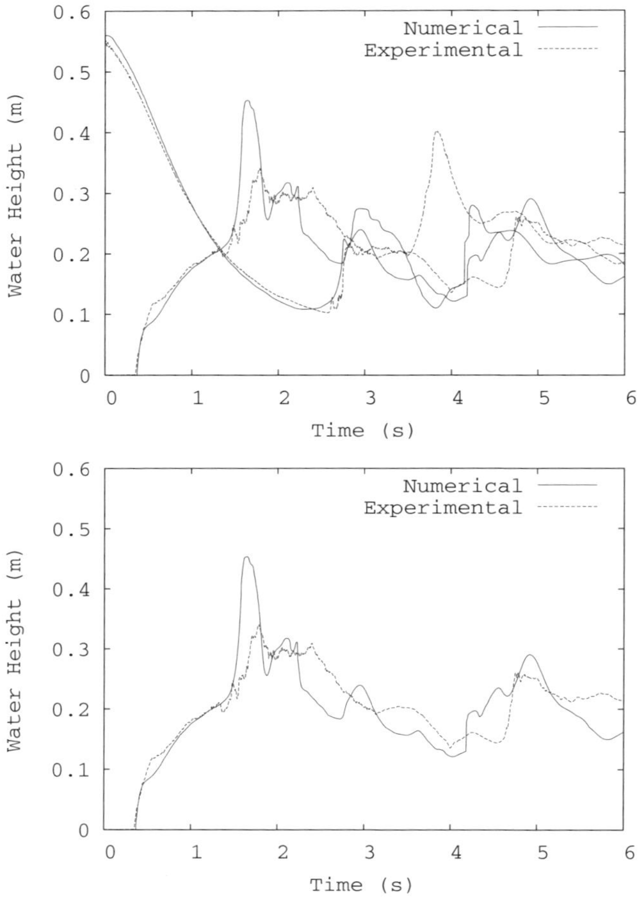

Fig. 20. Vertical water heights in the reservior (top) and the tank (bottom).

Godunov-type which capture the moving free surface accurately as part of the numerical solution without interface-tracking or other special provisions commonly used in volume-of-fluid (VoF) methods. Mesh generation in the conventional sense is eliminated in favour of defining local cut cell data on a stationary background Cartesian mesh. In cases where the boundaries are in motion or moving bodies are to be included, the only required changes to the meshing procedures consist of local updates to the cut cell data at the boundaries that are in motion for as long as the motion continues. These updates involve only a small percentage of the total number of cells in the

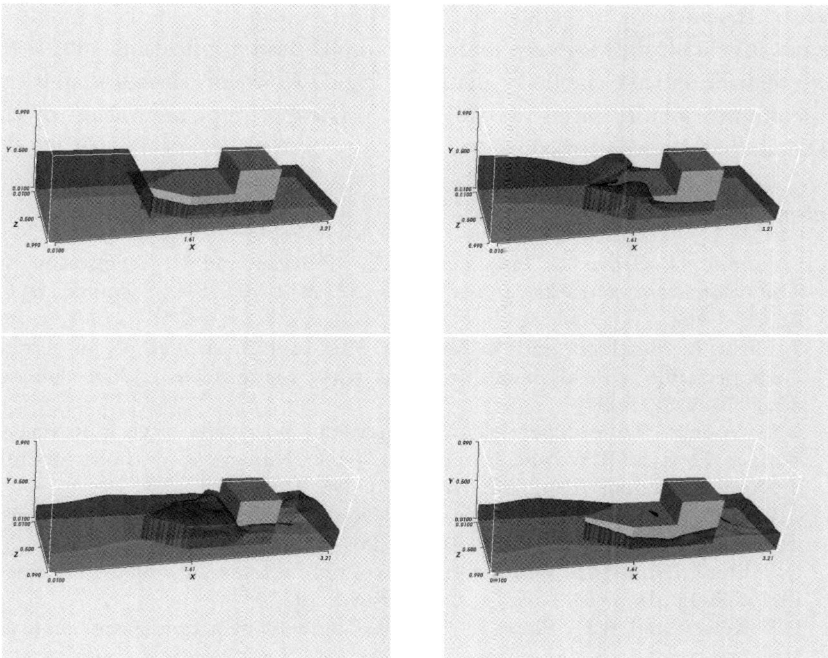

Fig. 21. Simulation of green water overtopping of a fixed container ship geometry at $t = 0.0$, 0.4, 1.4 and 2.5 seconds.

mesh, the majority being unaffected. This is in contrast to conventional mesh generation methods which must be re-meshed in the field at least locally.

The extension of the interface capturing two-fluid solver to 3-D flow problems is straightforward as shown by the initial results for sample 3-D test cases. The method is currently in use for numerically modelling wave attack at porous seawall structures and is also being integrated with an existing 3-D Cartesian cut cell library to study the hydrodynamic performance of several new wave energy devices. Although the reported cases are inviscid, these other applications have necessitated the incorporation into the new method of viscous effects including turbulent transport. A 3-D version of the present method will be applicable to any problem involving free surfaces, interface break-up and recombination and wave interactions with static or moving boundaries such as green water overtopping of vessels and offshore structures with appropriate wave climate boundary conditions.

Finally, the solver is not restricted to two fluids and the potential exists to extend it in a straightforward manner to multi-fluid applications and there is no restriction that the body should be rigid. However, these calculations are computationally intensive and thus a parallel implementation of the 3-D code has been initiated.

References

1. J. Glimm, J. Grove, B. Lingquist, O.A. McBryan and G. Tryggvason, G. The bifurcation of tracked scalar waves. *SIAM J. Sci. Stat. Comput.*, **9(1)**, 61–79, 1988.
2. J. Farm, L. Martinelli and A. Jameson, Fast multimesh method for solving incompressible hydrodynamic problems with free surfaces. *AIAA Journal*, **32**, 1175–1182, 1994.
3. D.L. Youngs, Time dependent multi-material flow with large fluid distortion. In Morton, K.W. and Baines M.J. (eds.) Numerical Methods for fluid dynamics. London: Academic Press, 273–285, 1982.
4. C.W. Hirt and B.D. Nichols, Volume of fluid (VOF) method for dynamics of free boundaries. *J. Comput. Phys.*, **39**, 201–225, 1981.
5. O. Ubbink, Numerical prediction of two fluid systems with sharp interface. Ph.D Thesis, Imperial College, unpublished, 1997.
6. F.J. Kelecy and R.H. Pletcher, The development of a free surface capturing approach for multidimensional free surface flows in closed containers. *J. Comput. Phys.*, **138**, 939–980, 1997.
7. D. Pan and H. Lomax, A new approximate LU factorization scheme for the Navier-Stokes equations. *AIAA Journal*, **26**, 163–171, 1988.
8. D.M. Causon, D.M. Ingram, C.G. Mingham, G. Yang and R. V. Pearson, Calculation of shallow water flows using a Cartesian cut cell approach. *Advances in Water Resources*, **23**, 545–562, 2000.
9. D.M. Causon, D.M. Ingram, C.G. Mingham, A Cartesian cut cell method for shallow water flows with moving boundaries. *Advances in Water Resources*, **24**, 899–911, 2001.
10. L. Qian, D.M. Causon, D.M. Ingram and C.G. Mingham, A Cartesian cut cell two fluid solver for hydraulic flow problems. *ASCE Journal of Hydraulic Engineering*, **9**, 688–696, 2003.
11. L. Qian, D.M. Causon, C.G. Mingham and D.M. Ingram, A free surface capturing method for two fluid flows with moving bodies, *Proceedings of the Royal Society A*, **462**, No. 2065, 21–41, 2006.
12. S.E. Rogers and D. Kwak, An upwind difference scheme for the time-accurate incompressible Navier-Stokes equations. *AIAA J.*, **28**, 113–134, 1990.
13. G. Yang, D.M. Causon, D.M. Ingram, R. Saunders and P.A. Batten, A Cartesian cut cell method for compressible flows - part B: moving body problems. *Aeronaut J*, **101**(1001), 57–65, 1997.
14. L. Qian, D.M. Causon, D.M. Ingram, J.G. Zhou and C.G. Mingham, A Cartesian cut cell method for incompressible viscous flows. *Proc. of ECCOMAS CFD 2001*, Southend-on Sea, UK, 2001.

15. G. Yang, D.M. Causon, D.M. Ingram, Calculation of compressible flows about complex moving geometries using a 3D Cartesian cut cell method. *Int J Num Meth Fluids*, **33**, 1121–1151, 2000.
16. C.G. Mingham, L. Qian, D.M. Causon and D.M. Ingram, M. Folly and T.J.T Whittaker, A two-fluid numerical model of the LIMPET OWC, in *Fifth European Wave Engergy Conference*, Lewis A. and Thomas G. (eds.), Hydraulic and Maritime Research Centre, Ireland, 119–123, 2005.
17. L. Qian, C.G. Mingham, D.M. Causon, D.M. Ingram, M. Folly and T.J.T Whittaker, Numerical simulation of wave power devices using a two-fluid free surface solver, *Modern Physics Letters B*, **19**, No. 28–29, 51–54, 2005.
18. F. Gao, D.M. Ingram, D.M. Causon and C.G. Mingham, The development of a Cartesian cut cell method for incompressible viscous flows, *Int. J. Numer. Meth. Fluids*, **54**, 1033–1053, 2007.
19. I. Tadjbakhsh and J.B. Keller, Standing surface waves of finite amplitude, *J. Fluid Mech.*, **8**, 442–451, 1960.
20. F. Gao, Ph.D. thesis, Brighton University, UK, 2003.
21. M. Greenhow and W.-M. Lin, Nonlinear free surface effects: experiments and theory. *Rep. No. 83–19, Department of Ocean Engineering, MIT*, Cambridge, Massachusetts, 1983.
22. M. Greenhow, Wedge entry into initially calm water. *Applied Ocean Research*, **9(4)**, 214–223, 1987.
23. M. Folley, T.J.T. Whittaker and M. Osterried, The oscillating wave surge converter, *Proceedings of the 14th International Offshore and Polar Engineering Conference (ISOPE 2004)*, CD-ROM, Paper No. JSC-379, Toulon, France, 2004.
24. R. Zhao and O. Faltinsen, Water entry of two-dimensional bodies. *J. Fluid Mech.*, **246**, 593–612, 1993.
25. X. Mei, Y. Liu and D.K.P Yue, On the water impact of general two dimensional sections. *Applied Ocean Research*, **21**, 1–15, 1999.
26. C. Hu and M. Kashiwagi, Experimental validation of the computation method for strongly nonlinear wave-body interactions, *21st International Workshop on Water Waves and Floating Bodies*, April 2nd-5th, Loughborough University, England, UK, 2006.
27. P. Troch, T. Li, J. De Rouke and D.M. Ingram, Wave interaction with a sea dike using a VOF finite volume method, *Proceedings of the 13th International Offshore and Polar Engineering Conference*, vol. 3, 325–332, 2003.
28. Y.A. Goda, A comparative review on the functional forms of directional wave spectrum. *Coastal Engineering Journal*, **41(1)**, 1–20, 1999.
29. K.M. Kleefsman, G. Fekken, A.E.P. Veldman, B. Iwanowski and B. Buchner, A volume-of-fluid based simulation method for wave impact problems, *J. Comput. Phys.*, **206**, 363–393, 2005.

CHAPTER 12

NUMERICAL COMPUTATION METHODS FOR STRONGLY NONLINEAR WAVE-BODY INTERACTIONS

Masashi Kashiwagi*, Changhong Hu and Makoto Sueyoshi

Research Institute for Applied Mechanics, Kyushu University
6-1 Kasuga-koen, Kasuga, Fukuoka 816-8580, Japan
**kashi@riam.kyushu-u.ac.jp*

This chapter introduces numerical computation methods recently developed at RIAM (Research Institute for Applied Mechanics), Kyushu University, for strongly nonlinear wave-body interaction problems, such as ship motions in large-amplitude waves, resulting green-water impact on deck, and so on. Introduced first is the CIP-based Cartesian grid method, in which the wave-body interaction is treated as a multi-phase flow with liquid, gas, and solid phases. Development of an efficient interface-capturing scheme is a key and a couple of new conservative schemes are tested. Introduced second is the MPS method, which is a so-called particle method and hence no grid is used. The features and calculation procedures of these methods are described somewhat at length, and their validation is performed through comparisons with some experiments conducted at RIAM.

1. Introduction

Strongly nonlinear seakeeping problems include ship motions in large-amplitude waves, resultant slamming on ship's bottom and flare, green-water impact on deck, and so on. Development of numerical calculation methods applicable to such strongly nonlinear problems is a subject of current interest. In strongly nonlinear phenomena, wave breaking and air entrainment may inevitably occur. For these cases, conventional calculation methods based on the potential-flow theory cannot be applied, and CFD (Computational Fluid Dynamics) techniques based on the finite difference or finite volume method must be used. However, since the free surface will be largely distorted and the splash of fluid may be generated, conventional CFD techniques break down or cannot avoid large numerical diffusion, which results in a serious failure in the interface capturing. To surmount these difficulties, we have been developing a number of CFD techniques; among those the CIP (Constrained Interpolation Profile) based Cartesian grid method and the MPS (Moving Particle Semi-implicit) method are introduced in this chapter.

The CIP scheme was initiated by Yabe et al.[1] for computing the advection equation with less numerical diffusion, and its application to strongly nonlinear problems in ship and ocean engineering has been mainly done at RIAM of Kyushu University.[2-7] Further modification and improvement of the method culminated in the code named RIAM-CMEN (Computation Method for Extremely Nonlinear hydrodynamics). However, the calculation method introduced in this chapter is referred to as the CIP-based method in general.

The main feature of this method is that the wave-body interaction problem is treated as a multi-phase problem and a stationary Cartesian grid is adopted. In this kind of method, the free surface must be determined by an interface-capturing scheme, for which we adopt basically the CIP method using a density function. Although the thickness of an interface is zero in real physical problems, a transient region of the density function across the interface exists in numerical computations. A good interface-capturing scheme is to have a very compact transient region, and it is believed that the accuracy of computed results depends largely on the efficiency of the interface-capturing scheme adopted. We have tested various schemes, such as CIP schemes with function transformation and the CIP-CSL3 (Conservative Semi-Lagrangian scheme with 3rd-order polynomial function) scheme.

In 2-D computations with sufficient grid resolution, a CIP scheme with a simple linear function transformation[5] works well. However, for 3-D computations, we are obliged to use a coarse grid with variable grid spacing of high aspect ratio. Then numerical problems such as interface smearing and deterioration of the mass conservation are unavoidable. The CIP-CSL3 scheme was applied to the numerical simulation of violent sloshing in a rectangular tank,[3] and we observed that the CIP-CSL3 scheme could reduce drastically the numerical diffusion by adjusting a free parameter included in the scheme. However, a drawback of the CIP-CSL3 scheme is that an overshoot or undershoot in the density function occurs at the interface and the scheme is rather complicated. Recently, in order to surmount these drawbacks, we have studied the THINC (Tangent of Hyperbola for Interface Capturing) scheme,[6,7] which uses the hyperbolic tangent function to represent a step-like sharp change of the density function across the interface. Performance of this new interface-capturing scheme is demonstrated in this chapter through some comparisons between measured and computed results.

No matter which scheme is used for interface capturing, the numerical diffusion originating from the computation of advection terms cannot be zero as long as a stationary Cartesian grid is used. In that respect, the MPS method is advantageous, because this method is a fully Lagrangian scheme. The MPS method was proposed and developed by Koshizuka and Oka[8] for studying extremely complicated free-surface problems of incompressible fluids in nuclear engineering. Sueyoshi and Naito[9] applied the MPS method to strongly nonlinear free-surface problems in ocean engineering and popularized this method. These days many other researchers seem to be working with this method, and the number of related papers is increasing.[10] In this chapter, the calculation procedure of the MPS method is briefly introduced and then an example of the comparison between computed and measured results is shown for a tank-sloshing problem, which suggests that the MPS method is a promising tool for strongly nonlinear hydrodynamic problems.

2. CIP-Based Cartesian Grid Method

2.1. *Flow solver*

The wave-body interaction problem is treated as a multi-phase flow with liquid phase (water), gas phase (air), and solid phase (floating body), and the flow solver in the numerical method under development is based on the CIP scheme for advection equations and the C-CUP (CIP Combined Unified Procedure) method[1] for the coupling of the velocities and the pressure. The C-CUP method was originally proposed for multi-phase compressible flows. However, we realized after a pilot study on a dam-breaking problem[2] that treating the flow as incompressible gives reasonable results and the calculation scheme becomes simpler than the original.

Therefore we consider here an unsteady, viscous and incompressible flow. Then the governing equations for the fluid velocity u_i in the ith direction ($i = 1,2,3$ for x, y, and z respectively) and the pressure p are as follows:

$$\frac{\partial u_i}{\partial x_i} = 0, \tag{1}$$

$$\frac{\partial u_i}{\partial t} + u_j \frac{\partial u_i}{\partial x_j} = -\frac{1}{\rho}\frac{\partial p}{\partial x_i} + \frac{1}{\rho}\frac{\partial}{\partial x_j}\left(\mu S_{ij}\right) + f_i\,, \tag{2}$$

where S_{ij} denotes the viscous stress written as $\partial u_i / \partial x_j + \partial u_j / \partial x_i$, ρ is the fluid density, μ is the viscosity coefficient, and f_i denotes a body force such as the gravity force, etc. The surface tension is neglected and no turbulence models are introduced here.

In order to apply the CIP scheme explained later, not only the fluid velocity u_i but also its derivatives with respect to the spatial coordinates must be computed to determine the interpolation profile with a cubic polynomial inside one cell. For this purpose, Eq. (2) is differentiated with respect to the spatial coordinates, which gives the following:

$$\frac{\partial (u_i)_\xi}{\partial t} + u_j \frac{\partial (u_i)_\xi}{\partial x_j} = -(u_j)_\xi \frac{\partial u_i}{\partial x_j} - \frac{\partial}{\partial \xi}\left(\frac{1}{\rho}\frac{\partial p}{\partial x_i}\right) + \frac{\partial H_i}{\partial \xi}, \tag{3}$$

where $(u_i)_\xi = \partial u_i / \partial \xi$ ($\xi = x_1,\ x_2,\ x_3$) and H_i represents the viscous and body force terms in Eq. (2).

We employ a Cartesian grid with staggered arrangement of the variables to be computed in the C-CUP method (the pressure is at the volume center of a cell and the velocity components are at the corresponding surface center of a cell). The calculation of time evolution in Eq. (2) is performed by a fractional step method, with the equation divided into three steps: 1) advection phase ($q^n \to q^*$), 2) first non-advection phase ($q^* \to q^{**}$), and 3) second non-advection phase

($q^{**} \rightarrow q^{n+1}$), where q represents the variables to be computed, and $\Delta t = t^{n+1} - t^n$ is the size of one time step.

2.1.1. *Advection phase*

In the advection phase, only the advection equation is solved by using the CIP scheme. The basic idea of the CIP scheme is to construct the profile of a function within a computational cell in terms of a higher-order polynomial and to preserve it during the advection. This can be accomplished by using some supplementary constraints, such as the spatial gradient of the function, in addition to the values at grid points.

Therefore the advection phase includes the solution of two equations of the following kind:

$$\left.\begin{aligned} \frac{\partial q}{\partial t} + u_j \frac{\partial q}{\partial x_j} = 0 \\ \frac{\partial q_\xi}{\partial t} + u_j \frac{\partial q_\xi}{\partial x_j} = 0 \end{aligned}\right\}. \tag{4}$$

For example, in the 1-D case, an interpolation function using the cubic polynomial can be constructed around a grid point x_m as follows:

$$\hat{q}(\eta) = a_m \eta^3 + b_m \eta^2 + c_m \eta + d_m, \tag{5}$$

where $\eta = x - x_m$ and x is located in an upwind computational cell, i.e. $[x_{m-1}, x_m]$, and the four unknown coefficients in Eq. (5) can be determined using q_m, q_{m-1}, $(q_x)_m$, and $(q_x)_{m-1}$.

In the same way, we can construct interpolation functions for 2-D and 3-D cases; for example in the 2-D case, we have

$$\begin{aligned} \hat{q}(\eta_1, \eta_2) = {} & C_{30}\eta_1^3 + C_{21}\eta_1^2\eta_2 + C_{12}\eta_1\eta_2^2 + C_{03}\eta_2^3 \\ & + C_{20}\eta_1^2 + C_{11}\eta_1\eta_2 + C_{02}\eta_2^2 + C_{10}\eta_1 + C_{01}\eta_2 + C_{00}, \end{aligned} \tag{6}$$

where $\eta_1 = x_1 - x_{1i}$ $\eta_2 = x_2 - x_{2j}$, and (x_{1i}, x_{2j}) are the coordinates for the grid point concerned. Eq. (6) contains 10 unknown coefficients C_{mn}, which can be determined by using the known values of q, q_{x_1}, and q_{x_2} at the surrounding grid points of a computational cell. For more details, the readers are referred to Hu and Kashiwagi.[2]

Once the interpolation function is constructed as described above, a semi-Lagrangian procedure is applied for the time evolution of Eq. (4) as follows:

$$\left.\begin{aligned} q^*(\boldsymbol{x}) &= \hat{q}^n(\boldsymbol{x} - \boldsymbol{u}\Delta t) \\ q^*_\xi(\boldsymbol{x}) &= \hat{q}^n_\xi(\boldsymbol{x} - \boldsymbol{u}\Delta t) \end{aligned}\right\}. \tag{7}$$

Here the superscript 'n' denotes the values at current time level and the superscript '$*$' denotes the values after advection calculation.

As is clear from the above procedure, Eq. (7) can be computed in terms of only the information at the grid points of one cell. Therefore we can see that the CIP scheme has both features of a sub-cell resolution and a compact structure. These features are effective in maintaining sharpness of the interfaces in a multi-phase computation including discontinuities or large gradients in physical quantities in a computational domain.

2.1.2. *Non-advection phase*

In the first non-advection phase, all terms on the right-hand side of Eqs. (2) and (3) except for the pressure-related terms are evaluated using an Euler explicit scheme with a central finite difference algorithm.

In the second non-advection phase, coupling between the velocity and the pressure is treated by an implicit scheme. Taking account of the continuity equation, Eq. (1), we can obtain the following Poisson-type equation for the pressure:

$$\frac{\partial}{\partial x_i}\left(\frac{1}{\rho}\frac{\partial p^{n+1}}{\partial x_i}\right) = \frac{1}{\Delta t}\frac{\partial u_i^{**}}{\partial x_i}. \tag{8}$$

Equation (8) is assumed valid for liquid, gas, and solid phases, thus giving the pressure distribution in the whole computation domain. The pressure obtained in this manner inside a rigid body is a fictitious one satisfying the divergence-free condition of the velocity field. In this treatment, no explicit boundary conditions are required on the free and body surfaces, because those surfaces are treated as inner interfaces. On outer boundaries, a Neumann condition is imposed. With a finite difference scheme, Eq. (8) can be rewritten in a linear system of simultaneous equations with large dimension, to which a fast solver (such as SOR,

ICCG or Multigrid method[11]) or a parallel computing technique may be easily applied.

2.2. *Free surface capturing method*

The interface between water and air is called the free surface, which is determined by an interface capturing method. To recognize different phases in a multi-phase flow, we define density functions ϕ_m for liquid ($m=1$), gas ($m=2$), and solid ($m=3$) phases, see Fig. 1.

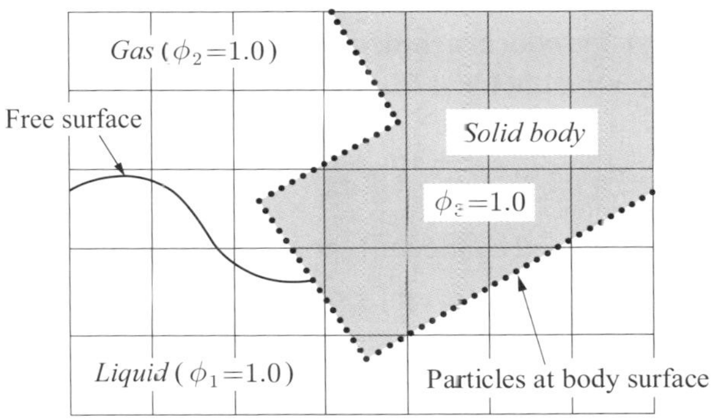

Fig. 1. Multi-phase computation model for wave-body interactions, composed of liquid, gas, and solid body. Artificial particles are used to represent the surface of a moving rigid body.

Whichever method is adopted for free-surface capturing, the density function for a solid body ϕ_3 will be computed first by integrating in time the equations of body motions, thereby determining the exact position of the rigid-body surface.

Next the density function for liquid ϕ_1 will be determined by solving the following advection equation:

$$\frac{\partial \phi_1}{\partial t} + u_j \frac{\partial \phi_1}{\partial x_j} = 0. \tag{9}$$

Then the remaining density function for gas ϕ_2 can be determined from a simple identity $\phi_1 + \phi_2 + \phi_3 = 1$.

The CIP method may be applied to solve Eq. (9), and the surface of $\phi_1 = 0.5$ is considered the free surface. As already noted, the CIP scheme has the sub-cell resolution property that is very useful to suppress the numerical diffusion at the free surface. However, there are some disadvantages even in the CIP method. First, since the CIP scheme is based on the non-conservative form of the advection equation, the mass conservation may deteriorate in long-term simulations. Furthermore, although the CIP scheme is featured with sub-cell resolution, the numerical diffusion may become prominent at the interface when the wave breaking and splash of water occur. To surmount these problems and enhance sharpness of the interface, a couple of improved schemes are proposed.

The first is to introduce a function transformation $\Phi = \Phi(\phi_1)$. The equation for Φ can be obtained from Eq. (9) in the form

$$\frac{\partial \Phi}{\partial t} + u_j \frac{\partial \Phi}{\partial x_j} = 0. \tag{10}$$

Yabe et al.[1] introduced a tangent-type transformation function:

$$\Phi(\phi_1) = \tan[\varepsilon\pi(\phi_1 - 0.5)], \tag{11}$$

where $0 < \varepsilon < 1$ is the parameter to control the sharpness.

A simpler linear transformation function given by

$$\Phi(\phi_1) = 0.5 + \alpha(\phi_1 - 0.5) \tag{12}$$

was used in Hu et al.[4] and Hu and Kashiwagi,[5] where $\alpha > 1$ is the parameter for sharpness enhancement. $\alpha = 1.2$ is adopted in numerical examples shown in those papers, and with this choice of α, it is found for a dam-breaking problem that the interface thickness (defined as $\phi_1 = 0.05 \rightarrow 0.95$) is kept within about 5 grid widths even after the returning jet plunges into the free surface.

Schemes which may be recommended for much better interface capturing are based on a conservative form of the advection equation:

$$\frac{\partial \phi_1}{\partial t} + \frac{\partial (u_j \phi_1)}{\partial x_j} = 0. \tag{13}$$

One of the schemes of this kind is the CIP-CSL3 (Conservative Semi-Lagrangian scheme with 3rd polynomial function), proposed by Xiao and Yabe.[12] The original CIP scheme adopts spatial gradients of a quantity of concern as the additional constraints to determine the interpolation function, while the CIP-CSL3 scheme uses cell-integrated values as the additional constraints.

The interpolation function used in the CIP-CSL3 scheme is also expressed with a cubic polynomial, which is given as Eq. (5) for the 1-D case. To determine four unknowns for an upwind computational cell $[x_{m-1}, x_m]$, the following constraints are used:

$$\left.\begin{aligned} &\hat{q}^n(x_m) = q_m^n, \quad \hat{q}^n(x_{m-1}) = q_{m-1}^n \\ &\int_{x_{m-1}}^{x_m} \hat{q}^n(x)dx = \overline{q}_{m-1/2}^n, \quad \frac{d\hat{q}^n(x)}{dx} = \lambda_{m-1/2}^n \end{aligned}\right\}. \tag{14}$$

Here $\overline{q}_{m-1/2}$ denotes the value of cell-integrated average and $\lambda_{m-1/2}$ is the slope of q at the center of the cell. The latter is used in fact as a free parameter to make numerical diffusion small at the phase interface; details of which are shown in Xiao and Ikebata.[13]

Although the scheme explained above is just for one dimension, it may be applied without modification to multi-dimensional problems with the idea of directional splitting technique. In fact, this CIP-CSL3 scheme was applied to a 2-D tank sloshing problem by Kishev et al.,[3] and much better results were confirmed through comparisons with measured results of the free-surface profile and the time history of the pressure impinging on the side wall of a tank.

As an alternative conservative scheme for free-surface capturing, Hu and Kashiwagi[6] incorporated the THINC (Tangent of Hyperbola for INterface Capturing) scheme in a CIP-based Cartesian grid method. The THINC scheme was originally proposed by Xiao et al.[14] In this scheme, the profile of the density function (ϕ_1) inside a computation cell is approximated with an interpolation function and the time evolution of ϕ_1 is performed by a semi-Lagrangian scheme, as in the original CIP scheme. However, differing from a cubic polynomial, a hyperbolic tangent function is used in the THINC scheme to reproduce a step-like sharp change of ϕ_1 near the free surface. Therefore no spurious oscillation appears in the THINC scheme. Furthermore, since the THINC scheme uses the cell-averaged density function with temporal integration carried out by the finite volume method, the conservation of mass flux may be guaranteed.

In order to explain the calculation procedure of the THINC scheme, a conservative form of 1-D advection equation for the density function $\phi(x,t)$, equal to Eq. (13) with $j=1$ only, is considered here. Extension to multi-dimensional problems may be achieved using the idea of directional splitting technique.

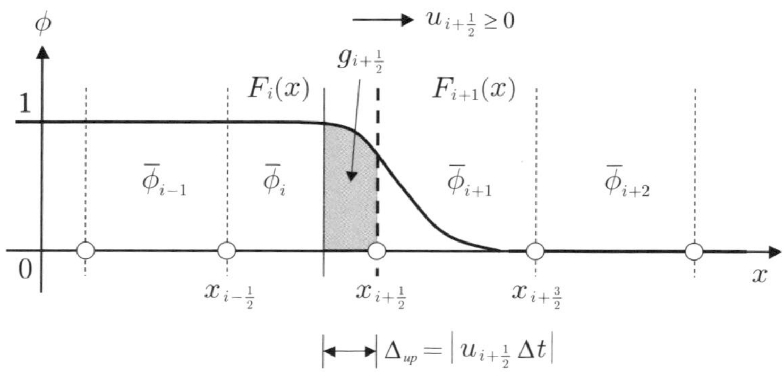

Fig. 2. Concept of the THINC scheme.

Let $\overline{\phi}_i^{\,n}$ denote the cell-averaged density function at the nth time step ($t=t^n$), defined over $[x_{i-1/2}, x_{i+1/2}]$. Then, with a finite-volume integration of Eq. (13), the density function ϕ may be updated as

$$\overline{\phi}_i^{\,n+1} = \overline{\phi}_i^{\,n} + \frac{1}{\Delta x_i}\left(g_{i-1/2} - g_{i+1/2}\right), \tag{15}$$

where

$$g_{i\pm 1/2} = \int_t^{t+\Delta t} (u\phi)_{i\pm 1/2}\, dt$$

is the flux and $\Delta x_i = x_{i+1/2} - x_{i-1/2}$ (see Fig. 2).

A step-like distribution of the density function within cell i is approximated using a modified hyperbolic tangent function of the form

$$F_i(x) = \frac{\alpha}{2}\left[1 + \gamma \tanh\left\{\beta\left(\frac{x - x_{i-1/2}}{\Delta x_i} - \delta\right)\right\}\right], \tag{16}$$

where α, β, γ and δ are the parameters to be determined as follows.

First α and γ can be explicitly determined as

$$\alpha = \begin{cases} \bar{\phi}_{i+1} & \text{if } \bar{\phi}_{i+1} \geq \bar{\phi}_{i-1} \\ \bar{\phi}_{i-1} & \text{otherwise} \end{cases} \tag{17}$$

$$\gamma = \begin{cases} 1 & \text{if } \bar{\phi}_{i+1} \geq \bar{\phi}_{i-1} \\ -1 & \text{otherwise.} \end{cases} \tag{18}$$

Parameter β is associated with the sharpness of a jump in the interpolation function. Following a result after several numerical experiments by Xiao et al.,[14] $\beta = 3.5$ is used in the results shown in this chapter and in Hu and Kashiwagi.[6]

Parameter δ is associated with the middle point of a transition jump and can be determined by the relation below:

$$\frac{1}{\Delta x_i} \int_{x_{i-1/2}}^{x_{i+1/2}} F_i(x)\, dx = \bar{\phi}_i . \tag{19}$$

Hence we have

$$\delta = \frac{1}{2\beta} \ln \left| \frac{e^A - e^{\beta}}{e^A - e^{-\beta}} \right|, \quad A = \frac{2\beta}{\alpha\gamma}\left(\bar{\phi}_i - \frac{\alpha}{2} \right). \tag{20}$$

Having determined a step-like profile, the flux $g_{i+1/2}$ across the cell boundary at $x_{i+1/2}$ can be evaluated in a semi-Lagrangian way as follows:

$$\begin{aligned} g_{i+1/2} &= \int_{x_{i+1/2}-\Delta_{up}}^{x_{i+1/2}} F_i(x)\, dx \\ &= \frac{\alpha}{2}\Delta_{up} + \frac{\alpha\gamma}{2\beta}\Delta x_i \ln \frac{\cosh \beta(1-\delta)}{\cosh \beta(1-\delta-\Delta_{up}/\Delta x_i)}, \end{aligned} \tag{21}$$

where, as shown in Fig. 2, $\Delta_{up} = |u_{i+1/2}\Delta t|$. Equation (21) is valid for $u_{i+1/2} \geq 0$, and a similar result may be obtained for $u_{i+1/2} \leq 0$ in the same manner.

In order to demonstrate the performance of the THINC scheme, we computed a 2-D dam-breaking problem, illustrated in Fig. 3. In the numerical computation, a variable grid is used, in which grid points are concentrated near the floor and the right-hand wall. The number of total grid points is 240×96 and the minimum grid spacing is 0.4 mm.

Two interface-capturing schemes, THINC and CIP, are applied and variation of the free-surface profile is compared in Fig. 4. Three lines show the values of the density function, equal to $\phi = 0.05$, 0.5, and 0.95, respectively. The free surface is defined as $\phi = 0.5$, which is indicated by the solid line. The thickness between the two dash lines roughly represents the transient region from liquid to gas, while this distance should be

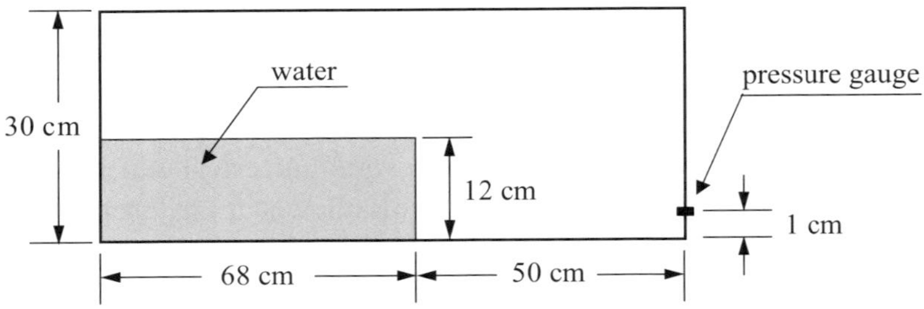

Fig. 3. Schematic view and dimensions in a dam-breaking problem.

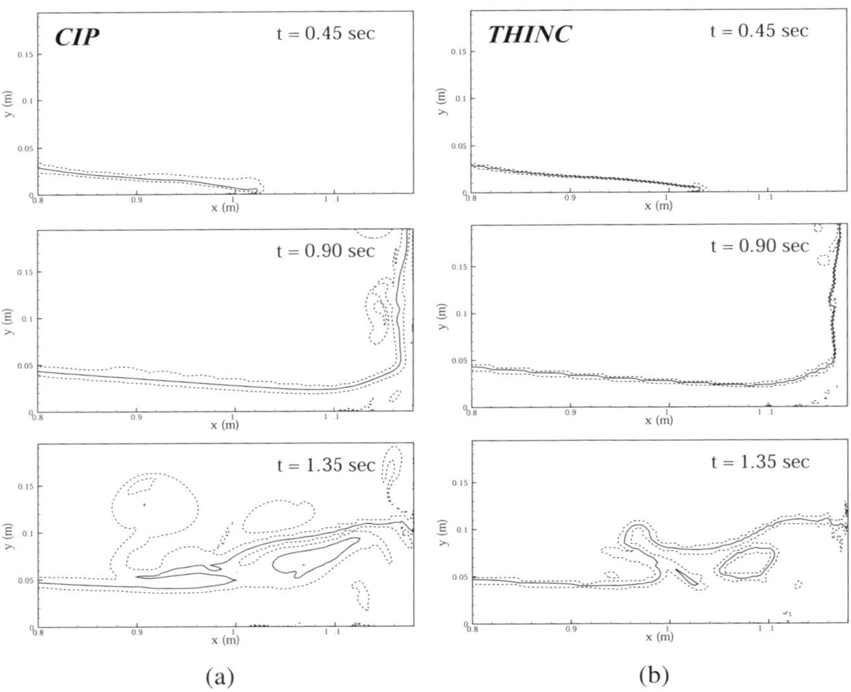

Fig. 4. Comparison of contours of the density function. Three lines indicate ϕ_1 = 0.05, 0.5, and 0.95. (a) CIP and (b) THINC schemes are used as the interface-capturing scheme.

zero in a real physical problem. We can see that the thickness of the free surface computed by the THINC scheme is very thin, while the free surface computed by the CIP scheme is diffusive after the overturning water plunges into the free surface. It is confirmed that time variation of the total water volume in the tank is virtually zero in the THINC scheme, whereas up to ±0.5 % variation exists in the CIP scheme.

2.3. *Virtual particle method for treating a rigid body*

A floating body is treated as a rigid body in this chapter. A Lagrangian method using a Cartesian grid is developed to calculate directly the density function for the solid phase ϕ_3 and the local velocity of the body.

As shown in Fig. 2, the body surface is approximated using the virtual particles which have the geometrical information of the local area and the normal vector to the body surface. The pressure at a particle can be evaluated by interpolation in terms of the pressures at surrounding grid points. Using the pressure thus obtained and the geometrical information at particles, the hydrodynamic forces on the body can be calculated by integrating the pressure at wet particles:

$$F_j = -\iint_{S_B} p\, n_j\, dS\,, \tag{22}$$

where S_B denotes the wetted surface of a body and n_j the *j*th component of the normal vector, which is defined positive when pointing from the body surface into the fluid.

Then, by integrating in time the motion equations of a rigid body, we can obtain the translational and rotational velocities at the center of gravity of the body, which gives the velocity at any particle at each new time step, say U_{pi}^{n+1}. The density function for a solid body ϕ_3 in a computation cell can easily be calculated by an approximate way using the local coordinates of the particles in the cell. The velocity at the body boundary cell denoted by $\hat{U}_i^{n+1}$ can be calculated using the velocities at nearby particles U_{pi}^{n+1}. Therefore this velocity $\hat{U}_i^{n+1}$ is the 'body velocity', with which the body boundary condition is specified. The velocity in the whole computational domain is updated by

$$U_i^{n+1} = \phi_3 \hat{U}_i^{n+1} + (1-\phi_3)\, u_i^{**}\,, \tag{23}$$

where u_i^{**} denotes the velocity to be obtained after the first non-advection step in the fractional step method adopted for computing Eq. (2).

In a boundary cell, Eq. (23) is a volume fraction weighting treatment for the velocity interpolation. We note that the method of imposing the velocity distribution inside and on the body boundary is equivalent to applying a forcing term to the momentum equation.

A non-slip boundary condition on the surface of a moving solid body can be imposed as described above. However, in order to apply a slip (or partial slip) boundary condition in the framework of Eq. (23), the velocity at ith particle U_{pi}^{n+1} needs to be determined differently. For example, the normal and tangential velocities at particles may be set equal to those of the neighboring flow. In order to realize this condition numerically, we introduce an outer layer of virtual particles, with the distance between the two layers chosen equal to the order of one cell. Then the velocity at outer-layer particles can be interpolated with the velocities at surrounding grid points.

Once the density functions for liquid and solid phases are determined, the remaining density function for the gas phase can be readily obtained by $\phi_2 = 1 - \phi_1 - \phi_3$. Then a physical quantity q at each cell in the whole computational domain can be evaluated by

$$q = \sum_{m=1}^{3} \phi_m q_m . \tag{24}$$

3. Validation of the CIP-Based Method

3.1. *2-D experiments*

2-D experiments were conducted in a wave channel (18 m long, 0.13 m wide, and 0.4 m deep) at RIAM of Kyushu University. The model of a floating body used in the experiments is, as shown in Fig. 5, of box-shaped with $B = 0.5$ m in breadth, $d = 0.1$ m in draft, and $f = 0.023$ m in freeboard. A box-shaped upstructure is installed on the deck.

Regular waves were generated by a plunger-type wave maker, and a wave-absorbing apparatus installed at the other end of the wave channel was actuated. This implies that the wave reflection may occur between the wave maker and the floating body, but transmitted waves past a body may be absorbed.

In this experiment, we measured the forced vertical motion of the wave maker, wave-induced motions in heave, roll, and sway of the floating body, and the wave elevation at a point between the wave maker and the floating body. The heave and roll motions of floating body were free, while the sway motion was either fixed (denoted as sway-fixed condition) or weekly restrained by a spring with spring constant of 7.6 N/m^2 converted for a 2-D problem. Figure 5 shows the position of a wave probe, the initial position of a floating body, and other principal dimensions of the wave maker and so on, including the damping zone used in the numerical computation.

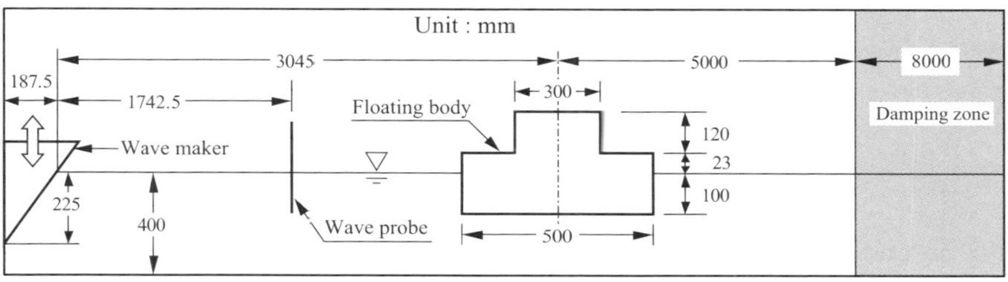

Fig. 5. Schematic view of a 2-D wave channel and a floating body used in the experiment and numerical computations. A damping zone in the downwave side does not exist in the real wave channel.

In addition to the above experiment, a forced oscillation test in heave in otherwise calm water was also carried out using the same floating body. In this test, the gap between the body and the side wall of the wave channel was set equal to 1.0 mm to reduce 3-D effects associated with a resonant flow in the gap. The hydrodynamic force (added mass and damping) coefficients obtained from this test are compared with corresponding results by the numerical computation to check the accuracy of the calculation method described in this chapter.

3.2. *2-D results and comparison*

3.2.1. *Hydrodynamic forces*

First we discuss the results of the forced oscillation test. In this test, the heave motion of floating body was given as

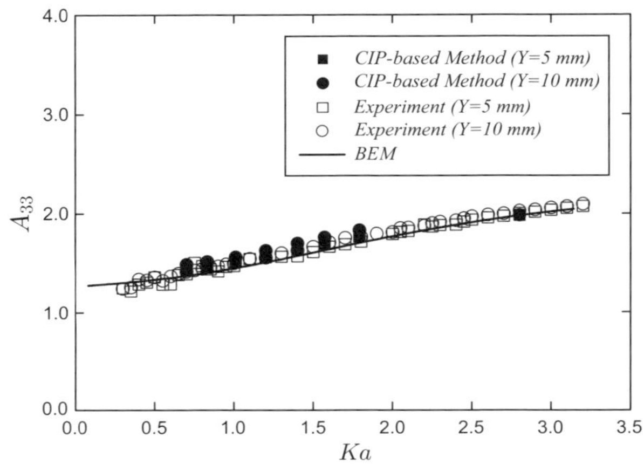

Fig. 6. Comparison of the added-mass coefficient in heave of a 2-D box-shaped floating body shown in Fig. 5. The results are shown in non-dimensional form with ρa^2.

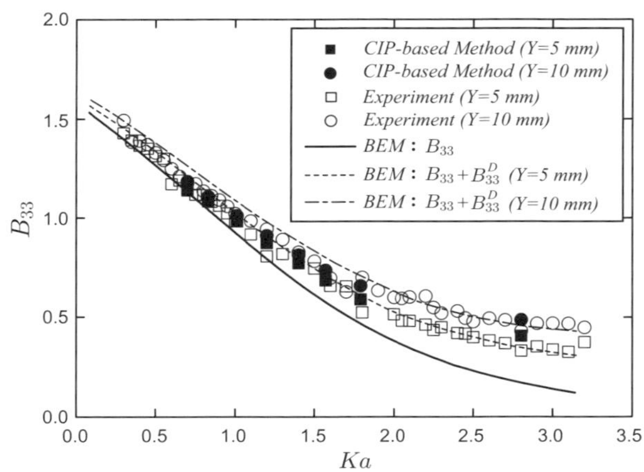

Fig. 7. Comparison of the damping coefficient in heave of a 2-D box-shaped floating body shown in Fig. 5. The results are shown in non-dimensional form with $\rho a^2 \sqrt{g/a}$.

$$y(t) = Y \sin(\omega t), \tag{25}$$

where Y is the amplitude of oscillation and $\omega = 2\pi / T$ with T being the period of oscillation.

Numerical computations were performed with a grid of 400×183 (in the horizontal (x) and vertical (y) directions, respectively), in which the minimum grid spacing was $\Delta x = \Delta y = 3$ mm around the body. The time step size was $\Delta t / T = 1/1000$. The total simulation time was $15T$ and the time history for the last $10T$ was used for the Fourier analysis to calculate the added-mass and damping coefficients. In this simulation for the forced heave oscillation, a numerical damping beach was installed on both sides of the body to prevent wave reflection.

Figures 6 and 7 show a comparison of the added-mass and damping coefficients respectively. The amplitude of forced oscillation was set to $Y = 5$ mm and 10 mm in the experiment and corresponding numerical computations by the CIP-based method. Computed results by a BEM (Boundary Element Method) using the free-surface Green's function are also shown. BEM results are for an inviscid fluid with irrotational motion. The added mass (A_{33}) and damping (B_{33}) coefficients are made non-dimensional with ρa^2 and $\rho a^2 \sqrt{g/a}$, respectively, where $a = B/2$ is the half breadth of floating body, g the gravitational acceleration, and ρ the density of water. The abscissa is taken as $Ka = \omega^2 a / g$.

The agreement in the added mass between the experiment and the computation is very good. For the damping coefficient, computed results by the present CIP-based method agree well with measured ones, whereas potential-flow results by BEM are slightly lower at higher frequencies. A reason is that the computation of BEM neglects viscous effects. The present computation of the CIP-based method is based on the Navier-Stokes equations and thus can simulate vortex shedding which is prominent from sharp corners.

The damping force due to vortex shedding may be roughly estimated using Morison's formula:

$$F_D = C_D \frac{1}{4} \rho B \dot{y} |\dot{y}|, \tag{26}$$

where the constant 1/4 is used instead of 1/2, for a floating body can be regarded as half a body in an unbounded fluid, if the free surface is assumed rigid.

The drag coefficient C_D is dependent on the KC number defined by $KC = 2\pi Y / B$. In the present experiment, the oscillation amplitude Y is 5 mm and 10 mm, which results in $KC = 0.0628$ and 0.1257, respectively. A rough estimation of the value of C_D for a square cylinder from the experiment by Vesgatesan et al.[15] may be $C_D = 25$ and 20 for the two KC numbers mentioned above, respectively.

In terms of the damping coefficient due to vortex shedding (denoted as B_{33}^D), let us write the viscous damping force in the form

$$\hat{F}_D = B_{33}^D \rho a^2 \sqrt{\frac{g}{a}}\, \dot{y}. \tag{27}$$

Calculating the work to be done by F_D and $\hat{F}_D$ over one period of oscillation and equating them, we obtain the following result:

$$B_{33}^D = C_D \frac{4Y}{3\pi a} \sqrt{Ka}. \tag{28}$$

The value estimated by Eq. (28) is added to the result of BEM and the total value is shown in Fig. 7 for each oscillation amplitude. From this estimation, the difference between the experiment and the computation by BEM can be accounted for by taking the damping force due to vortex shedding into consideration.

3.2.2. *Wave-induced motions*

Next we discuss the results of the body motions in waves. Numerical computations for this case were performed with a stationary Cartesian grid of 525×183 in the x- and y-directions, respectively; in which the minimum grid spacing was $\Delta x = \Delta y = 3$ mm around the body and the wave maker. The time step size was taken as $\Delta t / T_w = 1/1000$, where T_w is the wave period. Among a large amount of computations corresponding to the experiment, typical results for two different wave periods are shown in this subsection; both are for the case of sway-fixed but heave and roll are free.

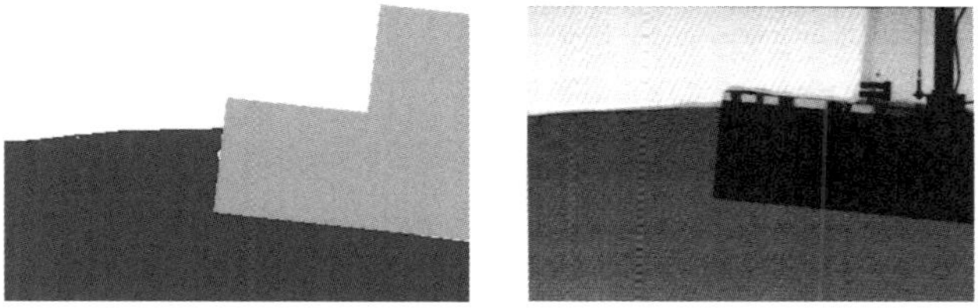

(a) Computation (b) Experiment

Fig. 8. Snap shots at $t/Tw = 17.3$ for the case of $Tw = 1.0$ sec and sway fixed.

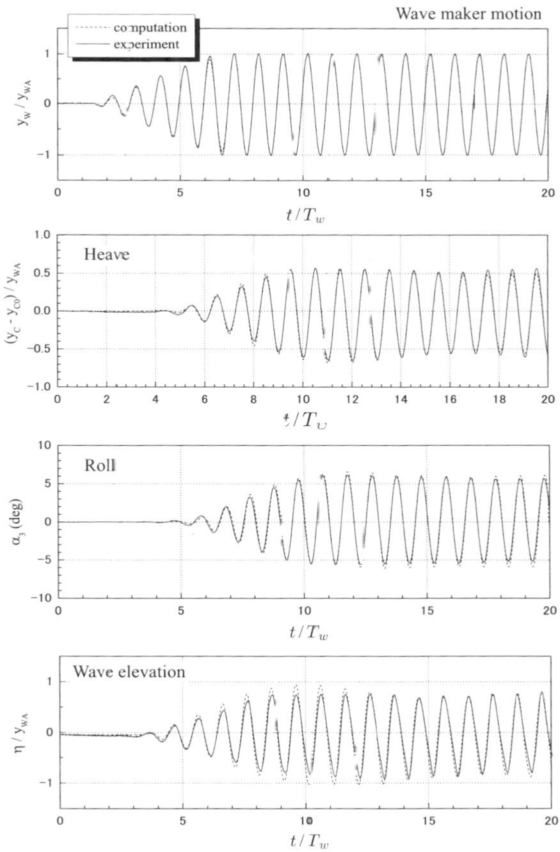

Fig. 9. Comparison of time histories of the body motions (heave roll) and the wave elevation for the case of $Tw = 1.0$ sec and sway-fixed condition.

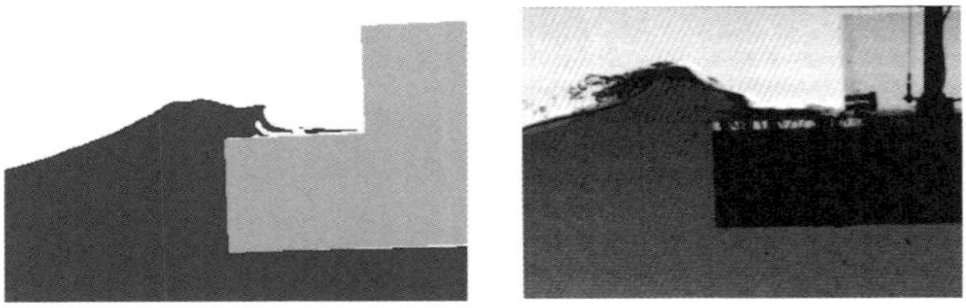

(a) Computation (b) Experiment

Fig. 10. Snap shots at $t/Tw = 17.3$ for the case of $Tw = 0.7$ sec and sway fixed.

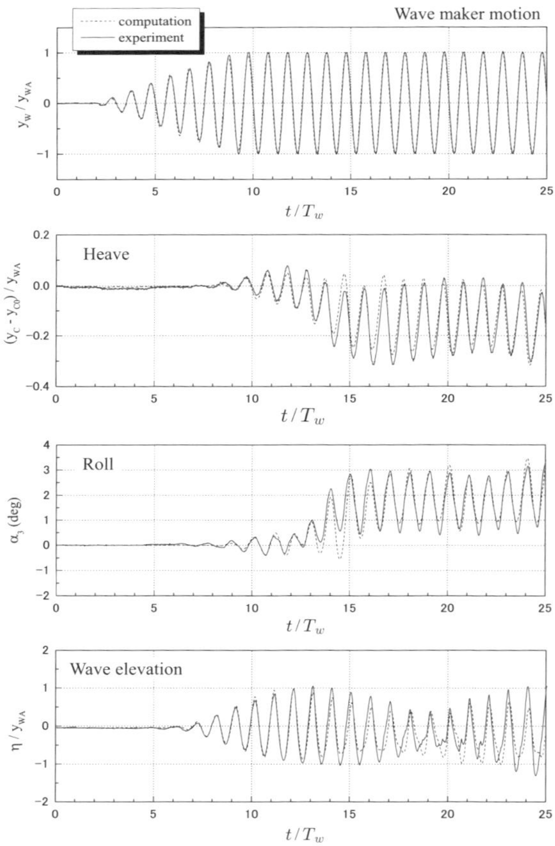

Fig. 11. Comparison of time histories of the body motions (heave roll) and the wave elevation for the case of $Tw = 0.7$ sec and sway-fixed condition.

In the numerical computations, the wave-maker motion measured in the experiment is the only input data. The amplitude of the wave-maker motion is 25 mm, and the wave period T_w is 1.0 sec for the first case and 0.7 sec for the second case. We note that the period of resonance in roll is about $T_r \approx 0.9$ sec for the set-up of a model used in the experiment, and that the body motion tends to be in phase with the incident wave for $T_w > T_r$ and reversely tends to be out of phase for $T_w < T_r$.

For the case of $T_w = 1.0$ sec, the results are shown in Figs. 8 and 9. In this case, no prominent water on deck was observed, and thus the body motions as well as the free-surface elevation are quite regular in variation. The agreement between the computation and the experiment is very good.

For the case of $T_w = 0.7$ sec, as shown in Figs. 10 and 11, strongly nonlinear phenomena appear in both motions and free-surface elevation. We could observe as in Fig. 10 that an incident wave comes onto the deck, impinges upon the upstructure, and flushes down with some water remaining on the deck. Consequently, as in Fig. 11, the heave becomes negative (i.e. sinkage) in average and the roll becomes positive (i.e. heel to the weather side) in average. It is confirmed that these non-zero values in average are due mainly to the weight of shipped water. We can see that these strongly nonlinear phenomena are successfully reproduced by the present calculation method.

3.3. *3-D experiments*

3-D experiments were conducted in the towing tank (65 m long, 5 m wide, and 7 m deep) at RIAM of Kyushu University. The ship model used in the experiments is a modified Wigley model which is of mathematical form represented by

$$\eta = (1-\zeta^2)(1-\xi^2)(1+a_2\xi^2+a_4\xi^4)+\zeta^2(1-\zeta^8)(1-\xi^2)^4, \qquad (29)$$

where $\xi = 2x/L$, $\eta = 2y/B$, and $\zeta = z/d$, with L, B, and d the length, breadth, and draft, respectively. The bluntness parameters, a_2 and a_4, are chosen as $a_2 = 0.6$ and $a_4 = 1.0$. Three different drafts (denoted as deep, standard, and shallow in Table 1) were considered in the experiment. The principal dimensions of the model are also shown in Table 1.

In the paper of Hu and Kashiwagi,[5] the experiments measuring the pressure and the motions of ship model in waves were carried out for the case of deep draft only, and the results are compared with computed ones by the CIP-based Cartesian grid method using the CIP scheme with a linear function transformation, Eq. (12), as the interface capturing method. Figure 12 shows a photo of the modified Wigley model with deep draft, set in the towing tank. As seen in this photo, a vertical wall is installed on the deck of the model to block the inflow of green water. Six pressure gauges are installed on the horizontal deck and vertical wall along the longitudinal centerline; their locations are indicated in Fig. 13.

Table 1. Principal dimensions of a modified Wigley model used in the experiment.

	deep	standard	shallow
Length: L (m)		2.500	
Breadth: B (m)		0.500	
Draft: d (m)	0.205	0.175	0.125
Freeboard: f (m)	0.045	0.075	0.125
Displacement volume: ∇ (m^3)	0.1688	0.1388	0.0892
Gyrational radius in pitch: κ_{yy}/L	0.244	0.258	0.278
Height of gravitational center: KG (m)	0.150	0.149	0.168

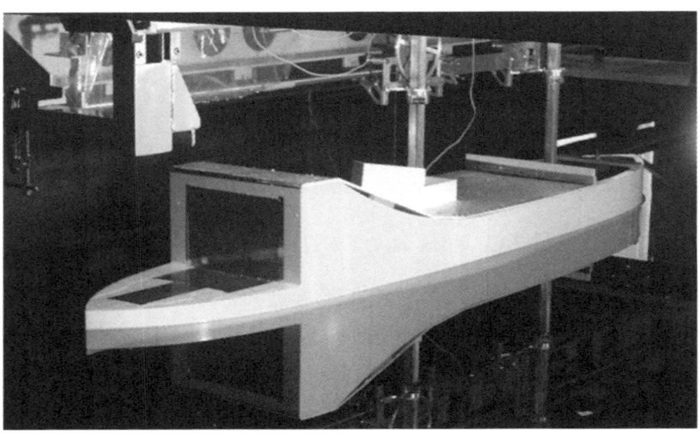

Fig. 12. Modified Wigley model set in the towing tank (for motion-free condition).

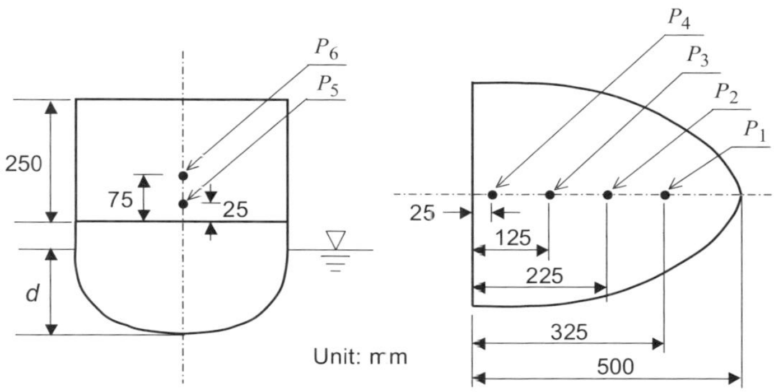

Fig. 13. Locations of pressure measurement on the deck and vertical wall.

Recently we also carried out the experiments measuring the same items for the standard and shallow draft cases. Some of the measured results are shown in Hu and Kashiwagi[6] together with a comparison with computed results by an improved CIP-based method using the THINC scheme for interface capturing. The experiments were performed for both motion-fixed and motion-free conditions, and three regular waves of different wave lengths ($\lambda / L = 0.75$, 1.0 and 1.25) and three ship speeds ($Fn = 0.0$, 0.15 and 0.20) were tested. Nevertheless, only the results for the case of $\lambda / L = 1.25$, $Fn = 0.20$ and of standard draft are shown in this chapter as a typical example.

3.4. *3-D results and comparison*

Numerical computations corresponding to the 3-D experiments were implemented in a numerical wave tank which moves at the same constant speed as that of a ship. The size of the numerical wave tank is taken as $-2.7L \sim 5.9L$ in the x-axis, $-1.1L \sim 1.1L$ in the y-axis, and $-0.8L \sim 1.3L$ in the z-axis, where the origin of the coordinate system is placed at the midship and undisturbed free surface. The positive x-axis is directed to the stern along the longitudinal centerline, and a numerical damping zone is installed at the end of downstream side. A non-uniform grid of $181 \times 80 \times 80$ in the x-, y-, and z-axes, respectively is employed

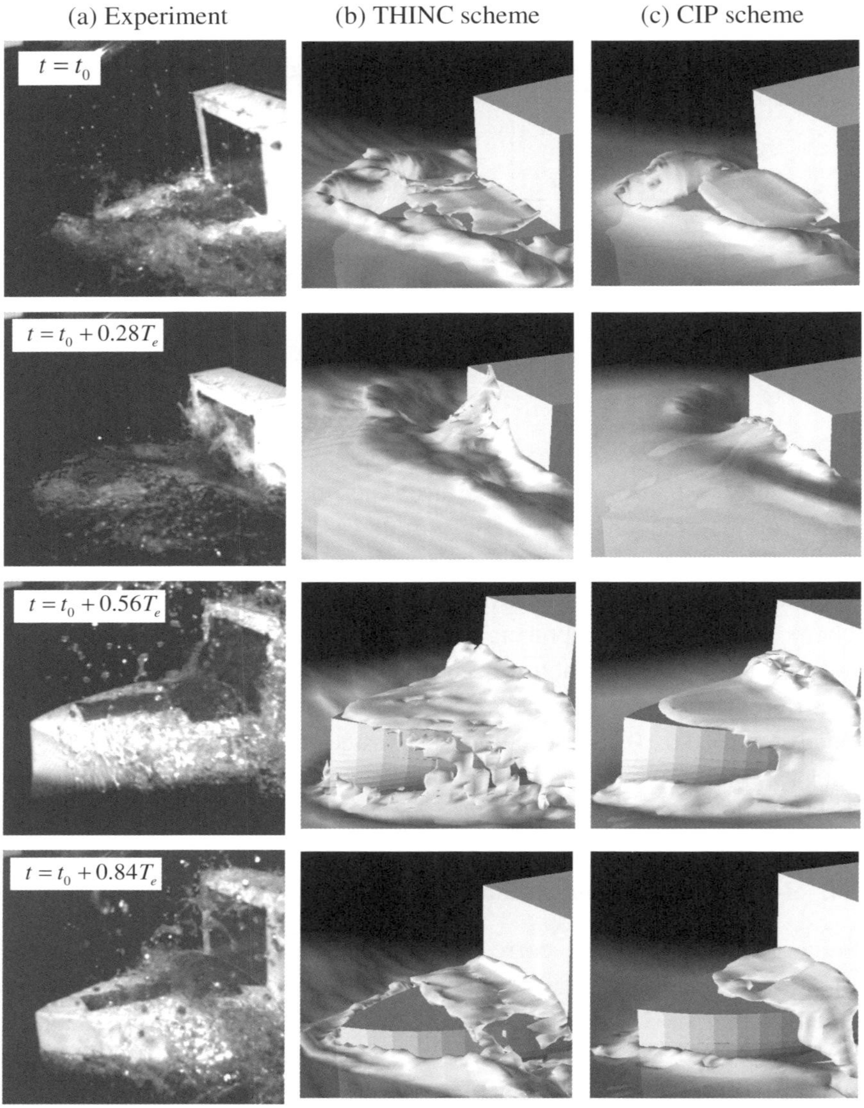

Fig. 14. Comparison of the free-surface profile at $\lambda / L = 1.25$ and $Fn = 0.20$. The numerical computation is performed by using (b) THINC and (c) CIP schemes for interface capturing.

for this numerical wave tank, where the minimum grid spacing was $\Delta x = 0.006L$ near the ship bow and $\Delta y = \Delta z = 0.005L$ near the ship hull and free surface. The time step size was taken equal to $\Delta t / T_w = 1/2000$, where T_w is the period of the incident wave. Numerical computations with these grids and time steps were performed using a PC workstation with CPU of Intel Xeon 3.4 GHz and the CPU time for the simulation of $6T_w$ was about 36 hours.

Deformation of the free surface around the fore deck is compared in Fig. 14 between the experiment and numerical computations at four time instants during one period of encounter T_e. Both THINC and CIP schemes are applied as the interface capturing method in the numerical computations, and the experimental images are taken from the record by a high-speed digital video camera. The computed free surfaces are iso-surfaces corresponding to $\phi_1 = 0.5$. The overall phenomena of strongly nonlinear flow on the deck are favorably simulated by both the interface-capturing schemes. However, looking at the details of the flow such as splash and fragmentation of water, we can see that the THINC scheme is much better than the CIP scheme. Specifically in the snapshot at $t = t_0 + 0.28T_e$, the result by the THINC scheme shows higher run-up of the green water on the vertical wall than that by the CIP scheme. When compared to the experiment, details of the flow are still much different even with the THINC scheme. However, this difference should be attributed to lack of resolution, and the number of grid points must be increased for further improvement.

Figure 15 shows time histories of the pressures measured at P_1, P_2, P_4, P_5 and P_6. (We note that the measurement at P_3 was unsuccessful due to some trouble.) From this result, the behavior of the pressure variation may be classified into non-impact type (P_1, P_2) and impact type (P_4, P_5 and P_6) with larger impulsive values at the beginning of a cycle. Such feature is well simulated by the present calculation method, and the agreement in the order of magnitude looks favorable. However, discrepancies can be seen in the maximum value of an impulsive pressure and in some other details. In particular, existence of a smaller second peak in the pressure during one period observed in the time history of P_2 is not properly reproduced by the numerical computation.

Figure 16 shows time histories of the pitch, surge, and heave motions. The agreement between the experiment and the numerical computation is satisfactory especially for pitch and heave. Meanwhile, slight discrepancy can be seen in surge. (We note that the surge motion was restrained by a strong spring with spring constant of 2,800 N/m in the experiment, and hydrodynamic forces in surge are relatively small in magnitude.)

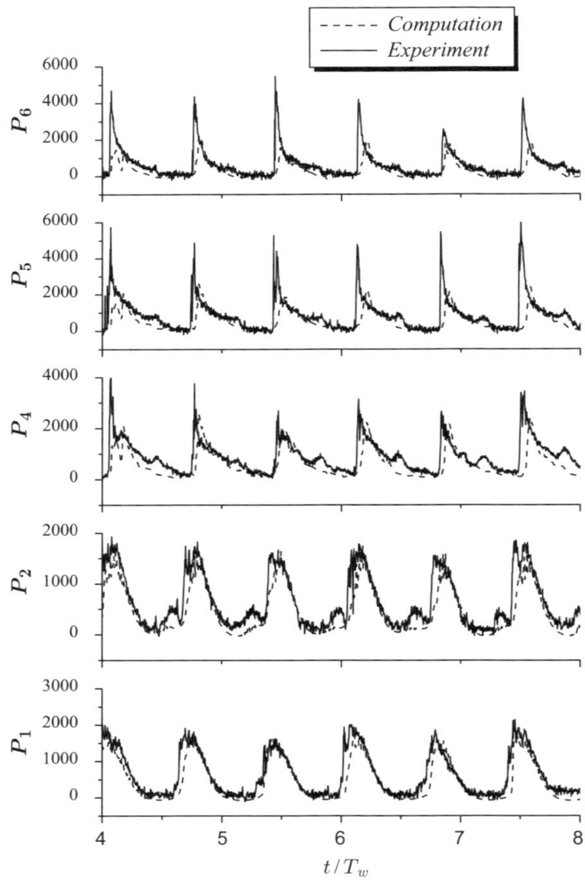

Fig. 15. Comparison of the pressure on the deck and vertical wall of a modified Wigley model at $\lambda/L = 1.25$ and $Fn = 0.20$ for the standard draft case. The numerical computation is performed using the THINC scheme for interface capturing.

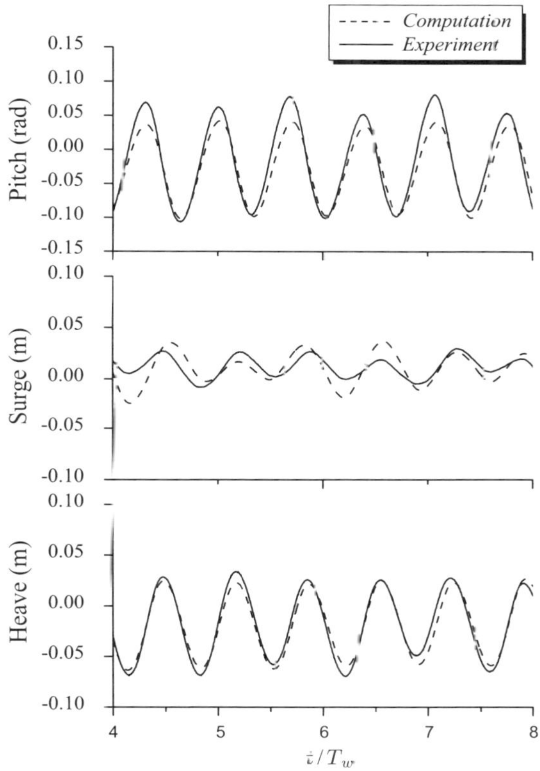

Fig. 16. Comparison of wave-induced motions of a modified Wigley model at $\lambda/L = 1.25$ and $Fn = 0.20$ for the standard craft case. The numerical computation is performed using the THINC scheme for interface capturing.

Nevertheless, the over-predicted value by the numerical computation may be reasonable, because a mechanical friction in the measuring system is not taken into account in the computation.

4. MPS Method

As demonstrated above, the CIP-based Cartesian grid method is robust and can deal with strongly nonlinear free-surface flows, and by using a conservative interface-capturing scheme such as the THINC or CIP-CSL3 scheme, sharpness of the free surface may be kept even for long-term simulations. However, as long as a stationary Eulerian mesh is used, the numerical diffusion originating from advection terms is unavoidable, and the resolution will be worse in 3-D computations due to limitation in the number of grid points.

The MPS (Moving Particle Semi-implicit) method, initiated by Koshizuka and Oka[8], is also considered to be a promising method for treating violent free-surface flows, because it has the following features:

- Gridless Lagrangian method, and thus
- No numerical diffusion due to convection and perfect conservation of mass
- Robust and simple because of no necessity of the geometric connectivity among particles

On the other hand, a demerit of the MPS method is that the computation time becomes generally long even for 2-D problems and maybe prohibitive for large-scale 3-D computations. It is also pointed out that the pressure computed by the original MPS method oscillates violently in time, which is obviously different from real phenomena. To overcome these problems, some continual studies have been made up to date, and at present the MPS method seems to attract much attention of researchers.[10]

4.1. *Overview of the MPS method*

The MPS method is a fully Lagrangian method, considering particles as discrete points. Then the velocity of fluid $\boldsymbol{u}$ corresponds to the velocity of particles. When the particle is numbered by subscript i, $\boldsymbol{r}_i$ and $\boldsymbol{u}_i$ mean the vectors of position and velocity of particle i, respectively, and p_i is the pressure exerted by the fluid at $\boldsymbol{r}_i$. From a Lagrangian viewpoint, they can be written as

$$\boldsymbol{u}_i = \boldsymbol{u}(t,\, \boldsymbol{r}_i), \quad p_i = p(t,\, \boldsymbol{r}_i). \tag{30}$$

In the MPS method, water is treated as incompressible. Thus the density of water ρ is constant from the continuity equation. The Navier-Stokes equations governing the fluid motion are described as

$$\frac{D\boldsymbol{u}}{Dt} = -\frac{1}{\rho}\nabla p + \nu \nabla^2 \boldsymbol{u} + \boldsymbol{f}\,, \tag{31}$$

where D/Dt denotes the substantial derivative, t is time, ν is the kinetic viscosity coefficient, and $\boldsymbol{f}$ denotes the vector of a body force. No turbulent models are considered here.

In order to solve Eq. (31) numerically, discrete models for the first- and second-order differential operators need to be introduced, for which Koshizuka[8] considered the particle interaction models. The first-order differential operator, the gradient of a scalar quantity ϕ, at the ith particle is described as

$$\nabla\phi_i = \frac{d}{n^0}\sum_{j\neq i}^{N} w(r_{ij})\frac{\phi_j-\phi_i}{r_{ij}}\frac{\boldsymbol{r}_j-\boldsymbol{r}_i}{r_{ij}}, \tag{32}$$

where $\boldsymbol{r}_i$ is the position vector of particle i, $r_{ij}=|\boldsymbol{r}_i-\boldsymbol{r}_j|$ is the distance between particle i and j, d is the number of dimensions of the space, and $w(r_{ij})$ is the weight function of the distance r_{ij}. The shape of this weight function can be arbitrary, but according to the study by Koshizuka, the following form is commonly used:

$$w(r)=\begin{cases}\dfrac{r_0}{r}-1 & \text{for } r\le r_0\\ 0 & \text{for } r>r_0\end{cases}. \tag{33}$$

Here r_0 is the cut-off radius which is constant and related to the influence area in the particle interaction model.

The particle number density is defined as

$$n_i = \sum_{j\neq i}^{N} w(r_{ij}) \equiv n^0, \tag{34}$$

which must be constant throughout the computational domain, because the fluid density is uniform in an incompressible fluid. This constant value is denoted as n^0 in Eq. (32). The treatment for the particle number density near the boundary surfaces will be described later.

The second-order differential operator, the diffusion of a quantity ϕ, at the ith particle is modeled as the distribution of ϕ and described as

$$\nabla^2\phi_i = \frac{2d}{\lambda}\sum_{j\neq i}^{N} w(r_{ij})\left(\phi_j-\phi_i\right), \tag{35}$$

where λ is a parameter to adjust a distributed quantity to the analytical result and given by

$$\lambda = \sum_{j \neq i}^{N} r_{ij}^2 \, w(r_{ij}). \tag{36}$$

The algorithm of the MPS method is a semi-implicit one using a fractional step method. The first step is an explicit computation of temporary velocity and position vectors, and the results are written as

$$\boldsymbol{u}_i^* = \boldsymbol{u}_i^n + \Delta t \left(\nu \nabla^2 \boldsymbol{u} + \boldsymbol{f} \right), \quad \boldsymbol{r}_i^* = \boldsymbol{r}_i^n + \Delta t \, \boldsymbol{u}_i^* , \tag{37}$$

where Δt is the time increment, and the viscosity term $\nu \nabla^2 \boldsymbol{u}$ may be discretized with the diffusion model given by Eq. (35).

With the temporary position vector $\boldsymbol{r}_i^*$ computed above, the particle number density is calculated according to Eq. (34), which is denoted as n_i^*. In the MPS method, the particle number density is required to be constant at each time step. To accomplish this requirement, Poisson's equation for the pressure in the second step may be derived, and the result is expressed in the form

$$\nabla^2 p_i^{n+1} = -\frac{\rho}{\Delta t^2} \frac{(n_i^* - n^0)}{n^0}. \tag{38}$$

From Eq. (35), Eq. (38) can be rewritten as

$$\sum_{j \neq i}^{N} w(r_{ij}) \, p_j^{n+1} - \left\{ \sum_{j \neq i}^{N} w(r_{ij}) \right\} p_i^{n+1} = -\frac{\rho}{\Delta t^2} \frac{\lambda}{2d} \frac{(n_i^* - n^0)}{n^0}. \tag{39}$$

This provides a linear system of simultaneous equations for the pressure p_i^{n+1}; the coefficient matrix of which is symmetric and sparse, because $r_{ij} = r_{ji}$ and the weight function $w(r_{ij})$ has a cut-off radius. Therefore, Eq. (39) can be solved efficiently by an iterative solution method.

In actual numerical computations, the boundary conditions must be given on the free surface and solid surfaces surrounding the fluid concerned. The free surface may be detected using only the value of particle number density; that is, when the particle number density is below $0.97 n^0$, the particle is regarded as on the free surface. Then a Dirichlet condition for the pressure, equal to the atmospheric pressure, is given to this particle on the free surface.

On the solid surface, particles are arranged to represent the boundary surface and the velocity at each particle must be equal to the velocity of the boundary surface. In the conventional MPS method, dummy particles are also arranged along the outside of a solid surface just to compute the particle number density for the solid surface. In this case, the relative position of dummy particles are unchanged. As an alternative treatment, Sueyoshi and Naito[9] proposed a method which does not use dummy particles. Depending on the curvature at each particle position, a constant compensation value corresponding to the dummy particles can be computed, which is added in computing the particle number density at a point on the boundary surface. Whichever method is used, generation of a grid system is not needed and only the arrangement of particles is enough for starting the computation. Therefore, there is no numerical difficulty even if fragmentation of a fluid or large deformation of the free surface occurs in real phenomena.

4.2. *Computed example*

Although the numerical computation has been implemented for various problems, shown here as an example is the simulation of a violent sloshing. To provide validation data for comparison, we also carried out a series of experiments with a horizontally oscillating rectangular tank. A schema of the tank is shown in Fig. 17. From a large number of experimental data, we select only the case of shallow water depth for comparison with numerical results. The amplitude and period of oscillation are set equal to $A = 0.06$ m and $T = 1.3$ sec, respectively. We note that the first resonant period for the present rectangular tank and water depth is estimated to be around $T = 1.3$ sec.

The number of particles in the simulation depends on the area of fluid. For the test case shown in Fig. 17, the number of particles used was 6,456, and the CPU time using a PC workstation with CPU of Intel Xeon 2.4 GHz was about 10 hours for the simulation of 3 oscillation periods.

A comparison is shown in Fig. 18 for the free-surface profile. We can see very good agreement between measured and computed results, in spite of occurrence of large deformation of the free surface and fragmentation of the fluid.

Figure 19 shows a comparison of time history of the pressure measured at point P_1 indicated in Fig. 17, and again good agreement exists between the experiment and numerical computation. We note that not only the first impulsive peak but also the second peak within one cycle of oscillation are well reproduced in the numerical simulation by the MPS method. It should be mentioned that the pressure computed from Eq. (38)

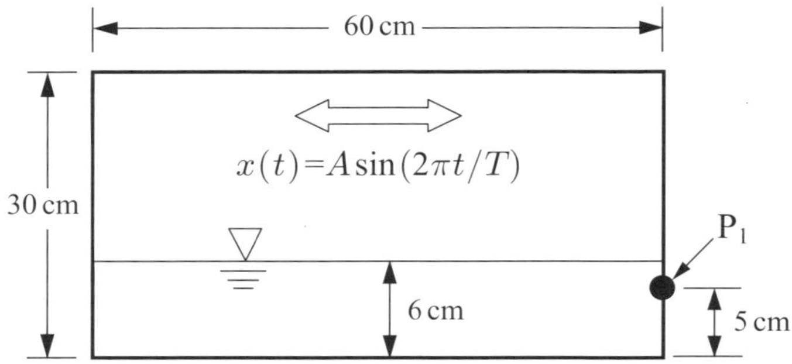

Fig. 17. Schema of the sloshing experiment with a rectangular tank. The tank is forced to oscillate harmonically in sway. P1 indicates the position where the pressure is measured.

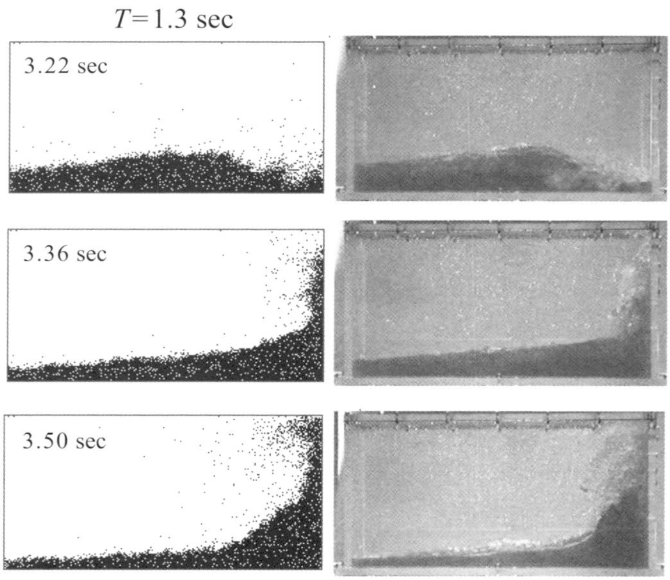

Fig. 18. Comparison of the free-surface profile in sloshing experiment at $T = 1.3$ sec between computed results by the MPS method (left) and observed ones (right).

oscillates violently in time, although the local mean value in the time history looks reasonable. A reason of this high-frequency oscillation in the pressure is that, as can be envisaged from Eq. (39), the pressures at particle points tend to be sensitive to the difference between temporal and initial constant values of the particle number density. To suppress this numerical oscillation, Sueyoshi and Naito[9] devised an auxiliary calculation method for the pressure, in which Poisson's equation for the pressure is driven not by Eq. (38) but by the divergence of the fluid velocity, like Eq. (8). The results shown in Fig. 19 are obtained by this auxiliary calculation method.

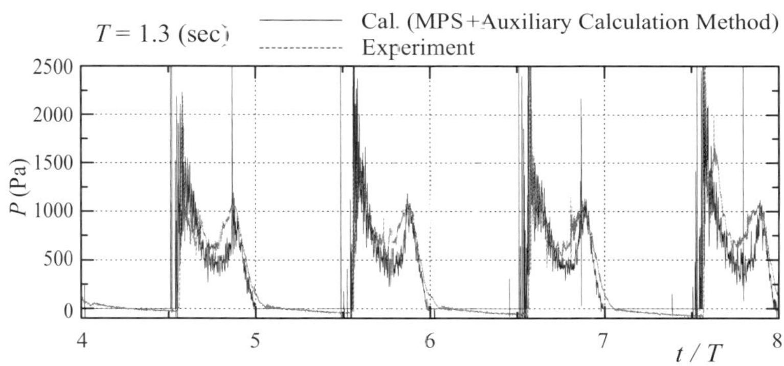

Fig. 19. Comparison of the pressure in sloshing experiment at $T = 1.3$ sec between computed results by the MPS method and observed results.

In the MPS method, 3-D computations can be performed without major modification of the computer code, but the computation time may become prohibitive in general. To make it possible to compute for particles more than 10^6 with practical computation time, Sueyoshi and Naito[16] devised a couple of numerical techniques, such as the cell portioning method and the parallel computing with optimum memory distribution. With these techniques, the MPS method can be performed at present for particles of the total number more than 10^6 by using personal computers.

Recently, for a purpose of shortening the computation time for large-scale computations based on the MPS method, a hybrid method using the MPS method and a time-domain boundary element method (BEM) is investigated by Kihara et al.[17] The idea is that the MPS method is applied to a near field around a floating body where strongly nonlinear phenomena are prominent and the BEM is applied to a far field where no breaking waves exist because of decay in amplitude of strongly nonlinear waves. Hybrid methods of this kind seem to be promising and should be studied for further development.

5. Conclusions and Outlook

In order to study strongly nonlinear wave-body interactions near the free surface, several numerical computation methods using CFD techniques are under development. Those methods are required to be robust and accurate even for a case where large deformation of a fluid occurs, such as wave breaking, splash, air entrainment, and so on. As the two promising calculation methods for fulfilling these requirements, the CIP-based Cartesian grid method and the MPS method are introduced in this chapter.

In the CIP-based Cartesian grid method, current efforts are mainly devoted to the development of efficient interface-capturing schemes and thus several schemes are described, such as the original CIP scheme with function transformation, the CIP-CSL3 scheme, and the THINC scheme which was recently developed. In particular, the THINC scheme looks promising in that the conservation of mass is virtually perfect and the numerical thickness of the free surface is very thin and no smearing is observed even in long-term simulations. Validation of the method has been done through comparison with 2-D and 3-D experiments. Overall agreement looks good, but further enhancement in the accuracy still need to be accomplished especially for enhancement of 3-D problems. Generally speaking, despite the use of CIP-based schemes, numerical accuracy depends on the number of grids (so-called grid resolution). Therefore for 3-D problems parallel computations with a massive number of unknowns must be performed or some techniques for the localized grid refinement should be developed.

Studies on the MPS method are also being made. A current problem to be resolved is to shorten the computation time, which is practically very important in applying the method to 3-D problems with large computation domain such as a flow around a real ship. The techniques being studied in this direction are development of a computer code for massive-scale parallel computing and a hybrid method that combines the MPS method for treating violent flows near a body with the BEM for treating an outer flow with no wave breaking. Experiments should also be conducted not only to validate the methods under development but also to understand the mechanism of complicated nonlinear phenomena.

Acknowledgments

The work introduced in this chapter was partially supported by the Ministry of Education, Culture, Sports, Science and Technology in Japan, providing us with the Grant-in-Aid for Scientific Research (A) from 2006 through 2008. The authors would like to thank M. Inada and M. Yasunaga for their support in the experiments described in this chapter.

References

1. Yabe, Y., Xiao, F. and Utsumi, T., "The Constrained Interpolation Profile Method for Multiphase Analysis", *J. Comp Physics*, Vol. 169, pp. 556-569 (2001).
2. Hu, C. and Kashiwagi, M., "A CIP-Based Method for Numerical Simulations of Violent Free Surface Flows", *J. Marine Science and Technology*, Vol. 9, No. 4, pp. 143-157 (2004).
3. Kishev, Z., Hu, C. and Kashiwagi, M., "Numerical Simulation of Violent Sloshing by a CIP-Based Method", *J. Marine Science and Technology*, Vol. 11, No. 2, pp. 111-122 (2006).
4. Hu, C., Kishev, Z., Kashiwagi, M., Sueyoshi, M. and Faltinsen, O. M., "Application of CIP Method for Strongly Nonlinear Marine Hydrodynamics", *Ship Technology Research*, Vol. 53, No. 2, pp. 74-87 (2006).
5. Hu, C. and Kashiwagi, M., "Validation of CIP-Based method for Strongly Nonlinear Wave-Body Interactions", *Proc. 26th Symposium on Naval Hydrodynamics (Rome)*, Vol. 4, pp. 247-258 (2006).
6. Hu, C. and Kashiwagi, M., "Numerical and Experimental Studies on Three-Dimensional Water on Deck with a Modified Wigley Model", *Proc. 9th Numerical Ship Hydrodynamics (Ann Arbor, Michigan)*, Vol. 1, pp. 159-169 (2007).
7. Kashiwagi, M., Hu, C., Miyake, R. and Zhu T., "A CIP-Based Cartesian Grid Method for Nonlinear Wave-Body Interactions", *Proc. 10th PRADS International Symposium (Houston, USA)*, Vol. 2, pp. 894-902 (2007).

8. Koshizuka, S. and Oka, Y., "Moving-Particle Semi-Implicit Method for Fragmentation of Incompressible Fluid", *Nuclear Science and Engineering*, Vol. 123, pp. 421-434 (1996).
9. Sueyoshi, M. and Naito, S., "A Numerical Study of Violent Free Surface Problems with Particle Method for Marine Engineering", *Proc. 8th Numerical Ship Hydrodynamics (Busan)*, Vol. 2, pp. 330-339 (2003).
10. Kashiwagi, M., Ed., *Proc. VF-2007 International Conference on Violent Flows (Fukuoka, Japan, 2007).*
11. Nishi, Y., Hu, C. and Kashiwagi, M., "Multigrid Technique for Numerical Simulations of Water Impact Phenomena", *Int. J. Offshore and Polar Engineering*, Vol. 17, No. 1, pp. 39-46 (2007).
12. Xiao, F. and Yabe, T., "Completely Conservative and Oscillation-less Semi-Lagrangian Schemes for Advection Transportation", *J. Computational Physics*, Vol. 170, pp. 489-522 (2001).
13. Xiao, F. and Ikebata, A., "An Efficient Method for Capturing Free Boundaries in Multi-Fluid Simulations", *Int. J. Numerical Method in Fluids*, Vol. 42, pp. 187-210 (2003).
14. Xiao, F., Honma, Y. and Kono, T., "A Simple Algebraic Interface Capturing Scheme Using Hyperbolic Tangent Function", *Int. J. Numerical Methods in Fluids*, Vol. 48, pp. 1023-1040 (2005).
15. Vengatesan, V., Varyani, K.S. and Barltrop, N.D.P., "Wave-Current Forces on Rectangular Cylinder at Low KC Numbers", *Int. J. Offshore and Polar Engineering*, Vol. 10, No. 4, pp. 276-284 (2000).
16. Sueyoshi, M. and Naito, S., "A 3-D Simulation of Nonlinear Fluid Problem by Particle Method – Over One Million Particles Parallel Computing on PC Cluster –", J. Kansai Soc. Nav. Arch. Japan, No. 241, pp. 133-142 (2004).
17. Kihara, H., Sueyoshi, M. and Kashiwagi, M., "A Hybrid Computational Method for Nonlinear Free Surface Problems", *Proc. Int. Conf. Violent Flows (Fukuoka, Japan)*, pp. 305-314 (2007).

CHAPTER 13

SMOOTHED PARTICLE HYDRODYNAMICS FOR WATER WAVES

Robert A. Dalrymple*, Moncho Gómez-Gesteira†, Benedict D. Rogers‡, Andrea Panizzo§, Shan Zou¶, Alejandro J.C. Crespo‖, Giovanni Cuomo** and Muthukumar Narayanaswamy††

Johns Hopkins University, 3400 No. Charles Street, Baltimore MD 21204, USA
rad@jhu.edu

†*Grupo de Fisica de la Atmósfera y del Océano, Facultad de Ciencias Universidad de Vigo, 32004 Ourense, Spain*
mggesteira@uvigo.es

‡*School of Mechanical, Aerospace and Civil Engineering University of Manchester, Manchester, UK*
benedict.rogers@manchester.ac.uk

§*DITS Department, University of Rome "Sapienza", Rome, Italy*
andrea.panizzo@uniroma1.it

¶*Everest International Consultants, 444 W. Ocean Blvd, Suite 1104 Long Beach, CA 90802, USA*
shan.zou@everestconsultants.com

‖*Grupo de Fisica de la Atmósfera y del Océano, Facultad de Ciencias Universidad de Vigo, 32004 Ourense, Spain*
alexbexe@uvigo es

***Department of Civil Engineering, University of Roma TRE Via Vito Volterra 62, 00146 Roma, Italy*

††*Johns Hopkins University, 3400 No. Charles Street Baltimore, MD 21204, USA*
muthu@jhu.edu

Smoothed Particle Hydrodynamics (SPH) is becoming a useful tool for examining hydrodynamic flows that are difficult to model with other existing numerical methods, such as time-dependent flows with convoluted free surfaces. This paper will describe the development of the method using interpolation, and then present some applications to free surface flows, primarily water waves. The ability of the method to easily handle large free surface deformations, including wave breaking, will be illustrated. Finally the ability of the SPH method to solve for other non-hydrodynamic variables is shown–here we solve for the suspended sediment concentration under waves.

1. Introduction

Smoothed Particle Hydrodynamics (SPH) was developed in 1977 by Lucy[1] and Gingold and Monaghan[2] for astrophysical studies. The method involves representing the fluid by particles, which follow the rules of hydrodynamics. The method is Lagrangian, rather than Eulerian, and the trajectory of each particle can be traced with time. Monaghan[3] applied the methodology to free surface flows for the first time, including a dam break problem and a breaking wave in a numerical wave tank. For reviews and background of the methodology, see Refs. 4, 5 and 6.

Here, we provide an introduction to the SPH method through interpolation, and then present some applications to water waves, showing how this novel method treats breaking waves and suspended sediments under waves.

2. Theoretical Background[a]

In SPH, the fluid flow is decomposed into parcels of fluid. Each of these parcels (particles) has an associated mass, density, velocity, pressure, etc., and these values are constants at any instant for that particle. Through interpolation we can determine the value of any of these variables at any location within the fluid. We can obtain derivatives of variables at any location as well. Due to the pressure gradients in the fluid, the particles move with time, carrying along their values of density, velocity, etc., which change with time. The SPH model then follows the trajectory of each of the (thousands of) particles. The collection of all the particles represents a flow. As an alternative conceptional framework, we can envision the SPH particles as irregularly spaced nodes of a numerical interpolation scheme that move with the flow.

Interpolation involves determining values of a quantity at a given point based on data known at other points. We will use Moving Least-Squares interpolants (MLS), which is an interpolating scheme for irregularly spaced

[a]Much of this section is taken from Ref. 7, presented at the Fourth Asian and Pacific Coasts Conference.

data, given as $u(\mathbf{x}_i)$, for $i = 1, \ldots, N$, in several dimensions, denoted by $\mathbf{x}$. MLS has the distinction of providing a fitting function $g(\mathbf{x})$ that does not explicitly go through the given data points as the interpolant at a given location is best fit only to nearby points. The method is similar to a Least-Squares fit to points, except that the fit is done locally; that is, many different least-squares fitting polynomials are found for different regions of the domain of $u(\mathbf{x}_i)$.

The interpolating function at any value of $\mathbf{x}$ (following Belytschko *et al.*[8] and Dilts[9]) is a low order polynomial

$$u_p(\mathbf{x}) = \sum_{j}^{M} a_j(\mathbf{x}) p_j(\mathbf{x}) \equiv \mathbf{p}^T(\mathbf{x})\, \mathbf{a}(\mathbf{x}) \tag{2.1}$$

where $p^T(\mathbf{x})$ are monomials in the coordinate $\mathbf{x}$ and $M < N$. For example, in one dimension, $p^T(\mathbf{x}) = [1, x]$ and $M = 2$ for a linear function (giving $u_p(x) = a_1(x) + a_2(x)\, x$), or, for a quadratric function, $p^T(\mathbf{x}) = [1, x, x^2]$ and $M = 3$. In two dimensions, the corresponding linear and quadratic bases are

$$p^T(\mathbf{x}) = [1, x, y], \quad \mathrm{M} = 3 \tag{2.2}$$
$$p^T(\mathbf{x}) = [1, x, y, x^2, xy, y^2], \quad \mathrm{M} = 6\,. \tag{2.3}$$

The coefficients of the interpolant, $\mathbf{a}(\mathbf{x})$ vary with $\mathbf{x}$, hence the name "moving," as the interpolant changes with location. To find these coefficients at a position $\mathbf{x}$, the sum of the weighted squared errors between the interpolating function for each given position $\mathbf{x}_i$ and data $u_i, i = 1, N$ is computed:

$$E(\mathbf{x}) = \sum_{i=1}^{N} W(\mathbf{x} - \mathbf{x}_i) \left(\mathbf{p}^T(\mathbf{x}_i)\, \mathbf{a}(\mathbf{x}) - u_i\right)^2 \tag{2.4}$$

where $W(\mathbf{x} - \mathbf{x}_i)$ is a weighting function for node i that decays with the distance from the node to position $\mathbf{x}$. Because of the weighting functions, contributions to the sum from distant nodes is either small or zero, giving then a local fit.

Minimizing the mean-squared error with respect to the coefficients, we obtain

$$\frac{\partial E}{\partial \mathbf{a}} = \mathbf{A}(\mathbf{x})\mathbf{a}(\mathbf{x}) - \mathbf{B}(\mathbf{x})\mathbf{u} = 0 \tag{2.5}$$

where

$$\mathbf{u}^T = (u_1, u_2, ...u_N) \tag{2.6}$$

$$\mathbf{P} = \begin{bmatrix} p_1(\mathbf{x}_1) & p_2(\mathbf{x}_1) & \dots & p_m(\mathbf{x}_1) \\ p_1(\mathbf{x}_2) & p_2(\mathbf{x}_2) & \dots & p_m(\mathbf{x}_2) \\ \dots & \dots & \dots & \dots \\ p_1(\mathbf{x}_n) & p_2(\mathbf{x}_n) & \dots & p_m(\mathbf{x}_N) \end{bmatrix} \tag{2.7}$$

$$\mathbf{W}(\mathbf{x}) = \begin{bmatrix} W(\mathbf{x}-\mathbf{x}_1) & 0 & \dots & 0 \\ 0 & W(\mathbf{x}-\mathbf{x}_2) & \dots & 0 \\ \dots & \dots & \dots & \dots \\ 0 & 0 & \dots & W(\mathbf{x}-\mathbf{x}_N) \end{bmatrix} \tag{2.8}$$

$$\begin{aligned} \mathbf{A} &= \mathbf{P}^T \mathbf{W}(\mathbf{x}) \mathbf{P} \\ \mathbf{B} &= \mathbf{P}^T \mathbf{W}(\mathbf{x}) \end{aligned} \tag{2.9}$$

solving,

$$\mathbf{a}(\mathbf{x}) = \mathbf{A}^{-1}(\mathbf{x})\, \mathbf{B}(\mathbf{x})\, \mathbf{u}\,. \tag{2.10}$$

The local approximation at $\mathbf{x}$ is then given by Eq. 2.1. The weighting function $W(\mathbf{x}-\mathbf{x}_i)$ could be chosen as a cubic spline of the following form:

$$W(q) = \frac{2}{3h} \begin{cases} 1 - \frac{3}{2}q^2 + \frac{3}{4}q^3, & \text{for } q \le 1 \\ \frac{1}{4}(2-q)^3, & \text{for } 1 \le q \le 2 \\ 0, & \text{for } q > 2\,. \end{cases} \tag{2.11}$$

The argument q is the normalized distance $|\mathbf{x}-\mathbf{x}_i|/h$, where h determines the size of the support for $W(q)$ as the value of $W(q)$ goes to zero for $q > 2$. This means that given data within a range of $\pm 2h$ of a given point will be used. So, again, the MLS is a local fit to the data. Figure 1 shows the shape of the cubic spline. This weighting function is used also in many Smoothed Particle Hydrodynamics models.

The choice of the weighting function, it turns out, is not that important, except that it must be positive everywhere and decrease monotonically in magnitude with distance. It also should be differentiable for as many times as needed.

The size h of the support for the weighting function is important. For rapidly varying data, large h values lead to excessive averaging. For a linear interpolant, only a region of data where the variation is linear should be included.

The results of a moving least squares interpolation are no longer a single polynomial, but a different polynomial fit for each point $\mathbf{x}$. However, it

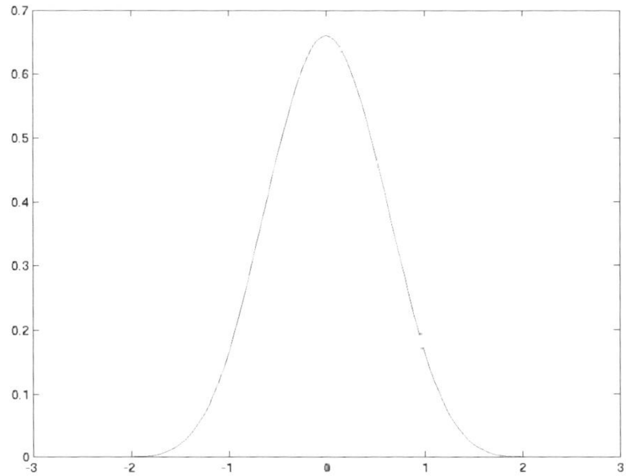

Fig. 1. Cubic spline weighting function (h=1 for this case), as defined in Eq. 2.11.

can be shown that data, provided by a polynomial of the same degree as the monomial basis function, is reproduced exactly by the method. Also, if the weight function is differentiable r times, then the resulting MLS polynomials are also differentiable r times.

The simplest interpolation that can be done is constant interpolation; that is, locally a constant is fit to the data. The polynomials in the moving least squares approach would be $\mathbf{p}^T = [1\ 1\ 1\ldots]$. Substituting into Eqs. 2.9 and 2.10, we find that

$$\mathbf{a} = \frac{\sum_j W_j u_j}{\sum_j W_j} \tag{2.12}$$

and

$$u(\mathbf{x}, t) = \sum_j \frac{W(\mathbf{x} - \mathbf{x}_j, h)}{\sum_j W(\mathbf{x} - \mathbf{x}_j, h)}\, u_j(t)\,. \tag{2.13}$$

This is known as a Shepard interpolant to the data.

Monaghan[3] derives SPH from the integral interpolant:

$$u(\mathbf{s}, t) = \int_v W(\mathbf{s} - \mathbf{x}, h)\, u(\mathbf{x}, t)\, dv \tag{2.14}$$

where the integral is over the domain v and dv is $dx\, dy\, dz$ in three dimensions, $dx\, dy$ in two, and dx in one dimension. The h is the parameter that determines the size of the support for the weighting function, $W(\mathbf{x}, h)$, which is a circle of radius $2h$ in 2D and a sphere of the same radius in 3D. In digitized form:

$$u(\mathbf{s},t) \approx \sum_j W(\mathbf{s}-\mathbf{x}_j,h)\, u_j\, \Delta v_j \tag{2.15}$$

$$\sum_j W(\mathbf{s}-\mathbf{x}_j,h)\, \Delta v_j = 1\,. \tag{2.16}$$

The second equation, a partition of unity, is found by interpolating a constant in Eq. 2.15; that is, setting u to unity. Here Δv_j is the incremental volume (or area in 2D) associated with particle j. This quantity Δv_j is replaced by the mass of the particle, m_i, divided by the local density of the fluid, ρ_i, which results in Monaghan's final form of the SPH interpolant:

$$u(\mathbf{s},t) = \sum_j W(\mathbf{s}-\mathbf{x}_j,h)\, u_j\, \frac{m_i}{\rho_i}\,. \tag{2.17}$$

Note that in the interior of the fluid, the condition (Eq. 2.16) shows that the Shepard interpolant and Monaghan's form are the same. However, near boundaries, Eq. 2.16 does not hold and the Shepard interpolant is better as it normalizes the kernel, accounting for the missing portion of the kernel's support.

A higher order interpolant could be used to derive SPH. Moving Least Squares approaches have been used by Dilts[9] and Colagrossi.[10] It is likely that going to much higher order than a linear polynomial is similar to the case in finite element or boundary element methods. Higher order elements lead to greater accuracy for a given size element, but the additional computational cost may not outweigh the use of smaller, lower order, elements.

2.1. *Gradients*

One of the powerful advantages of the SPH kernel approach is that the derivative of a function is calculated analytically, as compared to a method like finite differences, where the derivatives are calculated from neighboring points using the spacing between them. For the irregularly spaced SPH particles, this would be extremely complicated.

To take the gradient of $u(\mathbf{x},t)$ with respect to x, we use Eq. 2.17 to obtain

$$(\nabla \mathbf{u})_i = \sum_j \frac{m_j}{\rho_j}\, u_j\, \nabla W(x_j - x_i) \tag{2.18}$$

as the other terms in the summation are constant for particle j.

There are some accuracy problems with this gradient, however. The simplest correction is given by the method shown below after Eq. 2.27 to ensure that the gradient of a constant field is zero. Another correction occurs by taking the derivative of Eq. 2.13 and compare to Eq. 2.18. Refs. 11 and 12 show different gradient corrections.

2.2. *Governing hydrodynamics equations*

In Smoothed Particle Hydrodynamics, for simulations of fluid motions, the fluid has been traditionally considered compressible. The reason is that it is easier to calculate the pressure from an equation of state rather than having to solve for the pressure from a Poisson-type equation. For large changes in pressure, water is actually compressible, but for coastal applications we can consider it to be incompressible. We will get around the issue of compressibility by assuming that the water is only slightly compressible (in Section 2.2.3).

2.2.1. *Conservation of mass*

The conservation principle for a compressible fluid is, in Lagrangian form,

$$\frac{1}{\rho}\frac{d\rho}{dt} = -\nabla \cdot \mathbf{u}\,. \tag{2.19}$$

Following a parcel of fluid, we conserve the mass of the parcel j by assigning a constant mass to each particle, m_j. As long as we keep the same number of particles in our problem, mass is then conserved automatically.

To find a form of this equation for use in SPH, we multiply both sides of the equation by the kernel and integrate over the domain, υ:

$$\int_\upsilon W(\mathbf{s}-\mathbf{x},h)\left(\frac{1}{\rho}\frac{d\rho}{dt}\right)d\upsilon = -\int_\upsilon W(\mathbf{s}-\mathbf{x},h)\,\nabla\cdot\mathbf{u}\,d\upsilon\,. \tag{2.20}$$

By the reproducing nature of the kernel (Eq. 2.14), we find the left side easily as $\frac{d\rho}{dt}/\rho$ evaluated at position $\mathbf{s}$, and the right side is rewritten using an identity as

$$-\int_\upsilon \left[\nabla\cdot\left(W(\mathbf{s}-\mathbf{x},h)\,\mathbf{u}\right) - \mathbf{u}\quad \nabla W(\mathbf{s}-\mathbf{x},h)\right]d\upsilon \tag{2.21}$$

where the gradient is applied on the variable $\mathbf{x}$. Next we use Gauss's theorem, which is stated as

$$\int_\upsilon \nabla\cdot\mathbf{f}\,d\upsilon = \int_S \mathbf{f}\cdot\mathbf{n}\,dS \tag{2.22}$$

where the last integral is a surface integral over the surface that encloses the volume v and $\mathbf{n}$ is the outward surface normal.

$$\frac{1}{\rho}\frac{d\rho}{dt} = -\int_S W(\mathbf{s}-\mathbf{x},h)\,\mathbf{u}\cdot\mathbf{n}\,dS + \int_v \mathbf{u}\cdot\nabla W(\mathbf{s}-\mathbf{x},h)\,dv\,. \tag{2.23}$$

If the surface is far from the point s, then the integrand in the first integral is zero as the weighting function goes to zero rapidly. What remains is the last term.

$$\frac{1}{\rho}\frac{d\rho}{dt} = \int_v \mathbf{u}\cdot\nabla W(\mathbf{s}-\mathbf{x},h)\,dv\,. \tag{2.24}$$

The discrete form of this equation for use in SPH, when evaluated at particle x_i, is

$$\left(\frac{1}{\rho}\frac{d\rho}{dt}\right)_i = \sum_j \frac{m_j}{\rho_j}\,\mathbf{u}_j\cdot\nabla W_{ij} \tag{2.25}$$

where $W_{ij} \equiv W(\mathbf{x}_i - \mathbf{x}_j)$ and $\nabla W_{ij} = \nabla W(\mathbf{x}-\mathbf{x}_j)$ for $\mathbf{x} = \mathbf{x}_i$. Due to symmetry, the gradient of the weighting function with respect to the particle at $\mathbf{x}_j$ is $\nabla_j W_{ij} = -\nabla_i W_{ij}$, the gradient with respect to the first particle $\mathbf{x}_i$ in the argument of $W(\mathbf{x}_i - \mathbf{x}_j)$; therefore we can rewrite the above equation for the conservation of mass as

$$\left(\frac{1}{\rho}\frac{d\rho}{dt}\right)_i = -\sum_j \frac{m_j}{\rho_j}\,\mathbf{u}_j\cdot\nabla_i W_{ij}\,. \tag{2.26}$$

Another form of this equation is found by examining the expression for a gradient (Eq. 2.18). From that expression, we see that the gradient of a constant is zero, or

$$0 = \sum_j \frac{m_j}{\rho_j}\,\nabla_i W_{ij}\,. \tag{2.27}$$

Multiplying by $\mathbf{u}_i$, we can then add this zero sum to the previous form of the conservation of mass equation to get our final form.

$$\left(\frac{d\rho}{dt}\right)_i = \rho_i \sum_j \frac{m_j}{\rho_j}\,(\mathbf{u}_i - \mathbf{u}_j)\cdot\nabla_i W_{ij}\,. \tag{2.28}$$

As others have pointed out before, this form ensures that the gradient of a constant field is zero (when $\mathbf{u}_i = \mathbf{u}_j$).

Often it is necessary to obtain the fluid density at a location not corresponding with a particle. It can be calculated by the usual SPH definition:

$$\rho(\mathbf{s}) = \sum_j m_j W(\mathbf{s} - x_j),$$

which indicates that the density at a point is due to the contribution of the neighboring particles (those within the compact support of the kernel at $\mathbf{s}$). Another approach that is more accurate is due to Randles and Libersky:[13]

$$\rho(\mathbf{s}) = \frac{\sum_j m_i W(\mathbf{s} - x_j)}{\sum_j \left(\frac{m_j}{\rho_j}\right) W(\mathbf{s} - x_j)} . \tag{2.29}$$

SPH calculations can be tamed by using this equation periodically to smooth out density variations[10,14,15] due to acoustic waves in the fluid.

2.2.2. *Equations of motion*

The equations that govern the fluid motion are derived in particle form by the same method as used for the conservation of mass. We start with a reduced set of equations (viscosity will be added later):

$$\frac{d\mathbf{u}}{dt} = -\frac{1}{\rho}\nabla p - \mathbf{g} \tag{2.30}$$

where $\mathbf{g}$ is the acceleration of gravity and p is the fluid pressure. If we multiply through by ρ and multiply both sides by the kernel and integrate over the domain, we have

$$\left(\frac{d\mathbf{u}}{dt}\right)_i = -\frac{1}{\rho_i}\sum_j \frac{m_j}{\rho_j}\, p_j\, \nabla_i W_{ij} - \mathbf{g} . \tag{2.31}$$

To this equation, we can subtract the following zero sum:

$$\frac{p_i}{\rho_i}\left(\sum_j \frac{m_j}{\rho_j}\, \nabla_i W_{ij}\right) = 0$$

which gives us the final form

$$\left(\frac{d\mathbf{u}}{dt}\right)_i = -\sum_j m_j \left(\frac{p_i + p_j}{\rho_i \rho_j}\right) \nabla_i W_{ij} - \mathbf{g} . \tag{2.32}$$

Another form of the equation of motion very often used is

$$\left(\frac{d\mathbf{u}}{dt}\right)_j = -\sum_j m_j \left(\frac{p_i}{\rho_i^2} + \frac{p_j}{\rho_j^2}\right) \nabla_i W_{ij} - \mathbf{g} . \tag{2.33}$$

In addition, particles are moved in time with the following equation:

$$\frac{d\mathbf{r}_i}{dt} = \mathbf{u}_i + \epsilon \sum_j m_j \left(\frac{\mathbf{u}_i - \mathbf{u}_j}{\bar{\rho}_i}\right) W_{ij} . \tag{2.34}$$

The second term, including the parameter $\epsilon \approx 0.5$, is the so-called XSPH correction of Monaghan,[16] which ensures that neighboring particles move with approximately the same velocity.

2.2.3. *Pressure and sound speed*

For a compressible SPH model, the pressure for each particle is obtained from an equation of state. Monaghan[3] uses a form due to Batchelor:[17]

$$p = B\rho_s\big[\big(\frac{\rho}{\rho_s}\big)^{\gamma} - 1\big]\,. \tag{2.35}$$

The B is a constant, related to the bulk modulus of elasticity of the fluid, ρ_s is a reference density, usually taken as the density of the fluid at the free surface, so that the pressure is zero there, and γ is the polytropic constant, usually between 1 and 7. A value of 7 fits oceanic data, while Morris[18] suggests for low Reynold number flows that γ is best taken as unity. (The inclusion of the minus one term in the equation of state is to give zero pressure at a surface.)

The speed of sound is given by the derivative of this equation of state with respect to density:

$$C^2(\rho) = \frac{\partial p}{\partial \rho} = B\gamma\big(\frac{\rho}{\rho_s}\big)^{\gamma-1} = C_s^2\big(\frac{\rho}{\rho_s}\big)^{\gamma-1} \tag{2.36}$$

where B is now seen to correspond to C_s^2/γ, the speed of sound at the surface divided by the polytopic constant, γ. The choice of B is very important. If we use a value corresponding to the real value of the speed of sound in water, then for numerical modeling, we would have to chose a very small time step, based on such criteria as a Courant-Fredrich-Levy condition. Monaghan has found by experience that the sound speed could be artificially slowed significantly for fluids without affecting the fluid motion; however, he argues that the minimum sound speed should be about ten times greater than the maximum expected flow speeds. Dalrymple and Rogers[19] verify Monaghan's experience for the case of water waves: C_s should be taken as at least 10 times the speed of any water waves. Further they show that using an appropriately fast speed of sound makes the model results independent of the choice of γ. (Issa[20] shows that, in the absence of gravity, this is not necessarily true.)

Incompressible SPH models have been developed by Refs. 21,22, and 23. The first two of these papers solve an elliptic Poisson equation for pressure, which provides a considerable computational burden. The advantage is that the computational time steps are based on a Courant condition based on flow velocities rather than the speed of sound in the model (on the order of a factor of ten). So there is a trade off of more computation versus more time steps with less computation (using the equation of state). In Japan, an incompressible model derived from SPH is the Moving Particle

Semi-Implicit model of Koshizuka,[24] which has been used by Gotoh and colleagues in a large number of papers, *cf*, Refs. 25 and 26.

2.3. *Viscosity and Large Eddy Simulation*

Monaghan[5] uses an artificial viscosity term to represent viscosity and to prevent particle interpenetration. Morris[18] use a more realistic velocity representation for viscous flows. Rogers and Dalrymple[15,27] replace the Monaghan artificial viscosity term with a Large Eddy Simulation approach. For incompressible SPH, this had been done by Lo and Shao[22] and Shao and Ji;[28] however, for a compressible fluid, the procedure is slightly different and involves Favre averaging, which is a density-weighted time averaging scheme. The Favre-averaged equations are:

$$\frac{d\bar{\rho}}{dt} = -\bar{\rho}\,\nabla \cdot \tilde{\mathbf{u}} \tag{2.37}$$

$$\frac{d\tilde{\mathbf{u}}}{dt} = -\frac{1}{\bar{\rho}}\,\nabla \bar{p} + \mathbf{g} + \frac{1}{\bar{\rho}}\,(\nabla \cdot \overline{\rho \nu_o}\,\nabla)\,\tilde{\mathbf{u}} + \frac{1}{\bar{\rho}}\,\nabla \cdot \tau^* \tag{2.38}$$

where τ^* is the sub-particle scale (SPS) stress tensor with elements (in tensor notation):

$$\tau^*_{i,j} = \bar{\rho}\left(2\nu_t \tilde{S}_{ij} - \frac{2}{3}\tilde{S}_{kk}\delta_{ij}\right) - \frac{2}{3}\bar{\rho}\, C_I \Delta^2 \delta_{ij}\,. \tag{2.39}$$

C_I is taken as 0.00066 and Δ is the initial particle spacing. The Favre-filtered rate of strain tensor is

$$\tilde{S_{ij}} = \frac{1}{2}\left(\frac{\partial \tilde{u_i}}{\partial x_j} + \frac{\partial \tilde{u_j}}{\partial x_i}\right). \tag{2.40}$$

The standard Smagorinsky model[29] is used to determine the eddy viscosity, ν_t.

Additional discussions of turbulence and SPH is given in Ref. 30.

2.4. *Boundary conditions*

There have been a variety of boundary conditions proposed for use in SPH. Monaghan[3] proposed a Lennard-Jones boundary condition using fixed boundary particles that exerted a radial force on approaching particles. He revised this subsequently[31,32] to account for the fact that particles moving along the boundary were bumping along the radial boundary force as if the boundary were corrugated with the spacing of the boundary particles. The new boundary conditions permitted a particle to travel parallel with a boundary without disturbance from boundary particles. Dalrymple and Knio[33] alternatively used a double (staggered) row of fixed particles with no imposed Lennard-Jones types of forces. The advantage of this condition was that no special programming was required for the boundary particles. They just were not allowed to advance their position with time as the fluid particles did. Experience however shows that this boundary is sticky–often holding particles in place against gravity, when the fluid should drain away.

Other boundary conditions that have been used are "ghost particles," which are image particles outside the boundary that approach the boundary with the opposite normal velocities (but the same density) as the actual fluid particles approaching the wall.[10,34]

2.5. *Time stepping*

There are a variety of ways to march the solution of the SPH equations ahead in time. It is desirable to use at least a second order accurate scheme in time. Most schemes have been explicit, either multi-step or single step, such as Runge-Kutta. To illustrate some of these methods, we first digitize time for the computer: $t = 0, \Delta t, 2\Delta t, \ldots, n\Delta t, (n+1)\Delta t, \ldots$.

Monaghan and colleagues use a modified Euler predictor-corrector method for second order accuracy. The governing equation is taken in this form

$$\frac{d\mathbf{v}}{dt} = \mathbf{f} \tag{2.41}$$

where we will assume that $\mathbf{f}$ involves $\mathbf{v}$ in some way. The first step is to predict at the next time level using the Euler method:

$$\mathbf{v}_{n+1} = \mathbf{v}_n + \mathbf{f}_n \, \Delta t \, . \tag{2.42}$$

The value of $\mathbf{v}_{n+1/2}$ (at the time midpoint) is $(\mathbf{v}_{n+1} + \mathbf{v}_n)/2$, which is used to calculate $\mathbf{f}_{n+1/2}$. The correction step is then

$$\mathbf{v}_{n+1} = \mathbf{v}_n + \mathbf{f}_{n+1/2} \, \Delta t \, .$$

Monaghan[16] uses this development to show that the SPH method conserves both linear and angular momentum. In practice, they used the midpoint value of the previous time step instead of computing $\mathbf{f}_n$, which saves time and creates only a small error. The overall scheme is second order.

One of the time-marching methods we use is the Verlet[35] method, which can be derived from a Taylor series in time for the position $\mathbf{x}$:

$$\mathbf{x}(t+\Delta t)=\mathbf{x}(t)+\mathbf{x}'(t)\Delta t+\mathbf{x}''(t)\Delta t^2/2+\mathbf{x}'''(t)\Delta t^3/6+O(\Delta t^4) \quad (2.43)$$

$$\mathbf{x}(t-\Delta t)=\mathbf{x}(t)-\mathbf{x}'(t)\Delta t+\mathbf{x}''(t)\Delta t^2/2-\mathbf{x}'''(t)\Delta t^3/6+O(\Delta t^4) \quad (2.44)$$

where the primes are derivatives with respect to time. Adding these two equations we obtain a very accurate (fourth order in Δt) way to advance the position of a particle, based on knowing the position of the particle two times steps earlier and the acceleration at time level t:

$$\mathbf{x}(t+\Delta t) = 2\mathbf{x}(t) - \mathbf{x}(t-\Delta t) + \mathbf{a}(t)\,\Delta t^2 + O(\Delta t^4) \quad (2.45)$$

where $\mathbf{a}(t) = \mathbf{x}''(t)$. Subtracting the equations instead gives the simple Verlet equation for the velocity

$$\mathbf{v}(t) = \frac{\mathbf{x}(t+\Delta t) - \mathbf{x}(t-\Delta t)}{2\Delta t} + O(\Delta t^2), \quad (2.46)$$

which is two orders in Δt less accurate than the position expression. The so-called Velocity Verlet equation is found by expressing the velocity in a Taylor series:

$$\mathbf{v}(t+\Delta t)=\mathbf{v}(t)+\mathbf{v}'(t)\Delta t+\mathbf{v}''(t)\Delta t^2\ldots=\mathbf{v}(t)+\mathbf{a}(t)\Delta t+\mathbf{x}'''\Delta t^2+O(\Delta t^3). \quad (2.47)$$

We can obtain an expression for $\mathbf{x}'''(t)$ by differentiating the Taylor series for $\mathbf{x}(t+\Delta t)$ twice with respect to time, which gives us $\mathbf{a}(t+\Delta t) = \mathbf{a}(t) + \mathbf{x}'''(t)\Delta t + O(\Delta t^2)$. Solving for $\mathbf{x}'''(t)$ and substituting into our expression above, we obtain

$$\mathbf{v}(t+\Delta t) = \mathbf{v}(t) + \frac{1}{2}\Delta t\left(\mathbf{a}(t) + \mathbf{a}(t+\Delta t)\right) + O(\Delta t^3)\,. \quad (2.48)$$

These equations can then be rewritten as:

$$\mathbf{x}_{n+1} = 2\mathbf{x}_n - \mathbf{x}_{n-1} + \mathbf{a}_n\,\Delta t^2 \quad (2.49)$$

$$\mathbf{v}_{n+1} = \mathbf{v}_n + \frac{1}{2}\Delta t\left(\mathbf{a}_n + \mathbf{a}_{n+1}\right). \quad (2.50)$$

In practice, these equations are used with a half step in time so that the variables at three levels in time $(n+1), n$, and $(n-1)$ are not stored:

$$\mathbf{v}_{n+1/2} = \mathbf{v}_n + \mathbf{a}_n\,\Delta t/2 \quad (2.51)$$

$$\mathbf{x}_{n+1} = \mathbf{x}_n + \mathbf{v}_{n+1/2}\,\Delta t \quad (2.52)$$

$$\mathbf{v}_{n+1} = \mathbf{v}_{n-1/2} + \mathbf{a}_{n+1}\,\Delta t/2\,. \quad (2.53)$$

Another method, popular in molecular dynamics, is the Beeman[36] method, which is stated as

$$\mathbf{x}_{n+1} = \mathbf{x}_n + \mathbf{v}_n \Delta t + \big(\frac{2}{3}\mathbf{a}_n - \frac{1}{6}\mathbf{a}_{n-1}\big)\Delta t^2 + O(\Delta t^4) \quad (2.54)$$

$$\mathbf{v}_{n+1} = \mathbf{v}_n + \big(\frac{1}{3}\mathbf{a}_{n+1} + \frac{5}{6}\mathbf{a}_n - \frac{1}{6}\mathbf{a}_{n-1}\big)\Delta t + O(\Delta t^3) \quad (2.55)$$

which is more accurate for velocity than the Verlet method.

A predictor-corrector version of the Beeman method for velocity that is useful when the acceleration terms are dependent on the velocity and the position uses the same equation as above for position $\mathbf{x}_{n+1}$, but uses the following predictor-corrector equations for velocity.

$$\mathbf{v}_{p,n+1} = \mathbf{v}_n + \big(\frac{3}{2}\mathbf{a}_n - \frac{1}{2}\mathbf{a}_{n-1}\big)\Delta t \quad (2.56)$$

where everything on the righthand side is known. Then the corrected value for $\mathbf{v}_{n+1}$ is

$$\mathbf{v}_{c,n+1} = \mathbf{v}_n + \big(\frac{1}{3}\mathbf{a}_{n+1} + \frac{5}{6}\mathbf{a}_n - \frac{1}{6}\mathbf{a}_{n-1}\big)\Delta t\,. \quad (2.57)$$

This gives a scheme that is correct to order Δt^4 for both position and velocity.

2.6. *Accuracy*

Panizzo *et al.*[37,38] examine the accuracy of various choices of kernel and time-integration schemes based on analytical solutions for particle motion in a pressure field. For their two-dimensional case of a particle moving in a figure eight (ottovolante) motion, forced by an oscillatory pressure field: $p(x, y, t) = \rho\omega^2\big(x \sin \omega t + 2y \sin 2\omega t\big)$, where ω is the angular frequency of the motion, they computed net error in displacement between the computed results and the analytical solution. Figure 2 shows the normalized error for various kernels and time-integration schemes. What they found was that the higher the order of the kernel (quintic versus quadratic for example) and the higher the order of the time-stepping scheme, the greater the accuracy of the SPH scheme. However, once they corrected the gradient operator, using the method of Bonet and Lok,[11] then the accuracy improved markedly and was nearly independent of the kernel and integration scheme. However the computational time increases with the complexity of computing the kernel and the gradient corrections.

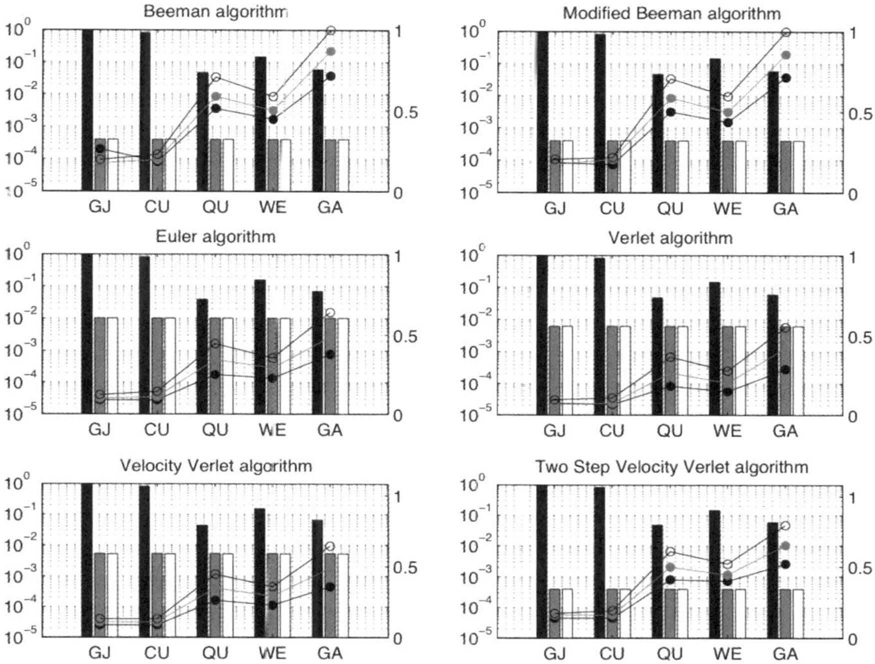

Fig. 2. Errors (bars) and computational time (lines) for the the ottovolante motion of a particle using various weighting kernels and time integration schemes. The black bar is for no kernel correction; the grey bar is for Bonet and Lok gradient correction; the white bar is both kernel and gradient corrections. The errors are normalized by the worst case and shown on the left ordinate. The right ordinate shows the normalized computational time. The polynomial kernels are quadratic, GJ;[39] cubic (CU) spline;[40] quartic, WE;[41] and quintic, QU.[42] GA is a Gaussian function, truncated at $3h$. From Panizzo *et al.*[37]

3. Examples

SPH has been applied to a number of examples of water waves, since (and including) Monaghan.[3] For example, Monaghan and Kos[31] examined tsunami waves. Due to its Lagrangian nature, the free surface of a breaking wave is part of the SPH solution. No special algorithms have to be developed, as in Eulerian models, to track the free surface. Here, several examples of the application of SPH to waves that we have done are shown, with an emphasis on the nearshore zone.

3.1. *Long waves*

The first case of free surface flow treated by SPH was that of a dam break. Monaghan[5] shows a dam break wave impinging onto a ramp and a vertical wall. Another dam break example was given recently by Crespo *et al.*,[43] who compared SPH to laboratory experiments of Janosi *et al.*[44] They released water from a vertical face dam into various depths of standing water in front of the dam (tail water). As the water leaves the dam, it collides with the standing water. Agreement is very good between the free surface elevation of experiment and the SPH model (2D). Crespo *et al.* also show that various depths of standing water in front of the dam would give rise to different types of waves–such as breaking waves or a non-breaking solitary wave (for deeper standing water). Figure 3 shows what happens when a dam (initial depth 0.15 m) breaks and releases water into standing water of depth 0.018m. Beyond the plunging jet and wave splash-up, the vorticity is also shown in the figure.

Landslides, either underwater or subaerial, can create tsunamis, *cf.* Ref. 46. Panizzo and Dalrymple[47] showed that a moving underwater slide creates a surface displacement using SPH particles that were larger than the slide thickness. (This can be done as the vertical motion of the SPH particles can be much smaller than the particle size.) Refs. 32 and 48 show the ability of SPH to model the impact of a sliding mass impacting a body of water, each comparing with experiments.

Figure 4 shows the comparison of breaking wave simulation between SPH model results and the experiment data of Synolakis[49] (see also his Ref. 50) of a solitary wave climbing a planar beach.[51] The Monaghan boundary condition was used along the beach.

Other long wave solutions have been given by Wang and Shen,[52] who used the St. Venant Equations and Ata and Soulaïmani,[53] who used the shallow water equations, and Del Guzzo *et al.*,[54] who used the nonlinear shallow water equations, instead of the Navier-Stokes equations for the equations of motion.

3.2. *Breaking waves*

Dalrymple and Rogers,[15] using SPH with large eddy simulation, discussed earlier, examined the coherent turbulence under breaking waves and showed that SPH is capable of depicting the wave roller, downbursting, splash-up, and probably obliquely descending eddies. Figure 5 shows one of their results for a breaker on a mild beach, with splash-up. The wave is created

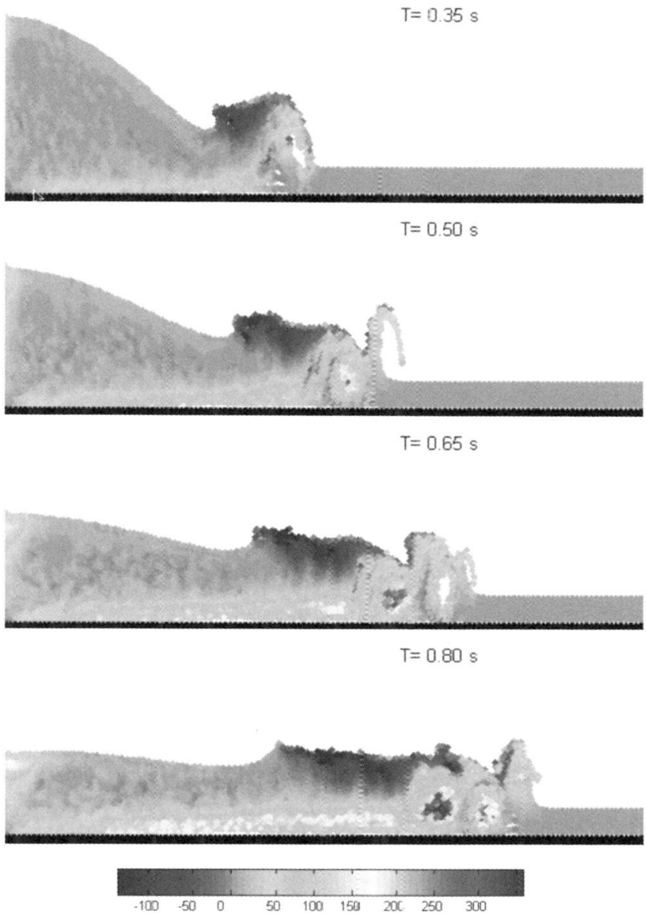

Fig. 3. The interaction of a flow from a dam break propagating into standing water in front of dam. Water depth behind dam, 0.15 m; standing water depth, 0.018 m. Shading represents vorticity, in units of s^{-1}. From Crespo *et al.*[45]

by a wavemaker offshore (not shown). The middle panel shows that the touchdown of a plunging breaker jet creates a large amount of vorticity, as expected, and the lower panel shows that the fluid in the plunger is both thrown forward in the splashup, but about half the fluid in the plunger is wrapped up in the recirculating flow at the face of the breaker, creating the roller.

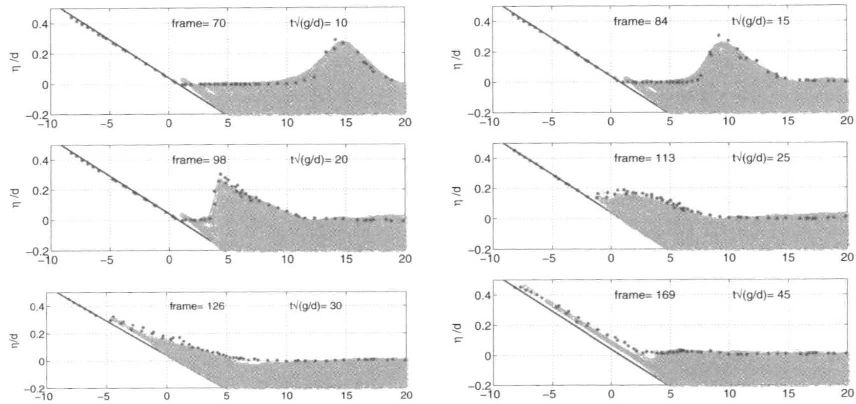

Fig. 4. Comparison of non-dimensional free surface profiles for the breaking case $H/d = 0.28$ solitary wave at non-dimensional time $t\sqrt{g/d} = 10,15, 20, 25, 30, 45$; solid fill, SPH model (Zou, 2007); dark dots, experiments (Synolakis,1987).

3.3. *Interaction with structures*

Gómez-Gesteira and Dalrymple[55] developed a 3D SPH model and examined the impact of a wave (generated by a dam break) onto a rectangular structure, representing a building. The results compare reasonably well with laboratory experiments performed by Arnason,[56] showing the interaction between the water from the dam interacting with still water that was pre-existing in front of the dam and around the building. Recently Crespo *et al.*[57] examined the behavior of the dam break wave with a small wall in front of a structure to determine the role of walls in moderating the impact forces on the structure. Figure 6 shows the wave hitting the wall from both the side and the top views. The low wall in this case serves to deflect the jet upwards and the water hits the building at a higher level than in the absence of the wall. An apparent example of this deflection during the Indian Ocean tsunami of 2004 was discussed by Dalrymple and Kriebel.[58] Shao and Gotoh[59] examine the interaction between waves and a floating curtain wall attached to the bottom using an SPH with LES. Lee *et al.*[60] examine the run-up of both Boussinesq and SPH waves on a coastal structure, while Shao *et al.*[61] use an incompressible SPH for run-up.

Other applications of SPH to wave-structure interaction includes green water overtopping of horizontal decks, such as Gómez-Gesteira *et al.*,[62] who compared their results to 2D laboratory experiments of Cox and Ortega.[63]

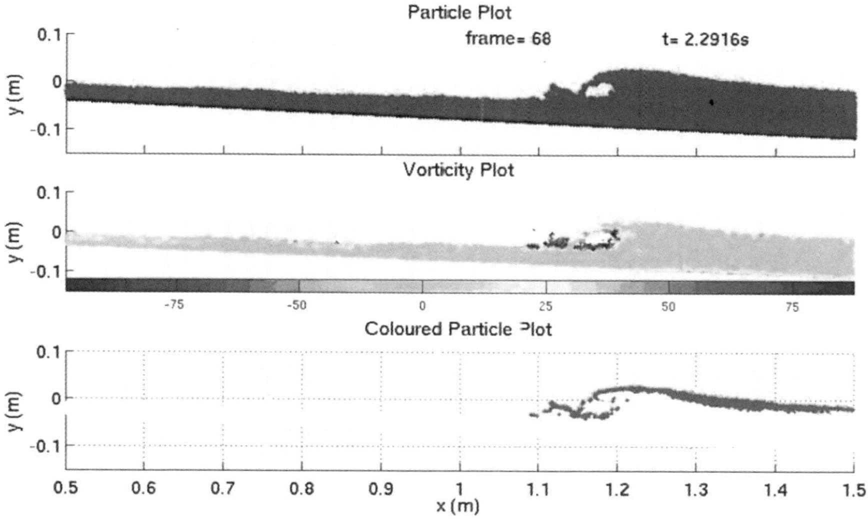

Fig. 5. Breaking wave on a beach.[15] Top image shows the plunging wave form and the beginning of splash-up. The middle image shows strong vorticity at both sides of the plunger touchdown, and the bottom image shows that the water from the plunging jet is both splashing forward in the splash-up but also caught up into the recirculating flow in the so-called roller. (Note: units of vorticity: 1/seconds).

The deck was represented by a horizontal fixed particles. The incoming waves, higher than the deck, are split, with part of the wave running across the top of the deck much like a dam break, and the other part of the wave underneath constrained by the plate–leading to uplift pressures. The two waves reunite downwave of the structure. For large waves, SPH predicted that the two parts of the wave would collide, creating a rooster tail splash.

The presence of air can be modeled with a two-phase SPH model. An early study of flows with fluids of different density (liquid/liquid) was done by Monaghan *et al.*[64] to examine stratified flows. For these cases, they treated two fluids with different densities (upwards of 30% difference) and comparisons were made successfully to lab experiments. May and Monaghan[65] and Colagrossi and Landrini[10] examined a rising bubble in a fluid. While their[10] bubble behaved similarly to a real rising bubble, the speed of sound in air in their model was much greater than that for water. They were unable to use the real equations of state due to strong instabilities that developed on the air/water interface.

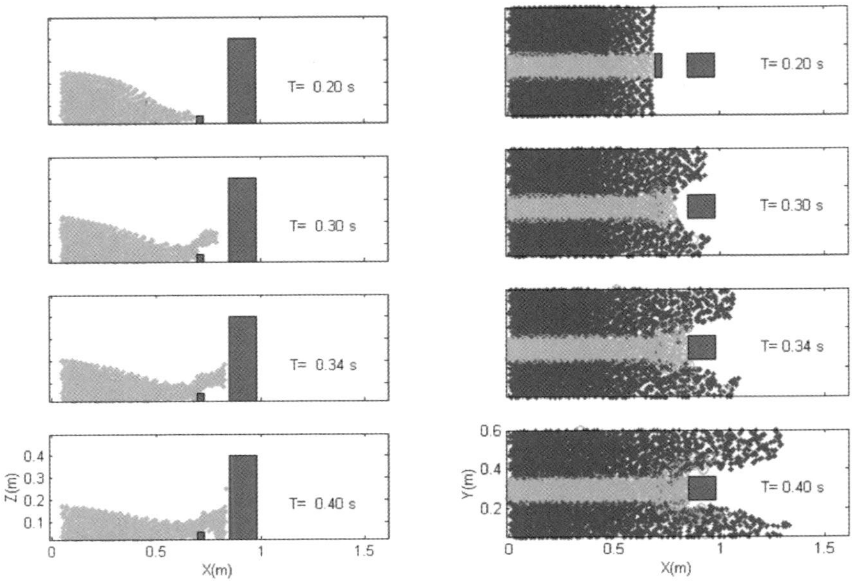

Fig. 6. Lateral and top view of dam-break wave striking a dike in front of a building. Here the low wall does deflect some of the flow around the building; however there is still wall overtopping.[57]

Cuomo *et al.*[66] also examined two-phase flows, taking the speed of sound in air and water to be the same. This is still not the correct value, but an improvement. The model has been developed further by Ref. 67 to allow any speed of sound ratio between air and water. They investigated the influence of this ratio (from 1:1 to 1:4) on dam break impact maxima, rise time, and impulse. One of the cases that they examined is the breaking of a solitary wave on a beach (see Figure 7), which compared well with the results of an SPH model with water particles only.

One problem of the current method is that the interface between the air and water, which is sharply defined, contains large (outsized) bubbles, due to an incorrect strong repellant force between the air and water. This work promises to allow the incorporation of the effect of air in the breaking process, which is likely to be important particularly when a wave hits a wall or the bottom of a bridge deck.

Recently, Hu and Adams[68] developed an incompressible multi-phase SPH model. They use a color index, which has a unit jump across an

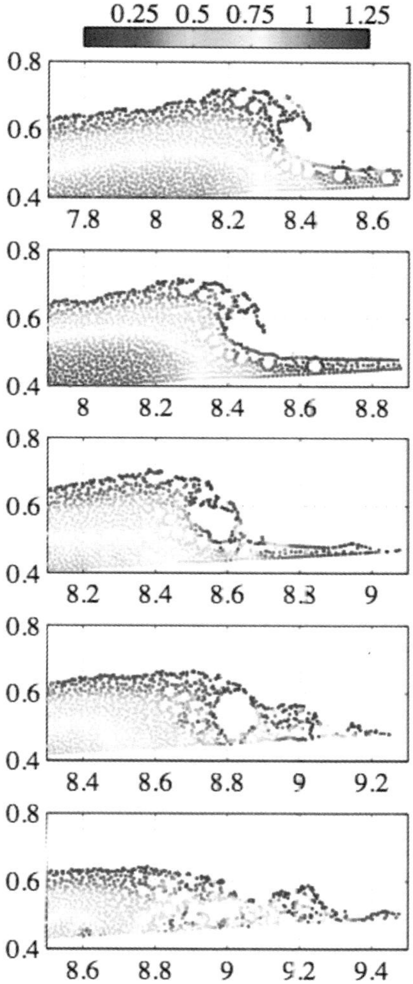

Fig. 7. Breaking solitary wave examined by two-phase SPH. Normalized pressure is shown by the color coding. From Ref. 67.

interface. The color index is used to determine a surface force between phases.

3.4. *Sediment suspension under breaking waves*

In addition to the equations that govern the fluid motion, the SPH particles can be used to solve any additional equations (in particle form, obtained as shown in Section 2.2). Zou and Dalrymple[69] and Zou[51] have examined suspended sediment under waves, using the Lagrangian form of convection-diffusion equation.

$$\frac{dC}{dt} = w_s \frac{\partial C}{\partial z} + \nabla \cdot (\epsilon_s \nabla C) \tag{3.1}$$

where C is sediment concentration, w_s is sediment settling velocity and ϵ_s is a diffusion coefficient. Here we use a constant value for $\epsilon_s = 0.000016$, which is referred to as Nielson's period averaged value.[70] The first term on the right hand side is the convection or settling term, due to the fall velocity of the sediment. The second is due to diffusive processes.

The response of the bed to settling or eroding sand is given by the morphological model:

$$(1-n)\frac{\partial z_b}{\partial t} = D - E \tag{3.2}$$

$$D = w_s C_{nb} \tag{3.3}$$

$$E = p \tag{3.4}$$

where n is the porosity, D is the downward flux due to settling, E is the upward flux by pick-up effect, C_{nb} is the near bed sediment concentration, z_b is the bed elevation, and p is the pick-up function.

Nielsen[71] suggested the modified Van Rijn's pick-up function as the vertical flux of sediment in unsteady flow. The following equations show his form of pick-up function and the net sediment vertical flux in convection-diffusion model.

$$p(t) = 0.00033(\frac{\theta(t) - \theta_c}{\theta_c})^{1.5} \frac{(s-1)^{0.6} g^{0.6} d^{0.8}}{\nu^{0.2}} \tag{3.5}$$

$$\text{where } \theta(t) = \frac{\tau_b}{\rho(s-1)gd} = \frac{u_*^2}{(s-1)gd} \tag{3.6}$$

where $p(t)$ is pick-up function, which is the instantaneous rate at which sand is picked up from the bed. It is a non-negative function with the dimensions of sediment flux in the unit of velocity, d is the grain size, $\theta(t)$ is the instantaneous Shields parameter, θ_c is the critical Shields parameter, here $\theta_c = 0.05$ is used, $s = \rho_s/\rho$ is the relative density of sediment, ν is the kinematic viscosity coefficient, and u_* is the friction velocity. Nguyen and Shibayama[72] also used this method.

The calculation of bottom shear stress is given by $\tau_b = \frac{1}{2}\rho f_w u_b |u_b|$. Following Cox *et al.*[73] and Karambas and Koutitas,[74] the friction factor

f_w is a constant 0.3 for $\frac{a_b}{k_s} \leq 1.57$, and is found from the following equation for larger values of $\frac{a_b}{k_s}$:

$$\frac{1}{4\sqrt{f_w}} + \log \frac{1}{4\sqrt{f_w}} = -0.08 + \log \frac{a_b}{k_s} \tag{3.7}$$

where u_b is the near-bed maximum horizontal velocity, a_b is the excursion amplitude, given as u_b/ω, where ω is the angular frequency of the waves, and k_s is the Nikuradse roughness, taken as $2d$.

3.4.1. *Numerical procedure*

At each time step, the hydrodynamics model computes the dynamic properties for each fluid particle, such as the position, velocity, pressure, and vorticity. The smoothed near-bed velocity can be used to estimate the Shields parameter and allows the determination of the bottom boundary condition for sediment concentration. The sediment distribution can be determined by solving the convection-diffusion equation, Eq. 3.1 above.

In our solution, an upwinding scheme is used; only the particles located above particle i are used when calculating the convection term. However, the upwind particle scheme introduces numerical diffusion, or artificial (numerical) diffusivity.[75] Calculating the convection term by this scheme should be corrected by normalization of the kernel with corrective SPH to eliminate the boundary effect due to partial integration, expressed as the following:[76]

$$(w_s \frac{\partial C}{\partial z})_i = w_s \frac{\sum_j \frac{m_j}{\rho_j}(C_j - C_i)\frac{\partial W_{ij}}{\partial z}}{\sum_j \frac{m_j}{\rho_j}(z_j - z_i)\frac{\partial W_{ij}}{\partial z}} \,. \tag{3.8}$$

The diffusion term is a second derivative, which in SPH is done with a scheme proposed by Brookshaw[77] and used for heat conduction problems by Cleary and Monaghan.[78]

$$(\epsilon_s \nabla^2 C)_i = \sum_j (\frac{m_j}{\rho_j})(\epsilon_{s,i} + \epsilon_{s,j})(C_i - C_j)\frac{\vec{r_i} - \vec{r_j}}{|\vec{r_i} - \vec{r_j}|^2 + \eta^2} \cdot \nabla_i W_{ij} \tag{3.9}$$

where j represents the neighboring particles for i, and η is the (constant) clip parameter to prevent a singularity.

3.4.2. *Model calibration and use*

In order to calibrate and verify the model, the above morphological model system was applied to reproduce the experiment by Ribberink and Al-Salem.[79] The vertical distribution of sediment concentration was compared with SPH model results, their Discrete Vortex Model results, and their experiment data in Figure 8.

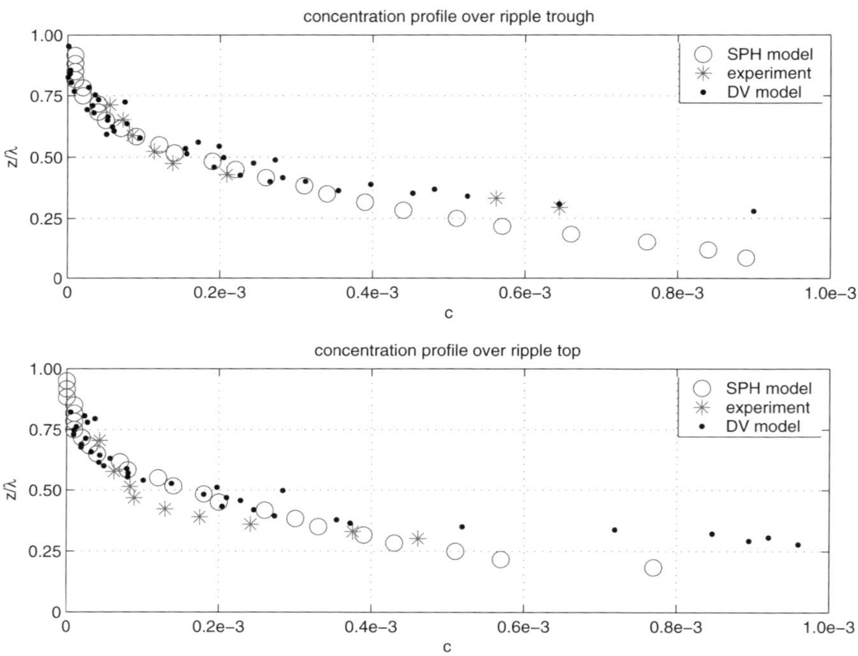

Fig. 8. Comparison of sediment concentration profile for experiment A of Ribberink and Al-Salem[79] using SPH, their Discrete Vortex Model and their experimental data.

Finally the model system was used to simulate the morphological change under the breaking waves. In Figure 9, plumes of sediment are shown rising from the bed under the breakers, created by the bottom shear stress under the breaking waves.

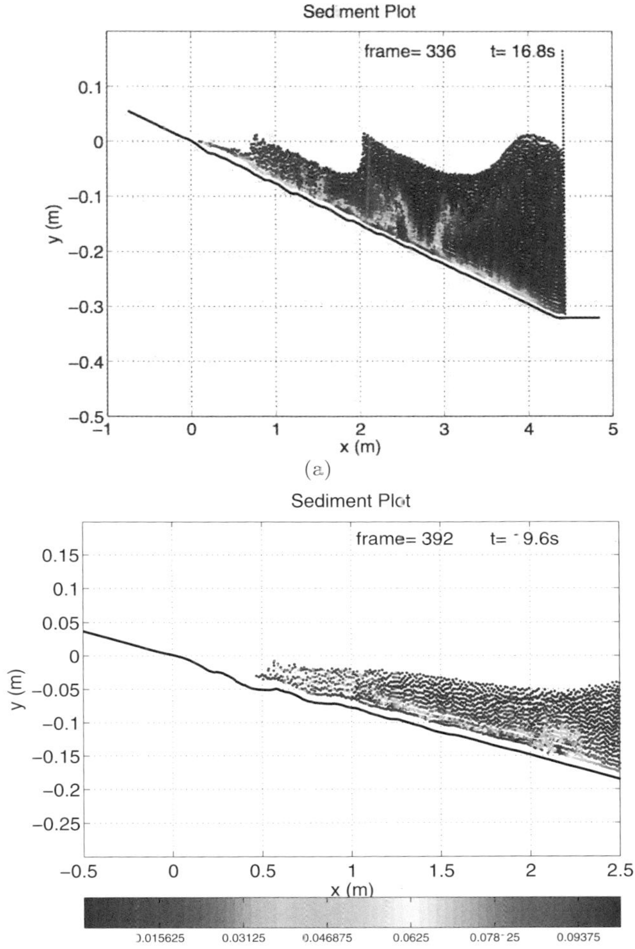

Fig. 9. (a) Suspended sediment under a breaking wave on an eroding beach in a wave tank. Wavemaker to the right. (b) Close-up of nearshore region of sediment volumetric concentration and morphological change of originally planar beach face. Here $T = 2.8s$, $H = 0.08m$, the slope is 1/13.5, with 11395 particles. Note that the vertical and horizontal scales are not the same.

4. Open Source Code

The authors of this paper have provided a generic source code in FORTRAN, with many options and modular construction, to the web. It is available for download at http://wiki.manchester.ac.uk/sphysics. The code is called *SPHysics* and comes with a users manual and test cases. It is hoped that this code will enable others to try SPH more easily, and will encourage others to contribute to the *SPHysics* project, by sending improvements in the code back to the developers.

5. Conclusions

Smoothed Particle Hydrodynamics is a powerful technique for treating deforming media in a variety of arenas, such as astrophysics, solid mechanics (impact and fracture), and fluid mechanics. The method excels when there is highly nonlinear deformation of the medium. For fluids problems such as water waves, wave breaking presents a very difficult test of a method. However, SPH easily handles breaking and other breaking related phenomena, such as roller formation and reverse breaking.

The examples presented here show that the SPH method is robust enough to do a variety of wave problems. From our experiences and those of others, the use of higher order kernels and the more elaborate time-stepping schemes gives more accuracy in the results. However, for many problems, simple kernels and time-stepping schemes are sufficient–particularly when the number of particles is small (thousands).

As shown with the suspended sediment problem, other equations can be solved in addition to the hydrodynamics equations. Tracking chemical species, temperature, or other variable is very easy to do, given the governing equation.

However, there are still important problems to be solved with the method. There are matters of code efficiency as high resolution runs in SPH require very small time steps $(O(10^{-5}\ s))$. Parallel computing is required to do meaningful problems.

There is much more to learn about the details of the wave breaking processes, such as coherent turbulent structures. A better turbulence model needs to be used. More work needs to be done with two-phase modeling to capture, for example, the energy transfers between the air and water.

Acknowledgements

Support for part of RAD's effort has been provided by the Office of Naval Research, Geosciences Program. M.G.G. and A.J.C.C. were partially supported by the Xunta de Galicia under project PGIDIT06PXIB383285PR.

A.P. was partially supported by PRIN2005 2005084953 "Tecnologie moderne per la riduzione dei costi nelle opere di difesa portuali." The authors would also like to thank Tatiana Capone for her contributions to this work.

References

1. L. Lucy, A numerical approach to testing of the fusion process, *Astronomical Journal.* **88**, 1013–1024, (1977).
2. R. Gingold and J. Monaghan, Smoothed Particle Hydrodynamics: theory and application to non-spherical stars, *Monthly Notices of the Royal Astronomical Society.* **181**, 375–389, (1977).
3. J. Monaghan, Simulating free surface flows with SPH, *Journal of Computational Physics.* **110**, 399–406, (1994).
4. J. Monaghan, Smoothed particle hydrodynamics, *Reports on Progress in Physics.* **68**, 1703–1759, (2005).
5. J. Monaghan, Smoothed particle hydrodynamics, *Annual Review of Astronomy and Astrophysics.* **30**, 543–574, (1992).
6. G. Liu and M. Liu, *Smoothed Particle Hydrodynamics: a meshfree particle method.* (World Scientific, 2003).
7. R. Dalrymple. Using Smoothed Particle Hydrodynamics for waves. In eds. Q. Zuo, X. Dou, and J. Ge, *Asian and Pacific Coasts 2007.* China Ocean Press, (2007).
8. T. Belytschko, Y. Liu, and L. Gu, Element-free Galerkin methods, *International Journal for Numerical Methods in Engineering.* **37**, 229–256, (1994).
9. G. Dilts, Moving-least-square-particle hydrodynamics-I. consistency and stability, *International Journal for Numerical Methods in Engineering.* **44**, 1115–1155, (1999).
10. A. Colagrossi and M. Landrini, Numerical simulation of interfacial flows by smoothed particle hydrodynamics, *Journal of Computational Physics.* **191** (2), 448–475, (2003).
11. J. Bonet and T.-S. Lok, Variational and momentum preservation aspects of Smooth Particle Hydrodynamics, *Computer Methods in Applied Mechanics and Engineering.* **180**, 97–115, (1999).
12. J. Chen and J. Beraun, A generalized smoothed particle hydrodynamics method for nonlinear dynamic problems, *Computer Methods in Applied Mechanics and Engineering.* **190**, 225–239, (2000).
13. P. Randles and L. Libersky, Smoothed Particle Hydrodynamics: some recent improvements and applications, *Computer Methods in Applied Mechanics and Engineering.* **138**, 375–408, (1996).
14. A. Panizzo. *Physical and numerical modelling of subaerial landslide generated waves.* PhD thesis, Universita degli Studi di L'Aquila, (2004).
15. R. Dalrymple and B. Rogers, Numerical modeling of water waves with the SPH method, *Coastal Engineering.* **53**(2/3), 141–147, (2006).
16. J. Monaghan, On the problem of penetration in particle methods, *Journal of Computational Physics.* **82**, 1–15, (1989).

17. G. Batchelor, *An Introduction to Fluid Mechanics.* (Cambridge University Press, 1973).
18. J. Morris, P. Fox, and Y. Zhu, Modeling low Reynolds number incompressible flows using SPH, *Journal of Computational Physics.* **136**, 214–226, (1997).
19. R. Dalrymple and B. Rogers. A note on wave celerities on a compressible fluid. In *Proc. 30th International Conference on Coastal Engineering.* World Scientific Press, (2006).
20. R. Issa, E. Lee, D. Violeau, and D. Laurence, Incompressible separated flows simulations with the Smoothed Particle Hydrodynamics gridless method, *International Journal for Numerical Methods in Fluids.* **47**(10-11), 1101–1106, (2004).
21. S. Cummins and M. Rudman, An SPH projection method, *Journal of Computational Physics.* **152**(2), 584–607, (1999).
22. E. Lo and S. Shao, Simulation of near-shore solitary wave mechanics by an incompressible SPH method, *Applied Ocean Research.* **24**, 275–286, (2002).
23. M. Ellero, M. Serrano, and P. Espanol, Incompressible smoothed particle hydrodynamics, *Journal of Computational Physics.* **226**, 1731–1752, (2007).
24. S. Koshizuka, H. Tamako, and Y. Oka, A particle method for incompressible viscous flow with fluid fragmentation, *Computational Fluid Dynamics Journal.* **4**(1), 29–46, (1995).
25. H. Gotoh and T. Sakai, Lagrangian simulation of breaking waves using particle method, *Coastal Engineering Journal.* **41**(3/4), 303–326, (1999).
26. H. Gotoh, T. Shibihara, and M. Hayashi, Sub-particle-scale model for the mps method-lagrangian flow model for hydraulic engineering, *Computational Fluid Dynamics Journal.* **9**(4), 339–247, (2001).
27. B. Rogers and R. Dalrymple. SPH modeling of breaking waves. In *Proc. 29th International Conference on Coastal Engineering*, pp. 415–427, (2004).
28. S. Shao and C.-M. Ji, SPH computation of plunging waves using a 2-d subparticle scale (SPS) turbulence model, *International Journal for Numerical Methods in Fluids.* **51**(913-936), (2006).
29. J. Smagorinsky, General circulation experiments with the primitive equations: I. the basic experiment, *Monthly Weather Review.* **91**, 99–164, (1963).
30. D. Violeau and R. Issa, Numerical modeling of complex turbulent free surface flows with the sph lagrangian method: an overview, *International Journal for Numerical Methods in Fluids.* **53**(2), 277–304, (2007).
31. J. Monaghan and A. Kos, Solitary waves on a Cretan beach, *Journal of Waterway, Port, Coastal, and Ocean Engineering.* **125**(3), 145–154, (1999).
32. J. Monaghan, A. Kos, and N. Issa, Fluid motion generated by impact, *Journal of Waterway, Port, Coastal, and Ocean Engineering.* **129**(6), 250–259, (2003).
33. R. Dalrymple and O. Knio. SPH modelling of waves. In *Proc. Coastal Dynamics, Lund, Sweden*, pp. 779–787. A.S.C.E., (2001).
34. L. Libersky and A. Petschek. Smoothed Particle Hydrodynamics with strength of materials. In *Lecture Notes in Physics*, vol. 365, pp. 248–257. Springer Verlag, (1991).

35. L. Verlet, Computer 'experiments' on classical fluids, I. thermodynamical properties of Lennard-Jones molecules, *Physical Review D.* **159**, 98–103, (1967).
36. D. Beeman, Some multistep methods for use in molecular dynamics calculations, *Journal of Computational Physics.* **20**, 130–139, (1976).
37. A. Panizzo, T. Capone, and R. Dalrymple. Accuracy of kernel derivatives and numerical schemes in SPH. Submitted to J.C.P., (2007).
38. T. Capone, A. Panizzo, C. Cecioni, and R. Dalrymple. Accuracy and stability of numerical schemes in sph. In ed. ERCOFTAC, *Second SPHERIC Workshop, Madrid*, (2007).
39. G. Johnson, R. Stryk, and S. Beissel, SPH for high velocity impact calculations, *Computer Methods in Applied Mechanics and Engineering.* **139**, 347–373, (1996).
40. J. Monaghan and J. Lattanzio, A refined method for astrophysical problems, *Astronomy and Astrophysics.* **149**(1), 135–143, (1985).
41. H. Wendland. Computational aspects of radial basis function approximation. In eds. K. Jetter et al., *Topics in Multivariate Approximation and Interpolation.* Elsevier, (2005).
42. J. Morris. *Analysis of Smoothed Particle Hydrodynamics with applications.* PhD thesis, Monash University, (1986).
43. A. Crespo, M. Gómez-Gesteira, and R. Dalrymple, Modeling dam break behavior over a wet bed by an SPH technique, *Journal of Waterway, Port, Coastal, and Ocean Engineering.* **revised**, (2007).
44. I. Janosi, D. Jan, K. Szabo, and T. Tel, Turbulent drag reduction in dam-break flows, *Experiments in Fluids.* **37**, 219–229, (2004).
45. A. Crespo, M. Gómez-Gesteira, and R. Dalrymple. Vorticity generated by a dam break over a wet bed modeled by Smoothed Particle Hydrodynamics. In *32nd Congress, International Association for Hydraulic Research and Engineering*, (2007).
46. P.-F. Liu, T.-R. Wu, F. Raichlen, C. Synolakis, and J. Borrero, Runup and rundown generated by three-dimensional sliding masses, *Journal of Fluid Mechanics.* **536**(1), 107–144, (2005).
47. A. Panizzo and R. Dalrymple. SPH modelling of underwater landslide generated waves. In *Proc. 29th International Conference on Coastal Engineering*, pp. 1147–1159, (2004).
48. A. Panizzo, G. Cuomo, and R. Dalrymple. 3D-SPH simulation of landslide generated waves. In *Proc. 30th International Conference on Coastal Engineering.* World Scientific Press, (2006).
49. C. Synolakis. *The run-up of long waves.* PhD thesis, California Institute of Technology, (1986).
50. C. Synolakis, The run-up of solitary waves, *Journal of Fluid Mechanics.* **185**, 523–545, (1987).
51. S. Zou. *Coastal Sediment Transport Simulation by Smoothed Particle Hydrodynamics.* PhD, Civil Engineering, Johns Hopkins University, (2007).
52. Z. Wang and H. Shen, Lagrangian simulation of one-dimensional dam-break flow, *Journal of Hydraulic Engineering.* **125**(11), 1217–1220, (1999).

53. R. Ata and A. Soulaïmani, A stabilized SPH method for inviscid shallow water flows, *International Journal for Numerical Methods in Fluids.* **47**, 139–159, (2005).
54. A. Del Guzzo, A. Panizzo, G. Bellotti, D. Longo, and P. D. Girolamo. Simulation of fluid-structure interaction using Smoothed Particle Hydrodynamics and non-linear shallow water equations. In *32nd Congress, International Association for Hydraulic Research and Engineering*, (2007).
55. R. Gómez-Gesteira and R. Dalrymple, Using a 3d SPH method for wave impact on a tall structure, *Journal of Waterway, Port, Coastal, and Ocean Engineering.* **130**(2), 63–69, (2004).
56. H. Arnason. *Interaction between an incident bore and a free-standing coastal structure*. PhD thesis, University of Washington, (2004).
57. A. Crespo, M. Gómez-Gesteira, and R. Dalrymple, 3D SPH simulation of large waves mitigation with a dike, *Journal of Hydraulic Research.* **45**(5), (2007).
58. R. Dalrymple and D. Kriebel, Lessons in engineering from the tsunami in thailand, *The Bridge.* **35**(2), (2005).
59. S. Shao and H. Gotoh, Simulating coupled motion of progressive wave and floating curtain wall by SPH-LES model, *Coastal Engineering Journal.* **46** (2), 171–202, (2004).
60. E.-S. Lee, D. Violeau, M. Benoit, R. Issa, D. Laurence, and P. Stansby. Prediction of wave overtopping on coastal structures by using extended Boussinesq and SPH models. In *30th International Conference on Coastal Engineering*, pp. 4727–4740, (2006).
61. S. Shao, C.-M. Ji, D. Graham, D. Reeve, P. James, and A. Chadwick, Simulation of wave overtopping by an incompressible SPH model, *Coastal Engineering.* **53**(723-735), (2006).
62. R. Gómez-Gesteira, D. Cerquiero, A. Crespo, and R. Dalrymple, Green water overtopping analyzed with an SPH model, *Ocean Engineering.* **32**(2), 223–238, (2005).
63. D. Cox and J. Ortega, Laboratory observations of green water overtopping a fixed deck, *Ocean Engineering.* **29**, 1827–1840, (2002).
64. J. Monaghan, R. Cas, A. Kos, and M. Hallsworth, Gravity currents descending a ramp in a stratified tank, *Journal of Fluid Mechanics.* **379**, 39–69, (1999).
65. D. May and J. J. Monaghan, Can a single bubble sink a ship?, *American Journal of Physics.* **71**(9), 843–849, (2003).
66. G. Cuomo, A. Panizzo, and R. Dalrymple. SPH-LES two phase simulation of wave breaking and wave-structure interaction. In *Proc. 30th International Conference on Coastal Engineering*, pp. 274–286. A.S.C.E., (2006).
67. G. Cuomo and R. Dalrymple. A compressible two-phase SPH-LES model for interfacial flows with fragmentation. Submitted manuscript, (2007).
68. X. Hu and N. Adams, An incompressible multi-phase SPH method, *Journal of Computational Physics.* **227**, 264–278, (2007).
69. S. Zou and R. Dalrymple. Sediment suspension simulation under oscillatory flow with SPH-SPS method. In *30th International Conference on Coastal Engineering.* World Scientific Press, (2006).

70. F. Marin, Eddy viscosity and Eulerian drift over ripples beds in waves, *Coastal Engineering.* pp. 139–159, (2004).
71. P. Nielsen, *Coastal Bottom Boundary Layers and Sediment Transport.* (World Scientific Press, 1992).
72. T. Nguyen and T. Shibayama, A convection-diffusion model for suspended sediment in the surf zone, *Journal of Geophysical Research.* **102**(C10), 23169–13186 (October, 1997).
73. D. Cox, N. Kobayashi, and A. Okayasu, Bottom shear stress in the surf zone, *Journal of Geophysical Research.* **101**(C6), 14337–14348, (1996).
74. T. Karambas and C. Koutitas, Surf and swash zone morphology evolution induced by nonlinear waves, *Journal of Waterway, Port, Coastal, and Ocean Engineering.* pp. 102–113, (2002).
75. C. Fletcher, *Computational techniques for fluid dynamics.* vol. I, (Springer Verlag, 1991).
76. J. Chen, J. Beraun, and C. Jih, Completeness of corrective smoothed particle hydrodynamics for linear elastodynamics, *Compuational Mechanics.* **24**, 273–285, (1999).
77. L. Brookshaw, Solving the heat diffusion equation in SPH, *Memorie della Societa Astronomia Italiana.* **65**(4), 1033–1042, (1994).
78. P. Cleary and J. J. Monaghan, Conduction modeling using Smoothed Particle Hydrodynamics, *Journal of Computational Physics.* **148**(1), 227–264, (1999).
79. J. Ribberink and A. Al-Salem, Sheet flow and suspension of sand in oscillatory boundary layers, *Coastal Engineering.* **25**, 205–225, (1995).

CHAPTER 14

MODELLING NONLINEAR WATER WAVES WITH RANS AND LES SPH MODELS

R. Issa*, D. Violeau, E.-S. Lee and H. Flament

EDF R&D, National Hydraulics and Environment Laboratory
Quai Watier, 6
Chatou, 78 400, France
**reza.issa@edf.fr*

Coastal and environmental phenomena involve elaborate physics such as breaking waves. The gridless method SPH could be an appropriate tool to numerically investigate these problems. Indeed, due to its Lagrangian feature, tracking a free surface is quite practical. This method is here applied to solitary wave generation, propagation and breaking. Results are satisfactory and are enhanced by modelling turbulent effects. Set up over a coral reef, as well as wave shoaling and overtopping, are also successfully simulated. Since coastal and environmental phenomena are mainly 3D, SPH LES has been considered and applied to the classical dam breaking problem. Free surface evolution simulated with SPH is very close to experimental results. Despite these promising results, the presented results suffer from unstable pressure fields and large computing time. Hence, trully incompressible SPH and massively parallel code (several thousand processors) will be considered in the future.

1. SPH and Coastal Engineering

The ocean waves generated by seismic events usually have long wavelengths and small wave heights. As they propagate shoreward, due to the offshore bathymetry, the wave heights can increase significantly, leading to breaking waves near the shoreline.[1] These breaking waves can run up at the shoreline, inundating coastal regions and causing large property damage and loss of life. In order to proceed to the design of sea defence structures and estimate the possible damage resulting from sea submersion due to a tsunami

*Corresponding author.

for instance, it is thus crucial to understand these phenomena. Due to the complexity of the problem and its theoretical intractability, most of investigations for breaking waves are experimental and numerical. Some methods were recently developed to perform such simulations, among them Volume Of Fluids (VOF) and Lagrangian methods. After its development for astrophysics applications, SPH has successfully been applied in fluid mechanics to simulate delicate free surface flows (dam breaking, wave flumes, etc.) which would require complex meshes with classical Eulerian codes. SPH is also very efficient for rapid, convection dominated flows. For the purpose of the presented simulations, since highly distorted flow are presented, turbulence modelling plays a crucial role. However, turbulence developments in SPH are rather recent and still not generalized, although some stochastic approaches have been tested.[2,3] Unfortunately, they are quite complex and time consuming. Some recent improvements have been achieved during the past few years: the authors developed a SPH mixing length model,[4] as well as k and k-ϵ models.[5] In the meanwhile, Lo and Shao developed 2D SPH Large Eddy Simulation,[6] also applied in Ref. 7. The authors[4] also investigated 3D Large Eddy Simulation with success. One should emphasize the fact that many authors still neglect turbulence modelling in SPH and consider a constant eddy viscosity. Contrary to this approach, the authors[5] have experienced that turbulence closures based on the concept of eddy viscosity (namely mixing length, k and $k-\epsilon$ models) provide both physical accuracy and numerical smoothness. We apply in this chapter several of these models to the simulation of nonlinear solitary water waves and reveal that turbulent effects must be taken into account for modelling such phenomena.

2. Turbulence Modelling

2.1. *Statistical modelling*

A way to investigate turbulent flows is to time-average unsteady fluctuations, hence achieving turbulence statistical modelling. The simplest statistical point of view consists of considering only averaged values, which are defined by ensemble, time or spatial averaging.[8] By applying the Reynolds average operator $\langle\ \rangle$ to the Navier-Stokes equations, one ends up with the Reynolds equations (1894).[9] The turbulence statistical models based on the Reynolds equations are thus called RANS models (Reynolds Averaged Navier-Stokes equations).

2.1.1. *Reynolds equations*

In an Eulerian formalism, the Reynolds continuity equation is written

$$\frac{\partial \rho}{\partial t} + \frac{\partial \rho \langle u_i \rangle}{\partial x_i} = 0\,. \tag{2.1}$$

When considering nearly incompressible flows (characterised by density fluctuations less than 1 %), terms due to fluid compressibility can be neglected. We hence consider that $\langle \rho \rangle = \rho$. For the same reason, compressibility terms can be neglected. Consequently, the Reynolds momentum equation reads

$$\frac{d\langle u_i \rangle}{dt} = \frac{\partial \langle u_i \rangle}{\partial t} + \langle u_j \rangle \frac{\partial \langle u_i \rangle}{\partial x_j} = -\frac{1}{\rho}\frac{\partial \langle p \rangle}{\partial x_i} + \nu \frac{\partial^2 \langle u_i \rangle}{\partial x_j \partial x_j} - \frac{\partial R_{ij}}{\partial x_j} + F_i^e \tag{2.2}$$

with a summation over j subscript (Einstein's conventions), where i varies from 1 to 3 in 3D and from 1 to 2 in 2D. $\langle u_i \rangle$ corresponds to the i-component of the averaged velocity and R_{ij} are the components of the Reynolds stress tensor defined by $R_{ij} = \langle u_i' u_j' \rangle$, $u_i' \equiv u_i - \langle u_i \rangle$, being the velocity fluctuating component satisfying $\langle u_i' \rangle = 0$. p, ρ and ν respectively denote pressure, density and kinematic viscosity while F_i^e corresponds to the i-component of an external volumetric force, *e.g.* the gravity. The Reynolds stress tensor incorporates the influence of the removed turbulence fluctuations on the mean flow and describes the influence of all scales of turbulent motion, including the anisotropic large scales.[10,11]

2.1.2. *The eddy viscosity assumption*

In order to close the equation set relative to (2.1) and (2.2), several techniques allow the Reynolds stress values to be expressed in terms of the resolved quantities: the more commonly used family of models is based on the eddy viscosity assumption, introduced by Boussinesq in 1880.[12] The eddy viscosity assumption consists in writing Reynolds stresses as

$$R_{ij} = \frac{2}{3} k \delta_{ij} - 2\nu_T \langle s_{ij} \rangle \tag{2.3}$$

where $k = \left\langle u_i' u_i' \right\rangle /2$, denotes the turbulent kinetic energy (per mass unit) and δ_{ij} is the Kronecker's symbol. ν_T is the eddy-viscosity, generally much higher than the fluid molecular viscosity. An important feature is that, conversely to the molecular viscosity, it is not a property of a given fluid but depends on the local turbulent characteristics. $\langle s_{ij}\rangle$ are the components of the rate of strain tensor based on averaged velocities

$$\langle s_{ij}\rangle = \frac{1}{2}\left(\frac{\partial \langle u_i\rangle}{\partial x_j} + \frac{\partial \langle u_j\rangle}{\partial x_i}\right). \tag{2.4}$$

With this model, the averaged momentum equation (2.2) is commonly written as

$$\frac{d\langle u_i\rangle}{dt} = -\frac{1}{\rho}\frac{\partial \langle p\rangle}{\partial x_i} + \frac{\partial}{\partial x_j}\left(2\left(\nu + \nu_T\right)\langle s_{ij}\rangle\right) + F_i^e \tag{2.5}$$

in which the turbulent kinetic energy k has been incorporated into the pressure term for convenience.

2.1.3. *The mixing length model*

The simplest (and historically the first) model to estimate the eddy viscosity ν_T introduces the "mixing length" assumption, such that the eddy viscosity can be written

$$\nu_T = L_m u_t \tag{2.6}$$

where u_t denotes the typical velocity of large eddies and L_m a mixing length representing the characteristic distance of large eddy diffusive action.[13] Considering energy balance,[13] (2.6) comes

$$\nu_T = L_m^2 S \tag{2.7}$$

where $S = \sqrt{2\langle s_{ij}\rangle\langle s_{ij}\rangle}$, is the scalar mean rate-of-strain. In its generalised form (2.7), the mixing length model is applicable to all turbulent flows,[12] provided the mixing length L_m is known, which is the major drawback of this model. For a complex flow such as a breaking wave, the specification of L_m requires a large measure of guesswork and consequently, one should have little confidence in the accuracy of the results.

2.1.4. *The k model*

The k model is also based on the mixing length assumption (2.6). However, the velocity u_t is here estimated through the turbulent kinetic energy k and gives

$$\nu_T = C_\mu \frac{k^2}{\epsilon} \tag{2.8}$$

where $C_\mu = 0.09$,[14] and ϵ is the energy dissipation rate (see below). In order to estimate the values of k, a transport equation is solved. The exact one for the turbulent kinetic energy is too complex to be directly solved and after modelling the diffusion term of this equation, it is commonly written as

$$\frac{\partial k}{\partial t} + \underbrace{\langle \mathbf{u} \rangle . \nabla k}_{advection} = \underbrace{\nabla . \left[\left(\nu + \frac{C_\mu}{\sigma_k} \frac{k^2}{\epsilon} \right) \nabla k \right]}_{diffusion} + \underbrace{P}_{production} - \underbrace{\epsilon}_{dissipation} . \tag{2.9}$$

The parameter σ_k corresponds to a Schmidt number and is equal here to 1.0. The production term P is usually modelled by

$$P = C_\mu \frac{k^2}{\epsilon} S^2 . \tag{2.10}$$

However, this relation overestimates turbulent kinetic energy k for highly distorted flows where production should become linear with respect to the rate-of-strain. To circumvent this weakness, one can use the following relation[15]

$$P = min \left(\sqrt{C_\mu}, C_\mu \frac{k}{\epsilon} S \right) kS . \tag{2.11}$$

The ϵ value is then obtained through the following dimensional relation

$$\epsilon = C_\mu^{3/4} \frac{k^{3/2}}{L_m} . \tag{2.12}$$

This model is well suited for unsteady flows, for which the energy production is characterised by a time scale $\tau \approx k/\epsilon$, comparable to the timescale of the mean flow, like in wave propagation. Once again, the drawback of this model is the mixing lenth definition requirement.

2.1.5. *Two-equation models*

The shortcoming of the k model can be sorted out by solving a behaviour equation for the dissipation. Although the true equation governing ϵ is rather complicated, one usually assumes that it can be approximated by an advection-diffusion equation similar to (2.9):

$$\frac{\partial \epsilon}{\partial t} + \langle \mathbf{u} \rangle . \nabla \epsilon = \nabla . \left[\left(\nu + \frac{C_\mu}{\sigma_\epsilon} \frac{k^2}{\epsilon} \right) \nabla \epsilon \right] + \frac{\epsilon}{k} \left(C_{\epsilon 1} P - C_{\epsilon 2} \epsilon \right) . \tag{2.13}$$

The so-called $k - \epsilon$ model is one of the most popular in industrial and environmental CFD. Launder and Spalding[14] propose the model constants $\sigma_\epsilon = 1,3$, $C_{\epsilon 1} = 1,44$, and $C_{\epsilon 2} = 1,92$.

Alternatively, the $k-\omega$ model consists of considering as a closure variable the quantity

$$\omega = \frac{\varepsilon}{C_\mu k} \tag{2.14}$$

representing a turbulent characteristic frequency. The eddy viscosity is then deduced from (2.8), which yields now

$$\nu_T = \frac{k}{\omega}\,. \tag{2.15}$$

Wilcox[16] proposed the following governing equation:

$$\frac{\partial \omega}{\partial t} + \langle \mathbf{u} \rangle.\nabla\omega = \nabla.\left[\left(\nu + \frac{1}{\sigma_\omega}\frac{k}{\omega}\right)\nabla\omega\right] + C_{\omega 1}\frac{\omega}{k}P - C_{\omega 2}\omega^2 \tag{2.16}$$

with $\sigma_k = \sigma_\omega = 2$, $C_{\omega 1} = 5/9$, and $C_{\omega 2} = 3/40$. Despite their ability to model correctly a large panel of turbulent flows, the above mentioned two-equation model suffers from several drawbacks. In particular they assume the Boussinesq model (2.3) which provides Reynolds stresses as linear functions of the rate-of-strain tensor, while a more complete model should define the R_{ij}'s as non-linear functions of $\langle s_{ij} \rangle$ and the mean vorticity tensor components, defined by

$$\langle \omega_{ij} \rangle = \frac{1}{2}\left(\frac{\partial \langle u_i \rangle}{\partial x_j} - \frac{\partial \langle u_j \rangle}{\partial x_i}\right). \tag{2.17}$$

Pope,[12] followed by Wallin and Johansson[17] proposed the following model for statistical two-dimensional flows:

$$R_{ij} = \frac{2}{3}k\delta_{ij} - 2C_\mu \frac{k^2}{\epsilon}\left(\langle s_{ij} \rangle + 2C_2\frac{k}{\varepsilon}\langle s_{ik} \rangle\langle \omega_{kj} \rangle\right). \tag{2.18}$$

The coefficients C_μ and C_2 are defined by

$$C_\mu = \frac{B_3}{2}\frac{P^* + B_0}{(P^* + B_0)^2 + (B_2\Omega^*)^2} \tag{2.19}$$

$$C_2 = \frac{B_2}{P^* + B_0} \tag{2.20}$$

where $P^* = P/\varepsilon$, is the production-to-dissipation ratio and $\Omega^* = k\Omega/\varepsilon$, where $\Omega = \sqrt{2\langle \omega_{ij} \rangle\langle \omega_{ij} \rangle}$, is the scalar mean rotation rate. The model constants are $B_0 = 0.8$, $B_2 = 4/9$, and $B_3 = 8/15$. The production is no longer defined by (2.10) or (2.11) but is the solution to a third-order polynomial equation:

$$P^{*3} + 2B_0 P^{*2} - \left(\frac{B_3}{2} S^{*2} - B_2^2 \Omega^{*2} - B_0^2\right) P^* - \frac{B_0 B_3}{2} S^{*2} = 0 \qquad (2.21)$$

with $S^* = kS/\varepsilon$. Wallin and Johansson[17] give an analytical solution to the above equation, based on Cardan's general solution. This model is often referred to as an Explicit Algebraic Reynolds Stress Model, but will be called in the following the 'non-linear $k - \epsilon$ model'. Boundary conditions of the one- and two-equation turbulent models are summarized in section (3.6.1).

2.2. *Large Eddy Simulation*

Large Eddy Simulation (LES) for turbulent flow modelling is motivated by the limitation of Direct Numerical Simulation (DNS) and RANS models;[18] the idea of LES is to simulate the large scales of motion of turbulence while approximating the scales smaller than the grid size. The justification of such a treatment is that large eddies contain most of the energy, do most of the transporting of conserved properties and vary most from flow to flow. On the other hand, small eddies are believed to be more universal and less important, except for their dissipative effects easily modelled by an enhanced viscosity. Moreover, the large structures are mostly anisotropic whereas the small eddies are closer to be isotropic.[19] When achieving LES, it is essential to define modified fields that only contain the large scale components of total fields. Each flow variable Φ is thus decomposed into a large scale (or grid scale) component $\overline{\Phi}$ and a small scale (or subgrid scale) component Φ' such as $\Phi = \overline{\Phi} + \Phi'$. The grid scale component is defined by the following filtering operator

$$\overline{\Phi}(\mathbf{x}, t) = \int_V G_\Delta (\mathbf{x} - \mathbf{x}')\, \Phi(\mathbf{x}', t)\, d\mathbf{x}' \qquad (2.22)$$

where $G_\Delta(\mathbf{x} - \mathbf{x}')$ is a spatial filter function depending on the separation between the position vectors $\mathbf{x}$ and $\mathbf{x}'$. Δ corresponds to the filter size and V denotes the volume of the computation domain. The large scale field is hence a local average of the complete field. Filter functions which have been applied in LES include for instance Gaussian function (see the left picture of figure 1), box function (see the right picture of figure 1). When applying the filtering operator to the Navier-Stokes momentum equation,

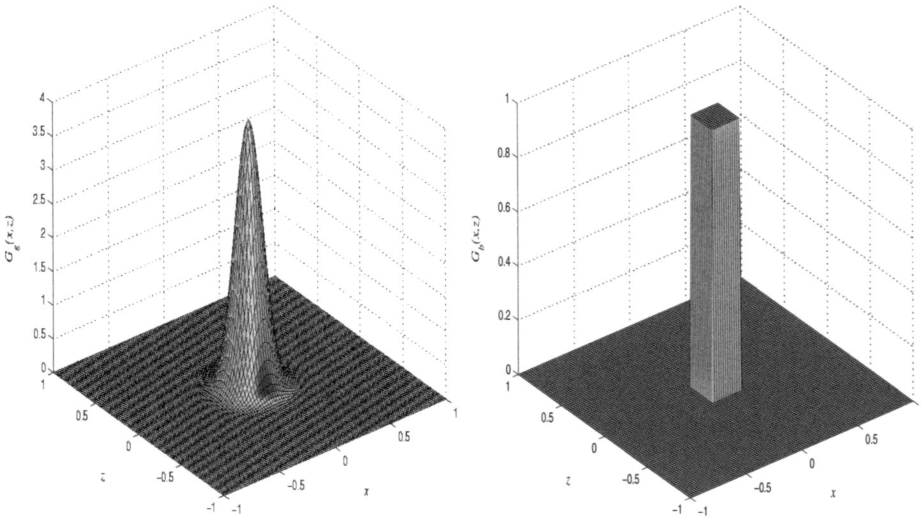

Fig. 1. Gaussian (left) and box (right) filters.

a new tensor τ_{ij}, namely the subgrid scale tensor, appears in the r.h.s. It is defined by

$$\tau_{ij} = \overline{u_i u_j} - \overline{u}_i \, \overline{u}_j \, . \tag{2.23}$$

In LES, the subgrid scale tensor plays a computational role similar to the role played by Reynolds stress tensor in RANS models but the physics that it contains is different. By far, the most commonly used subgrid scale model is the one introduced by Smagorinsky in 1963.[13] It is based on an eddy viscosity assumption that can be considered as an adaptation, to the subgrid scale, of the previously introduced Boussinesq assumption:

$$\tau_{ij} - \frac{1}{3}\tau_{kk}\delta_{ij} = -\nu_{Ts}\left(\frac{\partial \overline{u}_i}{\partial x_j} + \frac{\partial \overline{u}_j}{\partial x_i}\right) = -2\nu_{Ts}\overline{s}_{ij} \, . \tag{2.24}$$

δ_{ij} is Kronecker's symbol and ν_{Ts} corresponds to a subgrid (or subparticle) eddy viscosity. $\overline{s}_{ij}$ denotes the components of the rate of strain tensor based on filtered velocities $\overline{u}_i$, defined by

$$\overline{s}_{ij} = \frac{1}{2}\left(\frac{\partial \overline{u}_i}{\partial x_j} + \frac{\partial \overline{u}_j}{\partial x_i}\right) . \tag{2.25}$$

The Smagorinsky model expresses the subgrid eddy viscosity ν_{Ts} according to

$$\nu_{Ts} = (C_S \Delta)^2 \sqrt{2\overline{s}_{ij} \, \overline{s}_{ij}} \, . \tag{2.26}$$

The Smagorinsky constant C_S is generally close to 0.15[9] and the parameter Δ is often equal to $2\left(\Delta x \times \Delta y \times \Delta z\right)^{1/3}$, where Δx (respectively Δy, Δz) denotes a characteristic lengthscale of the grid in x (resp. y, z) direction. From a formalistic point of view, the Smagorinsky model is identical to a statistical mixing length model with a constant numerical mixing length. However, it has to be carried out in 3D to correctly simulate large eddies.

3. The SPH Method

3.1. *SPH formalism*

In SPH formalism, the fluid is discretized with a finite number of macroscopic fluid volumes called "particles". Each particle a is characterised by a mass m_a, a density ρ_a, a pressure p_a, a velocity vector $\mathbf{u}_a$ and a position vector $\mathbf{r}_a$ among other quantities. In the following, all notations A_a will refer to quantity A corresponding to particle a. At the heart of SPH is the exact formula (3.1) which evaluates the value of any flow property A at the position $\mathbf{r}$ by

$$A(\mathbf{r}) = \int_{\Omega} A(\mathbf{r}')\delta(\mathbf{r} - \mathbf{r}')dr' \,. \tag{3.1}$$

The previous summation is extended to the domain of interest Ω. For numerical reasons, the Dirac distribution δ is firstly approximated by a smooth kernel function w_h:

$$A(\mathbf{r}) = \int_{\Omega} A(\mathbf{r}')w_h(\mathbf{r} - \mathbf{r}')dr' + O\left(h^2\right). \tag{3.2}$$

The interpolating function w_h plays a central part in SPH: it depends on the distance between two particles (for a spherical kernel) and a parameter h called the smoothing length, proportional to the initial particle spacing. The transition to a discrete domain is achieved by approximating the integral of equation (3.2) by a Riemann summation:

$$A(\mathbf{r}) = \sum_b \frac{m_b}{\rho_b} A_b w_h(|\mathbf{r} - \mathbf{r}_b|) + O(h^2) \tag{3.3}$$

where A_b denotes the value of A at the point occupied by the particle b. The volume element dr' has been replaced by the particle volume m_b/ρ_b. The value of the function A relative to the particle a located at the point $\mathbf{r}$ is then expressed by

$$A_a = \sum_b \frac{m_b}{\rho_b} A_b w_h(r_{ab}) + O(h^2) \tag{3.4}$$

where r_{ab} denotes the distance between particles a and b. This summation is extended to all particles b that constitute the domain. In order to reduce the number of particles b involved in equation (3.3) (and thus to reduce the computing time), it is convenient to consider kernels characterised by a compact support of radius h_t, proportional to the smoothing length h. Consequently, only particles b located in the circle (sphere in 3D) of radius h_t and centred on a will contribute to the evaluation of function A relative to particle a (see the left picture of figure 2). Different types of kernel interpolation are provided in Ref. 20 as well as kernel improvement regarding consistency in Ref. 21. In most SPH codes, spline kernels are used and we consider here the fourth order spline kernel, represented on the right picture of figure 2. The interpolant of the function A established according to

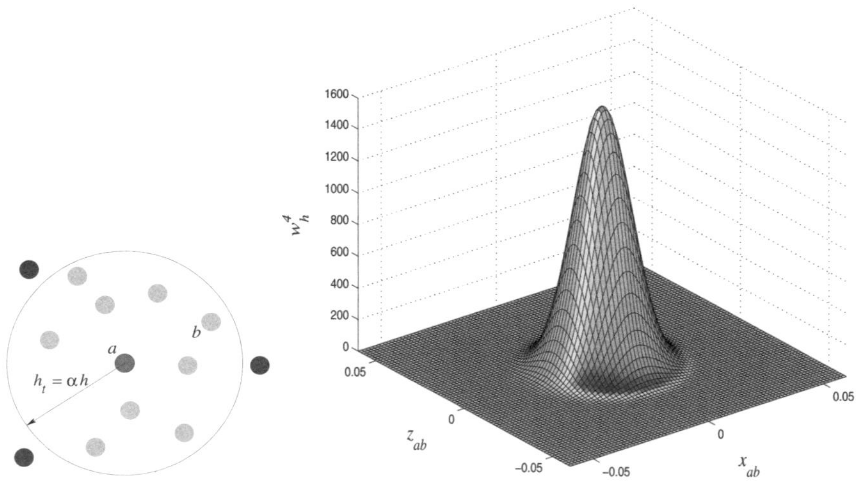

Fig. 2. Neighbours of particle a with a compact support kernel and representation of the fourth order spline kernel with $x_{ab} = x_a - x_b$ and $z_{ab} = z_a - z_b$.

equation (3.4) is differentiable provided the kernel function is also differentiable. Then, the gradient of a scalar field A relative to particle a can be written

$$(\nabla A)_a = \sum_b \frac{m_b}{\rho_b} A_b \nabla_a w_h(r_{ab}) \tag{3.5}$$

where $\nabla_a w_h\,(r_{ab})$ denotes the kernel gradient with respect to a-coordinates. One can notice that there is no need to use a grid to evaluate the gradient of a scalar field since it is a function of the kernel gradient which is analytically known. This SPH feature is essential and very attractive. As in finite-differences methods, the gradient of a scalar field A can be written in several ways in SPH formalism.[22] Among others, the following symmetric form

$$(\nabla A)_a = \frac{1}{\rho_a} \sum_b m_b (A_b - A_a) \nabla_a w_h (r_{ab}) \tag{3.6}$$

or the following asymmetric one

$$(\nabla A)_a = \rho_a \sum_b m_b \left(\frac{A_a}{\rho_a^2} + \frac{A_b}{\rho_b^2} \right) \nabla_a w_h (r_{ab}) \,. \tag{3.7}$$

Derivations of these expressions are provided in Ref. 22. In the same way, several forms of the divergence of a vector field could be established.[4]

Since we will mainly deal with nearly incompressible (characterised by density fluctuations less than 1 %) turbulent flows, a possible set of weakly compressible Reynolds equations and filtered Navier-Stokes equations in SPH formalism are herein presented.

3.2. *SPH weakly compressible Reynolds equations*

Modified equation (2.1) and equation (2.5) are used to derive an SPH form of Reynolds equations with the eddy viscosity assumption. From a formalistic point of view, they are identical to the classical SPH Navier-Stokes equations, except that

- pressure and velocities are considered as averaged values
- the kinematic viscosity is increased by the eddy viscosity

Consequently, for particle a, the averaged continuity equation can be written

$$\frac{d\rho_a}{dt} = \sum_b m_b \langle \mathbf{u} \rangle_{ab} . \nabla_a w_h (r_{ab}) \tag{3.8}$$

$\langle \mathbf{u} \rangle_a$ corresponds to the averaged velocity relative to particle a and $\langle \mathbf{u} \rangle_{ab}$ denotes the quantity $\langle \mathbf{u} \rangle_a - \langle \mathbf{u} \rangle_b$. The r.h.s. of equation (3.8) is an SPH form of the velocity divergence according to the general methodology presented in section 3.1. An SPH form of the averaged momentum equation (2.2) is

$$\frac{d\langle \mathbf{u} \rangle_a}{dt} = -\sum_b m_b \left(q_{a,b}^{pres} + q_{a,b}^{visc} \right) \nabla_a w_h (r_{ab}) + \mathbf{F}_a^e \tag{3.9}$$

with

$$q_{a,b}^{pres} = \frac{\langle p \rangle_a}{\rho_a^2} + \frac{\langle p \rangle_b}{\rho_b^2} \tag{3.10}$$

corresponding to the pressure gradient term and

$$q_{a,b}^{visc} = -8 \frac{\nu_{T,a} + \nu_{T,b}}{\rho_a + \rho_b} \frac{\langle \mathbf{u} \rangle_{ab} . \mathbf{r}_{ab}}{r_{ab}^2 + \eta^2} \tag{3.11}$$

corresponding to the viscous term,[22] which has here been modelled through the classical artificial term introduced by Monaghan[22] with $\nu_{T,a}$ corresponding to the eddy viscosity attached to particle a (the kinematic viscosity has here been neglected). $\mathbf{r}_{ab}$ denotes the vector $\mathbf{r}_a - \mathbf{r}_b$ while the parameter η (equals to $0.1h$) has been introduced in order to avoid a zero denominator. The averaged pressure $\langle p \rangle_a$ is defined by the following filtered state equation

$$\langle p \rangle_a = \frac{\rho_0 c_0^2}{\gamma} \left[\left(\frac{\rho_a}{\rho_0} \right)^{\gamma} - 1 \right] \tag{3.12}$$

where ρ_0 and c_0 respectively denote the reference density ($1000\ kg.m^{-3}$ for water) and a numerical speed of sound. The γ coefficient is equal to 7 for water,[22] which makes pressure strongly respond to density. Consequently, when particles are getting too close to each other, their pressure will highly increase and will repel these particles from each other through the pressure gradient term. Due to this equation, pressure automatically goes to zero when density equals the reference density (for instance near a free surface). In order to deal with a nearly incompressible flow (characterised by density fluctuation less than 1 %), the numerical speed of sound should be at least ten times higher than the maximum velocity of the flow.[23]

3.3. *SPH eddy viscosity models*

In this subsection, turbulent closure models presented in section 2.1 are applied to SPH. More details are available in Ref. 5.

3.3.1. *An SPH mixing length model*

Considering the mixing length model (2.7), the eddy viscosity $\nu_{T,a}$ for each fluid particle a is expressed by

$$\nu_{T,a} = L_{m,a}^2 S_a \tag{3.13}$$

$L_{m,a}$ corresponds to the value of the mixing length at the position occupied by particle a while the velocity gradients required to calculate the rate-of-strain S_a of particle a are estimated through the definition of $\langle s_{ij} \rangle$ (see equation (2.4)). The velocity gradients are then estimated by

$$\left. \frac{\partial \langle u_i \rangle}{\partial x_j} \right|_a = -\frac{1}{\rho_a} \sum_b m_b \langle \mathbf{u} \rangle_{ab} \otimes \nabla_a w_h \left(r_{ab} \right) . \tag{3.14}$$

3.3.2. *An SPH k model*

With an SPH k model, the eddy viscosity $\nu_{T,a}$ for each fluid particle a is expressed by

$$\nu_{T,a} = C_\mu \frac{k_a^2}{\epsilon_a} \tag{3.15}$$

where k_a and ϵ_a respectively correspond to the turbulent kinetic energy and the dissipation rate for particle a. The following k-transport equation (2.9) is written in SPH formalism as

$$\frac{dk_a}{dt} = \underbrace{\sum_b m_b \frac{\rho_a \nu_{T,a} + \rho_b \nu_{T,b}}{\rho_a \rho_b \sigma_k} \frac{k_{ab} \mathbf{r}_{ab}}{r_{ab}^2 + \eta^2} . \nabla_a w_h \left(r_{ab} \right)}_{diffusion} + \underbrace{P_a}_{production} - \underbrace{\epsilon_a}_{dissipation} \tag{3.16}$$

where $k_{ab} = k_a - k_b$. The diffusion term was written in the same form as the viscous diffusion term developed by Monaghan, as achieved for the temperature conductivity in Ref. 24. The production term P_a can be modelled according to

$$P_a = C_\mu \frac{k_a^2}{\epsilon_a} S_a^2 \tag{3.17}$$

or

$$P_a = min \left(\sqrt{C_\mu}, C_\mu \frac{k_a}{\epsilon_a} S_a^2 \right) k_a S_a^2 . \tag{3.18}$$

To close the system, the model (2.12) for ϵ_a is used, according to

$$\epsilon_a = C_\mu^{3/4} \frac{k_a^{3/2}}{L_{m,a}} . \tag{3.19}$$

3.3.3. *SPH two-equation models*

With an SPH $k-\epsilon$ model, the eddy viscosity $\nu_{T,a}$ relative to each fluid particle a is also expressed by (3.15). The previous k transport equation (3.16) is unchanged and the production term modelling (3.18) is also maintained. The transport equation for ϵ is expressed by

$$\begin{aligned}\frac{d\epsilon_a}{dt} &= \sum_b m_b \frac{\rho_a \nu_{T,a} + \rho_b \nu_{T,b}}{\rho_a \rho_b \sigma_\varepsilon} \frac{\epsilon_{ab} \mathbf{r}_{ab}}{r_{ab}^2 + \eta^2} . \nabla_a w_h (r_{ab}) \\ &+ \frac{\epsilon_a}{k_a} (C_{\epsilon 1} P_a - C_{\epsilon 2} \epsilon_a) \end{aligned} \quad (3.20)$$

with $\epsilon_{ab} = \epsilon_a - \epsilon_b$.

An SPH $k - \omega$ model consists of setting the eddy viscosity as

$$\nu_{T,a} = \frac{k_a^2}{\omega_a} \quad (3.21)$$

with a transport equation for ω_a:

$$\begin{aligned}\frac{d\omega_a}{dt} &= \sum_b m_b \frac{\rho_a \nu_{T,a} + \rho_b \nu_{T,b}}{\rho_a \rho_b \sigma_\omega} \frac{\omega_{ab} \mathbf{r}_{ab}}{r_{ab}^2 + \eta^2} . \nabla_a w_h (r_{ab}) \\ &+ C_{\omega 1} \frac{\omega_a}{k_a} P_a - C_{\omega 2} \omega_a^2 \end{aligned} \quad (3.22)$$

where $\omega_{ab} = \omega_a - \omega_b$.

An SPH non-linear $k - \epsilon$ model can be specified as follows (see Ref. 25). According to the continuous model (2.18) and considering the equation (2.8) of ν_T, the eddy viscosity is replaced by a tensor:

$$\mathbf{M}_{T,a} = C_\mu \frac{k_a^2}{\epsilon_a} \left(\mathbf{I} + 2C_{2,a} \frac{k_a \Omega_a}{\epsilon_a} \omega \right) \quad (3.23)$$

where Ω_a is the mean rotation rate of particle a and ω is a unit rotation tensor defined as

$$\omega = \begin{pmatrix} 0 & 1 \\ -1 & 0 \end{pmatrix}. \quad (3.24)$$

3.4. *SPH filtered equations*

The idea of modelling turbulence through LES in SPH was mentioned by Pope (2000) and based on the observation that the SPH integral interpolation (3.2) is mathematically similar to the LES filter (2.22). From a formalistic point of view, SPH filtered Navier-Stokes equations are identical to SPH Reynolds equations except that all averaged values are replaced by filtered ones and the eddy-viscosity by the subgrid eddy-viscosity. A main difference between LES and RANS models is that LES has to be achieved in 3D since large scales are explicitly resolved

3.5. *Temporal integration*

All previous SPH equations are integrated in time with a classical first order Euler scheme according to a time step estimated by

$$\delta t = \min\left(\delta t_{forces}, \delta t_{CFL}, \delta t_{visc}\right) \tag{3.25}$$

where

$$\delta t_{CFL} = 0.4\frac{h}{c_0} \tag{3.26}$$

corresponds to a CFL condition which imposes that the time step δt_{CFL} is less or equal to the convection time on the length h relative to the spatial discretisation.

$$\delta t_{forces} = 0.25 \min_a \sqrt{\frac{h}{|\mathbf{f}_a|}} \tag{3.27}$$

ensures that particles do not get too close to each other during the integration of their movement.[23] $\mathbf{f}_a$ denotes the internal and external forces associated to particle a (i.e. the magnitude of the r.h.s. of the momentum equation).

$$\delta t_{visc} = 0.125\frac{h^2}{\nu}\,. \tag{3.28}$$

This viscous criterion must be taken into account to make the time step inferior to the viscous phenomenon time scale.[23]

3.6. *Boundary conditions*

3.6.1. *Wall modelling*

Wall modelling herein described corresponds to the case where an SPH RANS model is considered. However, this formalism is identical when the LES model is used by replacing "averaged" by "filtered" values. In a more general view, these wall conditions work also in laminar cases, as shown in Ref. 26. Walls are here modelled with solid particles called edge particles. Moreover, in order to ensure the impermeability of the wall, three layers of dummy particles are added under the wall: the density of edge and dummy particles evolve thanks to the contribution of fluid particles through the averaged continuity equation (3.8). Indeed, when the fluid particle a is linked to the edge or dummy particle b, the b-contribution to the evolution of a-density is equal to (see equation (3.8))

$$m_b \langle \mathbf{u} \rangle_{ab} . \nabla_a w_h (r_{ab}) \, . \tag{3.29}$$

As particle b is also linked to fluid particle a, the contribution of particle a to the evolution of b-density is also given by equation (3.29), since the continuity equation (3.8) is symmetric with respect to a and b subscripts. The averaged pressure of edge and dummy particles are then computed with the averaged state equation and these particles are involved in the averaged pressure gradient relative to fluid particles in the momentum equation. These wall conditions also enable a perfect impermeability of the wall in rapid dynamic phenomena such as dam breaking. Contrary to the repulsive forces commonly used in SPH to represent walls,[22] the present formulation does not introduce any ad hoc coefficient. Moreover, in contrast to traditional mirror particles used in most SPH codes, the method proposed here considers fixed particles and can easily be implemented even for curved walls. Indeed, this has been achieved in Ref. 26 in the case of a hill channel. It is also practical to model a moving wall, as achieved in the following.

3.6.2. *Turbulent boundary conditions for SPH RANS models*

For turbulent flows, edge particles represent fluid particles immediately located at the bottom of the turbulent boundary layer. Their gravitational center is hence placed at $z = \delta$, where δ corresponds to a small distance to the wall larger than the viscous sublayer thickness, typically $\delta r/10$ where δr is the averaged initial particle spacing. In some Eulerian methods using cell vertex discretisation, although the first node is exactly located on the wall, the velocity at that point is non-zero and assumed to correspond to the value at the position δ. This means that the flow domain for a channel of a height H for instance is $[-\delta, H]$ and the computational domain $[0, H]$ (the wall is "pushed back"). The $-\delta$ offset is then negligible.[4] Prescribing turbulent wall boundary conditions with this approach requires wall functions, which give theoretical values of physical quantities in the vicinity of a wall. Although this method can be criticized for adverse pressure driven flows, it was shown to be efficient for simple shear flows in traditional Eulerian codes and was presented and used by the authors.[5] The following wall functions require the estimation of a friction velocity u_{*a} relative to each edge particle a, estimated through the following procedure:

(1) For each edge particle a, a fictitious point M located at a distance Δ from a on the normal to the wall (see figure 3) is defined.

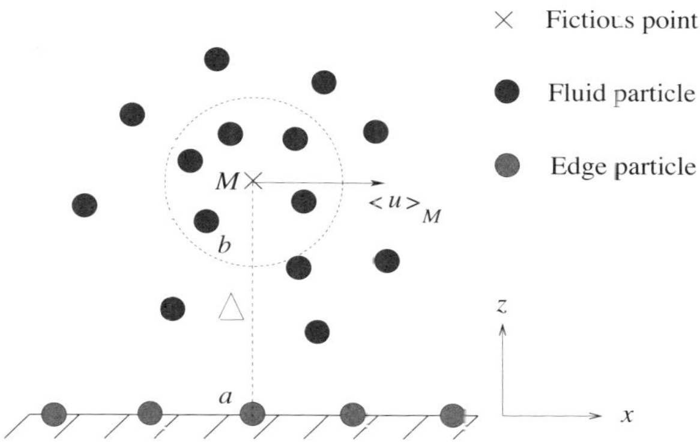

Fig. 3. Computation of the friction velocity u_*.

(2) The averaged axial velocity at M is computed with the classical following SPH relation

$$\langle u\rangle_M = \sum_b \frac{m_b}{\rho_b}\langle u\rangle_b w_h\left(r_{Mb}\right) \tag{3.30}$$

where r_{Mb} is the distance between the fictitious point M and the particle b. In principle,

(3) $\langle u\rangle_M$ should verify the following log-law for a rough wall

$$\langle u\rangle_M = u_{*a}\left[\frac{1}{\kappa}ln\frac{(\delta+\Delta)}{\xi}+D\right] \tag{3.31}$$

where ξ denotes a roughness scale. The constant D is equal to 8.5[10] and the Von Karman κ equals to 0.41. The friction velocity u_{*a} is then directly obtained from (3.31).

The edge particle averaged tangential velocity is then computed according to

$$\langle u\rangle_a = u_{*a}\left[\frac{1}{\kappa}ln\frac{\delta}{\xi}+D\right]. \tag{3.32}$$

It has been shown in Ref. 4 that the estimation of u_{*a} is very close to the theoretical value u_* for a turbulent steady open channel characterised by a constant slope. Near a wall, k-Dirichlet boundary conditions are prescribed according to

$$k_a = \frac{u_{*a}^2}{\sqrt{C_\mu}}\,. \tag{3.33}$$

The dissipation ε (respectively the frequency ω) obeys a Neumann boundary condition, which consists of setting the diffusion term appearing in (3.20) (respectively (3.22)) as $u_{*a}^4/\sigma_\varepsilon \delta^2$ (respectively $u_{*a}/\sqrt{C_\mu}\kappa\delta$) for all edge particles. At the surface, all turbulent quantities are assumed to tend naturally to zero; thus no particular boundary condition is prescribed.

3.6.3. *Turbulent boundary conditions for SPH LES models*

Previous wall functions have been used by the authors[27] to simulate a turbulent flow in a 3D channel with an SPH Smagorinsky model. However, for simplicity reasons, zero velocity will be considered for the test-case achieved with LES and presented here.

4. Applications

4.1. *The Spartacus-2D code*

The Spartacus-2D code has been developed since 1999 at EDF R&D mainly for coastal and environmental applications such as spillways, dam breaking, breaking waves,[28] etc. A first version has been commercialized in 2005 through the Telemac hydroinformatics system and reader can find all information about Spartacus-2D in Ref. 29. All 2D simulations herein presented have been achieved with the last version of Spartacus-2D (V1P2). Based on this 2D version, a 3D one, has been developped and has been used here for the 3D simulation shown at the end.

4.2. *Solitary wave propagation*

In order to generate a solitary wave characterised by the height $H = 0.2\ m$, and represented on figure 4, we consider a piston-type wave board moving on one side of a closed channel. The system is characterised by a flat bottom and a length $L = 15\ m$, long enough to avoid perturbations due to the right boundary. The tank is initially filled with water on a depth $d = 0.5\ m$. All equations presented in this part are derived from Ref. 30. The displacement $X_0(t)$ of the wave board has to evolve over the range

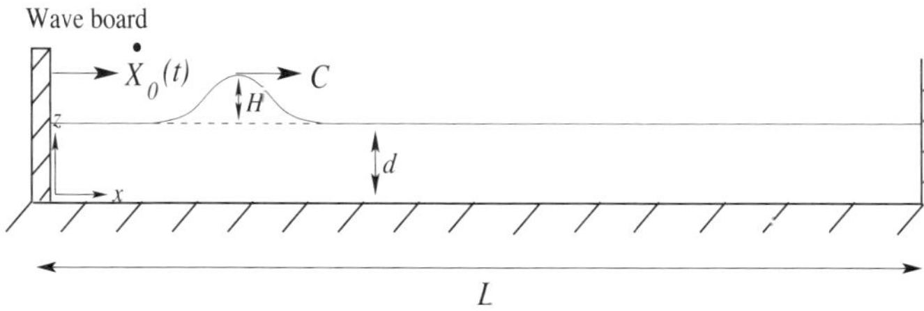

Fig. 4. Sketch of the considered solitary wave.

$-\infty < t < +\infty$ according to:

$$X_C(t) = \frac{H}{Kd} tanh\left[K\left(Ct - X_0\right)\right] \tag{4.1}$$

where

$$K = \sqrt{\frac{3H}{4d^3}} \tag{4.2}$$

and

$$C = \sqrt{g\left(d + H\right)}. \tag{4.3}$$

The velocity of the piston is thus evolving according to

$$\dot{X}_0(t) = \frac{\frac{H}{d}CA}{1 + \frac{H}{d}A} \tag{4.4}$$

with

$$A = \frac{1}{cosh^2\left[K\left(Ct - X_0\right)\right]}. \tag{4.5}$$

The total stroke S_s, corresponding to the distance between the maximum and minimum displacements of the wave board, is given by

$$S_s = \sqrt{\frac{16Hd}{3}}. \tag{4.6}$$

4.2.1. *Modelling*

The system was discretized with 315 000 particles, initially spaced by 5.10^{-3} m and distributed at the beginning of the calculation on a regular lattice. In order to generate the solitary wave, the velocity $\dot{X}_0(t)$ is imposed to particles constituting the wave paddle until they reach the distance S_s. It is well known that wave propagation is a inviscid phenomenon mainly driven by pressure forces: hence, no turbulence modelling has been considered and classical SPH parameters (fourth order spline kernel, $h/\delta r = 1.5$) and traditional SPH equations have been used. 90 000 iterations, for a time step of 6.10^{-5} s have been considered which enable the solitary wave to be conveyed without perturbation coming from the right boundary.

4.2.2. *Results*

127 hours were required to simulate this test-case on a 2.8 GHz Linux PC. Two investigation times have been considered: $t_1 = 3.42$ s (after the wave paddle stops) and $t_2 = 5.40$ s (before the solitary wave is perturbed by the right boundary) as showed on figure 5. Table 1 shows the errors obtained

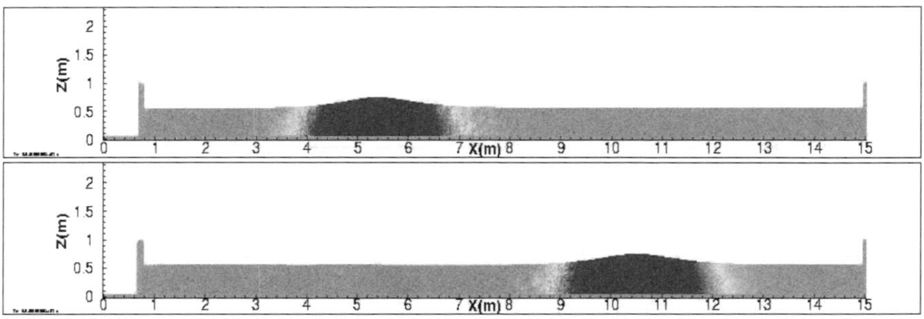

Fig. 5. Investigation times $t_1 = 3.42$ s (left) and $t_2 = 5.40$ s (right).

with Spartacus-2D regarding the estimation of two quantities: V_{inf} which represents the water volume located beneath the crest of the wave and L_{95} which is a pseudo wave length defining a zone where 95% of V_{inf} is located. At the beginning of the simulation, these two quantities are very well estimated by Spartacus-2D. At t_2, the agreement is still good but not

Table 1. Relative error relative to V_{inf} and L_{95} at t_1 and t_2.

	t_1	t_2
Relative error in % on V_{inf}	0.0	1.9
Relative error in % on L_{95}	1.0	2.8

as accurate as before. This lack of quality could be due to the simple time-scheme used. However, the errors are low enough to reveal the capability of the code to generate and propagate a solitary wave.

4.3. *Solitary breaking wave*

Several types of breaking waves are commonly described in the literature - spilling, surging, collapsing, and plunging breaking.[31] By using a simple plane beach, important characteristics of breaking waves can be studied in the laboratory. Indeed, the results for the simple two-dimensional case of a solitary wave propagating in a constant depth and impinging on a plane sloping beach can yield results that are useful for three-dimensional numerical modeling of coastal sites. A schematic of the solitary wave runup experiments achieved in Ref. 31 is presented on figure 6 with $L = 5\ m$, $d = 0.5\ m$, and $H = 0.2\ m$. Experiments were conducted with this arrangement for a slope $\beta = 1 : 15$ and waves were generated using a programmable vertical bulkhead wave generator which moved according to the classical solitary wave theory previously introduced.

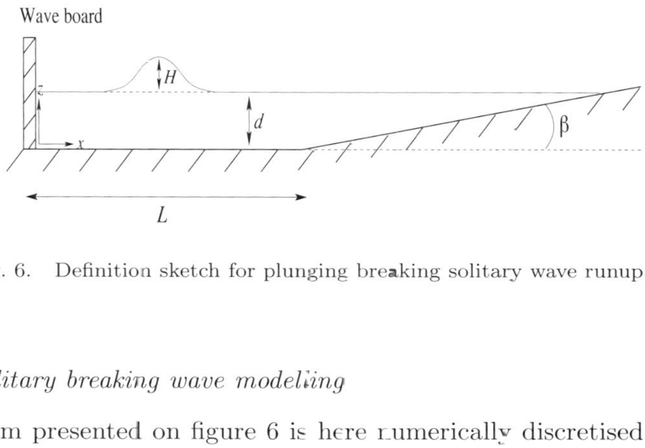

Fig. 6. Definition sketch for plunging breaking solitary wave runup.[31]

4.3.1. *Solitary breaking wave modeling*

The system presented on figure 6 is here numerically discretised with approximately 350 000 particles initially spaced by $2,5.10^{-3}$ m. In order to generate a plunging breaking solitary wave characterised by $H/d = 0.4$, the velocity of edge and dummy particles defining the wave board evolves according to equation (4.4) along a distance given by (4.6). Contrary to previous test case where turbulent effects were negligible, they are this time predominant during the breaking stage which involves high strains and surface deformation. All turbulent boundary conditions relative to edge particles, including velocity, are computed by considering a rough wall characterised by a roughness scale ξ of 0.01 m, according to the method presented in part 3.6.2. The previous turbulence models are then used in order to correctly capture the shape of the free surface and the wave velocity during and after breaking. We will examine the two models presented in section 3.3 (mixing length and k-model) and compare them to a model using a constant (in space and time) eddy viscosity. The value of the latter was estimated from simulations achieved with k-model. For all simulations herein presented, the time step was equal to 3.10^{-5} s and 363 hours were required on a 2.8 GHz Linux PC.

4.3.2. *Results*

Pictures (a) and (c) of figure 7 represent experimental results presented in Ref. 31, at two different times corresponding to the formation of the plunging wave. Unfortunately, no spatial or time indication are available in Ref. 31. The comparison between experimental and numerical results first reveal that Spartacus-2D is able to accurately reproduce breaking plunging wave. Moreover, by comparing Spartacus-2D results obtained with a constant eddy-viscosity, a mixing length model and a k-equation model (bottom pictures of figure 7), one can notice that there is no large discrepancy between the three models. The red solid line depicted on Spartacus-2D results of figures 7 to 8 corresponds to the experimental free surface shape. For this first step of plunging wave process, it seems as if turbulence modelling is not crucial, the rate of strain being still moderate. Conversely, on experimental pictures of figure 8, the plunging jet impacts the forward face of the wave with a shoreward directed jet generated at the impact point. This jet impact initiated the splashup/runup process. In this case, the jet is reflected at an angle with the bottom that is greater than the corresponding angle of the incident jet. It is postulated that this splashup is caused by the translation of the impact point of the jet up the slope and its interaction with the front face of the wave.[31] Spartacus-2D simulations of the splashup process initiation (see bottom pictures of figure 8) show that a constant eddy viscosity model does not correctly reproduce this phenomenon, while the mixing length model presents slightly better results. In contrast, results obtained with the k-equation model are in good agreement with the experiment. Up to this point, it can be established that turbulence modelling is important to simulate such phenomenon. The reason is that high shear stress generated in the vicinity of the impinging point leads

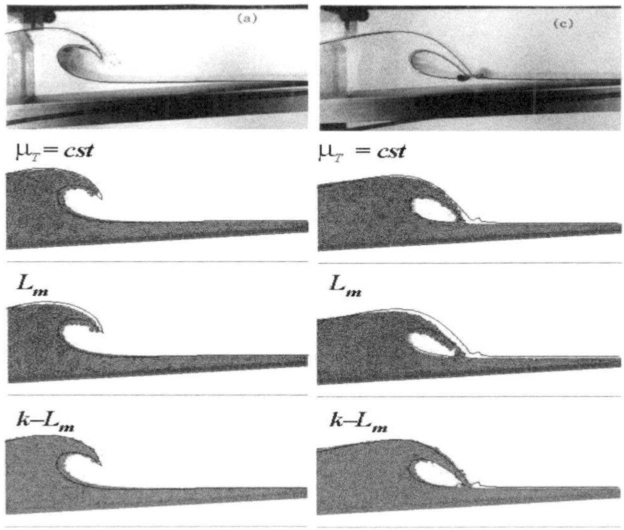

Fig. 7. Experimental plunging solitary wave[31] (top) and simulated one achieved by Spartacus-2D with various turbulent closure: constant eddy-viscosity, mixing length and k-model (from top to bottom). The solid lines mimic the experimental free surface.

to high turbulent kinetic energy production rate through (3.17) or (3.18). Thus, the turbulent kinetic energy - and hence the eddy viscosity - increases a lot and slows down the global motion, which is confirmed by figures 9 and 10 which represent the spatial distribution of k at two different stages (just before and during the splashup process initiation). Splashup development depicted on figure 11 reveals this time that all turbulence models give fairly good results. As the incident wave moves shoreward, the shape of the splashup (reflected) jet changes and curves back toward the incident wave. Finally, the incident jet disappears and the reflected jet ("counter-breaking") collapses. Once again, comparisons between experimental and numerical results of figure 12 prove that a constant eddy viscosity model is not accurate enough to reproduce the reflected wave. The other models are slightly better but have to be improved, specially regarding wall treatment. The reasons for which even the most sophisticated model presented here fails in predicting the last stages with accuracy may now be evoked. First of all, a two-equation model (*e.g.* the $k - \epsilon$ model) as presented in part 3.3 should be more appropriate to estimate the dissipation rate ϵ certainly crucial in this flow as we will see in the next section. Then, one could argue that the Boussinesq's assumption (2.3) is too poor to capture the

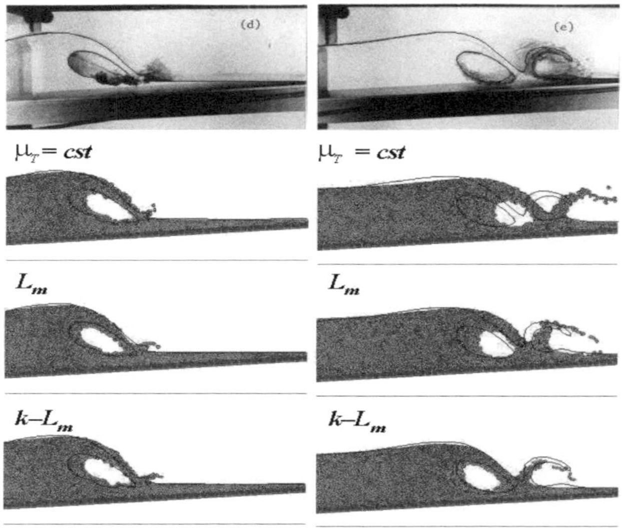

Fig. 8. Experimental splashup initiation. Same key as figure 7.

Fig. 9. Turbulent kinetic energy distribution before breaking.

complex turbulent stresses involved in the splash-up: high rate of strain and curvature effects would require a non-linear $k - \epsilon$ model (namely an Explicit Algebraic Reynolds Stress Model) as presented in part 3.3. Finally, one should mention the fact that no free-surface boundary condition regarding the turbulent kinetic energy was prescribed. Recent simulations have shown that imposing a zero-value of k on the free surface slightly improves the prediction of its shape. More generally, interactions between free-surfaces and turbulence (in terms of anisotropy of Reynolds stresses for

Fig. 10. Turbulent kinetic energy distribution after breaking.

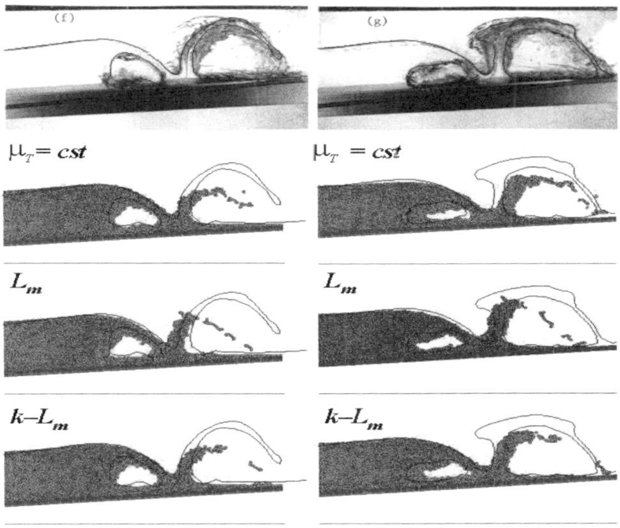

Fig. 11. Experimental splashup development. Same key as figure 7.

instance) remain an open field for which theoretical developments should be done in order to apply a method like SPH to very complicated flows.

4.4. *Collapse of a water column on a wet bed*

This case is described in Ref. 32. It consists of a 1.18 m long tank initially filled with a water layer of $d = 1.8$ cm, or 3.8 cm, at rest. A gate contains a water column of length 38 cm and depth 15 cm (see figure 13) and opens at time $t = 0$ s, to let the water collapse into the tank. We chose an initial

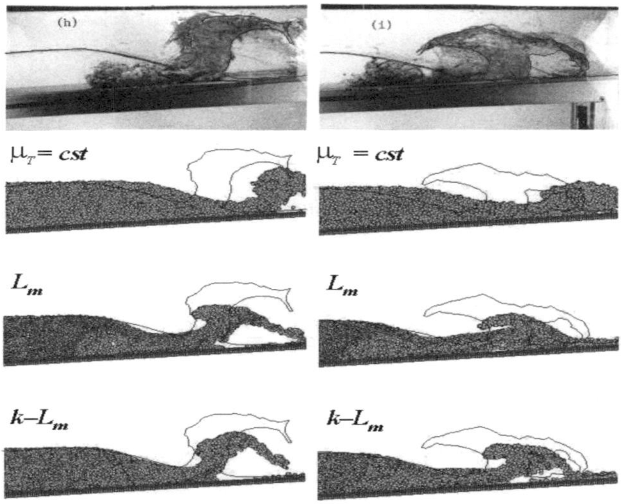

Fig. 12. Experimental reflected wave. Same key as figure 7.

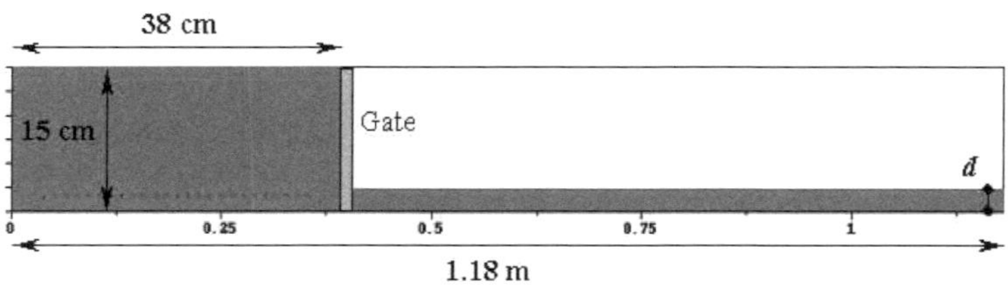

Fig. 13. Janosi et al's case: geometry.[32]

particle spacing of 1 mm, so that the total number of particles approaches 80,000. The maximum velocity in the flow was estimated (and checked) to be 2.4 m/s; hence the numerical speed of sound was set to 24 m/s. Results were output every 0.062 s to match experimental measurement times. 12 hours were required to simulate this test-case on a 2.8 GHz Linux PC. Experimental results concern the shape of the free surface. A small gap in time (about 43 ms) exists between the experiments and the computed results; for example the first presented graph on figure 14 was numerically obtained at t = 0.113 s, (instead of t = 0.156 s, as indicated on the experimental picture). We firstly tested the case with an initial water

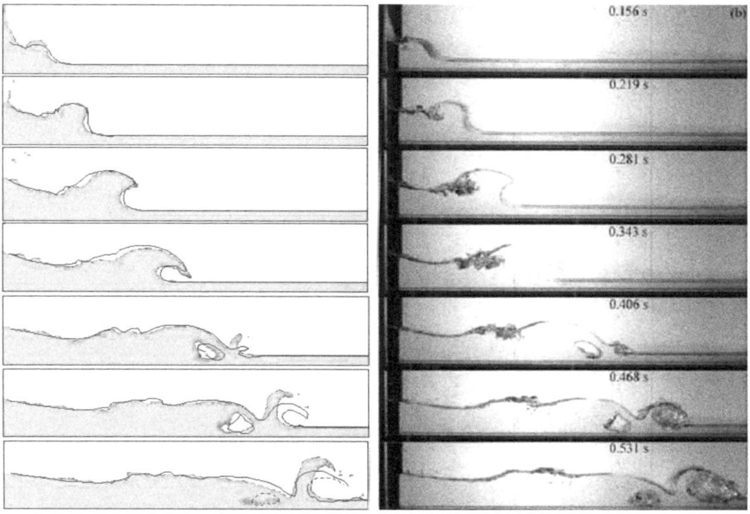

Fig. 14. Janosi et al's case: comparison of SPH (k-model) results with experiments. The solid red line on the left part mimics the experimental free surface.

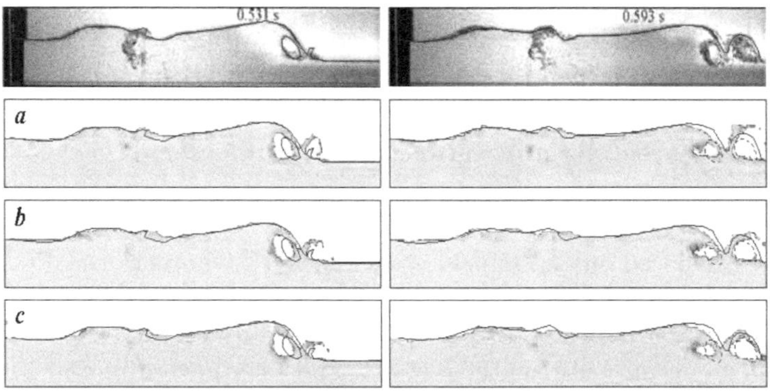

Fig. 15. Janosi et al's case with $d = 3.8cm$. From top to bottom: experiment, constant eddy-viscosity, k-model, non-linear $k-\epsilon$ model. The solid red line on the left part mimics the experimental free surface.

layer of thickness 1.8 cm. Generally speaking, the model performs well, as shown by figure 15 with the k-model for turbulence. However, during the last stages of the flow, discrepancies occur between observations and simulations: our model fails in representing correctly the splash-up after breaking. Generally speaking, in this case turbulence has a small effect; however, increasing the model complexity tends to improve the quality of predictions, especially in the vicinity of the jet just before breaking (the k-ω model is not presented here but gives results very close to the k-ε in this case). Similar conclusions concern the shape of the void bubble appearing after the first breaking. This is confirmed with the case $d = 3.8$ cm, presented on figures 15 and 16.

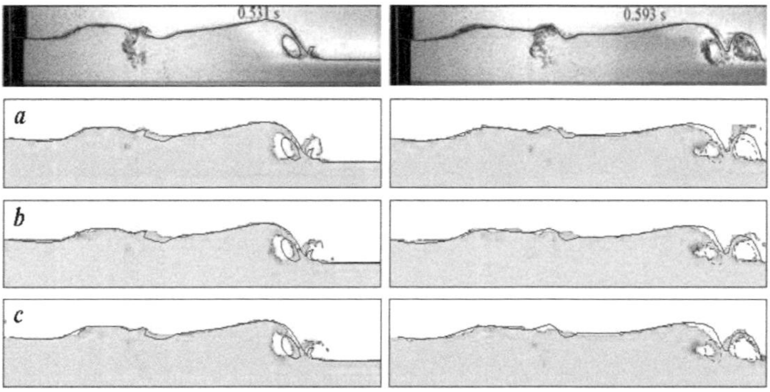

Fig. 16. Same key as picture 15.

4.5. *Wave breaking and setup over a coral reef*

4.5.1. *Model setup*

This case is based on experiments by Gourlay,[33] who performed tests with an idealised reef model in a wave flume of length 32 m and width 6 m with maximum water depth of 0.50 m. The height of the reef was 0.40 m with 8 m width of 'ocean' between the reef-face and wave generator and 5 m width of 'lagoon' behind it. The 'ocean' bottom had a slope of 1:88; therefore, the effective reef height was 0.32 m. The reef-face had a partially absorbing 1:1 slope by placing large pieces of gravel in order to simulate a natural rough steep reef-face. Gourlay[33] also examined the wave setup as a function of various still water depths on the reef-top, wave periods T and incident off-reef wave heights H_i. In this case the prediction of wave setup due to breaking on the coral platform is of highest importance. However, the main focus on this application is the maximum wave-setup. By keeping the water depth over the platform reef constant as $h_r = 0.1$ m, two different wave

Table 2. Physical parameters for Gourlays's case.[33]

Wave period (s)	1.10			1.48		
Incident wave height (m)	0.072	0.148	0.168	0.108	0.142	0.200
Deep water wave height (m)	0.077	0.158	0.181	0.118	0.154	0.218

periods and three different incident wave heights H_i in each wave period are examined (see table 2). For the SPH simulation, the original geometry, especially the length of platform reef and 'lagoon' is shortened as shown in figure 17 due to the computation limitations. The number of fluid particles used is different from case to case to maintain constant volume of water in 'ocean' during the simulations due to the propagation of water over the reef. The initial particle distance is 0.01 m. For some specific cases, these particles can move based on prescribed conditions. For instance, the wave paddle here moves according to wave-maker theory.[34]

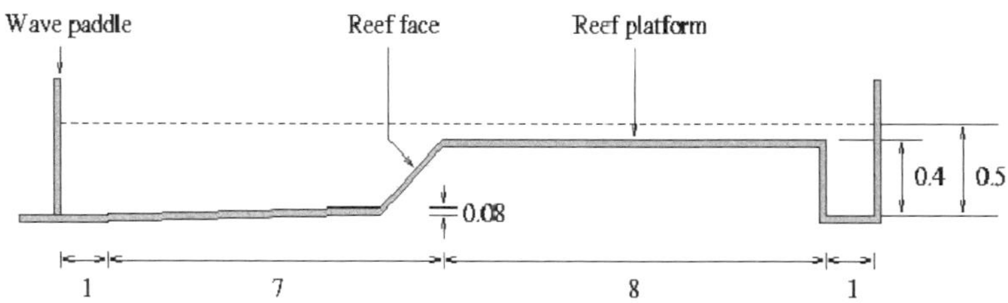

Fig. 17. Geometry of numerical configuration for Gourlay's reef case[33] (units in m).

4.5.2. *Results*

30 hours were required to simulate this test-case on a 2.8 GHz Linux PC. An example of the simulated flow is shown in figure 18. One can clearly see from these sample pictures the capabilities of SPH in modelling the wave breaking at the reef edge, with the formation on a breaking jet on the reef platform. The comparisons of measured and computed wave setup and wave height for the case $T = 1.10$ s, and $H_i = 0.148$ m, are shown in figures 19 and 20 respectively. The numerical models presented here are the refraction-diffraction equation by Massel and Gourlay,[35] two versions of a finite-difference model based on 1-D Boussinesq equations[36] and the SPH model, each of them respectively indicated in the legend as Gourlay

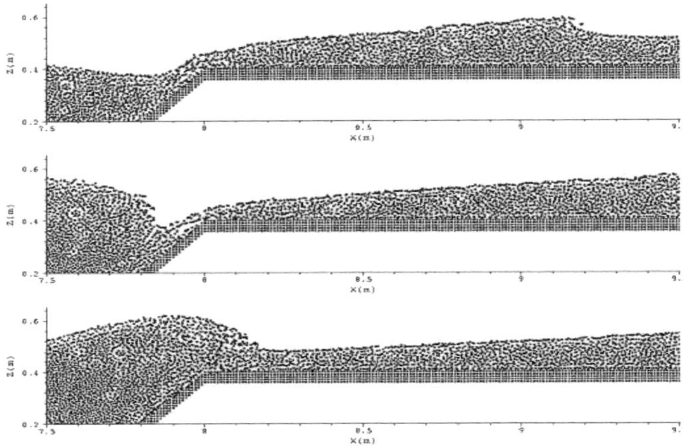

Fig. 18. Gourlay's case.[33] SPH results for the case $h_r = 0.1m$, $T = 148s$ and $H_i = 0.200m$ (top to bottom: $t = 39.3s$, $42.3s$, $42.6s$).

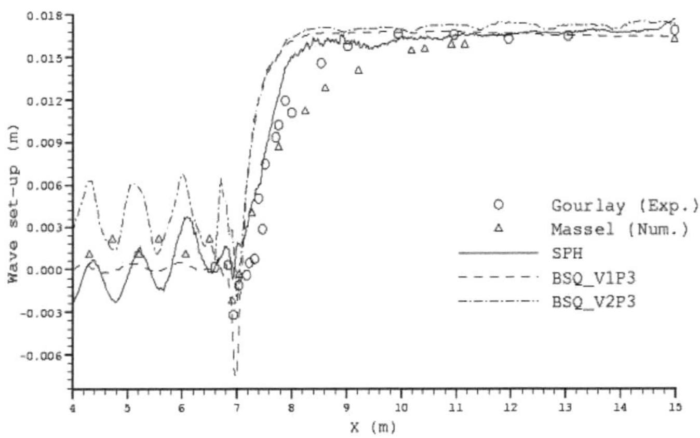

Fig. 19. Gourlay's case. Wave setup as a function of x ($h_r = 0.1m$, $T = 1.10s$, $H_i = 0.148m$). Experiments by Gourlay,[33] numerical simulation by Massel and Gourlay,[35] SPH simulation and finite-difference 1-D simulation based on the Boussinesq equations (BSQ_V1P3 and V2P3), see Ref. 36.

(Exp.), Massel (Num.), BSQ_V1P3, BSQ_V2P3 and SPH. Although the experimental data start just from the edge of the reef ($x = 7$ m), all the comparisons are made between the locations $x = 4$ m, and $x = 15$ m. The co-ordinate x is measured from the beginning of the slope. Although these

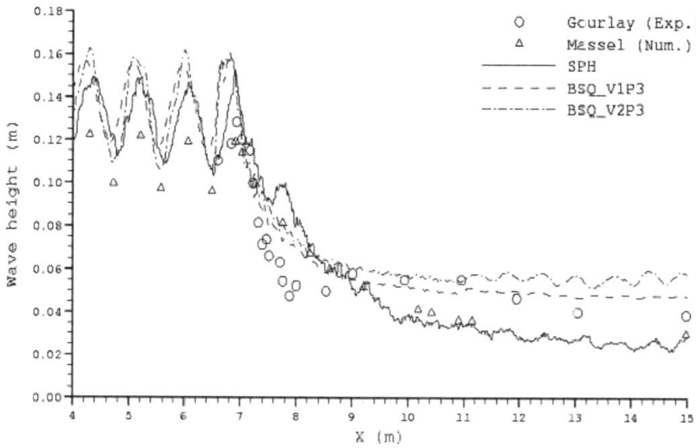

Fig. 20. Gourlay's case. Wave setup as a function of x ($h_r = 0.1m$, $T = 1.10s$, $H_i = 0.148m$). Experiments by Gourlay,[33] numerical simulation by Massel and Gourlay,[35] SPH simulation and finite-difference 1-D simulation based on the Boussinesq equations (BSQ_V1P3 and V2P3), see Ref. 36.

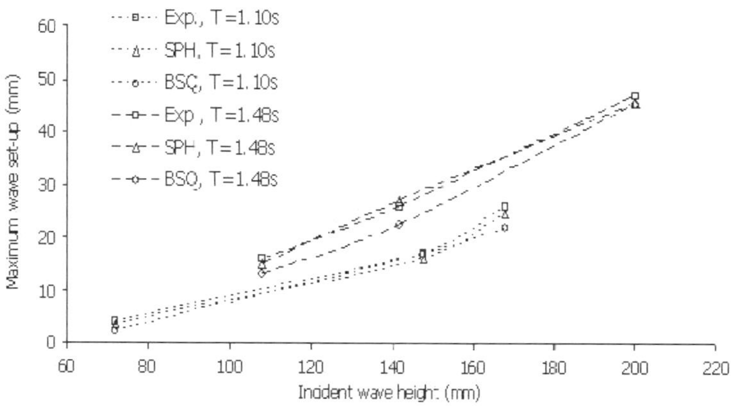

Fig. 21. Gourlay's case. Maximum wave setup over the reef top for the 6 cases ($h_r = 0.1m$). Experiments by Gourlay,[33] SPH simulation and finite-difference 1-D simulation based on the Boussinesq equations (see Ref. 36 for details).

four different numerical models show different behaviour of setup before the platform reef (from 4 m to 7 m), the predicted maximum wave setup on the reef agree quite well with each other. Especially, the SPH results are a bit closer to the experimental data. In terms of wave height, while SPH and Massel and Gourlay's results agree with each other, the Boussinesq models are closer to the experimental data. Before the reef, SPH and Boussinesq models wave heights are higher than Massel and Gourlay's. This undulating pattern of the wave height (and the setup) offshore of the reef face is a direct effect of the reflection of incident waves by the reef geometry. Massel and Gourlay[35] mentioned that the wave height is mostly constant after 3 m from the reef edge, as is the setup. The maximum wave setup is hence represented in figure 21 to compare models' results with experimental data, for the six test-cases presented herein. The increase of maximum setup with either increasing period or increasing incident wave height is remarkably well simulated by the two models (Boussinesq and SPH). In the SPH results, a small fluctuation of wave setup might be observed, due to the reflective waves. Note that the particle size in SPH is here 10 mm, which limits the accuracy of setup prediction.

4.6. *Wave shoaling and overtopping*

When approaching the coastline, the shape of water waves is modified by the change in bottom level. This phenomenon, referred to as "wave shoaling", is reproduced here following the experimental setup carried out at the University of Manchester by Stansby and Feng[37] in a wave flume of 13 m length, 0.5 m height and 0.3 m width, with a beach slope of 1:20 and a trapezoidal dyke (see figure 22). The flume is equipped with a piston-type wave paddle generating regular waves with period $T = 2.39$ s, and height $H = 0.10$ m, at the toe of the beach. Figure 23 presents some comparisons between SPH, simulations obtained by solving the Boussinesq equations (see Ref. 36 for more details) and experiments. 24 hours were required to simulate this test-case on a 2.8 GHz Linux PC. enerally speaking, the surface elevation is correctly predicted and SPH performs slightly better than the Boussinesq model considered. Estimating wave overtopping over

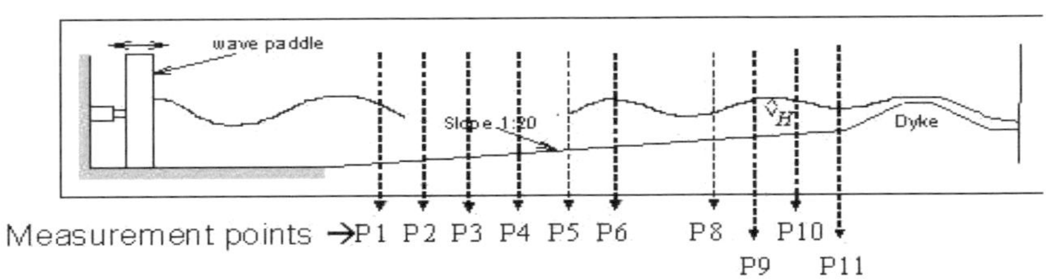

Fig. 22. Stansby and Feng's case.[37] Geometry and position of the measurements points.

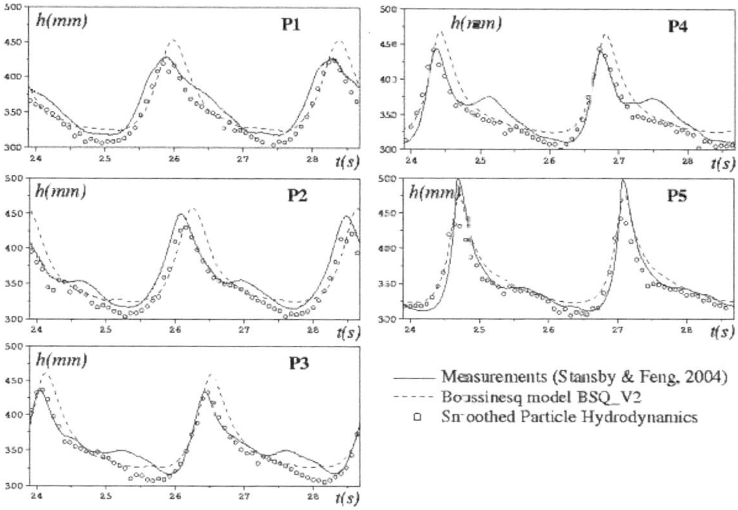

Fig. 23. Stansby and Feng's case.[37] Time series of the free surface elevation at the first 5 measurement points.

a dyke or a breakwater is a difficult task due to the complexity of the involved physical mechanisms. Empirical formulae were provided by several authors, giving the wave overtopping rate per unit length of the dyke Q (in l/s/m). We study here the prediction of wave overtopping rate over different types of dykes with varying slope, height, berm length, etc. and wave parameters (height and period). Non-dimensional wave overtopping rate $Q^* = Q/\sqrt{g.H^3}$ is used for comparisons, where H corresponds to the wave significant height. Figure 24 shows that our predictions are quantitatively in good agreement with the available experimental formulae.

4.7. *3D dam breaking*

A first attempt to apply the previous SPH Smagorinsky model to a complex case is herein presented: a 3D dam breaking problem, subject to gravity, is considered. The system, periodic in the spanwise direction z, is represented in figure 25. $30 \times 30 \times 60$ fluid particles according to x, y and z, are initially distributed on a Cartesian lattice, as shown on the top left picture of figure 26. Walls are modelled as previsouly described with wall and mirror particles. At $t = 0$, the fictitious dam is suddenly removed and one focuses

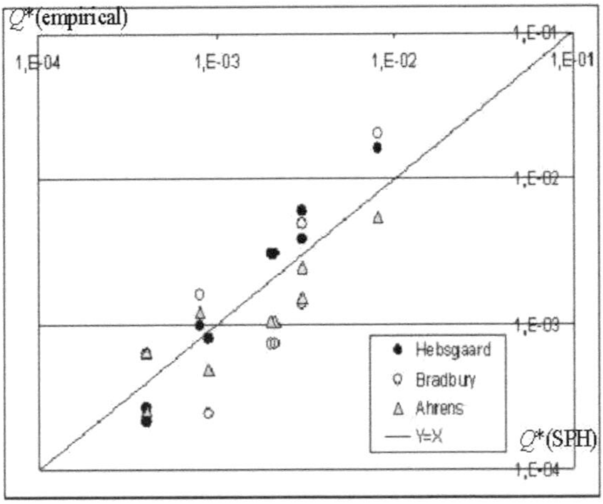

Fig. 24. Stansby and Feng's case.[37] Estimation of the wave overtopping rate over the dyke. Comparison of the non-dimensional SPH-computed overtopping rate Q^* (abscissa) with empirical formulae by Hebsgaard and other authors.[38–40]

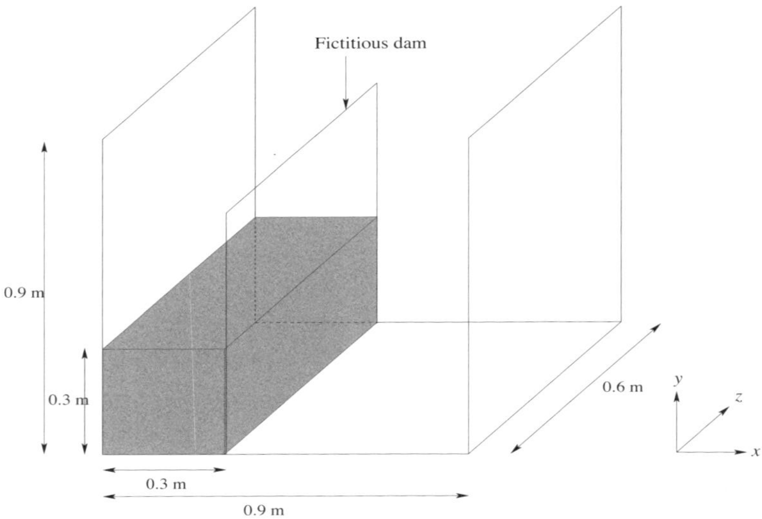

Fig. 25. Description of 3D dam breaking case.

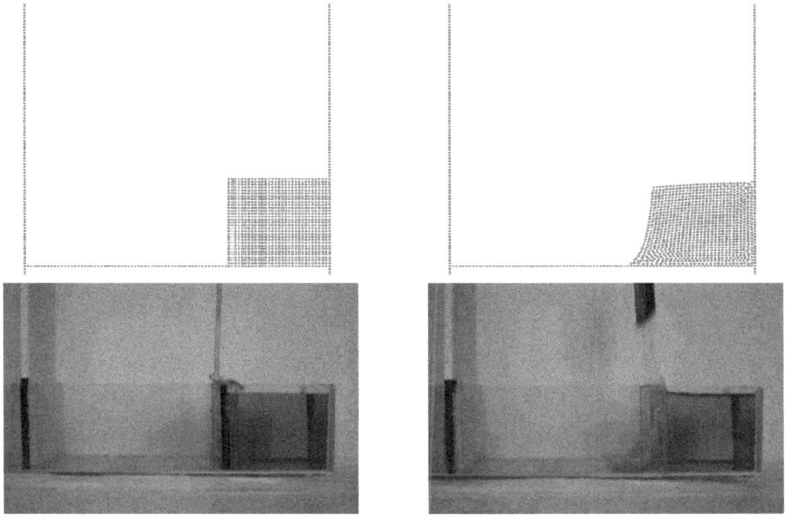

Fig. 26. Spartacus-3D results (top) compared to experimental results (bottom) around non-dimensional time 0 and $1.0.10^{-2}$ (from left to right).

on the instantaneous fluid motion. All the SPH equations considered here are identical to those introduced in part 3.4. In order to check the consistency of the simulation results, a basic experiment has been achieved at TU/Delft. A 3D channel, fitted with a sluice gate placed at a position consistent with simulation has been used. It is removed with a manual pulley system. A digital camera Sony DSC-P5 of 25 Hz has been used for the visualisation, and frames of size 320×240 are considered for the comparisons. Figures 26 to 31 represent the fluid motion simulated by Spartacus-3D and compared to experimental data, at similar non-dimensional times T_d defined by

$$T_d = \frac{t_i}{\sqrt{\frac{2g}{H_i}}} \tag{4.7}$$

where t_i denotes the physical time of the simulation (or of the experiment), H_i the initial height of the fluid considered by the SPH code (or in the experiment) and g the gravity. From a qualitative point of view, the results of SPH are on the whole quite satisfactory: pictures of figures 26, 27 and 28 firstly reveal that the evolution of the water front simulated by SPH is consistent with the experiment. Pictures of figure 29 show that the height

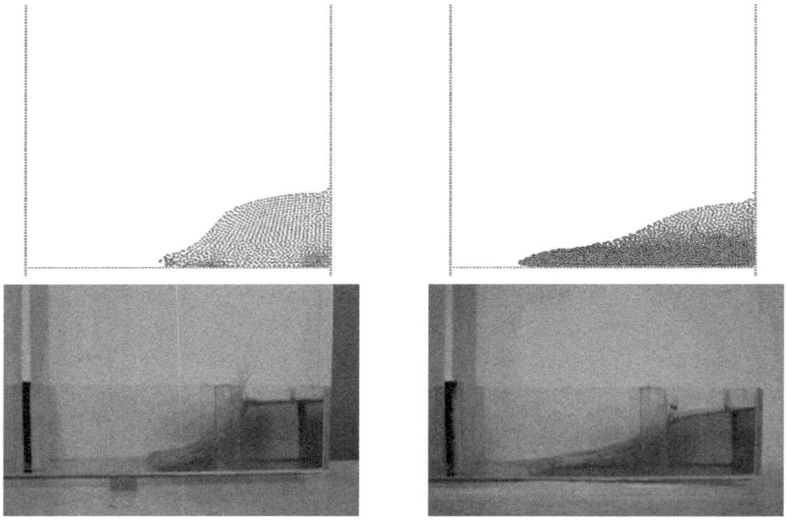

Fig. 27. Spartacus-3D results (top) compared to experimental results (bottom) around non-dimensional time $2.0.10^{-2}$ and $3.0.10^{-2}$ (from left to right).

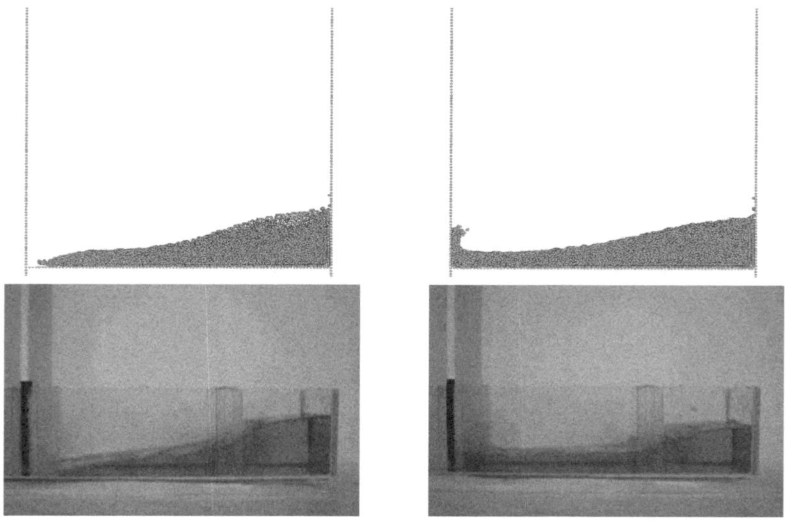

Fig. 28. Spartacus-3D results (top) compared to experimental results (bottom) around non-dimensional time $3.7.10^{-2}$ and $4.5.10^{-2}$ (from left to right).

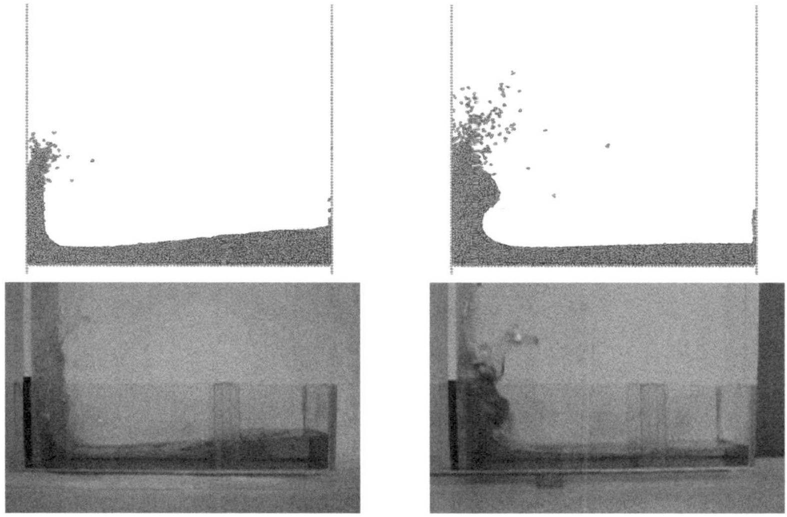

Fig. 29. Spartacus-3D results (top) compared to experimental results (bottom) around non-dimensional time $6.0.10^{-2}$ and $8.7.10^{-2}$ (from left to right).

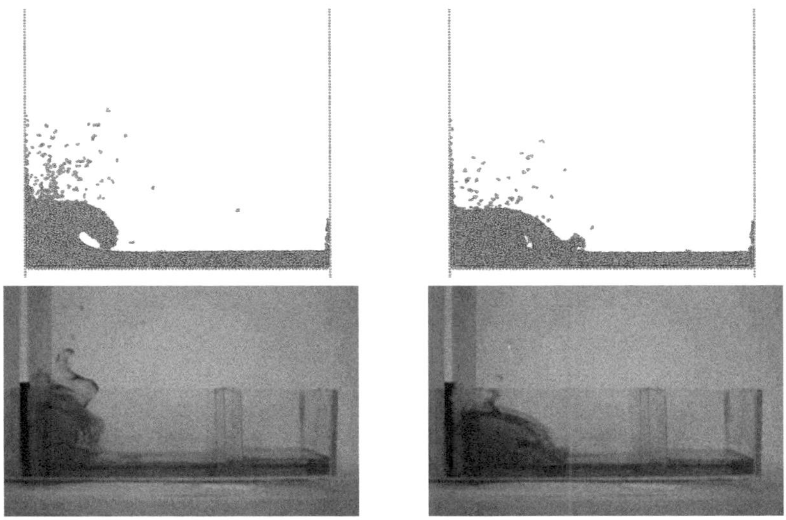

Fig. 30. Spartacus-3D results (top) compared to experimental results (bottom) around non-dimensional time $1.0.10^{-1}$ and $1.1.10^{-1}$ (from left to right).

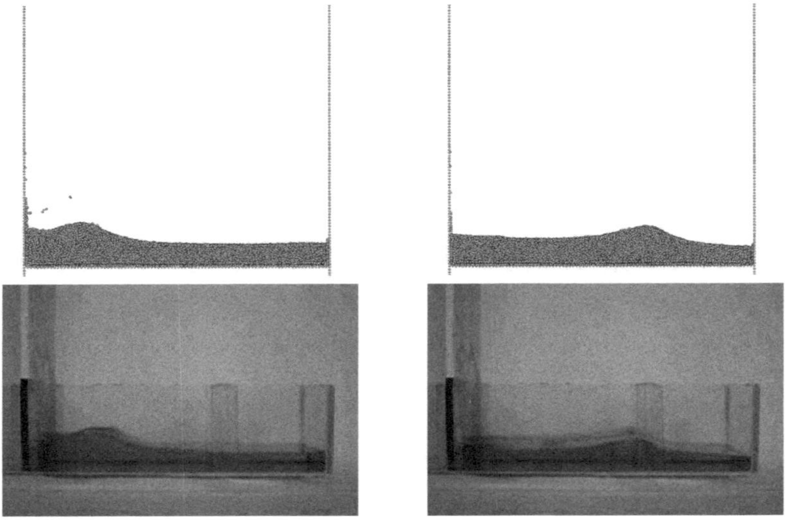

Fig. 31. Spartacus-3D results (top) compared to experimental results (bottom) around non-dimensional time $3.2.10^{-1}$ and $3.7.10^{-1}$ (from left to right).

of the fluid after impacting the wall is well predicted by SPH. The wave breaking depicted in figures 30 is also very similar to the experiment one. Moreover, the reflection wave motion (see figure 31) and the fluid behaviour near walls are also accurate.

5. Conclusion and Perspectives

It has been shown in this chapter that the SPH method (through Spartacus-2D code) is able to reproduce accurately solitary wave generation and propagation. Concerning plunging solitary wave simulation, it has been proved that turbulent effects have to be taken into account during this stage. Indeed, SPH statistical turbulence models developed by the authors here increase the accuracy of the results. Set up over a coral reef, as well as wave shoaling and overtopping, have also been successfully investigated by SPH: wave set-up and height match quite well with experimental results. Since coastal and environmental phenomena are mainly 3D, SPH LES has been considered and applied to the classical dam breaking problem. Free surface evolution simulated with SPH is very close to experimental results. However, all simulations presented here suffer from unstable pressure fields due to the state equation. In order to solve this problem, truly incompressible

SPH has been developed and successfully applied to concrete applications.[36] Incompressible SPH will hence be used in the future to improve presented work. Finally, since SPH is very time consuming, a massively parallel version (several thousand processor) of Spartacus (in 2D and in 3D) has been developed and will soon be applied to the previous test cases.

References

1. S. Li and W. K. Liu, Meshfree and particle methods and their applications, *Appl Mech Rev.* **55**, 1–34, (2002).
2. W. C. Welton, Two-dimensional PDF/SPH Simulations of Compressible turbulent flows, *Journal of Computational Physics.* **139**, (1998).
3. D. Violeau, S. Picon, and J.-P. Chabard, Two Attempts of Turbulence Modelling in Smoothed Particle Hydrodynamics, *Advances in Fluid Modelling and Turbulence Measurements.* (2002).
4. R. Issa, *Numerical assessment of the Smoothed Particle Hydrodynamics gridless method for incompressible flows and its extension to turbulent flows.* (University of Manchester, 2004).
5. D. Violeau and R. Issa, Numerical modelling of complex turbulent free surface flows with the SPH method: an overview, *International Journal for Numerical Methods in Fluids.* **in press**, (2006).
6. E. Y. M. Lo and S. Shao, Simulation of near-shore solitary wave mechanics by an incompressible SPH method, *Applied Ocean Research.* **24**, (2002).
7. R. A. Dalrymple and B. D. Rogers, Numerical modeling of water waves with the SPH method, *Coastal Engineering.* **53**, (2006).
8. J. H. Ferziger, Large eddy simulation. Turbulence course, Stanford University.
9. S. B. Pope, *Turbulent Flows.* (Cambridge University Press, 2000).
10. P. L. Viollet, J. P. Chabard, P. Esposito, and D. Laurence, *Mécanique des fluides appliquée. Ecoulements incompressibles dans les circuits, canaux, rivières, autour de structures et dans l'environnement.* (Presses de l'Ecole Nationale des Ponts et Chaussées, 1998).
11. J. G. M. Eggels, *Direct and Large Eddy Simulation of Turbulent Flow in a Cylindrical Pipe Geometry.* (TU/Delft, Aero en Hydrodynamica, 1994).
12. S. B. Pope, *Turbulent flows.* (Cambridge University Press, 2000).
13. M. Lesieur, P. Comte, E. Lamballais, O. Métais, and J. Silvestrini, Large-Eddy Simulations of shear flows, *Journal of Engineering Mathematics.* **32**, (1997).
14. B. E. Launder and D. B. Spalding, *Mathematical models of turbulence.* (Academic Press, London, 1972).
15. V. Guimet and D. R. Laurence, A linearised turbulent production in the $k-\epsilon$ model for engineering applications. Technical report, Proc. 5th International Symposium on Engineering Turbulence Modelling and Measurements, Mallorca, Spain, (2002).

16. D. C. Wilcox, *Turbulence modelling for CFD.* (DCW Industries Inc., 2004).
17. S. Wallin and A. V. Johansson, An explicit algebraic Reynolds stress model for incompressible and compressible turbulent flows, *Journal of Fluid Mechanics.* **403**, 89–132, (2000).
18. M. Lesieur and O. Métais, New trends in large-eddy simulations of turbulence, *Annu. Rev. Fluid Mech.* **28**, 45–98, (1999).
19. J. Eggels, *Direct and Large Eddy Simulation of Turbulent Flow in a Cylindrical Pipe Geometry.* PhD thesis, TU/Delft, Aero en Hydrodynamica, (1994).
20. R. Vignjevic, T. D. Vuyst, and M. Gourma, On interpolation in SPH, *Computer Modeling in Engineering and Sciences.* **2**, 319–336, (2001).
21. R. Vignjevic, J. R. Reveles, and J. Campbell, SPH in a Total Lagrangian Formalism, *Computer Modeling in Engineering and Sciences.* **14**, 181–198, (2006).
22. J. J. Monaghan, Smoothed Particle Hydrodynamics, *Annu. Rev. Astron. Astrophy.* **30**, (1992).
23. J. P. Morris, P. J. Fox, and Y. Zhu, Modelling low Reynolds Number Incompressible Flows Using SPH, *Journal of Computational Physics.* **136**, (1997).
24. P. W. Cleary and J. J. Monaghan, Conduction modelling using smoothed particle hydrodynamics, *Journal of Computational Physics.* **148**, (1999).
25. D. Violeau and R. Issa, Numerical modelling of complex turbulent free-surface flows with the sph method: an overview, *International Journal For Numerical Methods in Fluids.* **53**, 277–304, (2007).
26. R. Issa, E. S. Lee, D. Violeau, and D. Laurence, Incompressible separated flows simulations with the Smoothed Particle Hydrodynamics gridless method, *International Journal for Numerical Methods in Fluids.* **47**, 1101–1106, (2005).
27. R. Issa and D. Violeau, A first attempt to adapt 3d large eddy simulation to the smoothed particle hydrodynamics gridless method, (2006).
28. R. Issa and D. Violeau, Modelling a plunging breaking solitary wave with eddy-viscosity turbulent SPH models, *Computers, Materials and Continua.* To be published, (2007).
29. D. Violeau and R. Issa, Guide for the spartacus-2d v1p2 code : Lagrangian modelling of two-dimensional laminar and turbulent flows with SPH method, *EDF-R&D report.* (2007).
30. S. A. Hughes, *Physical models and laboratory techniques in coastal engineering.* (World Scientific, 1993).
31. Y. Li and F. Raichlen, Energy balance model for breaking solitary wave runup, *Journal of waterway, port, coastal and ocean engineering.* **47**(2), 47–59, (2003).
32. I. M. Jánosi, D. Jan, K. Gábor Szabó and T. Tél, Turbulent drag reduction in dam-break flows, *Experiments in Fluids.* **37**, 219–229, (2004).
33. M. R. Gourlay, Wave set-up on coral reefs. 1. Set-up and wave-generated flow on an idealised two dimensional horizontal reef, *Coastal Engineering.* **27**, 161–193, (1996).
34. R. G. Dean and R. A. Dalrymple, *Water wave mechanics for engineers and scientists.* (World Scientific, 1991).

35. S. Massel and M. Gourlay, On the modelling of wave breaking and set-up on coral reefs, *Coastal Engineering.* **39**, 1–27, (2000).
36. E.-S. Lee, D. Violeau, M. Benoit, R. Issa, D. Laurence, and P. Stansby, Prediction of wave overtopping on coastal structures by using extended Boussinesq and SPH models. Technical report, Proc. XXXth Int. Conf. Coastal Eng., (2006).
37. P. Stansby and T. Feng, Surf zone wave overtopping a trapezoidal structure: 1-D modelling and PIV comparison, *Coastal Engineering.* **51**, 483–500, (2004).
38. M. Hebsgaard, P. Sloth, and J. Juhl, Wave overtopping of rubble-mound breakwaters. Technical report, Proc. 26th Int. Conf. Coastal Eng., (1998).
39. A. P. Bradbury and N. W. H. Allsop, Hydraulics effects of breakwater crown walls. Technical report, Proc. Int. Conf. Breakwaters'88, (1988).
40. J. P. Ahrens and M. S. Heimbaugh, Seawall overtopping model. Technical report, Proc. 21st Int. Conf. Coastal Eng., (1988).

CHAPTER 15

MLPG_R METHOD AND ITS APPLICATION TO VARIOUS NONLINEAR WATER WAVES

Q.W. Ma

School of Engineering and Mathematical Sciences, City University
Northampton Square, London, EC1V 0HB, UK
q.ma@city.ac.uk

The author has extended the MLPG (Meshless Local Petrov-Galerkin) method developed by Prof. Atluri and his group (http://care.eng.uci.edu/atluri.htm) to model nonlinear water waves; and then made further development to produce a method called MLPG_R (Meshless Local Petrov-Galerkin method based on Rankine source solution) method. The MLPG_R method is much more efficient than the MLPG method when applied to water wave problems. This chapter will review the development of these methods and its application to various nonlinear water waves based on the author's previous work.

1. Introduction

In order to simulate nonlinear water waves, mesh-based methods, such as finite element (FEM), finite volume and finite difference methods, are widely used. These methods have provided many useful and satisfactory results. However, their successes largely rely on good quality meshes. The construction of such meshes, particularly unstructured ones, is usually a time-consuming task because it must be ensured that aspect ratios of all elements are not very large or very small and connectivity between nodes and elements must be carefully and accurately found and recorded. More problematically, elements can frequently become over-distorted during simulation of water wave evolutions unless fixed meshes are used. The over-distorted elements may be amended by remeshing.

However remeshing can be as expensive as generation of original meshes and may take a major proportion of computational time, considering that this has to be done every time step if unstructured mesh is necessarily used. In order to tackle the difficulty, the research group of the author devised a new method called quasi arbitrary Lagrangian-Eulerian finite element method (QALE-FEM)[1] which is reviewed by the author in another chapter (Ch 5) in this book. The main difference of the QALE-FEM from the conventional FEMs[41] is that the complex unstructured mesh is generated only once at the beginning and is moved by using a spring analogy method at all other time steps in order to conform to motion of the free surface. This feature allows one to use an unstructured mesh with any degree of complexity without the need of regenerating it at every time step and so the difficulty with regenerating meshes is bypassed. However, the QALE-FEM is so far suitable only for potential flow problems. As for using fixed meshes, although the over-distortion problems may not arise, numerical diffusions due to advection terms may become severe and a much larger computational domain than that for moving meshes may need to be considered.

A new class of methods has also been developed and applied to water wave problems, which do not pose the difficulties associated with meshes. Actually, these methods do not need use of any mesh to discretise computational domains and to construct approximate solutions. Alternatively, they are based only on randomly-ordered and -distributed nodes, implying that the problems associated with the shape and connectivity of elements and regeneration of mesh in mesh-based methods vanish automatically. Many Meshless (or particle) methods have been reported in the literature, such as element free Galerkin method,[2] diffusion element method,[3] reproducing kernel particle method,[4] smoothed particle hydrodynamic (SPH) method,[5] particle finite element method,[6] moving particle semi-implicit (MPS) method[7] and so on. Among them, the SPH and MPS methods have been applied to simulate water waves by many researchers. Reviews on them are given in other chapters (Chs 12, 13 and 14) of this book. Very recently, another meshless method, called Meshless Local Petrove-Galerkin (MLPG) method, has been proposed by Atluri and his colleagues in around 1998. Since then, this method has been rapidly developed, extended and applied to various problems. This chapter will focus mainly on its development related to nonlinear water waves but a brief review over a broader scope will be given in the next section.

Table 1. Development history of MLPG method.

Developments	Applications	Year
Proposed[8]	2D steady linear potential flow	1998
Extended to nonlinear problems[9]	Possion equation with nonlinear terms	1998
Solving 2D elasto-statics problems[10]	Plate with a circular or elliptical hole	2000
Extended to steady convection and diffusion flow[11]	2D steady problems	2000
Extended to steady Navier-Stokes equations[12]	Stokes flows and Lid-driven cavity flow	2001
Six MLPG methods[13]	Comparison of different MLPGs	2002
Solving 3D elasto-statics[14]	3D Boussinesq and Eshelby's Inclusion Problems	2003
Multiscale simulation[15]	2D grapheme sheet of a nano-system	2004
Extended for 2D Heat Conduction Problem[16]	Anisotropic square with homogeneous material properties	2004
Solving static problem of thick plates[17]	Clamped and simply supported square plates	2004
MLPG mixed Finite Volume Method (MFVM)[18]	Cube under uniform tension; cantilever and curved beams	2004
Solving 3D elasto-dynamic problems[19]	Cantilever beam under a sudden transverse load; a semi-infinite space under a pulse load	2004
Mixed MPLG for 4th order ODE[20]	No specific applications	2005
Extended to nonlinear water waves[21]	Various waves generated by wave makers	2005
MLPG_R method[22,23]	Waves generated by wave makers; sloshing waves and dam breaking	2005
Solving contact, impact, and penetration mechanics[24]	Rod-on-rod impact; plate-on-plate impact; ballistic impact	2006
MLPG mixed collocation method[25]	Cantilever and curved beams; a plate with a circular hole	2006
MLPG mixed finite difference method[26]	Curved beam under a end load a plate with a circular hole	2006
Non-singular MLPG(LBIE) method with free of MLS-derivatives[27]	Elastic problems	2007
SFDI interpolation[28]	For MLPG_R method	2008

2. Brief History of the MLPG Method

For completeness, the brief history of the MLPG method will be given in this section, which does not only cover those related to water waves but also other areas. In order to give readers an idea about the evolution of the method, the development stages of this method are outlined in Table 1. It should be noted that this table should not be considered as an extensive list and some developments may not be contained due to the limitation of the author's knowledge.

The MLPG method was first published by Atluri and Zhu in 1998[8]. After that, it has been developed along several lines, including (1) different forms; (2) different interpolations; (3) extension to different problems; and (4) different applications. Each of these is summarised below with the reference to Table 1.

Different Forms. The original MLPG method[8] was based on a local symmetric weak form. The shape function used to give relationship between unknown function and its nodal values was formulated by using a moving least square method (MLS). The derivative of an unknown function was found by direct differentiation of the shape function. Soon after the introduction of the concept of the MLPG method, many variants of the method were suggested[13]. They were named as MLPG1, MLPG2, MLPG3, MLPG4, MLPG5 and MLPG6, corresponding to symmetric or asymmetric local forms, different trial and test functions. Based on numerical tests, the MLPG5 seems to be more promising than others. It was not long before realising that the way for formulating the derivatives, in particular the second-order ones, by direct differentiation was very time consuming. To circumvent the drawback, various forms of mixed MLPG methods were developed by Atluri and his research group, which included the MLPG mixed finite volume method[18], the MLPG mixed collocation method[25] and the MLPG mixed finite different method[26]. The basic idea of these methods is that an unknown function and its derivative are interpolated, separately. Using these mixed techniques, the inefficient way to find the derivatives is bypassed. In addition to this, the requirement on completeness and continuity of shape functions is reduced by one order. The MLPG_R method[22,23] proposed by the author aimed also to avoid the estimation of derivative in the local weak form. However it was not through the change in the way formulating the shape function; instead, it was achieved by eliminating the derivatives in the local weak form. The method is particularly effective for nonlinear water wave problems. Due to these developments, the MLPG method becomes so general that many meshless methods may be considered as its special forms, for example the SPH and MPS methods may be considered as special cases of the MPLG method based on the collocation formulation.

Different Interpolations. So far, the MLS method has often been adopted to formulate the shape function or to interpolate unknown functions in terms of its nodal values in various forms of MLPG methods. Numerical tests[13] showed that the MLS method was superior compared with other well-known methods, such as Shepard function (SF), the partition of unity (PU), the reproducing kernel particle interpolation (RKPM) and radial basis function (RBF). Nevertheless, the MLS interpolation function has some drawbacks, as discussed in Section 4.2.1. One of them is related to the evaluation of the gradient by direct differentiation. To avoid the direct differentiation on the interpolation function, Atluri, Liu and Han (2006)[26] described a new formulation for gradients or derivatives, in which the unknown function is still evaluated using the MLS method but its gradient is estimated by a finite difference method and by minimising the norm of the unknown function with respect to the gradient components (similar to a least square method). Along this line, the author developed another method called simplified finite difference interpolation (SFDI)[28]. In this method, use of the MLS is eliminated not only for estimating the gradient but also for the function interpolation. Detail of this method will be given in Section 4.2 of this chapter.

Extension to Different Problems. Since the MLPG method was developed for linear and nonlinear Possion equations[8,9], it has been extended to deal with various problems in several fields. Most of them are related to solid mechanics, such as the 2D elasto-statics problems[10], 3D elasto-statics problems[14], static problem of thick plates[17], 3D elasto-dynamic problems[18] and problems about contact, impact, and penetration[24]. The significant extensions to other fields have also been made, which include the problems of steady convection and diffusion[11], the fluid flow governed by steady Navier-Stokes equations[12], 2D heat conduction problem[16] and multi-scale simulation of molecular motions[15]. It was extended to simulating nonlinear water waves by the author in 2005[21]. The MLPG has a potential to be applied to a wider range of areas and further extensions are expected in future.

Different Applications. With development of the MLPG method, its applications have become more and more extensive and sophisticated. In solid mechanics, it started with the application to beams[10], then to plates[10] and now to complicated contact, impact and even to penetration of projectiles against plates[24]. In fluid mechanics, its application was first made to simple linear potential flow around a 2D cylinder[8], then to the Stokes and Lid-driven cavity flows and then to heat conduction. It is now applied to solve various nonlinear water waves, such as mono- and bi-chromatic waves[21], sloshing waves[22], dam breaking[23] and freak waves[29].

To be consistent with the context of this book, the following discussions of this chapter will focus only on the development of the MLPG related to nonlinear water waves and will largely cover the author's work in this area over past several years.

3. Governing Equations and Numerical Procedure

3.1. *Governing equations*

In our work, fluids are assumed to be incompressible. The corresponding governing equations and boundary conditions are given by

$$\nabla \cdot \vec{u} = 0 \tag{1a}$$

$$\frac{D\vec{u}}{Dt} = -\frac{1}{\rho}\nabla p + \vec{g} + \nu\nabla^2\vec{u} \tag{1b}$$

$$\frac{D\vec{x}}{Dt} = \vec{u} \text{ and } p = p_{atm} \qquad \text{on the free surface} \tag{2a}$$

$$\vec{u}\cdot\vec{n} = \vec{U}\cdot\vec{n} \text{ and } \vec{n}.\nabla p = \rho\left(\vec{n}\cdot\vec{g} - \vec{n}\cdot\dot{\vec{U}}\right) \text{ on solitary boundary} \tag{2b}$$

where ρ is the density of fluid, ν its kinematic viscosity, $\vec{u}$ the velocity of fluids, $\vec{U}$ the velocity of solid boundaries, $\vec{g}$ the gravitational acceleration, p the pressure and p_{atm} the atmospheric pressure. To solve the fully nonlinear problem in time domain, the initial conditions are also required. We assume that the simulation starts from rest. As a result, the velocity is zero, the pressure is hydrostatic and the free surface is a horizontal straight line (2D case) or plane (3D cases).

The momentum equation (Eq. 1b) is given as the Lagrangian form. It can also be written in the Eulerian form that is suitable for mesh-fixed methods. The two forms are equivalent. However, the convective term does not explicitly appear in Eq. (1b) and will not be estimated during simulation process. Therefore, the numerical diffusion related to the convection term, which is of a great concern in mesh-fixed methods, is automatically eliminated. In addition, it is well known that the viscous term in Eq. (1b) may play a little role in water wave problems if there is no wave breaking. On this basis, this term has not been considered so far and so will be ignored hereafter. Without the viscous term, the flow can be formulated by a velocity potential theory. Nevertheless, the potential formulation is not adopted in our work related to the MLPG method even for non-breaking waves because our long term aim is to develop it into one suitable for all kinds of wave problems.

3.2. *Numerical procedure*

The above equations are solved using a time-step marching procedure. This starts from a particular instant when the velocity and the geometry of fluid flow are known and then evolves to next time step, at which all physical quantities are updated by solving the governing equations. During each time step, the problem is formulated using a well known time-split procedure. This formulation contains three sub-steps:

(1) Evaluating intermediate velocities and positions.

At start of a new time step, intermediate velocities ($\vec{u}^{(*)}$) and positions ($\vec{r}^{(*)}$) with ignoring viscosity are first evaluated by

$$\vec{u}^{(*)} = \vec{u}^{(n)} + \vec{g}\Delta t \tag{3}$$

$$\vec{r}^{(*)} = \vec{r}^{(n)} + \vec{u}^{(*)}\Delta t \tag{4}$$

where $\vec{r}$ is the position vector of a point, Δt is the time step and the superscripts (n) in the parenthesis indicate the quantities at time $t = t_n$. On this basis, the velocities at time $t = t_{n+1} = t_n + \Delta t$ can be expressed by

$$\vec{u}^{(n+1)} = \vec{u}^{(*)} + \vec{u}^{(**)}, \tag{5}$$

where $\vec{u}^{(**)}$ is a velocity correction.

(2) Estimating the pressure.

Integrating Eq. (1b) without considering viscosity over the time interval (t_n , t_{n+1}) results in

$$\vec{u}^{(n+1)} = \vec{u}^{(n)} + \vec{g}\Delta t - \int_{t_n}^{t_{n+1}} \left(\frac{1}{\rho} \nabla p \right) dt \,. \tag{6}$$

The integral with respect to time can be approximated by well-known methods, such as explicit, implicit or semi-implicit methods, i.e.

$$\int_{t_n}^{t_{n+1}}\left(\frac{1}{\rho}\nabla p\right)dt = \Delta t\left[\frac{\theta}{\rho}\nabla p^{(n+1)} + \frac{1-\theta}{\rho}\nabla p^{(n)}\right] \tag{7}$$

with $\theta=0$ for explicit, $\theta=1$ for implicit and $\theta=1/2$ for semi-implicit methods. Substituting Eq. (7) into Eq. (6) results in:

$$\begin{aligned}\vec{u}^{(n+1)} &= \vec{u}^{(n)} + \vec{g}\Delta t - \Delta t\left[\frac{\theta}{\rho}\nabla p^{(n+1)} + \frac{1-\theta}{\rho}\nabla p^{(n)}\right] \\ &= \vec{u}^{(*)} - \Delta t\left[\frac{\theta}{\rho}\nabla p^{(n+1)} + \frac{1-\theta}{\rho}\nabla p^{(n)}\right].\end{aligned} \tag{8}$$

From Eq. (5) it follows that

$$\vec{u}^{(**)} = -\frac{\Delta t}{\rho}\left[\theta\nabla p^{(n+1)} + (1-\theta)\nabla p^{(n)}\right]. \tag{9}$$

The velocity in Eq. (5) or (8) must also satisfy the continuity equation (1a), yielding

$$\nabla^2 p^{(n+1)} = \frac{\rho}{\theta\Delta t}\nabla\cdot\vec{u}^{(*)} - \left(\frac{1}{\theta}-1\right)\nabla^2 p^{(n)}. \tag{10}$$

It is clear that this formulation is not suitable for $\theta=0$. In fact, the explicit method corresponding to the zero value of θ is not stable and possesses low accuracy. Thus, it is suggested that the value of θ is always chosen in the range $0<\theta\le 1$. In our work, the fully implicit method ($\theta=1$) is employed, leading to

$$\vec{u}^{(n+1)} = \vec{u}^{(*)} - \frac{\Delta t}{\rho}\nabla p^{(n+1)} \tag{11}$$

and

$$\nabla^2 p^{(n+1)} = \frac{\rho}{\Delta t}\nabla\cdot\vec{u}^{(*)}. \tag{12}$$

Equation (10) or (12) governs the pressure at the new time level, from which the solution for the pressure at the new time level can be found.

(3) Computing the velocity and the position at time $t = t_{n+1}$.

After the solution for $p^{(n+1)}$ is found, the velocities can be estimated using Eq. (11) and the positions of fluids can be updated by numerically integrating the velocities.

The time split procedure was initially introduced for incompressible flow[30]. It has been extended to deal with various problems using a finite element method as described in the book of Zienkiewicz and Taylor[31]. The procedure has also been employed by a number of authors who use other meshless or particle methods, e.g., Lo and Shao[32] for the SPH method, Koshizuka and Oka[7] for the MPS method and Idelsohn et al.[33] for the particle FEM method. However, it is the publication by Ma[21] that combined this procedure with the MLPG method.

4. MLPG Formulation

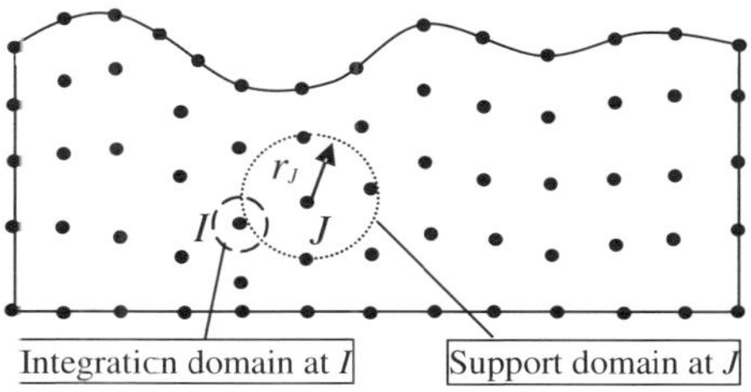

Fig. 1. Illustration of nodes, integration domain and support domain.

The key task in the above formulation is to find the solution for the pressure by solving Eq. (12). Various methods, such as finite element and finite different methods may be used. In this article, we will discuss how to employ the MLPG and MLPG_R methods to solve it. These two methods are composed of several numerical techniques, which will be summarised in the following sections.

4.1. *Weak forms*

These two methods are both based on a set of nodes, as illustrated in Fig. 1, which discretise the fluid domain. Some of these nodes are located on boundaries and others lie inside the fluid domain. Formulation around inner nodes is first discussed and formulation around boundary nodes is described in Section 4.4. At each of the inner nodes, a sub-domain is specified, which is a circle for two-dimensional (2D) and a sphere for three-dimensional (3D) cases. Equation (12), after multiplying by an arbitrary test function o, is integrated over each of the sub-domains, leading to the following weak form

$$\int_{\Omega_I}\left[\nabla^2 p-\frac{\rho}{\Delta t}\nabla\cdot\vec{u}^{(*)}\right]\varphi d\Omega=0\,, \tag{13}$$

where Ω_I is the area or volume of the sub-domain centred at node I. In Eq. (13), the second order derivative of unknown pressure and the gradient of the intermediate velocity are included. Direct calculation of the derivative and gradient requires not only much computational time but also degrades the accuracy. Equation (13) may be changed into a better form by choosing the test function. There are many options for the test function[13]. In Ma (2005a)[21], the Heaviside step function ($\varphi=1$ in Ω_I and $\varphi=0$ otherwise) was used and the following form was given:

$$\int_{\partial\Omega_I}\vec{n}\cdot\nabla p dS=\frac{\rho}{\Delta t}\int_{\partial\Omega_I}\vec{n}\cdot\vec{u}^{*}dS\,. \tag{14}$$

In Ma (2005b)[22], the solution for a Rankine source is taken as the test function. This solution satisfies that $\nabla^2\varphi=0$ in Ω_I except for its centre and $\varphi=0$ on $\partial\Omega_I$, the boundary of Ω_I, and can be expressed as

$$\varphi=\frac{1}{4\pi}\begin{cases}2\ln(r/R_I) & \text{for a two dimensional case}\\(1-R_I/r) & \text{for a three dimensional case}\end{cases}, \tag{15}$$

where r is the distance between a concerned point and the centre of Ω_I and R_I the radius of Ω_I. Using this solution, adding a zero term $p\nabla^2\varphi$ to Eq. (3) and applying the Gauss's theorem, it follows that

$$\int_{\partial\Omega_I+\partial\varepsilon}[\vec{n}\cdot(\varphi\nabla p)-\vec{n}\cdot(p\nabla\varphi)]dS=\int_{\partial\Omega_I+\partial\varepsilon}\frac{\rho}{\Delta t}\vec{n}\cdot\left(\varphi\vec{u}^{(*)}\right)d\Omega-\int_{\Omega_I}\frac{\rho}{\Delta t}\vec{u}^{(*)}\cdot\nabla\varphi d\Omega \tag{16}$$

where $\partial\varepsilon$ is a small closed surface surrounding the centre of Ω_I, which is a circle in 2D cases and a spherical surface in 3D cases with a radius of ε. The reason for adding $\partial\varepsilon$ is that the test function φ in Eq. (15) becomes infinite at $r = 0$ and so the Gauss's theorem can not be used otherwise. Considering the terms in Eq. (16) related to $\partial\varepsilon$, one can easily prove that taking $\varepsilon\to 0$ yields

$$\int_{\partial\Omega_I+\partial\varepsilon}\left[\vec{n}\cdot(\varphi\nabla p)\right]dS=0\,,\quad\int_{\partial\varepsilon}\left[\vec{n}\cdot(p\nabla\varphi)\right]dS=-p \text{ and } \int_{\partial\Omega_I+\partial\varepsilon}n\cdot\phi\vec{u}^{(*)}dS=0\,.$$

As a result, Eq. (13) becomes

$$\int_{\partial\Omega_I} \vec{n}\cdot(p\nabla\varphi)dS - p = \int_{\Omega_I} \frac{\rho}{\Delta t}\vec{u}^{(*)}\cdot\nabla\varphi d\Omega\ . \tag{17}$$

This equation is similar to Eq. (14) in that only boundary integral on the left hand side is involved but is distinct from the latter in that the pressure, rather than pressure gradient, is dealt with here. This is of great benefit because estimating pressure is much easier than estimating pressure gradient. In addition, although the left hand side of Eq. (17) is in a similar form to that of Eq. (9) in Zhu et al.,[34] the right hand side is different from the corresponding term in their paper. It is this difference that allows the term related to the intermediate velocity derivatives to be eliminated from the weak form. The improvement in these two aspects makes evaluation of the integrals and therefore the whole procedure more efficient. However, the right hand side in Eq. (17) is an integral over the local domain while the right hand in Eq. (14) is an integral over the boundary of the local domain. To avoid inefficiency of evaluating the right hand in Eq. (17), a semi-analytical technique[21] has been proposed, which will be presented in Section 4.3 of this chapter. The method based on the weak form in Eq. (17) is named as MLPG_R method in Ma (2005b)[22], where R in the name represents the Rankine source solution.

4.2. *Meshless interpolations*

In order to find solution for the pressure in Eq. (14) or (17) and to update the velocity in Eq. (11), the pressure or its gradient at a point (not necessary a node) must be expressed in term of discrete values of the pressure at nodes. The intermediate velocity needs also to be estimated by using nodal value of the velocity. To do so, one requires interpolation schemes.

Various available meshless interpolation schemes have been explored[13]. However, as indicated above, the MLS method is more popular than other interpolation schemes for the MLPG methods. It is the reason why it was utilised by the author at the beginning of development of the method[21,22]. However, the MLS method has some drawback as discussed in Section 2. The author very recently suggested a new scheme based on finite difference. Some details about these methods are given in the following two sub-sections.

4.2.1. *MLS method*

For the sake of convenience, the unknown function (pressure or velocity) is represented by $f(\vec{r})$ with $\vec{r}$ denoting the position vector of a point. In the MLS method, the unknown function $f(\vec{r})$ is written as

$$f(\vec{r}) \approx \sum_{J=1}^{N} \Phi_J(\vec{r}) \hat{f}_J , \tag{18a}$$

where N is the number of nodes that affect the function at Point $\vec{r}$; $\hat{f}_J$ are nodal variables but not necessarily equal to the nodal values of $f(\vec{r})$. In the equation, $\Phi_J(\vec{r})$ is called as the interpolation or shape function and is given as

$$\Phi_J(\vec{r}) = \sum_{m=1}^{M} \psi_m(\vec{r}) \left[\mathbf{A}^{-1}(\vec{r}) \mathbf{B}(\vec{r}) \right]_{mJ} = \boldsymbol{\psi}^T(\vec{r}) \mathbf{A}^{-1}(\vec{r}) \mathbf{B}_J(\vec{r}) \tag{18b}$$

with the base function, assumed to be linear, being

$$\boldsymbol{\psi}^T(\vec{r}) = [\psi_1, \psi_2, \psi_3] = [1, x, y] \ (M=3) \qquad \text{for 2D cases}$$

and

$$\boldsymbol{\psi}^T(\vec{r}) = [\psi_1, \psi_2, \psi_3, \psi_4] = [1, x, y, z] \, (M=4) \text{ for 3D cases;}$$

and with the matrixes $\mathbf{B}(\vec{x})$ and $\mathbf{A}(\vec{x})$ being defined as

$$\mathbf{B}(\vec{r}) = \boldsymbol{\Psi}^T \mathbf{W}(\vec{r}) = [w_1(\vec{r} - \vec{r}_1)\boldsymbol{\psi}(\vec{r}_1), w_2(\vec{r} - \vec{r}_2)\boldsymbol{\psi}(\vec{r}_2), \cdots] \tag{18c}$$

$$\mathbf{A}(\vec{r}) = \boldsymbol{\Psi}^T \mathbf{W}(\vec{r}) \boldsymbol{\Psi} = \mathbf{B}(\vec{r}) \boldsymbol{\Psi} \tag{18d}$$

where $\mathbf{W}(\vec{r})$ and $\boldsymbol{\Psi}$ are, respectively, expressed by

$$\mathbf{W}(\vec{r} - \vec{r}_J) = \begin{bmatrix} w(|\vec{r} - \vec{r}_J|) & 0 & \cdots & 0 \\ 0 & & & \\ \cdots & & & \\ 0 & & & w(|\vec{r} - \vec{r}_J|) \end{bmatrix} \tag{18e}$$

and

$$\mathbf{\Psi}^T = [\psi(\vec{r}_1), \psi(\vec{r}_2), \cdots, \psi(\vec{r}_N)], \tag{18f}$$

which shows each column of the matrix $\mathbf{\Psi}^T$ is the value of the base function ψ at a particular point. $w(|\vec{r} - \vec{r}_J|)$ in Eq. (18e) is a weight function. Its specific form will be given later. The gradient of the unknown function is estimated by

$$\nabla f(\vec{r}) \approx \sum_{J=1}^{N} \nabla \Phi_J(\vec{r}) \hat{f}_J . \tag{18g}$$

The partial derivatives (or the gradient) of the shape function, with respect to y for instance, are found by differentiating Eq. (18b), that is,

$$\Phi_{J,y} = \psi^T{}_{,y} \mathbf{A}^{-1} \mathbf{B}_J + \psi^T \mathbf{A}^{-1}{}_{,y} \mathbf{B}_J + \psi^T \mathbf{A}^{-1} \mathbf{B}_{J,y} \tag{18h}$$

where $\mathbf{A}^{-1}{}_{,y}$ is the partial derivative of $\mathbf{A}^{-1}$ (the inverse of $\mathbf{A}(\vec{r})$), with respect to y (x or z) and is evaluated by $\mathbf{A}^{-1}{}_{,y} = -\mathbf{A}^{-1} \mathbf{A}{}_{,y} \mathbf{A}^{-1}$; $\mathbf{B}_J$ is J-th column of Matrix $\mathbf{B}$ and its partial derivative is estimated by

$$\mathbf{B}_{J,y} = \frac{\partial w_J(\vec{r} - \vec{r}_J)}{\partial y} \psi(\vec{r}_J).$$

With the sufficient number of neighbour nodes, the interpolations given by Eqs. (18a) and (18g) are accurate for a linear function independent on the node distribution. This is a good feature, particularly for the problems, such as about nonlinear water waves, involving large deformation of computational domain. Nevertheless, they do have some drawbacks. (1) The interpolation function will not be well defined if the number of nodes (N) affecting the concerned point is not large enough. This implies that it may not work properly when modelling the problems associated with fragment phenomena, such as breaking waves. Although this may be overcome by enlarging the support domain, the larger support domain may lead to over-smoothing and so degrade the accuracy. (2) Interpolation of the function in Eq. (18a) needs the inverse of Matrix $\mathbf{A}$, which is the order of $m \times m$. Although m may be only 3 for 2D case and only 4 for 3D cases, the computational cost spent on it is not negligible due to the fact that the number of points at which the function $f(\vec{r})$ needs to be estimated is much larger (at least 8 times for

2D and 16 times for 3D) than the number of total nodes when discretising Eq. (16) and that the computational cost for the inverse of the matrix is proportional to m^3. (3) The evaluation of the gradient as shown in Eq. (18h) is even more expensive due not only to inverse of matrix but also to successive multiplication of several matrixes. To avoid the direct differentiation on the interpolation function, Prof. Atluri and his research group suggested a mixed approach in their several recent publications. In particular, Atluri, Liu and Han[26] describe the following new formulation. The unknown function is still evaluated using Eq. (18a) but its gradient is estimated by a different way. The components of the gradient are determined by minimising the norm of the unknown function with respect to the gradient components (similar to a least square method) and are expressed (different symbols used here to avoid some confusion with other equations) as

$$f_{,y}{}^{I}(\vec{r}_I) = \sum_{J=1}^{N} H_y^J f(\vec{r}_I + \varsigma\Delta\vec{r}_{IJ}) - \overline{H}_y f(\vec{r}_I)$$
$$\overline{H}_y = \sum_{J=1}^{N} H_y^J \qquad (19)$$

where $f_{,y}{}^{I}(\vec{r}_I)$ is the y-component (y can be changed into x or z to give other components) of the gradient of $f(\vec{r})$ at Node I, $\mathbf{H}=(\hat{\mathbf{h}}^{\mathrm{T}}\mathbf{W}\hat{\mathbf{h}})^{-1}\hat{\mathbf{h}}^{\mathrm{T}}\mathbf{W}$, $\hat{\mathbf{h}}^{\mathrm{T}}=\varsigma[\,\Delta\vec{r}_{I1},\ \Delta\vec{r}_{I2},\ \Delta\vec{r}_{I3}\ldots\ldots\ \Delta\vec{r}_{IN}\,]$ and **W** is the same matrix as in Eq. (18e). The ς is a shrink factor, controlling the position of sample points that are not necessarily the nodes. If the ς is not equal 1, the value of $f(\vec{r}_I + \varsigma\Delta\vec{r}_{IJ})$ must be found by Eq. (18a) before estimating the gradient.

4.2.2. *Simplified finite difference interpolation (SFDI) method*

To overcome drawbacks of the MLS method pointed above, a new method called simplified finite different interpolation (SFDI) method has been recently proposed by the author[28]. Its details for interpolating a function and its gradients are separately described below.

Interpolation of a function

A general function $f(\vec{r})$ is expanded into a Taylor series at Point $\vec{r}_0$:

$$f(\vec{r}) = f(\vec{r}_0) + (\nabla f)_{\vec{r}_0}\cdot(\vec{r}-\vec{r}_0) + O(|\vec{r}-\vec{r}_0|^2). \qquad (20)$$

Applying the expression to a set of nodes (J) near $\vec{r}_0$, multiplying a weight function $w(|\vec{r}_J - \vec{r}_0|)$ on both sides and taking the sum of equations at all relevant nodes yields

$$\sum_{J}^{N} f(\vec{r}_J) w(|\vec{r}_J - \vec{r}_0|) = \sum_{J}^{N} f(\vec{r}_0) w(|\vec{r}_J - \vec{r}_0|)$$
$$+ (\nabla f)_{\vec{r}_0} \cdot \sum_{J}^{N} (\vec{r}_J - \vec{r}_0) w(|\vec{r}_J - \vec{r}_0|) + O\left(\sum_{J}^{N} |\vec{r}_J - \vec{r}_0|^2 w(|\vec{r}_J - \vec{r}_0|)\right), \qquad (21)$$

where N is the total number of nodes with $w(|\vec{r}_J - \vec{r}_0|) \geq 0$; $(\nabla f)_{\vec{r}_0}$ is the gradient of $f(\vec{r})$ at Point $\vec{r}_0$. Ignoring the second or higher order terms in Eq. (21) gives:

$$f(\vec{r}_0) \approx \frac{\sum_{J}^{N} f(\vec{r}_J) w(|\vec{r}_J - \vec{r}_0|)}{\sum_{J}^{N} w(|\vec{r}_J - \vec{r}_0|)} - (\nabla f)_{\vec{r}_0} \cdot \vec{R}_0 \qquad (22)$$

where $$\vec{R}_0 = \sum_{J}^{N} (\vec{r}_J - \vec{r}_0) w(|\vec{r}_J - \vec{r}_0|) \Big/ \sum_{J}^{N} w(|\vec{r}_J - \vec{r}_0|).$$

It is obvious that if $f(\vec{r})$ is a linear function and $(\nabla f)_{\vec{r}_0}$ is its accurate gradient, the expression for $f(\vec{r}_0)$ in Eq. (22) is accurate. On the other hand, if $f(\vec{r})$ is a general function, the expression has a second order of accuracy. This is similar to the approximation of the MLS by using a linear basis. Nevertheless, the gradient $(\nabla f)_{\vec{r}_0}$ is generally unknown. Although we may derive another set of equations for $(\nabla f)_{\vec{r}_0}$ and solve them together with Eq. (10) as in Liszka et al.,[35] our purpose is to derive a shape function by avoiding matrix inverse or solution of linear algebraic equations. To do so, it is suggested that $(\nabla f)_{\vec{r}_0}$ is replaced by $(\nabla f)_{\vec{r}_I}$, where $\vec{r}_I$ is one of nodes among $\vec{r}_J$ (J=1,2, ……), that is,

$$f(\vec{r}_0) \approx \frac{\sum_{J}^{N} f(\vec{r}_J) w(|\vec{r}_J - \vec{r}_0|)}{\sum_{J}^{N} w(|\vec{r}_J - \vec{r}_0|)} - (\nabla f)_{\vec{r}_I} \cdot \vec{R}_0 . \qquad (23)$$

The alternation to the gradient in the above equation is a key point of the new scheme. Because of this, it is called as 'simplified' finite difference interpolation (SFDI). It is noted that this alternation does not affect the accuracy of the expression if $f(\vec{r}_0)$ is a linear function because $(\nabla f)_{\vec{r}_0} = (\nabla f)_{\vec{r}_I}$ for such a function. In other words, it does not affect the order of accuracy for a general function. It is also noted that $\vec{r}_I$ may be the nearest node to Point $\vec{r}_0$ but not necessary. Of course, the gradient $(\nabla f)_{\vec{r}_I}$ must be evaluated by using the model values $f(\vec{r}_J)$ of the function. There are variety ways to do so. For example, the method described in the next section may be used but it again needs the solution of a linear algebraic system. In our work, we estimate it by

$$\left(f_{,y}\right)_{\vec{r}_I} = \frac{1}{n_{I,y}} \sum_{J \neq I}^{N} [f(\vec{r}_J) - f(\vec{r}_I)] \frac{\left(\vec{r}_{J,y} - \vec{r}_{I,y}\right)}{\left|\vec{r}_J - \vec{r}_I\right|^2} w\left(\left|\vec{r}_J - \vec{r}_I\right|\right) \tag{24a}$$

where $$n_{I,y} = \sum_{J \neq I}^{N} \frac{\left(\vec{r}_{J,y} - \vec{r}_{I,y}\right)^2}{\left|\vec{r}_J - \vec{r}_I\right|^2} w\left(\left|\vec{r}_J - \vec{r}_I\right|\right) \tag{24b}$$

and y can be changed into x or z to give other components of the gradient. It should be noted that Eq. (24) is used to estimate the gradient in the MPS method[36]. Ma[28] pointed out that the gradient estimated in this way is not generally accurate to the second order and thus may produce a larger error for estimating the velocity directly or for discretising gradient in Eq. (14). However, in Eq. (23), the term $(\nabla f)_{\vec{r}_I} \cdot \vec{R}_0$ has higher order than the first term and so the error in the estimation of $(\nabla f)_{\vec{r}_I}$ may not significantly degrade the accuracy of $f(\vec{r}_0)$. This has been confirmed by numerical tests[28]. Substituting Eq. (24) into (23), it follows that

$$f(\vec{r}_0) = \sum_{J=1}^{N} \Phi_J(\vec{r}_0; \vec{r}_I) f(\vec{r}_J) \tag{25}$$

where $\Phi_J(\vec{r}_0)$ is the shape function in the SFDI interpolation and is defined by

$$\Phi_J(\vec{r}_0;\vec{r}_I)=\frac{w(|\vec{r}_J-\vec{r}_0|)}{\sum_K^N w(|\vec{r}_K-\vec{r}_0|)}-(1-\delta_{IJ})B_{0,J}(\vec{r}_I)+\delta_{IJ}\sum_{K\neq I}^N B_{0,K}(\vec{r}_I)$$

where $$\delta_{IJ}=\begin{cases}1 & I=J\\ 0 & I\neq J\end{cases},$$

$$B_{0,J}(\vec{r}_I)=\frac{w(|\vec{r}_J-\vec{r}_I|)}{|\vec{r}_J-\vec{r}_I|^2}\sum_{k=1}^{d}\frac{\vec{R}_{C,x_k}}{n_{I,x_k}}(\vec{r}_{J,x_k}-\vec{r}_{I,x_k}),$$

and $$\vec{R}_{0,x_k}=\sum_J^N(\vec{r}_{J,x_k}-\vec{r}_{0,x_k})w(|\vec{r}_J-\vec{r}_0|)\Big/\sum_J^N w(|\vec{r}_J-\vec{r}_0|),$$

where d is the number of spatial dimensions (e.g. d=3 for 3D cases) and x_k for k=1,2 and 3 is x, y and z, respectively. It may be noted that the methodology to formulate the above interpolation function is some what similar to that for formulating the consistent SPH approximation in Chen and Beraun.[37] However there are four differences. (1) Equation (20) is integrated over a domain in the SPH approximation rather than taken as the weighted sum. (2) The integration must be performed on a mesh while Eq. (22) or (23) is based only on the weight function. (3) Therefore, the consistent SPH approximation relies on a background mesh. Comparatively, Eq. (23) is a truly meshfree expression. (4) The consistent SPH approximation for a function did not have the second term, i.e., similar to the form used in the SF method.

Estimation of gradient

The gradient can be estimated by using the similar method. For this purpose, Eq. (20) is rewritten as

$$f(\vec{r}_J)-f(\vec{r}_0)=(\nabla f)_{\vec{r}_0}\cdot(\vec{r}_J-\vec{r}_0)+O(|\vec{r}_J-\vec{r}_0|^2).$$

Multiplying $\dfrac{\left(\vec{r}_{J,x_m}-\vec{r}_{0,x_m}\right)}{\left|\vec{r}_J-\vec{r}_0\right|^2}w\left(\left|\vec{r}_J-\vec{r}_0\right|\right)$ on both sides, taking the sum of equations at all the relevant nodes and ignoring the error term, it follows that

$$\sum_J^N\left[f(\vec{r}_J)-f(\vec{r}_0)\right]\frac{\left(\vec{r}_{J,x_m}-\vec{r}_{0,x_m}\right)}{\left|\vec{r}_J-\vec{r}_0\right|^2}w\left(\left|\vec{r}_J-\vec{r}_0\right|\right)=\sum_J^N\frac{\left(\vec{r}_{J,x_m}-\vec{r}_{0,x_m}\right)^2}{\left|\vec{r}_J-\vec{r}_0\right|^2}w\left(\left|\vec{r}_J-\vec{r}_0\right|\right)\left(f_{,x_m}\right)_{\vec{r}_0}$$

$$+\sum_J^N\sum_{k=1,k\neq m}^d\left(\vec{r}_{J,x_k}-\vec{r}_{0,k}\right)\left(f_{,x_k}\right)_{\vec{r}_0}\frac{\left(\vec{r}_{J,x_m}-\vec{r}_{0,x_m}\right)}{\left|\vec{r}_J-\vec{r}_0\right|^2}w\left(\left|\vec{r}_J-\vec{r}_0\right|\right).$$

Alternatively, we have

$$\left(f_{,x_m}\right)_{\vec{r}_0}+\sum_{k=1,k\neq m}^d a_{0,mk}\left(f_{,x_k}\right)_{\vec{r}_0}=C_{0,m}\quad (\text{m}=1,2,\ldots,d), \tag{26}$$

where $$C_{0,m}=\frac{1}{n_{0,x_m}}\sum_J^N\left[f(\vec{r}_J)-f(\vec{r}_0)\right]\frac{\left(\vec{r}_{J,x_m}-\vec{r}_{0,x_m}\right)}{\left|\vec{r}_J-\vec{r}_0\right|^2}w\left(\left|\vec{r}_J-\vec{r}_0\right|\right)$$

and $$a_{0,mk}=\frac{1}{n_{0,x_m}}\sum_J^N\frac{\left(\vec{r}_{J,x_m}-\vec{r}_{0,x_m}\right)\left(\vec{r}_{J,x_k}-\vec{r}_{0,x_k}\right)}{\left|\vec{r}_J-\vec{r}_0\right|^2}w\left(\left|\vec{r}_J-\vec{r}_0\right|\right).$$

Solving the system in Eq. (26), one can find the gradient. For example, in 2D cases, it is easy to obtain

$$\left(f_{,x}\right)_{\vec{r}_0}=\frac{C_{0,1}-a_{0,12}C_{0,2}}{1-a_{0,12}a_{0,21}}$$

$$\left(f_{,y}\right)_{\vec{r}_0}=\frac{C_{0,2}-a_{0,21}C_{0,2}}{1-a_{0,12}a_{0,21}}.$$

In 3D cases, Eq. (26) can be generally written as

$$[A]\{F\}=[C]\{\delta f\}$$

where $$\{F\}=\left[\left(f_{,x}\right)_{\vec{r}_0},\left(f_{,y}\right)_{\vec{r}_0},\left(f_{,z}\right)_{\vec{r}_0}\right]^T$$

$$\{\delta f\}=\left[f(\vec{r}_1)-f(\vec{r}_0),f(\vec{r}_2)-f(\vec{r}_0),\cdots,f(\vec{r}_J)-f(\vec{r}_0),\cdots\right]^T$$

$[C]$ is an 3×N matrix and its components defined as

$$C_{mJ} = \frac{1}{n_{0,x_m}} \frac{\left(\vec{r}_{J,x_m} - \vec{r}_{0,x_m}\right)}{\left|\vec{r}_J - \vec{r}_0\right|^2} w\left(\left|\vec{r}_J - \vec{r}_0\right|\right).$$

$[A]$ is an 3×3 matrix and its entries are defined as

$$A_{mk} = a_{0,mk} \quad (m \neq k) \text{ and } A_{mm} = 1 .$$

At last, we may write the gradient components as

$$\left(f_{,x_m}\right)_{\vec{r}_0} = \sum_{J=1}^{N} \tilde{\Gamma}_{mJ}(\vec{r}_0)\left[f(\vec{r}_J) - f(\vec{r}_0)\right]$$
$$\tilde{\Gamma}_{mJ}(\vec{r}_0) = \sum_{k=1}^{d} \tilde{A}_{mk} C_{kJ} \quad J = 1,2,3\cdots \tag{27}$$

where $\left[\tilde{A}\right]$ is the inverse matrix of $[A]$, i.e. $\left[\tilde{A}\right] = [A]^{-1}$. $\tilde{\Gamma}_{mj}(\vec{r}_0)$ is the shape function for gradient components. To obtain the shape function, one needs to solve a linear algebraic system. However, the algebraic system involved here contains a smaller number of equations than that for Eq. (18g) in the MLS method and therefore require less CPU time, making the method more efficient, particularly in 3D cases.

It is easy to deduce that if $f(\vec{r})$ is linear Eq. (27) gives exact value of gradient components, independent of the distribution of nodes. This equation may be used in Eq. (11) to formulate the shape function for the interpolation of unknown function with sacrifice to CPU time and with slight improvement to the accuracy. In addition, it may be interesting to point out that Eq. (27) become equivalent to Eq. (19) if the shrink factor in the later is taken as 1. Therefore the same expression of the gradient is obtained by using two different methods.

Based on numerical tests[28], it was showed that (1) the SFDI is as accurate as the MLS method generally but may be more accurate than the later when the number of neighbour nodes is small and (2) the SFDI needs considerably less CPU time than the MLS.

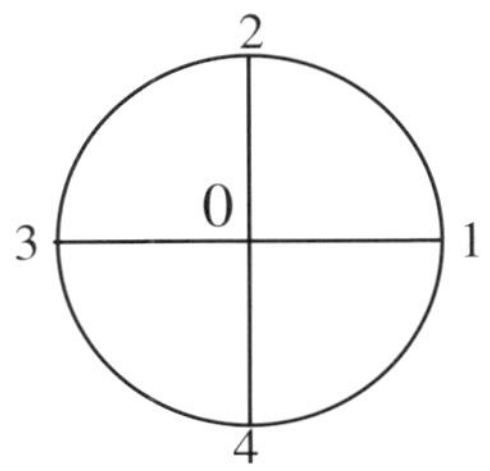

Fig. 2. Illustration of division of an integration domain.

4.3. *Semi-analytical techniques for numerical domain integrations*

In the weak form (Eq. 17) used by the MLPG_R method, the domain integration is involved on the right hand side. It may be evaluated by using the Gaussian quadrature. To do so, more than 16 Gaussian points for 2D cases and 64 Gaussian points for 3D cases may be required to obtain satisfactory results, at which the intermediate velocities are estimated. Evaluation of the velocities at so many points is time-consuming. In order to make the method more efficient, a semi-analytical technique was proposed[22]. The basic idea of the technique is

(1) Dividing an integration domain into several sub-domains;
(2) Assuming intermediate velocities to linearly vary over each subdomain;
(3) Performing the integration over each subdomain analytically.

The technique can be demonstrated by 2D cases. Let us consider a circular integration domain with a radius of R, centred at (x_0, y_0), as shown in Fig. 2, and divide it into four subdomains, 0-1-2, 0-2-3, 0-3-4 and 0-4-1. Over each subdomain, e.g., 0-1-2, the intermediate velocity components are assumed to be linear with respect to coordinates. Thus the velocity components are given by

$$u^{(*)} = u_0^{(*)} + c_{ux}(x - x_0)/R + c_{uy}(y - y_0)/R \tag{28a}$$

$$v^{(*)} = v_0^{(*)} + c_{vx}(x - x_0)/R + c_{vy}(y - y_0)/R \tag{28b}$$

where $\left(u^{(*)}, v^{(*)}\right)$ are the intermediate velocity components at any point (x,y) in the subdomain 0-1-2; and $\left(u_0^{(*)}, v_0^{(*)}\right)$ are that at its centre. c_{ux}, c_{uy}, c_{vx} and c_{vy} are constants, which are determined in such a way that the velocity components equal to those at Point 1 and 2. Taking the x-component as an example, one has

$$c_{ux}(x_1 - x_0) + c_{uy}(y_1 - y_0) = (u_1^{(*)} - u_0^{(*)})R$$

$$c_{ux}(x_2 - x_0) + c_{uy}(y_2 - y_0) = (u_2^{(*)} - u_0^{(*)})R$$

which gives

$$c_{ux} = \frac{\left(u_1^{(*)} - u_0^{(*)}\right)(y_2 - y_0) - \left(u_2^{(*)} - u_0^{(*)}\right)(y_1 - y_0)}{(x_1 - x_0)(y_2 - y_0) - (x_2 - x_0)(y_1 - y_0)} R \tag{29a}$$

$$c_{uy} = \frac{(x_1 - x_0)\left(u_2^{(*)} - u_0^{(*)}\right) - (x_2 - x_0)\left(u_1^{(*)} - u_0^{(*)}\right)}{(x_1 - x_0)(y_2 - y_0) - (x_2 - x_0)(y_1 - y_0)} R \qquad (29b)$$

c_{vx} and c_{vy} can be found similarly or obtained by replacing $\left(u_1^{(*)}, u_2^{(*)}\right)$ with $\left(v_1^{(*)}, v_2^{(*)}\right)$ in Eq. (29); and thus the velocities in this subdomain are determined by Eq. (28). The velocities in other subdomains can also be estimated in this way. The only difference is that they may be related to the velocities at Points 3 and 4, depending on which subdomain is concerned. Consequently, the velocities at any points in the circle are determined by those at only five points (0, 1, 2, 3 and 4). Based on this, let us consider the integral on the right hand side of Eq. (17) for a 2D case. It is first written as:

$$\int_{\Omega_I} \vec{u}^{(*)} \cdot \nabla \varphi \, d\Omega = \frac{1}{2\pi} \int_0^{2\pi} \int_0^{R_I} u_r^{(*)}(r,\theta) dr d\theta = \frac{1}{2\pi} \sum_{i=1}^{N_i} \int_{\vartheta_i}^{\vartheta_{i+1}} \int_0^{R_I} u_r^{(*)}(r,\theta) dr d\theta \qquad (30)$$

where $u_r^{(*)}$ is the radial component of $\vec{u}^{(*)}$ and $N_i = 4$ for 4 divisions and $\vartheta_5 = \vartheta_1$. The integral over each subdomain can be evaluated analytically using Eq. (28). For this purpose, Eq. (28) is rewritten as

$$u^{(*)} = u_0^{(*)} + c_{ux}\,\bar{r}\cos\theta + c_{uy}\,\bar{r}\sin\theta\,,$$

$$v^{(*)} = v_0^{(*)} + c_{vx}\,\bar{r}\cos\theta + c_{vx}\,\bar{r}\sin\theta\,,$$

and so

$$u_r^{(*)} = u_0^{(*)}\cos\theta + v_0^{(*)}\sin\theta + c_{ux}\bar{r}\cos^2\theta$$
$$+ c_{uy}\bar{r}\cos\theta\sin\theta + c_{vx}\bar{r}\cos\theta\sin\theta + c_{vy}\bar{r}\sin^2\theta \qquad (31)$$

where the transformation of $(x - x_0)/R = \bar{r}\cos\theta$ and $(y - y_0)/R = \bar{r}\sin\theta$ have been employed. It is easy to show that the integration of $u_0^{(*)}\cos\theta + v_0^{(*)}\sin\theta$ over the whole circle becomes zero and so can be omitted. The integration of other terms in Eq. (31) over Subdomain 0-1-2 is given by

$$\int_{\vartheta_1}^{\vartheta_2}\int_{0}^{R_I} u_r^{(*)}(r,\theta)\,drddrd\theta = \frac{1}{4}R_I\begin{bmatrix}(c_{ux}+c_{vy})(\vartheta_2-\vartheta_1)\\ +(c_{ux}-c_{vy})(\sin\vartheta_2\cos\vartheta_2-\sin\vartheta_1\cos\vartheta_1)\\ +(c_{uy}+c_{vx})(\sin^2\vartheta_2-\sin^2\vartheta_1)\end{bmatrix}.$$

Similar results can be obtained for other subdomains and the sum of these gives the results for the whole circular domain. Although the results depend on the number of subdomains, numerical tests show that four subdomains are good enough for 2D cases. It is envisaged that eight subdomains may be required for 3D cases.

With this semi-analytical technique, the velocities at only five points for a 2D circle with four divisions, instead of more than 16 points if the Gaussian quadrature would be used, need to be evaluated. In 3D cases, a spherical domain may be divided in to 8 subdomains as indicated above and the velocities at only 7 points need to be evaluated, instead of 64 points when using the Gaussian quadrature. As a result, the reduction in CPU time spent on the evaluation of the domain integration is considerable.

The similar domain division technique was used in other publications[38,39]. However, in those papers, the Gaussian quadrature is still used for the integration over each subdomain while we do not use the Gaussian quadrature at all, as demonstrated above.

4.4. *Imposing boundary conditions*

Generally, there are two kinds of boundary conditions for the pressure in water wave problems, as shown in Eqs. (2a) and (2b): the free surface condition specifying the pressure (Dirichlet condition, similar to essential boundary conditions in solid dynamics) and the solid boundary condition specifying the normal derivative of the pressure (Neumann condition). The condition on the solid boundary was generally suggested to be imposed by applying Eq. (13) to incomplete circular or spherical sub-domains of boundary nodes[8,13]. The implementation of an essential boundary condition is not very straightforward if the shape function obtained by the MLS method is used since the unknown nodal values in the MLS approach are not physical values as pointed out above. This issue has received a considerable amount of research[40]. The methods suggested therein include the penalty method, transformation method, collocation method and so on. These approaches work well for fixed (or with little movement) boundary problems [8,13,39]. However, the penalty method for the nonlinear water wave problems can cause big errors, particularly near the free surface [21]. The reason may be due to the fact that the addition of the penalty term inevitably produces some errors since it is not an exact representation of the free surface boundary condition. These errors may be negligible if the final solution could be found in one or a few time steps and if the boundary is not a part of solution but specified. However, they may accumulate to significant ones in several thousand time steps even though the whole scheme is stable in simulating nonlinear water waves. Although the accumulated errors may be reduced by shortening the length of the time step, it inevitably increases the computational costs. Therefore, based on numerical tests [21], a better way to impose the free surface and solid boundary conditions in water wave problems is to use the following scheme:

$$\sum_{J=1}^{N} \Phi_J(\vec{x})\hat{p}_J = p_{atm} \quad \text{for nodes on the free surface} \tag{32a}$$

and

$$\sum_{J=1}^{N} \vec{n}\cdot\nabla\Phi_J(\vec{x})\hat{p}_J = \vec{n}\cdot\left(\vec{g}-\dot{\vec{U}}\right) \quad \text{for nodes on the solid boundary}\,. \tag{32b}$$

These work well for the MLPG formulation described by Ma (2005a)[21]. For the MLPG_R method[22], numerical tests show that Eq. (32a) may be replaced by

$$p = p_{atm} \tag{32c}$$

with the pressure at nodes on the free surface being directly imposed.

4.5. *Discretised equations*

With use of interpolation functions discussed in Section 4.2, the weak form in Eqs. (14) or (17) can be transferred to a discretised algebraic system in term of the nodal values of the unknown pressure. Although we have discussed two methods for meshless interpolation in Section 4.2, the MLS scheme will be only implemented in this section. That is because the SFDI scheme has been proposed very recently and its implementation is still under investigation when the article is written. As mentioned above, the MLPG_R method uses Eq. (17) as its weak form while the MLPG method[21] is based on Eq (14). No matter which weak form is used, the resulting algebraic systems are similar, which can be given by

$$\mathbf{K}\cdot\hat{\mathbf{P}}=\mathbf{F} \tag{33}$$

where $\mathbf{K}$ is the coefficient matrix, $\hat{\mathbf{P}}$ is the vector of unknown nodal values and $\mathbf{F}$ is the vector of the values of the right hand terms in Eq. (14) or (17) after discretised. $\mathbf{K}$ and $\mathbf{F}$ are different for the MLPG_R and the MLPG methods, which are given separately in the following.

In the MLPG_R method, the weak form in Eq. (17) is employed and the boundary conditions in Eqs. (32b) and (32c) are imposed. The resulting expressions of each element in $\mathbf{K}$ and $\mathbf{F}$ are given by

$$K_{IJ}=\begin{cases}\int_{\partial\Omega_I}\Phi_J(\vec{x})\vec{n}\cdot\nabla\varphi ds-\Phi_J(\vec{x}_I) & \text{for inner nodes}\\ \vec{n}\cdot\nabla\Phi_J(\vec{x}_I) & \text{for nodes on solid boundaries}\end{cases} \tag{34a}$$

and

$$F_I=\begin{cases}\int_{\Omega_I}\frac{\rho}{\Delta t}\vec{u}^{(*)}\cdot\nabla\varphi d\Omega & \text{for inner nodes}\\ \vec{n}\cdot\left(\vec{g}-\dot{\vec{U}}\right) & \text{for nodes on solid boundaries}\end{cases} \tag{34b}$$

where Node I denotes nodes that do not lie on the free surface; and Nodes J are those influencing Node I.

In the MLPG method in Ma (2005a)[21], the weak form in Eq. (14) is used and the boundary conditions in Eqs. (32a) and (32b) are imposed. The resulting expressions of each element in $\mathbf{K}$ and $\mathbf{F}$ are written as

$$
K_{IJ} = \begin{cases} \int_{\partial\Omega_I} \vec{n}\cdot\nabla\Phi_J(\vec{x})ds & \text{for inner nodes} \\ \vec{n}\cdot\nabla\Phi_J(\vec{x}_I) & \text{for nodes on solid boundaries} \\ \Phi_J(\vec{x}_I) & \text{for nodes on the free surface} \end{cases}
$$

and

$$
F_I = \begin{cases} \frac{\rho}{\Delta t}\int_{\partial\Omega_I} \vec{n}\cdot\vec{u}^* dS & \text{for inner nodes} \\ \vec{n}\cdot\left(\vec{g}-\dot{\vec{U}}\right) & \text{for nodes on solid boundaries} \\ p_{atm}(x_I) & \text{for nodes on the free surface.} \end{cases}
$$

It was pointed in Ma (2005b)[22] that the MLPG_R method is twice as fast as the MLPG method[21] for nonlinear water waves. Therefore development in our current research has focused on the MLPG_R method.

It could be understood that the size of integration domain and the size of influence domain within which the weight function is not zero play an important role in the methods discussed in the article. In addition, the pressure gradient and velocity at and near the free surface may need special treatment. Discussions about these issues have been given in author's other papers[21,22] and will not be presented here.

5. Applications

The MLPG and the MLPG_R methods have been applied to various nonlinear water waves. Some examples will be given in this section. Unless mentioned otherwise, the results presented in this section are obtained by using the MLPG_R method.

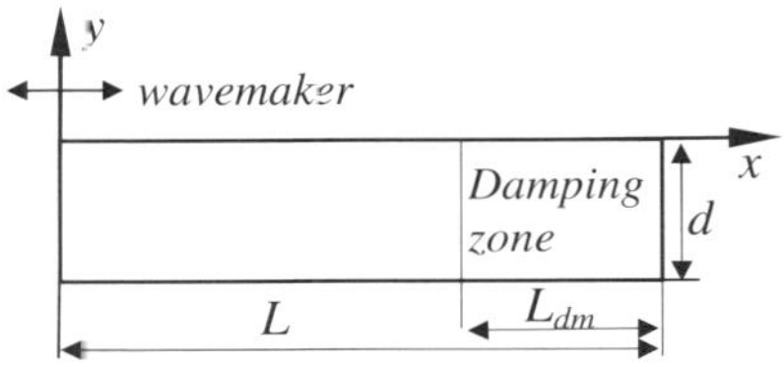

Fig. 3. Sketch of the wavemaker problem and coordinate system.

In the following discussion, all variables and parameters are non-dimensionalised using the water depth d and the gravitational acceleration g, i.e.

$$(x, y, L, a) \rightarrow (x, y, L, a)d \quad t \rightarrow \tau\sqrt{d/g} \quad \omega \rightarrow \omega\sqrt{g/d}$$

where L represents the length (e.g, the length of water domain), a the amplitude (e.g., the amplitude of the motion of the wavemaker), t the time and ω the frequency of oscillation.

5.1. *Freak waves generated by a wavemaker*

The sketch of wavemaker problems and the coordinate system is illustrated in Fig. 3. In order to reduce reflection, a damping zone is added at the far end opposite to the wavemaker, in which the velocity ($\vec{u}_{dm}$) is modified by adding an artificial damping term and computed by using

$$\vec{u}_{dm}^{(n+1)} = \vec{u}_m^{(n+1)} - \nu(x)\,\vec{u}_m^{(n+1)}, \tag{35a}$$

where $\nu(x)$ is an artificial damping coefficient defined as

$$\nu(x) = \frac{1}{2}\nu_0\left[1 - \cos\left(\frac{\pi(x - x_d)}{L_{dm}}\right)\right] \quad x \geq x_d\,, \tag{35b}$$

where L_{dm} is the length of the damping zone taken as $L_{dm}= 3d$; x_d is the x-coordinate of the left end of the damping zone; and ν_0 is the magnitude of the damping coefficient. A similar form of the damping coefficient was used in Ma, Wu and Eatock Taylor [41] for a numerical wave tank based on potential theory using a finite element method, but combined with a Sommerfeld condition. Extensive numerical tests were carried out in those papers to choose the optimum values for ν_0. Those values are not necessarily optimum here as the formulation is different. Nevertheless, similar numerical investigations have yet been carried out for the MLPG_R or MLPG method. Instead, only one value is chosen based on some numerical tests, which is $\nu_0 = 0.1$. With this value, the reflection wave may not always be reduced sufficiently, unless a long tank is used or simulations stop before reflection waves come into the area of interest.

Although various waves may be generated in numeral wave tanks by specifying different motions of the wavemaker, $S(t)$, two cases for generating freak waves are only presented here.

In the first case, comparison with experimental results is made to show the accuracy of the numerical method. For this purpose, the exact same motion of the wavemaker as in Nestegard (1999) is used, whose time history is plotted in Fig. 4. A scaling factor may multiply the time history to obtain different amplitude of waves. The case for the factor of 0.749 is given here. In order to simulate this case, the tank length is chosen as 20. Figure 5 shows the wave time histories recorded at $x = 13.436$ (where the largest wave amplitude is expected to occur) computed by the MLPG_R method together with the experimental data given by Nestegard[42]. In these figures, the relative error of the numerical results to those from experiments at the peak is less than 8%. The agreement of the numerical results with the experimental data is satisfactory.

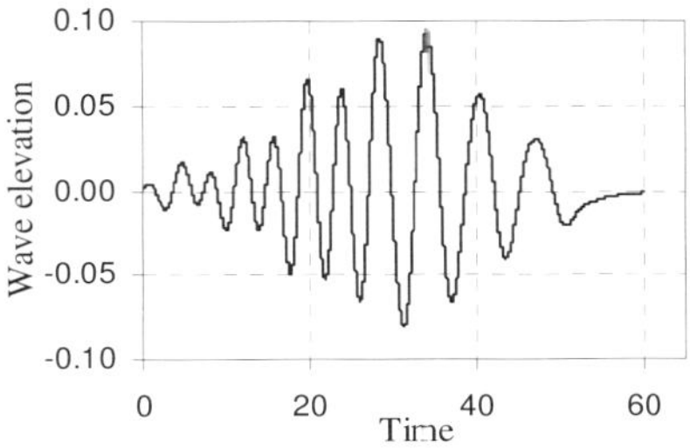

Fig. 4. Time history of wavemaker motion.

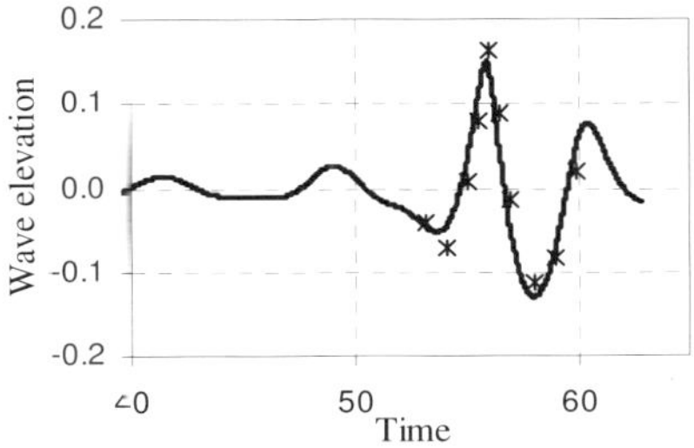

Fig. 5. Time histories (*: Nestegard, 1999).

The second case is to model the freak wave generated by the following equations

$$S(t) = \sum_{n=1}^{N} \frac{a_n}{F_n} \cos(\omega_n t + \varepsilon_n) \tag{36a}$$

with

$$F_n = \frac{2[\cosh(2k_n) - 1]}{\sinh(2k_n) + 2k_n}, \tag{36b}$$

where N is the total number of components; ε_n is the phase of the n-th component; ω_n are the dimensionless frequency of the n-th component, related to k_n by $\omega_n^2 = k_n \tanh(k_n)$ with k_n being the dimensionless wave number of the n-th component; and a_n is the amplitude of n-th component. As indicated by Dean and Dalrymple[43], each component of the waves generated by using Eq. (36) consists of two parts: one is the progressive wave and the other is the standing or local wave. The local wave decays with the increase in the distance from the wavemaker and becomes negligible after a point that is 3 or more water depths away from the wavemaker. The progressive wave consists of a transient wave profile in the front part of a harmonic wave train even though the motion of the wavemaker is purely harmonic. Therefore, the wave component at a certain point downstream of the wavemaker become harmonic only after the transient wave has passed this point. Under certain conditions, these harmonic components can focus at a point where all of them reach their peaks. Based on a linear theory[44] and choosing $\varepsilon_n = k_n x_f - \omega_n t_f$ with x_f and t_f being the focusing point and the focusing time, the waves in the tank generated by using Eq. (36) may be expressed at such a point as

$$\zeta(x,t) = \sum_{n=1}^{N} a_n \cos\left[k_n\left(x - x_f\right) - \omega_n\left(t - t_f\right)\right] \quad . \tag{37}$$

In Eqs. (36) and (37), the frequencies of the wave components are equally spaced over the band $[\omega_1, \omega_N]$ with ω_1 being the smallest and ω_N being the largest frequencies, respectively. In this work, they are selected as $\omega_1 = 0.8683$ and $\omega_N = 1.8865$. The number (N) of wave components affects the continuity of the wave spectrum and depends on the bandwidth ($\omega_N - \omega_1$) of spectrum. If it is not large enough, the wave spectrum is not smooth and the properties of generated waves depend on the number. Based on our numerical tests, the waves are almost the same when N>20 for the above frequencies specified. It is taken as 32 here. The amplitude of each component is chosen as the same to simplify the relationship between the target amplitude (A) of the freak wave and the amplitudes of the components, leading to $a_n = a = A/N$ or $Na_n = Na = A$. Two values of amplitudes (Na = 0.074 and 0.1488) are considered to look at the influence of different amplitudes. x_f must be large enough so that the local wave generated by the wavemaker should become negligible in the region where the target freak wave occurs. This requires that the nearest point of the region to the wavemaker should be 3 or more water depths away from the wavemaker as mentioned above. t_f must also be large enough so that the shortest wave components have passed point x_f when the freak wave occurs. The relationship for them and may be given as:

$$3 < x_f < t_f C_N$$

where C_N is determined by $C_N = 1/\omega_N$. For the conservative reason, the nondimensional focusing point coordinate and the focusing time are taken as $x_f = 14.65$ and $t_f = 83.68$ respectively. The tank length for this case is selected as 38, much longer than the distance from the wavemaker to the theoretical focusing point, and thus the focusing wave is hardly affected by the reflection even without the artificial damping zone. The results corresponding to the two amplitudes are presented in Fig. 6, where the wave elevations have been divided by the value of Na. One may see from the figure that the wave corresponding to the larger amplitude tends to travel faster, which implies that the focusing point takes place further downstream than those to the smaller amplitude, being consistent with that observed in physical experiments[44].

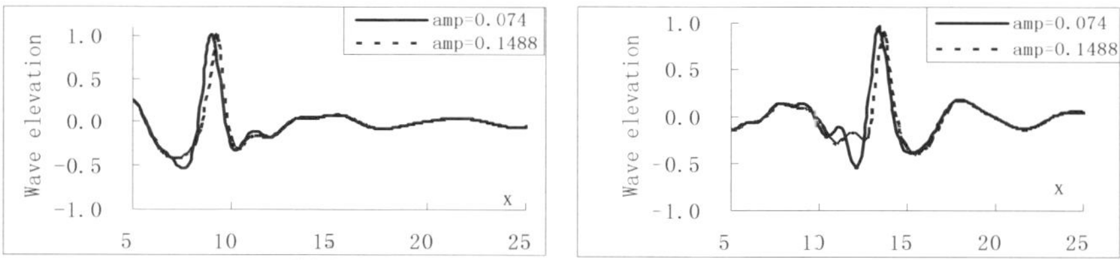

(a) t=66.0 (b) t=77.25

Fig. 6. Wave profiles for different amplitudes.

5.2. *Sloshing waves*

Sloshing waves are those generated in a moving tank, which are associated with various engineering problems, such as liquid oscillations in large storage tanks caused by earthquakes, motions of liquid fuel in aircrafts and spacecrafts, liquid motions in containers and the water flow on decks of ships. The loads produced by wave motions of this kind can cause structural damage and the loss of the motion stability of objects such as ships. There has been a considerable amount of experimental, numerical and analytical work on wave sloshing. Nevertheless, the detail review on sloshing waves will not be given here, which may be found in Wu, Ma and Eatock Taylor[45] and other papers cited therein.

Although one may here employ the same governing equations as in Eqs. (1) and (2) established for a fixed coordinate system, it is advantageous to use alternative ones established for a coordinate system moving together with the tank; and thus one can study the fluid flow relative to the tank, which is of primary interest. The sketch of the sloshing wave problem described in the moving coordinate system is still similar to that in Fig. 3 but without the wavemaker and damping zone. For such a system, the governing equations and boundary conditions for the problem are given as:

$$\nabla \cdot \vec{v} = 0 \tag{38a}$$

$$\frac{D\vec{v}}{Dt} = -\frac{1}{\rho}\nabla p + \vec{g} - \frac{D\vec{U}}{DT} \tag{38b}$$

$$\frac{D\vec{r}}{Dt} = \vec{v} \text{ and } p = p_{atm} \text{ on the free surface} \tag{38d}$$

$$\vec{n} \cdot \vec{v} = 0 \text{ and } \frac{\partial p}{\partial n} = \rho\left(n \cdot \vec{g} - n \cdot \frac{D\vec{U}}{DT}\right) \tag{38c}$$

on the walls and the bed of the tank.

where $\vec{v}$ is the relative velocity of fluid and $\vec{U}$ the velocity of the tank. These governing equations and boundary conditions are very similar to Eqs. (1) and (2). The only difference is that an extra term associated with the acceleration of the tank is contained in Eqs. (38b) and (38c), which presents the effects of the motion of the tank on the relative velocity of fluid. Although these equations can be used for general motions of the tank, we only consider a motion in x-direction defined by

$$\begin{aligned} X(t) &= a\sin(\omega t) \\ U(t) &= \dot{X}(t) = a\omega\cos(\omega t) \end{aligned} \tag{39}$$

where a and ω are the amplitude and frequency, respectively. For such a motion, a linearised analytical solution[46] can be found when wave amplitudes are small. Based on this solution, the dimensionless natural frequency of the tank is given by

$$\omega_{nj} = \sqrt{\frac{(2j+1)\pi}{L}\tanh\frac{(2j+1)\pi}{L}} \quad \text{for } j=0,1,2\ldots.$$

When the frequency of motions equals one of these natural frequencies, the amplitude of the wave in the tank can theoretically grow to infinity. Even when the frequency of the tank motion is not equal but near to the natural frequencies, the wave amplitudes can also grow to large ones. The numerical results from the MLPG_R method will be compared with this solution. For this purpose, the dimensionless length of the tank is taken as $L=2$, which gives

$$\omega_{n0} = \sqrt{\frac{\pi}{2}\tanh\frac{\pi}{2}}\ .$$

Two cases are presented here. In these cases, the amplitudes are the same (0.002) but the frequencies of the tank motion are different, i.e. $\omega = 0.9\ \omega_{n0}$, and $1.02\ \omega_{n0}$, respectively. Convergent tests have been carried out and found that $dx = dy = 0.08$ and 100 time steps ($\Delta t = T/100$, $T = 2\pi/\omega$) in each period of the tank motion yield satisfactory solutions. The numerical time histories at $x = 0$ (upper left corner of the tank) are depicted in Fig. 7 together with an analytical solution[46]. The agreement between them is quite good as can be seen from this figure.

The next example is for a case with an amplitude of $a=0.0186$, which is about nine times larger than that in the previous case. The frequency of the motion is taken as $\omega/\omega_{n0}=0.999$, very close to the first resonant frequency. The case with these parameters has been experimentally investigated by Okamoto and Kawahara[47]. To simulate this case, nodes are uniformly distributed initially with $dx=dy=0.06$ and the time step is also taken as $\Delta t=T/100$. Figure 8 plots nodal configurations at two time instants, $t=2.5T$ and $3T$, respectively. The experimental data of Okamoto and Kawahara[47] are also shown in this figure. The free surfaces obtained by numerical simulation agree well with their experimental data.

5.3. *Dam breaking*

The dam breaking problem is often used to validate numerical methods because it involves a large deformation of fluid domain. The result for the problem obtained by the MLPG_R method is also given here. Initially the dam is confined between two walls, with the width of $0.5d_0$, where d_0 is the initial depth. At the beginning of the simulation, the right wall (dam) is instantaneously removed and the water is allowed to flow out along a dry horizontal bed. The position of the lower-right corner of the water volume obtained by the MLPG_R method is plotted in Fig. 9, together with some experimental results from Martin[48]. The agreement between them is very good, as can be observed.

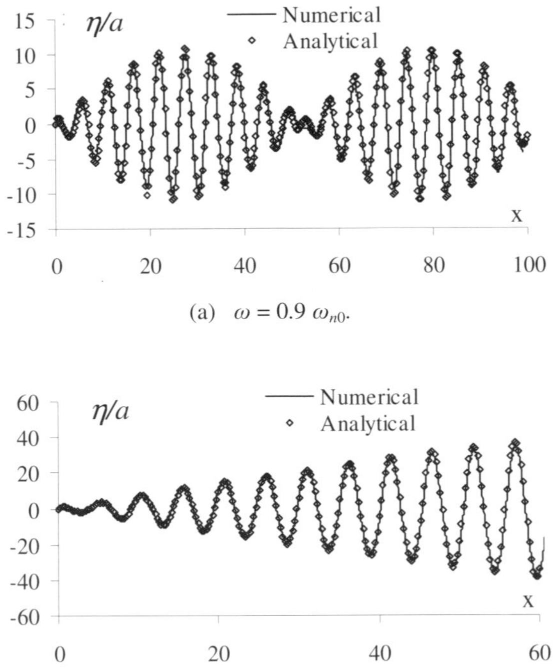

(a) $\omega = 0.9\ \omega_{n0}$.

(b) $\omega = 1.02\ \omega_{n0}$.

Fig. 7. Time histories of sloshing waves at $x = 0$ for different frequencies.

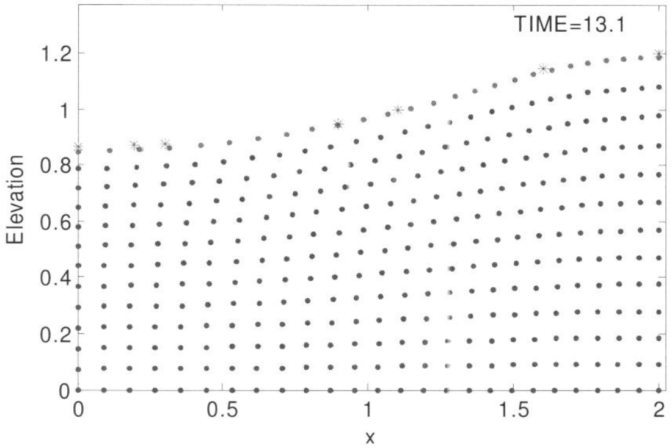

(a) t = 2.5T (T = 2π/ω).

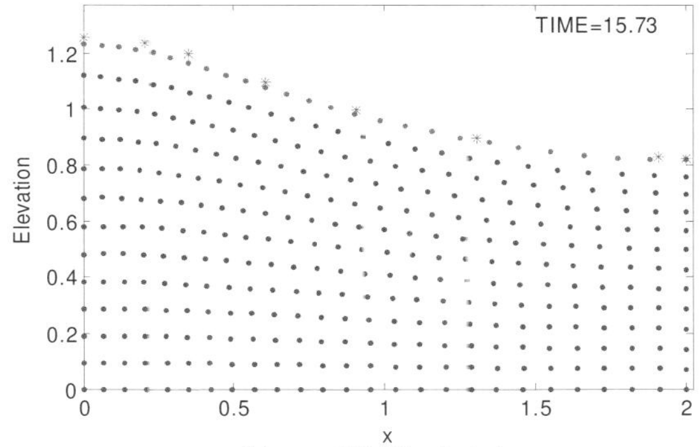

(b) t = 3T (T = 2π/ω).

Fig. 8. Configurations of nodes at two instants for $a = 0.0186$ and $\omega/\omega_{n0} = 0.999$ (*: experimental data from Okamoto and Kawahara (1990)).

6. Conclusions and Future Work

This article has discussed developments of the MLPG method related with nonlinear water waves. These include that (1) the MLPG was first extended to nonlinear waters in 2005; (2) it was then developed into the MLPG_R method in 2006; (3) the semi-analytical technique for domain integration was proposed also in 2006; and (4) a new interpolation schemed was devised in 2008. This method has been applied to various water wave problems, such as freak waves, sloshing waves, dam breaking and so on. For water wave problems, the MLPG_R method, rather than MLPG method, should be employed because it is more efficient.

It should be noted that the MLPG or MLPG_R method is relative young compared with other meshless methods (such as SPH) and so needs further developments: (1) considering viscous effects and turbulent flow; (2) applying it to model breaking waves; (3) applying it to model interaction between large waves and floating bodies and so on.

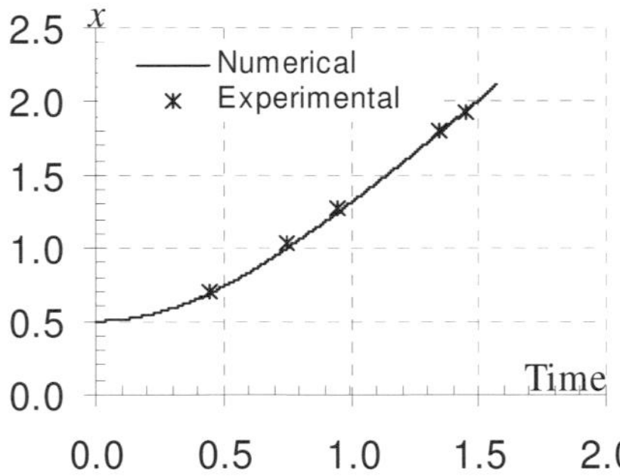

Fig. 9. The position of the lower corner of water volume for broken dam problem.

Acknowledgment

Some part of this work was supported by the Leverhulme Trust, UK.

References

1. Q.W. Ma and S. Yan, *J. Comput. Phys.*, Vol. 212, Issue 1, 52-72 (2006).
2. T. Belytschko, Y.Y. Lu and L. Gu, *Int. J. Numr. Meth. Eng.*, Vol. 37, 229-256 (1994).
3. B. Nayroles, G. Touzot and P. Villon P., *Computational Mechanics*, Vol. 10, 307-318 (1992).
4. W.K. Liu, Y. Chen, S. Jun, J.S. Chen, T. Belytschko, C. Pan, R.A. Uras and C. Chang, *Archives of Computational Methods in Engineering: State of the Art Reviews*, Vol. 3, 3-80 (1996).
5. J.J. Monaghan, *J. of Comput. Phys.*, Vol. 110, 399-406 (1994).
6. E. Onate, S.R. Idelsohn, F.D. Pin and R. Aubry, *Int. J. of Comput. Meth.*, Vol. 1 (2), 267-308 (2004).
7. S. Koshizuka and Y. Oka, *Nuclear Science and Engineering*, Vol. 123, 421-434 (1996).
8. S.N. Atluri and T. Zhu, *Computational Mechanics*, Vol. 22, 117-127 (1998).
9. S.N. Atluri and T. Zhu, *Computer Modeling and Simulation in Engg.*, Vol. 3, 187-196 (1998).
10. S.N. Atluri and T. Zhu, *Computational Mechanics*, 25 (2-3), 169-179, MAR (2000).
11. H. Lin and S.N. Atluri, *CMES-Computer Modelling in Engineering and Sciences*, Vol. 1 (2), 45–60 (2000).

12. H. Lin and S.N. Atluri, *Computer Modeling in Engineering and Sciences*, Vol. 2 (2), 117-142 (2001).
13. S.N. Atluri and S.P. Shen, *CMES-Computer Modeling in Engineering and Sciences*, 3 (1): 11-51, (2002).
14. Q. Li, S. Shen, Z.D. Han, S. N. Atluri, *CMES-Computer Modeling in Engineering and Sciences*, Vol. 4, No. 5, 571-585 (2003).
15. S.P. Shen and S.N. Atluri, *CMES-Computer Modeling in Engineering and Sciences*, Vol. 5, No. 3, 235 – 255 (2004).
16. J. Sladek, V. Sladek and S.N. Atluri, *CMES-Computer Modeling in Engineering and Sciences*, Vol. 6, No. 3, 309-318 (2004).
17. J. Soric, Q. Li, T. Jarak and S.N. Atluri, *CMES-Computer Modeling in Engineering and Sciences*, Vol. 6, No. 4, 349-357 (2004).
18. S.N. Atluri, Z.D. Han and A.M. Rajendran, *CMES-Computer Modeling in Engineering and Sciences*, Vol. 6, No. 6, 491 – 513 (2004).
19. Z.D. Han and S.N. Atluri, *Computers, Materials and Continua*, Vol. 1 (2), 129-140 (2004).
20. S.N. Atluri and S.P. Shen, *CMES-Computer Modeling in Engineering and Sciences*, Vol. 7, No. 3, 241-268 (2005).
21. Q.W. Ma, *Journal of Computational Physics*, Vol. 205, Issue 2, 611-625 (2005a).
22. Q.W. Ma, *CMES-Computer Modeling in Engineering and Sciences*, Vol. 9, No. 2, 193-210 (2005b).
23. Q.W. Ma, *Proceedings of the International Conference on Computational and Experimental Engineering and Sciences (ICCES)*, Chennai, India, 1st–6th Dec, 2005.
24. Z.D. Han, H.T. Liu, A.M. Rajendran and S.N. Atluri, *CMESS-Computer Modeling in Engineering and Sciences*, 14 (2): 119-128 (2006).
25. S.N. Atluri, H.T. Liu and Z.D. Han, *CMES-Computer Modeling In Engineering and Sciences*, 14 (3): 141-152 (2006a).
26. S.N. Atluri, H.T. Liu and Z.D. Han, *CMES-Computer Modeling In Engineering and Sciences*, 15 (1): 1-16 (2006b).
27. V. Vavourakis and D. Polyzos, *CMC-Computers, Materials and Continua* Vol. 5, No. 3, 185-196 (2007).
28. Q.W. Ma, *CMES-Computer Modeling in Engineering and Sciences*, Vol. 23, No. 2, 75-90 (2008).
29. Q.W. Ma, *CMES-Computer Modeling in Engineering and Sciences*, Vol. 18, No. 3, 223-234 (2007).
30. A.J. Chorin, *Math. Computation*, 22, 745-762 (1968).
31. O.C. Zienkiewicz and R.L. Taylor, *The Finite Element Method - Volume 3: Fluid Dynamics, 5th edition*, Butterworth-Heinemann (2000).
32. E.Y.M. Lo and S. Shao, *Applied Ocean Research*, 24, 275-286 (2002).
33. S.R. Idelsohn, M.A. Storti, E. Onate. *Comput. Methods. Appl. Mech. Engrg.*, 191 (2001), 583-593.

34. T. Zhu, J.D Zhang and S.N. Atluri, *Computational Mechanics*, Vol. 21, 223-235 (1998).
35. T.J. Liszka, C.A.M. Duarte and W.W. Tworzydlo, *Computer Methods in Applied Mechanics and Engineering*, Vol. 139, 263-288 (1996).
36. H.Y. Yoon, S. Koshizuka and Y. Oka, *International Journal of Multiphase Flow*, 27, 277-298 (2001).
37. J.K. Chen and J.E. Beraun, *Computer Methods in Applied Mechanics and Engineering*, Vol. 190, Issues 1-2, 225-239 (2000).
38. S.N. Atluri, H.G. Kim and J.Y. Cho, *Computational Mechanics*, 24:(5), 348-372 (1999).
39. E.J. Sellountos and D. Polyzos, *CMES-Computer Modeling in Engineering and Sciences*, Vol. 4, 619-636 (2003).
40. S.N. Atluri, The Meshless Local Petrov-Galerkin (MLPG) Method for Domain and Boundary Discretizations, Tech Science Press (2004).
41. Q.W. Ma, G.X. Wu and R. Eatock Taylor, *Int. J. Numer. Meth. Fluids*, Vol. 36, 287-308 (2001).
42. A. Nestegard, *Technical report*, DNV, Norway, (1999).
43. R.G. Dean and R.A. Dalrymple, R.A. Water Wave Mechanics for Engineers and Scientists, World Scientific, ISBN 981-02-0420-5(1991).
44. T.E. Baldock, C. Swan and P.H. Taylor, *Philos. Trans. R. Soc. London, Ser. A*, Vol. 354, 1-28 (1996).
45. G.X. Wu, Q.W. Ma and R. Eatock Taylor, *Applied Ocean Research*, Vol. 20, 337-355 (1998).
46. O.M. Faltinsen, *Journal of Ship Research*, Vol. 18, 224-241 (1978).
47. T. Okamoto and M. Kawahara, *International Journal for Numerical Methods in Fluids*, Vol. 11, 453-477 (1990).
48. J.C. Martin and W.J. Moyce, *Philos Trans R Soc London, Ser A*, Vol. 244, 312-324 (1952).

CHAPTER 16

LARGE EDDY SIMULATION OF THE HYDRODYNAMICS GENERATED BY BREAKING WAVES

Pierre Lubin* and Jean-Paul Caltagirone

Université de Bordeaux, TREFLE UMR CNRS 8508
16 avenue Pey-Berland, 33607 Pessac Cedex, France
**lubin@enscpb.fr*

Progresses in applied mathematics and computer architecture offer the possibility of developing numerical models for studying the nearshore processes. A very recent way of simulating turbulence in breaking waves is the Large Eddy Simulation (LES) method. Turbulence is taken into account in the Navier-Stokes equations thanks to a turbulence model for the subgrid scales of the flow. The LES method has been applied to breaking waves and showed very promising results compared with experimental measurements, considering two-dimensional configurations. Then, some numerical results of three-dimensional LES have been presented, for spilling, weak and strong plunging breakers. Some very interesting visualizations and encouraging results describing the internal velocity field perturbed by three-dimensional vortices were shown, describing the air entrainment phenomenon. The scope of this review is thus to detail the very recent progresses which have been made thanks to the LES technique. Great improvements have been brought to the description of the turbulent hydrodynamics, leading to the understanding of the complicated features which occur when waves break.

1. Introduction

The coastal zone is divided into four separate zones: the shoaling zone, the breaking zone, the surf zone and the swash zone. In the shoaling zone, as the waves approach to the coastline, they start to deform. The wave height increases as the wave celerity decreases. In the breaking zone, the incoming wave becomes unstable and starts breaking. The surf zone is divided in two separate areas: the outer and inner regions. The outer surf zone starts at the breaking point. The flow is violently perturbed, highly unstationary and turbulence is spreading in the whole water column, from the free surface to the bottom. The inner surf zone starts when the flow becomes quasi-stationary and keeps on propagating towards the shore line. Then, the broken wave runs up the beach in the swash zone. The wave breaking phenomenon can be described as the transition from an irrotational particular motion, during wave steepening, overturning and plunging jet formation, to a rotational motion following jet impingement leading to a turbulent, chaotic flow motion[37]. It can also be characterized as the complete and irreversible transformation of the organized motion of the incident waves into motions of different types and scales, including large-scale coherent vortical motions and small-scale turbulence[1]. Once waves broke, a large amount of energy is released and turned into turbulence[42].

A very recent way of simulating turbulence in breaking waves is the Large Eddy Simulation (LES) method. LES is now used not only to calculate academic test cases, but also to study the physical phenomena that occur in more complex flows such as breaking waves. In LES, the contribution of the large, energy-carrying structures is computed exactly, and the effect of the smallest scales of turbulence is modelled. The scope of this review is to detail the very recent progresses which have been made thanks to the LES technique. Great improvements have been brought to the description of the turbulent hydrodynamics, leading to the understanding of the complicated features which occur when waves break.

In the following sections, we will first describe the complexity of the flow induced by the breaking waves. Then, mathematical formulation of the problem will be given and subgrid scale models will be introduced. The example of a solitary wave propagating over a submerged reef will illustrate the LES method. The structure of the flow under broken waves investigated with LES will then be detailed and discussed.

2. Breaking Waves in the Surf Zone

2.1. *General flow description*

Important reviews described the general processes involved in the wave steepening and subsequent breaking[37], and detailed the features occurring when waves break, such as turbulence generation and air entrainment[1-42].

The flow in the surf zone can be divided into the upper, the middle and the lower regions[6]. In the upper region, the waves break and large-scale eddies associated with small-scale turbulence and entrained air bubbles are generated[34]. These structures spread down in the middle region, air bubbles rise to the surface and more turbulence is produced. The lower region is characterized by the coexistence of the small-scale turbulence originated from the upper zone and the bottom originated turbulence, as the undertow interacts with the bed. The velocity field under broken waves is then characterized by the existence of very active turbulence associated with air entrainment.

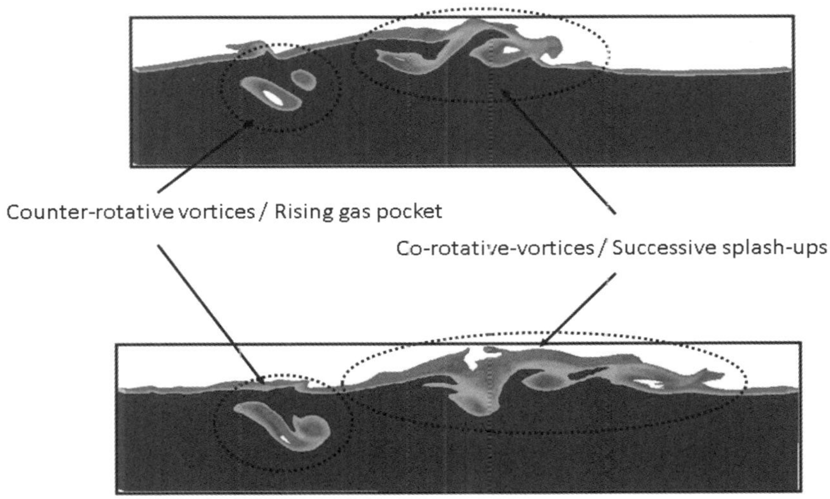

Fig. 1. A numerical illustration of jet-splash cycles[27]. The breaking wave propagates towards the right side.

The jet-splash cycles, occurring several times in a single breaker, are responsible for the generation of a sequence of large-scale coherent vortices (see Fig. 1). These cycles of plunges and splash-ups entrap pockets of air and create strong vortex-like motions in each cycle[32]. The gas pockets collapse and the trapped air escapes towards the free surface. The flow is then characterized with strong vortical motions and high concentrations of air bubbles rising gradually to the surface[1-4]. Lamarre and Melville[17] showed experimentally that up to *50 %* of the wave energy is spent in entraining air and resisting buoyancy forces. The rate of energy dissipation is increased by the bubble penetration depth[3].

Experimental measurements[21] identified that the main mechanism driving the motion in the bubble region was the vortex system generated during the jet-splash cycles. It appears that the entrained air bubbles are contained mostly in the large structures and diffused towards the bottom due to these eddies, which enhances the upwelling of sediment[34]. Nadaoka *et al.*[35] confirm the characteristic structures of large-scale horizontal eddies as found previously by some authors. Obliquely descending eddies are visualized and shown to have a significant role in the deformation of the mean flow field. These eddies are generally observed under spilling breakers or propagating bores, after the occurrence of plunging breakers, and responsible for a large amount of sediment to be put in suspension.

It is then a real challenge to be able to simulate accurately this very complicated flow, which induces turbulence and strong mixing of air and water. Adequate and robust numerical methods are then required to simulate this three-dimensional, unstationary, two-phase turbulent flow, with strong interface deformations, tearing and reconnections, from the largest to the smallest scales involved in the breaking process.

2.2. *Numerical description of turbulence*

The most direct way to describe numerically the complexity of the flow in the surf zone is to solve the Navier-Stokes equations, coupled with mathematical treatments for the free surface and turbulence descriptions. It has been proved to be efficient to treat aerodynamics and academic fluid mechanics problems, as well as complex engineering and environmental flows. The drawbacks are the numerical diffusion, induced by the numerical schemes used to discretize and solve the equations, and the large CPU time, which can be very time consuming when three-dimensional problems are tackled. Nevertheless, this method gives access to many reliable information, e.g. about the velocity, acceleration or pressure fields under the broken waves. Moreover, a considerable improvement in the numerical methods dedicated to the Navier-Stokes equations has been demonstrated, enabling the free surface and the general behavior of turbulent flow structures to be described with a very promising accuracy.

Numerical resolution of the Navier-Stokes equations applied to unstationary turbulent two-phase flow problems should take into account the dynamics existing at all the temporal and spatial scales. This requires very fine spatial and temporal discretizations of the governing equations, respectively smaller than the characteristic length and time associated with the smallest dynamically active scale in the flow. In general, three main distinct approaches exist for turbulence modeling.

The Direct Numerical Simulation (DNS) is the most straightforward method. The governing equations are solved without any modeling simplification or assumptions. All the scales of the flow are supposed to be resolved thanks to a sufficiently fine mesh grid. The numerical schemes are also assumed to be the less dispersive and dissipative possibly. Given the large number of discretizations points needed to simulate directly a turbulent flow (proportional to the *9/4* power of the Reynolds number *Re*), DNS has been limited to low *Re* numbers and simple geometries. The majority of the works found in the literature followed that method and gave very promising results for the overall flow description, including the shoaling and the breaking of the waves. The turbulence is not described nor analyzed, the main large-scale structures being shown as an illustration of the abilities of the numerical tools. A majority of these studies consider theoretical fluids, more viscous and lighter than water.

The Reynolds Averaged Navier-Stokes Equations (RANS) describe the evolution of mean quantities. The method is based on the splitting of the velocity field into a time or spatial averaged part and a fluctuating component. The RANS equations are found by applying the Reynolds' averaging to the Navier-Stokes equations. The effect of turbulence appears in the Reynolds stress term that has to be modelled to close the system of equations. The most used RANS models are the k-ε model, the RNG model, the v^2-f model or the k-l model, using an equation for k and an algebraic relation for the mixing length l. Lin and Liu[24] brought a great improvement in the understanding of the processes taking place in the surf zone, using again a k-ε model. The turbulence levels at breaking were found to be overestimated by *25 %* to *50 %*, as compared to the turbulence levels measured by Ting and Kirby[43]. The results of the plunging case showed a better agreement than the spilling case compared with the measurements reported by Ting and Kirby[44,45]. In the spilling case, the undertow was found to be sometimes in a completely wrong prediction, directing towards the shore line. Recent works[58] showing improvements still report discrepancies.

Large Eddy Simulation (LES) is an intermediate method, which allows one to reach unstationary information to a lower computational cost. This kind of simulation only solves the large scales which drive the flow dynamics, whereas the small scales are modeled thanks to a so-called subgrid-scale model. Simulations can be run with coarser mesh grids which cannot take into account the small dissipative scales of the flow. The separation between the resolved and the modeled scales is mathematically formalized by applying a convolution filter to the Navier-Stokes equations, which introduces the notion of cut-off length. LES should then be an efficient method to overcome the limitations of both DNS and RANS methods.

The interest of this paper is to present the major advances that have been made in the description of the complicated features occurring when waves break, thanks to the LES technique. Lin and Liu[23] gave an overview and discussion of the different numerical techniques which have been used for the interface tracking in numerical simulations of breaking waves. Lubin[27] also detailed this particular aspect of the numerical methods, which will not be treated in this paper, even if the accuracy of these numerical techniques is of great importance for the description of the flow.

3. Mathematical Formulation

Let $\boldsymbol{u}$ be the velocity field vector, p the pressure, $\boldsymbol{g}$ the gravity vector, t the time, ρ the density and μ the dynamic viscosity. The conservation equation system for incompressible flows is the classical Navier-Stokes equations supplemented by the condition of incompressibility. This system can be written as follows:

$$\nabla \cdot \boldsymbol{u} = 0 \tag{1}$$

$$\rho\left(\frac{\partial \boldsymbol{u}}{\partial t} + (\boldsymbol{u} \cdot \nabla)\boldsymbol{u}\right) = \rho \boldsymbol{g} - \nabla p + \nabla \cdot \left(\mu(\nabla \boldsymbol{u} + \nabla^{t} \boldsymbol{u})\right). \tag{2}$$

The selection between the large and small scales, which is the basic idea of LES, implies the definition of two categories subjected to the determination of a reference length, the cut-off length[38]. Are called the large scales, or resolved scales, those which characteristic length are greater than the cut-off length. Are called the small scales, or subgrid scales, the others.

These latest will be taken into account through a subgrid-scale model. The basic idea of the LES is then to deliberately separate the large scales from the rest of the flow, this being achieved thanks to a filtering technique detailed hereafter. Thus, the largest eddies which contain the energy of the flow are completely solved.

The extra terms, called subgrid terms, appearing during the filtering process of the governing equations and representing the interactions between small and large eddies, are modeled by introducing closure assumptions for the filtered system of equations.

3.1. *The filtering operation*

The filtering of a given space-time variable $f(x,t)$ is obtained in real space by applying a convolution product between the function f to be filtered (denoted by an overbar) and the defined filter function $G(x,t)$:

$$\overline{f}(x,t) = \int_{-\infty}^{+\infty}\int_{-\infty}^{+\infty} f(y,t')G(x-y,t-t')dydt'. \tag{3}$$

The cut-off scales associated to the filter function *G(x,t)* in space and time are respectively $\overline{\Delta}$ and $\overline{\tau}$. The non-resolved part of the variable *f(x,t)*, related to the subgrid scales, denoted *f '(x,t)*, is then defined in the physical domain as:

$$f'(x,t) = f(x,t) - \overline{f}(x,t)\,. \tag{4}$$

Among the classical filters used for LES in real space, the top-hat filter is defined as:

$$G(x) = \begin{cases} 1/\overline{\Delta} \text{ if } |x| \leq \overline{\Delta}/2 \\ 0 \text{ otherwise} \end{cases} \tag{5}$$

in a mono-dimensional case.

Applying a low-pass frequency filter to the governing equations, the filtering operation introduces exclusively the effect from the smallest eddies, which are not resolved due to the mesh grid size, to the larger eddies. The choice of the spatial filter is intrinsically linked to the method of simulation which is used. The Finite-Volume method in physical space is generally associated to a top-hat filter, *i.e.* the value in the center of the control volume is the result of integration over the elementary mesh grid of discretization. As a consequence, all the large eddies, which size is strictly greater than the mesh grid $\overline{\Delta}$, called cut-off length, are calculated thanks to the governing equations in every mesh grid points.

3.2. Classical filtered Navier-Stokes equations

The governing equations of the LES in real space are obtained by filtering the system of equations (1-2):

$$\nabla.\overline{\boldsymbol{u}} = \boldsymbol{0} \tag{6}$$

$$\rho\left(\frac{\partial \overline{\boldsymbol{u}}}{\partial t} + (\overline{\boldsymbol{u}}\cdot\nabla)\overline{\boldsymbol{u}}\right) = \rho\boldsymbol{g} - \nabla\overline{p} + \nabla\cdot\left(\mu(\nabla\overline{\boldsymbol{u}} + \nabla^t\overline{\boldsymbol{u}})\right) + \nabla.\tau\,. \tag{7}$$

The effect of the small scales appears in the momentum equation (Eq. 7) through the subgrid-scale tensor τ, which has to be expressed as a function of the filtered variable $\overline{\boldsymbol{u}}$ and a subgrid-scale velocity, defined as the decomposition $\boldsymbol{u}'=\boldsymbol{u}+\overline{\boldsymbol{u}}$. So, we introduce the subgrid-scale stress term τ, defined as:

$$\tau=\overline{\boldsymbol{u}}\otimes\overline{\boldsymbol{u}}-\boldsymbol{u}+\overline{\boldsymbol{u}\otimes\boldsymbol{u}}\,. \tag{8}$$

These equations govern the evolution of the large energy-carrying scales of motion, for a one-phase problem. The information coming from the small scales being lost, their effect is gathered in this subgrid-scale stress term τ, which contains the interactions between resolved and subgrid scales.

This subgrid-scale stress term has now to be modelled[38].

3.3. *Physical subgrid-scale models*

As already detailed previously, the main role of the subgrid-scale model is to remove energy from the resolved scales. Thus, most subgrid-scale models are eddy-viscosity models, the differences between the models found in the literature lying in the formulation of this eddy-viscosity. In most cases, the equilibrium assumption is made to simplify the problem: the small scales are in equilibrium, and dissipate entirely and instantaneously all the energy they receive from the resolved ones. So the forward energy cascade mechanism to the subgrid scales is modeled explicitly using the hypothesis stating that the energy transfer mechanism from the resolved to the subgrid scales is analogous to the molecular mechanisms represented by the diffusion term, in which the viscosity ν appears.

So the energy cascade mechanism will be modeled by a term having a mathematical formulation similar to that of molecular diffusion, in which the molecular viscosity μ in equation is completed with an eddy viscosity, or subgrid viscosity, denoted μ_T. We then have:

$$\rho\left(\frac{\partial\overline{\boldsymbol{u}}}{\partial t}+(\overline{\boldsymbol{u}}\cdot\nabla)\overline{\boldsymbol{u}}\right)=\rho\boldsymbol{g}-\nabla\overline{p^{*}}+\nabla\cdot\left((\mu+\mu_T)(\nabla\overline{\boldsymbol{u}}+\nabla^{t}\overline{\boldsymbol{u}})\right). \tag{9}$$

Assuming a linear relationship between the deviatoric part of the subgrid tensor τ^D and the resolved deformation rate tensor $\overline{S}$, the unknown term τ has then been replaced with:

$$\tau^D = \tau - \frac{1}{3}\boldsymbol{Tr}(\tau)Id = -2\mu_T\overline{S} \tag{10}$$

with $\overline{S} = 1/2\left(\nabla\overline{\boldsymbol{u}} + \nabla^T\overline{\boldsymbol{u}}\right)$. The isotropic part of the subgrid stress tensor being injected in the pressure term, which is now called the modified pressure and denoted $p^* = \overline{p} + 1/3\tau_{kk}$, where $\tau_{kk} = Tr(\tau)$. The eddy viscosity $\nu_T = \mu_T / \rho$ has now to be modeled, as a function of the resolved variables. In the following, the cut-off length $\overline{\Delta}$ is chosen to be calculated as the cube root of the cell volume[9]:

$$\overline{\Delta} = \left(\Delta x \Delta y \Delta z\right)^{1/3} \tag{11}$$

where Δx, Δy and Δz are the sizes of the Cartesian mesh grids in the respective directions X (length) , Y (width) and Z (height).

3.3.1. *Smagorinsky model*

The eddy viscosity, based on the resolved scales, is defined as[40]:

$$\upsilon_T(x,t) = \left(C_S\overline{\Delta}\right)^2\sqrt{\left(2\overline{S}_{ij}\overline{S}_{ij}\right)} \tag{12}$$

where C_S is the Smagorinsky constant, which takes values between *0.18* and *0.23* in the case of homogeneous isotropic turbulent flows. It must be decreased to *0.1* in the case of shear, near solid boundaries, or transitional flows (*e.g.* channel flow). This constant is determined thanks to the following relation:

$$C_S = \frac{1}{\pi}\left(\frac{3C_k}{2}\right)^{-3/4} \tag{13}$$

in the case of an infinite Kolmogorov energy cascade, in isotropic turbulence of kinetic energy spectrum *E(k)*. We then have $C_S \approx 0.18$ for $C_k = 1.4$, which is the usual value admitted for the Kolmogorov constant C_k.

The main problem of this model is its simplicity: it affects the flow as soon as the resolved velocity field contains spatial variations. This model is unable to describe the transition between a laminar to a turbulent state of a flow. Even worst, a turbulent flow can be relaminarized due to the eddy viscosity calculation[38]. The eddy viscosity does not vanish near solid walls, this causing the destruction of the coherent flow structures due to a too high dissipation rate. Subsequently, the model is often corrected with the Van Driest damping law:

$$\upsilon_T = \upsilon_T \left(1 - \exp\left(-z^+ / A\right)\right) \tag{14}$$

where A is a constant taken to be equal to *25*. z^+ is the distance from the solid boundaries in wall units, $z^+ = z\, u_\tau/v$, with u_τ being the wall shear velocity.

Yakhot and Orszag[51], based on a renormalization group theory, proposed a modification of the Smagorinsky model (Eq. 12):

$$\upsilon_T(x,t) = \upsilon \left[1 + H\left(\frac{C_S{}^2 \overline{\Delta}^4 \upsilon_T(x,t)}{\upsilon} \right) \left(2\overline{S}_{ij}\overline{S}_{ij} \right) - C \right]^{1/3} \tag{15}$$

where $C_S=0.0062$, $C=75$. $H(x)$ denotes the Heaviside function. When:

$$\frac{C_S{}^2 \overline{\Delta}^4 \upsilon_T(x,t)}{\upsilon} \gg C \tag{16}$$

the eddy viscosity asymptotically converges to the Smagorinsky model (Eq. 12). $v_T(x,,t)$ is supposed to reduce to the kinematic viscosity v in relatively low strains flows.

3.3.2. *The mixed scale model*

The eddy viscosity can be determined as a function of the cut-off length $\overline{\Delta}$ and the kinetic energy of the subgrid modes $q_{SGS}{}^2$. This subgrid kinetic energy is supposed to be equal to the kinetic energy at cut-off $q_c{}^2$, evaluated in real space as[38]:

$$q^2{}_{SGS}(\boldsymbol{x},t) \equiv q^2{}_c(\boldsymbol{x},t) = \frac{1}{2}\left(\overline{\boldsymbol{u}(\boldsymbol{x},t)}\right)' \left(\overline{\boldsymbol{u}(\boldsymbol{x},t)}\right)' \tag{17}$$

where the test field velocity $\overline{\boldsymbol{u}}'$ can be evaluated thanks to an explicit test filter, denoted ~, applied to the resolved scales:

$$\overline{\boldsymbol{u}}' = \overline{\boldsymbol{u}} + \tilde{\overline{\boldsymbol{u}}} . \tag{18}$$

The test filter is associated with the cut-off length $\tilde{\overline{\Delta}} \geq \overline{\Delta}$. This test field velocity represents the high frequency part of the resolved velocity field.

This model then exhibits a triple dependency on the large and small structures of the resolved field as a function of the cut-off length. The eddy viscosity is defined as:

$$\upsilon_T(x,t) = C_M \overline{\Delta}^{1+\alpha} \left(2\overline{S}_{ij}\overline{S}_{ij}\right)^{\alpha/2} \left(q^2{}_{SGS}(\boldsymbol{x},t)\right)^{\frac{1-\alpha}{2}} \tag{19}$$

with C_M a constant, a new parameter α which value varies between *0* and *1*. Generally, α is taken to be equal to *0.5*. If $\alpha=1$, we find the Smagorinsky model (Eq. 12).

The use of this model does not require any complementary wall model, as the Smagorinsky model does, because the eddy viscosity vanishes when the kinetic energy at cut-off tends to zero. So does it near solid walls.

3.4. *Improvements of subgrid-scale models*

Some works have been produced to improve the simulation results by adapting the subgrid models better to the local state of the flow.

3.4.1. *Structural sensor*

In order to improve the previously detailed subgrid-scale models, a sensor based on local information of the flow can be used[8]. This sensor, or selection function, is related to the local angular fluctuations of the vorticity $\boldsymbol{\omega}$. This sensor locally checks if the flow is of fully turbulent developed type. The selection criterion is based on an estimation of the angle θ between the instantaneous vorticity vector $\boldsymbol{\omega}$ and the local average $\tilde{\boldsymbol{\omega}}$, which is determined by applying a test filter to the instantaneous vorticity vector. The threshold angle θ_0 has been identified[8] to be around the value of *20°*.

The angle θ is given by:

$$\theta(\boldsymbol{x}) = \arcsin\left(\frac{\left\|\tilde{\omega}(x)\otimes\omega(x)\right\|}{\left\|\tilde{\omega}(x)\right\|.\left\|\omega(x)\right\|}\right). \tag{20}$$

A function is then developed to modulate the subgrid-scale model when the angle θ is less than the threshold angle θ_0 :

$$f_{\theta_0}(\theta) = \begin{cases} 1, \text{if } \theta \geq \theta_0 \\ r(\theta)^n = \dfrac{\tan^2(\theta/2))}{\tan^2(\theta_0/2)} \text{otherwise} \end{cases} \tag{21}$$

where n is in practice taken to be equal to 2.

The selection function is used as a multiplicative factor of the eddy viscosity, calculated through a chosen subgrid-scale model. This leads to what is called a selective model:

$$\upsilon^{slc}{}_T = \upsilon_T(\boldsymbol{x},t)\, f_{\theta_0}(\theta(x)) \tag{22}$$

where the constant that appears in the subgrid-scale model has to be multiplied by a factor *1.65*, evaluated on the basis of isotropic homogeneous turbulence simulations, in order to keep the same average eddy viscosity value over the whole domain[38].

3.4.2. *Dynamic procedure to compute the constants*

Another technique of better adapting subgrid-scale models to the local state of the flow, Germano *et al.*[11] developed an algorithm which can be used for any subgrid-scale models. It consists in computing automatically the constant at each point in space and at each time step, the constant being now denoted $C_d = C_d(\boldsymbol{x},t)$. A test filter is used, still denoted ~. Coarser than the previous one, denoted with an overbar, its characteristics length $\tilde{\bar{\Delta}}$ is greater than the characteristic length $\bar{\Delta}$ associated to the implicit filter.

Rewriting equation (10) under a symbolic form, two tensors, τ and T, are respectively associated to the first and second filtering levels, such that:

$$\begin{aligned} \tau_{ij} &= C_d f_{ij}\left(\overline{\Delta}, \overline{\boldsymbol{u}}\right) \\ T_{ij} &= C_d f_{ij}\left(\tilde{\overline{\Delta}}, \tilde{\overline{\boldsymbol{u}}}\right) \end{aligned} \tag{23}$$

under the assumption that the two subgrid tensors can be modeled by the same subgrid-scale model $f_{ij} = -2\mu_T \overline{S}$ with the same constant C_d, for both filtering levels.

This approach is based on the Germano identity[12], which states that the subgrid tensors corresponding to two different levels of filtering can be related through an exact relation. This can be formulated as:

$$L_{ij} = T_{ij} - \tilde{\tau}_{ij}\,. \tag{24}$$

Equation (24) can be written as:

$$L_{ij} = C_d f_{ij}\left(\tilde{\overline{\Delta}}, \tilde{\overline{\boldsymbol{u}}}_i \tilde{\overline{\boldsymbol{u}}}_j\right) - C_d \widetilde{f_{ij}}\left(\overline{\Delta}, \overline{\boldsymbol{u}}_i \overline{\boldsymbol{u}}_j\right). \tag{25}$$

If we assume that the constant does not vary over the test filter length, Lilly[20] proposes calculating the constant C_d by a least-squares method. To do so, we introduce the residual E_{ij}:

$$\begin{aligned} E_{ij} &= L_{ij} - \frac{1}{3} L_{ij}\delta_{ij} - C_d M_{ij} \\ M_{ij} &= f_{ij}\left(\tilde{\overline{\Delta}}, \tilde{\overline{\boldsymbol{u}}}_i \tilde{\overline{\boldsymbol{u}}}_j\right) - \widetilde{f_{ij}}\left(\overline{\Delta}, \overline{\boldsymbol{u}}_i \overline{\boldsymbol{u}}_j\right). \end{aligned} \tag{26}$$

This definition is made of six independent equations, which gives six values for the constant C_d.

To obtain a unique value of the constant, the constant C_d is a solution of the problem formulated as follows:

$$\frac{\partial E_{ij} E_{ij}}{\partial C_d} = 0\,. \tag{27}$$

A unique value is obtained for the dynamic constant C_d through the relation:

$$C_d = \frac{L_{ij} M_{ij}}{M_{kl} M_{kl}}. \tag{28}$$

This technique does not modify the subgrid-scale models, but only adjust locally the constant. The constant C_d can cancel out in some region of the flow and is not bounded. Some of these properties are interesting, but in order to solve some numerical instability problem, a treatment has to be done to the calculation of the constant to guaranty the good numerical behavior of the model. Some treatments are proposed:

- averaging the constant in space[11],
- averaging in space and time[52],
- bounding the constant and the eddy viscosity[52],
- combining these treatments[52].

The averaging procedures, denoted <.>, can be defined in two non-equivalent ways[56]: by averaging the numerator and denominator separately, or by averaging the constant itself:

$$C_d = \frac{\langle L_{ij} M_{ij} \rangle}{\langle M_{kl} M_{kl} \rangle}$$

or (29)

$$C_d = \langle C_d \rangle = \left\langle \frac{L_{ij} M_{ij}}{M_{kl} M_{kl}} \right\rangle.$$

Bounding the constant and the eddy viscosity is used to guaranty that the following conditions will be verified:

- $\nu + \nu_T \geq 0$ (dissipation positive or zero),
- $C_d \leq C_{max}$ (upper bound).

Practically, C_{max} is of the order of the theoretical value of the Smagorinky constant $C_{max} \approx (0.2)^2$.

3.5. *Explicit discrete filtering*

The subgrid-scale models imply two levels of filtering:

- the first level is a frequency filtering, denoted with an overbar, which associated characteristic length is $\bar{\Delta}$,
- the second level is an explicit discrete filtering, called test filtering, denoted ~, which associated characteristic length is $\tilde{\bar{\Delta}}$.

Some subgrid-scale models and improvement techniques detailed previously need the definition of the latest level of filtering:

- the Mixed Scale models,
- the structural sensor,
- the dynamic determination of the constant.

As already mentioned through the previous sections, the associated characteristic length of the test filter is defined as $\tilde{\bar{\Delta}} \geq \bar{\Delta}$. Best results are given with $\tilde{\bar{\Delta}} = 2\bar{\Delta}$. It has been shown that the results were very sensitive to the definition of this explicit discrete filter[38]. All the variables to be filtered are located on the middle of the computational cells. The explicit discrete filter is a linear combination of the neighboring values[38,39]. A three-dimensional test filter results from the tensorial product of the following mono-dimensional three-point filter:

$$\tilde{\bar{f}}_i = \frac{1}{6}\left(\overline{f_{i+1}} + 4\overline{f_i} + \overline{f_{i-1}}\right) \tag{30}$$

this weighted average being obtained by applying the Simpson rule to compute the average of the resolved variable $\bar{f}$ over the control cell surrounding the ith point. This formula is valid for a uniform Cartesian mesh grid, but most of the authors apply it to non-uniform mesh grids, considering the error as negligible. We do so, keeping in mind this approximation in the following.

3.6. *Application to the interaction between a solitary wave and a submerged reef*

The interaction between a solitary wave and a submerged reef has been investigated experimentally and numerically by Chang *et al.*[2] We performed the simulations with the characteristics of air and water and the numerical domain is discretized into *1000* x *170* non-uniform grids, in order to have the same Δx_{min} and Δz_{min} than Chang *et al.*[2] in the vicinity of the submerged obstacle ($1\ 10^{-3}\ m$). The solitary wave propagates towards the right side of the numerical domain. The main features of the interactions are observed in the vicinity of the obstacle: vortices are generated at the corners of the reef as the wave propagates towards and passes by the reef. The evolutions of the weatherside vortex is recorded and analyzed. We solved directly the full Navier-Stokes equations using the numerical model presented in the previous section (this simulation will be referred to as "DNS", as no turbulence model is added, so $\nu_T = 0$). With the same mesh grid configuration, we did LES simulation (Eq. 19). The point of this case is not to discuss about the validity of two-dimensional LES simulations of turbulent flows, but to illustrate the effect of a turbulent viscosity.

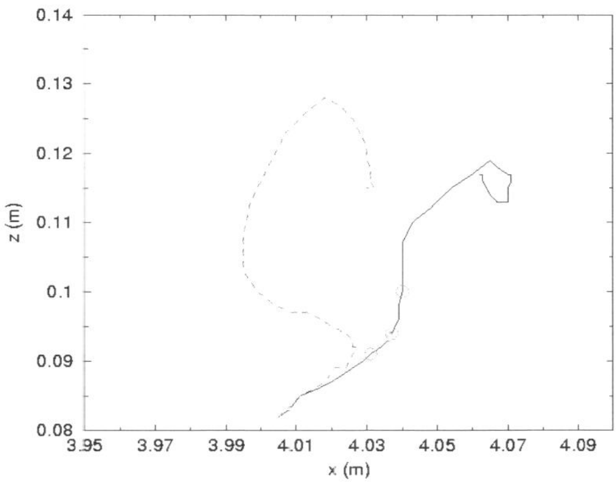

Fig. 2. Trajectories of the eye of the weatherside vortex – Experimental results[2]: circles – LES: solid line – DNS: dashed line.

Figure 2 presents the evolution of the trajectory of the weatherside vortex, which is very well reproduced by the LES simulation, the solid line following the circles showing the experimental results from Chang *et al.*[2] The DNS simulation (dashed line) shows a great discrepancy: the trajectory of the weatherside vortex seems, at the beginning, to follow the right tendency, but then diverges away from the experimental trajectory.

In order to analyze where the differences come from between our numerical simulations, we show two pictures (Fig. 3) of the weatherside vortex at $t = 2.1\ s$. On the left picture, corresponding to the LES simulation, the main structure can be easily seen, above the top left corner of the obstacle. The vortex is moving downstream and starts rising as the solitary wave propagates and passes by the corner of the reef, located at $x = 4\ m$. Some structures are generated near the corner of the reef (two smaller structures are observed at this instant) during the propagation of the wave, the main vortex moving downstream along the reef.

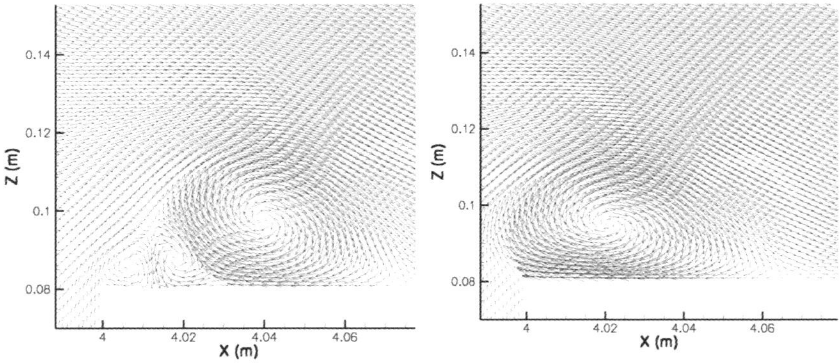

Fig. 3. Comparison at $t = 2.1\ s$ between LES (left) and DNS (right). One velocity vector over three is shown.

These are smaller and constrained by the main vortex, which is acting as a solid wall, the obstacle and the large amount of water increasing above them. These small vortices push the main vortex downstream. In the same way, the right picture corresponding to the DNS simulation shows the main structure, but no other vortex appears in the flow, near the corner. The vortex grows in size as the wave propagates. Once it passed by the reef, the vortex rises and then moves upstream, away from the obstacle. This seems to be due to the "extraction" of some energy from the main vortex thanks to the turbulent viscosity in the LES simulation. The smaller structures have then a chance to appear.

This feature is missing in the DNS simulation, which then should not be called DNS, because the energy cascade is not well simulated. The main vortex grows in size, no energy dissipation occurring, and no other vortex develops to block the main vortex to move upstream. A correct DNS simulation would have been obtained by refining the mesh grid, and so increasing the number of grid points, in order to resolve the smaller turbulent structures. Even if it is a two-dimensional simulation exercise, this test-case illustrates the effect of a turbulence model in the LES sense.

4. Large Eddy Simulation of Breaking Waves

Zhao and Tanimoto[57] first implemented the LES method to breaking waves over a submerged reef with a steep forward face. Smagorinsky model coupled with the van Driest damping function is used (Eq. 12-14), the constant C_S being set at *0.1*. Comparisons with experimental data are presented concerning the surface elevations and the vertical distributions of velocities. Very promising results were obtained, considering a two-dimensional configuration.

Watanabe and Saeki[46] first presented numerical results of three-dimensional plunging breakers over a sloping beach. The Navier-Stokes equations are filtered thanks to a Gaussian filter and the eddy viscosity is evaluated thanks to equations (15-16). The effect of air entrainment is neglected in this study. Watanabe and Saeki[46] investigated three-dimensional distribution and time variation of rotating components of velocity induced by wave breaking. They showed that two-dimensional large eddy motions generated by wave breaking rapidly broke down to fully three-dimensional flow structures during splash-up cycle as waves propagate on the sloping beach.

Mutsuda and Yasuda[33] presented numerical results of a three-dimensional solitary wave, describing the air entrainment phenomenon under the plunging breaking wave. A dynamic Smagorinsky model is used to compute the eddy viscosity (Eq. 12-29) and the numerical domain is discretized into *402* x *37* x *62* uniform grids ($\Delta x \approx 1.6\ 10^{-2}\ m$, $\Delta y \approx 1.6\ 10^{-2}\ m$ and $\Delta z \approx 8\ 10^{-3}\ m$). Entrained air pockets are generated while the plunging jet impacts the forward face of the wave and the subsequent splash-ups entrap air. The evolution of the air is visualized and described. The major role of the air entrainment on turbulence and vorticity generation is highlighted by a detailed study of time evolution of turbulence intensity.

Christensen and Deigaard[5] used a fully three-dimensional numerical tool to study spilling, weak and strong plunging breakers over a sloping beach, using a Smagorinsky model ($C_S = 0.1$) coupled with van Driest damping function (Eq. 12-14). In the breaking zone, the mesh grid sizes are $\Delta x \approx 9\ 10^{-3}\ m$, $\Delta y = 7.5\ 10^{-3}\ m$ and $\Delta z = 6\ 10^{-3}\ m$. The effect of air is not taken into account in the model. Some very interesting visualizations and encouraging results describing the internal velocity field perturbed by three-dimensional vortices are shown. They found similar results to Watanabe and Saeki[46], showing that three-dimensional flow structures spontaneously developed at the breaking point. They also identified the formation of the obliquely descending eddies. Analyzing the turbulent kinetic energy, they found that the obliquely descending eddies are located at local maximum in the intensity of the turbulence left by the propagating breaker. Nevertheless, these obliquely descending eddies were not always found, longitudinal eddies with horizontal axes being observed instead. Christensen *et al.*[6] developed the discussion about the mechanism responsible for the generation of the turbulent structures. They proposed that the obliquely descending eddies might be created by a similar mechanism to the one responsible for the Langmuir circulation cells. Christensen[6] later improved the numerical model to study spilling and plunging breaking waves in the configuration from Ting and Kirby[43–45]. Christensen[7] focused on the set-up, undertow and turbulence levels, using two different subgrid-scale models to evaluate the eddy viscosity. Smagorinsky model is compared to a *k-equation* subgrid-scale model (not detailed here). The outcome is not very clear as most comparisons led to the conclusion that the coarse resolution and the neglected effect of air in the model are probably responsible for the discrepancies. Indeed, the resolution was chosen to get acceptable computational times with a sufficient accuracy to resolve the effect of the wave boundary layer close to the bed. This resulted in missing the correct breaking point, which is very sensitive to any imperfections in the wave generation. Christensen *et al.*[6] recalled Nadaoka *et al.*[35] pointed out that the turbulence created in the surf zone by a broken wave will affect the exact position of the breaking point of the following wave.

Major discussion about the turbulent structures generation and their interactions is continued by Watanabe *et al.*[48,49] Spilling, plunging, strongly plunging and spilling/plunging breakers are simulated with the model developed by Watanabe and Saeki[46]. Strong surface distortions, surface tension and air-water interactions were omitted. The numerical tool was first validated compared to the experimental results from Ting and Kirby[44]. A general good agreement was found, some differences appearing when phase-averaged velocity and turbulent energy were examined in aerated regions. Sketches are proposed to illustrate the mechanism of formation of a vortex loop, which evolves into a rib structure (see Fig. 4).

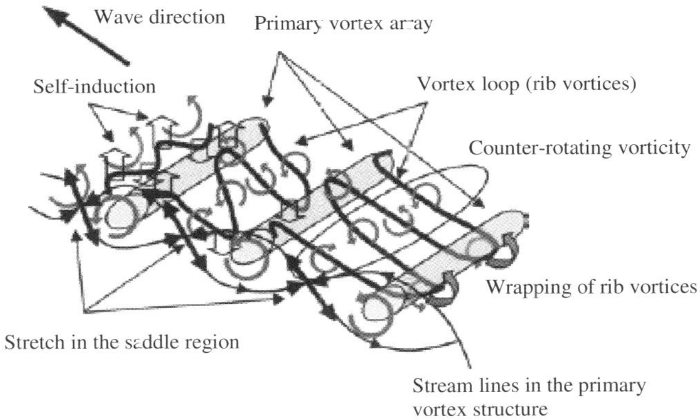

Fig. 4. A schematic representation of the mechanism to form the vortex loops evolving into the rib structure, from Watanabe *et al.*[47]

Vorticity is shown to be unstable in a saddle region localized between the rebounding plunging jet and the primary spanwise vortex formed by the initial free-falling tongue of water ejected from the crest of the breaking wave. The first splash-up rebounds and induces another main spanwise vortex. Between the first primary spanwise vortex and this latest one, a vortex loop takes place, undulates and wraps the adjacent primary vortices to form a rib structure. Watanabe *et al.*[49] speculate that this structure could be the obliquely descending eddies observed by Christensen and Deigaard[5]. The impact of these various structures on the free surface aspect is also illustrated. Finally, the obliquely descending eddies formation and the subsequent air entrainment is explained. The evolutions of these structures are found to be breaker dependent. The obliquely descending eddies are found to be much longer in spilling breaking wave cases, whereas the inclination angles are smaller in plunging breakers. This was explained to be due to the presence of the bottom which is closer to the free surface in the case of a plunging breaking wave, spilling breaker taking place in deeper water depth. The vortices are thus less constraint and free to develop behind the wave crest in spilling breaking waves. Trapped bubbles of air are proved to be entrained deep in the water when involved in obliquely descending eddies. The evolution of the eddies are found to be dependent to the wave shape, the rebounding height of the jet and the spacing between the primary vortices. However, concerning the detailed surface-vortex interactions, Watanabe *et al.*[49] recently indicated that this phenomenon would not be appropriately modeled in the computations since simplified free-surface boundary conditions were employed. Some experimental works would be necessary to confirm or not the numerical observations about the mechanism responsible for the obliquely descending eddies.

Hieu *et al.*[14] studied two-dimensional breaking waves, using Smagorinsky model (Eq. 12). The numerical domain is discretized into *1350* x *150* uniform grids ($\Delta x \approx 2\ 10^{-2}\ m$ and $\Delta z \approx 1\ 10^{-2}\ m$). Even if the surface tension was not implemented in the numerical simulations, the numerical model was shown to be able to reproduce the wave breaking processes in shallow water, compared to the experimental results from Ting and Kirby[43–45].

Zhao *et al.*[58] implemented a new multi-scale method where the two-dimensional flow structures are fully resolved with a *k-l* RANS approach, while the three-dimensional turbulence interactions are modeled with a subgrid-scale eddy viscosity model. Two-dimensional spilling and plunging breaking waves are studied compared to the experimental data from Ting and Kirby[43–45]. The numerical domain is discretized into *1000* x *100* uniform grids ($\Delta x \approx 2\ 10^{-2}\ m$ and $\Delta z \approx 8\ 10^{-3}\ m$). A general good agreement with experimental data is shown, improving significantly the RANS results found in the literature. In the spilling case, the numerical model slightly overestimates the wave heights near the breaking point and at the inner surf zone. In the plunging case, the wave heights are again overestimated near the breaking point, but are in a better agreement than in the spilling case. Some discrepancies also appear when simulated phase averaged surface elevations, horizontal and vertical velocities, in both cases, are compared to the experimental data. This has been mainly attributed to the numerical resolution being not high enough and to the fact that the air entrainment is not taken into account in the numerical model.

Hieu and Tanimoto[15] improved their numerical model[14] showing even better comparisons with experimental data concerning free surface elevations (with the same mesh grids $\Delta x \approx 2\ 10^{-2}\ m$ and $\Delta z \approx 1\ 10^{-2}\ m$). The numerical results are also compared to those from Zhao *et al.*[58] and a better accuracy is shown. They pointed out the major role of the air entrainment and the necessity to take it into account to get accurate numerical results. The numerical model is then used to study waves breaking over a submerged porous breakwater. Free surface profiles, velocity and dynamic pressure fields are carefully investigated, again compared to experimental measurements.

Lubin[28] presented the impact of LES (Eq. 19-22) compared with results obtained from DNS[26], considering three-dimensional unstable periodic waves (the mesh grid sizes are $\Delta x = \Delta y = \Delta z = 4\ 10^{-4}\ m$). Later, Lubin[27] investigated numerically several configurations. Numerical results for solitary waves colliding a vertical end-wall or head-on collisions between two solitary waves[29] are successfully compared to analytical and experimental data, illustrating the capacity of the numerical tool to take accurately into account highly complex nonlinear effects. The breaking of a stable solitary wave over a submerged rectangular reef is presented and compared to numerical and experimental data. This case was the occasion to lead a numerical exercise for comparing several numerical methods[13]. The agreement with the experimental gauges measurements was very satisfactory for the LES simulations. Lubin *et al.*[28-30] then discussed the results obtained from simulating three-dimensional plunging breaking waves by solving the Navier-Stokes equations (see Fig. 5), in air and water, coupled with the

Mixed Scale model and a structural sensor (Eq. 19-22). The splash-up mechanism is carefully detailed and the vortices generation and air entrainment process is described. The behavior of the gas pockets is analyzed and its impact on the energy dissipation is shown.

Fig. 5. Three-dimensional simulation of plunging breaking wave.

It can also be highlighted that, in the three cases detailed previously, the maximum of the total pre-breaking wave energy is dissipated during the splash-ups, an average value of *40 %* of the total pre-breaking wave energy being estimated from the numerical results. This observation is in a very good agreement with the results of Lamarre and Melville[17].

Based on these works, Lubin *et al.*[31] presented the results obtained for the LES of three-dimensional regular waves shoaling and breaking over a sloping beach, still taking air and water into account (Fig. 6).

Fig. 6. Three-dimensional simulation of wave breaking over a sloping beach (black line)[31].

A spilling/plunging breaking event was expected according to the experimental measurements, but the numerical results showed discrepancies, due to the coarse mesh grid resolution (the mesh grid sizes are $\Delta x = 2.8\ 10^{-2}\ m$, $\Delta y = 1.5\ 10^{-2}\ m$ and $\Delta z = 8.4\ 10^{-3}\ m$). Indeed, the main observed differences were that the first short spilling event was missed and the dislocation of the gas pockets into small bubbles could not be simulated, even if, in the numerical results, the gas pockets corresponded to some air-water mixing zones observed in the experimental pictures.

More recently, Liovic and Lakehal[24] modified the Smagorinsky model (Eq. 12) with a damping model function $f_{\mu Int}$ dependent on the normalized distance to the interface y^+_{Int} from the gas side:

$$\upsilon_T(x,t) = f_{\mu Int}\left(C_S\overline{\Delta}\right)^2\sqrt{\left(2\overline{S}_{ij}\overline{S}_{ij}\right)}$$
$$f_{\mu Int} = 1-\exp\left(-0.00013\left(y^+_{Int}\right)-0.00036\left(y^+_{Int}\right)^2-1.08.10^{-5}\left(y^+_{Int}\right)^3\right). \quad (31)$$

The coefficients and the exponential dependence have been obtained from the Direct Numerical Simulations from Fulgosi *et al.*,[10] considering low to moderate interface deformations. Two-phase flow studies of Fulgosi *et al.*[10] have shown the need for turbulence damping approaching deformable interfaces in the under-resolved simulation of such flows. Liovic and Lakehal[24] presented three-dimensional regular waves breaking over a sloping beach, setting the same parameters than Christensen and Deigaard[5], as an application of their numerical methods (the mesh grid sizes are $\Delta x = 4\ 10^{-2}\ m$, $\Delta y = 7.5\ 10^{-3}\ m$ and $\Delta z = 7.5\ 10^{-3}\ m$). Some qualitative results are presented about near-interface turbulence to prove the capacity of the numerical model to treat two-phase flows, more dedicated detailed works being announced for the future.

After some pioneering works studying LES of sediment transport over ripples[51–53], Suzuki *et al.*[41] introduced three-dimensional LES coupled with sediment pickup / advection model to study the spatial and temporal variations of sediment concentration on a sloping beach. Plunging breaking waves were generated on a uniform sloping beach. Smagorinsky model is used to calculate the eddy viscosity (Eq. 12). The sediment particles were assumed to have a spherical diameter of *0.15* mm and a density of $2.65\ g.cm^{-3}$. The sediment pickup rate per unit area is calculated, taking into account the density ratio between the fluid and the sand densities, the water viscosity, the bottom friction coefficient and the fluid velocity at the bottom. A fall velocity is also considered to evaluate the amount of sediment in the flow. Some good qualitative results are found compared to laboratory measurements. The figures presented confirm the major role of the large organized eddies in the sediment pickup. The study only focused on the development of the sediment transport modeling, so no comparisons have been presented concerning the flow field. Quantitative discrepancies between numerical and experimental results are also due to the neglected interactions between particles and the flow field.

LES has also been used to study tsunamis generated by sliding masses[25] or interactions between waves and submerged structures[18,19], proving the accuracy of the method for practical use in the coastal engineering management.

5. Conclusions

Some elements can be highlighted from this presentation of the existing works.

The authors proved a great improvement in the numerical methods allowing a description with a very promising accuracy both the free surface and the general behavior of the turbulent flow structures. Major contributions give a really interesting insight of the hydrodynamics under the broken waves. Based on the review, we find that the most widely used subgrid scale model is the Smagorinsky model. In spite of its negative aspects, its simplicity is widely appreciated. A general good agreement with experimental data is shown, improving significantly the RANS results found in the literature. Only few authors compared numerical models but the outcome is not very clear as most numerical LES results compared to experimental data still report discrepancies.

Three major points can be responsible for the observed differences. First, most of the published works are two-dimensional configurations, for reasons of simplicity or a first approximation. But turbulence and the mechanisms of dislocation of the entrained gas pockets into small bubbles are three-dimensional, so LES simulations should tend to be achieved in three-dimensions. Then, comparisons led to the conclusion that the use of coarse mesh grid resolutions, leading to bad free surface descriptions, is probably responsible for the differences. And finally, many authors neglected the effect of air in the numerical models, which has been widely identified to be an important feature to take into account.

In spite of progresses in applied mathematics and computer architecture, running three-dimensional two-phase Navier-Stokes simulations still remains very demanding in terms of CPU time and computer facilities. Numerical simulations involving turbulence investigations imply large number of mesh grid points to reach the small length-scales and thus to be reliable. Studies involving large sizes of numerical domains, with both small scales of turbulence and free surface deformations, are unreachable by DNS (in the strict sense of turbulence). Thus, the LES approach is the only alternative for dealing with breaking waves. Nevertheless, LES of breaking waves involve large numerical domains, to include the generation of regular waves propagating towards sloping beaches, and large CPU time of calculations are still required to simulate a sufficient number of wave periods to reach a steady state in the surf zone to perform correct averaging in time.

So, all the authors still agree to state that solving the Navier-Stokes equations in an air/water configuration is a real challenge, due to strong interface deformations and air entrainment. The bad description of air/water mixing has been widely proved to be responsible for the observed discrepancies. The choice for the mesh grid resolution should be first driven by the correct description of the free surface, otherwise the air entrainment process will not be correctly described. So, further advances in LES modeling of breaking waves first seem to be a matter of free surface tracking, even if LES allows the use of coarser mesh grid resolution compared to DNS.

Giving a final conclusion about which subgrid scale model should be used is not realistic in this paper. Our point is to highlight the benefit brought by LES to the numerical investigations of breaking waves. Comparisons between subgrid scale models, from simple to more sophisticated, will be further possible providing refined mesh grids are used to ensure an accurate free surface description. Parallel computing would certainly enhance the access to a better level of description of the turbulent behavior of the entrained and rising air bubbles.

Thus, the effect of air has not been studied yet in details in most of the cited two- and three-dimensional simulations. Taking the mixture of air and water into account is crucial and still remains one of the challenges of the coming years.

Another major point is that when simulating two-phase flows, the interface can become smaller than the mesh grid size (droplets or bubbles). These small inclusions are thus "lost" and their contribution to the flow can lead to a wrong description of the flow. It is then necessary to propose an appropriate modeling of these sugrid structures, also called subgrid interface. New subgrid scale models dedicated to two-phase flows have to be analytically developed. Some recent works heading in this direction[10–24] still have to be implemented or evaluated.

Acknowledgments

The authors gratefully acknowledge the financial and scientific support of the French INSU - CNRS (Institut National des Sciences de l'Univers - Centre National de la Recherche Scientifique) program IDAO (Interactions et Dynamique de l'Atmosphère et de l'Océan), and the computational facilities financially supported by the "Conseil Régional d'Aquitaine".

References

1. J. A. Battjes, *Annu. Rev. Fluid Mech.*, **20**, 257 (1988).
2. K.-A. Chang, T.-J. Hsu and P. L.-F. Liu, *Coast. Eng.*, **44**, 13 (2001).
3. H. Chanson and J.-F. Lee, *Coast. Eng.*, **31**, 125 (1997).
4. H. Chanson, S. Aoki and M. Maruyama, *Coast. Eng.*, **46**, 139 (2002).
5. E. D. Christensen and R. Deigaard, *Coast. Eng.*, **42**, 53 (2001).
6. E. D. Christensen, D.-J. Walstra and N. Emerat, *Coast. Eng.*, **45**, 169 (2002).
7. E. D. Christensen, *Coast. Eng.*, **53**, 463 (2006).
8. E. David, *PhD thesis*, Institut National Polytechnique de Grenoble (1993).
9. J. W. Deardorff, *J. Fluid Mech.*, **41**, 453 (1970).
10. M. Fulgosi, D. Lakehal, S. Banerjee and V. De Angelis, *J. Fluid Mech.*, **482**, 319 (2003).
11. M. Germano, U. Piomelli, P. Moin and W. H. Cabot, *Physics of Fluids*, **3**, 1760 (1991).
12. M. Germano, *J. Fluid Mech.*, **238**, 325 (1992).
13. P. Helluy, F. Gollay, S. T. Grilli, N. Seguin, P. Lubin, J.-P. Caltagirone, S. Vincent, D. Drevard and R. Marcer, *Math. Model. Numer. Anal.*, **39**, 591 (2005).
14. P. D. Hieu, K. Tanimoto and V. T. Ca, *Appl. Math. Model*, 983 (2004).
15. P. D. Hieu and K. Tanimoto, *Ocean Eng.*, 33 (2006).

16. E. Labourasse, D. Lacanette, A. Toutant, P. Lubin, S. Vincent, O. Lebaigue, J.-P. Caltagirone and P. Sagaut, *Int. J. Multiph. Flow*, **33**, 1 (2007).
17. E. Lamarre and W. K. Melville, *Nature*, **351**, 469 (1991).
18. T. Li, P. Troch and J. De Rouck, *J. Comp. Phys.*, **198**, 686 (2004).
19. T. Li, P. Troch and J. De Rouck, *J. Comp. Phys.*, **223**, 865 (2007).
20. D. K. Lilly, *Physics of Fluids*, **A 4**, 633 (1992).
21. C. Lin and H. H. Hwung, *Exp. Fluids*, **12**, 229 (1992).
22. P. Lin and P. L.-F. Liu, *J. Fluid Mech.*, **359**, 239 (1998).
23. P. Lin and P. L.-F. Liu, in *Advances in Coastal and Ocean Engineering*, **5** (World Scientific) 213 (1999).
24. P. Liovic and D. Lakehal, *J. Comp. Phys.*, **222**, 504 (2007).
25. P. L.-F. Liu, T.-R. Wu, F. Raichlen, C. E. Synolakis and J. C. Borrero, *J. Fluid Mech.*, **536**, 107 (2005).
26. P. Lubin, S. Vincent, J.-P. Caltagirone and S. Abadie, *C. R. Méc.*, **331**, 495 (2003).
27. P. Lubin, *PhD thesis (in English)*, Université Bordeaux I (2004).
28. P. Lubin, S. Vincent, J.-P. Caltagirone and S. Abadie, in *Proc. ASCE 29th Int. Conf. on Coastal Eng.*, 344 (2004).
29. P. Lubin, S. Vincent and J.-P. Caltagirone, *C. R. Méc.*, **333**, 351 (2005).
30. P. Lubin, S. Vincent, J.-P. Caltagirone and S. Abadie, *Coast. Eng.*, **53**, 631 (2006).
31. P. Lubin, H. Branger and O. Kimmoun, in *Proc. ASCE 30th Int. Conf. on Coastal Eng.*, 238 (2006).
32. R. L. Miller, *Chapter 24 In: Davis, R.A., Ethington, R.L. (Eds.), Spec. Publ.- Soc. Econ. Paleontol. Mineral*, 92 (1976).
33. H. Mutsuda and T. Yasuda, in *Proc. ASCE 27th Int. Conf. on Coastal Eng.*, 755 (2000).
34. K. Nadaoka and T. Kondoh, *Coast. Eng. J. Jpn*, **25**, 125 (1982).
35. K. Nadaoka, M. Hino and Y. Koyano, *J. Fluid Mech.*, **204**, 359 (1989).
36. A. Okayasu, T. Susuki and Y. Matsubayashi, *Coast. Eng. In Jpn*, 47 (2005).
37. D. H., Peregrine, *Annu. Rev. Fluid Mech.*, **15**, 149 (1983).
38. P. Sagaut, Large Eddy Simulation for incompressible flows — An introduction. Springer Verlag (1998).
39. P. Sagaut and R. Grohens. *Int. J. Numer. Methods Fluids*, **31**, 1195 (1999).
40. J. Smagorinsky, *Month. Weath. Rev.*, **91**, 99 (1963).
41. T. Suzuki, A. Okayasu and T Shibayama, *Coast. Eng.*, **54**, 433 (2007).
42. I. A. Svendsen and U. Putrevu, in *Advances in Coastal and Ocean Engineering*, **2**, (World Scientific) 1 (1996).
43. F. C. Ting and J. T. Kirby, *Coast. Eng.*, **24**, 51 (1994).
44. F. C. Ting and J. T. Kirby, *Coast. Eng.*, **24**, 177 (1995).
45. F. C. Ting and J. T. Kirby, *Coast. Eng.*, **27**, 131 (1996).
46. Y. Watanabe and H. Saeki, *Coast. Eng. J. Jpn*, 41 (1999).
47. Y. Watanabe, M. Yasuhara and H. Saeki, in *Proc. ASCE 27th Int. Conf. on Coastal Eng.*, 942 (2000).

48. Y. Watanabe and H. Saeki, *Int. J. Numer. Methods Fluids*, 39 (2002).
49. Y. Watanabe, H. Saeki and R. J. Hosking, *J. Fluid Mech.,* 545 (2005).
50. Y. Watanabe, A. Saruwatari and D. M. Ingram, *J. Comp. Phys.,* **227**, 2344 (2008).
51. V. Yakhot and S. Orszag, *J. Sci. Comp.*, 3 (1986).
52. Y. Zang, R. L. Street and J. R. Koseff, *Physics of Fluids*, **A 5**, 3186 (1993)
53. E. A. Zedler and R. L. Street, *J. Hydr. Engrg.*, **127**, 444 (2001).
54. E. A. Zedler and R. L. Street, in *Proc. ASCE 28th Int. Conf. on Coastal Eng.*, 1 (2002).
55. E. A. Zedler and R. L. Street, *J. Hydr. Engrg.*, **132**, 180 (2006).
56. H. Zhao and P. R. Voke, *Int. J. Numer. Meth. Fluids*, 19 (1996).
57. Q. Zhao and K. Tanimoto. *In Proc. ASCE 26th Int. Conf. Coastal Eng.*, 1 (1998).
58. Q. Zhao, S. Armfield and K. Tanimoto, *Coast. Eng.*, 51 (2004).

CHAPTER 17

RECENT ADVANCES IN TURBULENCE MODELING FOR UNSTEADY BREAKING WAVES

Qun Zhao

Halcrow Group Ltd, North-West Regional Office, Manchester
Deanway Technology Centre, Wilmslow Road, Handforth SK9 3FB, UK
zhaoqn@halcrow.com

Steve W. Armfield

School of Aerospace, Mechanical and Mechatronic Engineering
Sydney University, Sydney 2006, Australia
armfield@aeromech.usyd.edu.au

Recent researches have found that for high Reynolds number unsteady breaking wave problems, traditional turbulence models based on Reynolds-Averaged Navier-Stokes (RANS) simulations, such as the $k-\epsilon$, $k-\omega$ and Algebra Reynolds stress models (ARSM) may be inadequate. Alternative approaches such as Large Eddy Simulation (LES) and Direct Numerical Simulation (DNS) are being studied. In this paper, we review these recent advances in turbulence modeling for unsteady breaking waves.

1. Introduction

Breaking wave generated turbulence is of great interest for coastal engineers, because it modifies nearshore currents through momentum and energy transfer (see Refs. 1–2); enhances momentum mixing and alters the vertical structures of current velocities (Ref. 3). Moreover, it is the major force to put nearshore sediments into suspension. The understanding of breaking wave generated turbulence has been improved due to the advances of measuring technology. Numerical studies benefit from recent availability of fast computers, more efficient algorithms, as well as high quality data sets from laboratory measurements (Refs. 4–5). Several reviews are reported in the literature recently (Refs. 3–6), the present review focuses on turbulence modeling for this fast growing topic.

Numerical studies of breaking wave generated turbulence closely follow the advances of laboratory studies and theoretical analysis. Therefore we will first briefly summarize recent research progress in experimental and theoretical studies with special attention to those investigations that directly influence turbulence modeling. Then we will review recent progress made in numerical modeling in this subject including Reynolds-Averaged Navier-Stokes (RANS), Large Eddy Simulation (LES) and Direct Numerical Simulation (DNS).

2. Brief Review of Experimental and Theoretical Studies

Laboratory measurements and analysis of breaking wave generated turbulence aim at answering the following questions:

(1) Where is the origin of breaking wave generated turbulence?
(2) Which previously studied turbulent flow does this turbulence resemble?
(3) What is the length scale of this turbulence?
(4) What characteristic eddy structures does this turbulence have, particularly in the post breaking stage?

The first significant study of breaking wave generated turbulence was that of Peregrine and Svendsen.[7] They found that the turbulence generated by the breaking waves, while initiated at the toe of the turbulent front, spreads downwards and continues to do so long after the breaker has passed. The spreading of turbulence is similar to a mixing layer. However, based on their Laser Doppler Anemometry (LDA) measurements, Battjes and Sakai[8] argued that the turbulence generated by breaking waves is more like a wake rather than a mixing layer. Yeh[9] compared the initiation of vorticity of a mixing layer and a bore. He indicated the generation of the turbulence in a bore is caused by the baroclinic water surface, therefore, the turbulence in a bore cannot be similar to that in a mixing layer. Nadaoka[10] pointed out the existence of large-scale eddies under breaking waves is mainly responsible for the generation of Reynolds stress in the upper layer of the water. They were also the first to observe the oblique eddies. Tallent[11] showed that the incident wave condition and the bottom slope affect the eddy characteristics considerably, and the wave steepness plays a more important role than the bottom slope. Okayasu *et al.* (Refs. 12–13) performed extensive measurements of the velocity field under breaking waves using LDA. George[14] analysized turbulence generated by breaking waves from La Jolla beach California using a hotfilm anemometer. The hotfilm frequency spectra indicated the frequency f obeys a $f^{-5/3}$ inertial subrange following a f^{-3} regime of orbital motion. However, separating turbulence and orbital waves is not straightforward because of the existence of large-scale eddies,[15] especially in the field (see Refs. 16–17). In the laboratory, this problem is slightly less worrisome. Ting and Kirby's studies (Refs. 4–5) for typical spilling and plunging breaking waves show that the turbulence structure strongly depends on the type of the breaker. Turbulence kinetic energy is transported offshore under a spilling breaker but onshore under a plunging breaker. Their work also provides the main data sources for testing today's numerical models.

The above studies are all single point measurements and therefore cannot provide spatial pictures of the flow, which has been made possible in recent years by advances in particle image velocimetry (PIV) technology. Chang and Liu[18] using PIV investigated the particle velocities near the tip of the overturning jet of a breaking wave. They found the maximum velocity is 1.68 times the phase speed, and the maximum acceleration is 1.1 times the gravitational acceleration. They then examined the vertical vorticity post wave breaking and found the number and locations of these vortices are random. Their study also showed that the mean velocity in the cross-flume direction is not zero, which could be an indication of the existence of three-dimensional mean flow motions. But Chang and Liu[18] did not analysize the effect of side walls during the experiment.

Chang and Liu[19] carried out studies on spilling breaking waves to evaluate the magnitude of each term in the turbulence kinetic energy (TKE) equation. They show the turbulence intensities obtained from ensemble-average were 25% ∼ 50% smaller than those obtained from the moving-average method by Nadaoka and Kondoh[15] and Nadaoka *et al.*,[10] but larger than those obtained from the phase-averaging method of Ting and Kirby.[5] The mean values are also different for each separation method. Their study also showed that under the trough level, turbulent convection, production and dissipation are of the same order but at different phases and locations. Turbulent diffusion is negligible.

A review of theoretical and experimental work on spilling breakers by Duncan[20] draws attention to the surface tension effect. He showed the effect of surface tension on breaking waves increases as wavelength decreases. When the surface tension dominates, the free surface does not overturn during the breaking. Misra *et al.* (Refs. 21–22) pointed out the importance of including an air-water intermittency zone based on their PIV experiments of a hydraulic jump. They derived a surface boundary condition for turbulence models, and mean equations for water only single phase models for steady breakers. Their study showed that turbulence production dominates over dissipation near the edge of the breaker, but dissipation prevails further downstream. This is consistent with the results found by Chang and Liu.[19] Misra *et al.* (Refs. 21–22) also discussed the effects of the extra strain rate caused by surface curvature. They note that though the extra strain rates seem negligible in the mean and turbulence equations, they recommend considering this effect based on previous studies of curved flow by Gibson and Rodi[23] and Rumsey *et al.*[24]

In summary, some conclusions can be drawn from these experimental and theoretical studies:

(1) The Reynolds number for wave breaking problem can be defined as $Re = UL/\nu$, where $U \sim c$ and $L \sim$wave length are velocity and length scales, respectively, c the wave celerity and ν the kinetic viscosity.
(2) Breaking wave generated turbulence is a characteristic of unsteady high Reynolds number complicated flow, resemblance to basic turbulence flow types is difficult to find.
(3) Turbulent production, dissipation and convection are of the same order but dominate in different phases and locations.
(4) Separation of orbital wave motions and turbulence is not straightforward because of the existence of large-scale eddies.
(5) Air entrainment is important, especially at the post breaking stage of a plunging breaker.
(6) Surface tension plays an important role determing the shape of wave overturning for shorter waves.

3. Numerical Modeling

3.1. *The Navier-Stokes equations*

As noted above, breaking wave generated turbulence is an unsteady nonlinear complex high Reynolds number flow. The exact governing equations are the Navier-Stokes equations and the continuity equation,

$$\frac{\partial u_i}{\partial t} + u_j \frac{\partial u_i}{\partial x_j} = -\frac{1}{\rho}\frac{\partial p}{\partial x_i} + \frac{\partial}{\partial x_j}\nu\left(\frac{\partial u_i}{\partial x_j} + \frac{\partial u_j}{\partial x_i}\right) \tag{3.1}$$

$$\frac{\partial u_i}{\partial x_i} = 0 \tag{3.2}$$

where ρ is the water density, p the pressure and u_i the velocity components and ν kinematic viscosity.

The Navier-Stokes equations (3.1) are nonlinear equations, in which all scales are coupled through the nonlinear terms and they cannot be solved independently. Therefore, the exact Navier-Stokes equations are difficult to integrate for high Reynolds number turbulent flows, because of the wide range between the largest and the smallest scales involved. Approximations have to be made.

3.2. *The Reynolds Averaged Navier-Stokes equations (RANS)*

Depending on the problem to be solved and the accuracy required, there are several levels of approximations that can be made. For periodic wave problems, the most widely used approximation today is the phase-averaged Navier-Stokes equations, also referred as the *Reynolds-Averaged Navier-Stokes* (RANS) equations. In this method, the velocity u_i and the pressure p in the Navier-Stokes equations are split into the mean and the turbulent fluctuations parts,

$$u_i = \langle u_i \rangle + u_i' \tag{3.3}$$

$$p = \langle p \rangle + p' \tag{3.4}$$

where $'$ stands for the turbulent fluctuations. The phase-averaged quantities are defined as, for example for velocities,

$$\langle u_i \rangle(\theta) = \frac{1}{N} \sum_{n=1}^{n=N} u\left[\theta + 2\pi(n-1)\right] \tag{3.5}$$

where θ is the local fixed phase for n periods of waves, and $N \to \infty$.

Substituting (3.3), (3.4) into (3.1), the Reynolds-averaged equations read

$$\frac{\partial \langle u_i \rangle}{\partial t} + \langle u_j \rangle \frac{\partial \langle u_i \rangle}{\partial x_j} = -\frac{1}{\rho}\frac{\partial \langle p \rangle}{\partial x_i} + \frac{\partial}{\partial x_j}\nu\left(\frac{\partial \langle u_i \rangle}{\partial x_j} + \frac{\partial \langle u_j \rangle}{\partial x_i}\right) - \frac{\partial \langle u_i' u_j' \rangle}{\partial x_j} \tag{3.6}$$

in which the new term $\langle u_i' u_j' \rangle$ is called the "Reynolds stress term", and needs to be approximated by a turbulence model. This class of turbulence models are referred to as the RANS models.

Note the phase-averaging of (3.5) and (3.6) state that some large-scale turbulent structures, such as some coherent structures, are also solved by the RANS of (3.6). In this sense, the RANS (3.6) here can be called URANS (*Unsteady Reynolds Averaged Numerical Simulation*) or VLES (*Very Large Eddy Simulation*) comparing with traditional RANS,[25] which only solve the steady problem and model all the fluctuations. Figure 1 compares the decomposition of energy spectrum for RANS and URANS.

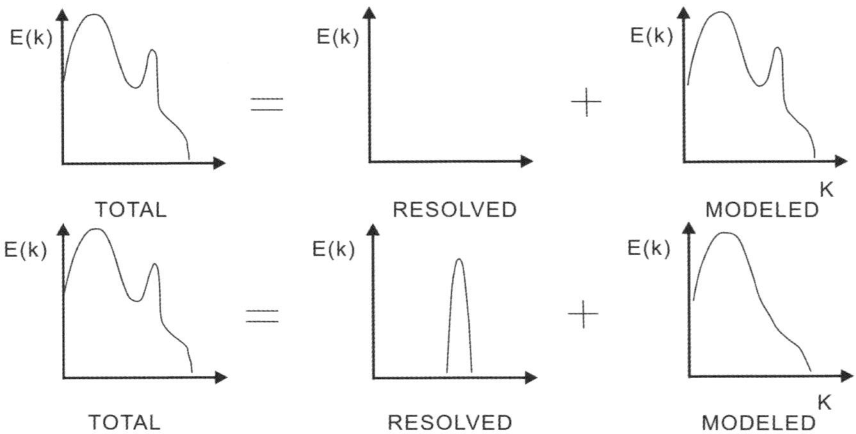

Fig. 1. Decomposition of the energy spectrum in the solution associated with RANS (top) and URANS (bottom) Simulation (after Sagaut,[26] with permission of use from author).

To solve (3.6), the Reynolds stress terms $\langle u'_i u'_j \rangle$ must be modeled in addition to the mean flow equations. Turbulent-viscosity models which are based on the turbulent-viscosity hypothesis relate the Reynolds stress $\langle u'_i u'_j \rangle$ to the mean flow through the shear strain rate,

$$\langle u'_i u'_j \rangle = \frac{2}{3} k \delta_{ij} - \nu_T \left(\frac{\partial \langle u_i \rangle}{\partial x_j} + \frac{\partial \langle u_j \rangle}{\partial x_i} \right) \tag{3.7}$$

in which $k = \frac{1}{2} u'_i u'_i$ is the turbulent kinetic energy, δ_{ij} is the identity matrix, and ν_T the eddy viscosity. Then the problem becomes how to obtain the eddy viscosity ν_T.

The intrinsic assumptions of the turbulent-viscosity models state:[27] 1) the Reynolds-stress anisotropy $a_{ij} = \langle u'_i u'_j \rangle - \frac{2}{3} k \delta_{ij}$ is determined by the mean velocity gradients; and 2) the relationship between the Reynolds-stress anisotropy a_{ij} and the mean velocity gradients follow (3.7). However, the wind tunnel experiments performed by Uberoi[28] and Warhaft[29] clearly showed that for flows where turbulence is subjected to a prior history of strain and when the mean strain rate changes fast, the turbulent-viscosity hypothesis is incorrect. But when the Reynolds-stress balance is dominated by local processes such as production and dissipation, and the nonlocal transport processes are small, the relationship between the mean shear strain rate and the Reynolds stresses are reasonable. An important observation for such flows is that the production and dissipation of turbulent kinetic energy are roughly in balance locally.

The most widely used turbulent-viscosity model today is the $k-\epsilon$ model. The $k-\epsilon$ model comprises transport equations for turbulent kinetic energy k and turbulent dissipation ϵ, respectively.

$$\frac{\partial k}{\partial t} + u_j \frac{\partial k}{\partial x_j} = \frac{\partial}{\partial x_j}\left(\frac{\nu_T}{\sigma_k}\frac{\partial k}{\partial x_j}\right) + Prod - \epsilon \tag{3.8}$$

$$\frac{\partial \epsilon}{\partial t} + u_j \frac{\partial \epsilon}{\partial x_j} = \frac{\partial}{\partial x_j}\left(\frac{\nu_T}{\sigma_\epsilon}\frac{\partial \epsilon}{\partial x_j}\right) + C_{\epsilon 1}\frac{Prod \cdot \epsilon}{k} - C_{\epsilon 2}\frac{\epsilon^2}{k} \tag{3.9}$$

$$\nu_T = C_\mu k^2/\epsilon\,. \tag{3.10}$$

Here $Prod$ represents the turbulent production from the mean flow, and σ_k, σ_ϵ, $C_{\epsilon 1}$, $C_{\epsilon 2}$ and C_μ are model coefficients. The "standard values" of these coefficients follows Launder and Spalding[30]

$$\sigma_k = 1.0, \sigma_\epsilon = 1.3, C_{\epsilon 1} = 1.44, C_{\epsilon 2} = 1.92, C_\mu = 0.09\,. \tag{3.11}$$

The k equation (3.8) states the rate of change of turbulent kinetic energy k is balanced by turbulent production $Prod$ and turbulent dissipation ϵ and smoothed by the turbulent diffusion $\frac{\partial}{\partial x_j}\left(\frac{\nu_T}{\sigma_k}\frac{\partial k}{\partial x_j}\right)$. Though the convection of the turbulent kinetic energy by the mean flow $\langle u_i \rangle \frac{\partial k}{\partial x_j}$ is included, the convection terms should not be important, that is local processes are supposed to dominate according to the turbulent-viscosity hypothesis. This is clearly not the case for breaking wave generated turbulence, where experimental analysis show that turbulent production dominates in front of the wave and turbulent dissipation dominates back of the wave. Turbulent convection is important and of the same order as production and dissipation.[19]

Lemos[31] first applied the standard $k-\epsilon$ model to simulate breaking waves on coastal structures. Only qualitative results were presented, and showed the kinetic energy is much higher than previously observed.

Lin and Liu (Refs. 32–33) recognized from previous measurements the eddy viscosity for breaking wave generated turbulence is generally anisotropic. Hoping to better describe the anisotropic eddy viscosity, they applied an Algebraic Reynolds Stress Model (ARSM), in which the turbulent shear stress is written as

$$\begin{aligned}\langle u_i' u_j' \rangle = & \frac{2}{3} k \delta_{ij} - C_\mu \frac{k^2}{\epsilon} \left(\frac{\partial \langle u_i \rangle}{\partial x_j} + \frac{\partial \langle u_j \rangle}{\partial x_i} \right) \\ & - \frac{k^3}{\epsilon^2} \left[C_1 \left(\frac{\partial \langle u_i \rangle}{\partial x_l} \frac{\partial \langle u_l \rangle}{\partial x_j} + \frac{\partial \langle u_j \rangle}{\partial x_l} \frac{\partial \langle u_l \rangle}{\partial x_i} - \frac{2}{3} \frac{\partial \langle u_l \rangle}{\partial x_k} \frac{\partial \langle u_k \rangle}{\partial x_l} \delta_{ij} \right) \right] \\ & + \frac{k^3}{\epsilon^2} \left[C_2 \left(\frac{\partial \langle u_i \rangle}{\partial x_k} \frac{\partial \langle u_j \rangle}{\partial x_k} - \frac{1}{3} \frac{\partial \langle u_l \rangle}{\partial x_k} \frac{\partial \langle u_l \rangle}{\partial x_k} \delta_{ij} \right) \right] \\ & + \frac{k^3}{\epsilon^2} \left[C_3 \left(\frac{\partial \langle u_k \rangle}{\partial x_i} \frac{\partial \langle u_k \rangle}{\partial x_j} - \frac{1}{3} \frac{\partial \langle u_l \rangle}{\partial x_k} \frac{\partial \langle u_l \rangle}{\partial x_k} \delta_{ij} \right) \right] \end{aligned} \tag{3.12}$$

where the coefficients C_1, C_2 and C_3 are determined following Refs. 34–35, and they ensured that the turbulent kinetic energy remains positive in all circumstances.[32] Note when $C_1 = C_2 = C_3 = 0$ is assumed in (3.12), the standard $k - \epsilon$ model of Launder and Spalding[30] with eddy viscosity (3.10) is recovered. Lin and Liu's ARSM did not solve the transport equations for the Reynolds stress directly, the Reynolds stress are solved using the $k - \epsilon$ model with an anisotropic eddy viscosity. So, the intrinsic assumptions of turbulent-viscosity models still apply.

Lin and Liu[32] first apply the ARSM to investigate the spilling breaking wave of Ting and Kirby.[4,5] The discrete equations are solved by the finite difference two-step projection method (Refs. 36–37). They use forward time difference for the time derivative, and a weighted upwind and second-order centered scheme for the convection terms. The free surface is tracked using the volume of fluid (VOF) method following Hirt and Nichols[38] and Kothe *et al.*[39] In Lin and Liu,[32] the numerical wave channel is 22.5m long with an impermeable 1/35 slope placed at the right end. Regular waves are generated at the inflow boundary using surface elevations. The water depth is 0.4m in front of the slope, and the incident wave height is 0.125m with wave period 2.0s. In total 900×77 cells are used with a uniform grid of 0.025m in the x direction, and a stretched grid with minimum 0.006m near the free surface in the vertical direction. Modeled results at $t = 16.6s \sim 18.6$s are used for the validation, where $t = 0$ represents the beginning of the computation. This data corresponds to the $8^{th} \sim 9^{th}$ wave generated at the inflow boundary, and roughly the 5th wave near the breaking point. The simulation takes about 16 hours using a single processor of supercomputer IBM SP2.

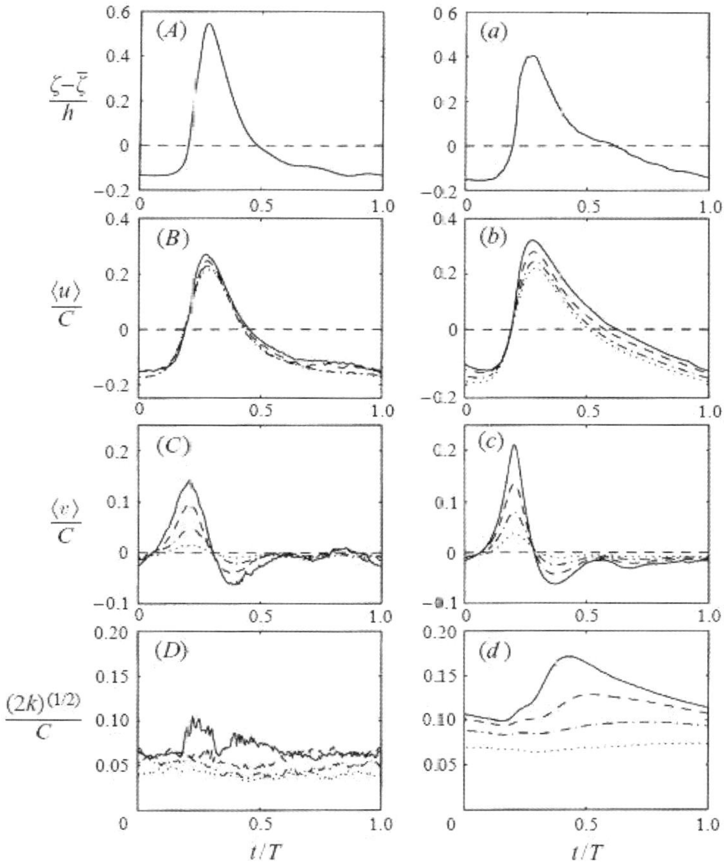

Fig. 2. Comparisons of experimental data (A-D) and numerical results (a-d) at $(x - xb)/h_b = 4.397$(measurement closest to the breaking point); $(y - \bar{\zeta})/h$ =-0.2623 (-),-0.4909(–),-0.7194(-.-.),-0.9080(...). From Ref. 32, with permission of use from publisher.

Lin and Liu's results show the surface elevations at the shoaling region and the inner surf zone are accurately modeled. But their numerical model predicts the breaking point nearly 2m earlier than the experimental observation (see Figure 2) and thus significantly underestimates surface elevations near the breaking point. Besides, Lin and Liu[32] found that their numerical model always overestimated the turbulence level near the breaking point from extensive numerical tests.

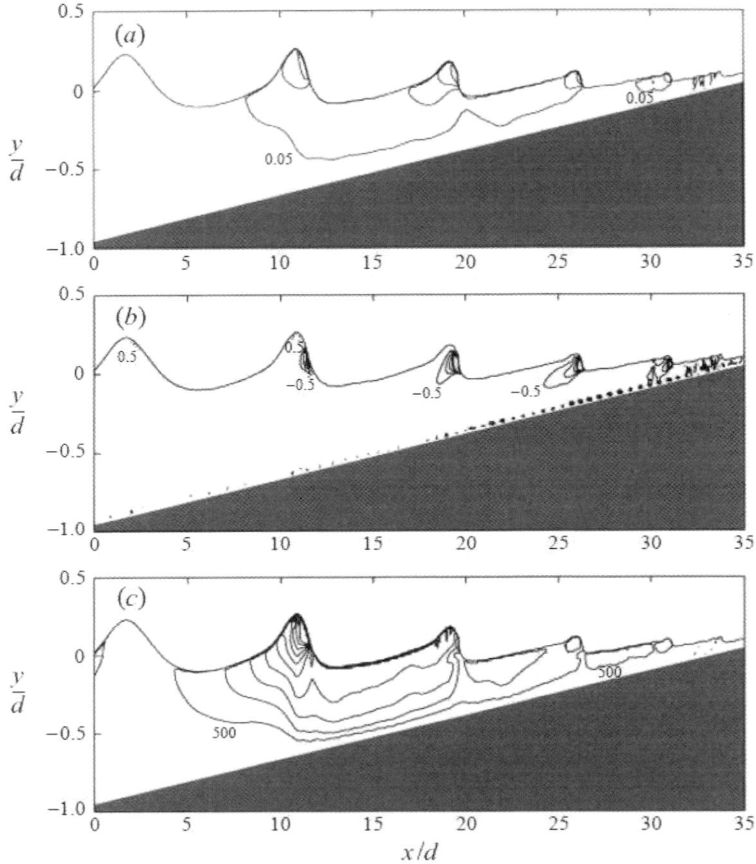

Fig. 3. Computed normalized (a) turbulence intensity $(2k)^{1/2}/[g(d_c + a)]^{1/2}$ (0.05 to 0.20 with interval of 0.05), (b) mean vorticity $\omega/[g(d_c+a)]^{1/2}$ [−0.5 to −2.5 with interval of −0.5 (solid line) and 0.5 (dashed line] (c) eddy viscosity ν_t/ν (500 to 2500 with interval of 500) at $t/T = 0.4$. Measured breaking point is at $x/d_c = 16$. From Ref. 32, with permission of use from publisher.

Using the same numerical model, Lin and Liu[33] then examined a plunging breaking wave. The performance of the numerical model is similar as that for the spilling breaking wave. The numerical model predicted the breaking point much earlier than the measurements and significantly underestimated the surface elevations and horizontal velocities near the breaking point (Figure 4). Lin and Liu[32] argue that the primary reason for this is that the turbulence model cannot accurately predict the initiation of the

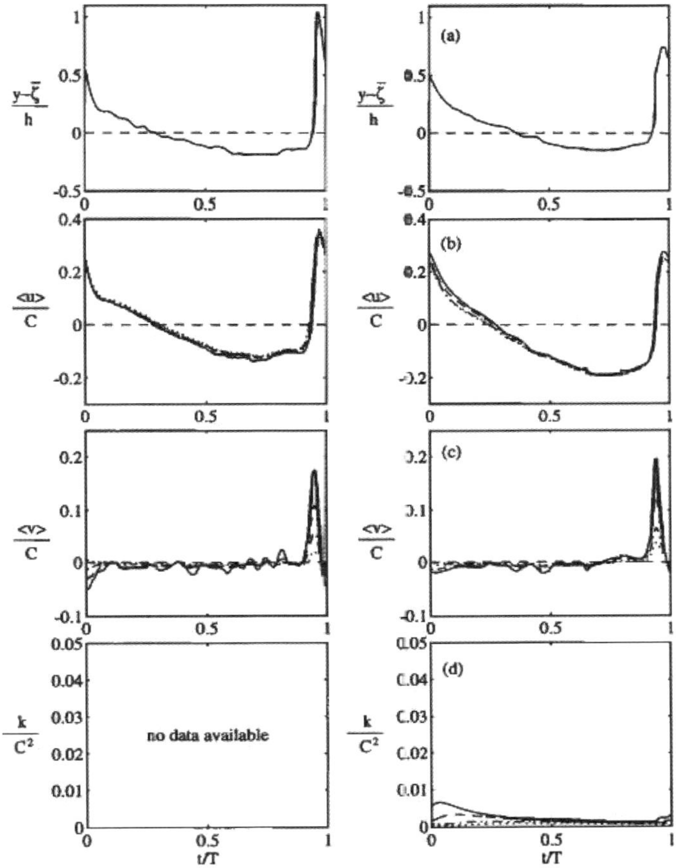

Fig. 4. Comparisons of experimental data (A-D) and numerical results (a-d) at $(x - xb)/h_b = 3.571$(measurement closest to the breaking point) $(y - \bar{\zeta})/h$ =-0.2967 (-),-0.4965(–),-0.7061(-.-.),-0.9161(...). From Ref. 33, with permission of use from publisher.

turbulence in a rapidly distorted flow region such as the initial stage of the breaking. Lin and Liu[32,33] did not compare their model results with isotropic eddy viscosity models such as the $k - \epsilon$ model in either of their studies, it is difficult to judge from the results how much the anisotropic eddy viscosity model affects the model performance.

Hsu *et al.*[40] applied a volume average to the RANS equations and solved the wave motions and turbulence flows in front of a porous composite breakwater using the same turbulence model as Lin and Liu.[32,33] The agreements between the numerical model and the experiments for surface elevations, particle velocities and turbulent intensities are reasonable. However, Hsu *et al.*[40] admitted that "the $k - \epsilon$ model tends to calculate unrealistically high turbulence in the supposedly low turbulence region after a long time computation". This observation is consistent with the findings of Christensen,[41] which we will review in a later section. Hsu *et al.*[40] suggested the reason for this unrealistic kinetic energy was the local dissipation of turbulent kinetic energy in the low turbulence region is not enough. To avoid this high unrealistic kinetic energy, they introduced a damping function following,[31]

$$\nu_{td} = \nu_t^2 \frac{1 - \exp(-\gamma\nu/\nu_t)}{\gamma\nu} \tag{3.13}$$

where ν_t is the eddy viscosity computed from their ARSM model, and ν_{td} the damped eddy viscosity. The damping coefficient $\gamma = 50$ is chosen in Hsu *et al.*,[40] which leads to a 2% decrease in eddy viscosity when ν_t is larger than 0.001, and an 80% reduction when $\nu_t \sim 10^{-5}(m^2/s)$. This means applying (3.13) will delay the turblulence development.

Based on the commercial code Flow 3D and using the same data set as Refs. 32–33, Bradford[42] compared the performance of a k model, the standard $k - \epsilon$ model and the Renormalized Group extension of the $k - \epsilon$ model (RNG). In the k model, the turbulent length scale is set to 0.1 local still water depth. The setup of Bradford[42] is similar to that of Refs. 32–33. The numerical wave channel is 16m long. In total 20s of waves are simulated for the spilling breaker, and 30s of waves for the plunging breaker. A 800×80 uniform grid was used, corresponding to a 0.02m increment in the cross-shore direction and 0.0075m in the vertical direction.

For the spilling breaking case, all models in Ref. 42 showed reasonable agreement between the measured and simulated surface elevations up to 2 meters offshore of the measured breaking point, and 2 meters onshore of the breaking point. In between, all models significantly underestimate data by up to 40%. All numerical models showed reasonable agreement for mean water levels and the trough levels except the k model(see Figure 5).

Bradford[42] also compared the undertow with data. In general, all models underestimate the undertow significantly (see Figure 6 left column). Since the model-data comparison were shown only up to the trough level, it is not clear if the numerical models conserve mass. Though the trough level is accurately predicted by most numerical models(except the k model), the undertows do not change directions at the right vertical level. This

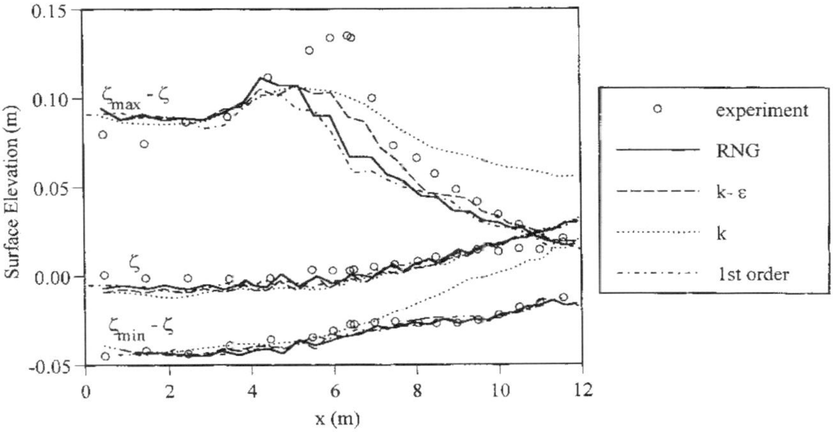

Fig. 5. Surface elevations of spilling breaker. From Ref. 42, with permission of use from ASCE.

maybe an indication that the numerical models have not reached a periodical steady stage. Comparison of the turbulent kinetic energy is also made for the spilling breaker case. All numerical models overestimate k at all locations except at $(x - x_b)/h_b = 4.937$, where the RNG model and the donor-cell RNG model underestimate k.

Model performance for the plunging breaker case is slightly better, especially for the RNG model and the $k-\epsilon$ model (see Figures 7, 6 right column). They only slightly underestimate the maximum surface elevations and the modeled breaking point is less than 1m earlier than the experimental measurements. But these models miss the trough level. Improved agreement can also be seen in undertow and turbulent kinetic energy, though in the most onshore position the undertow is overestimated.

In general, all models in Ref. 42 showed the same trend as that of the ARSM model of Lin and Liu[32] for the spilling breaker: they showed reasonable agreement in the shoaling zone and in the inner surf zone, but they predict the wave breaking far earlier than observed in the experiment. All models showed improved agreement with data for undertow for the plunging breaker case. Both Lin and Liu (Refs. 32–33) and Bradford[42] performed the numerical simulation in less than 10 wave periods. It is suspected the kinetic energy may not become periodic in such a short time; the mean water level may not become steady; the undertows may not conserve mass.

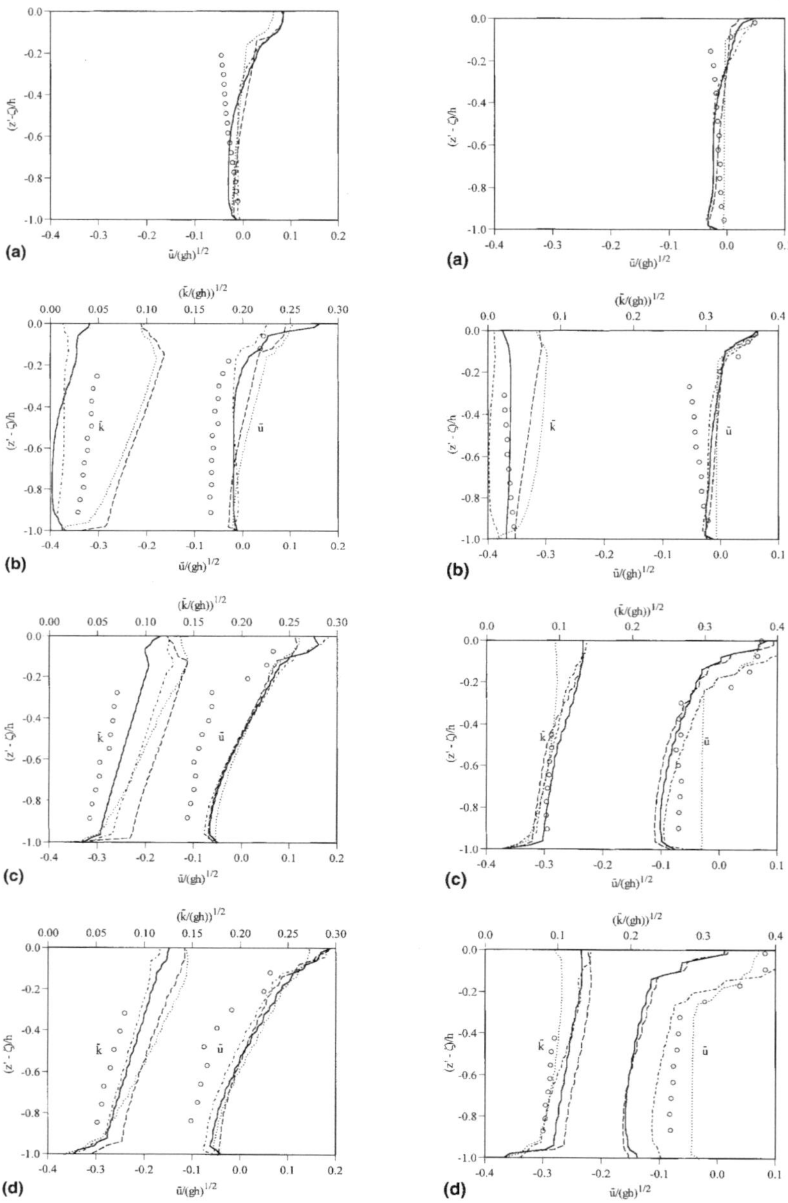

Fig. 6. Comparison of modeled and measured undertow. Left spilling breaker; right plunging breaker. Legends refer to previous Figure 5. Left (a) at $(x - x_b)/h_b = 1.332$; (b) at 4.397; (c) at 10.528; (d) at 16.709; Right at $(x - x_b)/h_b =$ -3.247; (b) at 3.571; (c) at 12.987; (d) at16.883. From Ref. 42, with permission of use from ASCE.

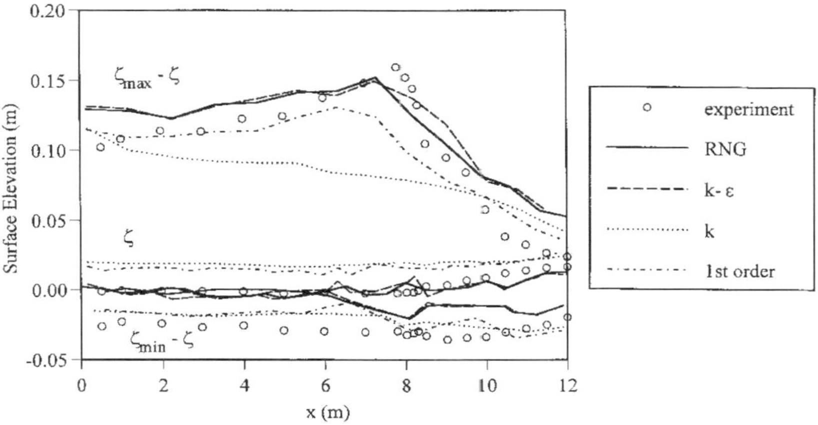

Fig. 7. Surface elevations of plunging breaker. From Ref. 42, with permission of use from ASCE.

Based on the $k - \omega$ model of Ref. 43, Christensen *et al.*[41] studied sediment transport under breaking waves using a non-orthogonal finite volume grid. Their numerical wave flume is 15m long, and two numerical resolutions with 688×32 and 1376×64 cells are used to test the model performance.

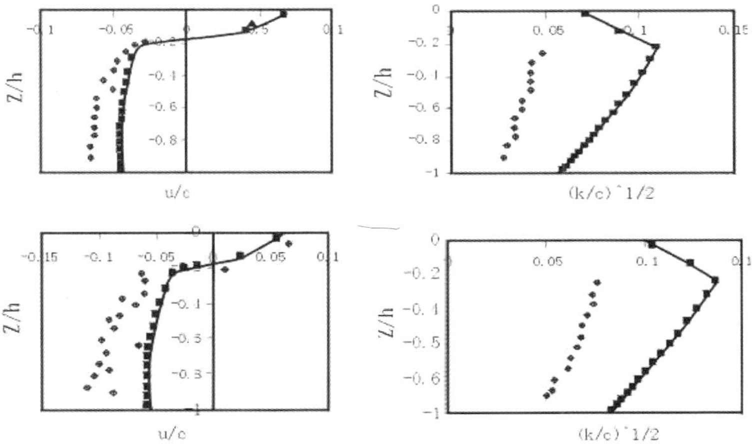

Fig. 8. The undertow and average of turbulence in the spilling breaker. Top: $(x - x_b)/h_b = 4.4$; Bottom: $(x - x_b)/h_b = 10.5$. ⊕: data from Ting and Kirby (1994); -■- numerical model results. After Ref. 41, with permission of use from ASCE.

Christensen *et al.*[41] focused their study on modeling the undertow and turbulent kinetic energy, and they compared the model results with data from Refs. 4–5. They show that their model underestimates the undertow for the spilling breaker case by 25 ∼ 30 % (see Figure 8) , but overestimates undertow for the plunging breaker case by 50 % near the breaking point. At positions close to the shoreline, the numerical model shows a perfect match to data (Figure 9). Christensen *et al.*[41] pointed out that they need at least 11 wave periods for the turbulence level to reach a periodically stationary state, and their $k-\omega$ model overpredicts the turbulent kinetic energy by two and three times for the spilling and plunging breaker cases, respectively (see Figure 10 and Figure 11).

In summary, these studies suggest the turbulent-viscosity models are inadequate when study breaking wave generated turbulence. More advanced models, especially those that can account for the rapid distortion of the fluid flow are required.

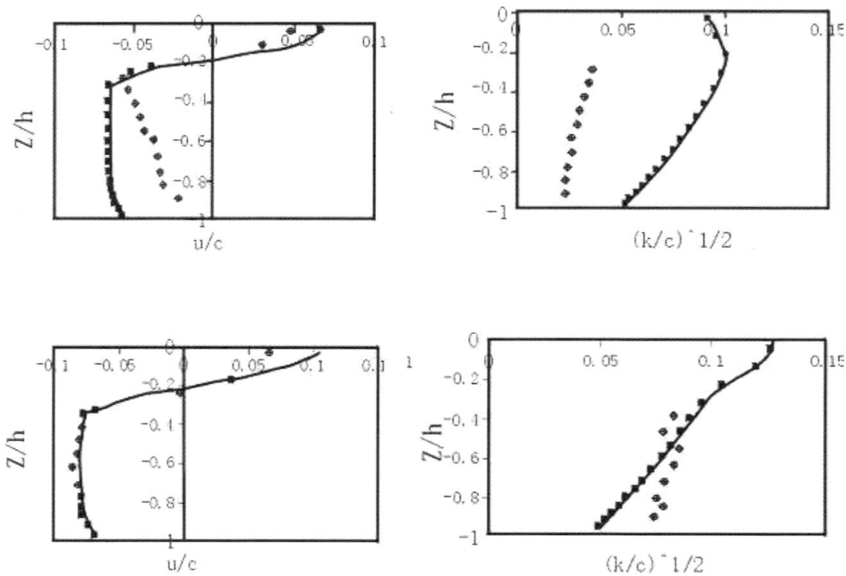

Fig. 9. The undertow and average of turbulence in the plunging breaker. Top: $(x - x_b)/h_b = 3.57$; Bottom: $(x - x_b)/h_b = 6.49$. ⊕: data from Ting and Kirby (1994); -■- Numerical model results. After Ref. 41, with permission of use from ASCE.

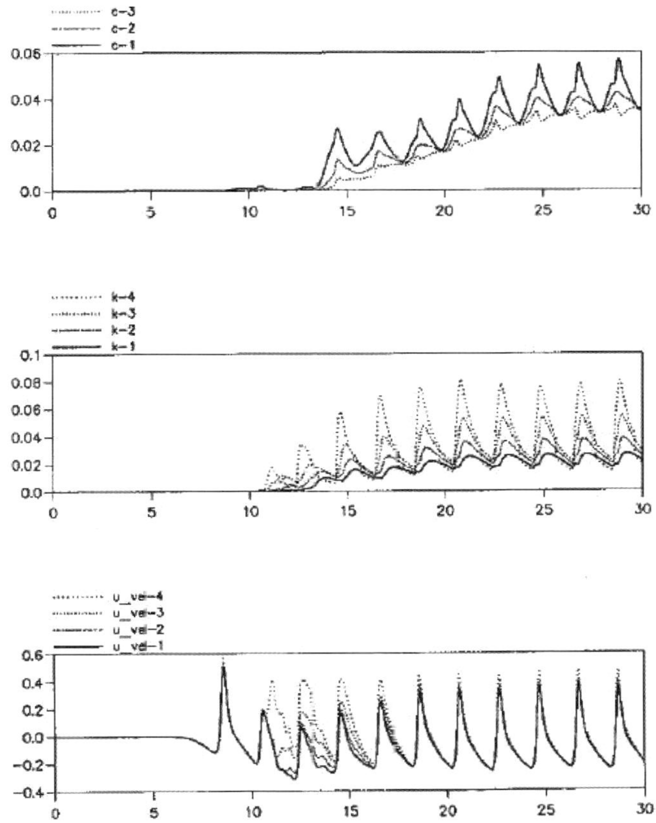

Fig. 10. The sediment concentrations, the non-dimensional turbulence levels and the non-dimensional velocity component at four levels above the bed under a spilling breaker. $(x - x_b)/h_b = 10.4$. Level-1: $z/h = -0.814$; level-2: $z/h = -0.68$; level-3: $z/h = -0.814$; level-4: $z/h = -0.48$. From Ref. 41, with permission of use from ASCE.

3.3. *Large Eddy Simulation — LES*

The next level of approximation is *Large Eddy Simulation*(LES), which was first established by Deardorff.[44] In LES, the Navier-Stokes equations are filtered in space. LES uses the full equations to compute the turbulent flow, but recognizes that it is very costly to compute the details of the small-scale turbulence. As a result, a closure model for scales below the selected cutoff length Δ is derived. To accomplish this, a filter is introduced to

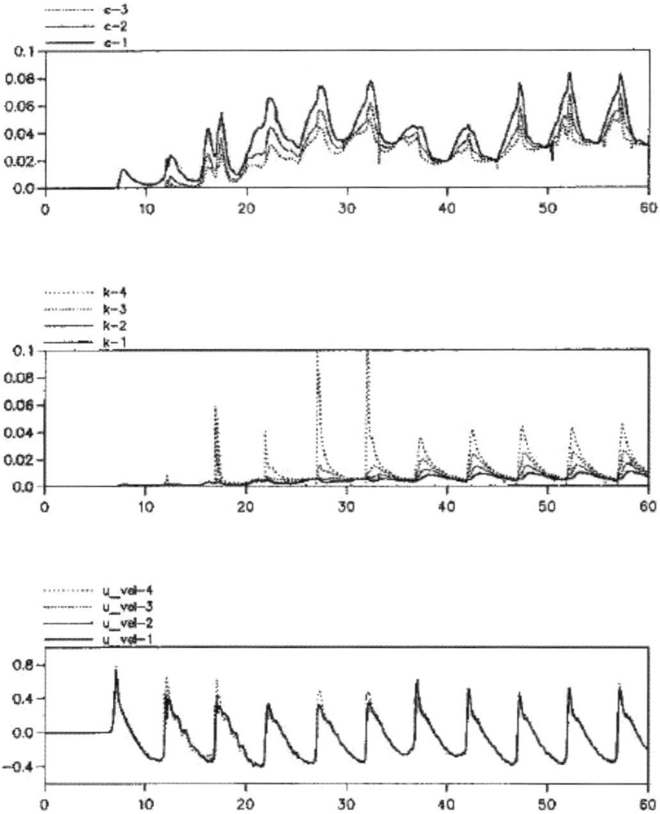

Fig. 11. The sediment concentrations, the non-dimensional turbulence levels and the non-dimensional velocity component at four levels above the bed under a plunging breaker. $(x - x_b)/h_b = 3.68$ level-1: $z/h = -0.814$; level-2: $z/h = -0.68$; level-3: $z/h = -0.814$; level-4: $z/h = -0.48$. After Ref. 41, with permission of use from ASCE.

the flow equations which generates a subgrid scale (SGS) Reynolds stress for the flow with scale length less than Δ. LES lies between the extremes of direct numerical simulation (DNS), in which all fluctuations are solved and no turbulence model is required; and the classical approach of RANS models, in which only mean values are calculated and all fluctuations are modeled. It is hoped the small-scales are more homogeneous and universal, and much less problem-dependent than the larger ones, so that they can be represented by simpler models than similar RANS equations. Furthermore, the details of the model are much less influential for the overall flow behavior than in closure models for the RANS approaches. This is because the main contribution to the turbulent transport stems from the large-scale motions that can be resolved on the numerical grid.

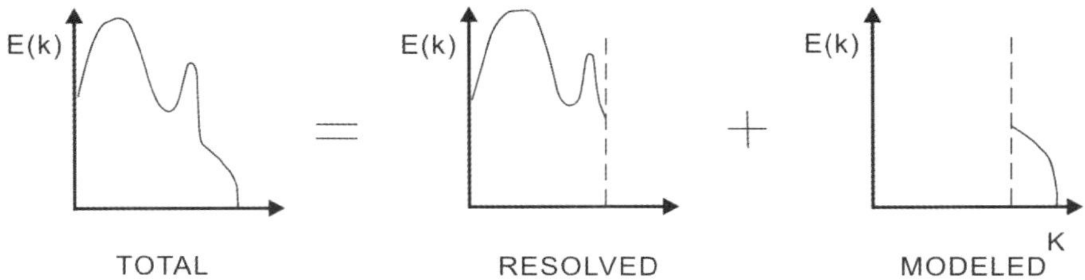

Fig. 12. Decomposition of the energy spectrum in the solution associated with LES Simulation (after Sagaut,[26] with permission to use from author.)

In LES, the unsteady Navier-Stokes and continuity equations are filtered in space,

$$\frac{\partial \bar{u}_i}{\partial t} + \bar{u}_j \frac{\partial \bar{u}_i}{\partial x_j} = -\frac{1}{p}\frac{\partial \bar{p}}{\partial x_i} + \nu \frac{\partial^2 \bar{u}_i}{\partial x_i^2} - \frac{\partial}{\partial x_j}(\overline{u_i u_j} - \bar{u}_i \bar{u}_j) \tag{3.14}$$

$$\frac{\partial \bar{u}_j}{\partial x_j} = 0 \tag{3.15}$$

where i=1,2,3 are the cross-shore (x), longshore (y) and the vertical direction (z), respectively. A space filter is applied in the above equations, and the overbars stand for the resolved filtered quantities (box filter for example),

$$\bar{u}_i(x, y, z, t) = \frac{1}{\Delta x \Delta y \Delta z} \int_{x-\frac{1}{2}\Delta x}^{x+\frac{1}{2}\Delta x} \int_{y-\frac{1}{2}\Delta y}^{y+\frac{1}{2}\Delta y} \int_{z-\frac{1}{2}\Delta z}^{z+\frac{1}{2}\Delta z} u_i(\zeta, \eta, \xi, t) d\zeta d\eta d\xi \tag{3.16}$$

where $\Delta x, \Delta y, \Delta z$ are taken as the grid size in the x, y, z directions, respectively. The new terms appearing in the filtered equations are

$$\overline{u_i u_j} - \bar{u}_i \bar{u}_j = L_{ij} + C_{ij} + R_{ij} \tag{3.17}$$

where L_{ij}, C_{ij}, R_{ij} are referred as the Leonard term, the cross term and the

SGS Reynolds term, respectively:

$$L_{ij} = \overline{\bar{u}_i\bar{u}_j} - \bar{u}_i\bar{u}_j \tag{3.18}$$

$$C_{ij} = \overline{\bar{u}_i u'_j} - \overline{\bar{u}'_i\bar{u}_j} \tag{3.19}$$

$$R_{ij} = \overline{u'_i u'_j}\,. \tag{3.20}$$

The prime represents a subgrid scale (SGS) quantity. Note that the sum of the Leonard term and the cross term $L_{ij} + C_{ij}$ is small comparing with the SGS Reynolds stress and thus can be neglected.[44] Finally we obtain the filtered Navier-Stokes equations in which the SGS Reynolds stress needs to be modeled.

3.3.1. *Two-dimensional mix models*

Zhao and Tanimoto[45] applied this method for waves incident on a submerged reef with a 1 : 2 slope. The model was first applied in two dimensions. To obtain stable meaningful results, they were careful with the inflow boundary condition, because strong reflective waves from the structure may spoil the whole computation. They applied an absorbing-generating piston-type wavemaker, which is essentially the same as is normally used in a physical experiment. The absorbing-generating wavemaker is effective and the wave field becomes stable after 15 wave periods.

Zhao and Tanimoto[45] used the two-dimensional Smagorinsky subgrid scale turbulence model. This model cannot provide three-dimensional vortex stretching, however, because of the dissipative nature of the Smagorinsky model, two-dimensional vortex stretching was approximated by the artificial viscosity. In that sense, the Smagorinsky model behaves like a mixing length turbulence model, with the turbulence length scale equal to the Smagorinsky length scale. The modeled surface elevations and wave heights were compared with experimental measurements and showed good agreement, see Figures 13 and 14.

Next, Zhao *et al.* (see Refs. 46–47) set up a multiscale $k - l$ model. In this model, a compromise of the RANS and three-dimensional LES approaches is achieved by setting up an artificial energy cascade explicitly using a multiscale subgrid approach that allows energy transfers between scales. Figure 15 shows a sketch of this idea. In Figure 15, the unresolved turbulence motions are parameterized into several length scales starting from the smallest resolved length scale κ_1, where κ is the wave number, to a cut-off length κ_N of the simulation. The nth length scale is $l_n = l_g/2^{n-1}$, and l_g is the grid scale $\sqrt{\Delta x \Delta z}$. The cutoff length scale is chosen to fall

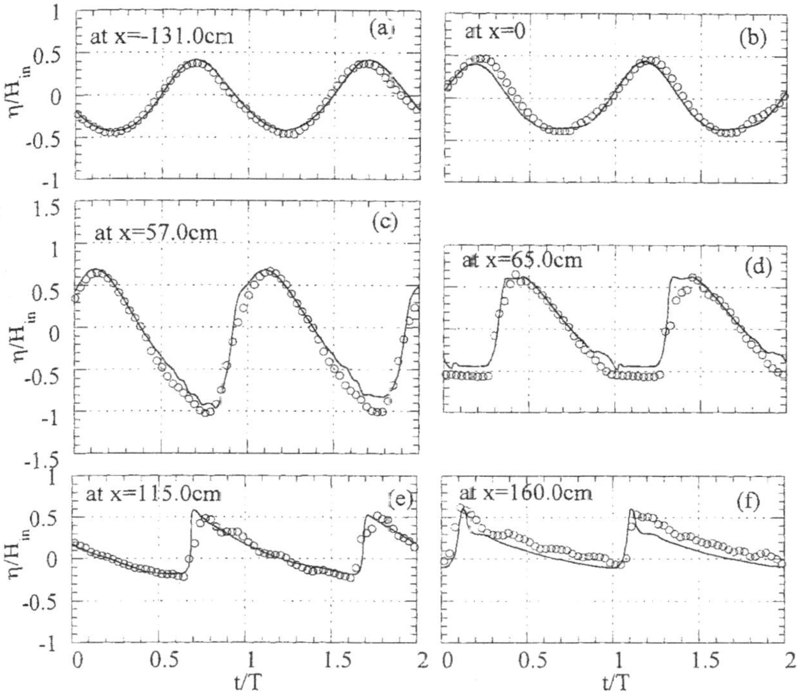

Fig. 13. Surface elevations at positions: before breaking (a, b); near the breaking point (c) and after breaking (d, e, f). circles: measurements; lines: model results. Refer Figure 14 for x-coordinates. From Ref. 45.

inside the inertial subrange, and thus can be assumed to be in local equilibrium where a simple subgrid model can be applied. Then N levels of $k-l$ equations are setup, in which turbulent kinetic energy can be transferred from larger length scales to smaller ones explicitly (see Figure 15).

The production term at the first level is dominated by the "large eddies", that is the smallest resolved scale (the grid scale, in this case),

$$PROD_1 = \nu_{t_1}\left[\frac{1}{2}\left(\frac{\partial \bar{u}_i}{\partial x_j}+\frac{\partial \bar{u}_j}{\partial x_i}\right)\left(\frac{\partial \bar{u}_i}{\partial x_j}+\frac{\partial \bar{u}_j}{\partial x_i}\right)\right]. \tag{3.21}$$

The dissipation term is the $k-l$ type dissipation,

$$\epsilon_n = C_d k_n^{\frac{3}{2}}/l_n, \qquad 1 \le n < N \tag{3.22}$$

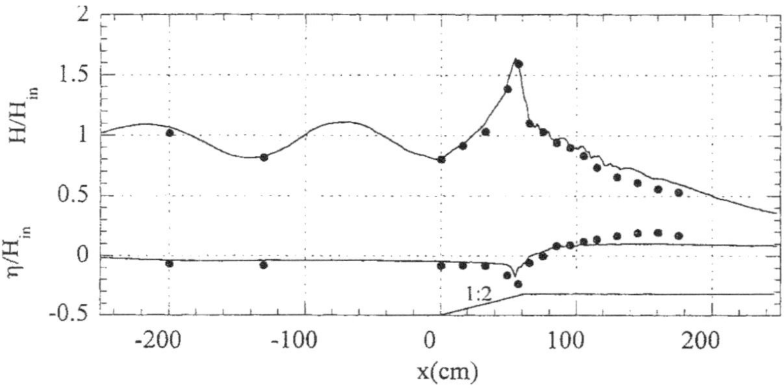

Fig. 14. Wave heights and mean water level distribution. Close circles: data; solid line: model results. From Ref. 45.

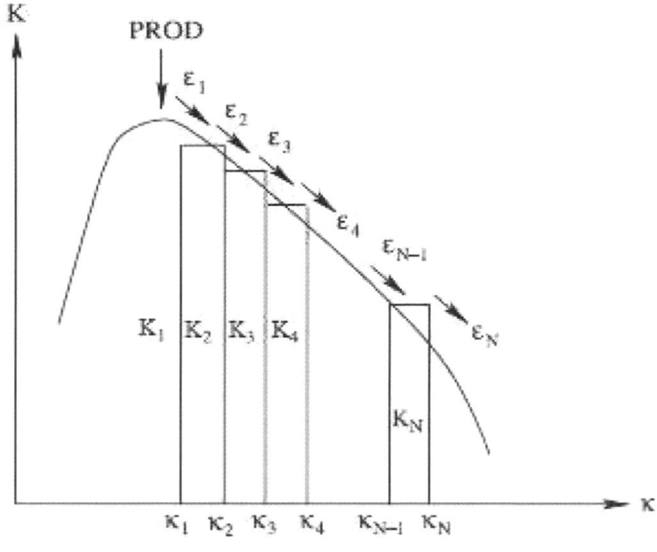

Fig. 15. Partition of the spectra density. From Ref. 47.

where the length scale for each partition is

$$l_n = \frac{\sqrt{\Delta x \Delta z}}{2^{n-1}} \qquad 1 \leq n < N\,. \tag{3.23}$$

For each partition $n > 1$, the production is set equal to dissipation at the next n level grid scale partition,

$$PROD_n = \epsilon_{n_1} \qquad 2 \leq n < N\,. \tag{3.24}$$

Then the total SGS turbulent kinetic energy is the sum of the kinetic energy in all partitions,

$$k = k_1 + k_2 + \ldots + k_N = \sum_{n=1}^{N} k_n\,. \tag{3.25}$$

The total dissipation of the system is characterized as

$$\epsilon = \epsilon_N \tag{3.26}$$

which means the turbulent kinetic energy leaves the system at the smallest scales.

Then, the nth level $k - l$ equation reads

$$\frac{\partial k_n}{\partial t} + u_j \frac{\partial k_n}{\partial x_j} = \frac{\partial}{\partial x_j}\left(\frac{\nu_{t_n}}{\sigma_k}\frac{\partial k_n}{\partial x_j}\right) + PROD_n - \epsilon_n\,. \tag{3.27}$$

Finally, the space-filtered Navier-Stokes equations read

$$\frac{\partial \bar{u}_i}{\partial t} + \bar{u}_j \frac{\partial \bar{u}_i}{\partial x_j} = -\frac{1}{\rho}\frac{\partial \bar{p}}{\partial x_i} + \frac{\partial}{\partial x_j}\left[(\nu + nu_T)\frac{\bar{\partial} u_i}{\partial x_j}\right] + g_i \tag{3.28}$$

$$\nu_T = \nu_{t_1} \tag{3.29}$$

where ν_{t_1} is the eddy viscosity for all the length scales smaller than the smallest resolved scales. $C_d = 0.17, C_s = 0.1$ and $\sigma_k = 1.0$ were used through out the calculation.

Numerical simulations of breaking waves on a sloping beach were conducted to evaluate the performance of this model. To facilitate comparisons with other models, the same set of data (Refs. 4–5) as that used by Refs. 32–33 and Ref. 42 for spilling and plunging breakers are used.

The computational domain for Ref. 47 is 20 m long and 0.8 m high with 1000 cells in the x direction and 100 cells in the z direction with the grid sizes set to $\Delta x = 0.02$ m and $\Delta z = 0.008$ m. The numerical scheme follows Hirt and Nichols[38] and Refs. 32–33 for time and diffusive terms, but used a fourth-order centered scheme for the convective terms. For the simulation, four levels of $k - l$ equations are solved, which gives the cut-off length scale as $l_4 = 0.0016$ m. The simulations were conducted for 40 and 60 seconds of waves for the spilling and plunging breaking waves, respectively. Therefore, the length of the simulation is similar to that presented by Christensen *et al.*,[41] but significantly longer than those presented by Lin and Liu[32,33]

18.6 and 22.4 seconds and Bradford,[42] 20 and 30 seconds for spilling and plunging breakers, respectively. Model results show that within the simulation time, the surface elevation, horizontal velocities, and especially the subgrid kinetic energy all become periodic, i.e., the mean value is almost steady. More importantly, this model does not need any damping function such as that used by Hsu *et al.*[40] for their ARSM model to obtain the right kinetic energy level.

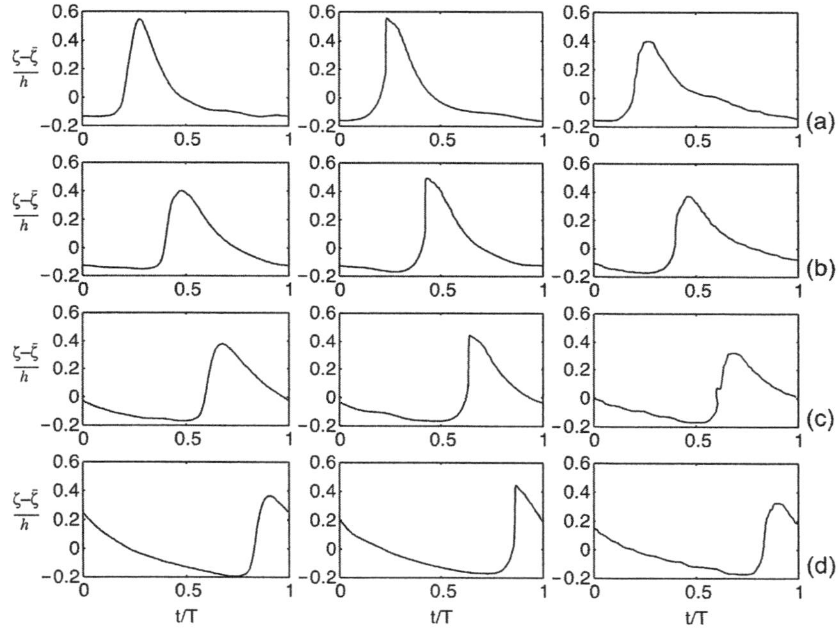

Fig. 16. Comparison of measured and simulated phase averaged surface elevations for the spilling breaker case. Left column: experimental data from Ting and Kirby (1996), middle column: results from Ref. 47; right column: results from the ARSM model.[32] From top to bottom (a-d) at locations $x = 7.28, 7.89, 8.50, 9.11$m (x locations refer to Figure 17). From Ref. 47.

Figure 16 presents the phase averaged surface elevations at four measured locations from nearest to the breaking point, Figure 16(a), to the bore region, Figure 16(d). The experimental data are shown in the left-hand column, model results from Ref. 47 are shown in the middle, and the ARSM results in Ref. 32 are shown in the right-hand column. In general, the Zhao *et al.*[47] model results agree well with the experimental measurements, but slightly overestimates the surface elevation in the inner surf zone. The ARSM model Lin and Liu,[32] on the other hand, significantly underestimates the surface elevation near the breaking point, but with improving accuracy in the inner surf zone.

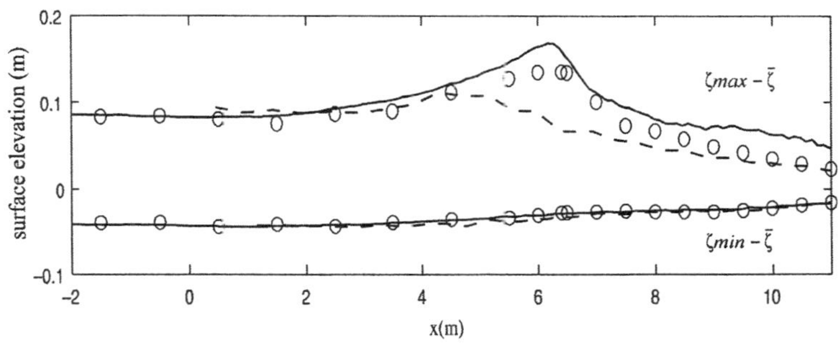

Fig. 17. Comparison of modeled and measured wave crest elevations and trough depressions of the spilling breaker case. Solid lines: results from Ref. 47; dashed lines: the RNG model;[42] circles: experimental data from Ref. 4. From Ref. 47.

The computed wave crest elevations and wave trough depressions are shown in Figure 17. The solid lines are the results from the Ref. 47 simulation, the dashed lines are the RNG model results from Ref. 42, and the circles are the experimental data of Ref. 4. Figure 17 shows that both models accurately simulate the wave trough, and the wave shoaling is also well predicted. The computed breaking point predicted by the multiscale $k-l$ model is at $x_b = 6.16$m, which is close to the measured breaking point of $x_b = 6.40$m. The RNG model, however, predicts the breaking point around $x = 4.25$ m, which is much earlier than the experimental measurement and significantly underpredicts the wave height in the outer surf zone. These results show the same trend as that obtained using the $k-\epsilon$ model, k model (Fig. 1 in Ref. 42) and the ARSM model.[32]

Figure 18 shows the measured and computed undertow profiles at six locations. The agreement between the multiscale $k-l$ model results and the experimental measurement is good with the structure and magnitude well represented. The RNG model, however, under-estimates the undertow in general and fails to capture the vertical structure.

The phase averaged surface elevations for the plunging breaker case at four measured locations are presented in Figure 19. Again, the multi-scale

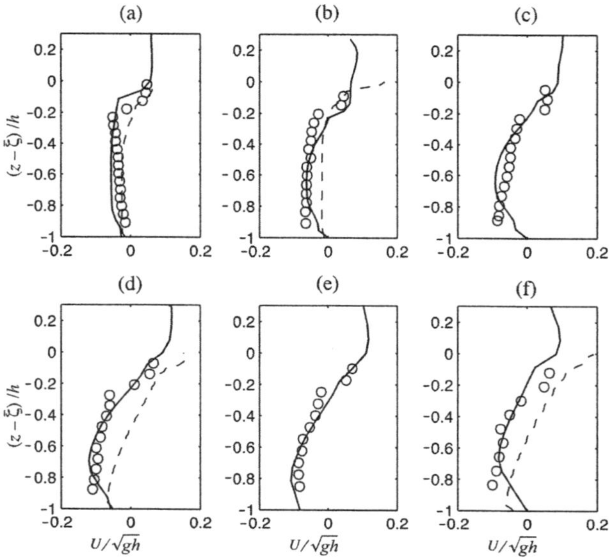

Fig. 18. Comparison of modeled and measured undertows for the spilling breaker case. Solid lines: results from Ref. 47; dashed lines: the RNG model;[42] circles: experimental data from Ref. 4. From (a) to (f) at locations x = 6.665; 7.275; 7.885; 8.495; 9.110; 9.725 m, coordinates refer to Figure 17. From Ref. 47

model shows good agreement with the experimental data. The ARSM still underestimates the surface elevation near the breaking point, as shown in Figure 19(a). However, comparing with Figure 16, we see the accuracy of the ARSM improved.

The comparison of wave crest elevations and wave trough depressions is presented in Figure 20. It is seen that the wave trough is accurately predicted by the multi-scale model, as well as the wave crest in the shoaling and the bore region. However, this model slightly overestimates the wave crest near the breaking point, while the RNG overestimates the wave trough in general and underestimates the wave crest. Comparing with Figure 17, it is seen the RNG model does a better job in predicting wave crest distribution for the plunging breaker case than for the spilling breaker case, consistent with the ARSM results.

The undertow profiles for the plunging breaker case are presented in Figure 21. The agreements between the measured data and the Zhao *et al.* model[47] prediction are reasonably good, showing this model can predict the

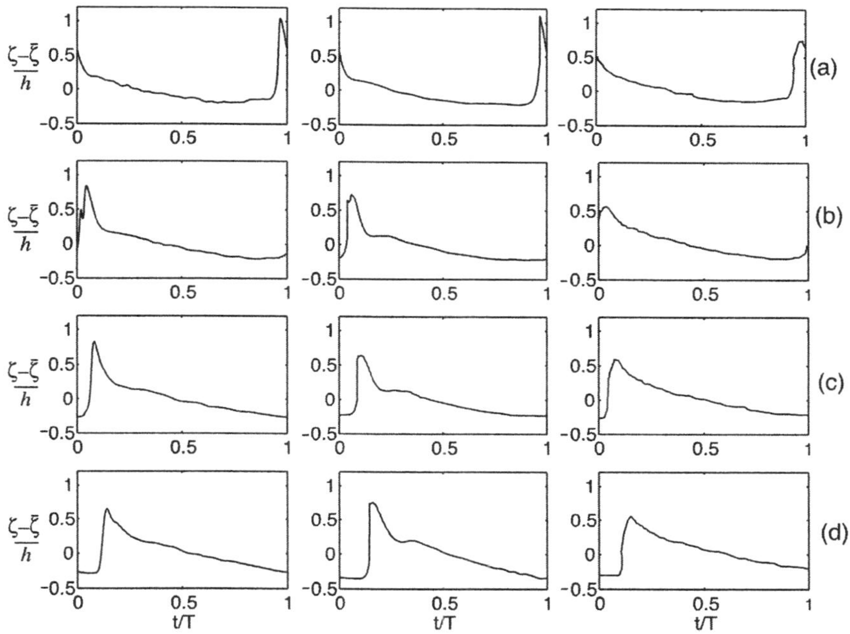

Fig. 19. Comparison of measured and simulated phase averaged surface elevations for the plunging breaker case. Left column: experimental data from Ref. 4; middle column: results from Ref. 47; right column: results from Reynolds stress model.[33] From Ref. 47.

undertow accurately in the outer surf zone. The general agreement of the undertow in the inner surf zone is also good, except the numerical model shows less vertical mixing than the experimental data. Zhao *et al.*[47] suggest this could be the lack of representation for buoyancy generated turbulence from the air entrainment post plunging in their numerical model.

3.3.2. *Three-dimensional models*

Christensen[48] applied three-dimensional LES to study the spilling and plunging breaking waves using the same dataset from Refs. 4–5, as Refs. 32–33, 42 and 47. Two subgrid scale turbulence models, the Smagorinsky model and the one equation k model, were investigated. The numerical wave channel is 20m long, and $320 \times 48 \times 32$ stretched grids were used for the resolved scales. Model results from 28 to 32 seconds are used for comparison with experimental measurements.

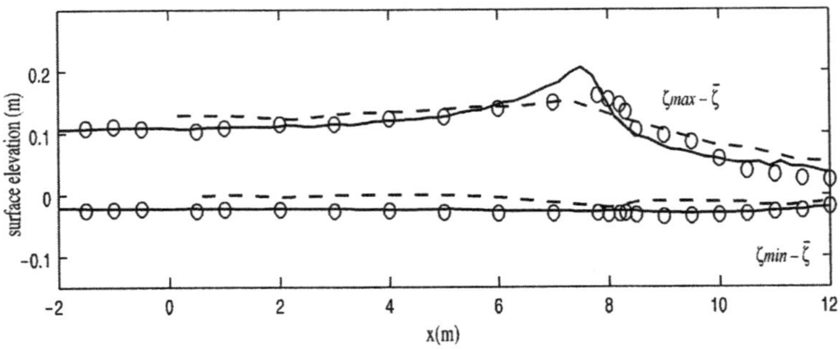

Fig. 20. Comparison of modeled and measured wave crest elevations and trough depressions for the plunging breaker case. Solid lines: results from Ref. 47; dashed lines: the RNG model;[42] circles: experimental data from Ref. 4. From Ref. 47.

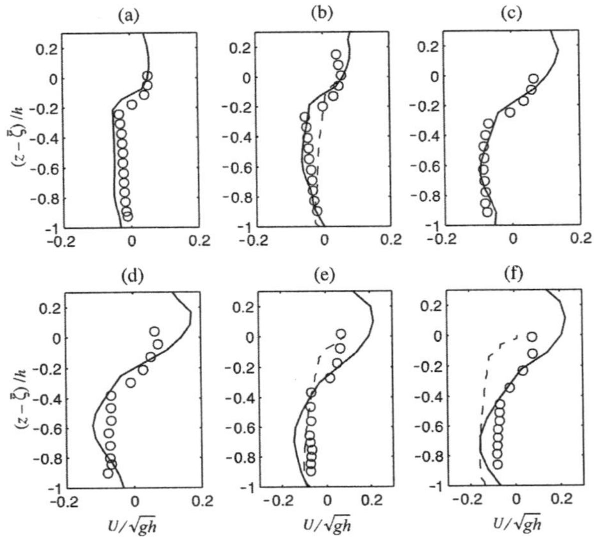

Fig. 21. Comparison of modeled and measured undertows for the plunging breaker case. Solid lines: results from Ref. 47; dashed lines: the RNG model;[42] circles: experimental data from Ref. 4. From (a) to (f) at locations x = 7.795; 8.345; 8.795; 9.295; 9.795; 10.395 m, coordinates refer to Figure 20. From Ref. 47.

The comparison of modeled and measured surface elevations for the spilling breaker showed the model results agree well with the data, especially in the shoaling region and the inner surf zone. But the initiation of the breaking point in the numerical model is slightly late, therefore the model slightly overestimates wave heights near the experimental breaking point. Christensen[48] argues that this is because the resolution near the free surface is not high enough due to the limited computer power. Results using the k equation SGS model show slightly better agreement with data than using the Smagorinsky SGS model. For the plunging breaker case, the numerical model predicts the breaking point earlier than the experimental measurements. In this case, the numerical model underestimates wave heights near the breaking point.

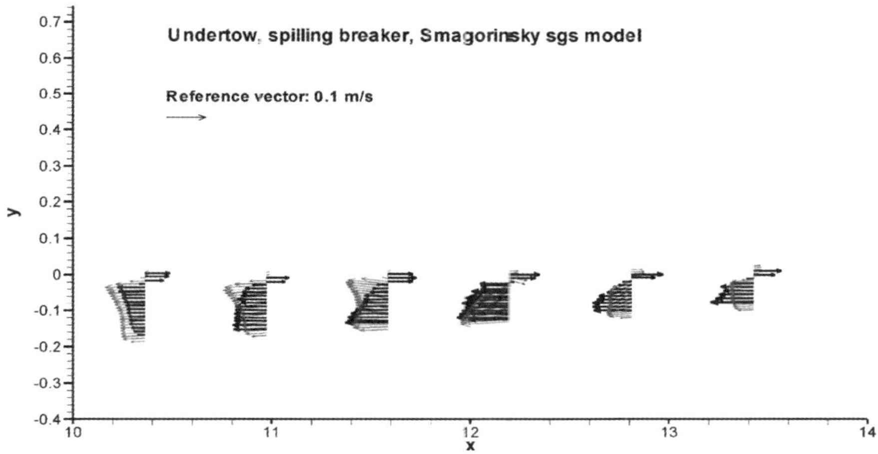

Fig. 22. Modeled and measured undertow spilling breaker. Thin: measurements of Ting and Kirby (1996); dark: model results using Smagorinsky SGS model. From Ref. 48 with permission of use from Elsevier. Refer Ref. 48 for color figure.

Christensen[48] also compared the modeled undertow with experimental measurements. Figures 22 and 23 show the results for the spilling breaker and the plunging breaker, respectively (Refer Ref. 48 for color figure). The thin arrows are data from Ref. 5 experiments, and the thick arrows are model results using the Smagorinsky SGS model. These results show the trough levels are accurately predicted with the undertows directed to the correct direction. However, the numerical model overestimates the undertow for the spilling breaker case near the bottom, and the curvature of the modeled undertow is also larger than the experimental measurements. Model performance for the plunging breaker seems slightly better, especially in the inner surf zone (Figure 23). Outside the surf zone, the

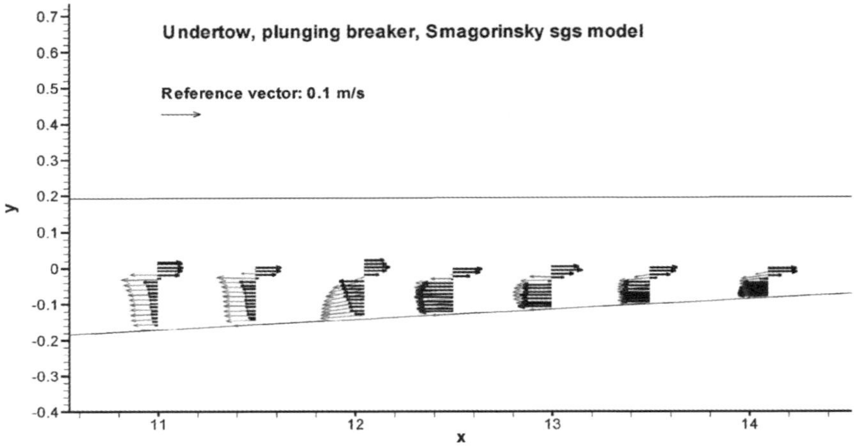

Fig. 23. Modeled and measured undertow plunging breaker. Thin: measurements of Ting and Kirby (1996); thick: model results using Smagorinsky SGS model. From Ref. 48 with permission of use from Elsevier. Refer Ref. 48 for color figure.

numerical model underestimates the undertow in general. Christensen[48] argued that the discrepancy between data and model results could be due to the stronger mixing in the simulation. No major differences are found between the two SGS turbulence models.

Finally, Christensen[48] compared modeled turbulent intensities with data. This is difficult to do for LES, because a long computation time is required to obtain meaningful results. To circumvent this, Christensen[48] define the "mean" quantities as the average over the transverse direction following by a phase-averaging. Any deviations from this "mean" are considered as "turbulence". Christensen[48] showed that his model over-estimates the turbulent kinetic energy and intensities by $50 \sim 100\%$. Because the definition for "turbulence" in experimental and numerical model are different, it is difficult to estimate how much of this discrepancy between data and model is from the different definition, and how much is from the modeling accuracy. It seems from these results, the three-dimensional LES model of Christensen[48] showed more details of the wave breaking process, but provided similar accuracy in terms of surface elevation and undertows as that of multi-scale two-dimensional model from Ref. 47 when compared with data.

More recently, Lubin *et al.*[49] used a three-dimensional two-phase LES model to study the air entrainment under a plunging breaker. This study is remarkable because the contribution of the air entrainment to the organization of coherent structures cannot be ignored for plunging breaking waves, particularly in the post plunging stage. Because of the demanding computer power required for this model, the computational domain in Lubin *et al.*[49] is only one wavelength long, and the modeling results can only be compared with experiments qualitatively. Therefore, Lubin *et al.*[49] focus their study on the generation of larger scale vortices, the air entrainment process and its implications for the turbulence generation and the energy dissipation during the breaking. They found that when breaking occurs, the jets do not penetrate deeply into the water. The tongues of water separate, one part feeds the upper part of the splash, while the other goes around the main pocket of entrapped gas where a topologically induced circulation takes place. This is consistent with previous laboratory observations. The circulation generated by air pockets, on the other hand, indicates the significance of air entrainment when examining the vortex structures of plunging breaking wave.

Lubin *et al.*[49] showed the major large vortices are first two-dimensional horizontal vortices under the wave crest, then they break up into three-dimensional flow structures. Lubin *et al.*[49] also observed some chaotic motions spreading towards the bottom, but they were unable to conclude that these motions are the "oblique eddies" discovered by Nadaoka *et al.*[10]

Lubin *et al.*[49] discussed the difference between the two-dimensional and three-dimensional numerical models for total energy dissipation. They show the two-dimensional turbulence is less dissipative. The discrepancy between the two numerical models starts at the impact of the jet on the forward face. When successive splash-ups entrain the gas pocket, 50% of pre-breaking total energy is dissipated in the two-dimensional model, compared with 45% dissipated in the three-dimensional model. Two and a half wave periods later, there remains 30% of the pre-breaking total wave energy in the two-dimensional model, but 20% in the three-dimensional model. Both curves of the total energy show the same decay rate at the end of the breaking process. The 10% difference in total energy between the two-dimensional and the three-dimensional simulations suggests the flow is mainly two-dimensional. However, Lubin *et al.*[49] note that this conclusion needs to be checked out more carefully with longer simulation time.

3.4. *Direct Numerical Simulation-DNS*

The approximation level on top of the pyramid is the *Direct Numerical Simulation* (DNS). In DNS, the Navier-Stokes equations are solved directly, resolving all the scales of the motions and involving no turbulence model.

Unlike RANS and LES, where the main focus is searching for better turbulence closure models, DNS pays more attention to the numerical methods used, because a wide range of length and time scales need to be accurately reproduced. The largest scale that needs to be resolved is determined by the physical parameters, such as the channel width, scale of the largest eddies and so on. The smallest scale that is required to be resolved is $O(\eta)$, where $\eta = (\nu^3/\varepsilon)^{1/4}$ is the Kolmogrov length scale, ν is the kinematic viscosity and ε the rate of dissipation of turbulent kinetic energy. Research has shown that to obtain reliable first and second order statistics, the resolution of a DNS model must be fine enough to capture most of the dissipation.[50] Therefore, DNS is restricted to low Reynolds number flow with simple geometries. A rough estimation for the breaking wave problem with Reynolds number of the order of 10^6 requires the grid size to be below 1mm. Few studies termed DNS modeling of breaking waves today satisfy these requirements.

Nonetheless, Wijayaratna and Okayasu[51] applied three-dimensional DNS and LES with the same resolution of 2mm to examine wave breaking on a sloping beach. The LES and DNS show similar results. The wave first breaks in two-dimensions, with three-dimensional turbulence slowly developing due to the two sidewalls. This three-dimensionality further developed following successive wave breaking. To save computational time, two-dimensional DNS and LES were used to compare model results with experimental data. Results from the $22^{th} \sim 23^{th}$ wave period were used for the comparison. The DNS and LES model showed minor differences with the breaking pattern, with the modeled breaking point and the bore propagation agreeing well with the experimental data. However, both numerical models showed a decrease in water surface behind the breaker, which is different from the experimental measurements. Wijayaratna and Okayasu[51] argue that this could be caused by the premature setup. They also show that though in two-dimensional simulation the stretching of eddies in the longshore direction is not allowed, the main features of the flow pattern closely follow the measurements.

Lubin *et al.*[52] also performed two-phase DNS for plunging breakers. To save computation time, Lubin *et al.*[52] generated wave overturning on a flat bottom by applying stability theory. They introduced a perturbation to the water depth in the longshore (y) direction to generate the three-dimensional breaking wave. They found the flow remains two-dimensional before plunging, the perturbation starts to grow when the second jet is falling down. A large amount of air is entrained into the water forming a topological vortex. This tube twisted along the y direction for the perturbed case, but stayed straight for the non-perturbed case. No model-data comparison is made using this model.

4. Conclusions and Future Work

Wave generated turbulence is an unsteady nonlinear complex high Reynolds number flow phenomenon. Resemblance to basic turbulence flow types is difficult to find. Numerical modeling is a useful tool to study this turbulent flow.

Independent studies show that turbulent-viscosity based RANS models show reasonable agreement with data in the shoaling zone and in the inner surf zone, but they predict the wave breaking far earlier than observed in experimental results. The two-dimensional multiscale $k - l$ model can be considered as an intermediate level between RANS models and three-dimensional LES. It provides reasonable agreement with experimental data for both spilling and plunging breakers with moderate computer requirements. This model should be further tested for second order statistics such as turbulent intensities and kinetic energy.

Vortex structures that develop post plunging are complex yet interesting. The effects of air entrainment can not be ignored when studying vorticity dynamics in this stage of the breaking process. Two-phase three-dimensional LES is a promising approach to include the effects of air entrainment, but is computationally very expensive. It will be interesting to see more investigations of the initiation, development and fate of these vortices; the role of air entrainment and the detection of the 'oblique eddies' using the two-phase LES approach.

References

1. M. S. Longuet-Higgins, Longshore currents generated by obliquely incident sea waves. 1, *J. Geophys. Res.* (75), 6783–6789, (1970).
2. F. Feddersen and J. H. Trowbridge, The effect of wave breaking on surf-zone turbuelcne and alongshore currents: a modeling study, *J. Phys. Oceanogr.* **35**, 2187–2203, (2005).
3. J. A. Battjes, Developemtns in coastal engineering research, *Coastal Engng.* **53**, 121–132, (2006).
4. F. C. K. Ting and J. T. Kirby, Observation of undertow and turbulence in a laboratory surf zone, *Coastal Engng.* **24**, 51–80, (1994).
5. F. C. K. Ting and J. T. Kirby, Dynamics of surf-zone turbulence in a spilling breaker, *Coastal Engng.* **27**(131-160), (1996).

6. S. Longo, M. Petti, and I. J. Losada, Turbulence in the swash and surf zones: a review, *Coastal Engng.* **45**, 129–147, (2002).
7. D. H. Peregrine and I. A. Svendsen. Spilling breakers, bores and hydraulic jumps. In *Proc. of the 16th Int. Conf. Coastal Eng.*, pp. 540–550., (1978).
8. J. A. Battjes and T. Sakai, Velocity field in a steady breaker, *J. Fluid Mech.* **111**, 421–437, (1981).
9. H. Yeh, Turbulent generation in a bore., *In Water Wave Kinematics.* pp. 525–534, (1990).
10. K. Nadaoka, M. Hino, and Y. Koyano, Structure of turbulent flow field under breaking waves in the surf zone., *J. Fluid Mech.* **204**, 359–387, (1989).
11. J. R. Tallent, T. Yamashita, and Y. Tsuchiya, Transformation characteristics of breaking water waves., *In Water Wave Kinematics.* pp. 509–524, (1990).
12. A. Okayasu, T. Shibayama, and N. Mimura. Velocity field under plunging waves. In ed. B. L. Edge, *Proc. of the 20th Int. Conf. Coastal Eng.*, pp. 660–674, (1986).
13. A. Okayasu, T. Shibayama, and K. Horikawa. Vertical variation of undertow in the surf zone. In ed. B. L. Edge, *Proc. of the 21st Int. Conf. Coastal Eng.*, pp. 478–491, (1988).
14. R. George, R. E. Flick, and R. T. Guza, Observation of turbulence in the surf zone, *J. Geophys. Res.* **99**(C1), 801–810, (1994).
15. K. Nadaoka and T. Kondoh, Laboratory measurements of velocity field structure in the surf zone by ldv., *Coastal Eng. Japan.* **25**, 125–, (1982).
16. J. Trowbridge and S. Elgar, Turbulence measurements in the suf zone, *J. Phys. Oceanogr.* **15**, 290–298, (1998).
17. J. Trowbridge, On a technique for measurement of turbulent shear stress in the presence of surface waves., *J. Phys. Oceanogr.* **31**, 2403–2417, (2001).
18. K.-A. Chang and P. L. F. Liu, Velocity, acceleration and vorticity under a breaking wave, *Phys. Fluids.* **10**(1), 327–329, (1998).
19. K.-A. Chang and P. L. F. Liu, Experimental investigation of turbulence generated by breaking waves in water of intermediate depth, *Phys. Fluids.* **11** (11), 3390–3400, (1999).
20. J. H. Duncan, Spilling breakers, *Annu. Rev. Fluid. Mech.* **33**, 519–547, (2001).
21. S. Misra, J. T. Kirby, and M. Brocchini, The turbulent dynamics of quasi-steady spilling breakers theory and experiments., *In Water Wave Kinematics.* (2005).
22. S. Misra, M. Brocchini, and J. T. Kirby. Turbulent interfacial boundary conditions for spilling breakers. In ed. J. Smith, *Proc. of the 30th Int. Conf. Coastal Eng.*, pp. 214–225, (2006).
23. M. M. Gibson and W. Rodi, A reynolds-stress closure model of turbulence applied to the calculation of a highly curved mixing layer., *J. Fluid Mech.* **103**, 161–182, (1981).
24. C. L. Rumsey, T. B. Gatski, and J. H. Morrison, Turbulence model predictions of extra strain rate effects in strongly-curved flows., *AIAA.* pp. 99–0157, (1999).

25. P. R. Spalart, Strategies for turbulence and simualtions., *Int. J. Heat and Fluid Flow.* **21**, 252–263, (2000).
26. P. Sagaut, *Large Eddy Simulation for Incompressible Flows.* (Springer, 2001).
27. S. B. Pope, *Turbulent flows.* (Cambridge University Press, 2000).
28. M. S. Uberoi, Effect of wind-tunnel contraction on free-stream turbulence., *J. Aero. Sci.* **23**, 754–764, (1956).
29. Z. Warhaft, An experimental study of the effect of uniform strain on thermal fluctuations in grid-generated turbulence., *J. Fluid Mech.* **99**, 545–573, (1980).
30. B. E. Launder and D. B. Spalding, The numerical computational of turbulent flows., *Comuter Methods in Applied Mechanics and Engineering.* **3**, 269–289, (1974).
31. C. M. Lemos, A simple numerical technique for turbulent flows with free surfaces., *Int. J. Numer. Methods Fluids.* **15**, 127–146, (1992).
32. P. Z. Lin and P. L.-F. Liu, A nuermical study of breaking waves in the surf zone, *J. Fluid Mech.* **359**, 239–264, (1998).
33. P. Z. Lin and P. L.-F. Liu, Turbulence transport, vorticity dynamics, and solute mixing under plunging breaking waves in surf zone, *J. Geophys. Res.* **103**(C8), 15677–15694, (1998).
34. F. H. Champagne, V. G. Harris, and S. Corrsin, Experiments on nearly homogeneous turbulent shear flow, *J. Fluid Mech.* **41**, 81–139, (1970).
35. T.-H. Shih, J. Zhu, and J. L. Lumley, Calculation of wall-bounded complex flows and free shear flows., *Intl. J. Numer. Meth. Fluids.* **23**, 1133–1144, (1996).
36. A. J. Chorin, Numerical solution of the navier-stokes equations., *Math. Comput.* **22**, 745–762, (1968).
37. A. J. Chorin, On the convergence of discrete approximations of the navier-stokes equations., *Math. Comput.* **23**, 341–353, (1969).
38. C. W. Hirt and B. D. Nichols, Volume of fluid (vof) method for the dynamics of free boundaries., *J. Comput. Phys.* **39**(1), 201–225, (1981).
39. D. B. Kothe, R. C. Mjolsness, and M. D. Torrey, Ripple: A computer program for incompressible flows with free surfaces *Los Alamos National Laboratory, LA-12007-MS.* (1991).
40. T. J. Hsu, T. Sakakiyama, and P. L.-F. Liu, A numerical model for wave motions and turbulence flows in front of a composite breakwater., *Coastal Eng.* **46**(1), 25–50, (2002).
41. E. D. Christensen, J. H. Jensen, and S. Mayer. Sediment transport under breaking waves. In ed. B. L. Edge, *Proc. of the 27th Int. Conf. Coastal Eng.*, pp. 2467–2480, (2000).
42. S. Bradford, Numerical simulation of surf zone dynamics, *J. Wtrwy., Port, Coast., and Oc. Engrg.* **126**(1), 1–13, (2000).
43. S. Meyer and P. A. Madsen. Simulation of breaking waves in the surf zone using a navier-stokes solver. In ed. B. L. Edge, *Proc. of the 27th Int. Conf. Coastal Eng.*, pp. 928–941, (2000).
44. J. W. Deardorff, A numerical study of three-dimensional turbulent channel flow at large reynolds numbers, *J. Fluid. Mech.* **41**(2), 453–480, (1970).

45. Q. Zhao and K. Tanimoto. Numerical simulation of breaking waves by large eddy simulation and vof method. In ed. B. L. Edge, *Proc. of the 26th Int. Conf. Coastal Eng.*, pp. 892–905, (1998).
46. Q. Zhao, S. W. Armfield, and K. Tanimoto. A two-dimensional multi-scale turbulence model for breaking waves. In ed. B. L. Edge, *Proc. of the 27th Int. Conf. Coastal Eng.*, pp. 80–93, (2000).
47. Q. Zhao, S. W. Armfield, and K. Tanimoto, Numerical simulation of breaking waves by a multi-scale turbulence model, *Coastal Engineering.* **51**, 53–80, (2004).
48. E. D. Christensen, Large eddy simulation of spilling and plunging breakers, *Coastal Engng.* **53**, 463–485, (2006).
49. P. Lubin, S. Vincent, S. Abadie, and J. P. Caltagirone, Three-dimensional large eddy simulation of air entrainment under plunging breaking waves, *Coastal Engng.* **53**, 631–635, (2006).
50. P. Moin and K. Mahesh, Direct numerical simulation: a tool in turbulence research., *Annu. Rev. Fluid. Mech.* **30**, 539–578, (1998).
51. N. Wijayaratna and A. Okayasu. Dns of wave transformation, breaking and run-up on sloping beds. In *Proc. 4th Int. Conf. on Hydrodynamics*, p. 5, Yokohama,Japan, (2000).
52. P. Lubin, S. Vincent, J. P. Caltagirone, and S. Abadie, Fully three-dimensional direct numerical simulation of a plunging breaker, *Comptes Rendus Mecanique.* **331**(7), 495–501, (2003).

CHAPTER 18

FREAK WAVES AND THEIR INTERACTION WITH SHIPS AND OFFSHORE STRUCTURES

Günther F. Clauss

Division of Naval Architecture and Ocean Engineering
Technical University Berlin, Germany
clauss@naoe.tu-berlin.de

Surviving a freak wave - what an experience. However, only scarce observations are available of such mystic disasters. For the design of safe and economic offshore structures and ships the knowledge of the extreme wave environment and related wave/structure interactions is inevitable. A stochastic analysis of these phenomena is insufficient as local characteristics in the wave pattern are of great importance for deriving appropriate design criteria. In general, extremely high rogue waves or critical wave groups are rare events embedded in a random seaway. The most efficient and economical procedure to simulate and generate such a specified wave scenario for a given design variance spectrum is based on the appropriate superposition of component waves or wavelets. As the method is linear, the wave train can be transformed down-stream and up-stream between wave board and target position. The desired characteristics like wave height and period as well as crest height and steepness are defined by an appropriate objective function. The subsequent optimization of the initially random phase spectrum is solved by a Sequential Quadratic Programming method (SQP). The linear synthesis of critical wave events is expanded to a fully nonlinear simulation by applying the subplex method. Improving the linear SQP-solution by the nonlinear subplex expansion results in realistic rogue waves embedded in random seas. This process of a deterministic evolution of tailored wave sequences is improved by combining the computer system with a wave tank forming a fully automized hybrid control loop for optimizing the selected wave train at target position. Nonlinear free surface effects are completely considered in the fitting process since the wave train propagates in the real wave tank, and thus the objective function is automatically validated. As an illustration of this technique a reported rogue wave - the Draupner "New Year Wave" is simulated and generated in a physical wave tank. Also a "Three Sisters" wave sequence with succeeding wave heights $H_s...2H_s...H_s$, embedded in an extreme sea, is synthesized. As wave evolution and propagation is governed by inertia and gravity forces the control signal for generating a tailored wave sequence which has been optimized in a small wave tank can easily be transferred to large seakeeping basins to investigate wave/structure interactions. For investigating the consequences of specific extreme sea conditions this paper analyses the seakeeping behaviour of a semisubmersible and a FPSO in the Draupner New Year Wave, embedded in extreme irregular seas. In addition, extreme roll motions and the capsizing of a RoRo vessel in a severe storm wave group are evaluated, and the procedures for safe ship design and operation are discussed.

1. Introduction

Freak waves are rare events, and hence only few observations are reported. The associated damages, however, are significant. Consequently, the design process has to consider these extreme scenarios. The most appropriate approach for evaluating the related wave/structure interactions is the targeted simulation of tailored wave sequences which are either documented (existing registrations) or are intentionally synthesized for investigating critical response characteristics of vessels, e.g. the capsizing of ships.

Figure 1 shows two exceptional white walls or rogue waves,[1] rare observations as it is generally not the dominant impulse to take a photo at such an occasion. Damages related to rogue waves are frequently reported, e.g. the destruction of the Norwegian tanker *WILSTAR* hit by a huge wave in the Agulhas current, South Africa in 1974 or the offshore oil production platform *EUGENE ISLAND 322* struck by Hurricane Lili in the Gulf of Mexico, in 2002. Figure 2(a) shows the supertanker *ESSO Languedoc* in a storm offshore Durban, South Africa (1980). Figure 2(b) illustrates the heavy damage of the tanker *Stolt Surf* in freak waves which even destroyed the bridge roof, 22m above sea level (20.10.1977). Similar disasters are reported by the *Norwegian Dawn* (16.4.2005) and *Queen Elisabeth II* (February 1995). The Hapag-Lloyd Lashcarrier *München* was lost in December 1978 as well as the British Oil-Bulk-Ore Carrier *Derbyshire* (1980). In the period 2001-2005 174 total losses for IACS vessels and 610 for non-IACS vessels were reported covering all seagoing merchant ships over this period of 100 gt and above, including fishing vessels (LR Fairplay). In total, more than 100 ships are lost every year - and in 60% of these disasters rogue waves are involved.

According to Murphy's law - the impossible will happen one day - we consider rogue waves as rare events beyond our present modelling

Fig. 1. Freak wave observations - and damages related to rogue waves, i.e. the offshore oil production platform *EUGENE ISLAND 322* struck by the Hurricane Lili in the Gulf of Mexico, in 2002 and the Norwegian tanker *Wilstar* hit by a huge wave in the Agulhas current, South Africa in 1974.

abilities. Haver and Anderson[2] suggest that freak waves (unexpected large crest height / wave height, unexpected severe combination of wave height and wave steepness or unexpected group pattern) should be defined as wave events which do not belong to the population defined by a Rayleigh model. To yield a sufficiently small contribution to the overall risk of structural collapse, the structure should withstand extreme waves corresponding to an annual probability of exceedance of say $10^{-5} - 10^{-4}$ as the Rayleigh model underpredicts the highest crest heights indicating that real processes may be strictly affected by higher order coefficients. In addition to the ULS (ultimate limit state) - based on a 100-year design wave - an ALS (accidental limit state) with a return period of 10000 years is suggested.

Based on observations Faulkner[3] proposes the freak or abnormal wave height for survival design $H_d > 2.5H_S$. He also recommends to characterize wave impact loads so they can be quantified for potentially critical seaways and operating conditions. Present design methods should be complemented by survival design procedures, i.e. two levels of design wave climates are proposed:

- The Operability Envelope which corresponds to the best present design practice
- The Survivability Envelope based on extreme wave spectra parameters which may lead to episodic waves or wave sequences (e.g. the *Three Sisters*) with extremely high and steep crests.

(a) (b)

Fig. 2. (a) Supertanker *ESSO Languedoc* in a storm offshore Durban, South Africa (1980). (b) Tanker *Stolt Surf* in heavy seas (1977).

Wave steepness, characterized by front and rear steepness as well as by horizontal and vertical wave asymmetries seems to be a parameter at least as important as wave height.[4]

A probability analysis of rogue wave data recorded at North Alwyn (160km East of the Shetland Islands in the central North Sea) from 1994 to 1998 (in total 345245 waves have been registered) reveals that these waves are generally 50% steeper than the significant steepness, with wave heights $H_{max} > 2.5H_S$.[5] The preceding and succeeding waves have steepness values around half the significant values while their heights are around the significant height. Figure 3 presents data of a five-days storm (Nov. 16-21, 1997) with 21 waves higher than $2H_S$. The highest wave $H_{max} = 22.03$m was registered at a 40 min registration, with $H_S = 8.64$m and $H_{max}/H_S = 2.55$. Based on these data Guedes Soares[6,7] analyses the characteristics of abnormal waves and the associated sea states in the North Sea (North Alwyn) as well as in the Gulf of Mexico (Hurricane Camille).

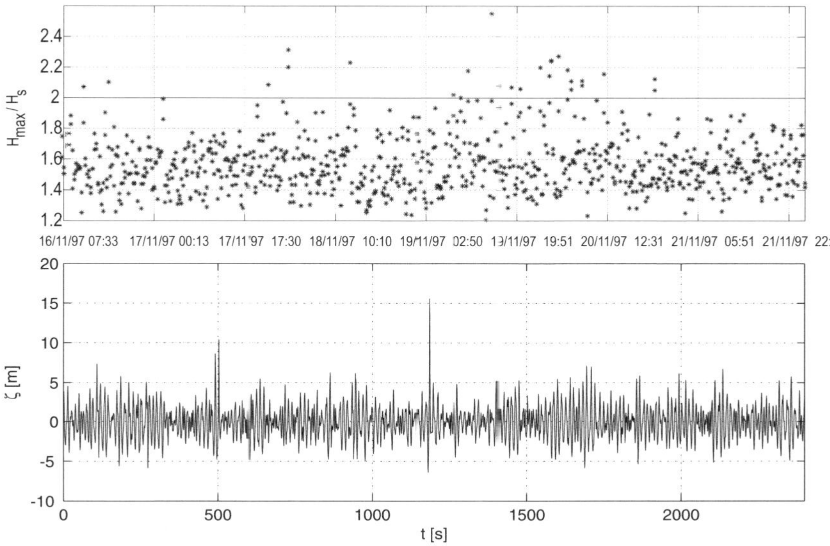

Fig. 3. North Alwyn storm data (Nov.16-21, 1997) as well as registration on November 19, 1997 - the 40 min record, with $H_S = 8.64$m and $T_P = 13.11$s (water depth 126m) includes a rogue wave with $H_{max} = 22.03$m ($H_{max}/H_S = 2.55$).[5]

Steep-fronted wave surface profiles with significant asymmetry in the horizontal direction excite extreme relative motions at the bow of a cruising ship with significant consequences on green water loading on the fore deck and hatch covers of a bulk carrier.[8] Heavy weather damages caused by giant waves are presented by Kjeldsen,[9] including the capsizing of the semisubmersible *Ocean Ranger*. Faulkner and Buckley[10] describe a number of episodes of massive damage to ships due to rogue waves, e.g. with the liners *Queen Elisabeth* and *Queen Mary*. Haver and Anderson[2] report on substantial damage of the jacket platform *Draupner* when a giant wave ($H_{max} = 25.63$m) with the crest height $\zeta_{crest} = 18.5$m hit the structure in 70m water depth on January 1, 1995 (Fig. 4). Related to the significant wave height $H_S = 11.92$m, the maximum wave rises to $H_{max} = 2.15 H_S$ with a crest of $\zeta_{crest} = 0.72 H_{max}$.

Not as spectacular but still exceptional are wave data from the Norwegian Frigg field - water depth 99.4m ($H_S = 8.49$m, $H_{max} = 19.98$m, $\zeta_{crest} = 12.24$m)[11] and the Danish Gorm field - water depth 40m ($H_S = 6.9$m, $H_{max} = 17.8$m, $\zeta_{crest} \approx 13$m).[12] Also remarkable are wave records

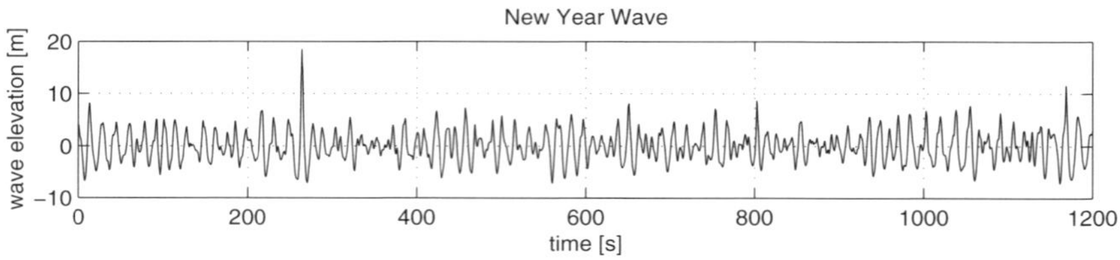

Fig. 4. Rogue wave registration - New Year Wave registered at the North Sea *Draupner* platform on 01.01.1995.[2]

of the Japanese National Maritime Institute measured off Yura harbour at a water depth of 43m ($H_S = 5.09$m, $H_{max} = 13.6$m, $\zeta_{crest} \approx 8.2$m).[13]

All these wave data (with $H_{max}/H_S > 2.15$ and $\zeta_{crest}/H_{max} > 0.6$) prove, that rogue waves are serious events which should be considered in the design process. Although their probability is very low they are physically possible. It is a challenging question which maximum wave and crest heights can develop in a certain sea state characterized by H_S and T_P, which extremely high wave groups are observed, and which consequences (motions, forces, bending moments) on structures and vessels are expected.

As the term "rogue" or "freak" wave is only defined in relation to the significant wave height H_s it is obvious that terrifying waves presume extremely rough sea states. This fact is illustrated by the above mentioned Yura-wave which is defined as a rogue wave, however, the related maximum wave height of $H_{max} = 13.6$m should not be critical for seagoing vessels. Consequently, the "ultimate limit state" (leading to the 100-year-design wave) has to be selected as the reference sea condition. Extended to the "accidental limit state"[2] (corresponding to a return period of 10000 years based on the - disputable - Rayleigh model) we then obtain extreme waves higher than 2...2.2 H_S which are termed "rogue" waves. As an example, a 33 m-wave undoubtly is an exotic monster in a sea state of $H_s = 15$m, and the related loads and motions of ocean structures may exceed the design limits.

In this context we should recall that rogue waves are often integrated in wave groups $H_s ... 2H_s ... H_s$[5] (see Fig. 3) - described as "Three Sisters" - which challenges the vessel performance also in relation to dynamic aspects. If natural frequencies of fixed or flexible structures (oscillations, vibrations) or of cruising vessels (roll or parametric roll resonance) are excited by wave sequences with relevant frequency characteristics the consequences may jeopardize.

In conclusion

- a single "rogue" wave hidden in a severe sea state is a highly dangerous event challenging the performance and safety of offshore structures.
- If a sea state contains wave groups which excite resonance motions, e.g. roll or parametric roll of cruising ships (dependent on course and speed) the dynamic behaviour of the vessel (or structure) may exaggerate the wave height effects.
- The worst case scenario is occurring if wave sequences with relevant frequency characteristics are combining with extreme wave heights.

In the following chapters a "standard model of ocean waves[14]" is used to generate tailored wave sequences with integrated "freak" waves by controlled superposition. An alternative approach for the description and explanation of rogue waves is the application of the (quantum physics) nonlinear Schrödinger equation (NLS) and the Benjamin-Feir instability[15] inducing the nonlinear self-focussing effect.[16–18] Even though the above results are well understood and robust from a physical[19] and mathematical[20,21] point of view, it is still unclear how freak waves are generated via the Benjamin-Feir instability in more realistic oceanic conditions.[22] This is addressed to formal limitations of the applicability of the NLS equation for three-dimensional ocean waves concerning the bandwidth.[23]

Zakharov[24] showed that the NLS equation can be deduced from the Zahkarov integral equation which is more general. This fact explains why the NLS equation and the modified nonlinear Schrödinger (MNLS) equation introduced by Dysthe[25] need less computation time to obtain numerical solutions. The MNLS for gravity waves is a fourth-order extension of the nonlinear Schrödinger equation for infinite depth. The application of the NLS equation and the Benjamin-Feir instability to rogue waves and holes in deep water wave trains is shown by Osborne.[26] Further studies use the JONSWAP power spectrum as initial condition.[22,27] A relation between spectral bandwidth and wave steepness is established by the Benjamin-Feir index (BFI)[28,29] which plays an important role in the appearance of large amplitude waves.[30] Comparisons to theory, experiments and real conditions are presented to validate this quantum physics model for oceanic rogue waves.[31,32]

2. Generation of Tailored Wave Sequences

2.1. *Linear optimization of target wave trains*

The method for generating linear wave groups is based on the wave focussing technique of Davis and Zarnick,[33] and its significant development by Takezawa and Hirayama.[34] Clauss and Bergmann[35] recommended a special type of transient waves, i.e. Gaussian wave packets, which have the advantage that their propagation behaviour can be predicted analytically. With increasing efficiency and capacity of computer the restriction to a Gaussian distribution of wave amplitudes has been abandoned, and the entire process is now performed numerically.[36] The shape and width of the wave spectrum can be selected individually for providing sufficient energy in the relevant frequency range. As a result the wave train is predictable at any instant and at any stationary or moving location. In addition, the wave orbital motions as well as the pressure distribution and the vector fields of velocity and acceleration can be calculated. According to its high accuracy the technique is capable of generating special purpose transient waves.

Extreme wave conditions in a 100-year design storm arise from unfavourable superpositions of component waves which represent the severe sea spectrum. Freak waves have been registered in standard irregular seas when component waves accidentally superimpose in phase. Extensive random time-domain simulation of the ocean surface for obtaining statistics of the extremes, however, is very time consuming. Generally, when generating irregular seas in a wave tank the phase shift is supposed to be random, however, it can be fixed by the control program on the basis of a pseudo-random process: Consequently, the phase spectrum is also given as a deterministic quantity and can be optimized. Why should we wait for rare events if we can achieve freak wave conditions by intentionally selecting a suitable phase shift, and generate a deterministic sequence of waves converging at a preset concentration point? At this position all waves are superimposed without phase shift resulting in a single high wave peak. Assuming linear wave theory, the synthesis and up-stream transformation of appropriate wave packets is developed from this concentration point, and the Fourier transform of the wave train (phase spectrum) is transformed back to the upstream position at the wave board.

In general, "rogue" waves or critical wave groups are rare events embedded in a random seaway. As long as linear wave theory is applied, the sea state can be regarded as superposition of independent harmonic "component" waves, each having a particular direction, amplitude, frequency and phase. For a given design variance spectrum of an unidirectional wave train, the phase spectrum is responsible for all local characteristics, e.g. the wave height and period distribution as well as the location of the highest wave crest in time and space.

2.2. *Nonlinear transient wave description*

The generation of higher and steeper wave sequences requires a more sophisticated approach as propagation velocity increases with height. Consequently, it is not possible to calculate the wave train linearly upstream back to the wave generator to determine the control signal of the wave board as the associated wave sequence - propagating due to nonlinear wave theory - will show substantial deviations at the target position. To solve this problem, Kühnlein[37] developed a semi-empirical procedure for the evolution of extremely high wave groups which is based on linear wave theory: The propagation of high and steep wave trains is calculated by iterative integration of coupled equations of particle positions. With this deterministic technique "freak" waves up to 3.2m high have been generated in a wave tank.[38] Figure 5 shows the genesis of this wave packet and presents registrations which have been measured at various locations including the position $x = 81.15$m close to the concentration point at 84m. The associated wave board motion which has been determined by the above semi-empirical procedure is the key input for the nonlinear analysis of wave propagation. Figure 6(a) shows the maximum (crest) and minimum (trough) surface elevation in the wave tank ζ_{max} and ζ_{min} as well as the difference, i.e. the wave height $\zeta_{max} - \zeta_{min}$. Note the sudden rise of water level (crest and trough) at the concentration point. Figure 6(b) illustrates numerically calculated orbital tracks of particles with starting locations at x=126m, which is very close to the concentration point. Generally, the orbital tracks are not closed. Particles with starting locations $z > -1$m are shifted in the x-direction, and due to mass conservation particles with lower z-coordinates are shifted in the opposite direction.

As has been generally observed - at wave groups as well as at irregular seas with embedded rogue wave sequences - substantial differences between the measured time series and the specified design wave train at target location are registered if a linearly synthesized control signal is used for the generation of higher and steeper waves. The main deviation, however, is localized within a small range. This promising observation proves that it is sufficient for only a short part of the control signal in the time-domain to be fitted.[39] Hence, only the wave environment close to the embedded rogue wave is modified. For these (temporally limited) local changes of the control signal the discrete wavelet transformation[40] is introduced into the optimization process. The discrete wavelet transform samples the signal into several decomposition levels and each resulting coefficient describes the wave in a specific time range and frequency band width. As a prerequisite, however, the optimization process should imply nonlinear wave theory and develop the wave evolution by using a numerical time-stepping method. Figure 7(a)

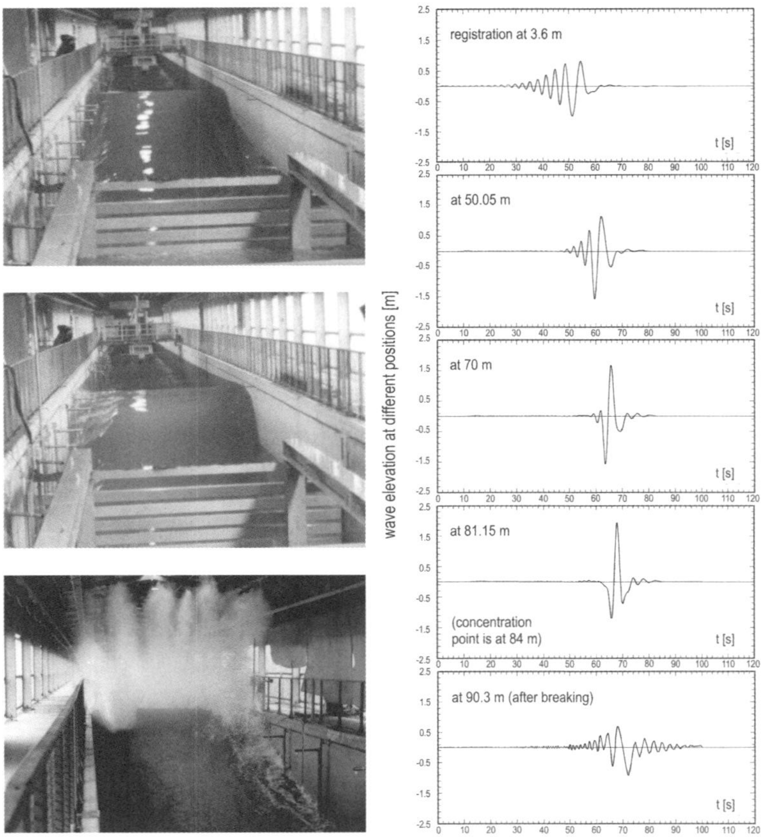

Fig. 5. Genesis of a 3.2m rogue wave by deterministic superposition of component waves (water depth d=4m) (right hand side). The effect of such a wave on a porous wall is shown at the left hand side.

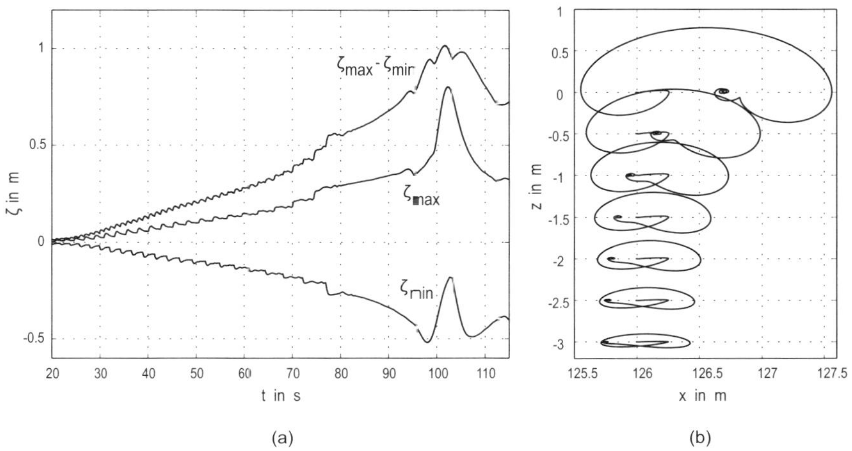

Fig. 6. (a) Maximum (crest) and minimum (trough) surface elevations (ζ_{max}, ζ_{min}) as well as wave height $\zeta_{max} - \zeta_{min}$. (b) Particle tracks with starting location at x=126m.

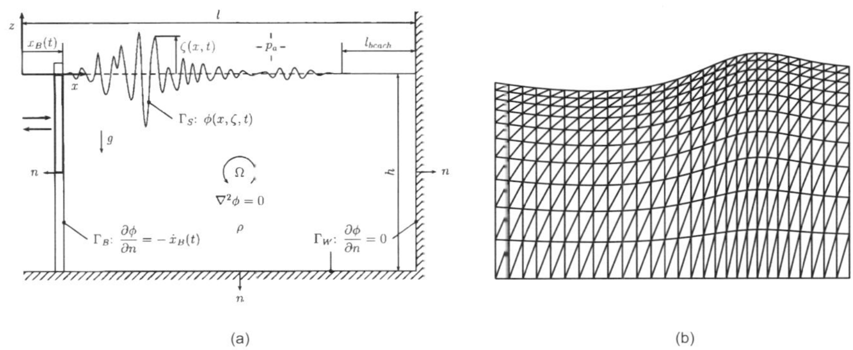

Fig. 7. (a) Numerical wave tank.[41] (b) Finite element mesh for nonlinear analysis.

summarises the basic equations and boundary conditions. A finite element method developed by Wu and Eatock Taylor[42,43] is used to determine the velocity potential, which satisfies the Laplace equation for Neumann and Dirichlet boundary conditions. The Neumann boundary condition at the wave generator is introduced in form of the first time-derivative of the measured wave board motion. To develop the solution in the time-domain the fourth order Runge-Kutta method is applied. Starting from a finite element mesh with 8000 triangular elements (401 nodes in x-direction, 11 nodes in z-direction, i.e. 4411 nodes - see Fig. 7(b)) a new boundary-fitted mesh is created at each time step. Lagrangian particles concentrate in regions of high velocity gradients, leading to a high resolution at the concentration point. This mixed Eulerian-Langrangian approach has proved its capability to handle the singularities at intersection points of the free surface and the wave board. Figure 8(a) presents registrations (numerical results and experimental data) at different positions to validate this nonlinear approach. Figure 8(b) shows wave profiles (snapshots) with associated velocity potential. Note how the faster long waves are catching up and finally swallow the slower short waves (in the registration diagrams $\zeta(t)$ the faster long waves are the last waves). Excellent agreement between numerical and experimental results is observed.

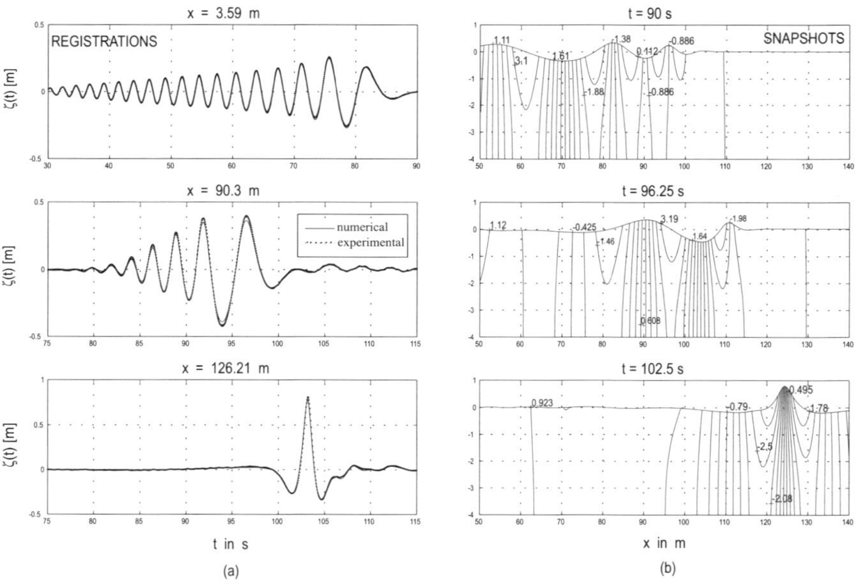

Fig. 8. Nonlinear numerical simulation of transient waves: (a) registrations at different positions and (b) surface elevation at different times with associated velocity potential.

2.3. *Integration of rogue wave sequences into irregular seas*

In the following it is shown how a tailored group of three successive waves is integrated into irregular seas using a Sequential Quadratic Programming (SQP) method. As a first approach the wave sequence with the integrated rogue wave group H_s ... $2H_s$ ... H_s is synthesized using linear wave theory. As illustrated in Fig. 9(a) all target features regarding global and local wave characteristics, including the rogue wave specification $H_{max} = 2H_S$ and $\zeta_{crest} = 0.6H_{max}$ are met.

The linear description of the wave train is a good starting point to further improve the wave board motion (i.e. time-dependent boundary conditions) required in the fully nonlinear numerical simulation. Using nonlinear wave theory, however, the same wave board motion results in the target registration shown in Fig. 9(b). Note that the design wave group deviates significantly. This is due to the fact that steep wave trains do not obey linear dispersion i.e. wave celerity is not only a function of frequency and water depth but also of wave height. Figure 10 illustrates the increase of wave celerity c with wave amplitude ζ_a for different frequencies - and steepness - using Stokes Third Order wave theory.[44] As a consequence, the nonlinear wave train at target location that originates from the first optimization process must be further improved. This is achieved by applying the subplex method developed by Rowan[45] for unconstrained minimization

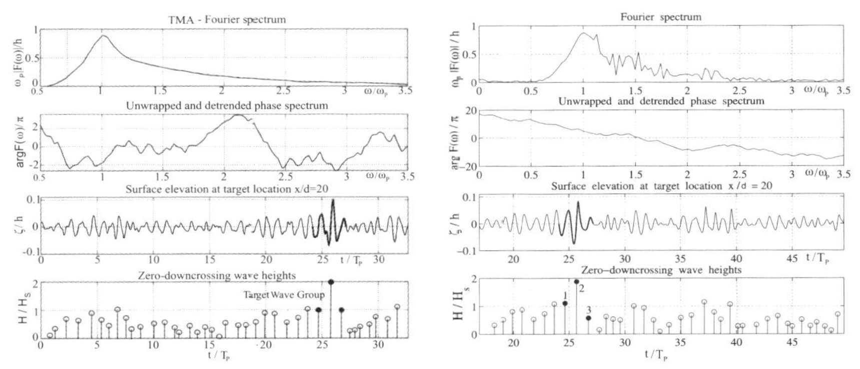

(a) (b)

Fig. 9. Wave evolution using the same board motion sequence: (a) Ideal wave train (linear wave propagation) at target position with predefined freak wave group $H_S...2H_S...H_s$ and (b) Real wave train (nonlinear wave propagation) at target position with deformation of freak wave group.

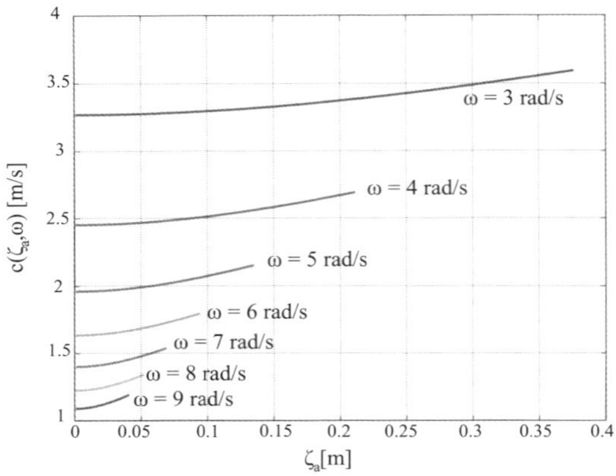

Fig. 10. Stokes III: wave celerity vs. wave amplitude ζ_a.

of noisy objective functions. The domain space of the optimization problem is decomposed into smaller subdomains which are minimized by the popular Nelder and Mead simplex method.[46] The subplex method is introduced because SQP cannot handle wave instability and breaking since the gradient of the objective function is difficult to determine in this case. Nonlinear free surface effects are included in the fitting procedure since the values of objective function and constraints are determined from the nonlinear simulation in the numerical wave tank. Figure 11 illustrates the results of this procedure, i.e. the evolution of the design wave sequence, with registrations at 5m, 50m and 100m (target position) behind the wave board (left side) as well as wave profiles (snapshots of the surface elevation) at $t = 75$s, 81s (target) and 87s (water depth $d = 5$m, $T_p = 3.13$s). The associated energy flux at the locations $x = 5$m, 50m and 100m is shown in Fig. 12(a). As has been expected the energy flux focuses at the target position. From the velocity potential which has been determined as a function of time and space all kinematic and dynamic characteristics of the wave sequence are deduced. Figure 12(b) presents associated velocity, acceleration and pressure fields.[41] Note that the effects of the three extremely high waves are reaching down to the bottom.

Summarizing the achievements of the numerical wave tank we can generate deterministic high and steep wave sequences at a selected position, and integrate them into irregular seas with defined significant wave height and period.

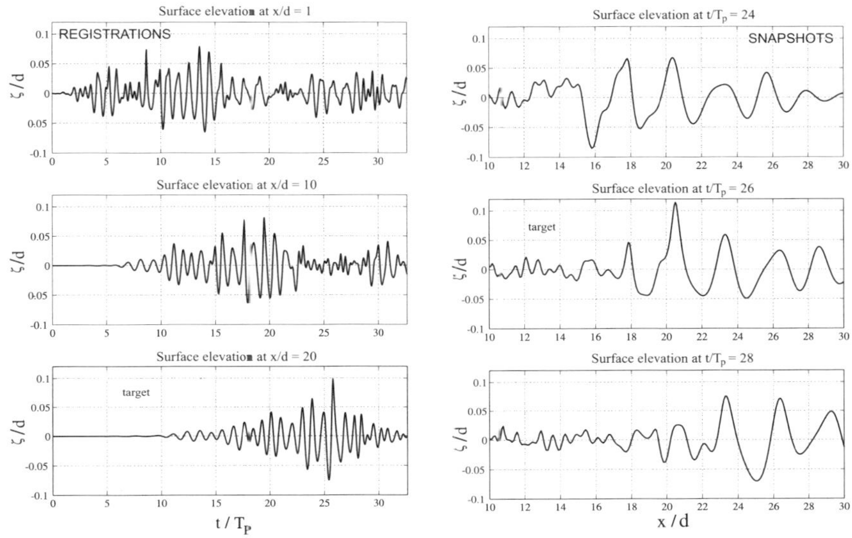

Fig. 11. Evolution of rogue wave sequence - registrations at x = 5m, 50m and 100m (left) as well as wave profiles (snapshots) at t=75s, 81s and 87s (right hand side) (water depth d = 5m, T_P = 3.13s).

2.4. *Self-validating procedure for optimizing tailored design wave sequences*

To further improve the process of the deterministic evolution of tailored wave sequences, a fully automized optimization technique with integrated validation has been developed. As shown in Fig. 13 the wave tank is combined with a computer system representing a hybrid control loop for optimizing the selected wave train at a target position. Nonlinear free surface effects are automatically considered in the fitting process since the wave train propagates in the real wave tank, and thus the objective function is automatically validated.[47] This self-validating optimization method has been applied to generate the New Year Wave at model scale (wave data see Fig. 4). Firstly, for the specified design variance spectrum, the SQP-method yields an optimized phase spectrum which corresponds to the desired wave characteristics at target position. The wave generator control signal is determined by transforming this wave train in terms of the complex Fourier transform to the location of the wave generator. The measured wave train at target position is then iteratively improved by systematic variation of

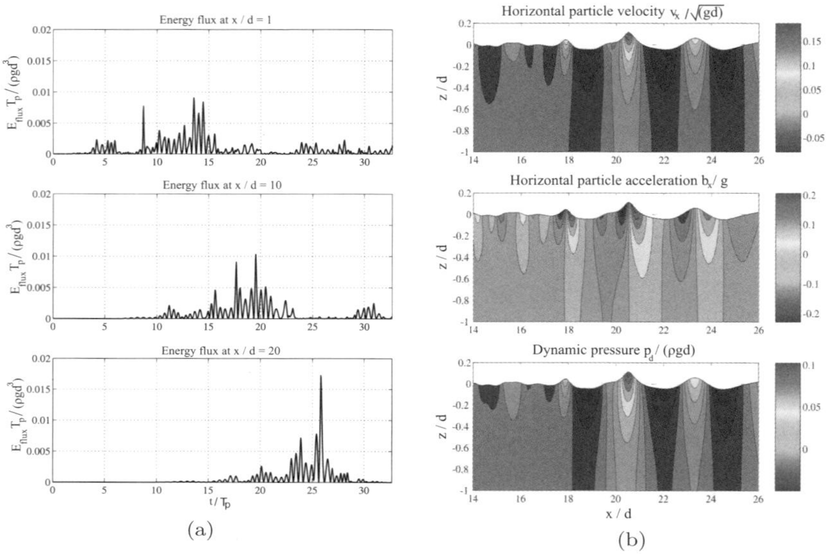

Fig. 12. (a) Energy flux of nonlinear wave at x = 5m, 50m and 100m (target), water depth d = 5m. (b) Kinematic and dynamic characteristics of rogue sequence $H_S...2H_S...H_S$ at target time t = 81s, water depth d = 5m.

the wave board control signal. To synthesize the control signal wavelet coefficients are used. The number of free variables is significantly reduced if this signal is compressed by low-pass discrete wavelet decomposition, concentrating on the high energy band. Based on deviations between the measured wave sequence and the design wave group at target location the control signal for generating the seaway is iteratively optimized in the fully automatic computer-controlled model test procedure (Fig. 13).

Figure 14(a) presents the evolution of the *Draupner* New Year Wave (full scale data deduced from model tests at a scale of 1:70). The registrations show how the extremely high wave develops on its way to the target position: Two kilometers ahead of the concentration point we observe a wave group with three high waves between 700s and 730s. The slightly longer waves (right hand side waves of the registration) are travelling faster catching up with the leading wave of the group. Just 500m ahead of the concentration point only two higher waves are visible, superimposing to the single New Year Wave at target. The registrations at 500m and 1000m behind the target document the dissipation of the wave

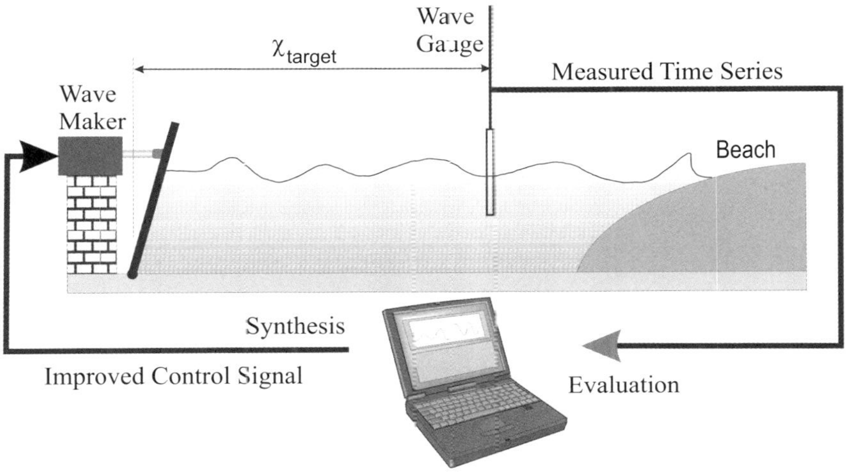

Fig. 13. Computer controlled experimental optimization of tailored design wave sequences.

sequence. Figure 14(b) presents the measured registration of the New Year Wave at target position. As compared to the registered New Year Wave the experimental simulation is quite satisfactory. Finally, Fig. 14(c) shows a snapshot of the freak wave at target time. Just to get a rough impression: The 14s - wave is approximately 300m long, and travels with a celerity of about 22m/s (42kn).

The hybrid procedure for optimizing tailored design wave sequences can also be used to adjust a target wave steepness. In this case the objective is to generate a wave sequence with a selected wave height and period, and in addition a certain crest front steepness $\varepsilon_t = 2\pi\zeta_{crest}/(gT_{rise}T_{zd})$, a defined vertical asymmetry factor $\lambda = T_{zd}/T_{rise}$ and a horizontal asymmetry factor $\mu = \zeta_{crest}/H$ (following the definition proposed by Kjeldsen[48]). Figure 15 illustrates these wave characteristics, and Fig. 16(a) shows the optimization process, i.e. the various stages of the synthesis of the target rogue wave, starting from lower values of crest front steepness to the requested near breaking limit of $\varepsilon = 0.49$. Figure 16(b) presents the target wave sequence as well as the associated crest front steepness and wave heights.[49] In addition the horizontal and vertical asymmetry factors are shown. Obviously, the embedded rogue wave is the steepest wave within this wave sequence. The third plot in Fig. 16(b) gives the relative wave height H/H_S. The height of the optimized rogue wave is $2.18H_S$. Thus, to generate such a

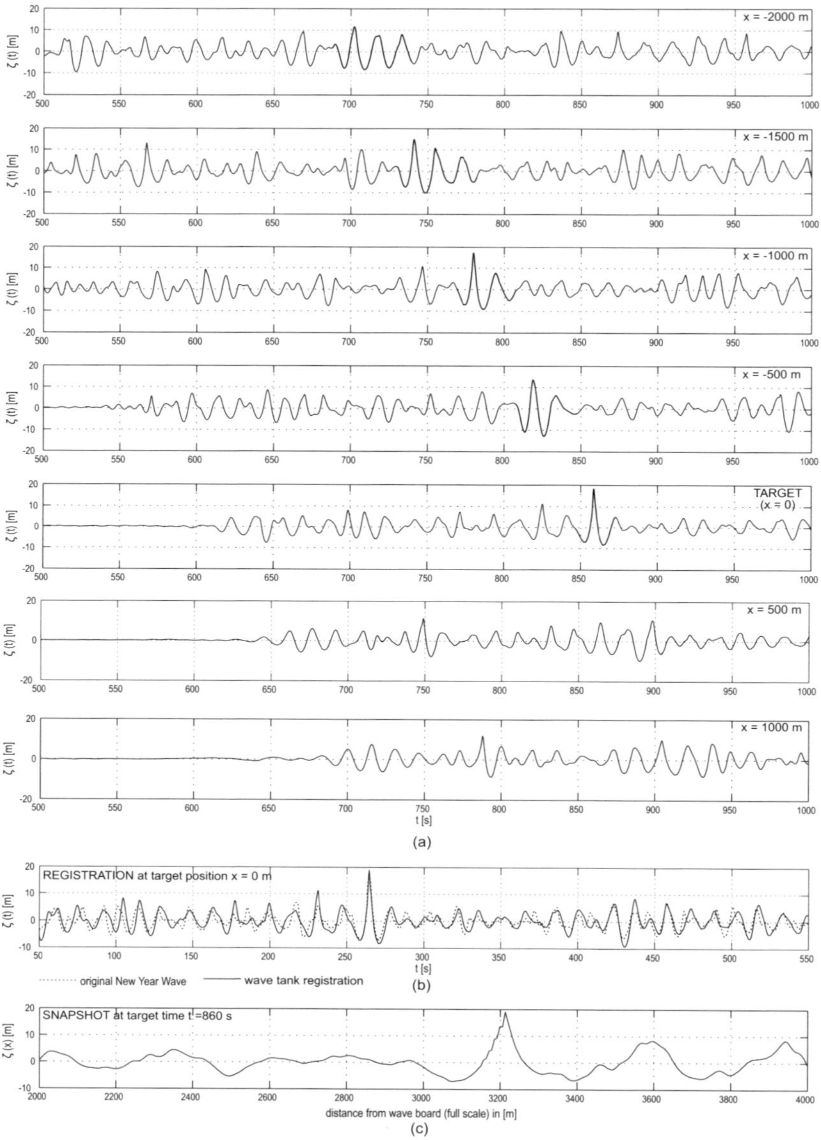

Fig. 14. New Year Wave: (a) Evolution of the New Year Wave (scale 1:70), (b) experimental simulation in comparison to the registered New Year Wave[2] (full scale wave data), (c) snapshot of the experimental simulation at target time.

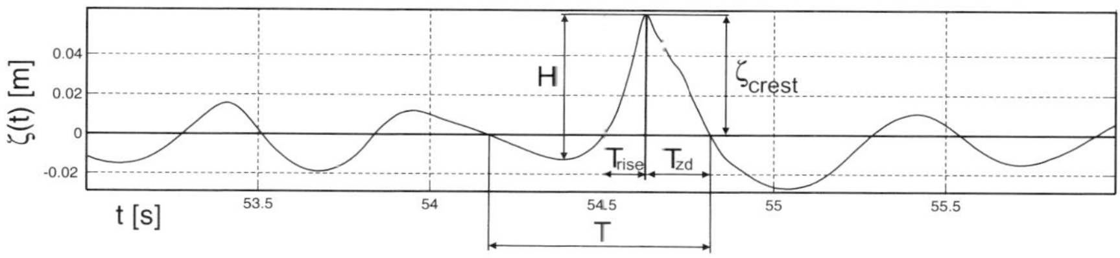

Fig. 15. Wave parameters for the definition of the wave crest front steepness, vertical asymmetry factor as proposed by Kjeldsen.[48]

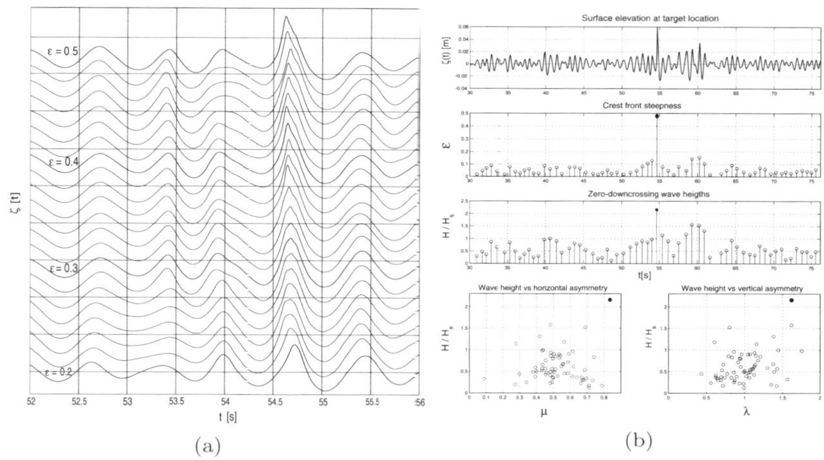

Fig. 16. (a) Synthesis of wave sequence showing 22 iteration steps. Wave train on top is the optimized wave sequence with a crest front steepness $\varepsilon = 0.49$. (b) Results of the experimental optimization in the wave tank: Registered wave sequence at target position; associated crest front steepness, zero-downcrossing wave height vs. horizontal asymmetry μ and dimensionless wave height H/H_S vs. vertical asymmetry λ. The corresponding properties of the optimized rogue wave are indicated by filled dots.

steep breaking wave the optimization algorithm increases the wave height more than demanded by the objective function ($2H_S$). As can be seen in the bottom diagram on the left hand side the horizontal asymmetry of the rogue wave $\mu = 0.83$, i.e. the crest height is $0.83 \cdot H_{max}$ while the trough is $0.17 \cdot H_{max}$ only. As the crest front steepness of the rogue wave is extremely high we also observe a large vertical asymmetry of $\lambda = 1.7$. For comparison, Kjeldsen[48] gives a range of $0.84 < \mu < 0.95$ for the horizontal asymmetry and $0.90 < \lambda < 2.18$ for the vertical asymmetry factor of breaking waves.

2.5. *Transfer of optimized wave scenarios to large seakeeping basins*

The advantage of the *hybrid optimization* process is the inherent inclusion of all nonlinear wave effects by using a physical wave tank. The basic idea of the wave generation method is to optimize wave sequences in a small wave basin, and to transfer the final signal to a large wave basin where tests with appropriate scaled models are conducted.

Small tanks are inexpensive and the decay time after each run is quite short, thus enabling (automatic and unattended) iteration cycles day and night. As inertia forces are dominant the results are reliable. For investigating wave/structure interactions large tanks are required as scale effects become increasingly important. To transfer the control signal from small to large tanks all transfer functions (electrical, hydraulic, hydrodynamic RAOs) must be known precisely. Figure 17 illustrates how the wave generator signal for the large wave tank is determined by an appropriate scaling-up procedure according to Froude, using FFT and IFFT routines repeatedly. The associated wave trains at target location agree perfectly, if compared

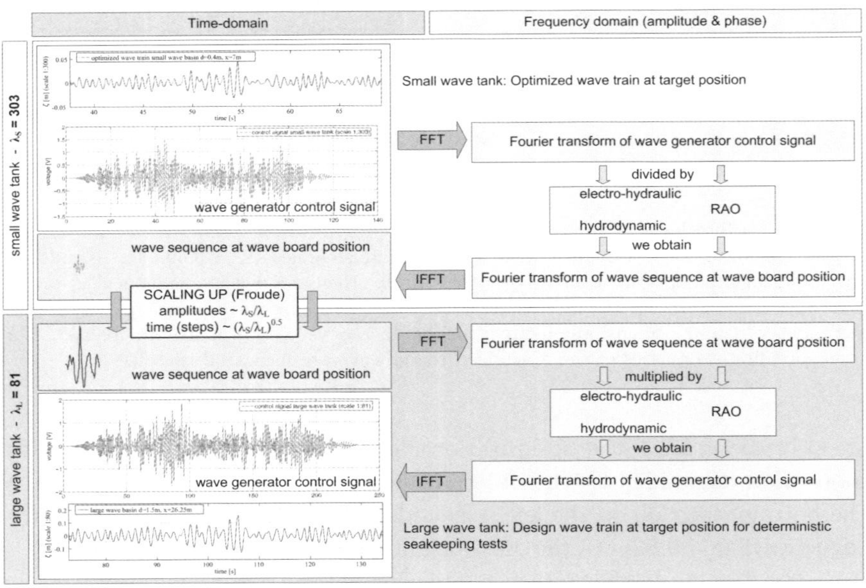

Fig. 17. Transfer of wave generator control signal from small (flap type) wave tank to large (piston-type) seakeeping basin (H_S = 13m, T_P = 13s, $H_{target} = 2H_S$, $\zeta_{crest} = 0.6H_{target}$).

at full scale. The electrical transfer functions are obtained efficiently by wave packet technology, and smoothed by regression. The hydrodynamic transfer functions are modelled using the Biésel function, relating the wave board stroke to the wave amplitude at the position of the wave generator.[50]

As a result, a "library" of relevant freak wave sequences is established which allows the accurate investigation of cause-effect relationship in time-domain. As all wave characteristics are known in space and time, the mechanisms of nonlinear structure dynamics can be evaluated. Finally, every model test is exactly repeatable.

3. Response Based Evaluation of Wave/Structure Interactions

In the previous sections the accurate generation of arbitrary extreme wave sequences embedded in irregular seas has been presented. Concerning wave/structure interactions, with respect to response based design the crucial question has to be answered: Is the highest wave with the steepest crest the most relevant design condition or should we identify wave sequences with critical frequency characteristics (e.g. close to the resonance frequency of a floating system) embedded in an irregular wave train? In addition to the global parameters H_S and T_P the wave effects on a structure depend on its dynamic behaviour as well as on superposition and interaction of wave components, i.e. on local wave characteristics.

If a cruising vessel is investigated the wave characteristics must be transformed to the moving reference frame, i.e. the characteristics of the encountering wave are relevant. Phase relations and nonlinear interactions are key parameters to specify the relevant surface profile at the (moving) structure. Only if wave kinematics and dynamics are known, cause-effect relationships can be detected.

The aiming target is the response of the structure which follows from the associated response amplitude operators (in frequency-domain) or impulse response functions (in time-domain). Provided that we know the wave field some miles ahead of our vessel we then can calculate the wave evolution up to the (moving) structure and evaluate the associated forces and motions. In other words - if we "see the future sea" we will hold an early warning system as we can prognosticate the expected behaviour of our system - or "see" its future which helps to decide on operations or manoeuvres.

In a first case study a semisubmersible in a freak wave is investigated, particularly with regard to motions and splitting forces. The second case study deals with a FPSO at various extreme wave sequences, and analyses the most critical bending moments. Finally, the capsizing of cruising ships is investigated using deterministic wave sequences which excite parametric rolling.

3.1. *Semisubmersible GVA 4000 in rogue waves*

The method of synthesizing extremely high waves integrated into severe irregular seas is applied to analyse the impact of reported rogue waves on semisubmersibles. As the procedure is strictly deterministic we can compare the numerical (time-domain) approach to model test results.[51]

For the numerical simulations the panel-method program for transient wave-body interactions TiMIT[52] (Time-domain investigations, developed at the Massachusetts Institute of Technology) is used, to evaluate the motions of the semisubmersible. TiMIT performs linear seakeeping analyses for bodies with or without forward speed. In a first module the transient radiation and diffraction problem is solved. The second module provides results like the steady force and moment, frequency-domain coefficients, response amplitude operators and time histories of body response in a prescribed wave sequence on the basis of impulse-response functions.

The drilling semisubmersible *GVA 4000* has been selected as a typical harsh weather offshore structure to investigate the seakeeping behaviour in rogue waves in time-domain. The wetted surface of the body is discretized into 760 panels (Fig. 18(b)). For validating TiMIT results of wave/structure interactions in extreme seas the Draupner New Year Wave (see Fig. 4) has

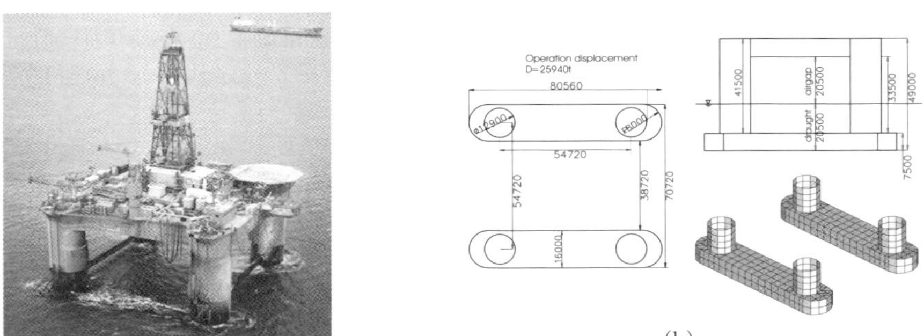

(a) (b)

Fig. 18. (a) Semisubmersible *GVA 4000*. (b) Main dimensions and discretization of the wetted surface using 760 panels.

been synthesized in a wave tank at a scale of 1:81. Using the proposed wave generation technique, the wave board signal is calculated from the target wave sequence at the selected wave tank location.[53]

Figure 19 presents the modelled wave train as well as splitting forces, heave motions and the associated airgap of the semisubmersible comparing numerical results and experimental data (scale 1:81). Note that the airgap is quite sufficient, even if the rogue wave passes the structure. However, wave run-up at the columns (observed in model tests) is quite dramatic, with the consequence that green water will splash up to the platform deck. As a general observation, the rogue wave is not dramatically boosting the motion response. The semisubmersible is rather oscillating at a period of about 14s with moderate amplitudes.

Related to the (modelled) maximum wave height of $H_{max} = 23$m we observe a maximum measured double heave amplitude of 7m. The corresponding peak value from numerical simulation is 8.6m. As a consequence,

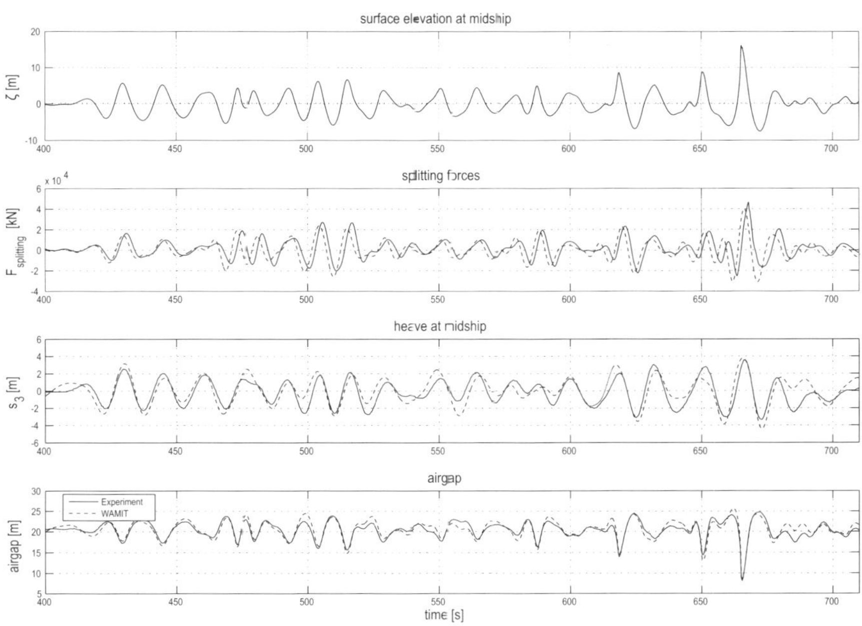

Fig. 19. Semisubmersible *GVA 4000* in the New Year Wave: Wave sequence and related splitting forces, heave motions and airgap records comparing numerical and experimental results.

the measured airgap is slightly smaller than the one from numerical simulation. Considering the complete registration it can be stated that the numerical approach gives reliable results. At rogue events the associated response is slightly overestimated due to the disregard of viscous effects in numerical calculations.

3.2. *FPSO — Heave, pitch and bending moment*

FPSOs are widely used in deep and harsh offshore areas. Therefore, a ship-type structure has also been selected for rogue wave investigations.[54] In Fig. 20 the model (scale 1:81) as well as the discretized wetted hull of the FPSO are presented. The vertical bending moment is the key parameter to ensure safe operation. If analyzed at water line level, however, the unknown influence of longitudinal forces may distort the results. Hence, the FPSO model is segmented, and force transducers are installed at midship at two levels to evaluate the vertical bending moment and the longitudinal forces (as difference and sum of the respective force transducer registrations). Investigations are performed in various deterministic wave sequences to identify the vertical bending moment and its associated neutral axis as well as the superimposing longitudinal forces. Both, frequency- and time-domain results are presented. With frequency-domain evaluation the profound data for the standard assessment of structures, concerning seakeeping behaviour, operational limitations and fatigue are obtained. In addition, time-domain analysis in real rogue waves gives indispensable data on extremes, i.e. motions and structural forces.

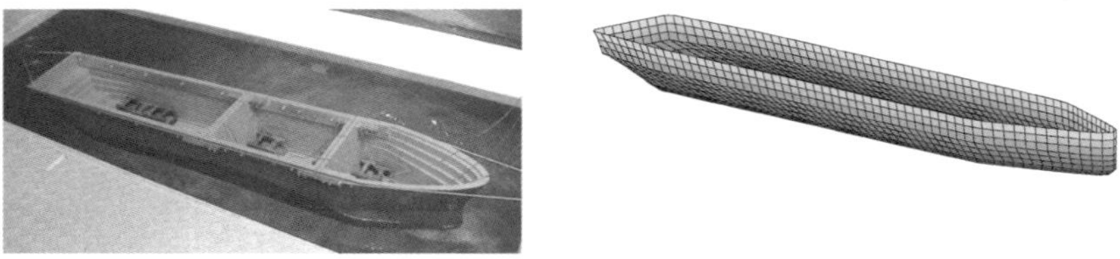

(a) (b)

Fig. 20. (a) Segmented model of the FPSO with two hull halfs connected by force transducers at different vertical levels to analyze the vertical bending moment and the longitudinal forces on the vessel. (b) Discretization of the FPSO with 1792 panels.

3.2.1. *Experimental program*

For the experimental investigations a wooden model of the selected FPSO design has been built at a scale 1:81 (Fig. 20). The model is cut into three segments with intersections at Lpp/4 and Lpp/2 measured from bow. Main dimensions of the FPSO are Lpp = 259.90m B = 46m and draft D = 16.67m (∇ = 174000t and block coefficient of $c_B = 0.87$). The model segments are connected by force transducers at two vertical levels to exactly determine the superposition of bending moments and associated longitudinal forces. As a hypothesis it is assumed that the neutral axis is quite below the still water level, and hence the horizontal forces generate a bending moment which counteracts the vertical bending moment caused by the vertical wave forces on the hull (Fig. 21).

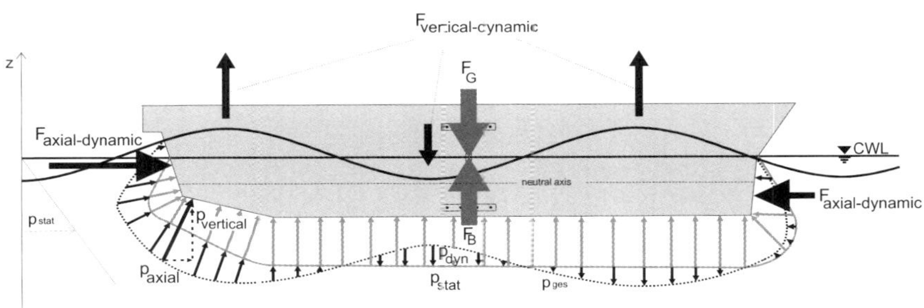

Fig. 21. Static and dynamic pressure distribution on a FPSO-hull in waves.

Again, the (modelled) New Year Wave has been used for comparative studies. Figure 22 presents this wave sequence as well as the associated heave and pitch motions of the FPSO. In addition the registrations of the horizontal forces in the mid-section are presented. These data are the basis for evaluating the vertical bending moment, the associated neutral axis and the longitudinal forces due to wave action. It should be noted that in this case the neutral axis relates to the dynamic pressure forces, i.e. it is not defined as the center of area of structural components. As the level of the neutral axis is about 7.5m above keel the longitudinal forces contribute to the bending moment if measured at waterline level or - even more pronounced - at deck level. Favourably, this additional moment is counteracting, and thus reduces the cyclic loads, in particular at deck level.

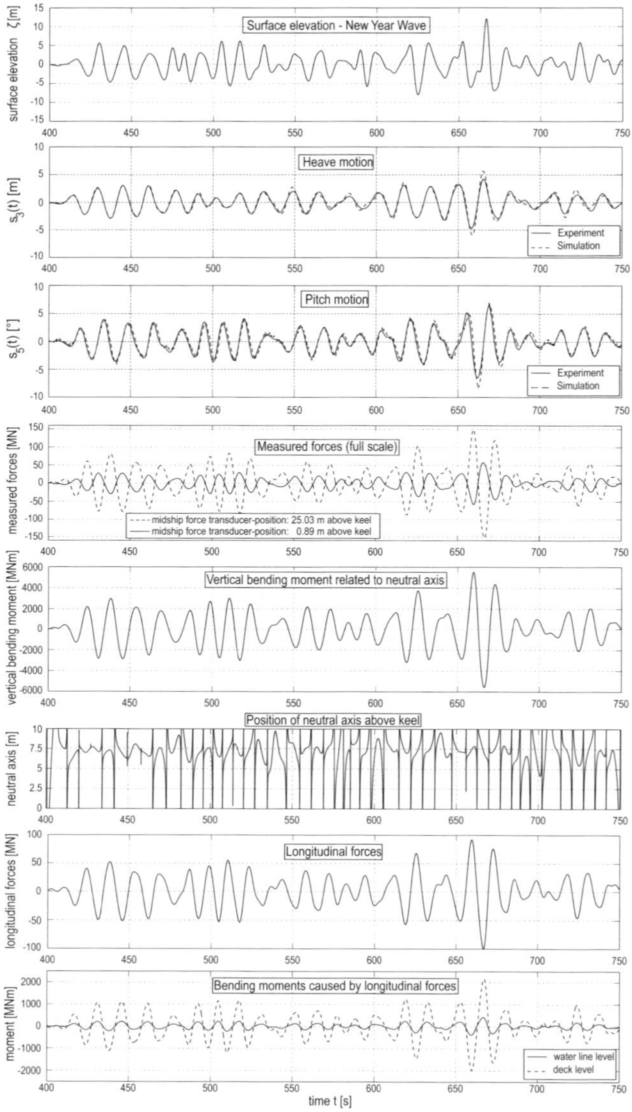

Fig. 22. Wave sequence (New Year Wave) and associated heave and pitch motions as well as related forces at midship: Based on these data the vertical bending moments and its associated neutral axis as well as the longitudinal forces and the associated (counteracting) bending moment (related to waterline and deck level) are presented at full scale.

3.2.2. *Numerical investigations*

An extensive evaluation of freak wave forces and moments requires detailed numerical investigations comparing frequency and time domain results. For the evaluation of motions, forces and bending moments the program system WAMIT[55] (Wave Analysis, developed at Massachusetts Institute of Technology) for wave/structure interaction at zero-speed is applied. WAMIT allows the analysis of generalized modes of flexural motions, in addition to the usual six degrees of rigid-body motions. By defining the ship bending modes the associated structural deformations can be calculated. Legendre polynomials $P_i(x), i = 2, 3, \ldots$[56] are found to approximate well the bending modes of a ship.[57] The deflection line for each bending mode is given by the product of the calculated amplitude and the corresponding Legendre polynomial.

$$w_i(x) = s_{ia} \cdot P_{i-5}(x),\ i = 7, 8, \ldots. \tag{3.1}$$

The indexing takes into account, that the first 6 indices are reserved for the conventional rigid body motions. For several bending modes, the total deflection results from complex addition of the individual deflection lines. Twice differentiation of the deflection line, multiplied with the flexural stiffness results in the bending moment of the ship:

$$M_b(x) = -w''(x) \cdot EI_y(x)\,. \tag{3.2}$$

The above WAMIT procedure has been successfully used to analyze vertical bending moments of stationary ships with high block coefficient.[58] Further simulations for comparative purposes have been carried out with SEAWAY, IST-code and F2T+, which are shortly described in the following.

The program SEAWAY (a frequency-domain ship motion code, based on linear strip theory) calculates wave induced loads, motions, added resistance and internal loads for six degrees of freedom of displacement ships, in regular and irregular waves. For analyzing the vertical bending moments the solid mass distribution is given to model the actual load case of the model. SEAWAY calculates the vertical bending moments from contributions of vertical forces as well as horizontal forces (see Theoretical Manual of Strip Theory Program "SEAWAY for Windows"[59]). Thus, SEAWAY can calculate the vertical bending moments with respect to an arbitrary vertical reference frame.

The IST-code using a time-domain nonlinear strip theory has been developed by Fonseca and Guedes-Soares.[60,61] The method assumes that the nonlinear contribution for the vertical bending moment is dominated by hydrostatic and Froude-Krylov forces, thus these components depend on the instantaneous wetted hull surface. Radiation and diffraction forces are linear. Additionally green water loads on the deck, which contribute to the calculation of motions and global loads, are represented by the momentum method. The exciting forces due to the incident waves are decomposed into a diffraction part and the Froude-Krylov part. The diffraction part, which is related to the scattering of the incident wave field due to the presence of the moving ship, is kept linear. Since this is a linear problem and the exciting waves are known a priori, it can be solved in frequency-domain and the resulting transfer functions are used to generate a time history of the diffraction heave force and pitch moment. The Froude-Krylov part is related to the incident wave potential and results from the integration at each time step of the associated pressure over the wetted surface of the hull under the undisturbed wave profile. The hydrostatic forces and moments are calculated at each time step by integration of the hydrostatic pressure over the wetted hull under the undisturbed wave profile. The radiation forces, which are calculated using a strip method, are represented in the time-domain by infinite frequency added masses, radiation restoring coefficients (which are zero for the zero speed case), and convolution integrals of memory functions. The convolution integrals represent the effects of the entire past history of the motion accounting for the memory effects due to the radiated waves. This code has been validated against measurements for the case of a containership[62] and of an FPSO hull[63] showing good results. The vertical forces associated with the green water on deck, which occurs when the relative motion is larger than the free board, are calculated using the momentum method.[64] The mass of water on the deck is proportional to the height of water on the deck, which is given by the difference between the relative motion and the free board of the ship. This approach to include these forces has been validated in.[65] The methodology to produce the correct wave field and the associated wave exciting forces which are consistent with a deterministic wave record (measured at one point fixed in space) is presented in.[66]

The F2T+ code - based on the transformation of response amplitude operators (frequency domain) into impulse response functions[67] (time domain) - has been developed by Jacobsen.[68]

- With frequency-domain results the motion behaviour of arbitrary structures in waves is investigated very fast and efficiently. The derived results, however, can be interpreted only statistically.

- If cause-reaction effects are of interest and wave/structure interactions are evaluated in detail a time-domain analysis in deterministic wave trains is required.

By Fourier-transforming frequency-domain results into impulse response functions and subsequent convolution with arbitrary wave trains a simple method is given to investigate wave/structure interactions in time-domain.[69,70] The response amplitude operators calculated by WAMIT are transformed into impulse response functions by Fourier transformation (Fig. 23):[67]

$$K_i(t) = \frac{1}{2\pi}\int_{-\infty}^{\infty} H_i(\omega)e^{i\omega t}d\omega\,. \tag{3.3}$$

For this purpose a Fortran routine F2T by J.N. Newman has been provided to the authors as beta-version. The Fourier Transforms are evaluated by Filon numerical integration. With known impulse-response functions $K_i(t)$ of the motions the time-dependent response in arbitrary wave trains $\zeta(t)$ are calculated by convolution:

$$s_i(t) = \int_{-\infty}^{\infty} K_i(t-\tau)\zeta(\tau)d\tau\,. \tag{3.4}$$

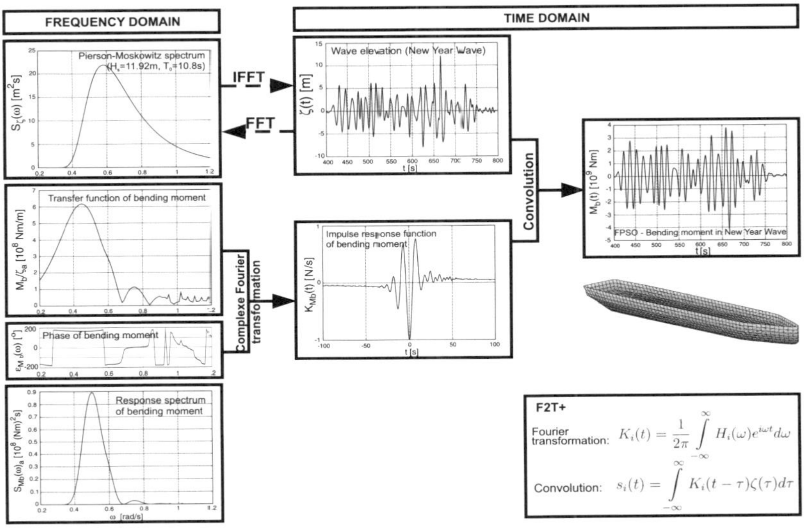

Fig. 23. F2T-code: Transformation of frequency-domain results into time-domain.

This extended F2T+ procedure combines the

- Transformation of the complex frequency-domain RAOs into time-domain impulse-response functions (F2T)
- Convolution of the impulse response function with arbitrary wave sequences to determine the behaviour of a structure in time-domain.[68]

Verification of the presented method F2T+ is carried out for single structures with TiMIT[52] (Time Domain Analysis Massachusetts Institute of Technology) which is based on the same theory as WAMIT.

3.2.3. *Validation of numerical codes*

As has been discussed in Section 3.2.1 the structural loads on a FPSO are caused by vertical and horizontal wave forces, and hence the vertical bending moment is influenced by longitudinal forces. Figure 24(a) presents the transfer functions of the vertical bending moment at midship comparing numerical programs to experimental results. Only WAMIT gives the same peak frequency as the experiments, whereas IST-code and SEAWAY yield slightly higher frequencies. Concerning the peak values, the model test data show good agreement with WAMIT and IST-code whereas SEAWAY predicts slightly lower values. As SEAWAY is the only program system that can account for the vertical position of the connecting elements at the model the sensitivity of this parameter is also investigated. Figure 24(b) presents a comparison of numerical and experimental results of the transfer

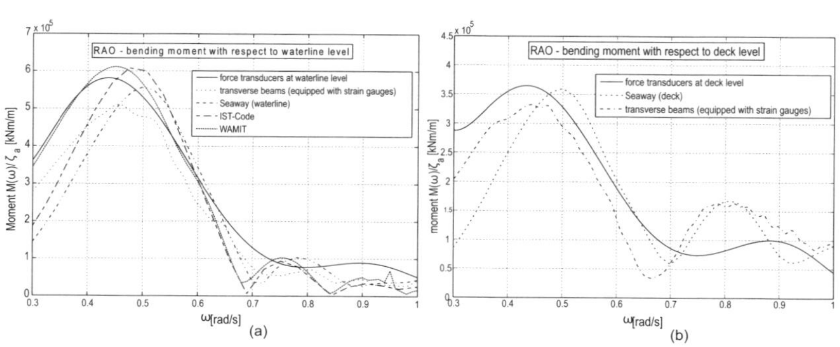

Fig. 24. Vertical bending moment at midship position, measured and calculated (a) at waterline and (b) at deck level.

function of the vertical bending moments related to a reference frame at deck level. As compared to the RAO-results at waterline level the maximum values match very good with SEAWAY.

Figure 25 presents F2T+ results compared to experimental data related to the waterline level. The agreement of the bending moment is excellent. The corresponding longitudinal forces show some discrepancies by phase and amplitude which are related to the fact that the distance of the two force transducers is to small. In conclusion it is stated that the magnitude of the longitudinal forces is quite significant. As the position of the neutral axis is far below the waterline level these forces induce an additional moment which is antiphase to the vertical bending moment. As a consequence, cyclic loads at the waterline level and especially at deck level are significantly reduced by the action of longitudinal forces.

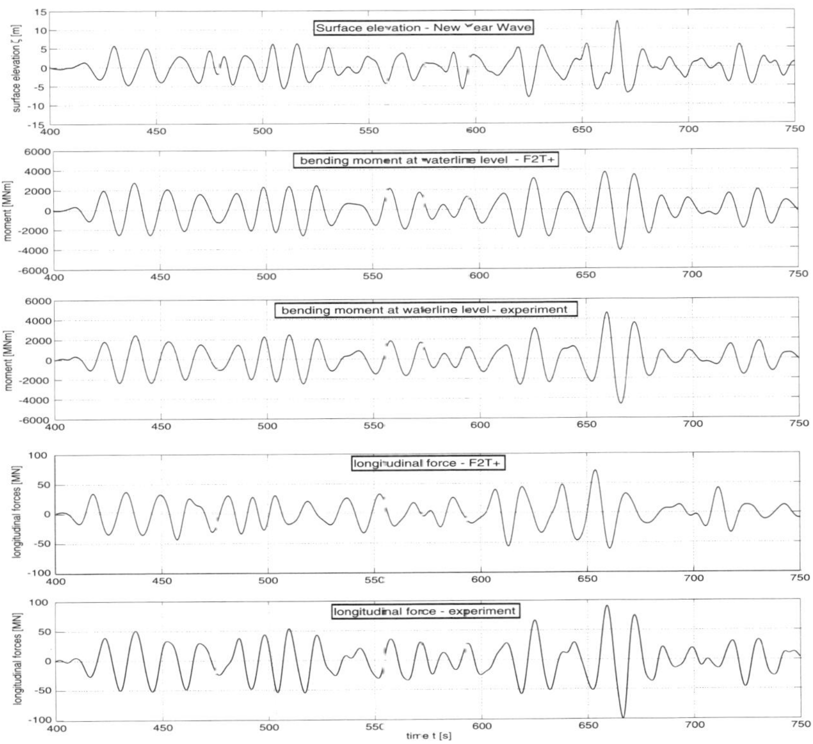

Fig. 25. FPSO - vertical bending moment at midship position (related to waterline level) as function of the New Year Wave sequence.

Finally it will be useful to compare the results in frequency- and time-domain with design bending moments. According to the IACS-Common Rules[71] the classification societies are considering a ship in design waves, and distinguish between hogging and sagging condition. The associated vertical wave bending moments

$$M_{WV} = L^2 \cdot B \cdot c_{wv} \cdot c_1 \cdot \underbrace{c_L \cdot c_M}_{=1} \tag{3.5}$$

depend on:

- c_{wv} - wave coefficient:

$$c_{wv} = \left[10.75 - \left(\frac{300 - L}{100}\right)^{1.5}\right] \cdot \underbrace{c_{RW}}_{=1}$$

- c_{1H} - hogging condition coefficient:

$$c_{1H} = 0.19 \cdot c_B$$

- c_{1S} - sagging condition coefficient:

$$c_{1S} = -0.11 \cdot (c_B + 0.7)$$

$L = L_{pp}$ is the ship length in [m], B its breadth, and c_B its block coefficient. Note that the length coefficient c_L (for L$\geq$90m) and the service range coefficient c_{RW} (unlimited service range) as well as the hogging and sagging distribution factors c_{MH} and c_{MS} (v_0 = 0m/s) are unity. With a wave coefficient $c_{wv} = 10.5$, the hogging and sagging condition coefficient $c_{1H} = 0.1653$ and $c_{1S} = -0.1727$, respectively, the design wave bending moment amounts to

$$\begin{aligned} M_{WV-hogging} &= 5.4 \cdot 10^6 kNm \\ M_{WV-sagging} &= -5.6 \cdot 10^6 kNm. \end{aligned}$$

Thus, the 25.6m freak wave (Fig. 25) which causes a bending moment of about $4 \cdot 10^6$ kNm is well covered by IACS-rules.

Similar results follow from the standard procedure for determining the significant (and maximum) vertical bending moment as a function of the sea state characteristics. As shown in Fig. 26 a variety of standard spectra (Pierson-Moskowitz - normalized for H_S = 1m) is multiplied by the squared RAO of the midship bending moment (related to waterline level - see Fig. 24(a)), and the resulting response spectra are evaluated to obtain the significant bending moment (double amplitude) as function of the zero-up-crossing period. Highest values of the significant RAO of bending moment (double amplitude) are expected in the range of $T_0 = 11$s (Fig. 26 bottom diagram). Even at this worst case ($T_0 = 11$s) - assuming a significant wave height of $H_S = 15$m (with a most probable maximum wave height of $H_{max} = 1.86 \cdot 15\text{m} = 28$m within a 3 hour storm) - the significant bending moment yields $M_{sign} = 3 \cdot 10^6$ kNm with a most probable maximum amplitude of $M_{WV_{max}} = 5.6 \cdot 10^6$ kNm which is still below IACS-limitations. Note that Fig. 26 also includes the (smoothed) energy spectrum of the New Year Wave with a zero-up-crossing period of $T_0 = 10.8$s (also normalized for $H_s = 1$m). Thus, this sea state ($H_S = 11.92$m, $H_{max} = 25.6$m) is quite critical.

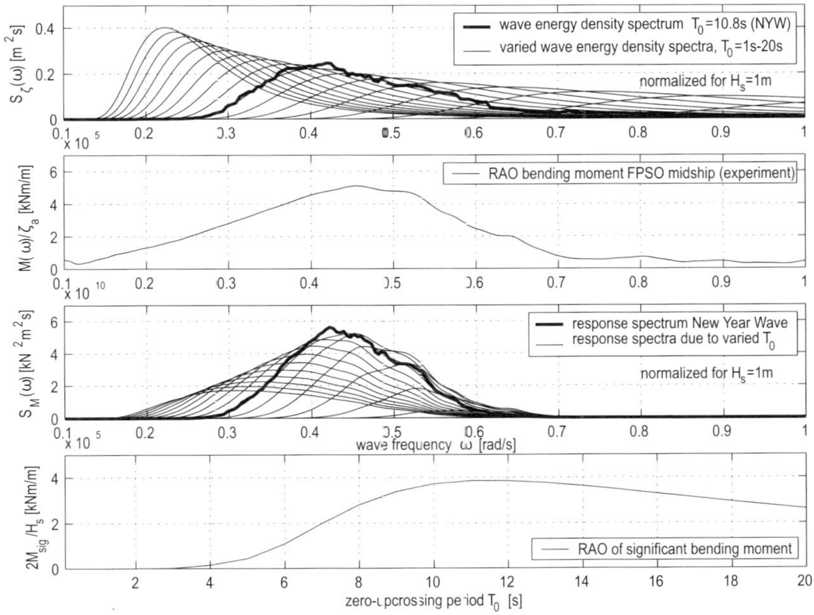

Fig. 26. Frequency-domain analysis of bending moments at midship section of FPSO: sea spectra - response amplitude operator - response spectra - significant RAO of bending moment double amplitude.

To achieve more information of freak wave impacts, and to prove the validity of numerical results as compared to experimental data a sensitivity study has been performed to investigate the worst position of the FPSO. Three locations have been selected ($x_{NYW} - L_{pp}$, x_{NYW}, $x_{NYW} + L_{pp}$).

Registrations are presented in Fig. 27. Before the seakeeping tests, the wave propagation at all locations has been recorded separately without the ship to obtain undisturbed wave registrations as input for the numerical simulation. Comparing numerical and experimental results the excellence of the numerical simulation (based on linear theory) is confirmed again. This is quite surprising, as nonlinear effects of wave/structure interactions seem to be less significant in this case. It should be noted, however, that the highest waves in the wave train are very steep, and these (real) wave kinematics are used in our analysis.

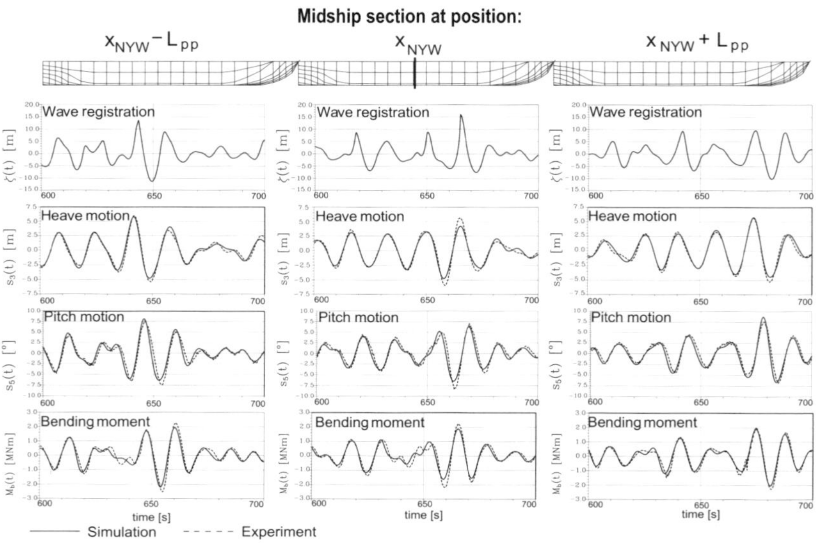

Fig. 27. Heave, pitch motions and bending moment of FPSO at varied positions of the midship section in the New Year Wave - numerical simulation and experiment (Note that the NYW - wave length roughly corresponds to the ship length).

The second surprise follows from the detailed study of the wave propagation: Our modelled reference freak wave sequence, see Fig. 27 (middle column), is not the highest wave. Just one ship length ahead (at $x_{NYW} - L_{pp}$) the maximum wave height is 26m, due to a very deep trough.

For the response based analysis of ships in freak wave sequences the dynamic behaviour is decisive. Consequently, for an improved design evaluation with regard to motion characteristics under severe weather conditions, not only wave height and steepness but also the frequency characteristics of the highest wave sequence are important parameters. Thus, the New Year Wave is modified with respect to height and period using the nonlinear optimization technique.[44,47] Figure 28 presents the respective registrations in the wave tank, i.e. the original wave train as well as the modified sequences with increased height (+25%) and elongated period (+20%). As a consequence, heave and pitch as well as the vertical bending moment at midship shows a nonlinear increase including significant asymmetry.[53] These results prove that the local wave pattern governs the maximum response. Consequently, investigations with deterministic wave sequences reveal the mechanism of wave/structure interactions and are the key tool for a response based design if worst case scenarios are evaluated. Fortunately, the comparison of time- and frequency-domain results

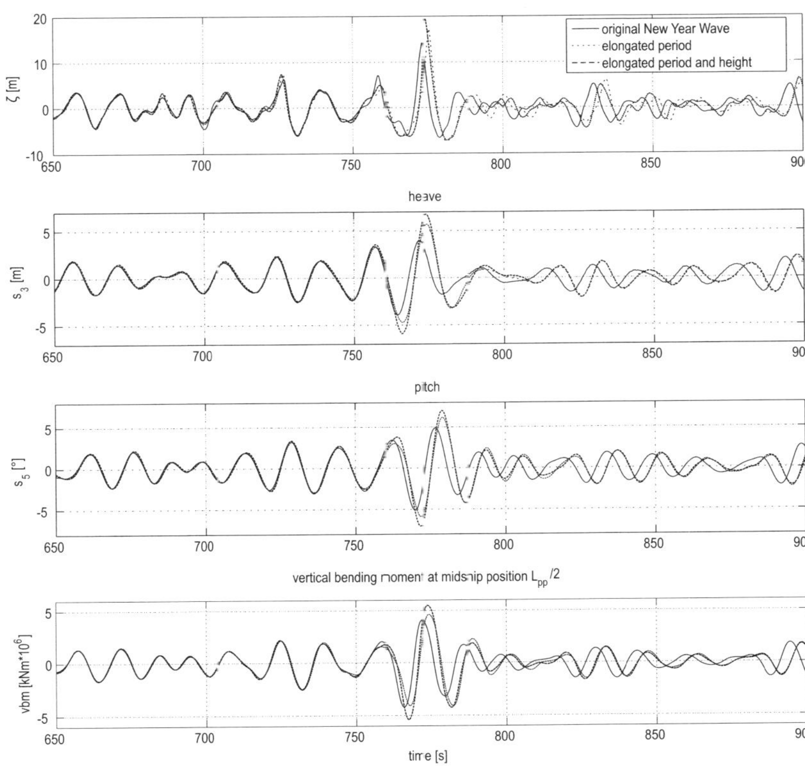

Fig. 28. Response based analysis of an FSPO - heave, pitch and vertical bending moment for the New Year Wave as well as related wave sequences with elongated period and increased height.

points out that the frequency-domain standard approach for investigating seakeeping characteristics and operational limitations as well as down-time evaluations appears satisfactory in this case, even if freak waves are hidden in the sea. The existing design rules seem to be still appropriate if impacts of rogue wave sequences[58] are considered.

3.3. *Computer controlled capsizing tests using tailored wave sequences*

The technique of generating deterministic wave sequences embedded in irregular seas is also used to analyse the mechanism of large roll motions with subsequent capsizing of cruising ships.[44]
Large roll motions follow from

- parametric excitation by encountering waves
- impact excitation, e.g. by rogue waves
- loss of stability at the wave crest
- broaching and associated loss of course stability

Figure 29 illustrates significant damages caused by parametric rolling. In a 12-hour storm this large post-panamax containership experienced rolling amplitudes up to 40 degrees combined with heavy pitching and yaw angles of 20 degrees. As a consequence the vessel lost one third of its 1300 deck containers, with another third wrecked, scattered about and smashed. For the targeted investigation of these phenomena with regard to

Fig. 29. Significant damage due to parametric rolling.

metacentric height, ship velocity and heading the parameters of the model seas (transient wave sequences consisting of random seas or regular wave trains with an embedded deterministic high transient wave) are systematically varied. The entire process including heading and velocity control of the freely cruising ship and the targeted evolution of tailored wave sequences is fully automated, i.e. no interaction of operators on the towing carriage is required. The wave elevation at the position of the ship model at any position in time and space is calculated (and controlled by registrations during model tests) in order to relate wave excitation to the resulting roll motion.[44] The ship's heading is controlled by the master computer by telemetry which commands a Z-manoeuvre at constant heading and model velocity. These test parameters as well as the model sea parameters are varied according to the metacentric height of the model, the expected rolling mode and occurrence of resonance. Ship motions in six degrees of freedom are registered precisely by computer controlled guidance of both, the towing and the horizontal carriage: During the entire test run, the ship model stays in the field of vision of the optical system line cameras. Additionally, the wave train is measured at several fixed positions in the wave tank. When the model reaches the critical safety limit at the wave maker or the absorbers at the opposite side of the tank, the ship and the carriage stop automatically. Thus, each test is realized by a deterministic course of test events which allow a reproducible correlation of wave excitation and ship motion. The wave generation process is illustrated in Fig. 30 for a two flap wave maker:

- The target position in time and space is selected - for example the position where the ship encounters the wave train at a given time. At this location, the target wave train is designed - either with chosen parameters or an existing wave registration measured in a storm.
- This wave train is transformed upstream to the position of the wave maker, and the corresponding control signals are calculated using adequate transfer functions of the wave generator.
- This control signal is used to generate the specified wave train which is measured at the selected position in the tank. The ship model should arrive at the target position at the corresponding target time (measured from the beginning of the wave generation). By registrations at the target position the compliance with the target wave parameters is validated.

- Finally, the wave train is converted to the moving reference frame of the model ship cruising at constant or non-constant speed. Again, the compliance of the target wave sequence and the measured wave train is validated by tests without ship model.

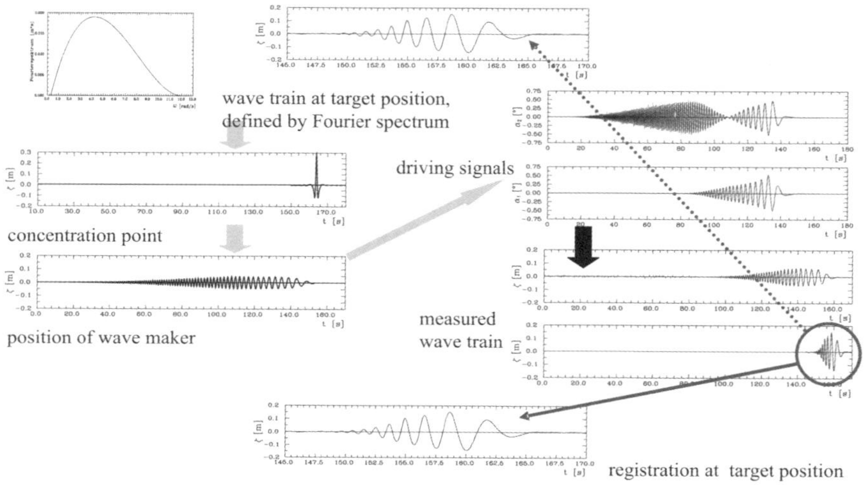

Fig. 30. Principle of the wave generation process. Calculation starts from the desired target wave train.

Figure 31(a) presents a model test with a RoRo vessel ($GM = 1.36$m, natural roll period $T_r = 19.2$s, $\nu = 15$kn) in extremely high seas from astern (ITTC spectrum with $H_S = 15.3$m, $T_P = 14.6$s, Z-manoeuvre: target course $\mu = \pm 10°$).[44] The upper diagram presents the registration at a stationary wave probe. As the waves are quite high the associated crests are short and steep followed by flat and long troughs. In contrast, the cruising ship - see wave elevation at ship center (moving frame) - apparently experiences extremely long crests and short troughs with periods well above 20s as the vessel is surfing on top of the waves. The ship loses stability while situated on a wave crest, broaches, and finally capsizes as the vessel roll exceeds 40° and the course becomes uncontrollable (Fig. 31(b)).

For parametric roll due to a high deterministic wave sequence from astern (significant wave height $H_S = 9.36$m, peak period $T_P = 11.66$s) Fig. 32 gives an example of reproducible test results.[72] The test conditions for the RoRo vessel (metacentric height $GM = 1.27$m, ship speed $\nu = 10.4$kn, ship heading $\mu = 3°$ - waves from astern) are simulated prior to

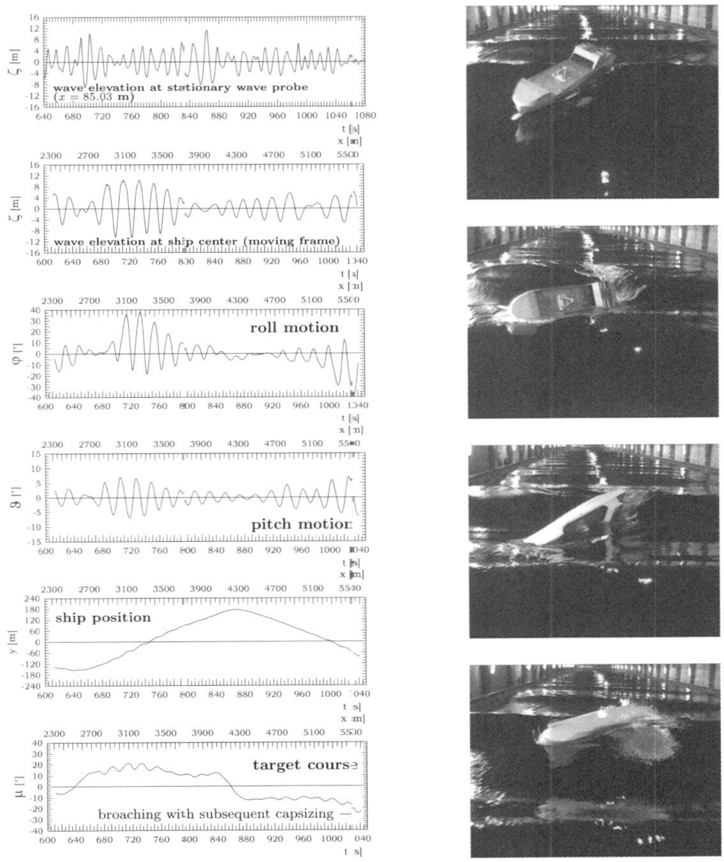

(a) (b)

Fig. 31. (a) Roll motion of the RoRo vessel in a severe storm wave train ($T_P = 14.6$s, $H_S = 15.3$m) at $GM = 1.36$m, $\nu = 15$kn. Z-manoeuvre with $\mu = \pm 10°$. (b) Capsizing of the RoRo vessel in a severe model storm.

the test. In both test runs, the ship encounters the wave train at almost identical conditions (top) and shows a similar roll response. However, in the first test (dark graph) the ship roll motions exceed 50° but the capsize occurs only in the second run (light graph). This is also an example for both the high reproducibility of capsizing tests with deterministic wave sequences and the sensitivity of the mechanism leading to capsizing. In reality, a capsize at a roll angle of more than 50° might have occurred due to shift of cargo, vehicles or equipment.

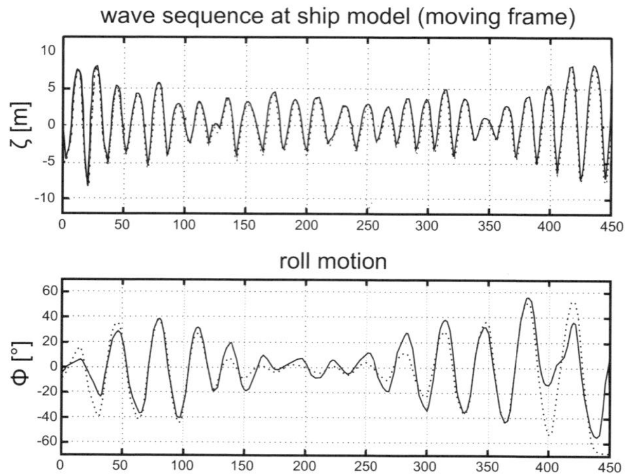

Fig. 32. RoRo vessel ($GM = 1.27$m) in extremely harsh seas ($H_s = 9.36$m, $T_P = 11.66$s) - Comparison of two capsizing test runs in deterministic wave sequences from astern inducing parametric roll ($\nu = 10.4$kn, $\mu = 3°$).

Apart from large container ships, also other modern ship designs are susceptible to parametric excitation such as RoRo, RoPax, ferries and cruise vessels - and the consequences can be dramatic[44,72] (see Fig. 29). Riding on the wave crest - the associated waterplane is small, and hence the stability extremely low (even negative lever arms are possible) - the ship abruptly tilts to high heeling angles. Just one half of the encountering wave period later, when the wave trough passes midship and the wave crest reaches the bow, an enormous uprighting moment - due to the large waterline area - is tipping the ship over to the other side. Thus, if the first wave throws the ship to starboard, the next wave tilts it over to port, and so on - with increasing heeling angles until capsize. Modern container ships with flat sterns and V-shaped frames in the forebody accompanied by large flare are highly susceptible to parametric rolling if the ship length is comparable to the length of the encountering wave. In head seas, wave encounter period is half the natural roll period of the ship, i.e. during one roll motion the ship passes two waves and hence pitches twice. In following seas, these instable conditions are also observed with roll resonance, i.e. if the natural roll period equals the wave encounter period. Surprisingly, ships in regular waves survive many parametric roll oscillations with incredibly high heeling amplitudes whereas they capsize rather fast in irregular seas.

4. Conclusions

What happens to a RoRo-ship in high seas from astern? How does a FPSO or a semisubmersible behave in a freak wave? Are motions and loads higher than predicted by standard stochastic evaluations?

For the analysis of wave/structure interactions the relation of cause and reaction is investigated deterministically to reveal the relevant physical mechanism. Based on the wave focussing technique for the generation of task-related wave packets a new technique is proposed for the synthesis of tailored design wave sequences in extreme seas.

The physical wave field is fitted to predetermined global and local target characteristics designed in terms of significant wave height, peak period as well as crest height and period of individual waves. The generation procedure is based on two steps: Firstly, a linear approximation of the desired wave train is computed by a sequential quadratic programming method which optimizes an initially random phase spectrum for a given variance spectrum. The wave board motion derived from this initial guess serves as starting point for directly fitting the physical wave train to the target parameters. The Subplex method is applied to systematically improve a certain time frame of the wave board motion which is responsible for the evolution of the response-related design wave sequence. The discrete wavelet transform is introduced to significantly reduce the number of free variables to be considered in the fitting process. Wavelet analysis allows the efficient localization of the relevant information on the electrical control signal of the wave maker in time and frequency-domain.

In general, a single rogue wave hidden in a severe sea state is a highly dangerous event. If a sea state contains wave groups which excite resonance motions, e.g. roll or parametric roll of cruising ships (dependent on course and speed) the dynamic behaviour of the vessel (or structure) may exaggerate the wave height effects. In addition memory effects play a major role for the seakeeping behaviour. The worst case scenario is occurring if wave sequences with relevant frequency characteristics are combining with extreme wave heights challenging the performance and safety of offshore structures. In this case the associated loads and motions may exceed the design limits.

As the presented technique permits the deterministic generation of design rogue wave sequences in extreme seas it is well suited for investigating the mechanism of arbitrary wave/structure interactions, including capsizing, slamming and green water as well as other survivability design aspects.

Even worst case wave sequences like the *Draupner* New Year Wave can be modelled in the wave tank to analyse the evolution of these events and to evaluate the response of offshore structures under abnormal conditions.

This procedure is illustrated at investigations of the behaviour of a semisubmersible and a FPSO in tailored freak waves as well as at the analysis of ship capsizing in deterministic wave sequences at selected target positions. Based on nonlinear methods the associated simulations - validated by identical seakeeping test - can be used for computer aided ship handling.[73] Consequently, when combined with an on-board radar system, we can "see the future seas", calculate the expected wave sequence at the ship's position (in the moving reference frame), and analyse the dynamics of the vessel well in advance. Thus we are able to "avoid the uncontrollable" and "control the unavoidable" by appropriate decisions.

Acknowledgements

The fundamentals of transient wave generation and optimization have been achieved in a research project funded by the German Science Foundation (DFG). Applications of this technique, i.e. the significant improvement of seakeeping tests and the analysis of wave breakers and artificial reefs in deterministic wave packets have been funded by the Federal Ministry of Education, Research and Development (BMBF). The technique is further developed to synthesize abnormal rogue wave sequences in extreme seas to evaluate the mechanism of large roll motions and capsizing of cruising ships (BMBF funded research projects ROLL-S and SINSEE as well as BMWi funded project LaSSe). The author wishes to thank the above research agencies for their generous support. He is also grateful for the invaluable contributions of Dipl.-Ing. Robert Stück and Dipl.-Ing. Marco Klein for finalizing this paper.

References

1. Nickerson, Freak waves!, *Mariners Weather Log, NOAA.* **37**(4), 14–19, (1993).
2. S. Haver and O. J. Anderson. Freak Waves: Rare Realization of a Typical Population or Typical Realization of a Rare Population? In *Proceedings of the 10th International Offshore and Polar Engineering Conference (ISOPE)*, pp. 123–130, Seattle, USA, (2000).
3. D. Faulkner. Rogue Waves – Defining their Characteristics for Marine Design. In *Rogue Waves 2000*, Brest, France, (2000).

4. S. P. Kjeldsen and D. Myrhaug. Breaking waves in deep water and resulting wave forces. Technical report, (1997).
5. J. Wolfram, B. Linfoot, and P. Stansell. Long- and short-term extreme wave statistics in the North Sea: 1994-1998. In *Rogue Waves 2000*, pp. 363–372, Brest, France, (2000).
6. C. Guedes Soares, Z. Cherneva, and E. M. Antão, Abnormal Waves during the Hurricane Camille, *Journal of Geophysical Research.* (2004).
7. C. Guedes Soares, Z. Cherneva, and E. M Antão, Characteristics of Abnormal Waves in North Sea Storm Sea States, *Applied Ocean Research.* (2004).
8. K. Drake. Wave profiles associated with extreme loading in random waves. In *RINA International Conference: Design and Operation for Abnormal Conditions*, Glasgow, Scotland, (1997).
9. S. P. Kjeldsen, Example of heavy weather damage caused by giant waves, *Bul. of the society of Naval Architects of Japan.* (820), (1996).
10. D. Faulkner and W. H. Buckley. Critical survival conditions for ship design. In *RINA International Conference, Design and Operation for Abnormal Conditions*, pp. 1–41, Glasgow, Scotland, (1997).
11. S. Kjeldsen. Breaking waves. In eds. A. Tørum and O. Gudmestad, *Proceedings of the NATO Advanced Research Workshop on Water Wave Kinematics*, pp. 453–473, Molde, Norway, (1990). Kluwer Academic Publishers, Dordrecht. NATO ASI Series, Series E - Volume 178.
12. S. Sand, H. Ottesen, P. Klinting, O. Gudmestadt, and M. Sterndorff, *Freak Wave Kinematics*, pp. 535–549. Kluwer Academic Publisher, NATO ASI Series edition, (1990).
13. N. Mori, T. Yasuda, and S. Nakayama. Statistical Properties of Freak Waves Observed in the Sea of Japan. In *Proceedings of the* 10th *International Offshore and Polar Engineering Conference (ISOPE)*, vol. 3, pp. 109–122, Seattle, USA, (2000).
14. G. Z. Forristall. Understanding rogue waves: Are new physics really necessary? In *Rogue Waves 14th 'Aha Hulikoʻa Hawaiian Winter Workshop, University of Hawaii*, Manoa, Honolulu: School of Ocean and Earth Science and Technology, (2005). `http://www.soest.hawaii.edu/PubServices/AhaHulikoa.html`.
15. T. B. Benjamin and J. E. Feir, The disintegration of wavetrains on deep water, *Journal of Fluid Mechanics.* **27**, 417–430, (1967).
16. K. Trulsen and K. B. Dysthe. Freak waves - a three-dimensional wave simulation. In *Proceedings of the 21st Symposion on Naval Hydrodynamics*, pp. 550–560, Washington, DC, (1997).
17. K. B. Dysthe and K. Trulsen, Note on breather type solutions of the NLS as model for freak waves, *Phys. Scr.* **T82**, 48–52, (1999).
18. K. L. Henderson, D. H. Peregrine, and J. W. Dold, Unsteady water wave modulations: fully nonlinear solutions and comparison with the NLS equation, *Wave motion.* **29**, 341–361, (1999).
19. H. Yuen. Recent advances in nonlinear water waves: an overview. In ed. A. R. Osborne, *Nonlinear Topics in Ocean Physics*, pp. 461–498, North-Holland, Amsterdam, (1991).

20. A. R. Its and V. P. Kotljarov, Explicit formulas for solutions of a nonlinear Schrödinger equation, *Dokl. Akad. Nauk. Ukr.* **SSR Ser. A. v1051**, 965–968, (1976).
21. E. R. Tracy and H. H. Chen, Nonlinear self-modulation: An exactly solvable model, *Phys. Rev. A.* **37**(3), 815–839 (Feb, 1988).
22. M. Onorato, A. R. Osborne, M. Serio, and S. Bertone, Freak waves in random oceanic sea states, *Phys. Rev. Lett.* **86**(25), 5831–5834 (Jun, 2001).
23. K. Trulsen and K. B. Dysthe, A modified nonlinear Schrödinger equation for broader bandwidth gravity waves on deep water, *Wave motion.* **24**, 281–289, (1996).
24. V. E. Zakharov, Stability of periodic waves on finite amplitude on the surface of a deep fluid, *J. Appl. Mech. Tech. Phys.* **9**, 190–194, (1968).
25. K. B. Dysthe, Note on a modification to the nonlinear Schrödinger equation for application to deep water waves, *Proc. Roy. Soc. Lond.* **Ser. A369**, 105–114, (1979).
26. A. R. Osborne, M. Onorato, and M. Serio, The nonlinear dynamics of rogue waves and holes in deep-water gravity wave trains, *Physics Letters A.* **275**, 386–393 (Oct., 2000).
27. M. Onorato, A. Osborne, M. Serio, and R. Fedele, Freak Waves and Wave Spectra Instability, *EGS XXVII General Assembly, Nice, 21-26 April 2002, abstract #4436.* **27**, 4436–, (2002).
28. P. Janssen, Nonlinear four-wave interactions and freak waves, *J. Phys. Oceanogr.* **33**, 863–884, (2003).
29. M. Onorato, A. R. Osborne, M. Serio, L. Cavaleri, C. Brandini, and C. T. Stansberg, Extreme waves and modulational instability: wave flume experiments on irregular waves, *ArXiv Nonlinear Sciences e-prints* (Nov. 2003).
30. M. Onorato, A. R. Osborne, M. Serio, L. Cavaleri, C. Brandini, and C. T. Stansberg, Observation of strongly non-Gaussian statistics for random sea surface gravity waves in wave flume experiments, *Phys. Rev.* **70**(6), 067302– (Dec., 2004).
31. M. Onorato, A. R. Osborne, M. Serio, L. Cavaleri, C. Brandini, and C. T. Stansberg, Extreme waves, modulational instability and second order theory: wave flume experiments on irregular waves, *European Journal of Mechanics B/Fluids.* **25**, 586–601, (2006).
32. L. Bertotti and L. Cavaleri, Analysis of the voyager storm, *Elsevier, Ocean Engineering.* **35**, 1–5, (2008).
33. M. C. Davis and E. E. Zarnick. Testing ship models in transient waves. In *5th Symposium on Naval Hydrodynamics*, (1964).
34. S. Takezawa and T. Hirayama. Advanced experimental techniques for testing ship models in transient water waves. Part II: The controlled transient water waves for using in ship motion tests. In eds. R. Bishop, A. Parkinson, and W. Price, *Proceedings of the 11th Symposium on Naval Hydrodynamics: Unsteady Hydrodynamics of Marine Vehicles*, pp. 37–54, (1976).
35. G. F. Clauss and J. Bergmann, Gaussian wave packets — a new approach to seakeeping tests of ocean structures, *Applied Ocean Research.* **8**(4), (1986).

36. G. Clauss and W. Kühnlein. Transient wave packets - an efficient technique for seakeeping tests of self-propelled models in oblique waves. In *Third International Conference on Fast Sea Transportation)*, Lübeck-Travemünde, Germany, (1995).
37. W. Kühnlein. *Seegangsversuchstechnik mit transienter Systemanregung.* PhD thesis, Technische Universität Berlin (D 83), (1997).
38. G. Clauss and W. Kühnlein. Simulation of design storm wave conditions with tailored wave groups. In 7th *International Offshore and Polar Engineering Conference (ISOPE)*, Honolulu, Hawaii, USA, (1997).
39. G. Clauss, C. Pakozdi, and W. Kühnlein. Experimental simulation of tailored design wave sequences in extrem seas. In 11th *International Offshore and Polar Engineering Conference (ISOPE), Vol.3*, Stavanger, Norway, (2001). ISBN 1-880653-54-0, JSC-MP-05.
40. S. *Gonzáles Sánches*, N. *Gonzáles* Prelic, and S. Garcia Galan. UviWave, wavelet toolbox for use with matlab. Technical report, (1996).
41. U. Steinhagen. *Synthesizing Nonlinear Transient Gravity Waves in Random Seas.* PhD thesis, Technische Universität Berlin (D 83), (2001).
42. G. Wu and R. Eatock-Taylor, Finite element analysis of two-dimensional non-linear transient water waves, *Applied Ocean Research.* (16), 363–372, (1994).
43. G. Wu and R. Eatock-Taylor, Time stepping solutions of two-dimensional non-linear wave radiation problem, *Ocean Engineering.* **22**(8), 785–798, (1995).
44. J. Hennig. *Generation and Analysis of Harsh Wave Environments.* PhD thesis, Technische Universität Berlin (D 83), (2005).
45. T. Rowan. *Functional Stability Analysis of Numerical Algorithms.* PhD thesis, University of Texas at Austin, (1990).
46. J. Nelder and R. Mead, A simplex method for function minimization, *Computer Journal.* (7), 308–313, (1965).
47. G. Clauss. Dramas of the sea: episodic waves and their impact on offshore structures. In *Applied Ocean Research, Vol. 24(3), ISSN 0141-1187*, (2002).
48. S. P. Kjeldsen. Determination of severe wave conditions for ocean systems in a 3-dimensional irregular seaway. In *Contribution to the VIII Congress of the Pan-American Institute of Naval Engineering*, pp. 1–35, (1983).
49. G. Clauss, C. Schmittner, and M. Klein. Generation of Rogue Waves with Predefined Steepness. In 25th *International Conference on Offshore Mechanics and Arctic Engineering (OMAE)*, pp. 1–7, Hamburg, Germany (June, 2006). OMAE2006-92272.
50. G. Clauss and C. Schmittner. Experimental Optimization of Extreme Wave Sequences for the deterministic Analysis of wave/structure interaction. In 24th *International Conference on Offshore Mechanics and Arctic Engineering (OMAE)*, Halkidiki, Greece, (2005). OMAE2005-67049.
51. G. Clauss, C. Schmittner, and K. Stutz. Time-domain investigations of a semisubmersible in rogue waves. In 21th *International Conference on Offshore Mechanics and Arctic Engineering (OMAE)*, Oslo, Norway, (2002). OMAE2002-28450, ISBN 0-7918-3599-5.

52. F. Korsmeyer, H. Bingham, and J. Newman. *TiMIT- A panel-method program for transient wave-body interactions.* Research Laboratory of Electronics, Massachusetts Institute of Technology, (1999).
53. C. Schmittner. *Rogue Wave Impact on Marine Structures.* PhD thesis, Technische Universität Berlin (D 83), (2005).
54. G. Clauss, A. Kauffeldt, and K. Jacobsen. Longitudinal forces and bending moments of a FPSO. In 26th *International Conference on Offshore Mechanics and Arctic Engineering (OMAE)*, San Diego, USA (June, 2007). OMAE2005-29091.
55. WAMIT. *WAMIT Version 5.1 – A Radiation-Diffraction Panel Program For Wave-Body Interactions.* Department of Ocean Engineering, MIT, (1994). user guide.
56. I. Bronstein, K. Semendjajew, and G. Musiol, *Taschenbuch der Mathematik.* (G. Grosche & V. Ziegler, 1. Edition, 1993).
57. J. Newman, Wave effects on deformable bodies, *Applied Ocean Research.* **16**, 47–59, (1994).
58. G. Clauss, K. Stutz, and C. Schmittner. Rogue Wave Impact on Offshore Structures. In *Proceedings of OTC. Offshore Technology Conference*, pp. 1–11, Houston, Texas (May, 2004). OTC2004-16180.
59. J. Journee. Theoretical Manual of Strip Theory Program SEAWAY for Windows. Shl-report no. 1370, Delft University of Technology, Ship Hydromechanics Laboratory, Mekelweg 2, 2628 CD Delft, Netherland, (2003).
60. N. Fonseca and C. Guedes Soares, Time-Domain Analysis of Large-Amplitude Vertical Motions and Wave Loads, *Journal of Ship Research.* **42** (2), 100–113, (1998).
61. N. Fonseca and C. Guedes Soares. Nonlinear Wave Induced Responses of Ships in Irregular Seas. In *OMAE 1998 - 17th International Conference on Offshore Mechanics and Arctic Engineering*, Lisbon, Portugal, (1998). OMAE 98-0446.
62. N. Fonseca and C. Guedes Soares, Comparison of Numerical and Experimental Results of Nonlinear Wave-Induced Vertical Ship Motions and Loads, *Journal of Marine Science and Technology.* **6**, 139–204, (2002).
63. C. Guedes Soares, N. Fonseca, and R. Pascoal. Experimental and Numerical Study of the Motions of a Turret Moored FPSO in Waves. In *OMAE 2001 - 20th Conference on Offshore Mechanics and Arctic Engineering*, Rio de Janeiro, Brasil, (2001). OMAE2001/OFT-1071.
64. B. Buchner. On the Impact of Green Water Loading on Ship and Offshore Unit Design. In *PRADS95 - 6th Symposium on Practical Design of Ships and Mobile Units*, Seoul, Korea, (1995).
65. N. Fonseca and C. Guedes Soares. Experimental Investigation of the Shipping of Water on the Bow of a Containership. In *OMAE 2003 - 22nd International Conference on Offshore Mechanics and Arctic Engineering*, Cancun, Mexico, (2003). OMAE2003-37456.
66. C. Guedes Soares, N. Fonseca, and R. Pascoal. An Approach for the Structural Design of Ships and Offshore Platforms in Abnormal Waves. In *MAXWAVE Final Meeting, 8-10, October*, Geneva, Switzerland, (2003).

67. W. Cummins, The impulse response function and ship motions, *Schiffstechnik, Band 9.* (Heft 47), 101–109 (June, 1962).
68. K. Jacobsen. *Hydrodynamisch gekoppelte Mehrkörpersysteme im Seegang Bewegungssimulationen im Frequenz- und Zeitbereich.* PhD thesis, Technische Universität Berlin (D 83), (2005).
69. K. Jacobsen and G. Clauss. Multi-body Systems in Waves Impact of Hydrodynamic Coupling on Motions. In 12th *International Congress of the International Maritime Association of the Mediterranean (IMAM)*, Lisboa, Portugal (September, 2005). ISBN 0 415 39036 2.
70. K. Jacobsen and G. Clauss. Time-Domain Simulations of Multi-Body Systems in Deterministic Wave Trains. In 25th *International Conference on Offshore Mechanics and Arctic Engineering (OMAE)*, pp. 1–10, Hamburg, Germany (June, 2006). OMAE2006-92348.
71. Germanischer Lloyd, IACS Common Structural Rules an Complementary Rules of Germanischer Lloyd. Hamburg, Germany, (2007).
72. G. F. Clauss, J. Hennig, H. Cramer, and E. Brink. Validation of numerical motion simulations by direct comparison with time series from ship model tests in deterministic wave sequences. In *OMAE 2005 - 24th International Conference on Offshore Mechanics and Arctic Engineering*, Halkidiki, Greece, (2005). OMAE2005-67123.
73. G. Clauss. Transient wave model testing for the evaluation of extreme motions and loads of ships and semisubmersibles. In *Sobena Vol.1 Nr. 3*, pp. 103–119 (December, 2005).

LIST OF CONTRIBUTORS

Name	Institution	Country
Steve W. Armfield	Sydney University	AUSTRALIA
Felicien Bonnefoy	Ecole Centrale de Nantes	FRANCE
J.P. Caltagirone	University of Bordeaux	FRANCE
D.M. Causon	Manchester Metropolitan University	UK
Günther F. Clauss	Technical University of Berlin	GERMANY
A.J.C. Crespo	University of Vigo	SPAIN
Giovanni Cuomo	University of Roma	ITALY
Robert A. Dalrymple	Johns Hopkins University	USA
Frederic Dias	ENS Cachans	FRANCE
Guillaume Ducrozet	Ecole Centrale de Nantes	FRANCE
Francois Enet	Alkyon Inc	NETHERLAND
K.Z. Fang	Dalian University of Technology	CHINA
Pierre Ferrant	Ecole Centrale de Nantes	FRANCE
H. Flament	Environmental and Hydraulic National Laboratory	FRANCE
C. Fochesato	ENS Cachans	FRANCE
Dorian Fructus	University of Oslo	NORWAY
David R. Fuhrman	Technical University of Denmark	DENMARK
M. Gomez-Gesteira	University of Vigo	SPAIN
Deborah Greaves	University of Plymouth	UK
Philippe Guyenne	University of Delaware	USA
Stéphan T. Grilli	University of Rhode Island	USA
John Grue	University of Oslo	NORWAY
Changhong Hu	Kyushu University	JAPAN
R. Issa	Environmental and Hydraulic National Laboratory	FRANCE

Name	Institution	Country
Masashi Kashiwagi	The Osaka University	JAPAN
Christian Kharif	Ecole Centrale de Marseille	FRANCE
Pierre Lubin	University of Bordeaux	FRANCE
E.-S. Lee	Environmental and Hydraulic National Laboratory	FRANCE
Z.B. Liu	Dalian University of Technology	CHINA
Q.W. Ma	City University	UK
Per A. Madsen	Technical University of Denmark	DENMARK
C.G. Mingham	Manchester Metropolitan University	UK
M. Narayanaswamy	Johns Hopkins University	USA
Andrea Panizzo	University of Roma La Sapienza	ITALY
L. Qian	Manchester Metropolitan University	UK
Benedict D. Rogers	University of Manchester	UK
S.A. Sannasiraj	Indian Institute of Technology Madras	INDIA
V. Sriram	Indian Institute of Technology Madras	INDIA
Makoto Sueyoshi	Kyushu University	JAPAN
V. Sundar	Indian Institute of Technology Madras	INDIA
Julien Touboul	Laboratoire de Sondages Électromagnétiques de l'Environnement Terrestre	FRANCE
David Le Touzé	Ecole Centrale de Nantes	FRANCE
Ting-Kuei Tsay	National Taiwan University	TAIWAN
D. Violeau	Environmental and Hydraulic National Laboratory	FRANCE
S. Yan	City University	UK
Der-Liang Young	National Taiwan University	TAIWAN
Qun Zhao	Halcrow Group Ltd	UK
Shan Zou	Johns Hopkins University	USA
Z.L. Zou	Dalian University of Technology	CHINA
Nan-Jing Wu	National Taiwan University	TAIWAN